Diagnostic Pathology of Ovarian Tumors

Robert A. Soslow • Carmen Tornos
Editors

Diagnostic Pathology of Ovarian Tumors

Springer

Editors
Robert A. Soslow, MD
Department of Pathology
Memorial Sloan-Kettering Cancer Center
1275 York Avenue
New York, NY 10065
USA

Carmen Tornos, MD
Department of Pathology
Stony Brook University Medical Center
SUNY Stony Brook
Hospital Level 2, Room 766
Stony Brook, NY 11794
USA

ISBN 978-1-4419-9750-0 e-ISBN 978-1-4419-9751-7
DOI 10.1007/978-1-4419-9751-7
Springer New York Dordrecht Heidelberg London

Library of Congress Control Number: 2011931785

Printed on acid-free paper

Springer is part of Springer Science+Business Media (www.springer.com)

Preface

By Robert A. Soslow, M.D. and Carmen Tornos, M.D.

This book attempts a comprehensive overview of ovarian neoplasia. The primary focus is diagnostic pathology, although the breadth of information provided here should be of interest to those who study or practice gynecologic oncology, clinical genetics, molecular genetics, oncogenesis, and epidemiology.

We have divided the book into four major parts. The introductory chapters, which are general in scope, review the importance of morphologic patterns in diagnosis, diagnostic discrepancies that arise in practice, and frozen section evaluation. The latter two chapters include information on quality assurance issues. Chapters about morphologic patterns and diagnostic discrepancies also serve as an index to guide the reader though later portions of the book. A special feature of the frozen section chapter is that it contains figures photographed from actual frozen sections and a short discussion of the use of telepathology.

The second part of the book focuses on the most common types of ovarian neoplasms, those usually referred to as "surface-epithelial" tumors. In addition to the expected chapters about serous, endometrioid, clear cell, mucinous, and metastatic neoplasms (included in this part because of frequent diagnostic problems distinguishing them from primary ovarian surface epithelial tumors), we also present here a chapter that documents the state of the art in clinical management of ovarian cancer and a special pathology oriented chapter that offers a compelling rationale supporting rigorous separation of different types of surface epithelial tumors. This chapter makes the important points that older data that rely on distinctions between ovarian carcinoma subtypes are likely to be inaccurate, and that clinical trials must make use of clinically validated pathologic criteria, instead of historical conventions that were empirically derived. Although the clinical chapter focuses primarily on the management of high grade serous carcinoma, by far the most common lethal malignancy of its type, a unifying theme emphasized throughout the pathology oriented chapters is that each type of ovarian carcinoma represents a distinct disease entity, with unique morphologic, immunophenotypic, genotypic, and clinical features. Several recent concepts are presented in each chapter. Examples include: the existence of low- and high-grade serous carcinomas; that most undifferentiated and transitional cell carcinomas represent high grade serous carcinomas; that architectural features are more important than the presence of clear cytoplasm for a diagnosis of clear cell carcinoma; the importance of "confirmatory endometrioid features" in the diagnosis of endometrioid carcinoma; the indolent clinical course of endometrioid and mucinous carcinoma when high grade serous carcinoma and metastases have been excluded; the importance of immunohistochemistry as a validated diagnostic adjunct; and the presence of type-specific abnormalities that may be targeted with new therapeutic agents in the near future. These include PARP in high grade serous carcinoma, *BRAF* and *KRAS* in low grade serous carcinoma, *PIK3CA* in clear cell carcinoma, and *Her2neu* in mucinous carcinoma.

Although relatively much less important than the aforementioned substantive issues, the problem of accommodating different preferences in terminology remained a concern of the editors throughout. We encouraged each of the authors to use terms with which they felt most comfortable, provided familiar synonymous terms were also referenced. For example, the terms "atypical proliferative tumor" and "borderline tumor," and "disseminated peritoneal adenomucinosis" and "low grade mucinous neoplasm" are used interchangeably.

The next part of the text details some of the most unusual ovarian neoplasms, those assigned to the sex cord stromal, germ cell, and mesenchymal categories, and those of uncertain histogenesis. There is an inherent paradox here; these chapters are longer and in some cases even more detailed than chapters in the second part. That mostly reflects that these neoplasms are extraordinarily complex and that pathologists tend to be relatively unfamiliar with a variety of the entities. It is legitimate to ask why we chose to include certain entities in one chapter and not in another. When such a question arose, we asked ourselves where in the book a pathology neophyte would search for a given entity. An example is ovarian adenosarcoma, a tumor that has historically been categorized with the endometrioid neoplasms; we chose, instead, to include it in the mesenchymal chapter. A very nice clinical chapter also accompanies the pathology chapters in this part.

The last part deals with subjects that were previously on the periphery of discussions about ovarian cancer, yet are included here because a grasp of these concepts is essential to accurate diagnosis and clinical management of "ovarian cancer" patients. Chapter 17 deals with tumors of the fallopian tube and peritoneum. It makes the case that many high grade serous ovarian carcinomas arise in the tubal fimbria, as noted earlier in Chaps. 4 and 6, and that what we currently recognize as "high grade ovarian serous carcinoma" may represent tubal carcinoma with disproportionate presentation in the ovary. This is then the perfect lead in to Chap. 18, which provides a theoretical and practical approach to understanding, processing, and diagnosing lesions in patients suspected of having hereditary breast/ovarian cancer syndrome. Last, but not least, is a chapter that covers the cytologic evaluation of peritoneal fluids from patients suspected of having ovarian cancer.

Acknowledgment

The publication of this book would not have been possible without the invaluable contributions of many individuals, particularly pathologists and gynecologists who have taken time from their busy lives to produce beautifully illustrated chapters containing the newest information about ovarian cancer. Jen Grady and Alex MacDonald, members of the Gynecology Service's editorial team at Memorial Sloan-Kettering Cancer, have been absolutely essential, hardworking team constituents, always helpful, supportive and knowledgeable. Thanks also go to the expert photography staff in Memorial Sloan-Kettering Cancer Center's Department of Pathology, Kin Kong and Allyne Manzo. Our spouses, Roman Perez-Soler and Michael Ogborn, deserve recognition and thanks for their support and encouragement. Finally, Carmen and I would like to acknowledge the power of mentorship, collaborations, and friendship. We were introduced by mutual colleagues, subsequently had the great fortune of working together and now count the other as the best of friends. We hope the reader finds that this book represents a composition that is greater than the sum of its parts. If so, it will be because of the spirit of cooperation and teamwork that has colored this work from its inception.

Contents

Contributors

Katherine M. Bell-McGuinn, M.D., Ph.D. Department of Medicine, Gynecologic Medical Oncology Service, Memorial Sloan-Kettering Cancer Center, New York, NY, USA

Martin C. Chang, M.D., Ph.D. Pathology and Laboratory Medicine, Mount Sinai Hospital, Toronto, ON, Canada

Terence J. Colgan, M.D. Pathology and Laboratory Medicine, Mount Sinai Hospital, Toronto, ON, Canada

Deborah F. DeLair, M.D. Department of Pathology, Memorial Sloan-Kettering Cancer Center, New York, NY, USA

Ginger J. Gardner, M.D. Department of Surgery, Gynecology Service, Memorial Sloan-Kettering Cancer Center, New York, NY, USA

C. Blake Gilks, M.D., F.R.C.P.C. Department of Pathology and Laboratory Medicine, University of British Columbia/Vancouver General Hospital, Vancouver, BC, Canada

Jason A. Konner, M.D. Department of Medicine, Gynecologic Medical Oncology Service, Memorial Sloan-Kettering Cancer Center, New York, NY, USA

Mario M. Leitao Jr., M.D. Department of Surgery, Gynecology Service, Memorial Sloan-Kettering Cancer Center, New York, NY, USA

Gkeok Stzuan Diana Lim, M.B.B.S., F.R.C.Path., F.R.C.P.A. Department of Pathology, National University Health Service, Singapore, Singapore and
Yong Loo Lin School of Medicine, National University of Singapore, Singapore

Oscar Lin, M.D., Ph.D. Department of Pathology, Memorial Sloan-Kettering Cancer Center, New York, NY, USA

Andre L. Moreira, M.D., Ph.D. Department of Pathology, Memorial Sloan-Kettering Cancer Center, New York, NY, USA

Melissa P. Murray, D.O. Department of Pathology, Memorial Sloan-Kettering Cancer Center, New York, NY, USA

Esther Oliva, M.D. Department of Pathology, Massachusetts General Hospital, Boston, MA, USA and
Harvard Medical School, Boston, MA, USA

Kay J. Park, M.D. Department of Pathology, Memorial Sloan-Kettering Cancer Center, New York, NY, USA

Jeffrey D. Seidman, M.D. Department of Pathology, Washington Hospital Center, Washington, DC, USA

Robert A. Soslow, M.D. Department of Pathology, Memorial Sloan-Kettering Cancer Center, New York, NY, USA

Carmen Tornos, M.D. Department of Pathology, Stony Brook University Medical Center, Stony Brook, NY, USA

Anna Yemelyanova, M.D. Department of Pathology, The Johns Hopkins University School of Medicine, Baltimore, MD, USA

Charles J. Zaloudek, M.D. Deparment of Pathology, University of California San Francisco, San Francisco, CA, USA

Morphologic Patterns

1

Robert A. Soslow

Keywords
Cribriform • Morphology • Ovarian cancer • Papillary • Solid

This initial chapter is meant to enable the reader to visualize the landscape of ovarian cancer, organized by the tumor's predominant morphologic pattern. This is not the only approach to diagnosing and grading ovarian cancer, but it is an option that may be attractive to trainees and practitioners who do not handle large volumes of ovarian tumors.

It is obvious that pathologists have a mandate to recognize which ovarian tumors are primary ovarian neoplasms and to distinguish epithelial, germ cell, and sex cord stromal tumors, but only recently has the importance of accurate and reproducible ovarian carcinoma subclassification and grading been emphasized. The rationale is provided in Chap. 4. To summarize, ovarian carcinoma types represent distinct disease entities, instead of just morphologic variations of a single disease type. Talking about ovarian carcinoma is like talking about "heart disease," a vague term with a myriad of etiologies, pathologies, prognoses, and treatments. Rigorous application of diagnostic and grading criteria bestows on each carcinoma type a unique epidemiologic, biologic, immunohistochemical, prognostic, and therapeutic profile that was previously unrecognized because of subjective diagnostic practices that were, in many cases, not meaningful.

We recommend that the reader use this chapter to familiarize him/herself with the various entities discussed elsewhere in the text. Schematic diagrams organized by morphologic pattern are the focus of this chapter (Figs. 1.1–1.4). Also presented are examples of diagnostically problematic cases with accompanying figure legends (Figs. 1.5–1.7) that guide the reader through application of the paradigm illustrated in the decision tree diagrams. The chapter concludes with a series of tables summarizing the key diagnostic features of each ovarian carcinoma subtype (Tables 1.1–1.11). These tables can be used as an adjunct to the schematic decision tree diagrams. The criteria tabulated here represent the basis of a revision to the Gynecologic Oncology Group's criteria for subtyping and grading ovarian carcinoma.

R.A. Soslow
Department of Pathology, Memorial Sloan-Kettering Cancer Center, 1275 York Avenue, New York, NY 10065, USA

R.A. Soslow and C. Tornos (eds.), *Diagnostic Pathology of Ovarian Tumors*,
DOI: 10.1007/978-1-4419-9751-7_1, © Springer Science+Business Media, LLC 2011

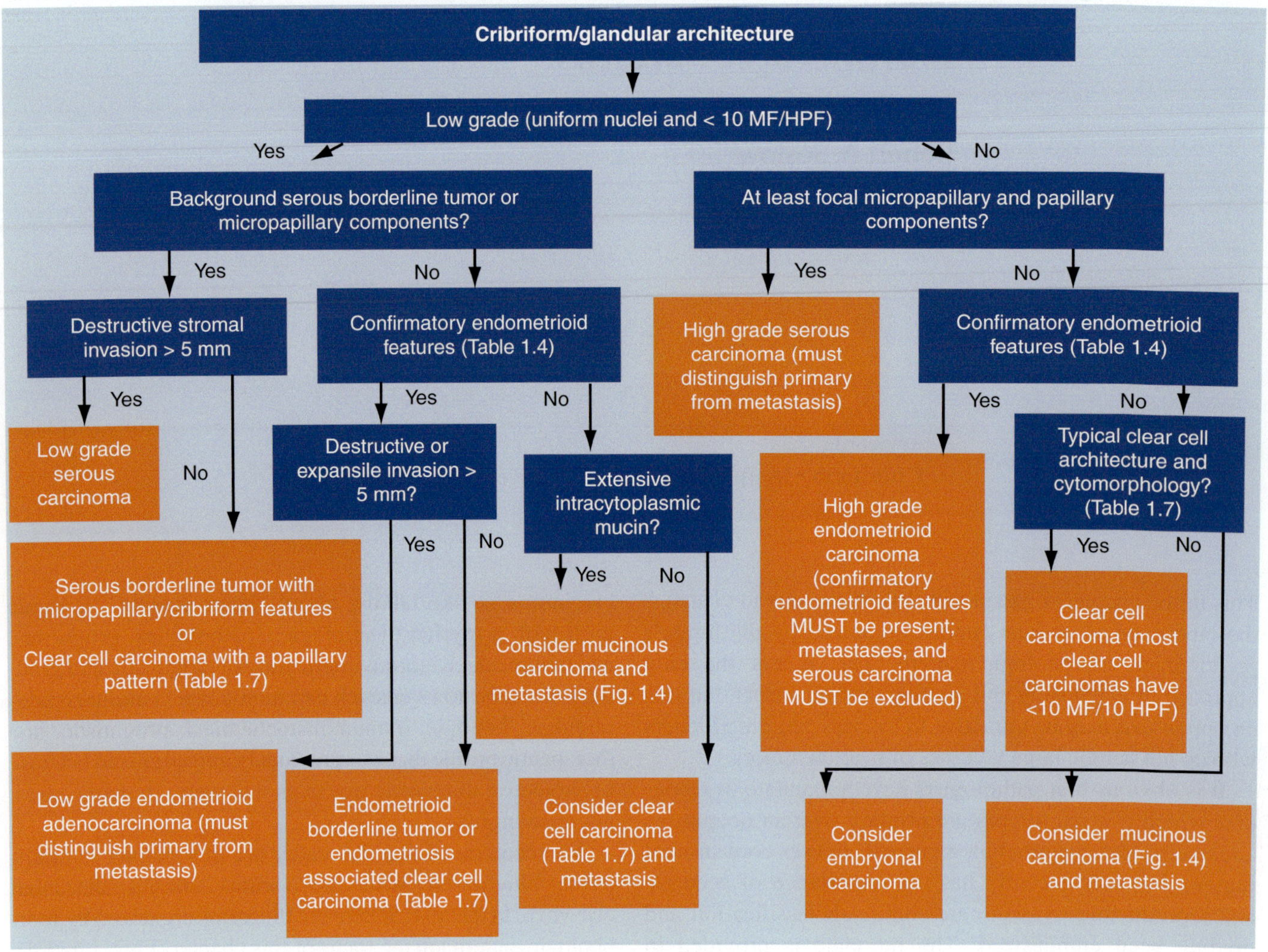

Fig. 1.1 Differential diagnosis of ovarian tumors with cribriform and glandular architecture

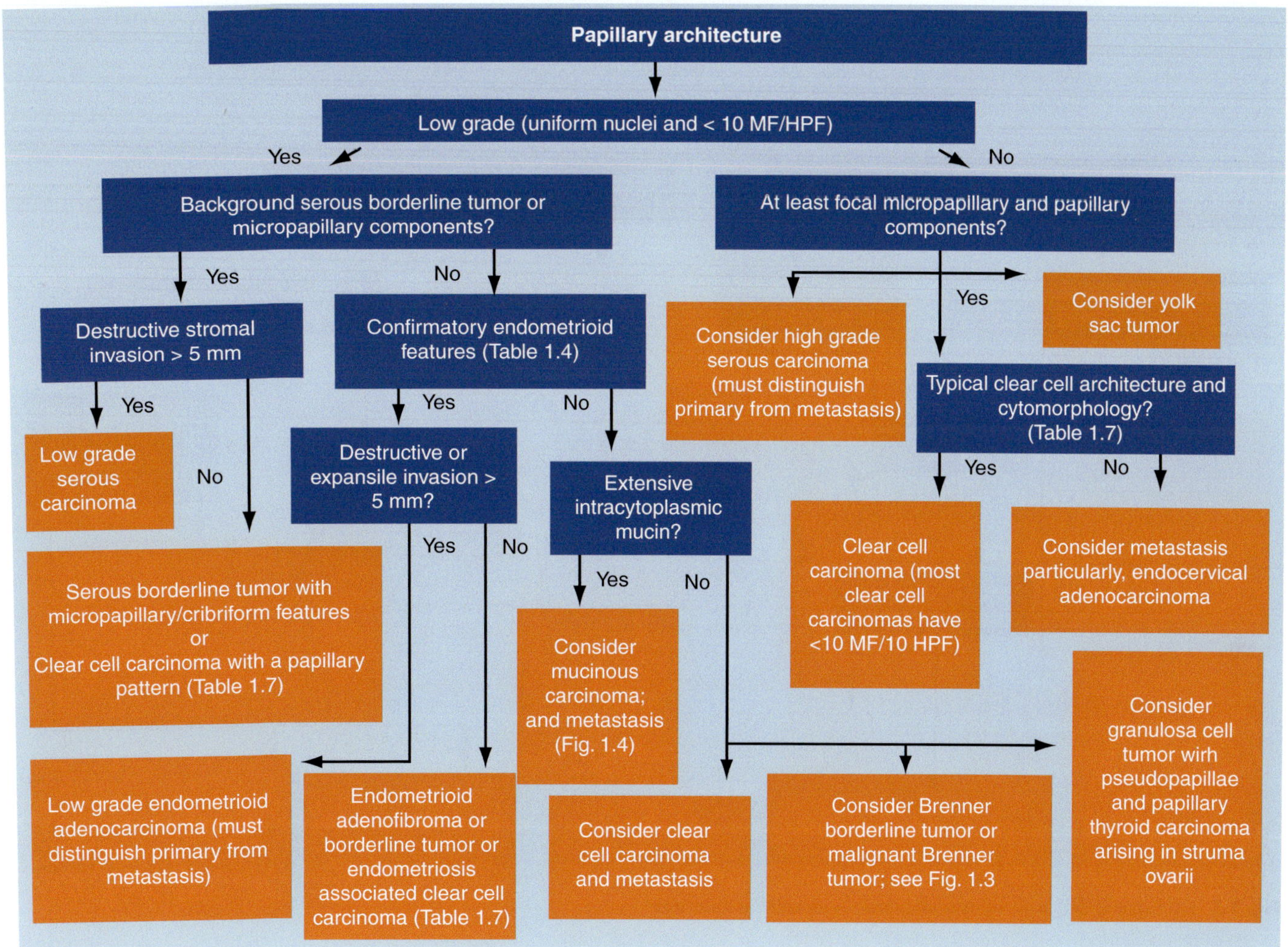

Fig. 1.2 Differential diagnosis of ovarian tumors with papillary architecture

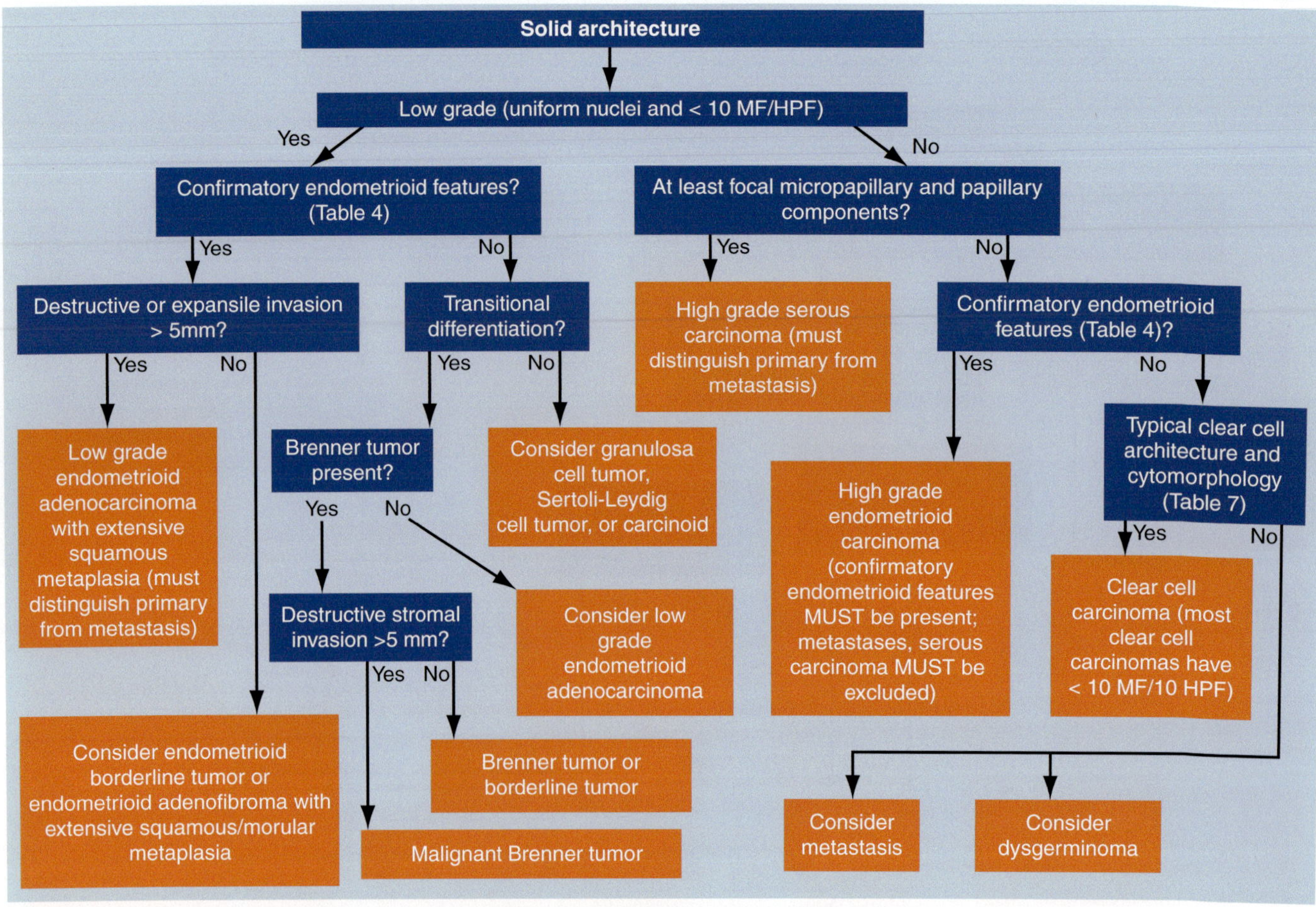

Fig. 1.3 Differential diagnosis of ovarian tumors with solid architecture

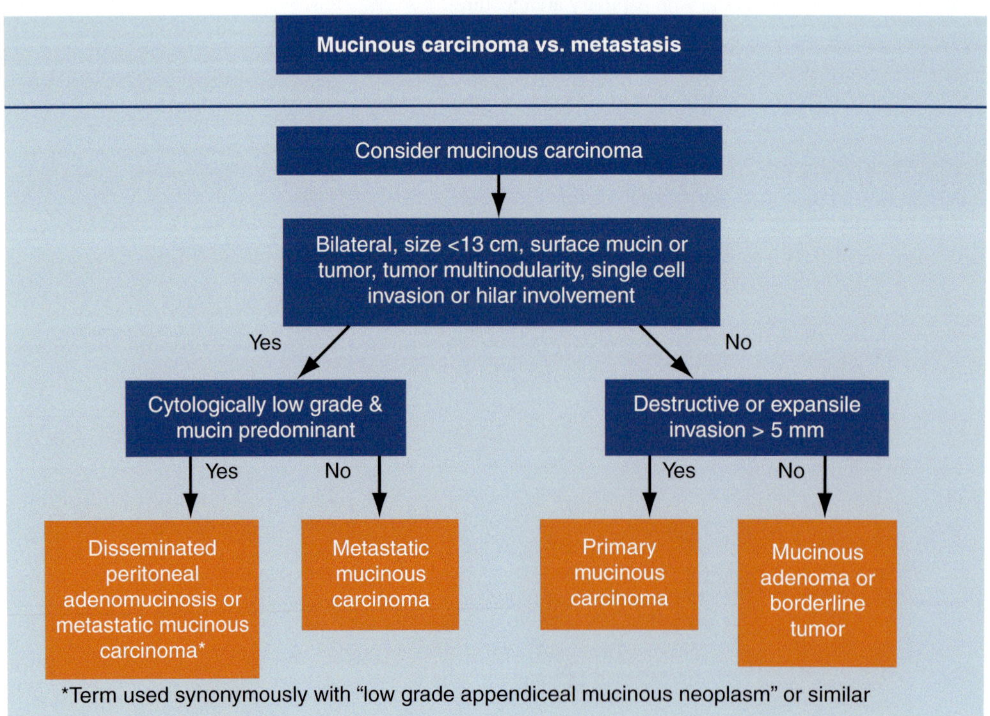

Fig. 1.4 Differential diagnosis of primary and metastatic mucinous carcinomas

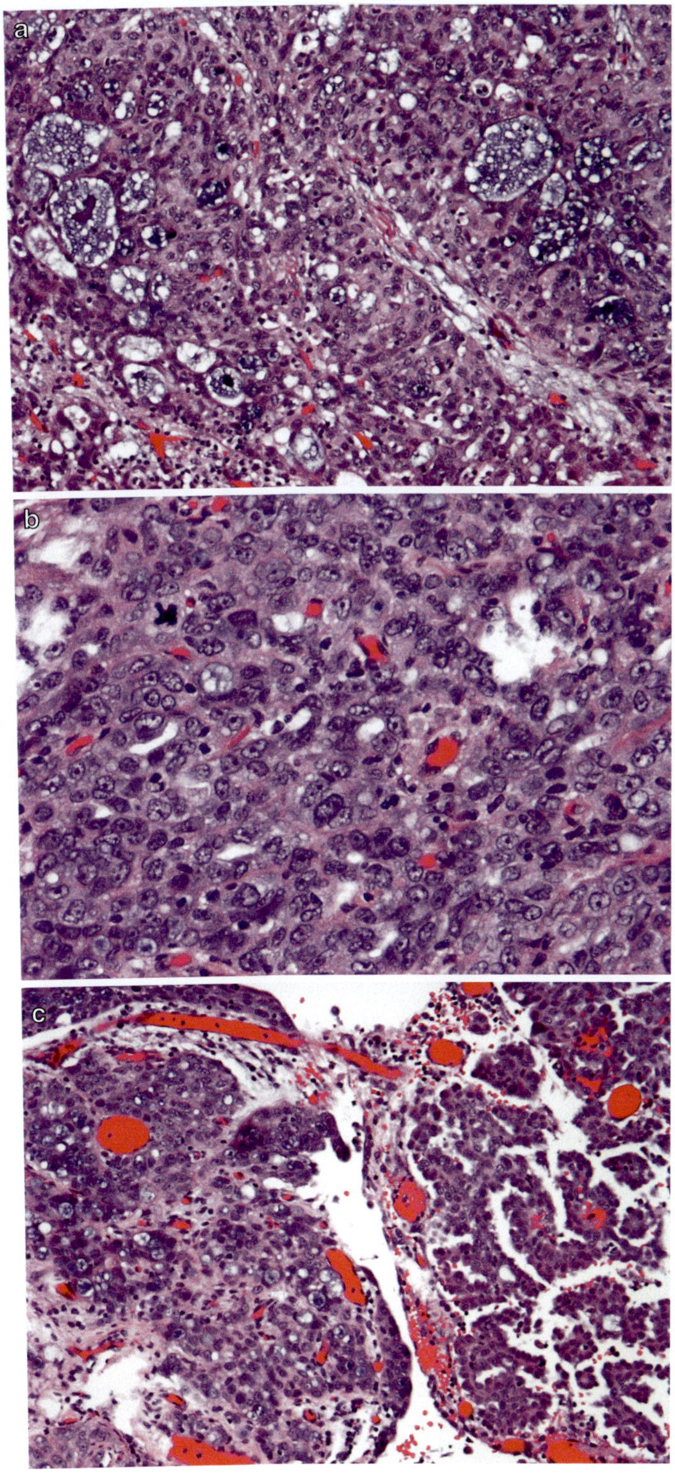

Fig. 1.5 High grade serous carcinoma mimicking high grade endometrioid carcinoma. (**a**) Adenocarcinoma with cribriform architecture (see schematic decision tree diagram Fig. 1.1). The tumor is not low grade. There is evident variability in nuclear size and mitotic figures are found easily in this field. The mitotic index overall was 23 mitotic figures per 10 high power fields. (**b**) Adenocarcinoma with solid architecture (see schematic decision tree diagram Fig. 1.3). Photograph taken from the same tumor illustrated in panels **a** and **c**. (**c**) Adenocarcinoma with papillary architecture (see schematic decision tree diagram Fig. 1.1). Another section from the same tumor illustrated in panel **a** shows focal micropapillary architecture. This high grade tumor therefore qualifies as high grade serous carcinoma. Confirmatory endometrioid features (Table 1.4) are lacking. Hysterectomy and salpingo-oophorectomy revealed tumor involving both ovaries and fallopian tubes; the endometrium was benign

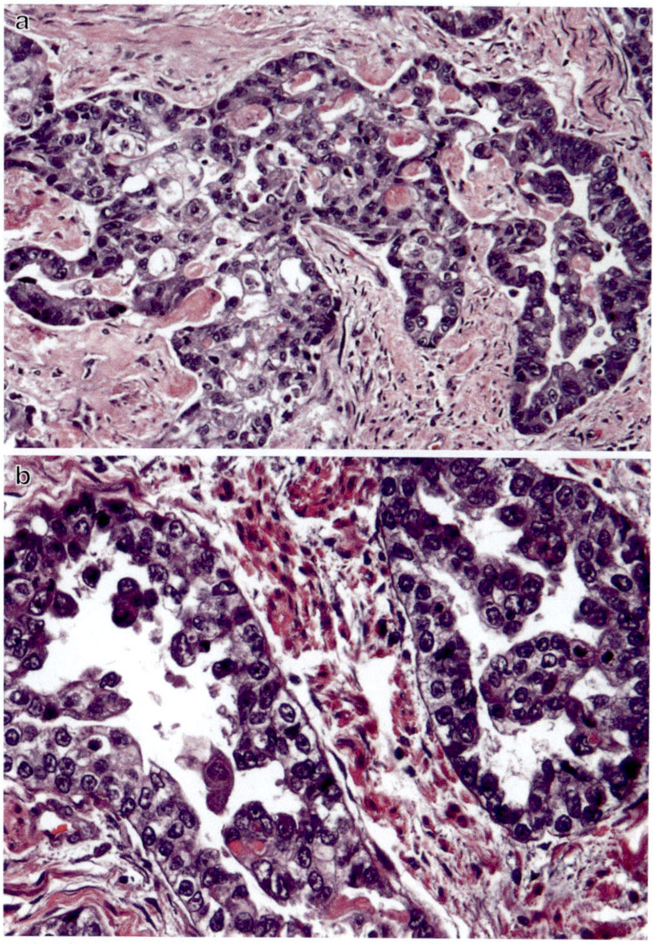

Fig. 1.6 High grade serous carcinoma mimicking clear cell carcinoma. (**a**) Adenocarcinoma with glandular architecture and clear cells (see schematic decision tree diagram Fig. 1.1). There is evident variability in nuclear size and a mitotic figure is present in this field. The mitotic index overall was 28 mitotic figures per 10 high power fields. There is also a suggestion of a micropapilla. (**b**) Adenocarcinoma with glandular architecture and clear cells (see schematic decision tree diagram Fig. 1.1). Another section from the same tumor illustrated in panel 6**a** shows micropapillary architecture along with other architectural and nuclear features that are more compatible with high grade serous carcinoma than with clear cell carcinoma (Table 1.7). The carcinoma expressed WT1 diffusely (not shown), support for serous differentiation in this context

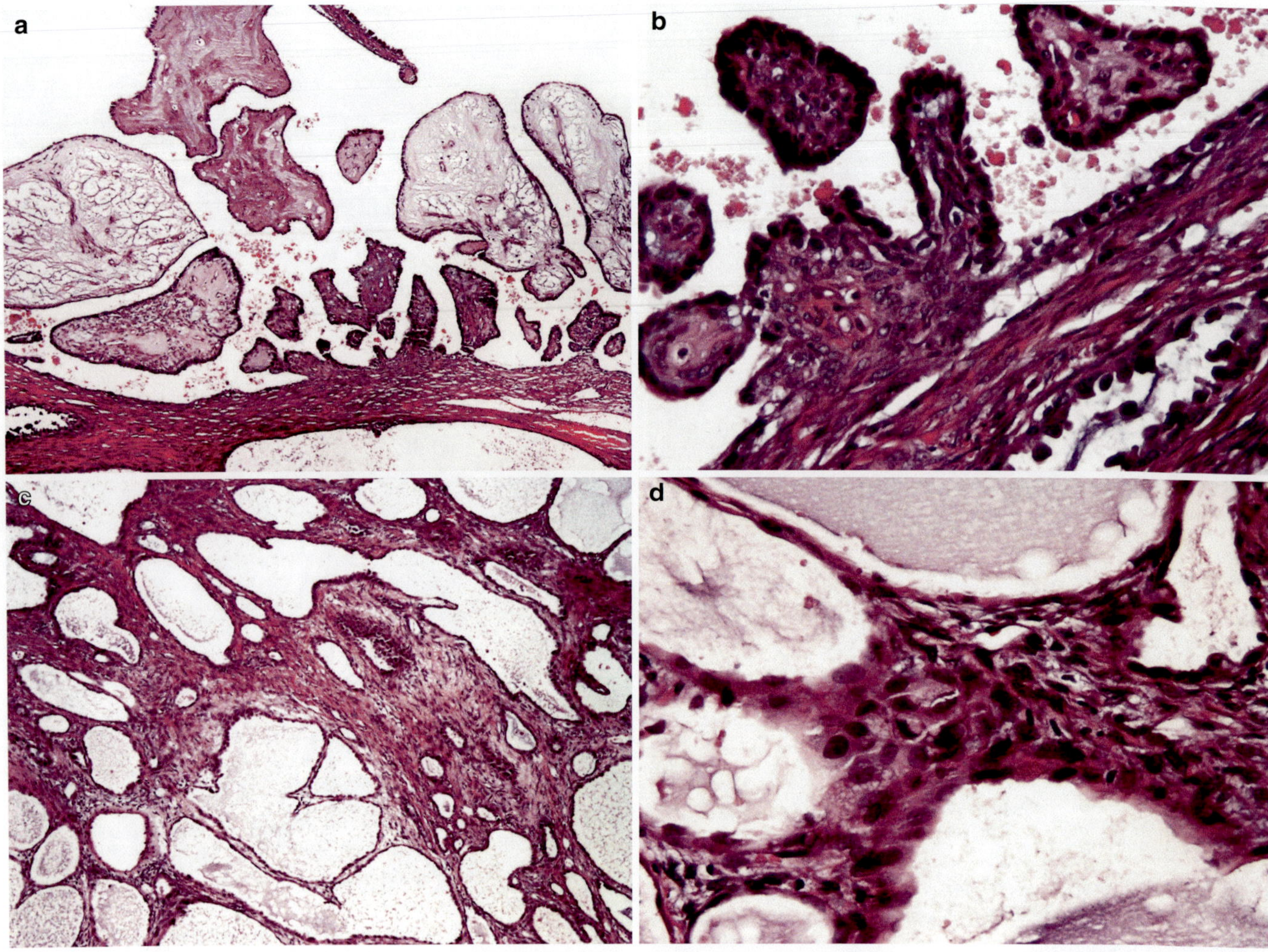

Fig. 1.7 Clear cell carcinoma mimicking serous borderline tumor and serous cystadenofibroma. (**a**) Papillary tumor lined by a single layer of epithelial cells (see schematic decision tree diagram Fig. 1.2). (**b**) The presence of a single layer of epithelial cells is not compatible with serous borderline tumor, although these features are typical of serous cystadenofibroma and clear cell carcinoma. (**c**) Tubulocystic clear cell carcinoma. Additional sections from the tumor illustrated in panel **a** reveal small cysts lined by flattened epithelial cells embedded in altered stroma. (**d**) Tubulocystic clear cell carcinoma at high power. Although most of the tumor cells are flattened and have small, dark nuclei, a rare tumor cell has an enlarged, vesicular nucleus with a nucleolus

Table 1.1 High grade serous carcinoma (Chap. 6)

A.	A common tumor
B.	Can also be diagnosed as Grade 2 or Grade 3 serous carcinoma using Shimizu-Silverberg grading scheme. Whether distinguishing these two sub-grades is clinically relevant is controversial
C.	Wide spectrum of architectural patterns, including solid, glandular, and cribriform patterns, and patterns resembling transitional cell carcinoma. At least focal papillae and micropapillae with gaping and slit-like architectural features are present
D.	Several histologic variants have been described. One is transitional cell carcinoma and another is serous carcinoma with microcystic features
E.	High nuclear grade, with extreme nuclear size variability (>5X)
F.	More than 10 mitotic figures per 10 high power fields
G.	Typically disseminated at presentation. WT1 expression should be sought for Stage I tumors
H.	WT1, p53, and/or p16 overexpresssion may be sought if the differential diagnosis includes low grade serous carcinoma, endometrioid carcinoma, or clear cell carcinoma
I.	Is distinguished from serous borderline tumor by the presence of high nuclear grade if obvious stromal invasion is not identified after examination of multiple sections

Table 1.2 Low grade serous carcinoma (Chap. 6)

A. A less common tumor than high grade serous carcinoma

B. "Low grade micropapillary serous carcinoma" is a synonym

C. Papillary, micropapillary, and cribriform patterns are typical. Intraluminal mucin frequently present

D. Frequent association with serous borderline tumor (serous tumor of low malignant potential)

E. Distinguished from serous borderline tumor by destructive stromal invasion exceeding 5 mm in any dimension in the ovary

F. Distinguished from high grade serous carcinoma by the presence of uniform nuclei and a mitotic rate that does not exceed 10 mitotic figures per 10 high power fields. Atypical mitotic forms are not seen. Nucleoli may be present

G. Noninvasive variant typified by well-developed micropapillary or cribriform architecture exceeding 5 mm in any dimension. This may be referred to as "serous borderline tumor (or serous tumor of low malignant potential) with micropapillary (or cribriform) features" or "noninvasive low grade micropapillary (or cribriform) serous carcinoma." Since this tumor is frequently associated with ovarian or peritoneal invasion, (i.e., invasive low grade serous carcinoma), every effort should be made (by meticulous sampling and staging) to exclude the presence of invasion

H. Metastatic low grade serous carcinoma, or synchronous or metachronous low grade serous carcinoma (i.e., extraovarian invasive implants) is distinguished from desmoplastic noninvasive implants of serous borderline tumor by epithelial predominant deposits lacking granulation tissue-like stroma, preferably infiltrating normal connective tissue including fibrous or adipose tissue or smooth muscle

I. The following issues are unsettled:

1. Whether invasive implants in peritoneum or lymph nodes should mandate diagnosis as "low grade serous carcinoma" when the ovarian tumor is borderline
2. Whether noninvasive peritoneal implants in the setting of an invasive low grade ovarian serous carcinoma has the same clinical significance as invasive peritoneal disease

Table 1.3 Low grade endometrioid carcinoma (Chap. 7)

A. A common tumor, particularly among low stage tumors

B. Can also be diagnosed as Grade 1 or Grade 2 endometrioid carcinoma using Shimizu-Silverberg grading scheme

C. Cribriform, glandular, and papillary architecture predominates. Foci of solid architecture lacking overt squamous differentiation (< 50% of tumor) are allowed

D. At least one confirmatory endometrioid feature is almost always present (see Table 1.4)

E. Nuclei show mild or moderate atypia with size variations that do not exceed 5X. Small nucleoli are allowed. Mitotic rate does not usually exceed 10 per 10 high power fields

F. Bilateral presentation and/or extrapelvic spread should be met with suspicion

G. High grade serous carcinoma is excluded, specifically, either by careful attention to morphologic features (see Table 1.1) or with WT1, p53, and p16 immunohistochemical stains

H. Metastatic endometrial carcinoma is excluded

I. Mucinous carcinoma and clear cell carcinoma are excluded. ER and PR stains may be used for this purpose, as mucinous and clear cell carcinomas are negative or only very weakly positive

J. Metastatic colorectal carcinoma is excluded

K. Distinguished from endometrioid borderline tumor by the presence of complex papillary or glandular architecture (i.e., resembling well-differentiated endometrioid adenocarcinoma in the endometrium, also referred to as expansile invasion) or destructive stromal invasion (resembling myometrial invasion) measuring more than 5 mm in any dimension. More evidence is needed that the expansile pattern has a significantly worse clinical course than tumors uniformly accepted as borderline

Table 1.4 Confirmatory endometrioid features (Chap. 7)

- Differentiated endometrioid neoplasm of endometrium (atypical hyperplasia and well-moderately differentiated endometrioid adenocarcinoma)
- Endometriosis
- Endometrioid adenofibroma or borderline tumor
- Squamous or morular metaplasia
- Mucinous metaplasia
- Secretory change

Note: Gland formation and cribriform structures alone are insufficient support for an endometrioid neoplasm

Table 1.5 High grade endometrioid carcinoma (Chap. 7)

A.	A rare tumor
B.	Typical examples show destructive stromal invasion by histologically high grade tumor with solid architecture that arises in the setting of a low grade endometrioid adenocarcinoma or adenofibroma (see Table 1.3)
C.	Bilateral presentation and/or extrapelvic spread should be met with suspicion
D.	High grade serous carcinoma is excluded, specifically, either by careful attention to morphologic features (see Table 1.1) or with a WT1 immunohistochemical stain
E.	At least one confirmatory endometrioid feature is present: squamous, morular, mucinous, or tubal metaplasia, secretory change, endometrioid adenofibroma, endometriosis, endometrial adenocarcinoma
F.	Nuclei show moderate or severe atypia with size variations that exceed 5X. Nucleoli are present and are sometimes large. Mitotic rates exceed 10 per 10 high power fields.
G.	Metastatic endometrial carcinoma is excluded
H.	Mucinous carcinoma and clear cell carcinoma are excluded.
I.	Metastatic colorectal carcinoma is excluded

Table 1.6 Clear cell carcinoma (Chap. 8)

A.	Architectural, cytoplasmic, and nuclear features *together* should be used for diagnosis; clear cytoplasm by itself is *insufficient* for diagnosis (and sometime it is lacking) see Table 1.7
B.	Most tumors have papillary patterns that are present at least focally. Papillary patterns have round, non-hierarchically branched papillae lined by only one or two layers of cells. Stromal hyalin is sometimes prominent
C.	Tumor cells are typically cuboidal, not high columnar. Flattened and low columnar cells may occasionally be seen
D.	Cytoplasm may be clear or eosinophilic (oxyphilic)
E.	Nuclei in papillary tumors are round, tend to be uniform in size, and may have prominent nucleoli. Occasional, scattered larger nuclei are present
F.	Tubulocystic tumors have flattened or cuboidal cells, with small nuclei and/or nuclei that are typical of the papillary tumors (see above)
G.	Tumors with solid architecture are composed of cuboidal cells arranged in a sheet resembling cobblestones. Nuclei may be round and regular or resemble koilocytes
H.	Mitotic rates exceeding 8–10 mitotic figures per 10 high power fields are exceptional. When present, tumors are solid, high stage, or recurrent and high grade serous carcinoma is excluded.
I.	Serous and endometrioid tumors are specifically excluded (see Table 1.1–1.4). Immunohistochemistry, using WT1 and ER/PR, may be necessary
J.	Distinguished from clear cell adenofibroma and clear cell borderline tumor by either papillary architecture or destructive stromal invasion
K.	Tumors may be graded using the Shimizu-Silverberg scheme, although clinical relevance is uncertain

Table 1.7 Clear cell carcinoma morphology (Chap. 8)

- Polygonal or cuboidal cells; NOT columnar
- Preserved N:C ratios; NOT increased
- Low mitotic index (mean=5 MF/10 HPF)
- Papillary architecture
  - Round papillae lacking hierarchical branching, stratification, and tufting
- Tubulocystic and solid architecture may also be present
- Round nuclei with a prominent nucleolus or wrinkled nuclei
  - Only occasional pleomorphic cells
- Tumor cells may be clear or oxyphilic

Table 1.8 Intestinal (or gastrointestinal) mucinous carcinoma (Chap. 9)

A.	A rare tumor
B.	Primary ovarian adenocarcinoma with intracytoplasmic mucin demonstrable by H and E staining and may be confirmed with mucin stains
C.	Bilateral presentation, tumor size less than 13 cm, and/or extrapelvic spread should be met with suspicion
D.	Extensive mucin dissection, hilar invasion, extensive lymphovascular invasion, nodular growth pattern, and signet ring cells suggest metastatic mucinous carcinoma and should also be met with suspicion
E.	Metastatic mucinous carcinoma is excluded
F.	Adenocarcinoma arising in teratoma should be specifically noted
G.	Endometrioid adenocarcinoma is excluded (see Table 1.3)
H.	Mucinous borderline tumor (tumor of low malignant potential) is frequently present (but patterns resembling this may also be seen in metastatic adenocarcinomas)
I.	Mucinous carcinomas may be graded using the Shimizu-Silverberg grading scheme. Grade 3 examples should be met with suspicion. Some represent metastatic adenocarcinoma
J.	Distinguished from intestinal mucinous borderline tumor by the presence of complex papillary or glandular architecture (i.e., resembling well-differentiated endometrioid adenocarcinoma in the endometrium, also referred to as expansile invasion) or destructive stromal invasion (resembling myometrial invasion) measuring more than 5 mm in any dimension. More evidence is needed that the expansile pattern has a significantly worse clinical course than tumors uniformly accepted as borderline

Table 1.9 Malignant Brenner tumor (Chap. 10)

A.	A rare tumor
B.	An invasive carcinoma that arises in the setting of benign or borderline Brenner tumor
C.	The invasive component usually resembles invasive squamous or urothelial carcinoma and measures more than 5 mm in any dimension (tumors with smaller invasive components are referred to as "Brenner borderline tumor [with or without intraepithelial carcinoma] with microinvasion")
D.	The immunophenotype is usually similar to that of urothelial carcinoma, not serous or ovarian "transitional cell carcinoma"
E.	Grading is unresolved

Table 1.10 Undifferentiated carcinoma (Chap. 10)

A.	A rare tumor
B.	Does not show ANY features of a differentiated carcinoma, as defined in Tables 1.1–1.9 above. This category specifically excludes serous carcinomas (and those with transitional carcinoma-like features) with solid architecture. Neuroendocrine carcinomas and mural nodules within mucinous cystic tumors are to be classified separately
C.	The only two tumor types remaining after exclusion of the above entities are highly anaplastic tumors, with or without solid architecture, and monomorphic tumors composed of epithelial cells resembling lymphocytes or plasma cells
D.	Immunohistochemistry may be needed to ensure that a tumor is not a solid variant of a differentiated ovarian carcinoma (i.e., solid serous carcinoma or high grade endometrioid carcinoma). It is also recommended to exclude undifferentiated sarcoma, melanoma, neuroendocrine carcinoma, hypercalcemic small cell carcinoma, and lymphoma. EMA and/or CK18 expression is sufficient support for epithelial differentiation. WT1 positivity should prompt consideration for a diagnosis of high grade serous carcinoma with a solid growth pattern
E.	Tumors are, by definition, high grade

Table 1.11 Mixed epithelial carcinomas (Chap. 10)

A.	Rather uncommon tumors
B.	Most ovarian carcinomas are morphologically heterogeneous, so criteria for establishing the presence of a mixed epithelial carcinoma should be stringent
C.	The minor component must be starkly different in appearance from the major component and immunohistochemistry may be used to highlight two separate tumor types
D.	Gradual transitions between areas, especially when the areas are only subtly different, should prompt reconsideration of a mixed epithelial carcinoma
E.	The minor component should represent at least 10% of the tumor, but it is recognized that low grade tumors that harbor even small foci of high grade tumors should be handled very cautiously

Common Diagnostic Discrepancies

2

Robert A. Soslow

Keywords
Ovarian cancer • Quality assurance • Diagnostic discrepancy • Metastatic carcinoma • Primary ovarian carcinoma

This chapter is more an index of entities that feature as diagnostic discrepancies than it is a traditional textbook chapter. References are provided to the chapter in which each entity is discussed. The most common types of discrepancies are found in Tables 2.1–2.4, whereas less common ones are found in the subsequent tables (Tables 2.5–2.14). As an example, mistaking metastatic colorectal carcinoma for primary endometrioid carcinoma is rather common (Table 2.2), whereas mistaking heterologous Sertoli-Leydig cell tumor for immature teratoma (Table 2.12) is much less common.

Published data regarding diagnostic discrepancies in gynecologic pathology are limited. Most studies distinguish between major and minor discrepancies. Major discrepancies are mostly defined as those that affect patient care, whereas minor discrepancies do not. Overall discrepancy rates range from 7% to 17% [1–5], with minor discrepancies accounting for between 2% and 20% of all cases and major discrepancies ranging from 5% to 8%. Some studies have looked at discrepancies uncovered only during tumor board meetings, whereas others have looked only at consultation cases. Cytologic diagnostic discrepancies were captured in some studies, but not all. Nevertheless, even a 7% discrepancy rate is significant and troubling and requires attention.

Most diagnostic errors described in the literature involve uterine, not ovarian, pathology. An informal study of gynecologic quality assurance records at Memorial Sloan-Kettering Cancer Center in New York, NY revealed that only one of the five most common types of major diagnostic discrepancies involved ovarian pathology. The most common type of major diagnostic discrepancy in ovarian pathology was mistaking metastatic carcinoma to ovary as primary ovarian carcinoma or borderline tumor. This situation contrasts with that of discrepancies encountered at the time of frozen section (Chap. 3), where problems involving ovarian pathology were relatively more common.

Reported major diagnostic discrepancies involving ovarian pathology are not well described in the literature; they constitute between 14% and 21% of major discrepancies in gynecologic specimens [1, 3, 5]. The most common type of error described involves "overcalling" ovarian carcinoma at the expense of failing to diagnose ovarian borderline tumor. Other rare reported errors involve undercalling ovarian carcinoma (in favor of borderline tumor) and mistaking adult granulosa cell tumor for carcinoma. Of interest is the observation that none of the studies reviewed highlighted the problem posed by distinguishing primary ovarian neoplasms and metastasis. Most of these studies reviewed cases diagnosed more than 10 years ago, a period when these types of diagnostic problems were not scrutinized.

Arguably the most clinically important diagnostic problems involve distinctions between primary and metastatic tumors (Chap. 11). Of these, the most common problem concerns tumors thought to represent primary ovarian mucinous carcinomas that are instead metastatic carcinomas (and appendiceal mucinous tumors) secondarily involving ovary. A related problem, and one that is nearly as frequently encountered, is a tumor with an endometrioid appearance that proves to be metastatic colorectal carcinoma. Pathologists are also expected to distinguish synchronous tumors of ovary and endometrium and primary tumors of either organ with metastasis to the other.

R.A. Soslow
Department of Pathology, Memorial Sloan-Kettering Cancer Center, 1275 York Avenue, New York, NY 10065, USA

R.A. Soslow and C. Tornos (eds.), *Diagnostic Pathology of Ovarian Tumors*,
DOI: 10.1007/978-1-4419-9751-7_2,

Table 2.1 Mistaken diagnosis: primary mucinous carcinoma

Correct diagnoses:
Major discrepancies:
Metastatic mucinous carcinoma and appendiceal mucinous neoplasm (Chap. 11)
Metastatic endocervical adenocarcinoma (Chap. 11)
Discrepancy of equivocal nature:
Mucinous borderline tumor[a] (Chap. 9)

[a]Discrepancies involving grade I, stage IA primary mucinous carcinoma and mucinous borderline tumor are minor because the expected treatment and survival are similar. Discrepancies involving other grades and stages are major

Table 2.2 Mistaken diagnosis: primary endometrioid carcinoma

Correct diagnoses:
Major discrepancies:
Metastatic colorectal adenocarcinoma (Chap. 11)
Metastatic endocervical adenocarcinoma (Chap. 11)
Synchronous or metastatic endometrioid carcinoma (Chap. 7)
Female adnexal tumor of probable Wolffian origin (FATWO) (Chap. 16)
Primitive neuroectodermal tumor (central type) (Chap. 13)
Endometriosis, including atypical endometriosis (can also be mistaken for adenocarcinoma in a peritoneal washing) (Chap. 7)
Benign Müllerian inclusions in lymph node (i.e., endosalpingiosis) (Chap. 6)
Adult granulosa cell tumor (Chap. 14)
Well-differentiated Sertoli-Leydig cell tumor (Chap. 14)
Discrepancies of equivocal nature:
Endometrioid borderline tumor[a] (Chap. 7)
High-grade serous carcinoma[a] (Chap. 6)

[a]Discrepancies involving grade I, stage IA primary endometrioid carcinoma and endometrioid borderline tumor are minor because the expected treatment and survival are similar. Discrepancies involving other grades and stages are major. Discrepancies involving high-grade serous carcinoma and high-grade endometrioid carcinoma are minor, although diagnosing the latter may lead to omission of surveillance for *BRCA1*- and 2-associated carcinomas. Discrepancies involving low-grade endometrioid carcinoma and high-grade serous carcinoma are major

The significance of this type of diagnostic error ranges from almost inconsequential to major. A metastatic high-grade serous carcinoma, clinically thought to represent an ovarian primary and treated with neoadjuvant chemotherapy, is sometimes found to be endometrial in origin after interval debulking and hysterectomy. This type of discrepancy is minor. However, the distinction between synchronous endometrioid carcinomas of endometrium and ovary versus metastatic endometrial carcinoma to ovary remains consequential.

Discrepancies that involve benign and malignant tumors are uncommon in the surface epithelial category (although they remain extremely clinically important), while they are slightly more commonly encountered in the sex cord stromal category. An example would be mistaking a cystic granulosa cell tumor for a large follicle cyst. Disagreements concerning borderline tumors (low malignant potential/atypical proliferative)

Table 2.3 Mistaken diagnosis: high-grade serous carcinoma

Correct diagnoses:
Major discrepancies:
Low-grade serous carcinoma and serous borderline tumor (Chap. 6)
Endometriosis, including atypical endometriosis (can also be mistaken for adenocarcinoma in a peritoneal washing) (Chap. 7)
Endosalpingiosis, including in lymph node (can also be mistaken for adenocarcinoma in a peritoneal washing) (Chap. 6)
Desmoplastic small round cell tumor (Chap. 16)
Adult granulosa cell tumor (Chap. 14)
Mesothelioma (Chap. 17)
Discrepancies of equivocal nature:
Metastatic endometrial serous carcinoma[a] (Chap. 6)
Clear cell carcinoma[a] (Chap. 8)
Minor discrepancy:
Malignant mixed mesodermal tumor (carcinosarcoma) (Chap. 10)

[a]The distinction of endometrial and ovarian primary is clinically relevant because treatment protocols differ depending on primary site in some centers. Discrepancies involving high-grade serous carcinoma and clear cell carcinoma are minor when the tumor is low stage, although diagnosing the latter may lead to omission of surveillance for *BRCA1*- and 2-associated carcinomas. For high-stage tumors, there are data suggesting that clear cell carcinomas are relatively chemoresistant and cannot be cured at present

Table 2.4 Mistaken diagnosis: mixed epithelial carcinoma

Correct diagnosis:
Minor discrepancy:
High-grade serous carcinoma (Chap. 6)

Table 2.5 Mistaken diagnosis: serous borderline tumor

Correct diagnoses:
Major discrepancies:
Mesothelioma (Chap. 17)
Clear cell carcinoma (Chap. 8)
High-grade serous carcinoma (Chap. 6)
Discrepancies of equivocal nature:
Endometriosis, including atypical endometriosis (can also be mistaken for adenocarcinoma in a peritoneal washing)[a] (Chap. 7)
Endosalpingiosis, including in lymph node (can also be mistaken for adenocarcinoma in a peritoneal washing)[a] (Chap. 6)
Endometrioid borderline tumor[a] (Chap. 7)
Serous cystadenofibroma[a](Chap. 6)
Minor discrepancy:
Seromucinous borderline tumor (Chap. 9)

[a]Mistaking endometriosis or endosalpingiosis for a noninvasive implant of serous borderline tumor is of indeterminate significance because the prognosis is different, but neither diagnosis mandates chemotherapy. However, when the discrepancy involves endometriosis or endosalpingiosis and invasive implant, the magnitude is major. Endometrioid borderline tumor versus unstaged serous borderline tumor is a discrepancy of equivocal significance because the prognosis is theoretically different. In centers where staging surgery is performed solely in the setting of serous borderline tumor (and not for other types of borderline tumor), the discrepancy is major. Clinical outcomes for stage I serous cystadenofibroma and borderline tumor are similar, although when the borderline tumor is unstaged or has spread beyond the confines of the ovary, the discrepancy is at least of indeterminate significance

Table 2.6 Mistaken diagnosis: low-grade serous carcinoma

Correct diagnoses:
Major discrepancies:
Endometriosis, including atypical endometriosis (can also be mistaken for adenocarcinoma in a peritoneal washing) (Chap. 7)
Endosalpingiosis, including in lymph node (can also be mistaken for adenocarcinoma in a peritoneal washing) (Chap. 6)
Mesothelioma (Chap. 17)
High-grade serous carcinoma (Chap. 6)
Desmoplastic noninvasive implant (can be mistaken for invasive implant of serous borderline tumor/low-grade serous carcinoma)
Clear cell carcinoma (Chap. 8)
Discrepancy of equivocal nature:
Serous borderline tumor, including in lymph node[a] (Chap. 6)

[a]Some pathologists do not consider invasive implants of serous borderline tumor and metastatic, low-grade serous carcinoma to represent different entities. However, the prognosis and therapy for metastatic low-grade serous carcinoma versus borderline tumor with or without *non*invasive implants differ. These are major discrepancies. Borderline tumor and low-grade serous carcinoma proven to be confined to the ovary have roughly similar outcomes and are treated similarly, provided the ovarian surface is uninvolved and the peritoneal washing is benign. This discrepancy is therefore minor.

Table 2.7 Mistaken diagnosis: metastatic gynecologic carcinoma (both in surgical and cytologic specimens)

Correct diagnoses:
Major discrepancies:
Endometriosis, including atypical endometriosis (Chap. 7)
Implant of serous borderline tumor (Chap. 6)
Endosalpingiosis, including in lymph node (Chap. 6)
Serous borderline tumor in lymph node (Chap. 6)
Synchronous tumors (Chap. 7)
Desmoplastic small round cell tumor (Chap. 16)
Mesothelioma (Chap. 17)
High-grade serous carcinoma (Chap. 6)

Table 2.8 Mistaken diagnosis: clear cell carcinoma

Correct diagnoses:
Major discrepancies:
Low-grade serous carcinoma and serous borderline tumor (Chap.6)
Mesothelioma (Chap. 17)
Endometrioid carcinoma (Chap. 7)
Endometriosis, including atypical endometriosis (can also be mistaken for adenocarcinoma in a peritoneal washing) (Chap. 7)
Endosalpingiosis, including in lymph node (can also be mistaken for adenocarcinoma in a peritoneal washing) (Chap. 6)
Metastatic renal cell carcinoma (Chap. 11)
Seromucinous borderline tumor (Chap. 9)
Discrepancy of equivocal nature:
High-grade serous carcinoma[a] (Chap. 6)

[a]Discrepancies involving high-grade serous carcinoma and clear cell carcinoma are minor when the tumor is low stage, although diagnosing the latter may lead to omission of surveillance for *BRCA-1* and *2*-associated carcinomas. For high stage tumors, there are data suggesting that clear cell carcinomas are relatively chemoresistant and cannot be cured at present.

Table 2.9 Mistaken diagnosis: undifferentiated carcinoma

Correct diagnoses:
Major discrepancies:
Lymphoma (Chap. 15)
Primitive neuroectodermal tumor (Chap. 13)
Small cell carcinoma (Chap. 10)
Rhabdomyosarcoma (Chap. 15)
Dysgerminoma (Chap. 13)
Adult granulosa cell tumor (Chap. 14)
Undifferentiated sarcoma (Chap. 15)
Metastatic melanoma (Chap. 11)
Unclassified sex cord stromal tumor (Chap. 14)
Metastatic poorly differentiated ductal carcinoma of breast (Chap. 11)
Minor discrepancy:
High-grade serous carcinoma (Chap. 6)

Table 2.10 Mistaken diagnosis: metastatic breast carcinoma

Correct diagnoses:
Major discrepancies:
High-grade serous carcinoma (Chap. 6)
Hilus and theca cells

Table 2.11 Mistaken diagnosis: malignant mixed mesodermal tumor (carcinosarcoma)

Correct diagnoses:
Major discrepancies:
Endometrioid carcinoma with spindle cell features (Chap. 7)
Heterologous and/or poorly differentiated Sertoli-Leydig cell tumor (Chap. 14)
Immature teratoma (Chap. 13)
Discrepancies of equivocal nature:
De-differentiated endometrioid carcinoma (i.e., combined endometrioid and undifferentiated carcinoma)[a] (Chap. 7)
Combined adenocarcinoma and neuroendocrine carcinoma, including small cell carcinoma[a] (Chap. 10)

[a]Although these tumors have different histologic appearances and arise via different pathogenetic pathways, their outcomes are broadly similar and are currently treated more or less the same way.

Table 2.12 Mistaken diagnosis: immature teratoma

Correct diagnoses:
Major discrepancies:
Heterologous Sertoli-Leydig cell tumor (Chap. 14)
Gliomatosis peritonei (Chap. 13)
Malignant mixed mesodermal tumor (carcinosarcoma) (Chap. 10)

Table 2.13 Mistaken diagnosis: mullerian adenosarcoma

Correct diagnoses:
Major discrepancies:
Polypoid endometriosis (Chap. 7)
Malignant mixed mesodermal tumor (carcinosarcoma) (Chap. 10)

Table 2.14 Mistaken diagnosis: juvenile granulosa cell tumor

Correct diagnoses:
Major discrepancies:
Small cell carcinoma of hypercalcemic type (Chap. 16)
Metastatic melanoma (Chap. 11)
Germ cell tumor (Chap. 13)
Desmoplastic small round cell tumor Chap. 16

and carcinomas can be very important, although in some specific cases, separation of these entities is relatively clinically unimportant. When the differential diagnosis involves borderline tumor and a surgically staged, stage IA, well-differentiated carcinoma, the distinction has little relevance. However, when the discrepancy concerns carcinoma versus stage III borderline tumor or a borderline tumor with intraepithelial carcinoma, for example, there are important clinical implications.

Other major diagnostic discrepancies, thankfully, are less common, but these problems remain crucial as they involve clinically distinct entities. Mistaking a surface epithelial tumor for a sex cord stromal tumor or germ cell tumor falls into this category. Even less common but equally substantial are mistakes that involve very rare entities such as desmoplastic small round cell tumor, small cell hypercalcemic carcinoma, mesothelioma, ependymoma, lymphoma, melanoma and primitive neuroectodermal tumor.

The last type of commonly encountered discrepancy involves the distinction of ovarian carcinoma type. With only several exceptions, none of these mistakes has the same profound significance as mistaking a lymphoma for a carcinoma, for example. Diagnoses mandating treatment with chemotherapy should be rigorously separated from those that are managed conservatively, however. The separation of stage IA and IB carcinomas from IC carcinomas (whether the ovarian surface is involved, for example) is important as is the distinction of low- and high-grade organ-confined ovarian carcinoma (stages IA and IB). Despite the fact that other distinctions are not presently highly clinically relevant, there is a compelling rationale to rigorously and reliably subclassify and grade ovarian carcinomas accurately, as summarized in Chap. 4.

References

1. Santoso JT, Coleman RL, Voet RL, et al. Pathology slide review in gynecologic oncology. Obstet Gynecol. 1998;91(5):730–4.
2. Chan YM, Cheung AN, Cheng DK, et al. Pathology slide review in gynecologic oncology: routine or selective? Gynecol Oncol. 1999;75(2):267–71.
3. Selman AE, Niemann TH, Fowler JM, Copeland LJ. Quality assurance of second opinion pathology in gynecologic oncology. Obstet Gynecol. 1999;94(2):302–6.
4. Santoso JT, Schwertner B, Coleman RL, Hannigan EV. Tumor board in gynecologic oncology. Int J Gynecol Cancer. 2004;14(2):206–9.
5. Khalifa MA, Dodge J, Covens A, Osborne R, Ackerman I. Slide review in gynecologic oncology ensures completeness of reporting and diagnostic accuracy. Gynecol Oncol. 2003;90(2):425–30.

Frozen Section of Ovarian Lesions

3

Carmen Tornos and Robert A. Soslow

Keywords
Frozen section • Intraoperative • Ovarian cancer • Quality assurance • Telepathology

Indications

Frozen sections (FS) of ovarian lesions are only indicated if the surgeon plans to do immediate additional surgery, or in rare occasions if there is a need to confirm the diagnosis before starting emergency therapy. Inappropriate requests for frozen section include satisfying clinical curiosity or desiring solely to communicate results to the family. Unnecessary frozen sections should be discouraged since they are a waste of time and resources and a source of possible errors. Pathologists need to monitor frozen section requests to identify inappropriate utilization and provide opportunities for educating surgeons with the aim of improving patient care, avoiding misuse of fresh tissue, and encouraging rational expenditure of resources.

Most general surgeons and general gynecologists are not trained to perform definitive surgery for borderline or malignant ovarian tumors, such as formal staging and/or tumor debulking. Specialized surgery such as this is performed by gynecological oncologists, surgeons who have completed a residency in gynecology and obstetrics followed by a fellowship in gynecological oncology. Pathologists should engage nonspecialized surgeons in a dialogue when a frozen section is requested in order to promote optimal patient care and discourage unindicated frozen sections. Such a dialogue might result in cancelling the frozen section or consultation between the surgeon and a gynecologic oncologist.

R.A. Soslow (✉)
Department of Pathology, Memorial Sloan-Kettering Cancer Center, 1275 York Avenue, New York, NY 10065, USA

Preparation for Frozen Section

When called to do a FS of an ovarian lesion there are some steps that the pathologist must follow (Table 3.1):

1. Review the clinical history, previous laboratory results, and previous pathology reports (and slides if available).
2. Review current radiology reports, if available.
3. Examine the specimen grossly for evidence of external surface nodules, papillary growth, adhesions, etc. The entire surface should be inked.
4. Measure and weigh the specimen.
5. If the tumor is cystic, open it carefully to drain any contents. Examine the quality of the contents: watery fluid (serous), thick, sticky, gelatinous material (mucinous), fatty with hairs (teratoma), etc.
6. Empty and clean the contents of the cyst very thoroughly to find any intramural nodules, papillary excrescences, trabecular areas, or other mass lesions. This intraluminal growth may harbor areas that need to be sampled at the time of FS. Some endometrioid and clear cell carcinomas arising in endometriosis may appear as small mural

R.A. Soslow and C. Tornos (eds.), *Diagnostic Pathology of Ovarian Tumors*,
DOI: 10.1007/978-1-4419-9751-7_3,

Table 3.1 Required information when doing a frozen section of an ovarian lesion

- Patient's age
- Relevant clinical/family history
- Serum markers (AFP, HCG,CA125), hormonal levels (estrogens, androgens)
- Imaging studies
- Previous history of malignancy
- Previous pathology reports and slides

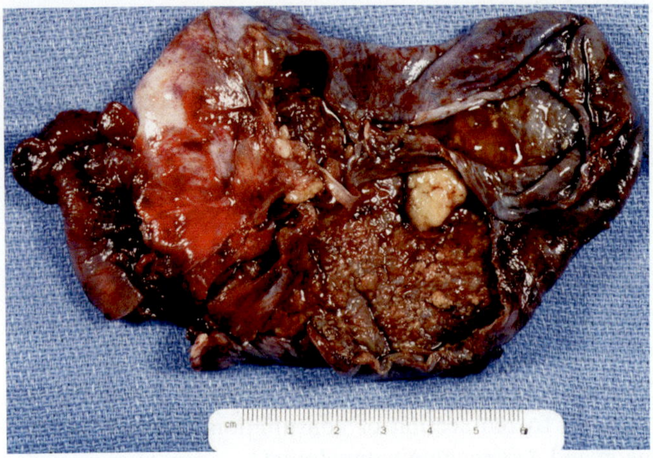

Fig. 3.1 Endometrioma with a small endometrioid carcinoma (white nodular mass on *center right*)

Table 3.2 Ovarian tumors with grossly papillary growth

- Cystadenofibroma
- Serous borderline tumor
- Seromucinous borderline tumor (also known as endocervical mucinous or müllerian mucinous tumor)
- Serous carcinoma

Table 3.3 Ovarian tumors involving the ovarian surface

- Serous cystadenofibroma
- Serous borderline tumor
- Serous carcinoma
- Clear cell carcinoma (rarely)
- Endometrioid carcinoma (usually metastasis from endometrium)
- Any metastatic tumor

Table 3.4 Gross appearance of primary malignant ovarian tumors

- Heterogeneous gross appearance
- Mixture of solid, cystic, with necrosis
- Clear cell carcinomas and endometrioid carcinomas can arise from benign cysts (endometriosis) and appear as small nodular areas in the cyst wall

Table 3.5 Gross appearance of ovarian metastases

- Bilateral
- Multiple nodules preserving normal ovarian parenchyma in between
- Involvement of ovarian surface

nodules within a large hemorrhagic cyst (endometrioma) (Fig. 3.1). Seromucinous borderline tumors may have a similar appearance. In mature cystic teratomas, the intracystic white nodular area (Rokitansky protuberance) is the most likely to harbor any malignant component, and the likelihood of malignant transformation increases with age.

It is obvious that the gross appearance of the lesion can be diagnostically helpful. A variety of gross features with a list of differential diagnosis appear in Tables 3.2–3.5 and Figs. 3.1–3.6. Ideally, the "best-" and "worst-" looking areas should be sampled from any given tumor regardless of the cell type, although the morphologically heterogeneous nature of mucinous tumors should encourage even more extensive sampling. Sampling a smooth lined cystic component helps in classification of cell type (i.e., mucinous, serous, endometrioid), while sampling a more solid or papillary area can determine the degree of malignancy (i.e., borderline tumor or carcinoma).

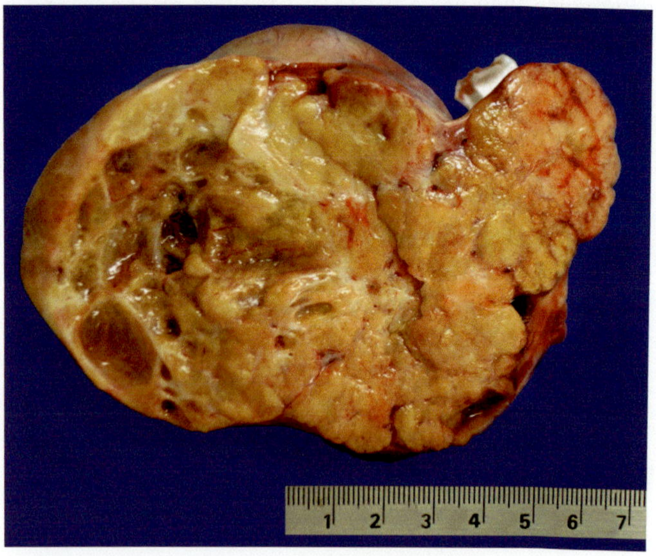

Fig. 3.2 Ovarian carcinoid tumor with classic yellow appearance

The ultimate FS diagnosis should be based on a combination of careful gross examination, microscopic evaluation, and correlation with age, clinical history, radiologic evaluation, and discussion with the surgeon. The surgeon may have valuable information that is unknown to the pathologist at the time of the FS. When in doubt, it is always useful to discuss the case with the surgical team performing the

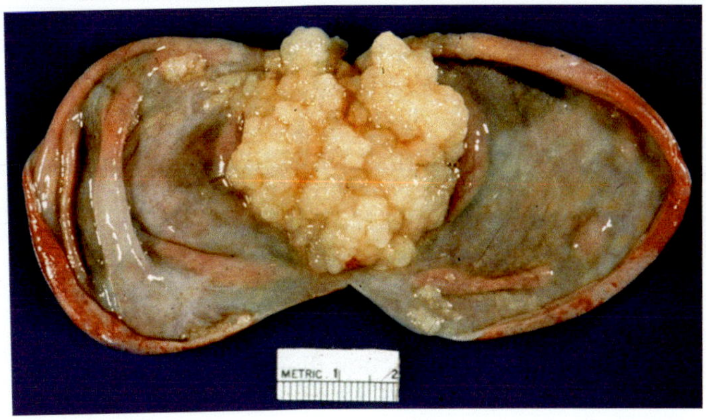

Fig. 3.3 Gross picture of a serous cystadenofibroma. The intracystic papillae are firm on palpation

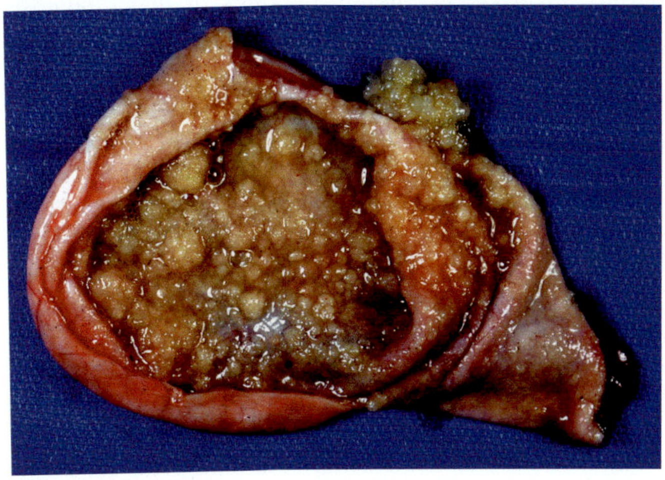

Fig. 3.4 Gross picture from a serous borderline ovarian tumor with classic papillary growth

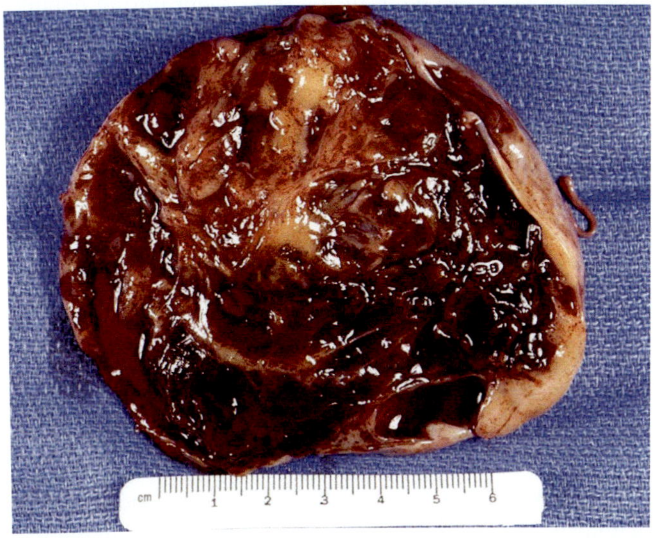

Fig. 3.5 Adult granulosa cell tumor showing a characteristic beefy and hemorrhagic gross appearance

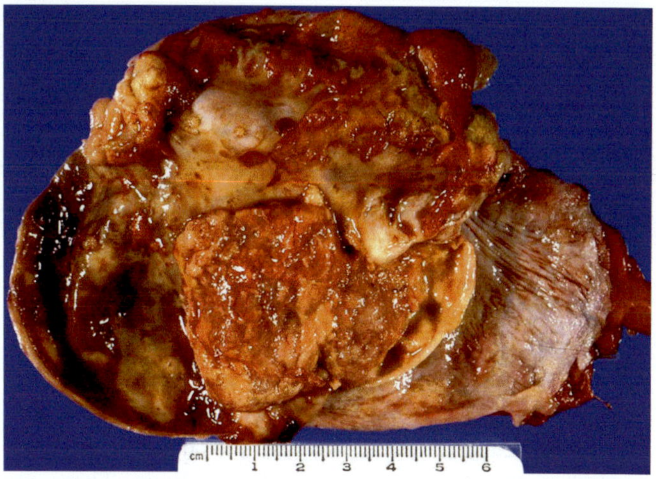

Fig. 3.6 High grade serous ovarian carcinoma with solid, cystic, and necrotic areas

Table 3.6 Malignant ovarian tumors in young girls/women (mean age in years)

- Germ cell tumors (18 years)
- Sex-cord stromal tumors
  - Juvenile granulosa cell tumor (13 years)
  - Retiform Sertoli-Leydig cell tumor (17 years)
  - Sertoli-Leydig cell tumor NOS (28 years)
- Small cell carcinoma hypercalcemic type (23 years)
- Desmoplastic small round cell tumor of peritoneum (19 years)
- Metastases from tumors seen in children/young adults (i.e., sarcoma)
- *Epithelial tumors are rare (most are borderline, not malignant)*

operation. The surgeons may be able to provide relevant clinical history, imaging results, or even a description of tumor dissemination as visualized intraoperatively. Experienced gynecological oncologists can often distinguish between primary and metastatic tumors by the way the tumor has metastasized.

The age of the patient is very important. There are tumors typically found in young patients that are very rare in older women. A list of these tumors and the mean age at presentation are found in Table 3.6. On the other hand carcinomas are very rare in young patients. Before rendering such a diagnosis in a patient 20 years or younger it is worth seeking another opinion, and consider the possibility of one of the diagnosis listed in Table 3.6 instead.

The quality of the FS slides is very important to render an accurate diagnosis. Pathologists working with inexperienced staff need to monitor this very carefully. If the frozen section is suboptimal, a new section needs to be cut, ideally by a senior staff member with more experience. Performing a touch prep or scraping the tumor in addition to the frozen section is often very useful. It allows the pathologist to better examine the cytologic features, and can be almost diagnostic in cases such as granulosa cell tumor, carcinoid tumor, Sertoli-Leydig cell tumor, and luteoma.

Information that the Surgeon Needs at the Time of Frozen Section

Specific information should be conveyed to the surgeon (Table 3.7) when the request for a frozen section is reasonable. If the lesion is benign, no further surgery is needed. Borderline ovarian tumors (tumors of low malignant potential/atypical proliferative tumors) are treated differently depending on the surgeon, the age of the patient, the type of borderline tumor, and co- morbid conditions. There are three theoretical reasons to perform additional surgery after a frozen section diagnosis of borderline tumor (BT): (1) debulk high stage disease, (2) discover invasive implants, and (3) anticipate an under-called frozen section diagnosis (i.e., expect a permanent section diagnosis of carcinoma) and perform a cancer-staging surgery. When the tumor involves the ovarian surface and/or grossly involves extraovarian structures, it is reasonable to remove all visible tumor (usually including omentectomy at least) and, depending on the surgeon's and patient's preferences, total abdominal hysterectomy and bilateral salpingo-oophorectomy (TAHBSO) and lymphadenectomy in addition. When the tumor appears confined to the ovary, the only reason to do additional surgery is anticipating an inaccurate frozen section diagnosis. To determine whether additional surgery is needed in this situation, it becomes important to understand the extent to which frozen section diagnoses are overturned on review of permanent sections. A recent review of the frozen section records at Memorial Sloan-Kettering Cancer Center in New York, NY reveals that under-calling carcinoma as borderline tumor on frozen section varies considerable by histologic type [1]. Five percent of tumors diagnosed as serous BTs, 0% of seromucinous BTs, 14% of intestinal mucinous BTs, and 80% of small numbers of endometrioid BTs were reclassified as carcinomas on permanent sections. Most under calls were due to sampling error, not interpretive error. Further study of the serous tumors revealed differences related to the presence of micropapillary architecture. Twenty-one percent of tumors diagnosed as micropapillary serous BTs were reclassified as carcinomas on permanent section compared to 1% of non-micropapillary serous BTs. Therefore, one can anticipate an accurate frozen section diagnosis of BT when the tumor is serous or seromucinous and non-micropapillary. This does not provide unmitigated support for performing staging surgeries for these types of borderline tumors when disease appears clinically confined to the ovary, without surface involvement. Significant rates of underdiagnosis can be anticipated with micropapillary serous, intestinal mucinous, and endometrioid tumors when diagnosed as BT on frozen section. This provides sufficient support for performing cancer-type staging surgeries at the discretion of the surgeon in these circumstances.

Table 3.7 Further surgical therapy depending on frozen section diagnosis

Benign: no further surgery
Borderline tumors:
• TAH-BSO in postmenopausal +/– staging[a]
• Unilateral SO in young patients +/– staging[a]
Ovarian carcinoma:
• TAH-BSO + debulking/staging
Malignant sex-cord stromal tumors and germ cell tumors:
• Unilateral SO in young patients + staging
• TAH-BSO + staging in older patients

TAH total abdominal hysterectomy, *BSO* bilateral salpingo-oophorectomy
[a]In general, the extent of staging surgeries for borderline tumors is less than for carcinomas. Staging may be omitted for certain types of borderline tumors, confirmed on permanent sections, that are clinically benign (i.e., endometrioid and intestinal mucinous borderline tumors) and when there are significant medical comorbidities

If the lesion is a primary malignant ovarian tumor, the operative approach depends chiefly on whether the tumor is a carcinoma, the tumor stage, patient age, and patient preferences about fertility preservation. When a carcinoma is diagnosed at frozen, reporting the type of carcinoma (serous, mucinous, endometrioid, clear cell, etc.) is much less clinically important than the stage. In many centers, surgeons attempt to debulk high stage ovarian carcinomas instead of using neoadjuvant chemotherapy. An intraperitoneal catheter, used to administer intra-peritoneal (IP) chemotherapy, will be placed during an optimal debulking surgery since both IP and intravenous chemotherapy is administered to these patients. Placement of the IP catheter during debulking surgery triggered by a FS diagnosis of high stage ovarian carcinoma spares the patient an additional surgical procedure.

If the tumor is a metastasis from an extra-gynecologic source, no further surgery is needed, but the surgeon may try to identify the primary source in collaboration with the pathologist, and in some cases attempt to debulk the tumor. For example, if the lesion is suggestive of metastatic colon cancer and the patient has no previous history, the surgeon may "run the bowel" to identify the primary tumor, and sometimes resect it to avoid further complications such as bowel obstruction. A diagnosis of primary malignant or borderline ovarian tumor should never be rendered if there are any doubts. The result would be extensive unnecessary surgery. In doubt it is best to be conservative and defer the diagnosis to review of permanent sections. Completion surgery and staging can be done at a later time if a primary ovarian malignancy is confirmed with the permanent sections.

Accuracy

The majority (56%) of ovarian lesions evaluated at frozen section in general hospitals are benign (Table 3.8). Most recent studies that include cases diagnosed after 1999 report accuracy rates of 95–97% [2–8]. The sensitivity and specificity vary with the type of diagnosis (benign, borderline, and malignant) as seen in Table 3.9. The majority of discrepancies are sampling errors seen in borderline tumors and mucinous tumors [2–8]. Other discrepancies can be attributed to misinterpretations by the pathologist reading the FS, or suboptimal quality of the frozen section. Some studies report lesser discrepancy rates when the FS is interpreted by a gynecological pathologist rather than a general surgical pathologist [4, 5, 9].

Most studies attribute the majority of discrepant diagnoses to problems posed by examining mucinous tumors [2–5, 7]. There are two principle reasons:

1. Mucinous ovarian tumors are morphologically heterogeneous. Such tumors can harbor areas with benign, borderline, and malignant appearances, such that limited sampling at the time of frozen section can provide misleading diagnostic information. It is therefore a good practice to submit frozen sections from two or three grossly different areas in a mucinous tumor. If clearly malignant areas are still not recognized, it is best to render a preliminary diagnosis of "mucinous neoplasm, final diagnosis deferred (to examination of multiple permanent sections)". The only exception is a unilateral and unilocular mucinous cyst with completely smooth internal and external surfaces. Such tumors are nearly always benign.
2. Metastatic mucinous tumors to the ovary often mimic primary benign, borderline, or malignant ovarian tumors (see Chap. 11). The most common metastatic mucinous tumors to the ovary are listed in Table 3.10. The differential diagnosis at the time of frozen section can be very difficult due to limited sampling and there are some features of primary and metastatic tumors that make this differential diagnosis even harder. Metastatic mucinous tumors to the ovary can display a "maturation phenomenon," in which the metastasis looks more differentiated than the primary tumor. In some cases, these metastatic carcinomas can have the deceptive deceiving appearance of a primary benign or borderline ovarian tumor (Fig. 3.7b). On the other hand, primary ovarian mucinous tumors can be relatively mucin-depleted and therefore resemble other primary or metastatic malignancies. In these cases the diagnosis of a mucinous carcinoma can be suspected by the presence of a benign or borderline mucinous component (Fig. 3.8a, b).

Table 3.8 Frozen section diagnosis of ovarian lesions

Diagnosis	% of cases
Benign tumor	56–60%
Malignant tumor	20–30%
Non-neoplastic	15–20%
Borderline tumor	6–10%

Table 3.9 Sensitivity and specificity of ovarian frozen section diagnosis by final diagnosis

Final diagnosis	Sensitivity%	Specificity%
Benign	97–100%	81–98%
Borderline tumor	57–99%	94–98%
Malignant	86–92%	97–100%

Table 3.10 Most common metastatic mucinous tumors to the ovary

- Colon
- Appendix
- Pancreas
- Gallbladder
- Cervix
- Bile ducts
- Small intestine
- Stomach
- Lung

Common Pitfalls, Major Discrepancies

Differential Diagnosis Between Primary and Metastatic Mucinous Ovarian Tumors on Frozen Section

As a general rule, any malignant bilateral mucinous tumor in the ovary is metastatic until proven otherwise [10–12]. It has been reported that approximately 1–2% of primary ovarian mucinous carcinomas are bilateral, but this estimate is based on studies performed before new pathologic criteria from the mid 1990s enabled us to discriminate more accurately between primary and metastatic mucinous carcinomas (see Chap. 9). Most primary ovarian carcinomas are unilateral and low stage.

Some studies report that unilateral mucinous tumors that measure less than 13 cm are most likely metastases, and most unilateral mucinous ovarian tumors larger than 13 cm are most likely primary [10–12]. This algorithm works well for smaller tumors, since most primary ovarian mucinous borderline and malignant tumors are 13 cm or larger, but is not as accurate for larger lesions, since a fair number of metastatic tumors are larger than 13 cm (Table 3.11). Using the 13-cm rule alone, 98% of primary ovarian mucinous carcinomas and 82% of metastases can be diagnosed correctly (Table 3.11). Tumors that do not comply with this rule include some large metastases (Table 3.11), and other primary tumor types with extensive mucinous metaplasia like

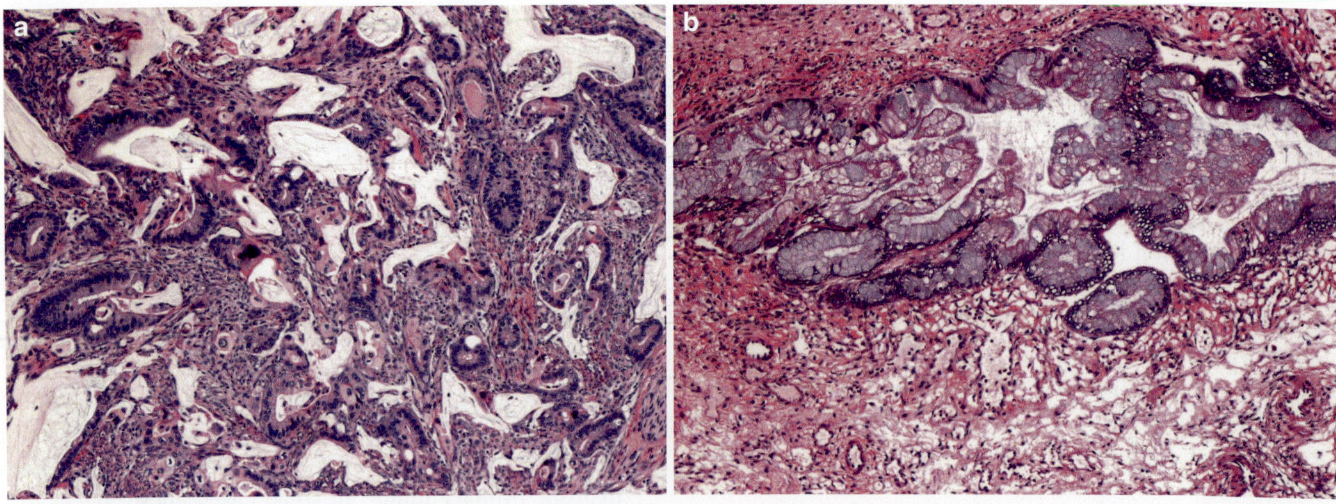

Fig. 3.7 (**a**) Colonic adenocarcinoma with mucinous features and focal marked cytologic atypia. (**b**) Ovarian metastasis from the same tumor with "maturation effect"

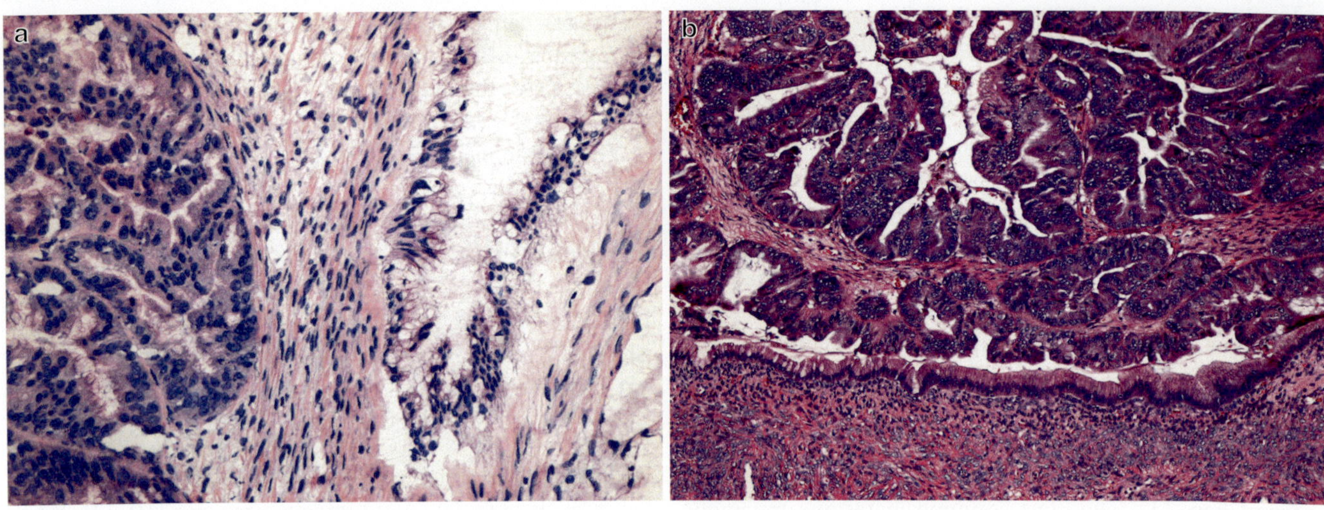

Fig. 3.8 (**a**) Frozen section of an ovarian mucinous carcinoma with mucin depletion (*left*) and a benign mucinous component (*right*). (**b**) Permanent section from the same area showing similar components, carcinoma (*top*), benign (*bottom*)

Table 3.11 Size and bilaterality of ovarian mucinous tumors (primary and metastatic)

	Range (cm)	Median (cm)	Bilateral (%)
Borderline	12–35	22	6% I, 40% E
Primary carcinoma	13–36	20	0 ?
Colorectal	2–45	13	67
Appendix	3–27	14	84
Pancreas	2–21	9	90
Small intestine	12–20	15	33
Gastric	6–15	9	40
Endocervix	2–30	12	35

I intestinal type of borderline tumor, *E* endocervical type of borderline tumor

mucin-rich endometrioid carcinoma and seromucinous borderline tumors that are often smaller than 13 cm. Other features useful for the differential diagnosis of primary and metastatic mucinous carcinoma at the time of FS are listed in Tables 3.12 and 3.13 and are discussed below (see also Table 11.1 and text in Chap. 11) [13].

Most primary ovarian mucinous carcinomas are low stage (i.e., FIGO stage I or II). The presence of a mucinous tumor

Table 3.12 Features that favor metastasis if the tumor is mucinous

- Bilaterality
- Size <13 cm
- Extensive intra-abdominal spread
- Ovarian surface involvement
- Hilar involvement
- Nodular invasive growth
- Abundant infiltrative pattern with desmoplastic stroma
- Extensive lymphovascular invasion
- Pseudomyxoma peritonei
- Extensive signet ring cell histology
- Colloid carcinoma pattern
- Small glands with bland cytology but invasive pattern

Table 3.13 Features that favor primary ovarian mucinous carcinoma

- Size >13 cm
- Smooth external surface
- Benign/adenofibroma/borderline areas
- Endometriosis
- Complex papillary architecture
- Multiple histologic patterns
- Association with teratoma, Brenner, Sertoli-Leydig cell tumor

Table 3.14 Entities associated with pseudomyxoma peritonei

- Mucinous lesions of the appendix (99%)
- Urachal mucinous carcinoma
- Mucinous adenocarcinoma of renal pelvis
- Intraductal papillary mucinous tumor of the pancreas
- Colonic polyps (tubulovillous adenoma)
- Mucinous ovarian tumors associated with mature teratomas

in the ovary associated with extensive intra-abdominal spread is suspicious for metastasis, as is the presence of mucinous ascites (i.e., *pseudomyxoma peritonei* = *PP*). PP is often secondary to a primary appendiceal tumor, including histologically benign-appearing tumors, although it rarely can be associated with other lesions listed in Table 3.14. Appendectomy should be recommended when PP is seen at the time of FS.

The presence of ovarian surface involvement by a mucinous tumor is also most consistent with metastatic disease. Ovarian surface involvement is often seen in primary ovarian serous tumors (borderline or malignant) but not in primary mucinous ovarian tumors. The surface involvement can on gross examination be seen as minute nodules, which microscopically, appear as "mushroom-like" deposits of tumor projecting above the ovarian surface. They are composed of sparse mucinous glands or clusters of tumor cells embedded in a desmoplastic or hyalinized stroma and associated with reactive mesothelium (Fig. 3.9). They are secondary to transcoelomic spread of the tumor, and are often seen in cases of metastatic colon carcinoma.

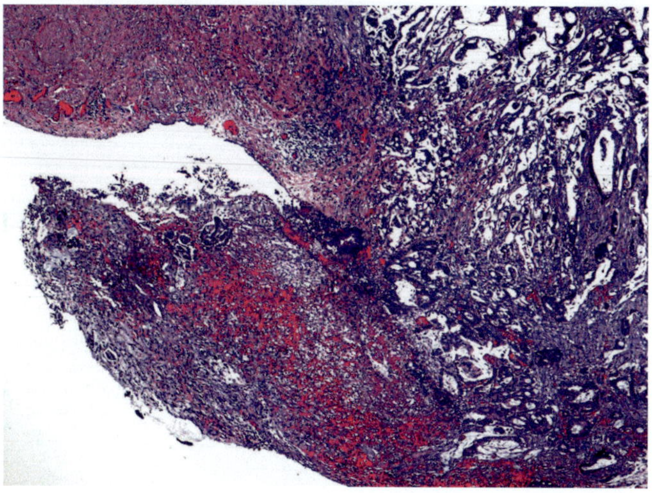

Fig. 3.9 Metastatic colonic carcinoma to ovary with an ovarian surface deposit

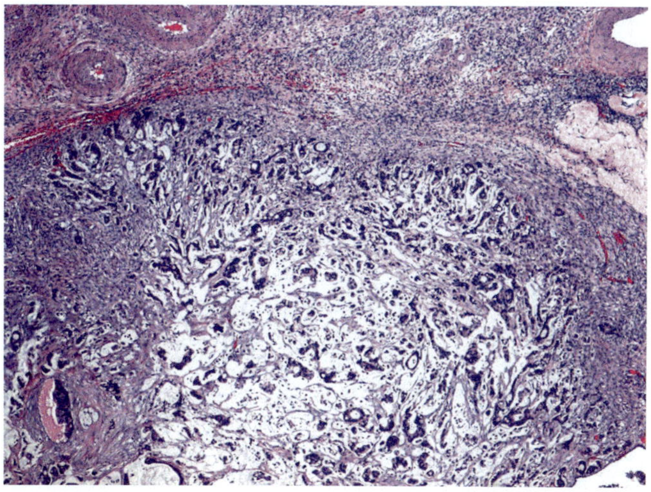

Fig. 3.10 Metastatic colon cancer with a nodular invasive growth pattern and preservation of normal structures (corpus albicans on the *right* and large hilar vessels at *top left*)

Microscopic features that support a diagnosis of a metastatic mucinous tumor (because of their rarity in primary lesions) include:

- Involvement of the ovarian hilum (Fig. 3.10)
- Nodular invasive growth pattern (Fig. 3.10)
- Extensive destructive stromal invasion with desmoplasia
- Preservation of normal structures (follicles, corpora albicans) with tumor infiltrating around them (Fig. 3.10)
- Extensive lymphovascular invasion
- Extensive pseudomyxoma ovarii (large pools of mucin dissecting the ovarian stroma, Fig. 3.11). This is often secondary to metastatic appendiceal tumor, and often associated with PP.
- Extensive colloid carcinoma morphology with nests of tumor cells swimming in pools of mucin (Fig. 3.12). This may be found in metastases from colorectal cancers.

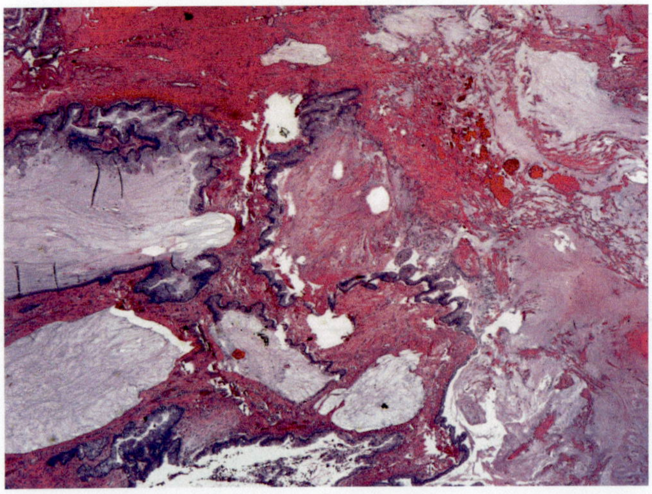

Fig. 3.11 Metastatic colon cancer with extensive pseudomyxoma ovarii (mucin dissection through ovarian stroma)

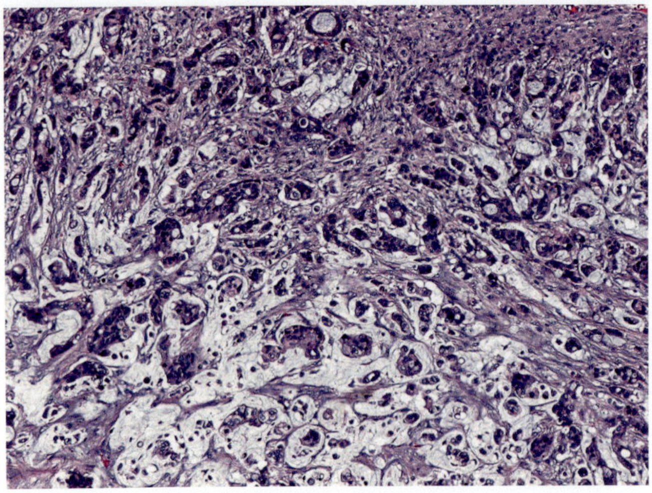

Fig. 3.12 Metastatic colon cancer with a "colloid carcinoma" pattern

- Raggedly infiltrating mucinous glands with low-grade cytology (Fig. 3.13a, b) are usually metastatic from pancreatico-biliary origin, or rarely endocervix.
- Signet ring cell morphology that sometimes can be associated with an extensive spindle cell component masking the presence of the mucinous cells (Fig. 3.14a, b), in Krukenberg tumors [14–16].

The majority of tumors with abundant signet ring cells are metastases from the stomach, although other primary sites may give rise to tumors with a similar appearance (listed in Table 3.15). Some primary ovarian tumors can have signet ring cells but they are usually a small component of the tumor (Table 3.16) [17, 18]. *Serous carcinomas* can have foci of signet ring cells that contain mucin, but they are always associated with more typical areas of serous carcinoma as seen in Fig 3.15a–c. The *signet ring cell stromal tumor* has an extensive component of signet-ring-like cells, but the cells do not contain mucin (Chap. 14). The tumor is always unilateral and often has hyaline globules. The presence of signet ring cells may cause confusion with a metastasis on FS, but since this tumor is benign, no further surgery is needed.

Primary ovarian mucinous tumors can rarely be associated with PP, pseudomyxoma ovarii, signet ring cells, and/or a colloid carcinoma component if the tumor is arising within an ovarian mature cystic teratoma [19, 20]. In these cases finding the teratomatous component is the only feature that unequivocally supports the diagnosis of a primary ovarian lesion.

Besides the presence of a mature cystic teratoma, other components that support the diagnosis of ovarian primary include Sertoli-Leydig cell tumor and Brenner tumor. Sertoli-Leydig cell tumors can have heterologous mucinous elements that can be benign, borderline, or malignant (Chap. 14), and Brenner tumors (Chap. 10) can have an extensive mucinous component that is usually benign.

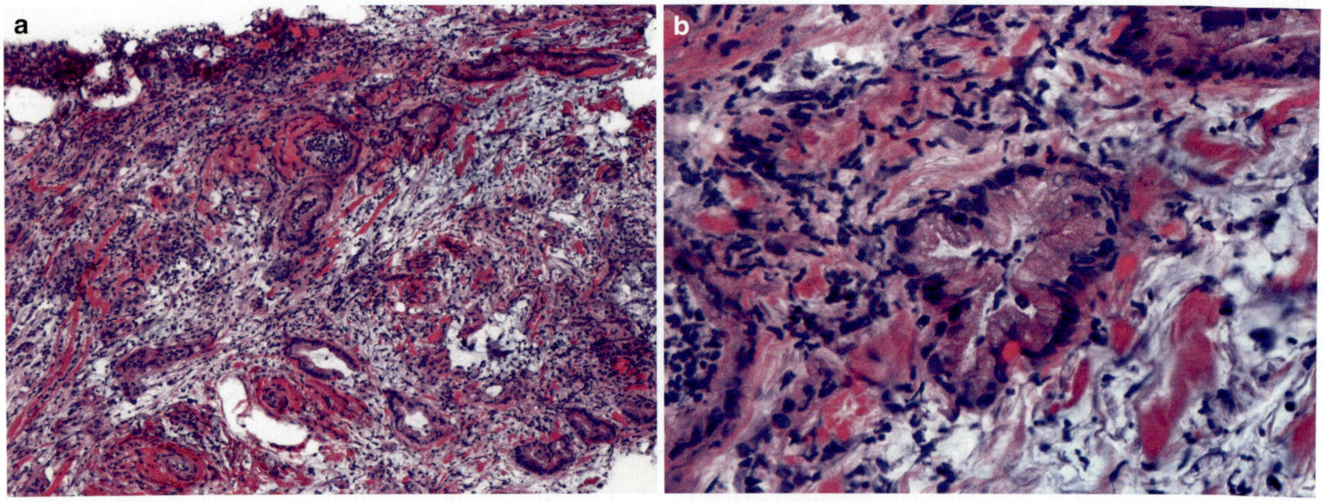

Fig. 3.13 (**a**) Frozen section of a metastatic gallbladder carcinoma with infiltration by mucinous glands with low grade nuclear features. (**b**) High power of the frozen section showing low grade cytologic features

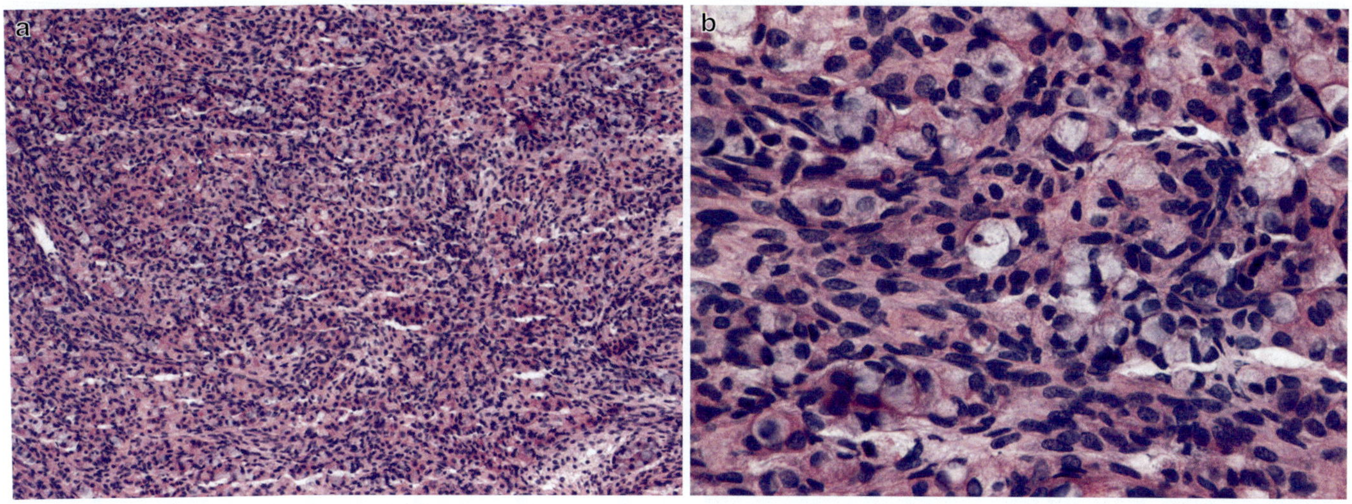

Fig. 3.14 (a) Frozen section of a metastatic signet ring cell carcinoma from the appendix showing solid architecture and spindle cells. (b) High power of the same frozen section shows signet ring cells interspersed within the spindle cell proliferation

Table 3.15 Possible primary sites of metastatic signet ring cell carcinoma to the ovary

- 76% stomach
- 11% colon-rectum
- 4% breast
- 3% biliary tract
- 3% appendix
- 3% others (small intestine, pancreas, cervix, lung, urinary bladder, urachus, renal pelvis)

Table 3.16 Primary ovarian tumors with signet ring or signet-ring-like cells

- Signet ring cell stromal tumor
- Serous carcinoma
- Primary ovarian mucinous carcinoma with signet ring cells (all stage 1A)
- Goblet cell carcinoid or signet ring cell-carcinoid (arising in a teratoma)
- Mucinous tumor associated with ovarian teratoma

Table 3.13 lists features that support the diagnosis of primary ovarian tumor. Large size (>13 cm) and smooth external surface are more commonly seen in primary tumors, as indicated previously, but are the least specific on the list. The presence of a benign and or borderline component is often seen in primary ovarian carcinomas, but can be seen in some metastases. The association with a mucinous adenofibroma is a more reliable feature since metastases do not have such components. Tumors associated with endometriosis are usually primary ovarian. Most carcinomas arising in or associated with endometriosis are endometrioid or clear cell tumors, but rarely, they can be mucinous. The most common mucinous ovarian tumor associated with endometriosis is a borderline seromucinous tumor. Most primary mucinous carcinomas have a complex papillary architecture with expansile type of invasion (Fig. 3.16a, b, and Chap. 9). Mucin depletion in the carcinomatous component contrasts with more abundant mucin in the benign and borderline areas of the same tumor (Fig. 3.8a, b). If at the time of the frozen section only the carcinomatous component is sampled, the lack of mucin may cause diagnostic difficulties. Sampling more than one area is always recommended in large lesions and it may be helpful to identify a more benign mucinous component to aid in the diagnosis and classification.

Other common and problematic differential diagnosis encountered during FS diagnosis include:

- Endometrioid carcinoma vs. metastases
- Ovarian endometrioid carcinoma vs. other primary ovarian tumors
- Serous tumors, benign, borderline vs. malignant
- High-grade ovarian cancer vs. granulosa cell tumor
- Diagnosis of clear cell carcinoma of ovary
- Primary high grade ovarian cancer vs. metastatic breast cancer

Differential Diagnosis of Endometrioid Carcinoma of Ovary and Metastases on Frozen Section

There are many metastatic tumors that can mimic primary ovarian endometrioid carcinoma (Table 3.17). Features that favor metastases are listed in Table 3.18. Some features are similar to those seen in metastatic mucinous tumors: ovarian surface involvement (Fig. 3.17), nodular invasive growth pattern (Fig. 3.18), tumor bilaterality, raggedly infiltrative pattern of invasion with abundant desmoplastic stroma,

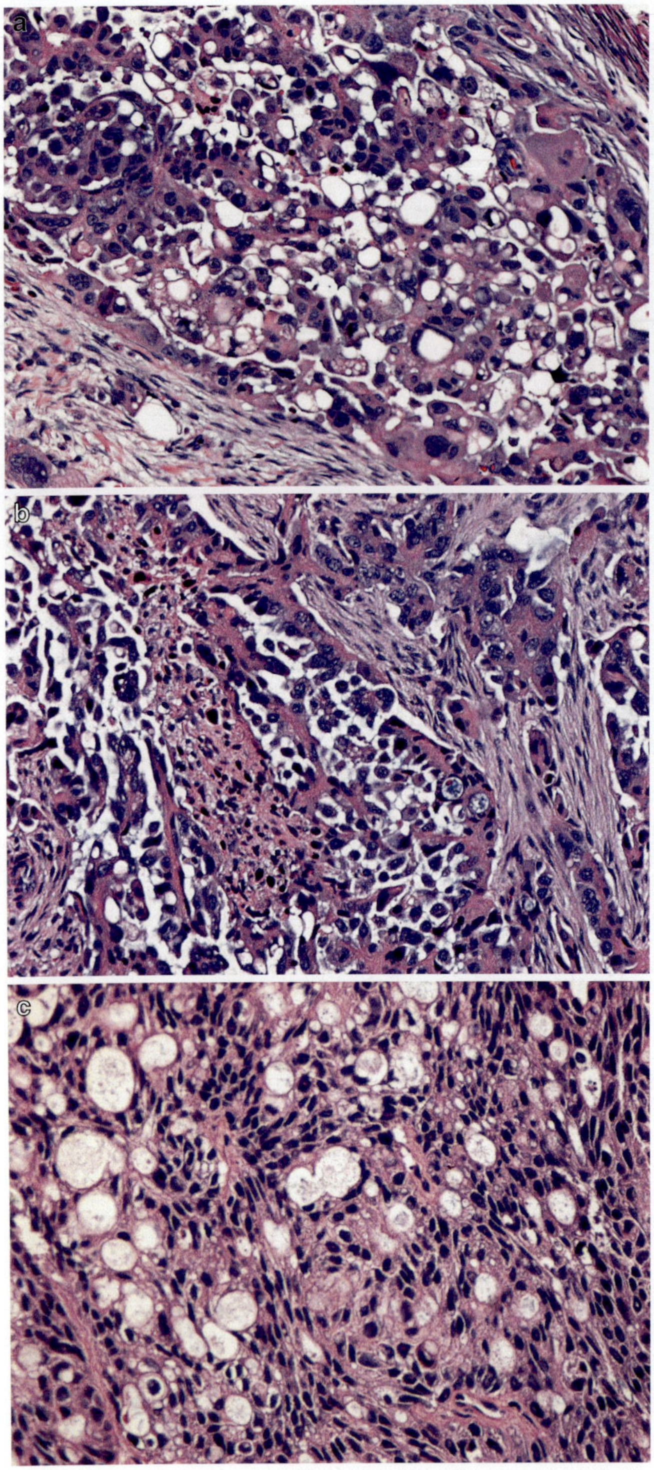

Fig. 3.15 (**a**) Serous carcinoma with signet ring cells, some of which were positive for mucicarmine. (**b**) The same tumor had more prominent papillary architecture, as is expected in serous carcinoma. (**c**) Frozen section of the same tumor mimicking a metastasis

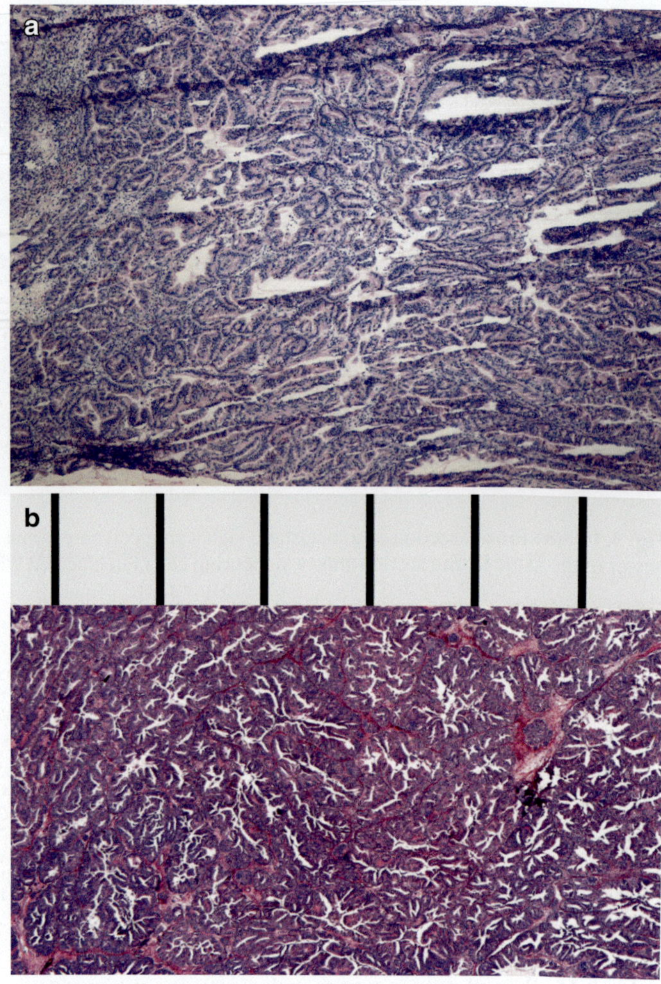

Fig. 3.16 (**a**) Frozen section of a primary ovarian mucinous carcinoma showing a characteristic complex glandular papillary pattern with mucin depletion. (**b**) Permanent section of the same block showing that the complex papillary architecture expands for more than 5 mm

Table 3.17 Metastatic tumors in the ovary that mimic endometrioid carcinomas

- Colon
- Endometrium
- Gallbladder
- Bile ducts
- Cervix
- Stomach
- Appendix
- Pancreas (rarely)
- Breast (rarely)
- Lung (rarely)

extensive lymphovascular invasion, preservation of normal structures with invasion in between them (Fig. 3.18). Some other features are more specific for metastases with an endometrioid-like appearance:

- Extensive necrosis (which is rare in primary ovarian tumors other than clear cell carcinomas, and some high-grade serous carcinomas) (Fig. 3.19)
- Segmental destruction of glands (Fig. 3.19)
- Cribriform pattern with dirty central necrosis (especially common in colon cancer), (Fig. 3.20a, b).

Table 3.18 Features that favor metastatic carcinoma to the ovary if the tumor has endometrioid features

- Bilateral multinodular solid masses
- Involvement of ovarian surface
- Nodular invasive growth
- Extensive lymphovascular invasion
- Extensive desmoplastic stroma
- Preservation of normal structures
- Uniform histology
- Extensive necrosis
- Segmental destruction of glands
- Garland pattern with cribriform architecture + dirty necrosis
- Low grade architecture with high nuclear grade
- Occasional goblet cells
- Metastases to mesentery and/or liver

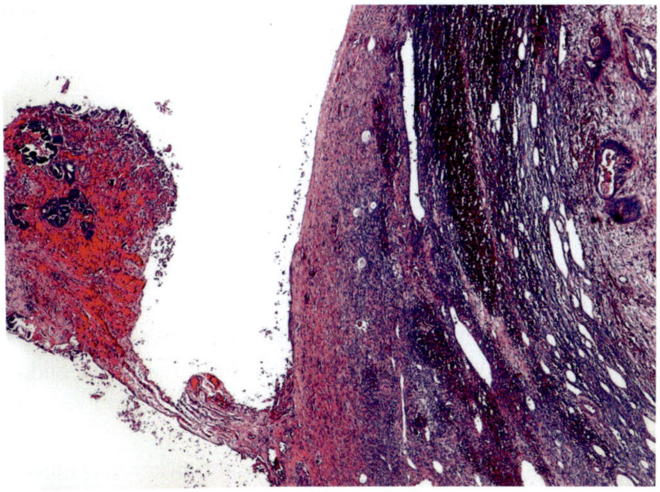

Fig. 3.17 Metastatic colon cancer with endometrioid features showing ovarian surface involvement

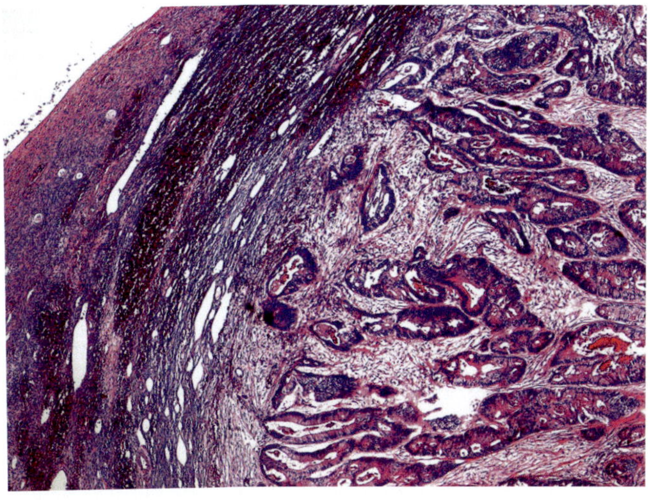

Fig. 3.18 Metastatic colon cancer (same tumor seen in Fig. 3.17) showing a nodular invasive growth pattern

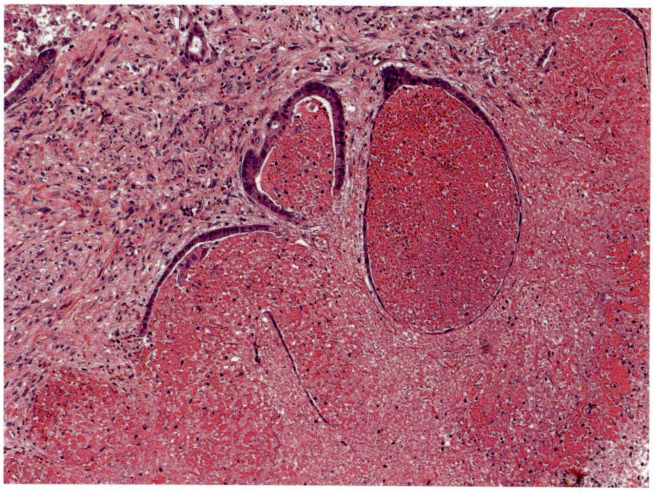

Fig. 3.19 Metastatic colon cancer with extensive necrosis and segmental destruction of glands

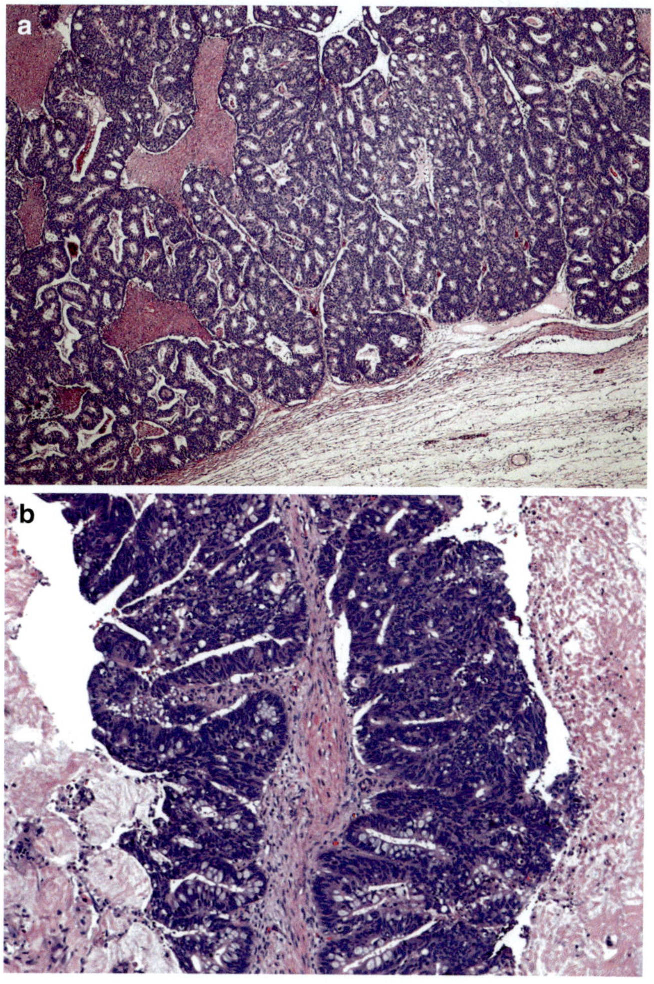

Fig. 3.20 (**a**) Metastatic rectal carcinoma with extensive cribriform architecture and abundant central dirty necrosis. (**b**) High power of same tumor

- Glandular architecture with very high nuclear grade. Most endometrioid ovarian carcinomas that are architecturally low grade are also low nuclear grade. In contrast, gastrointestinal adenocarcinomas tend to have more atypical nuclei despite a low-grade architectural pattern (Fig. 3.21).
- Presence of goblet cells (rarely seen in endometrioid carcinomas unless they are "mucin rich," an unusual variant) (Fig. 3.21)
- Metastases to mesentery and/or liver (common in colon cancer, but rare in ovarian cancer). Most ovarian carcinomas spread first to peritoneal surfaces, uterine serosa, omentum, and colonic serosa (Table 3.19).

Features that favor the diagnosis of primary ovarian endometrioid carcinoma are listed in Table 3.20. Most primary ovarian tumors show more than one histological pattern including solid, glandular, papillary, macro-, and microcystic components. In contrast most metastases are rather uniform and show the same pattern in many sections. There are some histologic features that are almost exclusively seen in endometrioid carcinomas but not in non-gynecological metastases. These include areas of squamous and/or morular metaplasia (Fig. 3.22a–d), sex-cord-like areas (like those seen in the sertoliform variant of endometrioid carcinoma), (Fig. 3.23) [21–27], and spindle cell change (like that seen in the spindle cell variant of endometrioid carcinoma) (Fig. 3.24) [28] (see Chap. 7).The presence of any of these three patterns in association with malignant endometrioid glands should prompt a diagnosis of endometrioid carcinoma of either ovarian or endometrial origin. *Endocervical adenocarcinomas* can have endometrioid features but are not associated with squamous/morular metaplasia, sex-cord-like areas or spindle cell metaplasia. Endocervical adenocarcinomas can sometimes be metastatic to the ovary even with minimal invasion in the endocervix. The metastases can be seen concomitantly with the cervical primary or present before the cervical carcinoma has been discovered [29, 30]. These metastases often have a mixture of endometrioid and mucinous features; they are cytologically low grade and are usually unilateral. They do not have a raggedly infiltrative pattern of invasion with desmoplasia, but instead exhibit expansile invasion, which often leads to confusion with borderline ovarian tumors (Fig. 3.25a, b). If there is doubt at the time of FS the pathologist can inquiry about a prior history of endocervical adenocarcinoma or request the surgeon to examine the cervix.

Metastatic colorectal carcinomas to the ovary that mimic the appearance of endometrioid carcinoma are not uncommon. They can be associated with an elevated CA125, can present at almost any age, including in premenopausal patients and are sometimes diagnosed before the colon carcinoma is discovered [31, 32]. Therefore, special effort should be exerted to rule out a metastasis before diagnosing an endometrioid carcinoma of ovary. The features listed in Tables 3.18 and 3.20 are helpful in the differential diagnosis. Other causes of elevated CA125 appear in Table 3.21 (see also Chap. 5).

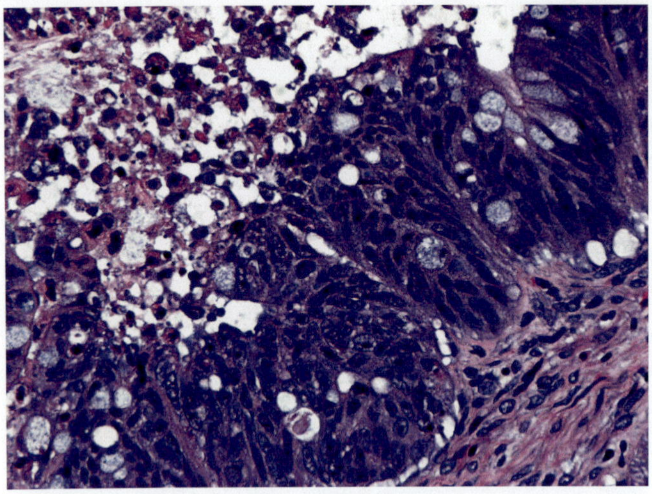

Fig. 3.21 Metastatic rectal carcinoma (same tumor seen in Fig. 3.20b) (high power) showing goblet cells, marked cytologic atypia, and central dirty necrosis (*top left*)

Table 3.19 Metastatic sites of ovarian and colonic carcinomas

Ovarian cancer	Colon cancer
Peritoneum	Mesenteric nodes
Uterine serosa	Liver
Omentum	
Intestinal serosa	

Table 3.20 Features that favor primary ovarian endometrioid carcinoma

- Intratumoral morphologic heterogeneity
- Areas resembling low-grade endometrioid carcinoma of endometrium
- Adenofibroma component
- Squamous metaplasia
- Morular metaplasia
- Sertoliform areas (sex-cord-like areas)
- Spindle cell features
- Endometriosis

Differential Diagnosis of Endometrioid Carcinoma of Ovary and Other Ovarian Tumors

Primary ovarian tumors with endometrioid-like glands include endometrioid carcinomas, well-differentiated Sertoli-Leydig cell tumors, endometrioid variant of yolk sac tumor, and carcinoid tumor (Table 3.22). The immediate surgical treatment is different with each of these diagnoses, so the differential diagnosis at the time of FS is very important (Table 3.23).

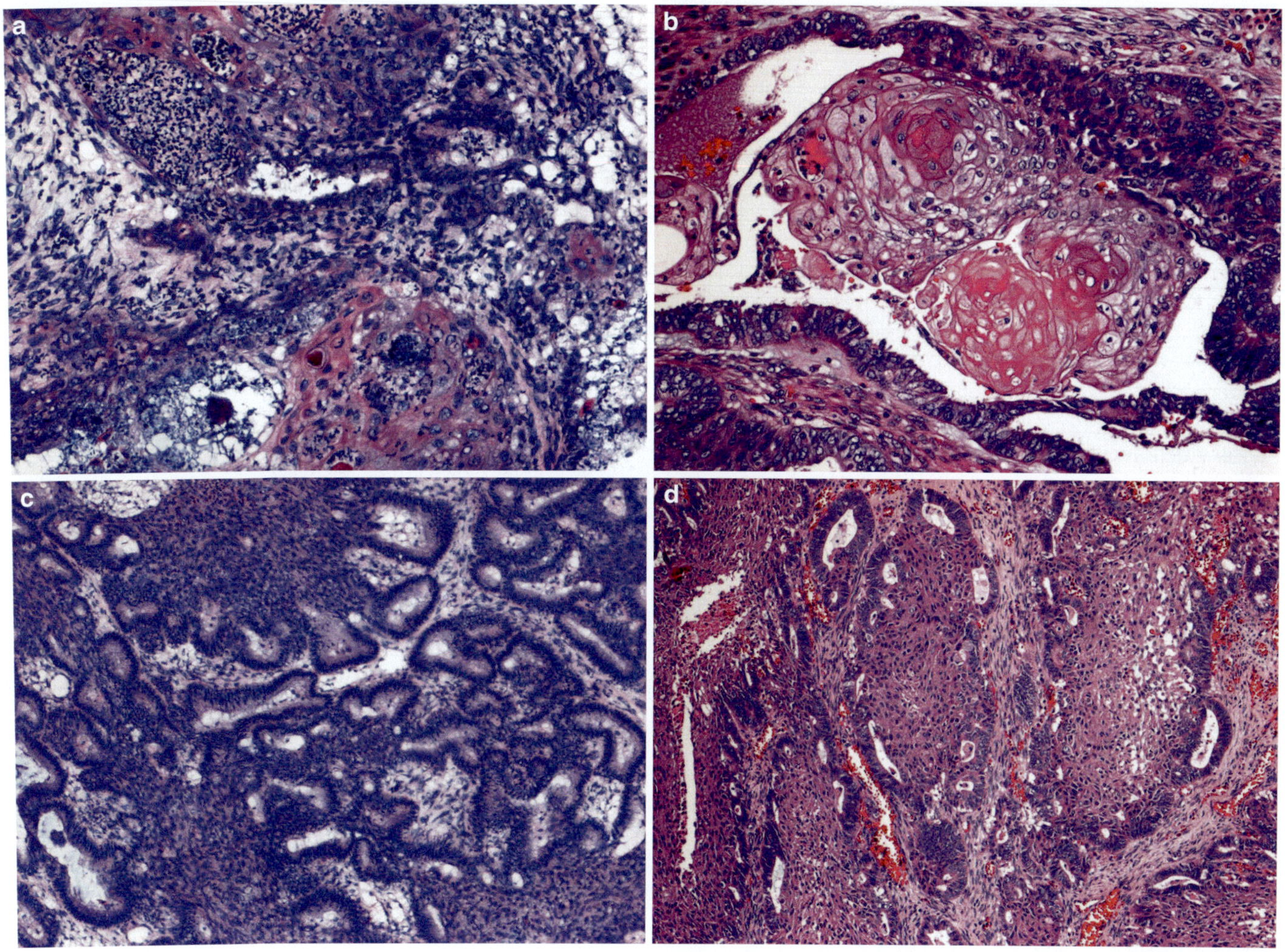

Fig. 3.22 Primary endometrioid carcinoma of ovary with squamous metaplasia. (**a**) Frozen section. (**b**) Permanent section. (**c**, **d**) Primary endometrioid carcinoma of ovary with morular metaplasia, permanent section

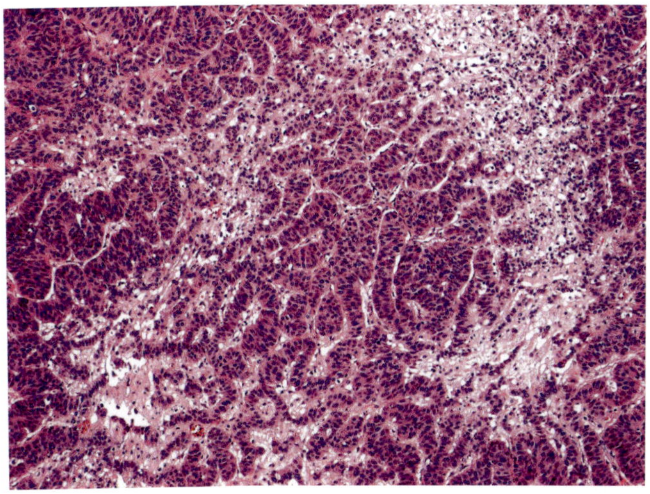

Fig. 3.23 Frozen section from a primary ovarian endometrioid carcinoma with sex-cord-like areas

Features that help in the differential diagnosis of sex-cord stromal tumor and endometrioid carcinoma are listed in Tables 3.24 and 3.25.

Sex-cord-like areas can be seen in endometrioid carcinomas, SLCT, and carcinoid tumors (Table 3.26). Since the cytologic appearance of these three tumors is different, an appreciation of these features can help in the differential diagnosis. A scraping at the time of FS can be very helpful. Sertoliform endometrioid carcinomas have stratified tall columnar cells with eosinophilic cytoplasm and round hyperchromatic nuclei with prominent nucleoli (Fig. 3.26). SLCT have cuboidal to columnar cells with round to oval basally placed nuclei, small nucleoli, and abundant eosinophilic or pale vacuolated cytoplasm (Fig. 3.27). Carcinoid tumors have cells with abundant granular eosinophilic cytoplasm and very classic "salt and pepper" nuclear chromatin (Fig. 3.28a–c).

Endometrioid variant of yolk sac tumor is rare and is seen in young patients, age 11–34 (mean, 22 years). The age is a helpful diagnostic feature. Features that favor the diagnosis of yolk sac

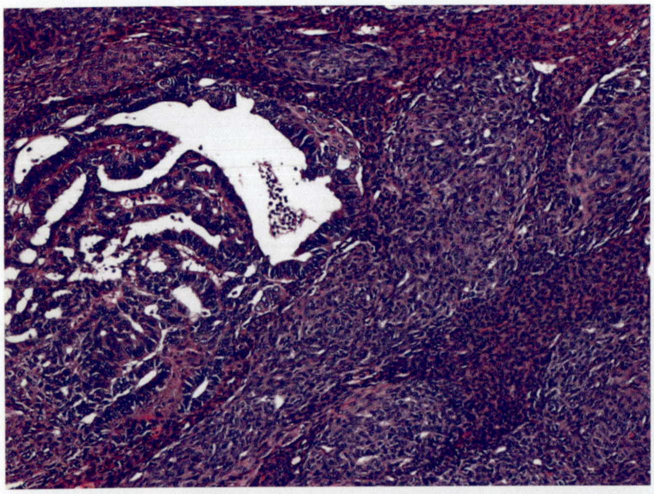

Fig. 3.24 Endometrioid ovarian carcinoma with spindle cell features

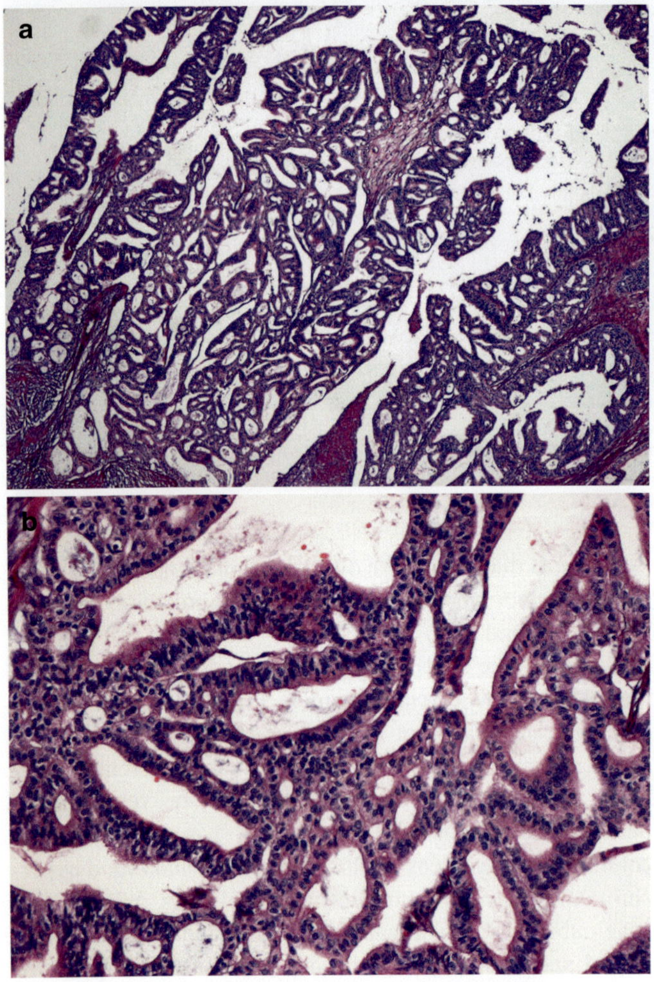

Fig. 3.25 (**a**, **b**) Metastatic endocervical adenocarcinoma to ovary

Table 3.21 Causes of elevations in serum CA125

- Ovarian cancer
- Colorectal cancer
- Pancreatic cancer
- Gastric cancer
- Lung cancer
- Breast cancer
- Benign: cirrhosis, chronic lung disease, pancreatitis

Table 3.22 Ovarian tumors with endometrioid-like glands

- Endometrioid carcinoma (primary ovarian)
- Metastatic endometrial carcinoma
- Metastatic colon cancer
- Metastatic endocervical adenocarcinoma
- Sertoli-Leydig cell tumor
- Endometrioid-variant of yolk sac tumor
- Carcinoid (primary or metastatic)

Table 3.23 Treatment of ovarian carcinoma versus Sertoli-Leydig cell tumor

Diagnosis	Treatment
Ovarian carcinoma	TAH-BSO and staging
Well-differentiated Sertoli-Leydig	Oophorectomy alone (all are benign)
Intermediate and poorly differentiated Sertoli-Leydig	USO and staging
Endometrioid variant of yolk sac	USO and staging
Carcinoid tumor	BSO if metastatic
	USO +/− staging if primary

TAH total abdominal hysterectomy, *BSO* bilateral salpingo-oophorectomy, *USO* unilateral salpingo-oophorectomy

Table 3.24 Features that support the diagnosis of Sertoli-Leydig cell tumor

- Small uniform glands
- Clear cytoplasm
- Solid tubules
- Leydig cells
- Presence of other patterns (sex-cord, trabecular, sarcomatoid, retiform, heterologous elements)
- Hypercellular nodules separated by hypocellular areas

Table 3.25 Differential diagnosis between sex-cord stromal tumor and endometrioid carcinoma of ovary

	Endometrioid carcinoma	Sex-cord stromal tumor
Age	Postmenopausal	<40 years
Bilaterality	~10%[a]	<2%
Mucin	Present	Absent
Squamous met	Present	Absent
Adenofibroma	Present	Absent
Nuclei	Nucleoli	Grooved, non atypical
Endometriosis	Present	Absent

[a]Unpublished data from Memorial Sloan-Kettering Cancer Center

tumor are listed in Table 3.27 [33] (Fig. 3.29a, b). Increased serum level of alpha-fetoprotein is also helpful. Table 3.28 has a list of all causes of increased serum level of alpha-fetoprotein.

Carcinoid tumors in the ovary can be primary or metastatic. The histology is similar in both cases. Features that support these diagnoses are listed in Table 3.29. Carcinoid tumors tend to be grossly yellow (Fig. 3.2). Table 3.30 has a list of ovarian tumors with a yellow gross appearance.

Table 3.26 Ovarian tumors with sex-cord-like areas

- Sertoli-Leydig cell tumor
- Adult granulosa cell tumor
- Sertoliform endometrioid carcinoma
- Carcinoid tumor (primary or metastatic)

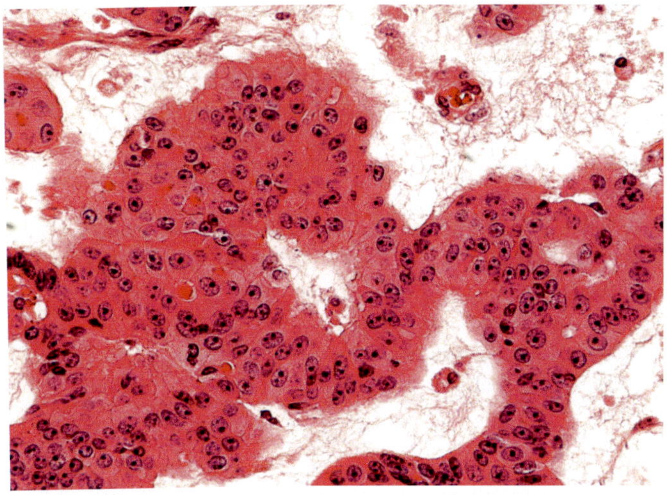

Fig. 3.26 Frozen section from a sertoliform variant of endometrioid carcinoma showing typical cytology with prominent nucleoli

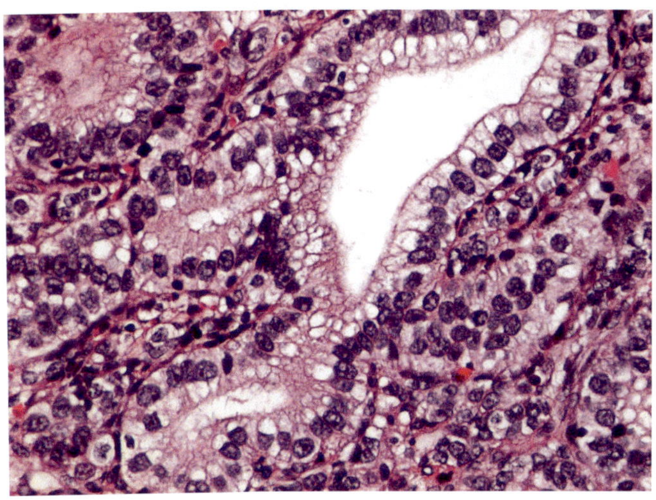

Fig. 3.27 Typical cytologic appearance of a Sertoli-Leydig cell tumor with vacuolated cytoplasm

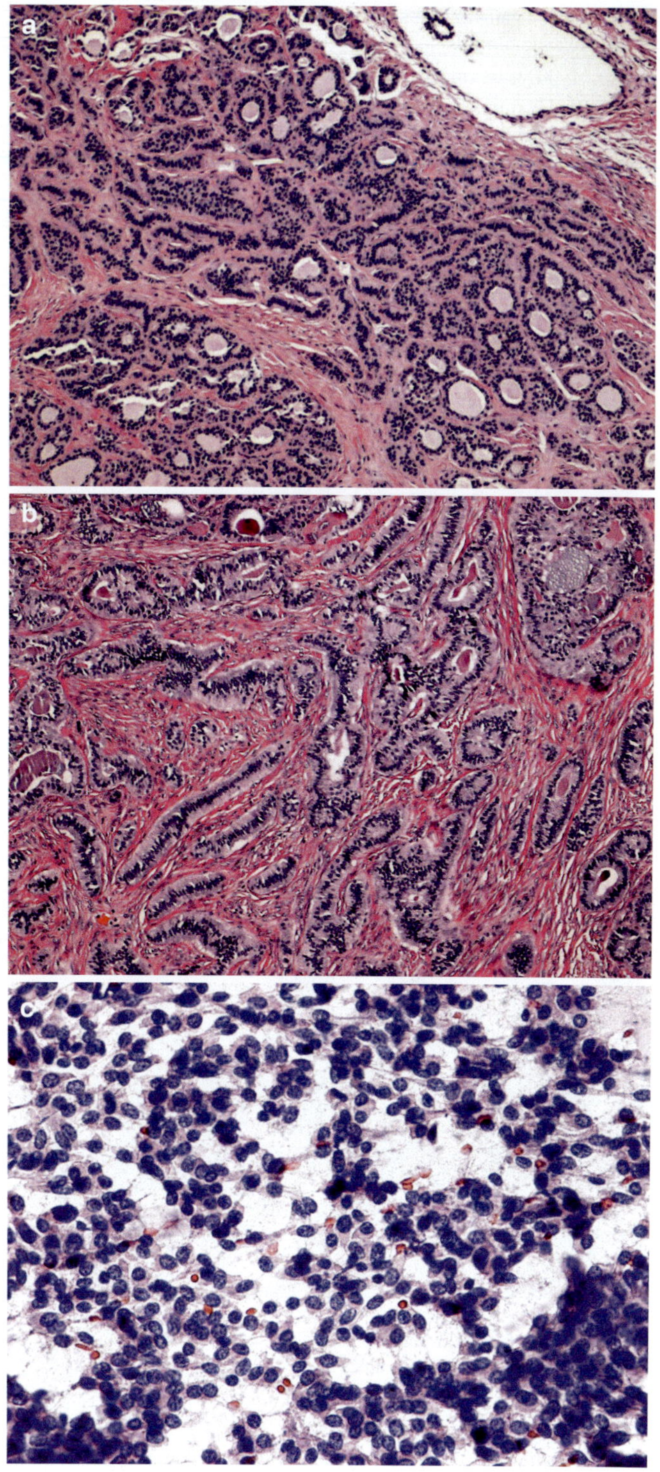

Fig. 3.28 (**a**) Frozen section from an ovarian carcinoid tumor with sex-cord-like areas. (**b**) Permanent section from the same tumor. (**c**) Cytologic appearance of a scrape preparation from the same tumor showing the classic salt and pepper nuclei

Differential Diagnosis Between Benign, Borderline, and Malignant Ovarian Serous Tumors

Cystadenomas are usually easy to diagnose because they are unilocular cysts with smooth internal linings and flat tubal-type epithelium. Somewhat more difficult is the differential diagnosis between a cystadenofibroma and a serous borderline tumor. Both tend to be cystic and have intraluminal papillary growth. In cystadenofibromas the papillae are firm and "crunchy," (Fig. 3.3) but in borderline tumors the papillae are friable and fall apart easily when cut (Fig. 3.4). Microscopically, cystadenofibromas have thick, broad-base papillae composed of densely fibrotic stroma and lined by a single layer of tubal-type epithelium. Borderline tumors have

Table 3.27 Features that support the diagnosis of endometrioid variant of yolk sac tumor

- Age 11–34 years (mean 22 years)
- Increased serum levels of alpha-fetoprotein
- Other classic components of yolk sac tumor
- Primitive nuclei
- Secretory vacuoles

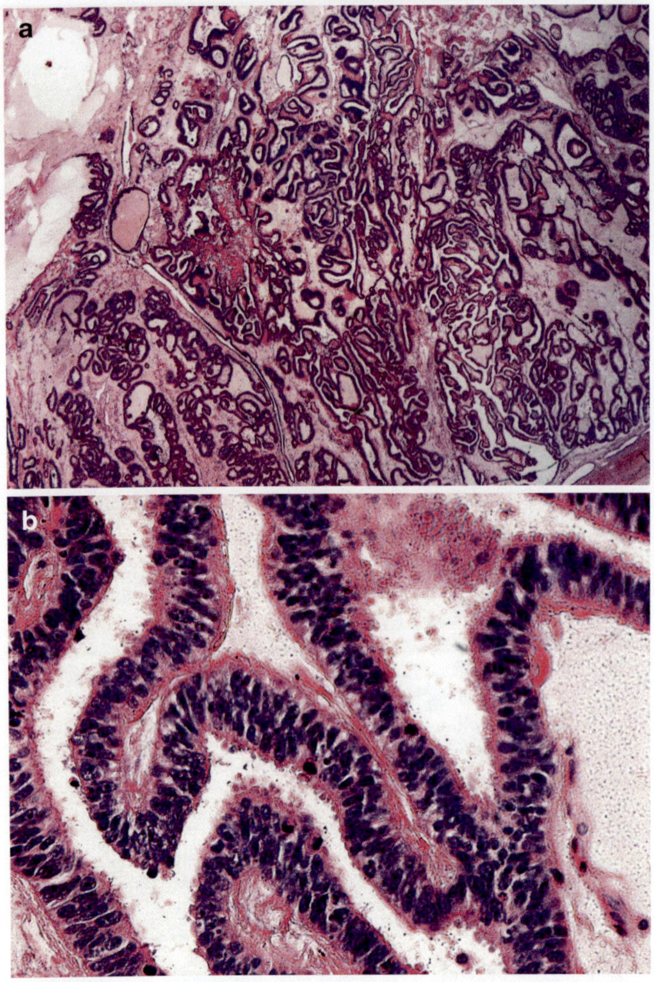

Fig. 3.29 (**a**, **b**) Endometrioid variant of yolk sac tumor with endometrioid glands, and primitive cytology on high power

Table 3.28 Causes of increased serum level of alpha-fetoprotein

- Hepatocellular carcinoma
- Non-seminomatous germ cell tumors
- Pancreatic cancer
- Colon cancer
- Lung cancer
- Granulosa cell tumors
- Sertoli-Leydig cell tumor
- Benign: pregnancy, hepatitis, cirrhosis

Table 3.29 Differential diagnosis of primary and metastatic ovarian carcinoid tumors

Primary	Metastasis
• Other tertomatous components	• Bilateral • Spread outside ovary in 90% cases • Carcinoid syndrome
• Sertoli-Leydig	• Vascular invasion
• Mucinous tumor	• Multinodular growth • Prominent stromal hyalinization

Table 3.30 Ovarian tumors with yellow gross appearance

- Carcinoid
- Sertoli-Leydig cell tumors, and Sertoli cell tumors
- Steroid cell tumors
- Thecoma
- Luteinized granulosa cell tumor
- Metastatic renal cell carcinoma
- Metastatic adrenal cortical carcinoma

a complex papillary growth with hierarchical branching, lined by tufted and stratified cells with cell detachment. These tumors often have two cell types in the same lesion: ciliated serous type cells and non-ciliated cells with a more abundant and very eosinophilic cytoplasm. They are differentiated from low-grade serous carcinomas by the lack of stromal invasion. Tangential sections of some borderline tumor papillae can mimic invasion, but do not have desmoplastic stroma and are in continuity with the surface papillae (recuts and levels can help in these cases). In addition to invasion, low-grade serous carcinomas tend to have more epithelial proliferation than serous borderline tumors.

Low-grade serous carcinomas are rare. The majority of serous carcinomas are high grade and have marked nuclear atypia and more than 10 mitoses/10HPF. They often have papillary architecture that, along with the high-grade cytology, is very characteristic (Fig. 3.30a, b). A micropapillary type of invasion is also fairly common. It should be noted that papillary and micropapillary architecture is easier to identify in formalin-fixed frozen sections. Laboratories that fix frozen sections with ethanol often see more solid serous carcinomas on frozen section than those where formalin is used (Fig. 3.31a, b). The cytologic appearance can help in the correct diagnosis when papillary features cannot be appreciated. High-grade serous carcinomas often have great variation in the size of the nuclei with occasional large nuclei (> 15 μm) interspersed with smaller, more abundant nuclei, and some large multinucleated cells (mean size, 28 μm) (Fig. 3.32a, b). Eosinophilic macronucleoli are also commonly seen in > 75% of cases. Psammomatous calcifications on the other hand are rare and are encountered in only 20% of cases [34].

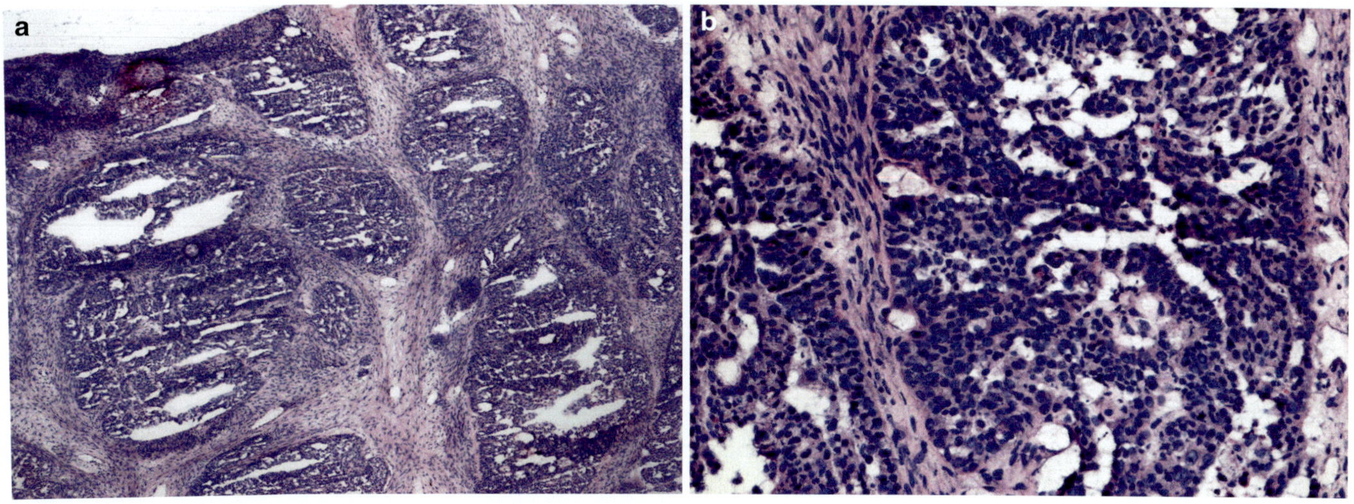

Fig. 3.30 (**a**, **b**) Frozen section from a high-grade serous carcinoma with characteristic papillary architecture and high nuclear grade

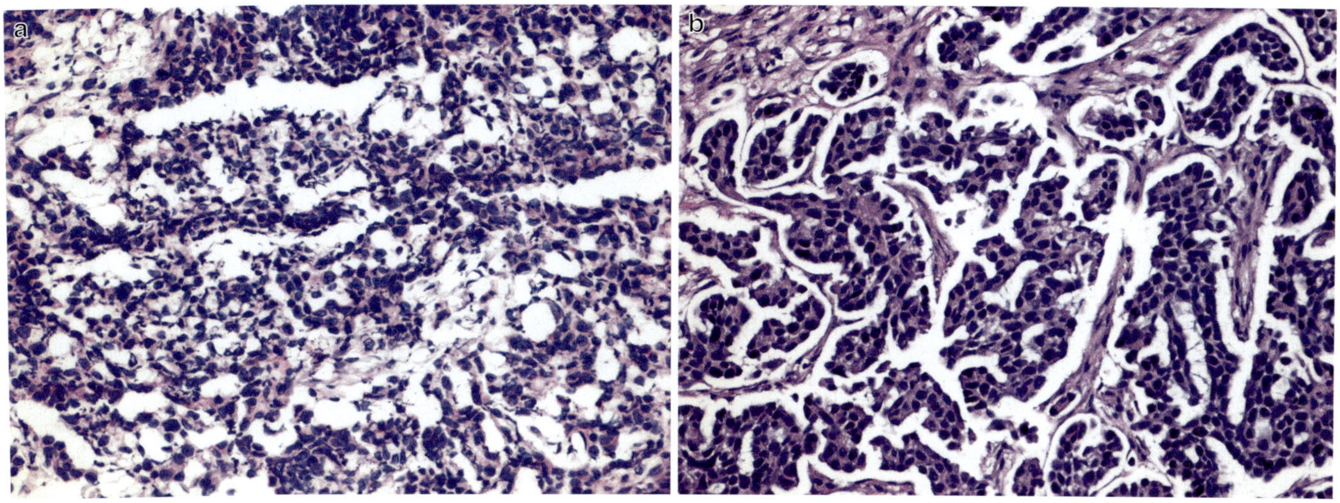

Fig. 3.31 (**a**) Frozen section from a serous carcinoma fixed in ethanol showing solid tumor. (**b**) Section from same block after formalin fixation showing micropapillary architecture

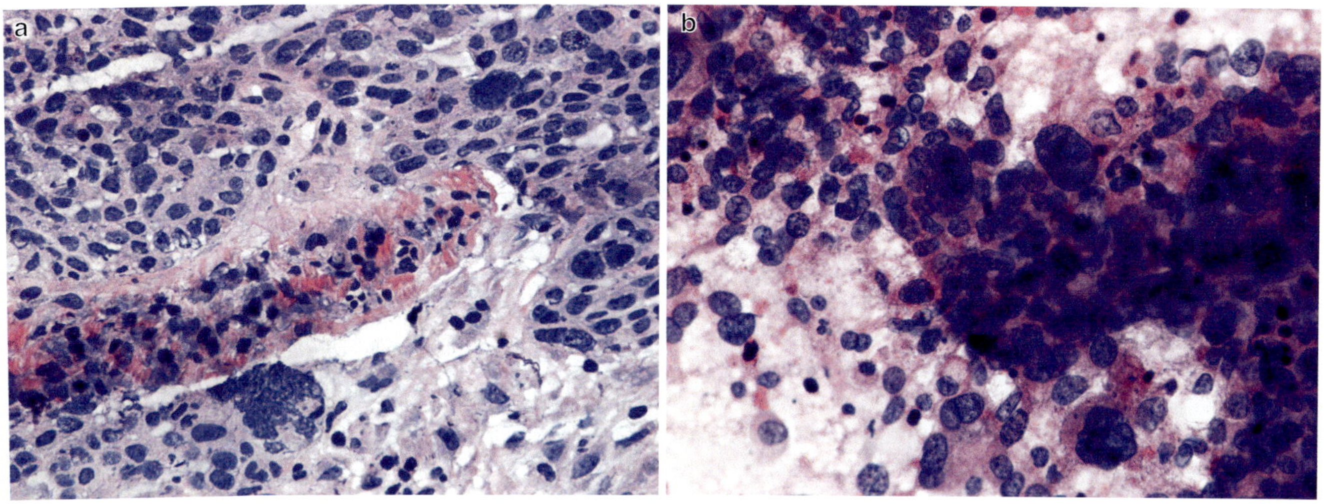

Fig. 3.32 (**a**) Frozen section from a high grade serous carcinoma showing significant variation in nuclear size and shape, with isolated large cells, and multinucleated giant cells. (**b**) Cytologic scraping from same tumor showing similar cytologic features

Differential Diagnosis Between High Grade Ovarian Carcinoma and Granulosa Cell Tumor (Adult Type)

Granulosa cell tumors (GCT) often have a solid growth that can mimic a primary high-grade ovarian carcinoma on FS. GCT are often beefy and very hemorrhagic on gross examination (Fig. 3.5), whereas serous carcinomas tend to be more heterogenous with cystic, solid, and papillary areas with a tan and necrotic appearance (Fig. 3.6). The differential diagnosis relies on the cytology and finding other architectural patterns that favor one or the other. GCT has very characteristic nuclei that can be seen in a scraping of the tumor at the time of FS. The nuclei are bland, uniform in size, and oval, round, or angular in shape, often pale and grooved (coffee bean appearance), and can have one inconspicuous nucleolus (Fig. 3.33a–c). The tumor cell arrangement appears disorganized in solid tumors. Architectural patterns found in GCT are listed in Table 3.31. The cytology of high-grade ovarian serous carcinomas was described above in the differential diagnosis of serous tumors (Table 3.32). Serous carcinomas often have areas of papillary and/or micropapillary type of invasion that can help in the diagnosis, although it should be noted that granulosa cell tumors can have a pseudopapillary appearance. As stated previously, the appearance of papillae and micropapillae is enhanced by formalin fixation. Therefore, laboratories that fix frozen sections with ethanol may benefit from fixing one frozen section slide with formalin whenever faced with a differential diagnosis of serous carcinoma [35].

Differential Diagnosis Between Ovarian Carcinoma and Metastatic Breast Cancer

Metastatic lobular carcinoma is usually not a diagnostic problem because it has a very characteristic pattern with rows of single cells. Ductal carcinomas, on the other hand, can have a solid pattern that can be confused with a primary ovarian carcinoma. The gross examination may be helpful since metastases often present as bilateral multinodular solid masses that often preserve normal structures like corpus luteum or corpora albicans (Fig. 3.34). Primary ovarian cancer in contrast tends to be more heterogenous with cystic, solid, and papillary areas with tan and necrotic appearance (Fig. 3.6). Microscopically primary ovarian carcinomas have pleomorphic features as mentioned above in the differential diagnosis of serous tumors, with cells of different sizes, with some large cells and multinucleated cells, and macronucleoli. Most ductal carcinomas tend to have a more uniform cytology (Fig. 3.35) with less atypia. Psammomatous calcifications are more commonly seen in ovarian carcinomas, but a small percentage of breast carcinomas can have calcifications that on FS can mimic psammoma bodies [34],

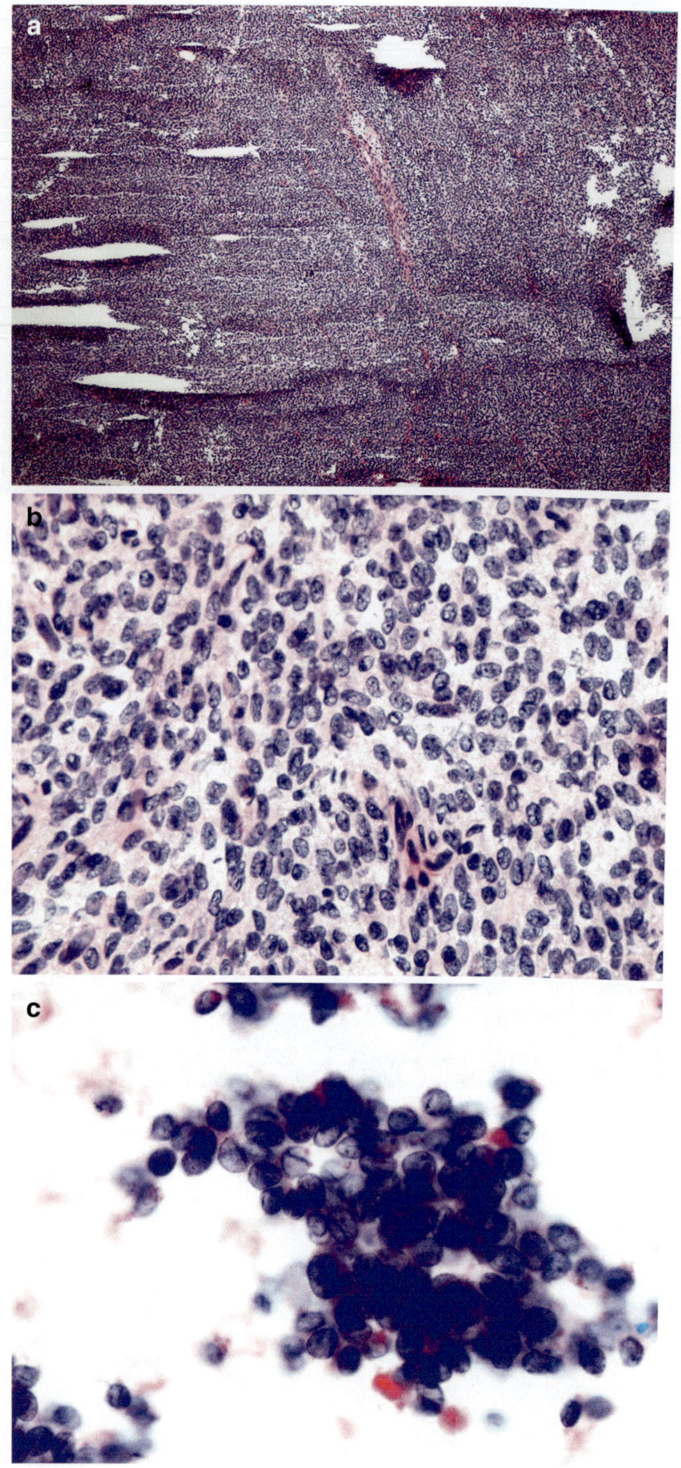

Fig. 3.33 (**a**) Frozen section from an adult granulosa cell tumor with solid architecture. (**b**) High power magnification of same tumor showing uniform cells with bland nuclei and few grooves. (**c**) Scraping from same tumor showing characteristic cytologic appearance with abundant grooves

Table 3.32. Comparison with the previous tumor can be very helpful if the patient has a history of breast cancer. The differential diagosis between primary solid high grade ovarian carcinoma and metastatic breast carcinoma can be impossi-

Table 3.31 Architectural patterns of adult granulosa cell tumors

- Microfollicular (Call-Exner bodies)
- Macrofollicular
- Trabecular
- Insular
- Hollow-tubular
- Solid tubular
- Solid
- Watered-silk

Table 3.32 Features of ovarian serous carcinomas and breast metastasis on frozen section

	Serous carcinoma	Metastases from breast
Large nuclei (15 μm)	96%	0
Multinucleated cells (28 μm)	89%	6%
Macronucleoli (>3 μm)	78%	45%
Psammomatous calcifications	21%	3%
Uniform nuclei	1.7%	80%

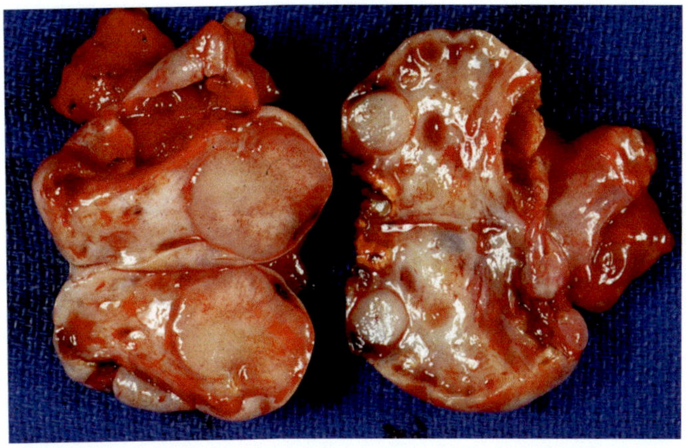

Fig. 3.34 Bilateral ovarian metastatic breast cancer with characteristic multinodular solid appearance, and preservation of normal structures (corpus luteum to the right of center)

ble in some cases at the time of FS. When in doubt, it is always best to be cautious and defer the diagnosis. If the surgeon reports a history of breast cancer, it is important not to assume that it was invasive (it could have been lobular carcinoma in situ or ductal carcinoma in situ) and to remember that most patients with a history of breast cancer who develop an ovarian tumor will have a new, primary ovarian, Fallopian tube or peritoneal carcinoma rather than a metastasis.

Diagnosis of Clear Cell Carcinoma of Ovary

Clear cell carcinoma has many different histological patterns and two of them are particularly deceiving and can cause diagnostic problems at the time of frozen section [36]. Most

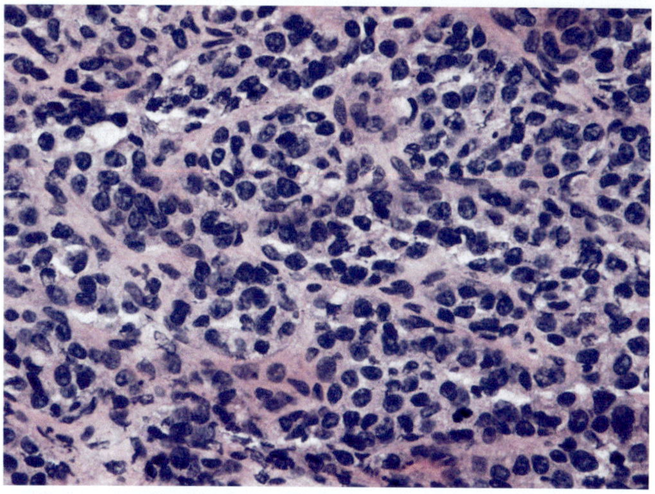

Fig. 3.35 Frozen section of metastatic ductal carcinoma to ovary showing rather uniform cytology

Table 3.33 Architectural patterns of ovarian clear cell carcinoma

- Tubulocystic
- Papillary
- Solid
- Adenofibromatous-like
- DON'T FORGET: Deceiving cystic pattern with flat lining cells! (Fig. 1.7c, d)

pathologists are familiar with the papillary, solid, and tubular patterns. Two additional less well-known patterns include the cystic pattern with flat epithelium, and the "adenofibromatous" pattern (Table 3.33). The cystic pattern with flat epithelium is composed of cysts of variable sizes, with little intervening stroma, and some intraluminal secretion, and are lined by predominantly flat cells, deceivingly benign in appearance. The diagnosis can be suspected in the presence of isolated plumper atypical nuclei, which may be scanty (Fig. 3.36a–c). Additional sections from other parts of the tumor may be diagnostically helpful if they contain patterns that are characteristic of clear cell carcinoma, like solid nests of clear cells or papillae with hob-nail cells and hyalinized fibrovascular cores. The adenofibromatous pattern has isolated clusters of tumor cells in a fibromatous stroma without an obvious desmoplastic reaction. At low power the tumor looks like an adenofibroma (Fig. 3.37). At higher magnification the clusters are composed of cells with abundant clear cytoplasm and large atypical nuclei, often with focal central necrosis. The clusters can be papillary and have the typical hyalinized fibrovascular cores, and hob-nail cells. Clearing of the cytoplasm is more readily seen in frozen sections fixed with formalin [35] (Fig. 3.38a, b). In laboratories that fix frozen sections with ethanol, the cells have eosinophilic cytoplasm and the diagnosis of a clear cell tumor may be difficult. Borderline clear cell adenofibromas un-associated with

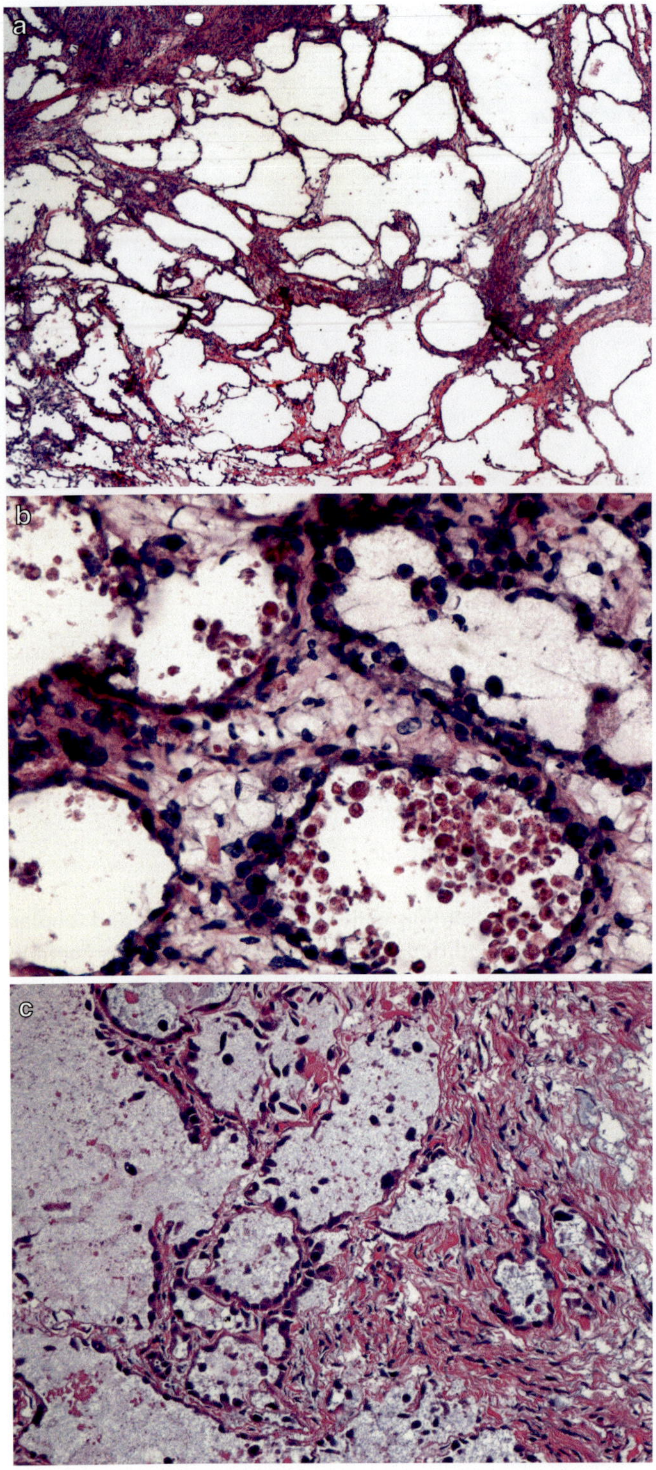

Fig. 3.36 (**a**, **b**) Frozen section of an ovarian clear cell carcinoma with a cystic pattern and flat epithelium (low and high power). (**c**) Same tumor with isolated atypical cells within the cyst, clues to the correct diagnosis

carcinoma are so rare that they should not be diagnosed at frozen section. The presence of a borderline clear cell pattern on FS should prompt additional sampling to rule out a concomitant carcinoma.

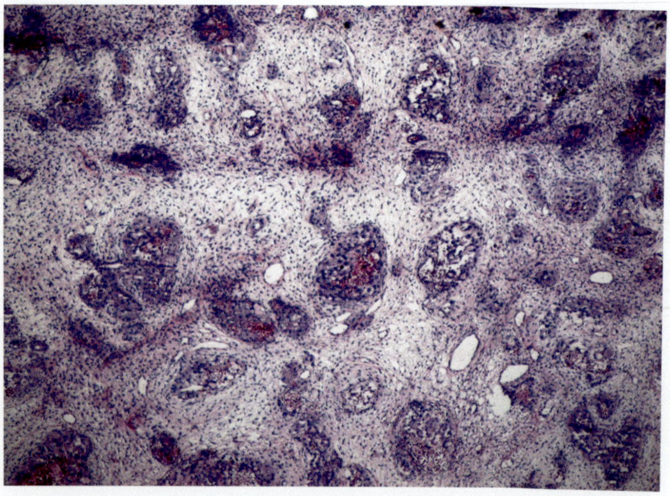

Fig. 3.37 Frozen section of an ovarian clear cell carcinoma with an adenofibromatous pattern (low and high power)

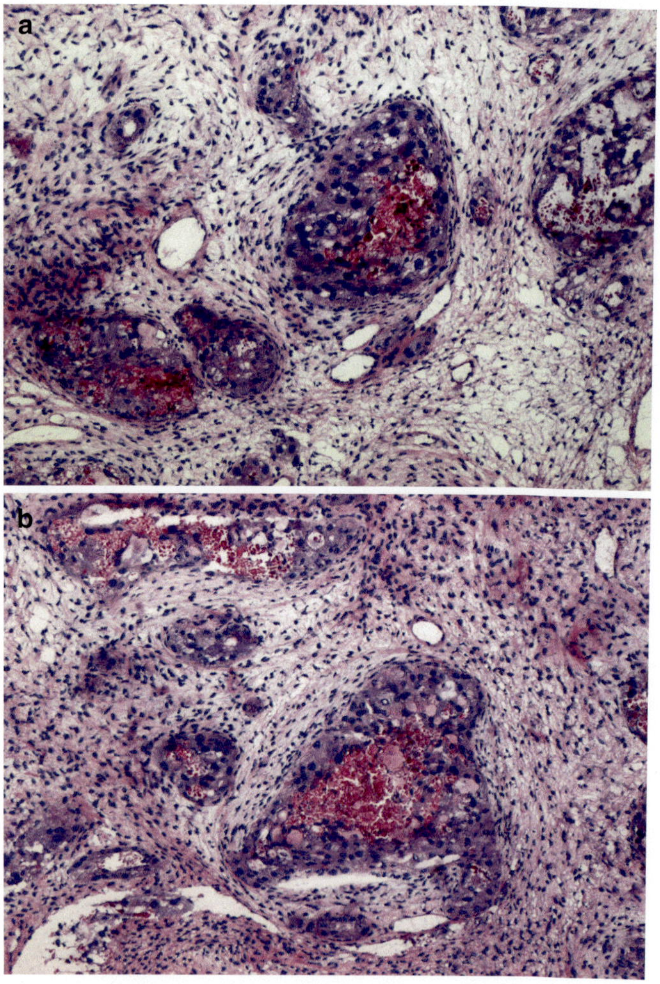

Fig. 3.38 (**a**) Frozen section of clear cell carcinoma fixed in ethanol shows eosinophilic cytoplasm. (**b**) Frozen section from the same block fixed in formalin shows slightly more evident cytoplasmic clearing

Table 3.34 Ovarian tumors with clear cells

- Clear cell carcinoma of ovary
- Dysgerminoma
- Yolk sac tumor
- Steroid cell tumor
- Metastatic renal cell carcinoma
- Metastatic colon cancer
- Metastatic hepatocellular carcinoma
- Metastatic clear cell carcinoma from endometrium, cervix, vagina
- Other metastatic carcinomas with clear cells (urachal, thyroid, adrenal, etc.)

Other tumors with clear cells that can be found in the ovary are rare; they are listed in Table 3.34. The differential diagnosis should be based on clinical history, age, and other features suggestive of metastasis like bilaterality, multinodular solid masses, and ovarian surface involvement. Dysgerminomas and yolk sac tumors are seen in young patients with a mean age of 19–22 years. A diagnosis of malignant germ cell tumor should never be rendered in a postmenopausal patient at the time of frozen section (see Chaps. 12 and 14).

Telepathology

Telepathology is the use of telecommunications technology to render pathology services over a distance. The use of telepathology is becoming more common and can be applied to different purposes including; frozen section diagnosis, consultation services, quality assurance, teaching, and research.

Telepathology systems can be divided into three categories: static microscopy (transmission of selected images), dynamic robotic microscopy (viewing pathologist has remote control over the microscope), and virtual slide technology. Virtual slide technology is the newer system and consists of slide scanners that can rapidly digitize whole glass slides that can be viewed over the Internet. The receiving pathologist can move these digitized images and change the magnification. The advantage of virtual slide telepathology (TP) is that the receiving pathologist does not need any technical equipment other than a computer, and the same images can be viewed by multiple pathologists at the same time.

There have been numerous studies validating the use of telepathology for frozen section diagnosis, and some include ovarian lesions [37–39]. Most of these studies have used static or dynamic robotic microscopy, and report accuracy rates that range from 92% to 100%. Virtual slide scanners are newer and fewer studies have been published using them for frozen section diagnosis [40–42]. Nevertheless, these studies report diagnostic accuracy rates that range from 94% to 100% [40–42]. Two of these studies include ovarian frozen section diagnosis [41, 42] and report similar accuracy rates as FS with routine glass slides. The time to scan each slide is dependent on several variables including: the size of the tissue, the maximum magnification attainable, and the technical system used. Modern systems are faster than older ones. Scanning times vary from 1 to 30 min per slide, with an average of 9 min [40, 41].

In summary, telepathology offers good results for interpretation of ovarian frozen sections regardless of the methodology used. Telepathology can be used to obtain consultations with more experienced gynecological pathologists in any practice setting.

Conclusion

Frozen sections of ovarian lesions can be hard to interpret for a variety of reasons. Important rules to remember include:

1. Frozen section should not be performed if it is not indicated.
2. Unless the pathologist is 100% certain of a primary malignant or borderline diagnosis, it is better to defer the diagnosis. Otherwise the patient will undergo unnecessary surgery.
3. Required information that is diagnostically helpful includes age, history, radiologic findings, laboratory results, and extent of the disease.
4. If uncertain about a diagnosis, ask for help from a more experienced pathologist.

References

1. Garg K, Shih K, Barakat R, Abu-Rustum N, Soslow RA. The accuracy of an intraoperative diagnosis of ovarian borderline tumor varies by histologic subtype. Mod Pathol. 2011;24:247.
2. Gol M, Baloglu A, Yigit S, et al. Accuracy of frozen section diagnosis in ovarian tumors: is there a change in the course of time? Int J Gynecol Cancer. 2003;13:593–7.
3. Wootipom V, Dechsukhum C, Hanprasertpong J, et al. Accuracy of intraoperative frozen section in diagnosis of ovarian tumors. J Med Assoc Thai. 2006;89:577–82.
4. Stewart CJR, Brennan BA, Hammond IG, et al. Intraoperative assessment of ovarian tumors: a 5 year review with assessment of discrepant diagnostic cases. Int J Gynecol Pathol. 2006;25: 216–22.
5. Brun JL, Cortez A, Rouzier R, et al. Factors influencing the use and accuracy of frozen section diagnosis of epithelial ovarian tumors. Am J Obstet Gynecol. 2008;199(244):e1–7.
6. Wasinghon P, Suthippintawong C, Tuipae S. The accuracy of intraoperative frozen sections in the diagnosis of ovarian tumors. J Med Assoc Thai. 2008;91:1791–5.
7. Suprasert P, Khunamornpong S, Phusong A, et al. Accuracy of intra-operative frozen sections in the diagnosis of ovarian masses. Asian Pac J Cancer Prev. 2008;9:737–7340.
8. Pavlakis K, Messini I, Vrekoussis T, et al. Intraoperative assessment of epithelial and non-epithelial ovarian tumors: a 7 year review. Eur J Gynaecol Oncol. 2009;30:657–60.
9. Fanfani F, Zannoni GF, Fagotti A, et al. Importance of a specialized pathologist for the examination of frozen sections of adnexal masses. Int J Gynecol Cancer. 2007;17:1034–9.

10. Seidman JD, Kurman RJ, Ronnett BM. Primary and metastatic mucinous adenocarcinomas in the ovaries. Incidence in routine practice with a new approach to improve intraoperative diagnosis. Am J Surg Pathol. 2003;27:985–93.
11. Stewart CJ, Brennan BA, Hammond IG, et al. Accuracy of frozen section in distinguishing primary ovarian neoplasia from tumors metastatic to the ovary. Int J Gynecol Pathol. 2005;24:356–62.
12. Yemelyanova AV, Vang R, Judson K, et al. Distinction of primary and metastatic mucinous tumors involving the ovary: analysis of size and laterality data by primary site with reevaluation of an algorithm for tumor classification. Am J Surg Pathol. 2008;32:128–38.
13. Lee KR, Young RH. The distinction between primary and metastatic mucinous carcinomas of the ovary: gross and histologic findings in 50 cases. Am J Surg Pathol. 2003;27:281–92.
14. Kiyokawa T, Young RH, Scully RE. Krukenberg tumors of the ovary: a clinicopathologic analysis of 120 cases with emphasis on their variable pathologic manifestations. Am J Surg Pathol. 2006;30:277–99.
15. Young RH. From Krukenberg to today: the ever present problems posed by metastatic tumors in the ovary: part I. Historical perspective, general principles, mucinous tumors including the Krukenberg tumor. Adv Anat Pathol. 2006;13:205–27.
16. Young RH. From Krukenberg to today; the ever present problems posed by metastatic tumors in the ovary. Part II. Adv Anat Pathol. 2007;14:149–77.
17. Che M, Tornos C, Deavers MT, et al. Ovarian mixed epithelial carcinomas with microcystic pattern and signet ring cells. Int J Gynecol Pathol. 2001;20:323–8.
18. McCluggage WG, Young RH. Primary ovarian mucinous tumors with signet ring cells. Report of 3 cases with discussion of so-called primary Krukenberg tumor. Am J Surg Pathol. 2008;32:1373–9.
19. Vang R, Gown AM, Zhao C, et al. Ovarian mucinous tumors associated with mature cystic teratomas: morphologic and immunohistochemical analysis identifies a subset of potential teratomatous origin that shares features of lower gastrointestinal tract mucinous tumors more commonly encountered as secondary tumors in the ovary. Am J Surg Pathol. 2007;31:854–69.
20. McKenney JK, Soslow RA, Longacre TA. Ovarian mature teratomas with mucinous epithelial neoplasms: morphologic heterogeneity and association with pseudomyxoma peritonei. Am J Surg Pathol. 2008;32:645–55.
21. Roth LM, Liban E, Czernobilski B. Ovarian endometrioid carcinomas mimicking Sertoli and Sertoli-Leydig cell tumors. Cancer. 1982;50:1322–31.
22. Young RH, Prat J, Scully RE, et al. Ovarian endometrioid carcinomas resembling sex-cord stromal tumors. A clinicopathological analysis of 13 cases. Am J Surg Pathol. 1982;6:513–22.
23. Guerrieri C, Franlum B, Malstrom H. Ovarian endometrioid carcinomas simulating sex-cord stromal tumors. Int J Gynecol Pathol. 1998;17:266–71.
24. Ordi J, Schammel DP, Rasekh L, et al. Sertoliform endometrioid carcinomas of the ovary: a clinicopathological and immunohistochemical study of 13 cases. Mod Pathol. 1999;12:933–40.
25. Zhao C, Bratthauer GL, Barner R, et al. Comparative analysis of alternative and traditional immunohistochemical markers for the distinction of ovarian Sertoli cell tumors, from endometrioid tumors, and carcinoid tumor. Am J Surg Pathol. 2007;31:255–66.
26. McCluggage WG, Young RH. Ovarian Sertoli-Leydig cell tumors with pseudo-endometrioid tubules. Am J Surg Pathol. 2007;31:592–7.
27. Misir A, Sur M. Sertoliform endometrioid carcinoma of the ovary: a potential diagnostic pitfall. Arch Pathol Lab Med. 2007;131:979–81.
28. Tornos C, Silva EG, Ordonez NG, et al. Endometrioid carcinoma of the ovary with a prominent spindle cell component, a source of diagnostic confusion. A report of 14 cases. Am J Surg Pathol. 1995;19:1343–53.
29. Elishaev E, Gilks CB, Miller D, et al. Synchronous and metachronous endocervical and ovarian neoplasms; evidence supporting interpretation of the ovarian neoplasms as metastatic endocervical adenocarcinomas simulating primary ovarian surface epithelial neoplasms. Am J Surg Pathol. 2005;29:281294.
30. Ronnett BM, Yemelyanova AV, Vang R, et al. Endocervical adenocarcinomas with ovarian metastases: analysis of 29 cases with emphasis on minimally invasive cervical tumors and the ability of the metastases to simulate primary ovarian neoplasms. Am J Surg Pathol. 2008;32:1835–53.
31. McCormicki JK, Vang R, Yemelanova AV, et al. Women with undiagnosed colorectal adenocarcinomas presenting with ovarian metastases: clinicopathologic features and comparison with women having known colorectal adenocarcinoma and ovarian involvement. Int J Gynecol Pathol. 2008;27:182–90.
32. Lewis MR, Deavers MT, Silva EG, et al. Ovarian involvement by metastatic colorectal adenocarcinoma: still a diagnostic challenge. Am J Surg Pathol. 2006;30:177–84.
33. Clemet PB, Young RH, Scully RE. Endometrioid-like variant of yolk sac tumor: a clinicopathologic analysis of 9 cases. Am J Surg Pathol. 1987;11:767–78.
34. Schreiber Z, Liang S, Pearl M, et al. Helpful histologic features of serous carcinoma during intraoperative diagnosis of ovarian tumors. Mod Pathol. 2008;21 suppl 1:224A.
35. Patel K, Schreiber Z, Kane PB, et al. Intraoperative frozen section fixation with formalin improves diagnostic accuracy of tumors with clear cells, papillary architecture, or micropapillary type of invasion. Mod Pathol. 2008;21(suppl1):368A.
36. Stewart CJR, Brennan BA, Hammond IG, et al. Intraoperative assessment of clear cell carcinoma of the ovary. Int J Gynecol Pathol. 2008;27:475–82.
37. Kaplan KJ, Burgess JR, Dandberg GD, et al. Use of robotic telepathology for frozen section diagnosis: a retrospective trial of a telepathology system fro intraoperative consultation. Mod Pathol. 2002;15:1197–204.
38. Frierson HF, Galgano MT. Frozen section diagnosis by wireless telepathology and ultraportable computer: use in pathology resident/faculty consultation. Hum Pathol. 2007;38:1330–4.
39. Liang WY, Hsu CY, Lai CR, et al. Low-cost telepathology system for intraoperative frozen-section consultation: our experience and review of the literature. Hum Pathol. 2008;39:56–62.
40. Evans AJ, Chetty R, Clarke MB, et al. Primary frozen section diagnosis by robotic microscopy and virtual slide telepathology: the University Health Network experience. Hum Pathol. 2009;40:1070–81.
41. Fallon MA, Wilbur DC, Prasad M. Ovarian frozen section diagnosis. Use of whole-slide imaging shows excellent correlation between virtual slide and original interpretation in a large series of cases. Arch Pathol Lab Med. 2010;134:1020–3.
42. Park KJ, Reuter VE. The use of digital microscopy for frozen section diagnosis; a validation study from Memorial Sloan Kettering Cancer Center, New York. Poster presentation at Pathology Visions 2009 Conference, San Diego, CA. 13–15 Sept. 2009.

Surface Epithelial Tumors: Pathology Introduction

4

C. Blake Gilks

Keywords
Surface epithelial-stromal tumors • Surface epithelial tumors • Borderline tumors • Carcinoma • Tumor cell type

Definition

Surface epithelial-stromal tumors (more commonly referred to as simply "surface epithelial tumors") are defined by the World Health Organization as being those ovarian tumors that "originate from the ovarian surface epithelium or its derivatives and occur in women of reproductive age and beyond" [1]. This is an inadequate definition, as in recent years it has become clear that many of these tumors do not originate from the ovarian surface epithelium. Furthermore, although most common in women of reproductive age, they are encountered at all ages. A working definition of this category is: all primary epithelial ovarian tumors except teratomas and primitive germ cell tumors. Even this definition is problematic as some mucinous tumors of intestinal type are associated with a teratomatous component and may be of germ cell origin. It should be noted that epithelial differentiation can also be seen in sex cord-stromal tumors (e.g., retiform Sertoli-Leydig cell tumors) and small cell carcinoma, hypercalcemic type.

C.B. Gilks
Department of Pathology and Laboratory Medicine, University of British Columbia/Vancouver General Hospital, Vancouver, BC V5Z 4E3, Canada

Subclassification of Surface Epithelial Tumors

Surface epithelial tumors are classified based on tumor cell type (serous, mucinous, endometrioid, clear cell, transitional) and are then further subclassified as benign borderline or malignant (carcinoma) (Fig. 4.1a–f). In the past, the emphasis was primarily on whether tumors were benign, borderline, or malignant. More specific criteria for distinction between benign, borderline, and malignant are cell-type specific and will be described and illustrated in the following chapters where different cell types are considered individually. It is now recognized that tumor cell type also carries significant information, and accurate cell type assignment is important in clinical practice. Five subtypes of ovarian surface epithelial carcinoma account for as much as 98% of cases [2]. These subtypes are high-grade serous, low-grade serous, endometrioid, clear cell, and mucinous (Fig. 4.2a–e), and they can be very reproducibly diagnosed based on examination of routine hematoxylin and eosin-stained sections, with use of immunohistochemistry in problematic cases [3]. The importance of accurate cell type diagnosis is reviewed in detail below.

R.A. Soslow and C. Tornos (eds.), *Diagnostic Pathology of Ovarian Tumors*,
DOI: 10.1007/978-1-4419-9751-7_4, © Springer Science+Business Media, LLC 2011

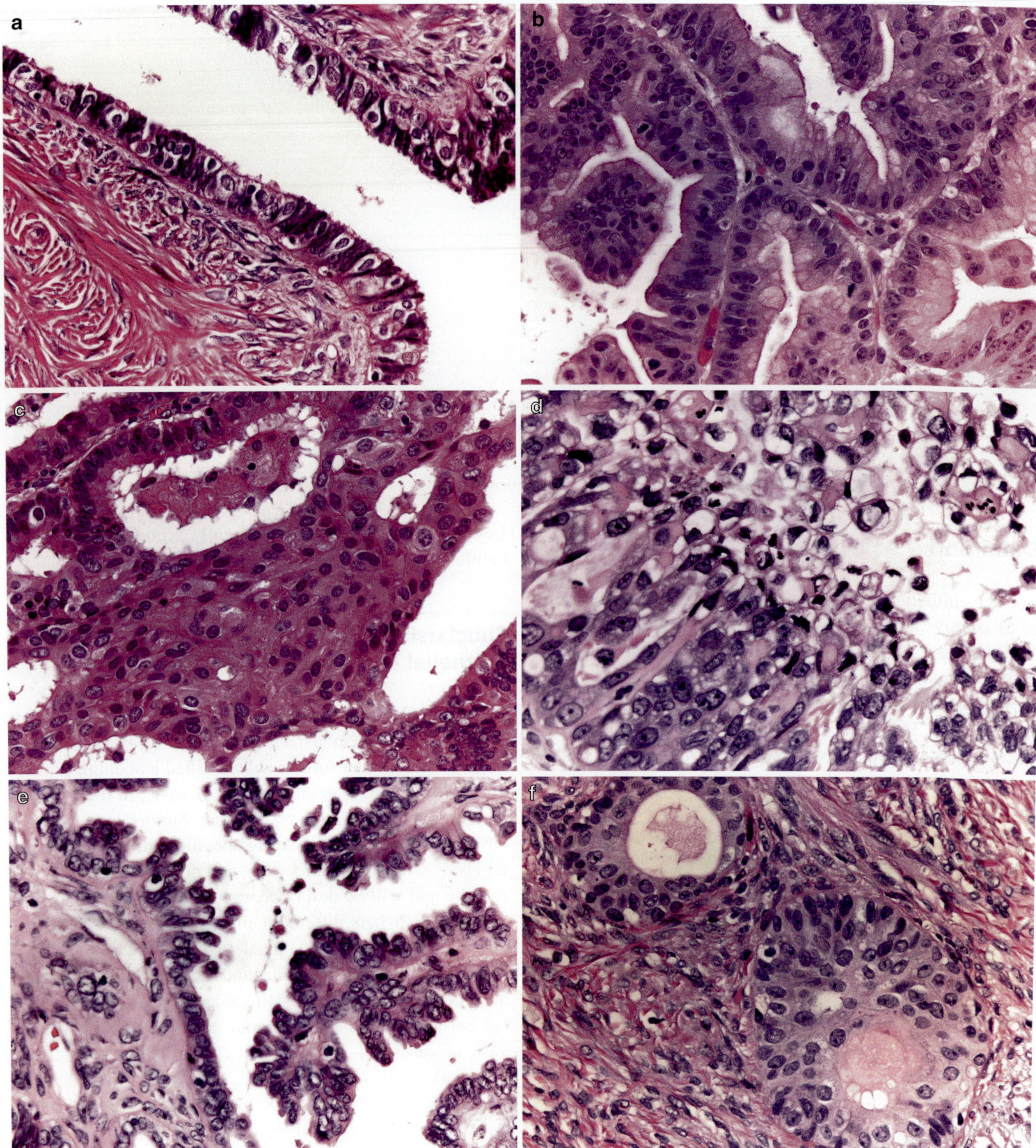

Fig. 4.1 The category of surface epithelial-stromal tumors encompasses a wide range of epithelial cell morphologies, including: (**a**) benign serous cystadenoma; (**b**) mucinous borderline tumor of intestinal type; (**c**) endometrioid carcinoma; (**d**) clear cell carcinoma; (**e**) serous borderline tumor; and (**f**) Brenner tumor

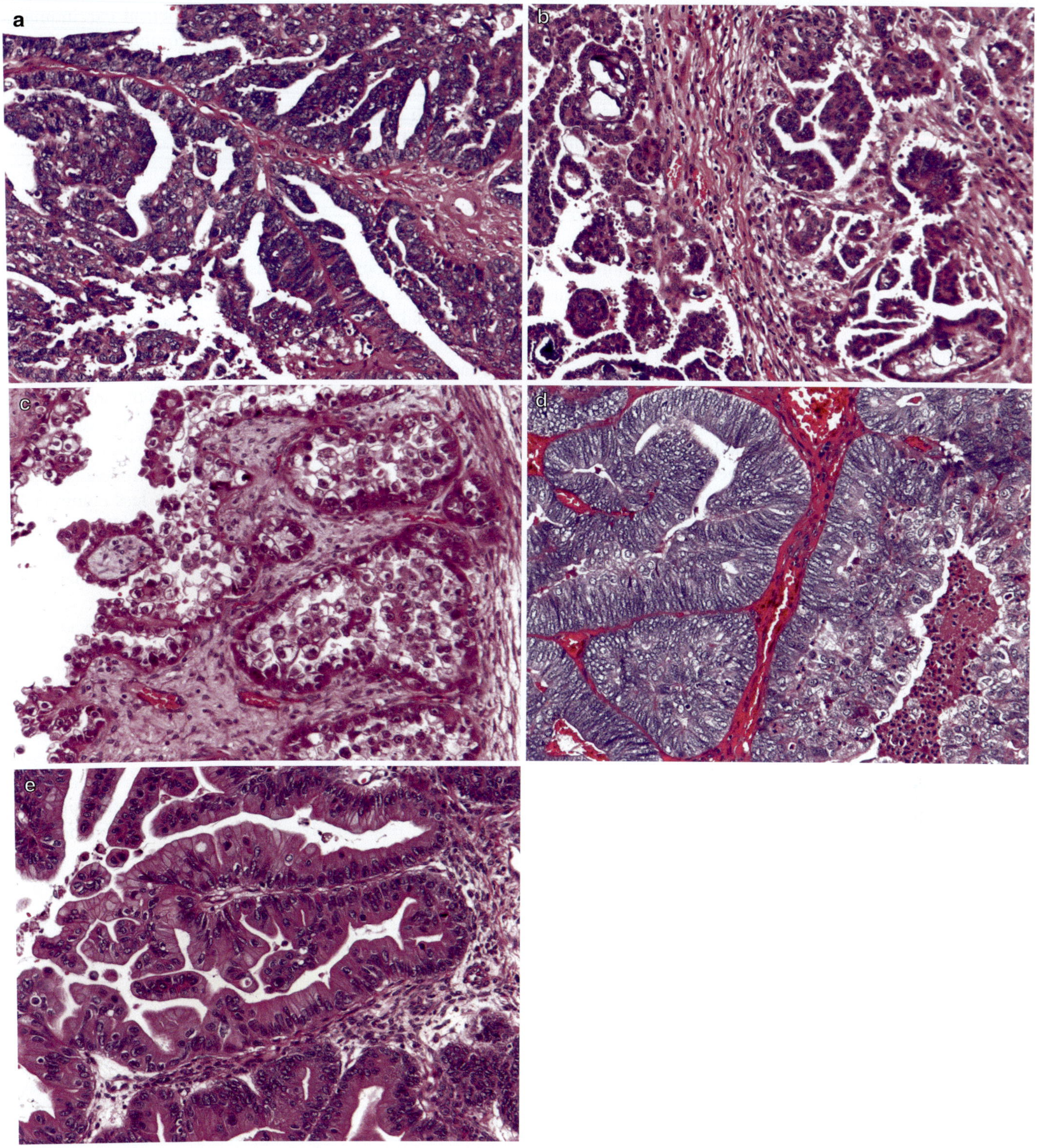

Fig. 4.2 The five main subtypes of ovarian surface epithelial carcinoma are: (**a**) high-grade serous carcinoma; (**b**) low-grade serous carcinoma; (**c**) clear cell carcinoma; (**d**) endometrioid carcinoma; (**e**) mucinous carcinoma. Together, these five subtypes account for more than 97% of cases of ovarian surface epithelial carcinoma

Why the Borderline (Low Malignant Potential/Atypical Proliferative) Category?

The ovarian surface epithelial tumors are distinguished by a category unique to ovarian surface epithelial tumors; borderline tumors (also referred to as low malignant potential or atypical proliferative tumors). Use of this category has resulted in confusion and controversy over the years, which continues to this day. This category was proposed in recognition that there are ovarian epithelial neoplasms that both morphologically and in terms of natural history seem to fall between usual carcinoma and completely benign. Morphologically, the defining feature has been absence of invasion as a basis for distinction between borderline tumor and carcinoma. The ovary does not have a mucosa or exocrine glandular architecture, and thus there is no clear definition of in situ disease. Arguably, the category of borderline has not proven useful for the categories of endometrioid, clear cell, or transitional, as there are no reported cases of borderline tumors of these cell types (when well sampled to exclude invasion, i.e., carcinoma) following a malignant course. These tumors show atypical or "pre-malignant" change, but are in no way a distinct tumor type that is intermediate between benign and carcinoma. They are best considered to be tumors at an early stage of progression toward full-blown carcinoma. Mucinous borderline tumors of intestinal type similarly appear to be entirely benign unless there is a component of intraepithelial carcinoma or microinvasion [4]. Mucinous borderline tumors of intestinal type with intraepithelial carcinoma, when recurrent or presenting at advanced stage, have a natural history indistinguishable from mucinous carcinoma. Thus, while examples of tumors in each of the non-serous categories with epithelial atypia are undoubtedly encountered in practice, they appear to lack malignant potential if the features of carcinoma (or intraepithelial carcinoma) are not present; there is no convincing evidence of a borderline category with a distinct natural history in these non-serous subtypes. This leaves serous and seromucinous borderline tumors as the only members of this group designated as "borderline" that do have a distinct morphology and natural history.

Magnitude of the Clinical Problem Posed by Surface Epithelial Tumors

Surface epithelial carcinomas are the most common ovarian malignancies, and ovarian carcinoma is the most common cause of death due to cancer arising in the female reproductive tract, even though endometrial carcinoma is more common, because of the unfavorable prognosis of ovarian carcinoma. There is considerable worldwide variation in the incidence of ovarian surface epithelial carcinoma, with higher incidence rates reported from North America and Europe, compared to Asia or Africa.

Prevalence

High-grade serous carcinomas account for approximately 70% of cases of ovarian carcinoma, while clear cell and endometrioid each account for 10% of cases; mucinous and low-grade serous carcinomas are relatively less common, and each comprise approximately 3.5% of cases, respectively (Fig. 4.3). There remain a few cases (<3%) of rare subtypes (undifferentiated, squamous, transitional, or malignant Brenner) or where the tumor morphology does not allow assignment to a specific cell type (adenocarcinoma N.O.S.). High-grade serous carcinoma is relatively less common in Japan, with relatively more cases of clear cell carcinoma reported. It is not clear whether this is due to differences in the incidence of high-grade serous carcinoma, clear cell carcinoma, or both subtypes.

Ovarian surface epithelial carcinomas of mixed type have decreased dramatically over recent years. This change is largely attributable to recognition that tumors with high-grade serous differentiation frequently also show glandular differentiation, solid growth, or both patterns, in addition to the typical serous architecture (Fig. 4.4a–c). These tumors, in the past, were often diagnosed as mixed serous and endometrioid, or serous and undifferentiated carcinoma, respectively. Based on immunophenotypic

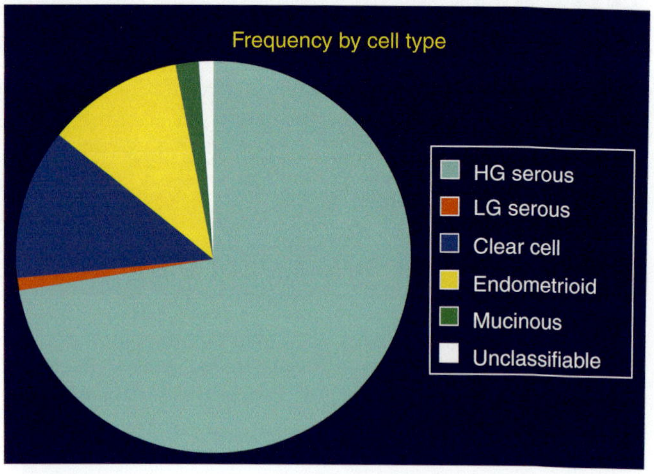

Fig. 4.3 The frequency of different subtypes of ovarian surface epithelial carcinoma in British Columbia, Canada. See also Table 4.2. (Adapted from Gilks and Prat [5])

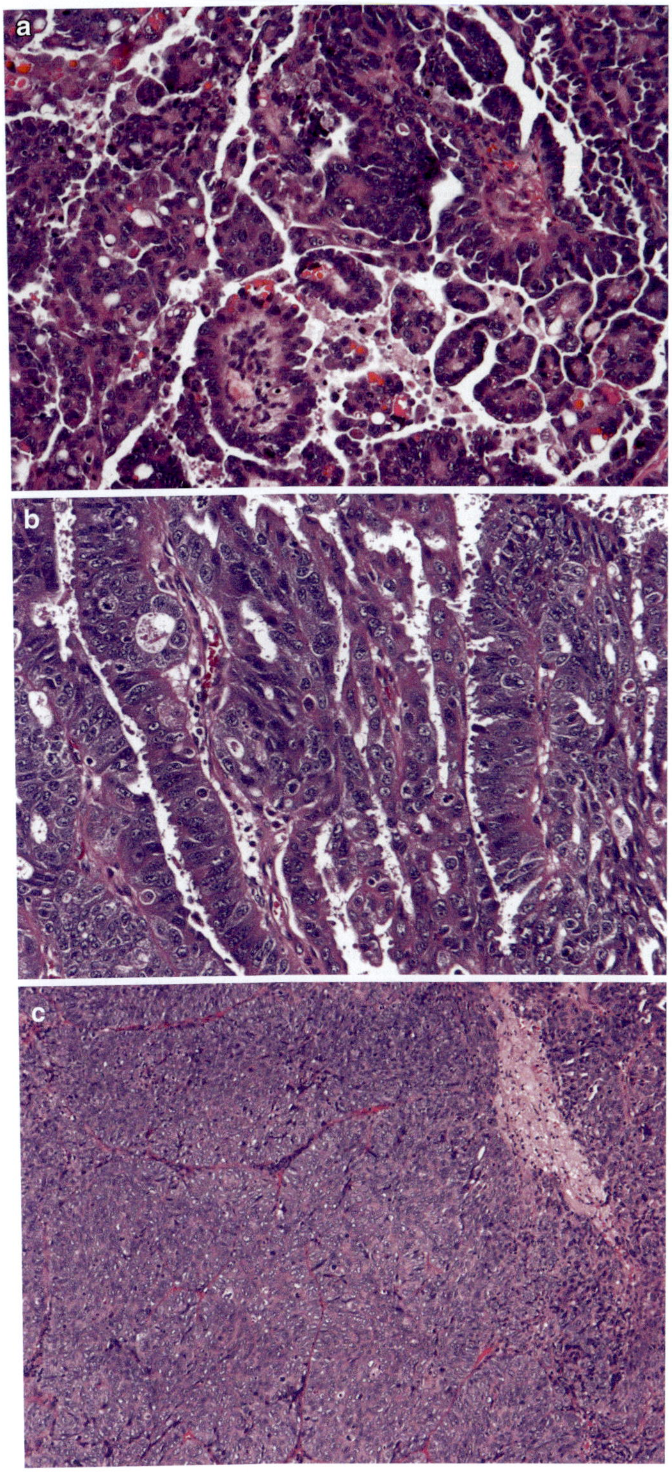

Fig. 4.4 The most common architectural patterns of high-grade serous carcinoma are: (**a**) papillary; (**b**) glandular; and (**c**) solid. These patterns often coexist in a tumor

and molecular data, it is now appreciated that these tumors are indistinguishable from typical high-grade serous carcinomas and are best diagnosed as such. This has the effect of dramatically improving the reproducibility of ovarian carcinoma subtype assignment and better reflects underlying disease biology [2, 3].

There remain ovarian surface epithelial carcinomas of mixed type, defined as tumors comprised of two or more tumor cell types, in which the minority component(s) account for at least 10% of the tumor, but they are relatively uncommon. The most common is an admixture of endometrioid and clear cell carcinoma.

There is relatively less data on the relative frequency of borderline tumors, based on cell type, but the vast majority are of serous, seromucinous, or mucinous, intestinal type. Borderline tumors are less common than carcinomas but will still be encountered regularly in practice, especially serous borderline tumors and mucinous borderline tumors of intestinal type.

For practical purposes, benign surface epithelial tumors are serous, mucinous, or transitional (Brenner) type, when cell type can be recognized. Benign clear cell and endometrioid tumors are rare. One regularly encounters in practice simple cysts of the ovary lined by non-descript cuboidal or flat epithelium without diagnostic features of specific cell type differentiation. These are diagnosed as "simple cysts," and it is likely that many of these are serous cystadenomas, given the frequency with which focal serous differentiation, with ciliated cells, is seen in such cases.

Importance of Accurate Diagnosis of Ovarian Carcinoma Subtypes

Different subtypes of ovarian carcinoma differ with respect to risk factors, precursor lesions, molecular abnormalities (including biomarker expression), pattern of spread and natural history, and response to treatment (summarized in Table 4.1) [5]. In effect, they are different diseases, which have in common epithelial lineage and manifestation as an ovarian mass. This is similar to how lymphomas or breast carcinomas are now understood, i.e., as heterogeneous groups of diseases, where subtype information is necessary to understand patient prognosis and plan treatment. As the characteristics of the different epithelial carcinoma subtypes will be covered in more detail in the following chapters, only a brief discussion will be provided here, addressing reasons why accurate histopathological diagnosis of cell type is important.

The first reason subtype is critically important is that prognosis and response to chemotherapy differ between the different subtypes. The stage at diagnosis is significantly different for the different subtypes of ovarian carcinoma, with high-grade serous carcinoma more likely to present with high-stage (Stage III or IV) disease compared to clear cell,

Table 4.1 Differences between ovarian carcinoma subtypes [5]

	HGSC	LGSC	MC	EC	CC
Risk factors	*BRCA1/2*	?	?	Lynch syndrome	Lynch syndrome
Precursor lesions	Tubal intraepithelial carcinoma	Serous borderline tumor	Cystadenoma/ borderline tumor?	Endometriosis	Endometriosis
Pattern of spread	Very early transcoelomic spread	Transcoelomic spread	Usually confined to ovary	Usually confined to pelvis	Usually confined to pelvis
Molecular abnormalities	*BRCA, p53*	*BRAF, KRAS*	*KRAS, HER2*	*PTEN*	HNF1
Chemosensitivity	High	Intermediate	Low	High	Low
Prognosis	Poor	Intermediate	Favorable	Favorable	Intermediate

HGSC high-grade serous carcinoma, *LGSC* low-grade serous carcinoma, *MC* mucinous carcinoma, *EC* endometrioid carcinoma, *HNPCC* hereditary non-polyposis colorectal carcinoma

Table 4.2 Stage distribution of ovarian carcinoma subtypes [6]

Stage	High-grade serous	Clear cell	Endometrioid	Mucinous	Low-grade serous	Other	All
I/II	35.5%	26.2%	26.6%	7.5%	1.9%	3.2%	37.7%
III/IV	87.7%	4.5%	2.5%	1.2%	5.3%	1.2%	62.3%
All	68.1%	12.2%	11.3%	3.4%	3.4%	1.6%	100%

endometrioid, or mucinous carcinomas, which typically present at stage I or II (Table 4.2) [6]. Even within the low-stage group (stage I or II), there are significant differences in prognosis between tumor cell types (Fig. 4.5) [7]. With regards to chemotherapy, it is appreciated that while most high-grade serous carcinomas respond, at least initially, to platinum/taxane chemotherapy, response rates for patients with clear cell or mucinous carcinoma are much lower, and a consensus conference sponsored by the United States National Cancer Institute identified subtype-specific clinical trials for patients with clear cell and mucinous carcinomas as a priority [8]. It is clear that the treatment offered to patients with ovarian carcinoma is based on the best treatment for the most common subtype of ovarian carcinoma, the high-grade serous subtype. To move forward and be able to offer patients more appropriate, subtype-specific treatment, there will have to be clinical trials; and for this to take place, patients must be diagnosed accurately by pathologists. Ultimately, progress in understanding and treating patients with ovarian carcinoma will depend on accurate pathological diagnosis. Recently it has been observed that clear cell carcinomas are more likely to be cured by radiotherapy [9], and it is this type of cell type-specific study that will become the norm in the future.

Subtype assignment in ovarian cancer research is also critically important, given the molecular and biological differences between the subtypes. Most biomarkers of prognostic significance are markers of tumor cell type (e.g., the more aggressive serous carcinomas) and do not have prognostic significance that is independent of tumor cell type [10]. An example is p53 overexpression, which in a meta-analysis of a large number of studies was shown to be associated with a worse prognosis in ovarian carcinoma [11]; however, p53 overexpression is associated with high-grade serous carcinoma, which in turn has an unfavorable prognosis. In these studies consisting of patients with a variety of ovarian carcinoma subtypes, p53 overexpression was serving as a marker of high-grade serous type. Looking specifically within the high-grade serous subtype, p53 overexpression is associated with a favorable prognosis [12]. Another example, where an important finding was obscured when tumor cell type was not considered, is in studies of tumor-infiltrating lymphocytes. The presence of tumor-infiltrating lymphocytes is of prognostic significance in the high-grade serous subtype, and this effect can be missed if a group of ovarian carcinomas consisting of a mixture of cell types is studied [13]. These examples illustrate how failure to appreciate the unique features of ovarian carcinoma subtypes and consider them as different diseases in study design can lead to misleading results. Finally, given that tumor cell type assignment has progressed in recent years, it is completely inappropriate to do retrospective cell type-specific studies on ovarian cancer without pathology review, as review using current diagnostic criteria will result in reclassification of a significant number of cases [2].

Accurate cell type classification is also necessary for borderline tumors, as pointed out in a recent consensus conference [4, 14]. Serous and mucinous (intestinal type) borderline tumors have different natural histories, molecular abnormalities, and prognoses. Accurate distinction between these two tumor types is straightforward because they are so differentiated. Arguably, for benign tumors, tumor cell type is not of clinical significance; in these tumors, however, the appropriate classification is usually obvious.

a

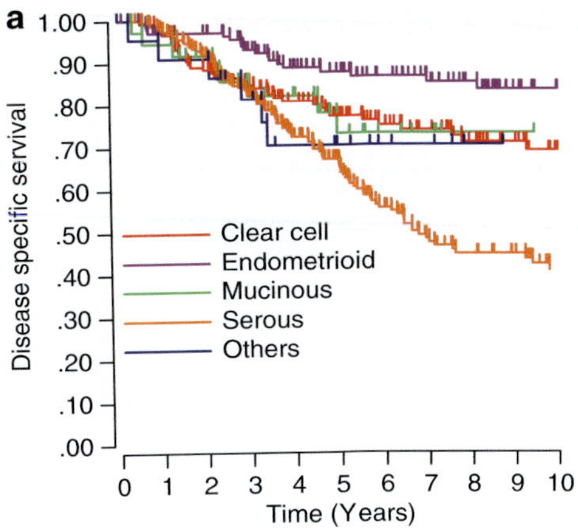

Logrank: Hazard ratio					
	Clear cell	Endometrioid	Mucinous	Serous	Others
Clear cell	1.00	0.50	0.99	1.96	1.20
Endometrioid	2.02	1.00	2.02	4.10	2.47
Mucinous	1.01	0.50	1.00	1.91	1.20
Serous	0.51	0.24	0.52	1.00	0.64
Others	0.83	0.40	0.83	1.56	1.00

Logrank: Chi-Square *p*					
	Clear cell	Endometrioid	Mucinous	Serous	Others
Clear cell		0.02	0.97	< 0.001	0.40
Endometrioid	0.02		0.10	< 0.001	0.01
Mucinous	0.97	0.10		0.06	0.49
Serous	< 0.001	< 0.001	0.06		0.41
Others	0.40	0.01	0.49	0.41	

b

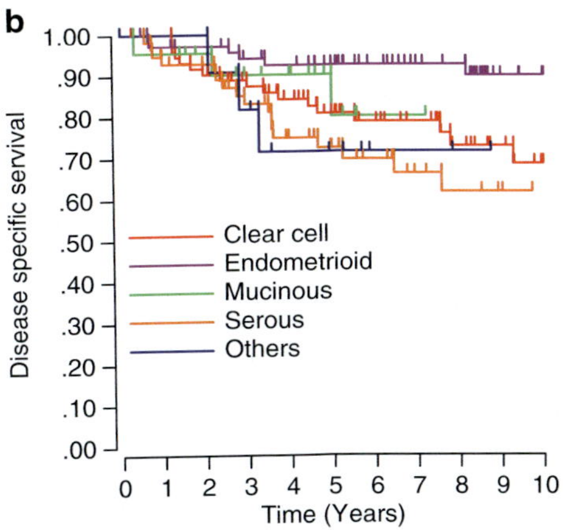

Logrank: Hazard ratio					
	Clear cell	Endometrioid	Mucinous	Serous	Others
Clear cell	1.00	0.33	0.71	1.45	1.35
Endometrioid	3.02	1.00	2.04	4.25	3.72
Mucinous	1.41	0.49	1.00	1.98	1.78
Serous	0.69	0.24	0.51	1.00	0.94
Others	0.74	0.27	0.56	1.07	1.00

Logrank: Chi-Square *p*					
	Clear cell	Endometrioid	Mucinous	Serous	Others
Clear cell		0.01	0.58	0.28	0.63
Endometrioid	0.01		0.30	< 0.001	0.05
Mucinous	0.58	0.30		0.27	0.47
Serous	0.28	< 0.001	0.27		0.92
Others	0.63	0.05	0.47	0.92	

Fig. 4.5 Outcomes (disease-specific survival) for patients with different subtypes of ovarian surface epithelial carcinoma in a cohort of patients who had no macroscopic residual disease after primary debulking surgery: (**a**) all patients; and (**b**) patients with stage I tumors. The outcomes for patients with different cell types differ, even when corrected for residual disease (**a**) and stage (**b**). The hazard ratios (HR) and *P*-values are shown to the right of the survival curves for each pair-wise comparison between cell types (Reproduced from Gilks et al. [2])

Problems in Pathological Assessment of Subtype

There has been significant progress in accurate cell type diagnosis of ovarian carcinoma (reviewed in Reference 5). This has been driven primarily by four completely separate lines of investigation: (1) the emergence of immunomarkers such as WT-1, and by global gene expression profiling studies, which demonstrated the clear separation between tumor cell types, based on gene expression signatures, and also showed that tumors diagnosed as "high-grade endometrioid" were indistinguishable from high-grade serous, and distinct from low-grade endometrioid (Fig. 4.6a–b); (2) the compelling body of work establishing that low-grade serous and high-grade serous carcinomas are different diseases (Fig. 4.7a–b); (3) demonstration that serous carcinomas with clear cell change are simply a morphological variant of high-grade serous carcinoma and are unrelated to clear cell carcinoma (Fig. 4.8a–c); and (4) evidence that cases diagnosed as "transitional cell carcinoma" are high-grade carcinomas that are indistinguishable immunophenotypically from high-grade serous carcinoma (Fig. 4.9a–c), and differ from transitional cell carcinomas arising from Brenner tumors (malignant Brenner tumors), leading to the dropping of the ill-defined category of high-grade transitional cell carcinoma, except for those rare tumors arising from Brenner tumors. As a result, there is now very reproducible diagnosis of ovarian cancer cell types. In a study involving pathologists from across Canada, it was possible to achieve greater than 95% inter-observer agreement in cell type diagnosis [3]. This was possible after a training exercise that typically takes less than 1 h, and is available online at http://www.gpecimage.ubc.ca/aperio/images/transcanadian/, to reproducibly diagnose ovarian carcinoma subtype. In most cases, this can be done based on hematoxylin and eosin-stained sections, with use of immunostaining in occasional cases. These ovarian carcinoma subtypes reflect underlying molecular abnormalities and

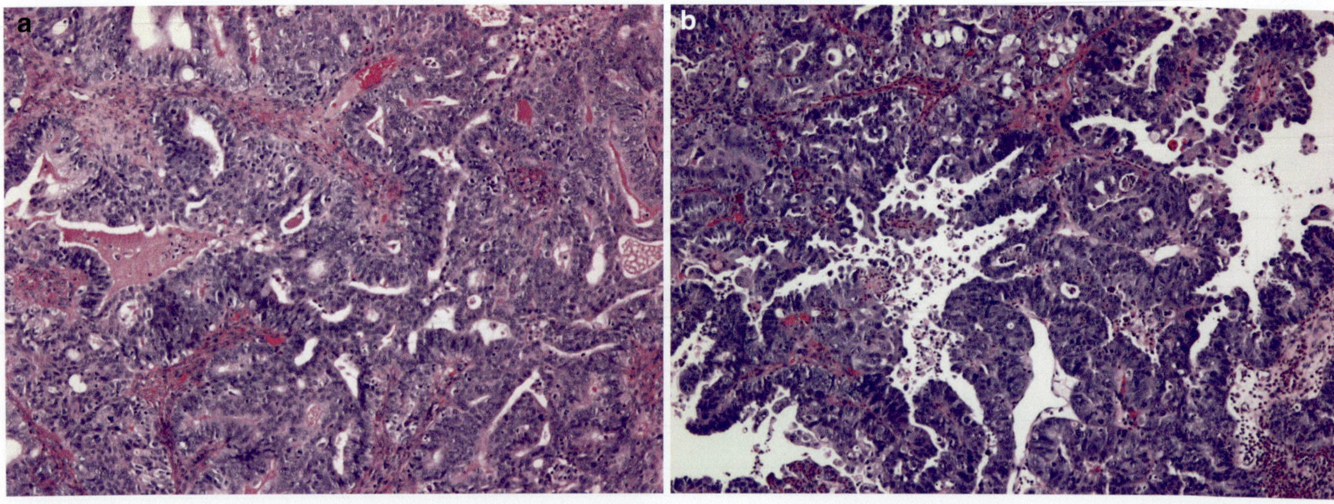

Fig. 4.6 A tumor reported as "high-grade endometrioid" carcinoma (**a**). Other areas of this cytologically high-grade tumor showed typical high-grade serous morphology (**b**). The tumor cells were strongly WT-1 positive on immunostaining. This tumor should be diagnosed as high-grade serous carcinoma

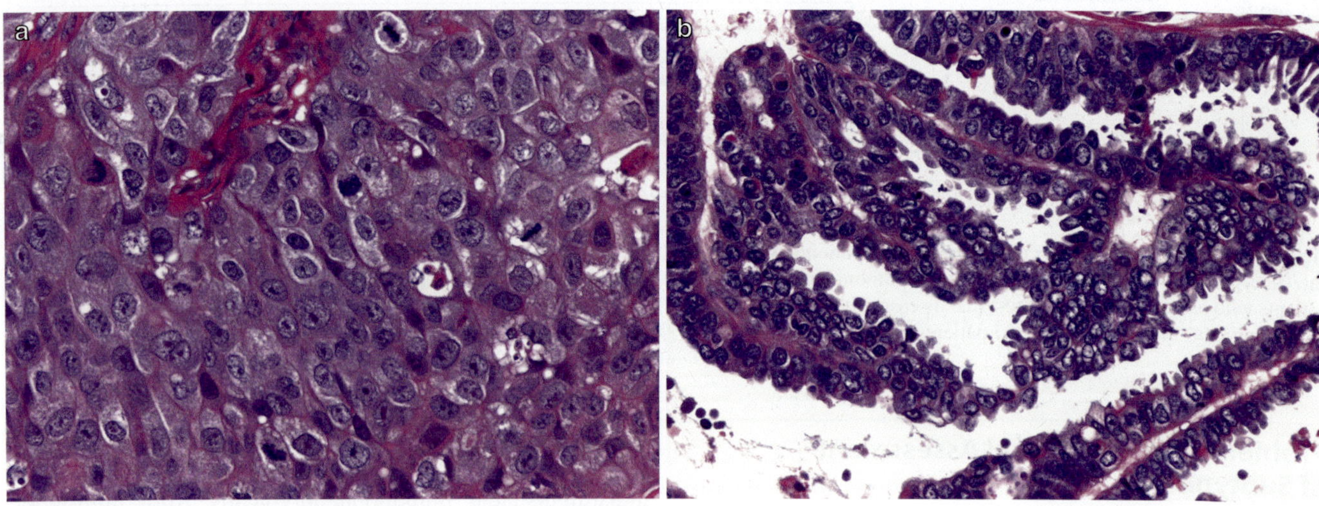

Fig. 4.7 Cytological features of high-grade serous (**a**) and low-grade serous (**b**) carcinoma, showing the greater variation in nuclear size and higher mitotic rate in the former

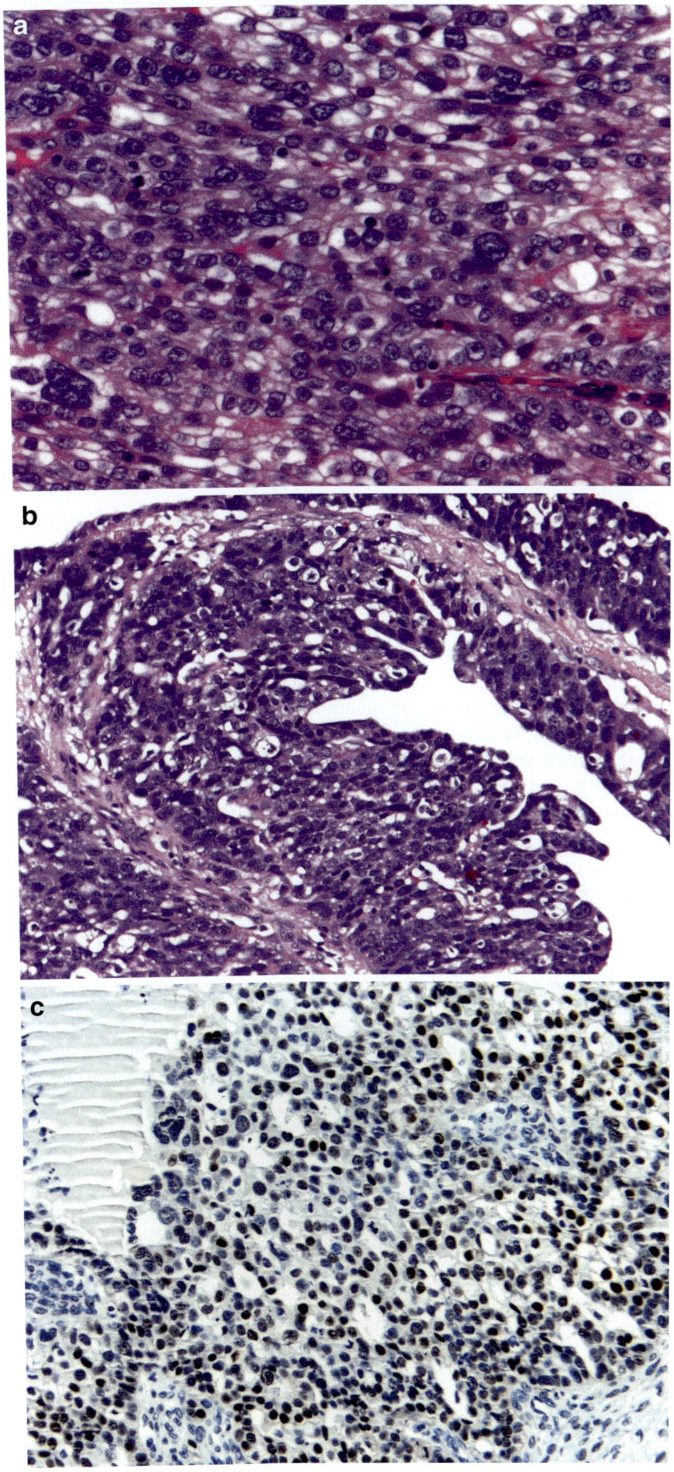

Fig. 4.8 Clear cell change in a high-grade serous carcinoma (**a**). Such tumors can be recognized as high-grade serous carcinoma because of the presence of typical serous morphology in other areas (**b**), and based on the immunophenotype: ER (**c**) and WT-1 positive, and HNF-1beta negative

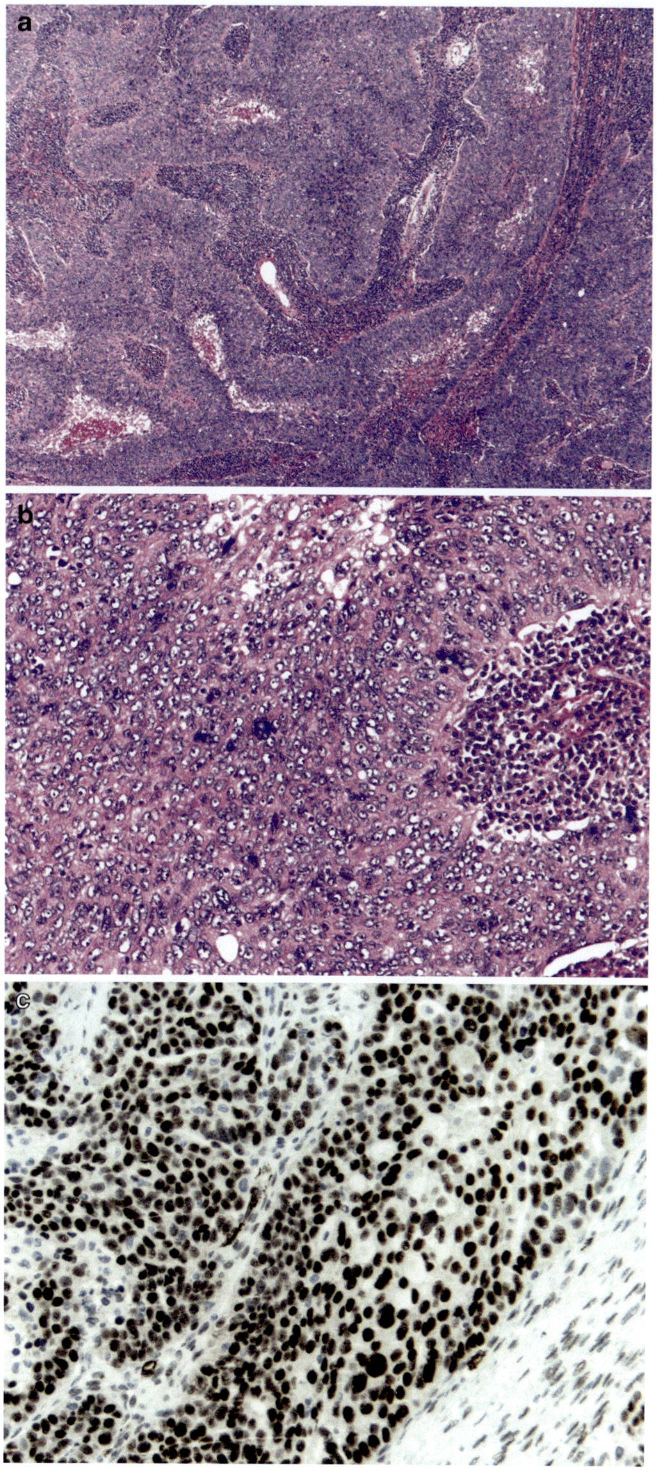

Fig. 4.9 Transitional-like growth of high-grade serous carcinoma (**a**). Higher power magnification shows high-grade cytological features with considerable variation in nuclear size, typical of high-grade serous carcinoma (**b**), and the tumor cells stain positively for WT-1 (**c**). This tumor is best considered a morphological variant of high-grade serous carcinoma

Table 4.3 Diagnoses that are overused and only to be made with trepidation

High-grade endometrioid carcinoma (in the absence of a low-grade endometrioid component or squamous differentiation): Likely Dx is high-grade serous carcinoma
Transitional cell carcinoma (in the absence of a benign or borderline Brenner tumor): Likely Dx is high-grade serous carcinoma
Mixed carcinomas composed of admixtures of high-grade serous, high-grade endometrioid, and/or undifferentiated carcinoma: Likely Dx is high-grade serous carcinoma
Mixed carcinomas composed of admixtures of high-grade serous and clear cell carcinoma: Likely Dx is high-grade serous carcinoma (with clear cell change)
Mixed serous borderline tumor and high-grade serous carcinoma: Likely Dx is low-grade serous carcinoma
Advanced stage mucinous borderline tumor of intestinal type: Likely Dx is mucinous carcinoma

are of increasing importance as there is movement toward subtype-specific treatments for ovarian carcinoma.

The shift in patterns of diagnosis of ovarian epithelial carcinoma cell type is ongoing. Table 4.3 has a list of diagnoses that should be made rarely, if ever, in practice, as well as providing a more likely diagnosis for consideration. More detailed discussion of differential diagnosis will be provided in subsequent chapters.

References

1. Tavassoli FA, Devilee P. World Health Organization classification of tumors: pathology and genetics. Tumours of the breast and female genital organs. Lyon: IARC Press; 2003.
2. Gilks CB, Ionescu D, Kalloger S, et al. Tumor cell type can be reproducibly diagnosed and is of independent significance in patients with maximally debulked ovarian carcinoma. Hum Pathol. 2008;39:1239–51.
3. Köbel M, Kalloger SE, Baker PM, et al. Diagnosis of ovarian carcinoma subtype is highly reproducible: a transcanadian study. Am J Surg Pathol. 2010;34:984–93.
4. Ronnett BM, Kajdacsy-Balla A, Gilks CB, et al. Mucinous borderline tumors: points of general agreement and persistent controversies regarding nomenclature, diagnostic criteria, and behavior. Hum Pathol. 2004;35:949–60.
5. Gilks B, Prat J. Ovarian carcinoma pathology and genetics: recent advances. Hum Pathol. 2009;40:1213–23.
6. Köbel M, Kalloger SE, Huntsman DG, et al. Differences in tumor type in low-stage versus high-stage ovarian carcinomas. Int J Gynecol Pathol. 2010;29:203–11.
7. Köbel M, Kalloger SE, Santos JE, et al. Tumor type and substage predict survival in stage I and II ovarian carcinoma: insights and implications. Gynecol Oncol. 2010;116:50–6.
8. Trimble EL, Fountain J, Birrer MJ. Recommendations of the 2005 ovarian cancer state of the science meeting. Gynecol Oncol. 2006;103:S26.
9. Nagai Y, Inamine M, Hirakawa M, et al. Postoperative whole abdominal radiotherapy in clear cell adenocarcinoma of the ovary. Gynecol Oncol. 2007;107:469–73.
10. Köbel M, Kalloger SE, Boyd N, et al. Ovarian carcinoma subtypes are different diseases: implications for biomarker studies. PLoS Med. 2008;5:e232.
11. de Graeff P, Crijns AP, de Jong S, et al. Modest effect of p53, EGFR, and HER-2/neu on prognosis in epithelial ovarian cancer: a meta-analysis. Br J Cancer. 2009;101:149–59.
12. Köbel M, Reuss A, du Bois A, et al. The biological and clinical value of p53 expression in pelvic high-grade serous carcinomas. J Pathol. 2010;222((2):191–8. epub ahead of print.
13. Clarke B, Tinker AV, Lee CH, et al. Intraepithelial T cells and prognosis in ovarian carcinoma: novel associations with stage, tumor type, and BRCA1 loss. Mod Pathol. 2009;22:393–402.
14. Bell DA, Longacre TA, Prat J, et al. Serous borderline (low malignant potential, atypical proliferative) ovarian tumors: workshop perspectives. Hum Pathol. 2004;35:934–48.

Surface Epithelial Tumors: Clinical Introduction

5

Katherine M. Bell-McGuinn and Mario M. Leitao Jr.

Keywords
Adnexal mass • Frozen section • Adjuvant chemotherapy • Neoadjuvant chemotherapy • Recurrent ovarian cancer, Ovarian cancer

Introduction

Adnexal masses are common in females of all ages, but epithelial neoplasms of the ovary are most often diagnosed after adolescence. The distribution of causes of adnexal masses in women of various ages must be kept in mind during the clinical and pathologic assessment of adnexal masses. Adnexal masses are relatively uncommon in premenarchal females, but large follicular cysts may sometimes be seen in newborns secondary to maternal estrogen exposure. Although they are rare, malignant germ cell tumors are the most common adnexal neoplasm in premenarchal girls. The vast majority of adnexal masses in adolescents are normal functional cysts. The most common neoplastic tumors in this age group are mature teratomas, and the most common malignant tumors are malignant germ cell tumors. In premenopausal women, the most common cause of an adnexal mass is still a benign functional cyst, but endometriomas and ectopic pregnancies can also have similar presentations. The most common neoplasms are mature teratomas, but ovarian serous carcinomas can sometimes arise in this age group. The most common adnexal masses in peri- and postmenopausal females again are still benign cysts, functional or not. Serous cystadenomas and carcinomas are the most common type of epithelial neoplasms in this age group.

This chapter will be limited to important clinical aspects of epithelial ovarian neoplasms. Ovarian cancer affects over 200,000 women worldwide, and nearly 125,000 women die annually from ovarian cancer [1]. Ovarian cancer is the fifth leading cause of cancer in the United States, and an estimated 21,550 women will be diagnosed with ovarian cancer in 2009. An estimated 14,600 among them will succumb to the disease [2]. The most common ovarian cancers are epithelial, with serous carcinomas accounting for the majority.

Protective and Risk Factors

The patient's past medical history and family history are important to pathologists as this may help focus the differential diagnosis. Age is extremely important as discussed above. There are also other patient factors and exposures that have been associated with either a decreased or increased risk of ovarian carcinoma. The baseline estimated lifetime risk of developing ovarian cancer is 1.4%. Infertility, nulliparity, late menopause, and early menopause have all been associated with modest increased risks of developing ovarian cancer (1.5–2.0-fold increase), whereas multiparity, oral contraceptive use, hysterectomy, and tubal ligation have all been associated with decreased risks.

The greatest known risk factor for the development of ovarian cancer is the presence of germline *BRCA1* or *BRCA2* mutations. Women from Lynch Syndfrome (HNPCC) kindreds are also at an increased risk of ovarian cancer.

M.M. Leitao Jr. (✉)
Department of Surgery, Gynecology Service, Memorial Sloan-Kettering Cancer Center, New York, NY, USA

R.A. Soslow and C. Tornos (eds.), *Diagnostic Pathology of Ovarian Tumors*,
DOI: 10.1007/978-1-4419-9751-7_5, © Springer Science+Business Media, LLC 2011

Hereditary Syndromes Associated with Ovarian Cancers

Hereditary syndromes are discussed in detail in Chap. 18. The estimated lifetime risk of developing ovarian cancer in patients with *BRCA1* or *BRCA2* mutations is 39% and 11%, respectively, based on pooled estimates [3]. The risk appears to start to increase at the age of 40 for patients with *BRCA1* mutations and the age of 50 for patients with *BRCA2* mutations. It is recommended that women with these mutations strongly consider prophylactic removal of both tubes and ovaries by these ages or once child-bearing is completed. There is no clear association with uterine malignancies and *BRCA* mutations and, therefore, no clear proven benefit to prophylactic hysterectomy. Fallopian tube carcinomas are increased in these patients. Very careful processing of the ovaries and tubes is imperative and requires extensive sectioning, much more than that done for ovaries removed at the time of hysterectomy for benign conditions, as occult malignancies can be diagnosed at the time of prophylactic surgery. Additionally, peritoneal washings are important and must be performed at the time of prophylactic surgery.

Aside from the necessity to identify which families have inherited predispositions to develop ovarian and breast carcinoma, there are also theoretical and practical reasons why it is important to determine which ovarian carcinomas have intact *BRCA* function. Numerous studies have now reported that *BRCA*-associated ovarian carcinomas are relatively prognostically favorable as compared to sporadic ovarian carcinoma [4–6]. Much of this survival benefit is likely provided by increased sensitivity to platinum-containing chemotherapeutic agents [7, 8]. Another factor that could account for increased survivals are the presence of tumor infiltrating lymphocytes in most *BRCA*-associated ovarian carcinomas [9, 10].

Ovarian cancer risk is also increased in women from Lynch Syndrome kindreds. The lifetime risk of ovarian cancer with Lynch Syndrome is 10% [11]. Since the risk of endometrial cancer is much greater than that of ovarian cancer prophylactic hysterectomy and removal of both tubes and ovaries is recommended once child-bearing is complete. Careful processing of the uterus, fallopian tubes, and ovaries is also extremely important.

Preoperative Evaluation of Adnexal Masses

Serum CA-125 is a tumor marker commonly assessed in patients with ovarian carcinomas. However, its role in screening and diagnostic evaluation of adnexal masses is extremely limited. An elevated serum CA-125 should not inappropriately influence the pathologic diagnosis. There are multiple causes of serum CA-125 elevations (Table 5.1). Serum CA-125 elevations in pre- and perimenopausal women are particularly problematic as the most common causes are not malignant. Additionally, a normal serum CA-125 does not eliminate the possibility of malignancy as nearly 50% of early stage ovarian carcinomas are not associated with elevated serum CA-125.

Table 5.1 Select possible causes of serum CA-125 elevations

Benign conditions	Malignant conditions
Leiomyomata uteri (fibroids)	Ovarian cancer
Endometriosis	Lung cancer
Benign ovarian cysts/masses	Breast cancer
Any condition leading to peritoneal, pleural, or pericardial inflammation	Pancreatic cancer
Intraperitoneal adhesions	Gastrointestinal cancer
Pregnancy	Metastatic endometrial and cervical cancer
Adenomyosis	Metastatic cancer to the ovary
Peritoneal tuberculosis	Lymphoma
Liver disease (especially cirrhosis with ascites)	Plasma cell dyscrasias
Heart failure	

The appropriate management of adnexal masses diagnosed preoperatively must be individualized. Each patient's history, as well as the imaging characteristics, must all be considered. In general, simple asymptomatic cysts most often can be followed conservatively in any age group. Solid masses are probably best managed with surgical intervention in most females, regardless of age. The management of complex masses is challenging and will depend on the individual case. Percutaneous biopsies or drainage of isolated adnexal masses is not generally recommended. Percutaneous biopsies are associated with technical difficulties as well as false-negative results. Most importantly, percutaneous biopsy of an unknown malignancy will upstage patients automatically and will lead to the use of adjuvant chemotherapy in a patient who may have otherwise not required it.

The best method of obtaining a definitive diagnosis is surgical. Ideally, removal of the entire ovary will result in the lowest risk of cyst or mass rupture and often facilitates the surgical procedure. Intraoperative rupture will upstage patients and will mandate additional chemotherapy. However, there is concern about possible impaired ovarian function and fertility by removing ovaries in young women. An attempt at merely excising the mass from the ovary (cystectomy) is reasonable in young women. Care must be taken to not rupture the cyst or mass. Cystectomy or removal of the ovary can be safely performed via laparoscopy by experienced surgeons.

Intraoperative Decision Making

Masses Confined to the Adnexa

A comprehensive discussion of frozen section is provided in Chap. 3. Accurate intraoperative frozen section (FSC) assessment of adnexal masses is crucial. The results of FSC assessment often determine the extent of the surgical procedure and must not be taken lightly. If the result is benign, then nothing further is needed. If only a cystectomy or salpingo-oophorectomy was completed laparoscopically or robotically, then the patient would be discharged the same day. If the result is malignancy, then a comprehensive staging procedure is required which would require at least a 1-day hospital stay if completed in a minimally invasive fashion. However, conversion from a minimally invasive procedure to a large vertical laparotomy incision may be necessary if the surgeon cannot perform the required staging procedures in a minimally invasive fashion. A diagnosis of malignancy rendered postoperatively when the frozen section had been interpreted as benign will often necessitate a second surgical procedure. Some of the most difficult surgical decisions arise when there is an FSC diagnosis of a borderline tumor (tumor of low malignant potential).

Table 5.2 describes the most important questions that should be addressed by FSC assessment and will have the greatest impact on the extent of surgery. The obvious first question is whether an adnexal mass is benign. If so, then often the surgical procedure is complete; hysterectomy, removal of the ipsilateral adnexa (if only excision of the mass was performed), removal of the contralateral adnexa, and comprehensive surgical staging are not required. These additional procedures may not be needed if the adnexal mass is felt to be metastatic from a non-gynecologic site.

Comprehensive surgical staging must be performed if the FSC assessment is a frankly invasive malignancy primary to the gynecologic tract. The following defines a comprehensive staging surgery: total abdominal hysterectomy; bilateral salpingo-oophorectomy; bilateral pelvic and para-aortic lymphadenectomy to the renal vessels; omentectomy; and multiple peritoneal biopsies. Preservation of the contralateral adnexa, if normal, and uterus is reasonable in young women who desire future fertility or when the patient's desires regarding fertility had not been discussed preoperatively [12]. The other procedures listed above must still be performed. Comprehensive surgical staging is of utmost importance as nearly 30% of patients who appear to have disease grossly confined to the adnexae and/or pelvis will be upstaged [13]. This will have significant implications in terms of prognosis and treatment recommendations. Appendectomy is required when the FSC reveals a mucinous carcinoma.

The most difficult intraoperative decision is when the FSC assessment is borderline tumor of the ovary. Comprehensive surgical staging is not necessary in cases in which the final pathologic assessment is a borderline ovarian tumor because comprehensive staging does not alter treatment or outcomes [14]. In younger women, the uterus and unaffected adnexa should not be routinely removed if a borderline tumor is confirmed after final pathologic assessment [15]. However, an FSC diagnosis of borderline tumor poses some difficulty. This difficulty is primarily due to the fact that an invasive malignancy may be diagnosed after thorough pathologic sectioning and assessment, which is not feasible during frozen section intraoperatively. Adenocarcinoma may be seen in approximately 20% of cases in which the FSC was interpreted as borderline tumor [16], but these discrepancy rates vary with histologic subtype. Micropapillary serous, intestinal mucinous, endometrioid, and clear cell borderline tumors have the highest discrepancy rates, whereas non-micropapillary serous and seromucinous borderline tumors have the lowest. The final pathology may also be entirely benign in another 6–7% [16]. These rates vary and ideally one should determine the individual institution's accuracy of FSC diagnosis of borderline tumor. Comprehensive surgical staging based on FSC diagnosis of borderline tumor may result in unnecessary additional surgery in approximately 80% of cases. Similarly, not performing a comprehensive staging at the time will result in 20% of patients possibly necessitating additional surgery at a later date.

Table 5.2 Key questions to be addressed at the time of frozen section assessment during surgical procedures for masses confined to the adnexa or with presumed metastatic gynecologic malignancy

Masses confined to the adnexa
1. Is the mass benign, a borderline tumor, or carcinoma?
2. If a borderline tumor or carcinoma, is it mucinous?
3. If carcinoma, is it possibly metastatic from another site, especially in patients with known non-gynecologic malignancies?
Biopsies from patients with grossly metastatic disease
1. Is it carcinoma?
2. Is it compatible with a gynecologic primary?

Frozen Section Assessment in Cases with Gross Extraovarian Disease

The majority of patients diagnosed with ovarian carcinoma will present with advanced extraovarian disease. The goal of surgery for advanced ovarian carcinoma is to achieve an optimal resection of all visible disease. The Gynecologic Oncology Group (GOG) defines this as no residual tumor measuring more than 1 cm. However, the best survival outcomes are seen

Table 5.3 FIGO staging of ovarian cancer [18]

Stage I	Growth limited to the ovaries IA: Limited to one ovary IB: Limited to both ovaries IC: Involving one or both ovaries, but with tumor on the ovarian surface, or with capsule rupture, or with malignant ascites present, or with positive peritoneal cytology
Stage II	Growth on one or both ovaries with pelvic extension IIA: Involvement of the uterus and/or fallopian tubes IIB: Involvement of other pelvic tissue IIC: Either stage IIA or IIB but with tumor on the ovarian surface, or with capsule rupture, or with malignant ascites present, or with positive peritoneal cytology
Stage III	Histologically confirmed peritoneal implants outside the pelvis and/or positive regional lymph nodes. Superficial liver metastases equals stage III. Tumor limited to true pelvis but with direct extension to omentum or small bowel is also stage III IIIA: Microscopic seeding; nodes negative IIIB: Gross implants ≤2 cm IIIC: Gross implants >2 cm and/or positive regional lymph nodes
Stage IV	Distant extraperitoneal metastases. If pleural effusion is present, there must be positive cytology. Parenchymal liver metastasis equals stage IV

in those in whom complete resection of all visible tumor is achieved [17]. Extensive multi-organ resections are often necessary in order to achieve a complete surgical cytoreduction. Ovarian carcinoma is unique compared to other solid tumors. In general, aggressive surgical cytoreductions are not performed for other grossly metastatic solid tumors.

Patients who present with presumed grossly metastatic ovarian carcinoma do not routinely have preoperative biopsies performed. The majority of these patients have large adnexal or pelvic masses with gross intra-abdominal disease with or without ascites. Additionally, other organs appear to be normal, so a different primary is not suspected. Therefore, the most likely diagnosis is a carcinoma of the ovary or Müllerian tract and surgery is indicated. However, in some instances these patients may actually have a extragenital that was not appreciated preoperatively. The most common other possible non-gynecologic sites include the colon, appendix, pancreas, stomach, and breast. Except for appendiceal carcinomas, these are not treated with aggressive surgical cytoreduction. A portion of omentum (usually) or other site of visible disease is obtained as soon as possible in the surgery and sent for frozen section assessment. The information that is necessary from frozen section is whether it is gynecologic in origin. The surgeon will often wait to proceed with anything further until the frozen section returns. Since borderline tumors with extensive extraovarian disease are also treated with aggressive surgery, the importance of distinguishing borderline tumor from carcinoma in this specific setting is minimized.

Staging

Comprehensive surgical staging of ovarian carcinoma, but not borderline tumor, is extremely important. Gynecologic tumors are most commonly staged according to International Federation of Gynecology and Obstetrics (FIGO) criteria and not American Joint Committee on Cancer systems. The FIGO staging system for ovarian cancer is presented in Table 5.3 [18]. Staging has tremendous implications in terms of prognosis as shown in Table 5.4 [19]. Most importantly, adjuvant therapies will vary significantly based on tumor stage, FIGO grade, and histology. Several important details should be mentioned. Cytologic confirmation of malignancy must be obtained before upstaging a patient based on the presence of ascites or an effusion. Patients are also not upstaged solely based on the adherence of an ovary to adjacent pelvic tissue – there must be biopsies of the adjacent tissue that confirm disease separate from the ovary. Metastasis

Table 5.4 Relative survival rates for patients ≥20 years of age with adenocarcinoma of the ovary, stratified by stage (excluding borderline tumors) [19]

FIGO stage	Relative survival rate (%)	
	5-Year	10-Year
I	89.3	84.1
IA	94.0	88.9
IB	91.1	78.7
IC	79.8	76.0
II	65.5	55.7
IIA	76.4	66.8
IIB	66.9	57.4
IIC	57.0	45.9
III	33.5	22.2
IIIA	45.3	31.4
IIIB	38.6	26.1
IIIC	35.2	22.6
IV	17.9	10.4

NB: In addition to stage, the following factors are currently considered to be prognostically important: patient age, performance status, debulking to minimal residual disease, histologic subtype, grade, and response to chemotherapy to the inguinal lymph nodes is still considered stage IIIC, but nodal metastases to the pleural, mediastinal, or supraclavicular regions are considered stage IV.

Chemotherapy

Adjuvant Chemotherapy

The decision whether or not to give adjuvant chemotherapy is directly related to stage at diagnosis and the grade of the tumor. Table 5.5 outlines key questions to be answered by the pathologist for patients with low-stage disease and for those requiring neoadjuvant chemotherapy. Patients with stage IC (ovarian surface involvement, tumor rupture intraoperatively, or positive peritoneal washings) or higher stages all receive adjuvant chemotherapy. For tumors confined to the ovary (stages IA and IB), adjuvant chemotherapy is only considered if the histology shows a serous, clear cell, or grade 3 (high-grade) tumor. The delivery method of chemotherapy also depends upon tumor stage and surgical outcome. For patients with stage III tumors who have optimal residual disease (implants ≤ 1 cm) after surgery, intraperitoneal chemotherapy is recommended as part of the adjuvant regimen.

The standard adjuvant chemotherapy combination for all histologic subtypes of ovarian cancer consists of a platinum and taxane drug, most commonly carboplatin and paclitaxel [20]. For stage III patients with optimally debulked disease, a combination of intravenous paclitaxel with intraperitoneal paclitaxel and cisplatin is used. Three large randomized studies using the intraperitoneal approach with cisplatin have all shown a survival benefit compared to intravenous chemotherapy alone in this population [21–23]. Current large randomized trials are evaluating whether bevacizumab should be added to the initial chemotherapy for ovarian cancer.

While large randomized studies that have defined the chemotherapeutic standard of care have included all histologic subtypes of ovarian cancer (serous; endometrioid; clear cell; mucinous; undifferentiated; adenocarcinoma not otherwise specified), there is increasing awareness that differences exist amongst the histologic subtypes with regard to chemosensitivity. For example, clear cell and mucinous ovarian cancers are more resistant to the standard carboplatin and paclitaxel combination [24, 25]. This has prompted newer trial designs to evaluate other chemotherapy combinations in the less common ovarian tumors such as clear cell and mucinous carcinomas and low-grade serous carcinomas. This paradigm shift is currently evolving, but will depend even more on the appropriate pathologic diagnosis of histologic subtype to determine therapy choices in the future [26].

With increasing recognition that each ovarian carcinoma subtype actually represents a different disease entity, each of which arises and progresses via a distinctive pathogenetic pathway, there is interest in determining which pathways can be targeted with existing and novel agents. Inhibition of poly ADP ribose polymerase (PARP) in high-grade serous carcinomas, particularly those with defective *BRCA* function, has been shown to be beneficial [27–29]. Trials studying the inhibition of MEK, a component that is downstream from RAF and RAS (mutations in which are implicated in serous borderline tumor and low grade serous carcinoma) are currently underway [30]. Endometrioid adenocarcinomas are increasingly being targeted with inhibitors of constituents of the PIK3CA pathway, which includes PTEN and AKT. Approximately 20% of mucinous carcinomas have been found to harbor amplification of HER2/neu [31], which provides yet another target. Since ovarian clear cell carcinomas share many characteristics with renal clear cell carcinomas, including activation of the hypoxia inducible factor (HIF) and mammalian target of rapamycin (mTOR) pathways, inhibitors of vascular endothelial growth factor (VEGF) and rapamycin derivatives are being studied. VEGF inhibitors are now frequently offered to patients with high-grade serous carcinoma and, increasingly, to patients with low-grade serous carcinoma as well. Given the way in which radiotherapy targets the microvasculature, it is tempting to hypothesize that the presence of a distinctive vasculature in ovarian clear cell carcinoma accounts for its preferential susceptibility to radiation [32].

Table 5.5 Key questions to be answered in the pathology report to assist in the determination of the need for adjuvant or neoadjuvant chemotherapy

Stage I tumors
1. Is the ovarian surface involved or was there rupture of the mass intra-operatively[a]?
2. Is the peritoneal washing positive for adenocarcinoma?
3. For tumors confined to the ovary, is the histology high grade serous, clear cell, or "grade 3"?
Neoadjuvant patients
1. Can malignancy (adenocarcinoma) be confirmed from ascites, FNA, or core biopsy?
2. If adequate tissue is available, can a non-Mullerian primary be excluded (ex. gastrointestinal or breast)?

[a]Information supplied by surgeon

Neoadjuvant Chemotherapy

Not all ovarian cancer patients are able to have initial debulking surgery given the extent of disease or comorbidities. In this setting, patients are treated with chemotherapy prior to surgery with possible surgical management after three or six cycles of chemotherapy. For these patients, the pathologist and cytopathologist play a key role in establishing the diagnosis. Either ascites fluid, fine needle aspiration (FNA), or core biopsy is usually obtained for diagnostic purposes.

Table 5.6 Response rates to single agent non-platinum chemotherapeutic agents

Agent	Platinum sensitive (recurrence >6 months from last platinum therapy)	Platinum resistant (recurrence <6 months from last platinum therapy)
Paclitaxel	20–41%	12–33%
Docetaxel	38%	27%
Etoposide(oral)	35%	25%
Topotecan	33%	12%
Liposomal doxorubicin	28%	14%
Vinorelbine	29%	21%
Gemcitabine		15%
Altretamine	27%	10%
Ifosfamide		10%

Ovarian cancer is often suspected based upon the pattern of disease spread and/or the presence of an abnormal ovarian mass on radiologic imaging. However, it is important to recognize that other primary tumor types can present with diffuse peritoneal carcinomatosis and ascites. In particular, it is important to try to distinguish whether a pancreatic, biliary, appendiceal, or colon cancer is likely, as these would be better treated with other chemotherapy combinations. Breast cancer is another tumor type best subjected to a different therapeutic regimen than ovarian carcinoma that can present or recur with ovarian masses or peritoneal disease. Even within gynecologic malignancies, it is often difficult to distinguish between an ovarian and endometrial primary as both tumors can metastasize to the other organ and cause diffuse peritoneal disease. Distinguishing these two tumor types is not as critical for chemotherapy choices but does influence overall prognosis. Patients with endometrial cancer with diffuse abdominal disease have a median overall survival of a year or less compared to 4–5 years in the ovarian cancer population with the same disease pattern [19].

Chemotherapy for Recurrent Disease

The majority of patients enter a clinical remission following surgery and adjuvant chemotherapy treatment. Unfortunately, long-term cure rates are low, ranging from 10–20% for stage III and IV patients. Thus, recurrent disease develops in the majority of patients. When the recurrence is within the first 5 years and involves characteristic sites of disease for ovarian cancer (peritoneal surfaces, lymph nodes, pleura) or a rising CA-125 tumor marker, biopsy proof of recurrent ovarian cancer may not be required. However, if the interval to the time of recurrence is especially long, or if the disease pattern is not consistent with that of recurrent ovarian cancer, a biopsy may be indicated prior to starting chemotherapy. In patients with a history of more than one malignancy (e.g., *BRCA* patients with breast and ovarian cancer), a biopsy can help to determine the etiology of the recurrent disease which will then guide therapy choices.

Choosing chemotherapy in this setting is initially based upon the interval from last platinum therapy. Patients who recur within a few months of completing initial therapy are less likely to respond to retreatment with platinum [33, 34]. An arbitrary window of 6 months has commonly been used in chemotherapy trial design to distinguish patients as "platinum-sensitive" or "platinum-resistant" based upon whether recurrence is greater or less than 6 months from the completion of their adjuvant platinum-based therapy, respectively. For the platinum-sensitive group, combination therapy with carboplatin and either paclitaxel, gemcitabine, or liposomal doxorubicin is standard. For the platinum-resistant group, United States Food and Drug Administration-approved options include liposomal doxorubicin or topotecan. Other chemotherapeutic agents have also been tested and can be used in this setting. However, the response rates are commonly quite low in the platinum-resistant population as shown in Table 5.6.

References

1. Parkin DM, Bray F, Ferlay J, Pisani P. Global cancer statistics, 2002. CA Cancer J Clin. 2005;55:74–108.
2. Jemal A, Siegel R, Ward E, Hao Y, Xu J, Thun MJ. Cancer statistics, 2009. CA Cancer J Clin. 2009;59:225–49.
3. Antoniou A, Pharaoh PDP, Narod S, Risch HA, Eyfjord JE, Hopper JL, et al. Average risks of breast and ovarian cancer associated with BRCA1 or BRCA2 mutations detected in case series unselected for family history: a combined analysis of 22 studies. Am J Hum Genet. 2003;72:1117–30.
4. Chetrit A, Hirsh-Yechezkel G, Ben-David Y, Lubin F, Friedman E, Sadetzki S. Effect of BRCA1/2 mutations on long-term survival of patients with invasive ovarian cancer: the national Israeli study of ovarian cancer. J Clin Oncol. 2008;26(1):20–5.
5. Ben David Y, Chetrit A, Hirsh-Yechezkel G, Friedman E, Beck BD, Beller U, et al. Effect of BRCA mutations on the length of survival in epithelial ovarian tumors. J Clin Oncol. 2002;20(2):463–6.
6. Boyd J, Sonoda Y, Federici MG, Bogomolniy F, Rhei E, Maresco DL, et al. Clinicopathologic features of BRCA-linked and sporadic ovarian cancer. JAMA. 2000;283(17):2260–5.
7. Hussain SP, Harris CC. Molecular epidemiology of human cancer: contribution of mutation spectra studies of tumor suppressor genes. Cancer Res. 1998;58:4023–37.

8. Yuan SS, Lee SY, Chen G, Song M, Tomlinson GE, Lee E. BRCA2 is required for ionizing radiation-induced assembly of Rad51 complex in vivo. Cancer Res. 1999;59:3547–51.
9. Clarke B, Tinker AV, Lee CH, Subramanian S, van de Rijn M, Turbin D, et al. Intraepithelial T cells and prognosis in ovarian carcinoma: novel associations with stage, tumor type, and BRCA1 loss. Mod Pathol. 2009;22(3):393–402.
10. Zhang L, Conejo-Garcia JR, Katsaros D, Gimotty PA, Massobrio M, Regnani G, et al. Intratumoral T cells, recurrence, and survival in epithelial ovarian cancer. N Engl J Med. 2003;348(3):203–13.
11. Vasen HFA, Watson P, Mecklin J, Lynch HT. New clinical criteria for hereditary nonpolyposis colorectal cancer (HNPCC, Lynch Syndrome) proposed by the International Collaborative Group on HNPCC. Gastroenterology. 1999;116:1453–6.
12. Leitao Jr MM, Chi DS. Fertility-sparing options for patients with gynecologic malignancies. Oncologist. 2005;10:613–22.
13. Leitao Jr MM, Chi DS. Operative management of primary epithelial ovarian cancer. Curr Oncol Rep. 2007;9:478–84.
14. Winter III WE, Kucera PR, Rodgers W, McBroom JW, Olsen C, Maxwell GL. Surgical staging in patients with ovarian tumors of low malignant potential. Obstet Gynecol. 2002;100:671–6.
15. Trimble CL, Kosary C, Trimble EL. Long-term survival and patterns of care in women with ovarian tumors of low malignant potential. Gynecol Oncol. 2002;86:34–7.
16. Geomini P, Bremer G, Kruitwagen R, Mol BWJ. Diagnostic accuracy of frozen section diagnosis of the adnexal mass: a metaanalysis. Gynecol Oncol. 2005;96:1–9.
17. Chi DS, Eisenhauer EL, Lang J, Huh J, Haddad L, Abu-Rustum NR, et al. What is the optimal goal of primary cytoreductive surgery for bulky stage IIIC epithelial ovarian carcinoma (EOC)? Gynecol Oncol. 2006;103:559–64.
18. FIGO. Current FIGO staging for cancer of the vagina, fallopian tube, ovary, and gestational trophoblastic neoplasia. Int J Gynaecol Obstet. 2009;105:3–4.
19. Ries LAG, Young JL, Keel GE, Eisner MP, Lin YD, Horner M-J (editors). SEER Survival Monograph: Cancer Survival Among Adults: U.S. SEER Program, 1988–2001, Patient and Tumor Characteristics. National Cancer Institute, SEER Program, NIH Pub. No. 07–6215, Bethesda, MD, 2007.
20. Ozols RF, Bundy BN, Greer BE, Fowler JM, Clarke-Pearson D, Burger RA, et al. Phase III trial of carboplatin and paclitaxel compared with cisplatin and paclitaxel in patients with optimally resected stage III ovarian cancer: a Gynecologic Oncology Group study. J Clin Oncol. 2003;21:3194–200.
21. Alberts DS, Liu PY, Hannigan EV, O'Toole R, Williams SD, Young JA, et al. Intraperitoneal cisplatin plus intravenous cyclophosphamide versus intravenous cisplatin plus intravenous cyclophosphamide for stage III ovarian cancer. N Engl J Med. 1996;335:1950–5.
22. Markman M, Bundy BN, Alberts DS, Fowler JM, Clark-Pearson DL, Carson LF, et al. Phase III trial of standard-dose intravenous cisplatin plus paclitaxel versus moderately high-dose carboplatin followed by intravenous paclitaxel and intraperitoneal cisplatin in small-volume stage III ovarian carcinoma: an Intergroup study of the Gynecologic Oncology Group, Southwestern Oncology Group, and Eastern Cooperative Oncology Group. J Clin Oncol. 2001;19:1001–7.
23. Armstrong DK, Bundy B, Wenzel L, Huang HQ, Baergen R, Lele S, et al. Intraperitoneal cisplatin and paclitaxel in ovarian cancer. N Engl J Med. 2006;354:34–43.
24. Pectasides D, Fountzilas G, Aravantinos G, Kalofonos C, Efstathiou H, Farmakis D, et al. Advanced stage clear-cell epithelial ovarian cancer: the Hellenic Cooperative Oncology Group experience. Gynecol Oncol. 2006;102:285–91.
25. Hess V, A'Hern RA, Nasiri N, King DM, Blake PR, Barton DPJ, et al. Mucinous epithelial ovarian cancer: a separate entity requiring specific treatment. J Clin Oncol. 2004;22:1040–4.
26. Gilks CB, Ionescu DN, Kalloger SE, Kobel M, Irving J, Clarke B, et al. Tumor cell type can be reproducibly diagnosed and is of independent prognostic significance in patients with maximally debulked ovarian carcinoma. Hum Pathol. 2008;39:1239–51.
27. Audeh MW, Carmichael J, Penson RT, Friedlander M, Powell B, Bell-McGuinn KM, et al. Oral poly(ADP-ribose) polymerase inhibitor olaparib in patients with BRCA1 or BRCA2 mutations and recurrent ovarian cancer: a proof-of-concept trial. Lancet. 2010;376(9737):245–51.
28. Tutt A, Robson M, Garber JE, Domchek SM, Audeh MW, Weitzel JN, et al. Oral poly(ADP-ribose) polymerase inhibitor olaparib in patients with BRCA1 or BRCA2 mutations and advanced breast cancer: a proof-of-concept trial. Lancet. 2010;376(9737):235–44.
29. Fong PC, Yap TA, Boss DS, Carden CP, Mergui-Roelvink M, Gourley C, et al. Poly(ADP)-ribose polymerase inhibition: frequent durable responses in BRCA carrier ovarian cancer correlating with platinum-free interval. J Clin Oncol. 2010;28(15):2512–19.
30. Nakayama N, Nakayama K, Yeasmin S, Ishibashi M, Katagiri A, Iida K, et al. KRAS or BRAF mutation status is a useful predictor of sensitivity to MEK inhibition in ovarian cancer. Br J Cancer. 2008;99(12):2020–8.
31. McAlpine JN, Wiegand KC, Vang R, Ronnett BM, Adamiak A, Köbel M, et al. HER2 overexpression and amplification is present in a subset of ovarian mucinous carcinomas and can be targeted with trastuzumab therapy. BMC Cancer. 2009;9:433.
32. Swenerton KD, Santos JL, Gilks CB, Köbel M, Hoskins PJ, Wong F, et al. Histotype predicts the curative potential of radiotherapy: the example of ovarian cancers. Ann Oncol. 2010;22(2): 341–7 [Epub ahead of print].
33. Gore ME, Fryatt I, Wiltshaw E, Dawson T. Treatment of relapsed carcinoma of the ovary with cisplatin or carboplatin following initial treatment with these compounds. Gynecol Oncol. 1990;36:207–11.
34. Markman M, Rothman R, Hakes T, Reichman B, Hoskins W, Rubin S, et al. Second-line platinum therapy in patients with ovarian cancer previously treated with cisplatin. J Clin Oncol. 1991;9:389–93.

Pathology of Serous Tumors

6

C. Blake Gilks

Keywords
Surface epithelial-stromal tumors • Surface epithelial tumors • Borderline tumors • Carcinoma • Tumor cell type

Introduction

The recognition that serous carcinomas are separable into high grade serous carcinoma (HGSC) and low grade serous carcinoma (LGSC) represents a critically important recent advance in our understanding of serous carcinoma. HGSC and LGSC are now understood to be different diseases rather than being two ends of a continuum of a single disease entity [1–3]. The major molecular and clinical differences are summarized in Table 6.1. The practice of lumping together HGSC and LGSC in studies of "serous carcinoma" has diminished our understanding of the natural history and optimal treatment of the much less common LGSC, as these cases have been overshadowed by HGSC in case series. Although the differences between HGSC and LGSC were first demonstrated at the molecular level, it turns out that the distinction between them can be achieved reproducibly in a large majority of cases by examination of routine hematoxylin and eosin (H&E)-stained slides alone [4].

In this chapter, HGSC and LGSC will be treated as separate diseases. Because LGSC share molecular abnormalities with serous borderline tumor (SBT), and frequently coexist with and may arise from SBT [1–3], LGSC and SBT will be considered together.

C.B. Gilks
Department of Pathology and Laboratory Medicine,
University of British Columbia/Vancouver General Hospital,
Vancouver, BC V5Z 4E3, Canada

Benign Serous Tumors

Prevalence

Benign serous tumors are very common; however, accurate estimates of incidence are difficult to obtain. They can appear at any age, but are most commonly encountered in adults.

Macroscopic Features

Benign serous tumors consist of uni-or multilocular cysts filled with watery fluid (Fig. 6.1). The cyst lining is smooth, although there may be papillary stromal projections into the cyst or from the surface (Fig. 6.2). There may be a solid component composed of firm white tissue. Predominantly cystic tumors are diagnosed as cystadenomas, tumors with mixed cystic and solid components are cystadenofibromas, and predominantly solid tumors are adenofibromas. Distinction between these categories is arbitrary and is not of clinical significance.

Microscopic Features

Benign serous tumors show spaces lined by bland tubal-type epithelium and variable amounts of benign ovarian-type

R.A. Soslow and C. Tornos (eds.), *Diagnostic Pathology of Ovarian Tumors*,
DOI: 10.1007/978-1-4419-9751-7_6, © Springer Science+Business Media, LLC 2011

Table 6.1 Low grade serous carcinoma versus high grade serous carcinoma

	Low-grade serous	High-grade serous
Molecular abnormalities	*BRAF* or *KRAS* mutations	*p53* or *BRCA* mutations
Proliferative index	Low	High
Apoptotic index	Low	High
Tumor cell DNA content	Diploid or near diploid	Aneuploid
Stage at diagnosis	Usually advanced	Often advanced
Response to chemotherapy	Poor	Good initial response

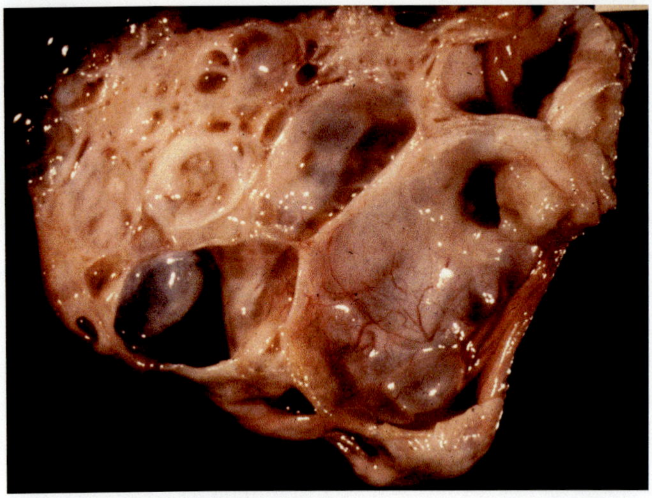

Fig. 6.1 Serous cystadenofibroma

stroma (Fig. 6.3). It is this tubal, ciliated epithelium that defines this subtype of benign ovarian surface epithelial tumor, and although much of the epithelium may be of a flattened, nonspecific type, there must be at least focal recognizable tubal differentiation to diagnose a benign serous tumor.

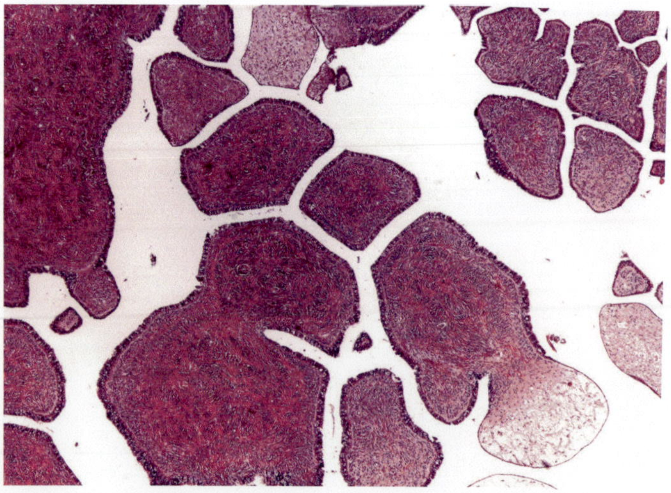

Fig. 6.2 A benign serous tumor with prominent stromal papillae. Grossly this may mimic SBT, but microscopically the papillae are covered by a single layer of epithelial cells, without stratification, atypia, or detached cell clusters

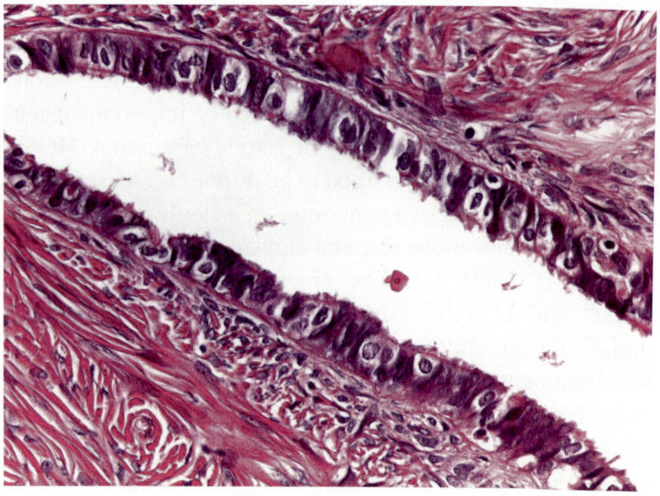

Fig. 6.3 Epithelium of tubal type lining a benign serous cystadenoma

Differential Diagnosis

1. *Benign serous tumor versus SBT*. This differential diagnosis will be discussed subsequently, in the section on SBT. The stromal projections that may be seen in benign serous tumors and that mimic SBT grossly (Fig. 6.2) are seen on microscopic examination to be stroma covered by a single layer of bland epithelial cells of tubal type (Fig. 6.3); that is, they lack stratification, atypia, or mitoses, the hallmarks of SBT.
2. *Benign serous tumor versus other subtypes of benign surface epithelial tumor*. This distinction is made based on the nature of the epithelial cells. Occasional benign surface epithelial tumors are mixed, with more than one epithelial cell type.
3. *Benign serous tumor versus cortical inclusion cyst*. By definition, a cortical inclusion cyst is less than 1 cm, while a serous cystadenoma is larger.
4. *Benign serous tumor versus struma ovarii*. *Struma ovarii* can occasionally contain cystic spaces lined by cuboidal or flattened epithelial cells [5]. Consideration of the possibility of *Struma ovarii* will lead to the correct diagnosis, as the presence of other more characteristic areas of *Struma*, absence of ciliated cells, and, in particularly challenging

cases, immunoreactivity for thyroglobulin, will allow a correct diagnosis of *Struma ovarii*.

Low-Grade Serous Carcinoma (LGSC) and Serous Borderline Tumor (SBT)

Prevalence

LGSC and SBT are less common than either HGSC or benign serous tumors. LGSCs account for approximately 3% of ovarian surface epithelial carcinomas. Serous borderline tumors are more common than LGSC and are encountered on a regular basis in a general surgical pathology practice.

Treatment and Clinical Outcomes

Our understanding of LGSC is limited, as there are relatively few studies that have specifically examined this uncommon and recently described subtype of ovarian carcinoma. When LGSC is confined to the ovary at diagnosis without ovarian surface involvement (stage Ia), the prognosis is excellent with surgical treatment alone (greater than 95% disease-free survival at 5 years), with no need for adjuvant therapy. Many LGSC are advanced stage at diagnosis, and the prognosis for these patients is poor. On one hand, they have low mitotic rates and relatively indolent growth; however, on the other hand, they do not respond well to conventional platinum-based chemotherapy [6]. The cornerstone of therapy is surgery. Because of the limited effectiveness of platinum-based chemotherapy, there are clinical trials of MAP-kinase inhibitors, given the frequent disruption of this signaling pathway in LGSC through mutations in *BRAF* or *KRAS*.

SBT are confined to the ovary at presentation, without ovarian surface involvement, in most cases, and these patients have a very favorable prognosis, although occasional recurrences are documented, sometimes after many years. Adjuvant therapy is not indicated for stage Ia SBT because of the very favorable prognosis. Cystectomy alone for SBT is associated with a significant likelihood of recurrence in the ipsilateral ovary [7] and is not recommended.

A significant minority of patients with serous borderline tumor have extra-ovarian implants (stage II–IV disease), estimated at 20–40% in different case series. These patients are at significant risk of recurrence and progression to invasive serous carcinoma. The presence of ovarian surface involvement in patients with serous borderline tumor is associated with a markedly increased likelihood of extra-ovarian spread [8]. At the very least, it should prompt a careful search for such extra-ovarian disease, as advanced stage at diagnosis is the most important prognostic indicator in patients with serous borderline tumor.

In patients with advanced-stage disease, the most important prognostic indicator is the presence of invasive implants (defined and illustrated below), present in approximately 10% of patients with advanced-stage SBT [9]. These are, both morphologically and biologically, foci of invasive LGSC at extra-ovarian sites. The convention is that tumors considered to be primary ovarian tumors are classified based on the appearance of the ovarian tumor. Somewhat confusingly, tumors showing SBT in the ovary and "LGSC" at extra-ovarian sites are considered SBTs (with invasive implants), but their natural history is that of advanced-stage LGSC, and outcomes are, accordingly, much worse than for patients with SBT without invasive implants. Most patients will die of disease. SBT with invasive implants are treated in an identical manner to LGSC. Although there is a need for adjuvant treatment, the best treatment option is not clear, as these tumors frequently are unresponsive to platinum-based chemotherapy. Note that classification of extra-ovarian implants as invasive or noninvasive is ONLY done for SBT and not for serous carcinomas.

SBT without invasive implants is typically treated by surgery alone. Recurrences are relatively common and can occur after many years. In a series of cases of advanced-stage SBT with noninvasive implants reported by Silva and colleagues, 44% experienced a recurrence, with 10% occurring within 5 years, 19% between 5 and 10 years, and 15% at greater than 10 years [10]. These recurrences can be typical SBT or there can be progression to LGSC, which is more common. It is important to document whether there has been progression in the recurrences as deaths due to disease are uncommon unless there is progression to invasive carcinoma.

Some SBT progress to LGSC, which is recognized either at the time of diagnosis (approximately 50% of LGSC have an associated component of SBT from which they are presumed to have arisen), or with recurrence, as noted above. The behavior of LGSC arising from SBT is identical to that of LGSC without associated SBT [11]. There can also be progression to high-grade carcinoma, but this is rare.

Micropapillary architecture in an SBT (defined and illustrated below) is also associated with increased likelihood of extra-ovarian implants, but, in the absence of extra-ovarian disease, is not of prognostic significance [12]. Similarly, microinvasion (also defined and illustrated below) within an ovarian SBT co-occurs with micropapillarity and the presence of extra-ovarian implants, but has not been consistently shown to be an adverse prognostic indicator, independent of stage and (for advanced-stage tumors) invasive implants. Nonetheless, the presence of micropapillarity or microinvasion is a marker of potentially more aggressive disease and should prompt thorough evaluation of stage/invasiveness of extra-ovarian implants. An unusual association exists

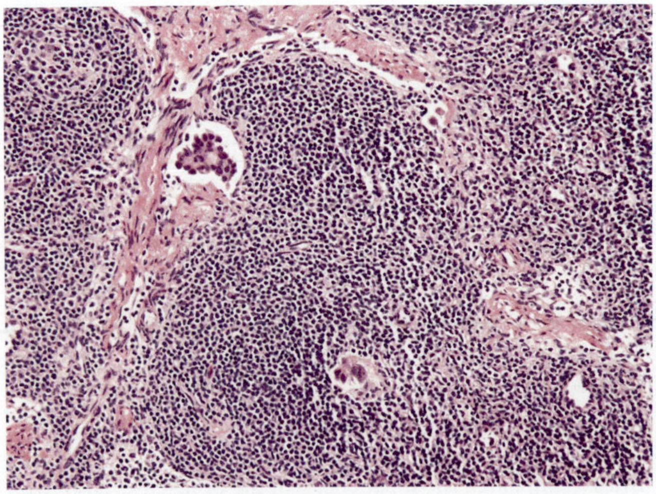

Fig. 6.4 Lymph node with SBT

between the presence of microinvasion and pregnancy; in pregnant patients there are no adverse associations with microinvasion [13].

Lymph node involvement (Fig. 6.4) in SBT is relatively common and is present in up to 42% of cases. Somewhat surprisingly, nodal involvement has consistently been shown to have no significant association with prognosis [12], and nodal sampling of clinically uninvolved nodes is not performed in many centers. One study showed an adverse prognosis in the subset of node-positive patients with nodular aggregates of SBT measuring greater than 1 mm [14], but this observation has not yet been independently validated.

LGSC Macroscopic Features

LGSC may be indistinguishable from SBT (see below for macroscopic features of SBT), but frequently show more solid areas. Compared to HGSC, there is less necrosis and hemorrhage. The extra-ovarian tumor implants are typically firm and may be gritty if there is abundant psammoma body formation within the tumor.

LGSC Microscopic Features

LGSC is distinguished from SBT by the presence of destructive stromal invasion. The three main patterns of invasion are: (1) infiltration of stroma by single cells and small clusters of cells; (2) infiltration of stroma by small nests of epithelial cells; the epithelial cells typically face outward, and have a clear space surrounding them, which may be a result of secretion by the epithelial cells; and (3) large papillae with a central, broad fibrovascular core infiltrating stroma, surrounded by a clear cleft (Fig. 6.5). This last pattern, of macropapillary invasion, is by far the least common. Single cell invasion (pattern 1) without micropapillary invasion (pattern 2) is unusual in LGSC, although it may be the predominant pattern in SBT with microinvasion (see following discussion). Papillary architecture is well-developed in LGSC, and psammoma body formation is often prominent. Psammocarcinoma is a rare variant of LGSC with psammoma bodies in greater than 75% of the epithelial nests. The individual tumor cells of LGSC are uniform in size and shape and show fewer than 12 mitotic figures/10 high power fields. They may show a single prominent nucleolus.

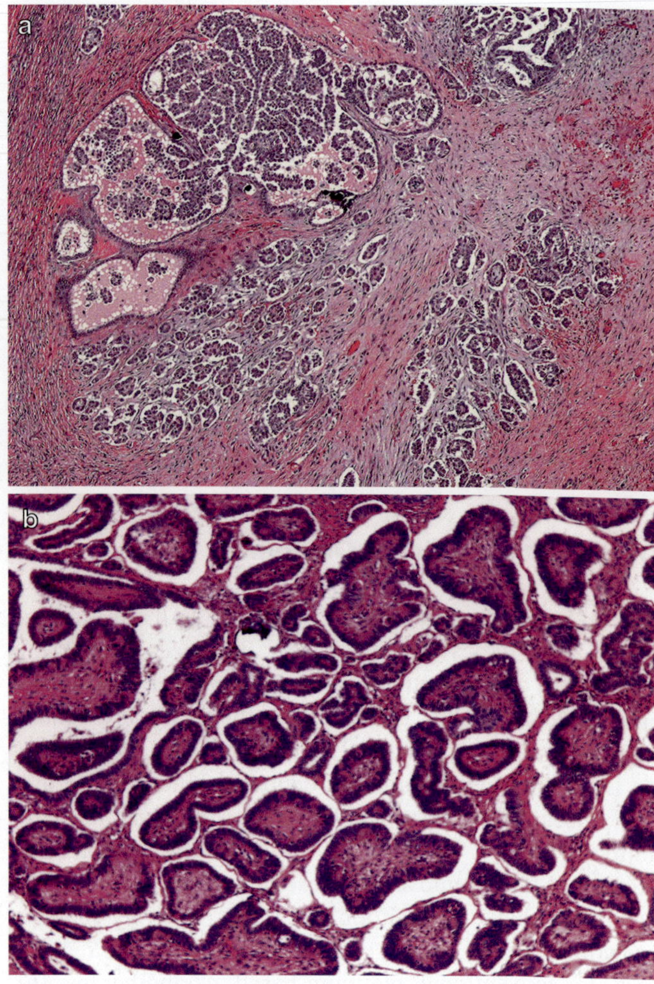

Fig. 6.5 Patterns of invasion of LGSC. (**a**) Infiltration of ovarian stroma by small nests of tumor cells (micropapillary pattern) and single cells in a background of SBT with micropapillary features. (**b**) Macropapillary pattern of invasion, with large papillae with a broad fibrovascular core invading stroma and surrounded by a clear space

SBT Macroscopic Features

The ovarian tumors are cystic, with papillary excrescences projecting into the cyst lumens. Similar papillary growths may be present on the ovarian surface (Fig. 6.6). These papillary projections have a relatively uniform appearance, from case to case, and the diagnosis of SBT may be suspected based on macroscopic examination. Extra-ovarian implants, if noninvasive,

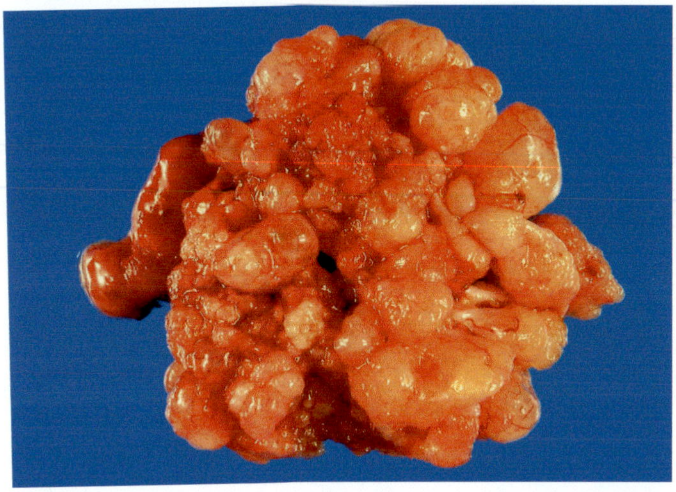

Fig. 6.6 SBT of the ovary, with ovarian surface involvement

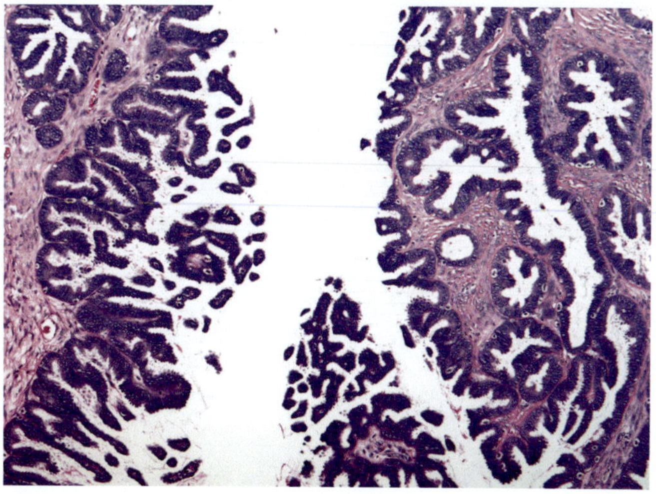

Fig. 6.7 Typical SBT architecture, with hierarchical branching and stratification and tufting of cells. Hierarchical branching refers to papillae that become progressively smaller with successive branching, while non-hierarchical papillae remain the same caliber

are typically tiny (less than 5 mm) and may be entirely incidental findings detected on submission of routine sections of macroscopically normal omentum. Invasive implants, on the other hand, are more commonly associated with macroscopic disease; if the omentum is visibly involved by tumor nodules, the presence of invasive implants is highly likely.

SBT Microscopic Features

The low-power architecture of serous borderline tumors is very consistent. It consists of well-formed papillae with stratification, and mild-to-moderate atypia of the epithelial cells (Fig. 6.7). Epithelial cells typically show some heterogeneity in the appearance of tumor cell cytoplasm, ranging from columnar and eosinophilic, and sometimes ciliated, to cuboidal with clear cytoplasm. Despite this, the nuclear size is uniform, and there may be a small nucleolus. Occasional mitotic figures are present. The stratification of the epithelium is variable and when very prominent, with non-hierarchical branching or cribriform architecture, the designation of micropapillary variant of serous borderline tumor can be used; by definition, this micropapillary area must occupy an area measuring at least 5 mm in greatest dimension. Micropapillary SBT (Fig. 6.8) tends to be composed of cells that are more uniform in appearance as compared to examples without well-developed micropapillary architecture. Recognizing micropapillary SBT is important as this tumor is more likely to be associated with extra-ovarian implants and invasive implants as compared to typical SBT. Cribriform architecture (Fig. 6.9) may co-exist with micropapillary architecture.

Microinvasion (Fig. 6.10) is relatively common within serous borderline tumors. Microinvasive foci in SBT are variably defined as being foci of destructive stromal invasion either smaller than 3 mm in greatest dimension, or smaller than 10 mm^2. These small foci of invasion show one or more

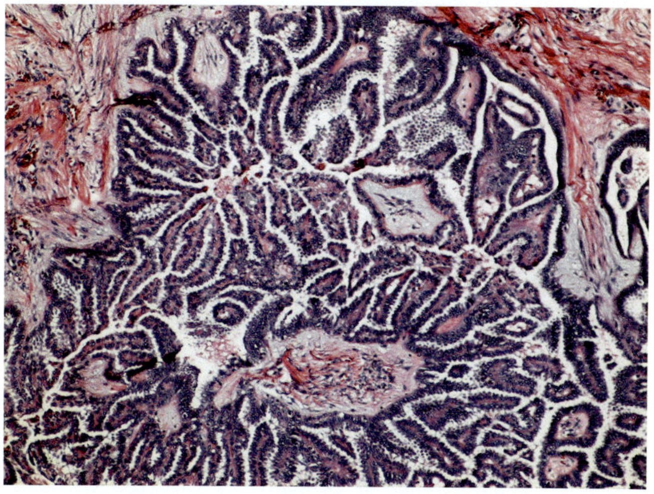

Fig. 6.8 Micropapillary pattern of SBT, with non-hierarchical branching of epithelial papillary structures. Note the more exuberant epithelial proliferation compared to typical SBT (Fig. 6.7)

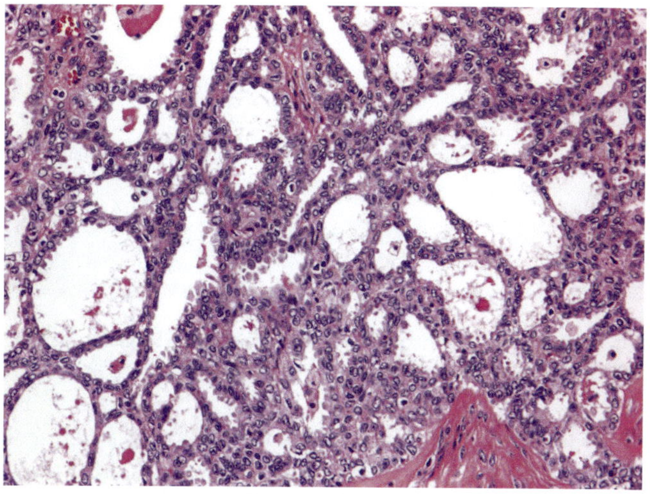

Fig. 6.9 SBT with cribriform architecture

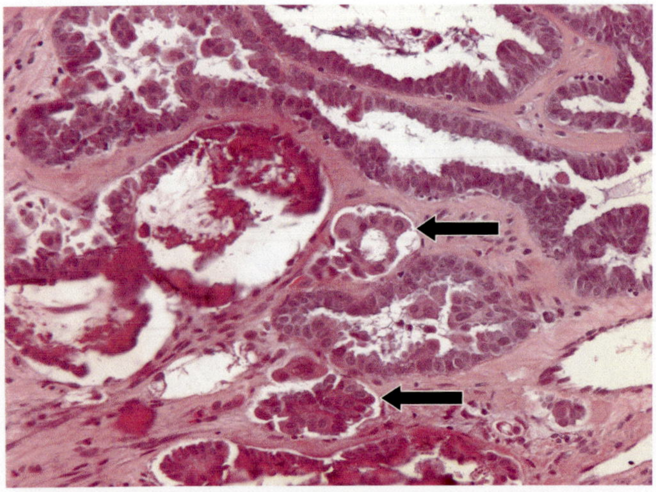

Fig. 6.10 Microinvasion in a SBT, with single cells and small nests of cells, surrounded by a clear space

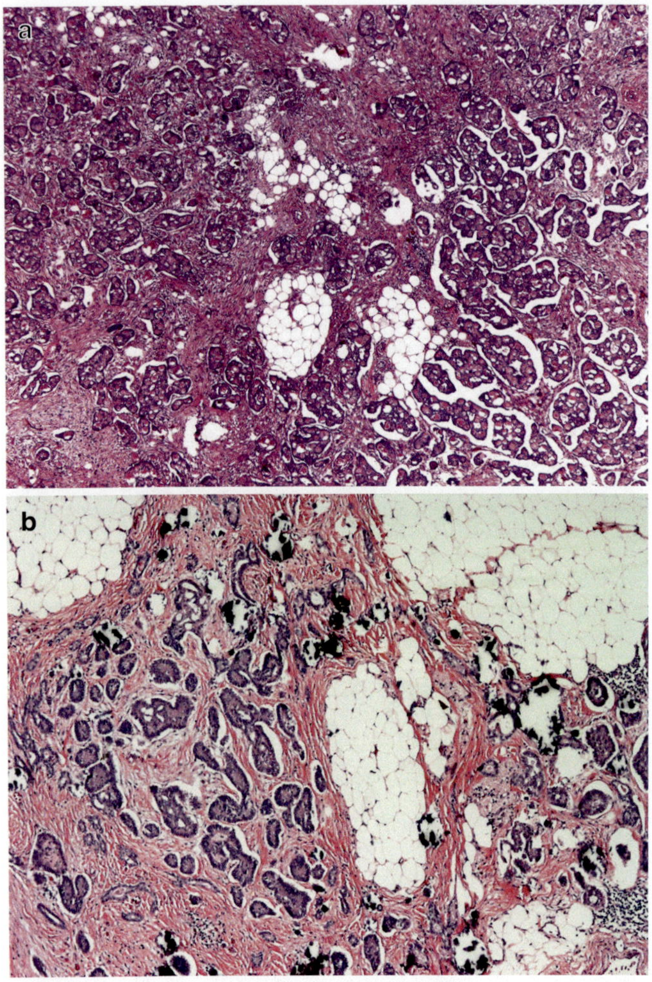

Fig. 6.11 (**a**) and (**b**). Examples of invasive omental implants of SBT. Note extensive infiltration of omental adipose tissue

of the three patterns of invasion described above in the section on morphology of LGSC. The invasive cells often show abundant eosinophilic cytoplasm. Microinvasiveness per se has not been consistently shown to be of prognostic significance, independent of tumor stage and invasive implants, but, like micropapillarity, it is more commonly associated with these features. Microinvasion is relatively more common in pregnancy; in this setting it shows the pattern of invasion that consists of individual large eosinophilic cells within the stroma.

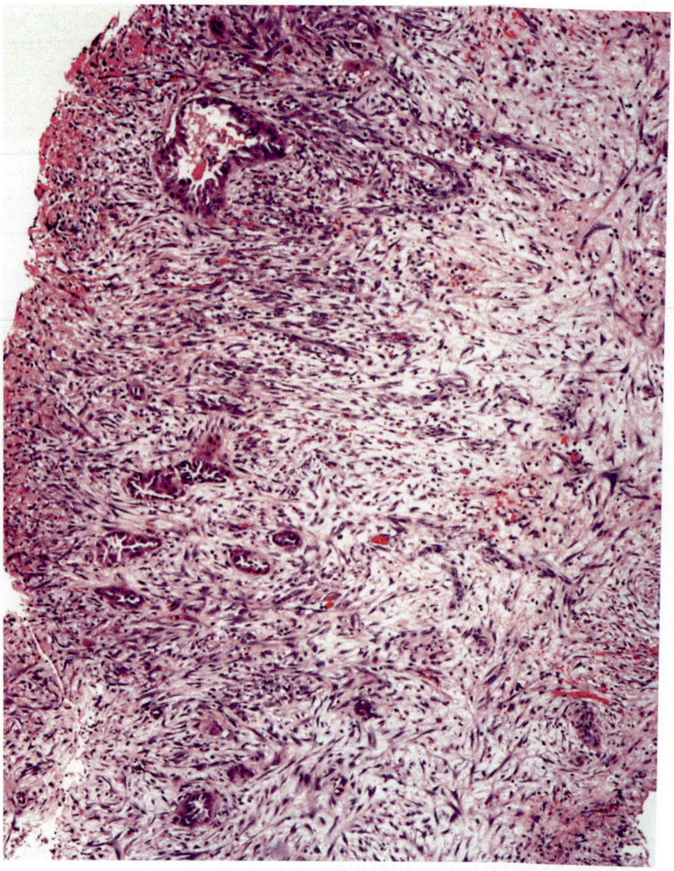

Fig. 6.12 A superficial biopsy of a noninvasive desmoplastic peritoneal implant of SBT, from the serosa of the uterus. Underlying myometrial tissue has not been sampled and the assumption is that this implant separated easily from the underlying tissue and is therefore noninvasive. Note low gland-to-stroma ratio and the appearance of granulation tissue, features that are typical of non-invasive desmoplastic implants

Recognition of invasive implants is critical in diagnosis of serous borderline tumor with extra-ovarian implants. The definition of invasive implants is as follows: invasive implants show destructive invasion, recognized by irregular, aggressive-appearing infiltration into underlying tissue, with tumor replacing or destroying associated non-neoplastic tissues (Fig. 6.11). Their appearance is indistinguishable from LGSC. To recognize invasion in omental biopsies, there must be obliteration of omental tissue. The presence of small papillae or single-tumor cells within abundant desmoplastic stroma is not interpreted as stromal invasion, and such implants are considered noninvasive desmoplastic implants (described below). Small superficial biopsy specimens of desmoplastic implants in which underlying tissue is absent are classified as noninvasive on the assumption that they have been easily stripped away from underlying tissue (Fig. 6.12) [15].

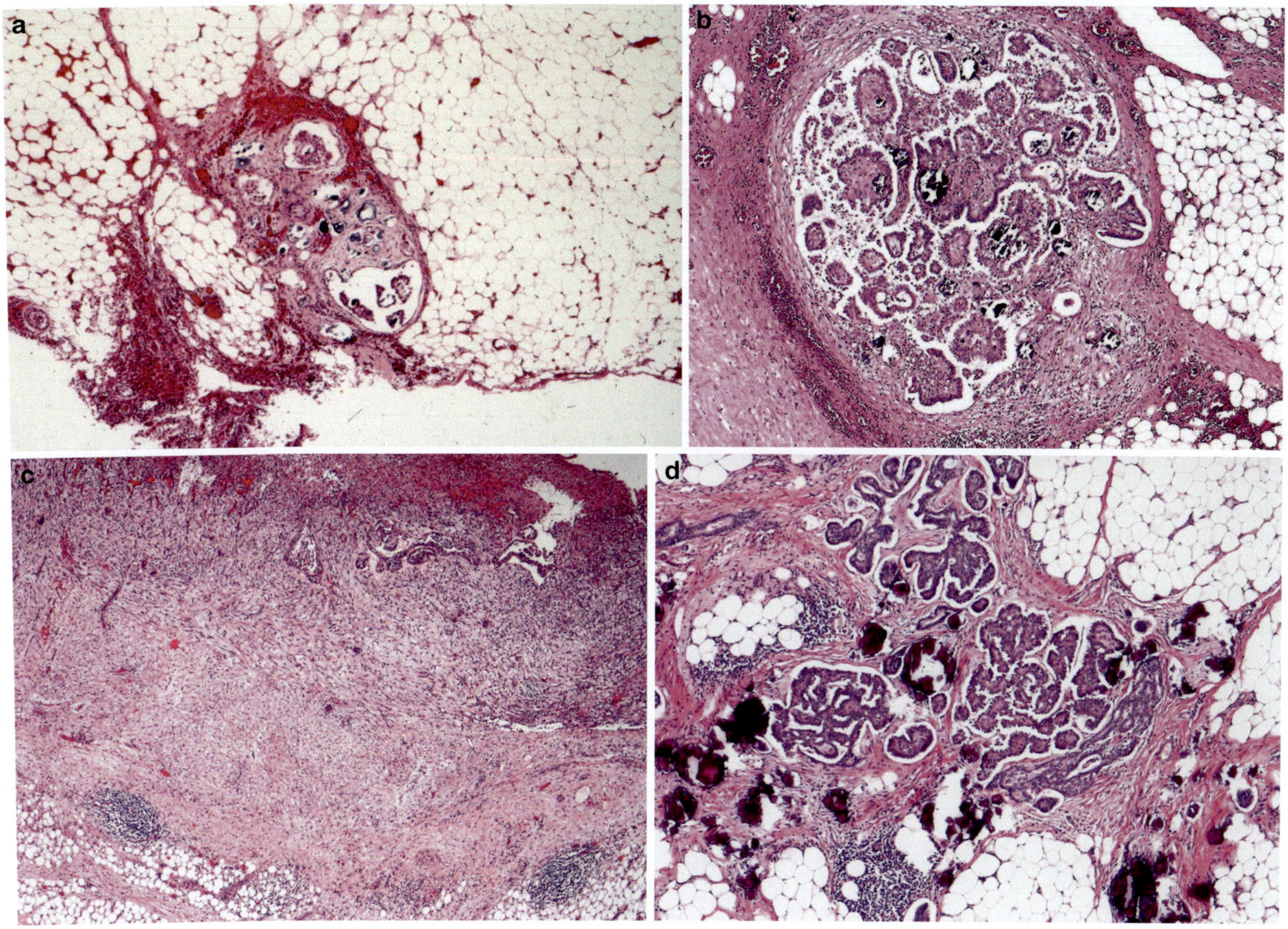

Fig. 6.13 Examples of omental implants of SBT. (**a**) and (**b**) are noninvasive epithelial implants, which resemble the intracystic growth of ovarian SBT. In both panels, the implant involves a mesothelial invagination, with fat surrounding the implant on three sides. Despite this, the implants' contours are rounded and irregular infiltration of adipose tissue is not observed. (**c**) is a noninvasive desmoplastic implant. Note the presence of a virtual plane that separates the implant from the underlying uninvolved adipose tissue. (**d**) is an invasive implant. In contrast to the desmoplastic noninvasive implant, this is epithelial-predominant, it lacks stroma resembling granulation tissue and, most importantly, it irregularly infiltrates adipose tissue

Invasive implants must be distinguished from noninvasive implants (Fig. 6.13). Two types of noninvasive implants are recognized: epithelial noninvasive implants and desmoplastic noninvasive implants. Epithelial noninvasive implants are easy to recognize as noninvasive lesions because they resemble typical serous borderline tumor. They show exophytic, hierarchically branched papillary architecture and the papillae are covered by bland-appearing cells. They can be found on the peritoneal surface or within mesothelial invaginations. Desmoplastic noninvasive implants are more common and can simulate the appearance of invasive carcinoma to the neophyte (Fig. 6.13). Low-power microscopy reveals a lesion in which stroma predominates over epithelium. Higher-power microscopy shows angulated glands, and single cells are embedded in a stroma that resembles granulation tissue or fasciitis. The epithelium may appear attenuated and frequently merges with spindle cells in the stroma. Despite this alarming appearance, the lesion usually forms a well-circumscribed plaque that can be easily separated (either manually at the time of surgery, or in the "mind's eye" on slide review) from underlying uninvolved fat or muscle. The histologic appearance of desmoplastic noninvasive implants, even when devoid of underlying fat or muscle, is so distinctive that its diagnosis can be established even in the absence of such underlying tissue. Plaque-like lesions on the ovarian surface or in septae between tumor nodules within SBTs, termed "autoimplants," have the same morphologic appearance as desmoplastic noninvasive implants and are similarly considered noninvasive (i.e., no evidence of LGSC) (Fig. 6.14). Support for classifying desmoplastic implants and autoimplants as "noninvasive" derives from clinical outcome data that are similar to that of SBT with epithelial noninvasive implants, and different (and significantly more favorable) than for patients with LGSC or SBT with invasive implants. Three main patterns of nodal involvement are seen in patients with ovarian SBT: papillary tumors within cystic spaces (equivalent to the noninvasive epithelial implants described below), a pattern usually associated with nodal endosalpingiosis; bland eosinophilic tumor cells in small cell

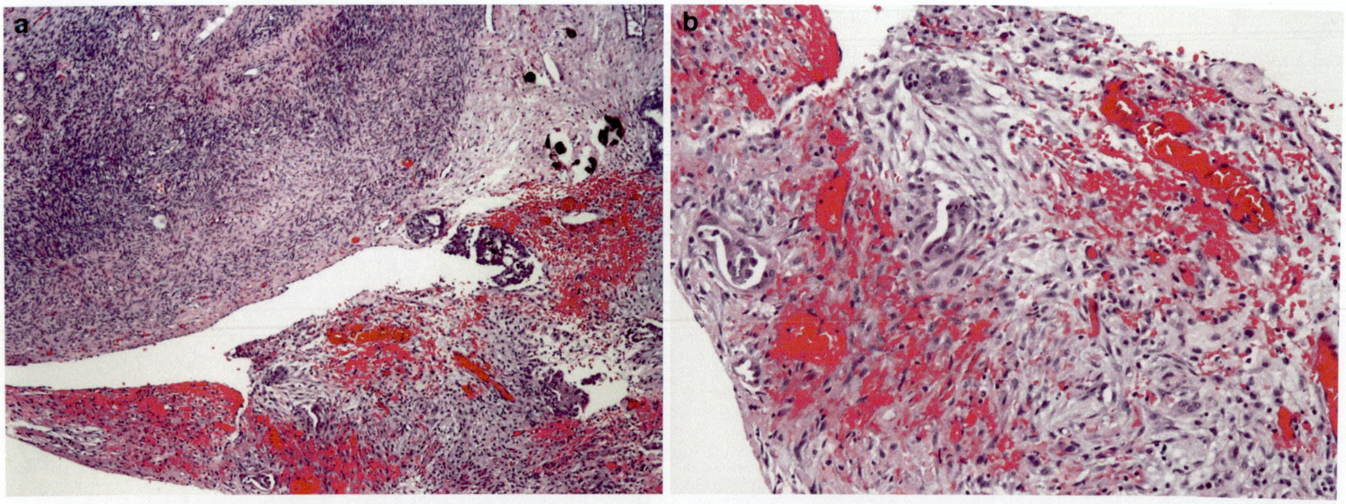

Fig. 6.14 "Autoimplant" of SBT (desmoplastic noninvasive implant of SBT on the ovarian surface) at low (a) and high (b) magnification. This is *not* evidence of invasive LGSC

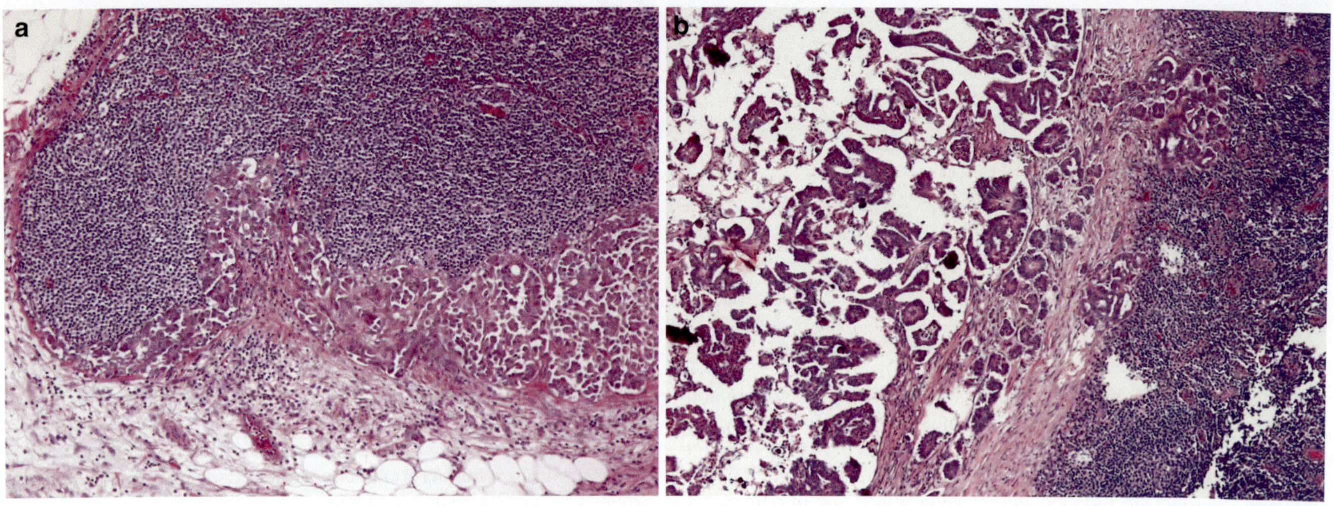

Fig. 6.15 Lymph node involvement. (**a**) Sinusoidal distribution of serous borderline tumor within the node is typical. (**b**) A nodule of invasive tumor showing features of low-grade serous carcinoma

clusters, typically in a subcapsular location; and nodules of tumor with features of LGSC (Fig. 6.15). This latter appearance may be associated with a worse prognosis, while the former are of no prognostic significance [14].

Some authors consider the presence of micropapillarity within extra-ovarian implants to be evidence of invasiveness, but this view is held by a minority of the gynecological pathology community.

Genotype

SBT and LGSC have in common frequent mutations in *KRAF* and *BRAF* [16], suggesting that SBT can progress to LGSC. They both lack *p53* and *BRCA* mutations and are not associated with any known ovarian cancer inherited susceptibility syndromes.

Immunophenotype

The tumor cells in SBT and LGSC have similar immunotypes and express many of the markers seen in HGSC. They are positive for WT1, ER, and PR in a large majority of cases. They are negative for HNF-1beta and the mesothelial marker calretinin.

Differential Diagnosis

1. *SBT versus LGSC*. SBT is distinguished from LGSC by the absence of invasion. Patterns of invasion are described above and illustrated in Fig. 6.5. This is a purely morphological distinction, and there are no special stains that can assist in recognition of invasion. If the invasive foci are less than 3 mm in greatest dimension, or less than 10 mm^2, a designation of "SBT with microinvasion" is appropriate.

2. *SBT versus HGSC.* Occasionally, especially in a small core biopsy consisting predominantly or exclusively of tumor cells, with minimal stroma, the absence of stromal invasion by itself does not allow distinction of these entities. In such a case, high nuclear grade, as defined subsequently in this chapter, is sufficient to diagnose HGSC even when demonstrable stromal invasion is lacking.
3. *LGSC versus HGSC.* This differential diagnosis will be discussed subsequently in the section on HGSC.
4. *SBT versus benign serous tumor.* SBT are distinguished from benign serous tumors by the presence of cellular stratification, with detached cell clusters, mild cytological atypia, and mitotic activity. The last feature, mitotic activity, is the least important of these criteria. As with the distinction between SBT and LGSC, this is based entirely on morphology, with no special stains that can help in differential diagnosis. By convention, when the borderline component comprises less than 10% of a tumor and the remainder is benign, the designation of "benign serous tumor with focal proliferation" (or "focal atypia") is used, as the observation has been that such tumors are benign on follow-up [17]. This only applies for stage Ia tumors; if there are extra-ovarian implants, a diagnosis of SBT is appropriate.
5. *SBT versus clear cell carcinoma.* Clear cell carcinomas with predominantly papillary architecture may closely mimic SBT, especially on frozen section [18]. The presence of other more typical patterns of clear cell carcinoma is helpful, if present. Within the papillary areas, there is typically less stratification of epithelial cells in the papillary areas of clear cell carcinoma, and absence of micropapillary architecture. The presence of more atypical cytological features is also in favor of clear cell carcinoma. In problematic cases, immunostaining for WT1, ER, and HNF-1beta can be used, as SBT show the same immunophenotype as serous carcinoma (WT1 and ER positive, HNF-1beta negative, the opposite staining pattern to that seen in clear cell carcinoma).
6. *SBT versus mucinous borderline tumor, endocervical-like (seromucinous borderline tumor).* Seromucinous borderline tumors are distinguished from SBT by the presence of conspicuous intracytoplasmic mucin within the tumor cells. The presence of this mucin frequently is associated with a prominent infiltrate of polymorphonuclear leukocytes. These two tumor types are otherwise indistinguishable and frequently coexist as mixed tumors. The presence of mucinous differentiation, as a focal finding in SBT, is not of clinical significance.
7. *SBT versus mesothelioma or mesothelial hyperplasia.* SBT, when confined to the ovary, are unlikely to be confused with mesothelioma, but primary peritoneal SBT can closely resemble well-differentiated papillary mesothelioma. The cells of well-differentiated papillary mesothelioma (Fig. 6.16) are typically cuboidal and form a single layer over the basement membrane, in contrast to the more cuboidal and stratified cells of SBT. In problematic cases, the same immunopanel used for HGSC versus mesothelioma can be used (calretinin, as a positive mesothelioma marker, and epithelial markers such as Ber-EP4, B72.3, and MOC31). Mesothelial hyperplasia can also mimic desmoplastic implants of SBT, which can become a diagnostic issue in patients with a primary ovarian SBT (Fig. 6.17). Again, immunostaining can be assistance in problematic cases.

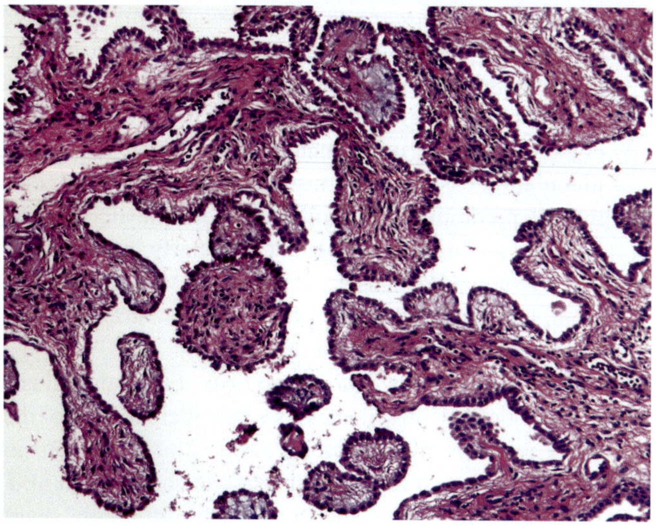

Fig. 6.16 Well-differentiated papillary mesothelioma. Note the lack of epithelial cell stratification compared to SBT

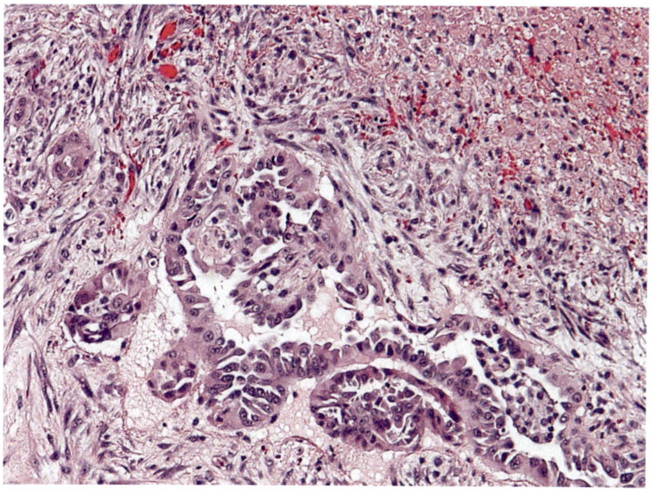

Fig. 6.17 Reactive mesothelial hyperplasia secondary to peritoneal irritation

8. *SBT versus retiform Sertoli-Leydig cell tumor.* Most retiform Sertoli-Leydig cell tumors have more typical patterns of Sertoli-Leydig cell tumor present. There may also be androgenic manifestations. Sertoli-Leydig cell tumors typically show inhibin immunoreactivity and are negative for CK7, a pattern opposite to that seen in SBT.

9. *Desmoplastic noninvasive implant versus invasive implant*. Desmoplastic noninvasive implant is a stromal predominant lesion containing angulated glands and single cells, some of which have attenuated cytoplasm, embedded in an inflammatory stroma that resembles granulation tissue. In contrast, invasive implants are epithelial predominant lesions with neoplastic glands, papillae, and micropapillae that irregularly infiltrate adipose tissue and/or muscle. If there is a stromal reaction at all, it is mild and appears loose or edematous, but not overtly desmoplastic. See Fig. 6.11.
10. *SBT implant versus endosalpingiosis*. Endosalpingiosis is commonly identified in cases where there is SBT of the ovary and can involve serosal surfaces or be present in lymph nodes, and endosalpingiosis can coexist with SBT implants in the same tissue section. This differential diagnosis is based on the same criteria outlined in point 4 above (SBT versus benign serous tumor) of stratification and atypia of the epithelial cells. There are instances where there is an indeterminate degree of atypia present in the epithelial cells lining a glandular space – insufficient for an unequivocal diagnosis of an implant of SBT – and the designation of "atypical endosalpingiosis" can be used in these cases. The clinical significance of such a finding, in a patient with stage I SBT, is not entirely clear, but this finding would typically not influence treatment.

High-Grade Serous Carcinoma (HGSC)

Prevalence

HGSCs account for 70% of ovarian surface epithelial carcinomas diagnosed in North America [19]. If one considers only those surface epithelial carcinomas that present with stage III or stage IV disease, they account for more than 90% of cases [19]. This is the ovarian carcinoma subtype that most people have in mind when they think of "ovarian cancer," and HGSC accounts for most deaths due to ovarian cancer. Among ovarian surface epithelial carcinomas presenting with disease confined to the pelvis (stage I or stage II), a minority (approximately 40%) will be of high-grade serous type. HGSCs are almost never confined to the ovary. Worldwide, there is considerable variability in the frequency of HGSC and, in general, HGSC are less common outside of Europe and North America, relative to other ovarian neoplasms.

Tissue of Origin and Nomenclature for Designation of Primary Site

A significant number (probably a majority) of high-grade serous carcinomas arise from the distal or fimbriated end of the fallopian tube, with a tendency to early transcoelomic spread [20–22]. Evidence in support of this is the frequency with which tubal intraepithelial carcinoma is seen in both hereditary and sporadic cases of HGSC, while early or precursor lesions in the ovary are vanishingly rare. HGSC designated in the past as being primary from fallopian tube or peritoneum are biologically the same disease as those cases considered, based on conventional criteria, to have arisen from the ovary, and it has been proposed that all such tumors should be designated as "pelvic HGSC" [20–22]. In practice, it is often not possible to be certain where a case of advanced-stage HGSC originates, and some uncertainty about primary site may be expressed in a diagnostic report for a tumor involving fallopian tube mucosa, ovary, and peritoneum (a not uncommon circumstance). The criteria used in the past to designate HGSC as being primary peritoneal, fallopian tube, or ovary (Table 6.2) were arbitrary and based on the assumption that most such tumors originated in the ovary. This assumption has since been shown to be incorrect, and these criteria should be abandoned. The move to a single inclusive designation of "pelvic HGSC" for such tumors is a sensible alternative and avoids clinically irrelevant speculation about primary site (ovary versus fallopian tube versus peritoneum) in a given case of HGSC. It remains important to separate pelvic HGSC from serous carcinomas arising in the uterus (endometrium or cervix); in practice, any disseminated serous carcinoma with involvement of the endometrial surface epithelium is most probably a primary serous carcinoma of endometrium; however, in individual cases there can remain uncertainty about primary site.

Table 6.2 Definitions formerly used to identify "primary site" in cases of pelvic high-grade serous carcinoma

Site	Definition
Fallopian tube	Fallopian tube tumor with mucosal involvement; no or minimal tumor involving ovary, endometrium and endocervix
Peritoneum	Extensive peritoneal involvement by tumor; ovaries and fallopian tubes of normal size with no or minimal involvement by tumor; no primary endometrial or endocervical tumor
Ovary	Any case with a significant ovarian mass and no primary endometrial or endocervical tumor

These definitions are arbitrary and are no longer considered to be able to accurately identify primary site in cases of high-grade serous carcinoma with involvement of multiple sites

Treatment and Clinical Outcomes

The early transcoelomic spread of HGSC, with rapid dissemination, has made screening for this disease incredibly difficult and, at present, there are no screening protocols of proven efficacy. There are reliable reports of patients who presented with advanced-stage HGSC less than 6 months after a completely negative laparoscopic examination, and also instances where patients with incidentally detected, microscopic, and apparently *in situ* tumors in the fallopian tube have developed recurrences, illustrating the challenges in screening for HGSC. With regards to treatment, all patients with HGSC (even stage Ia tumors) are at sufficient risk of recurrence that adjuvant chemotherapy is recommended [23]. This therapy consists of platinum and taxane combination chemotherapy. The main prognostic factor in a patient with an advanced-stage HGSC is whether the tumor has been completely resected at the time of primary surgery, with no macroscopic residual disease. Patients whose tumors can be resected completely have a 50% or greater 5-year recurrence-free survival, while the recurrence-free survival of those patients with macroscopic residual disease is closer to 15–20%. Patients whose tumors cannot be completely resected surgically have the same outcome with neoadjuvant chemotherapy and surgery performed after several cycles of chemotherapy, and with this neoadjuvant treatment approach, there are fewer complications secondary to the surgery [24]. As a result, there has been a tendency over the last few years for patients to either have more extensive surgery, including extensive upper abdominal surgery, to achieve complete resection of all macroscopic tumor, on one hand [25], or to have deferral of surgery until after the tumor has been shrunk by the chemotherapy, particularly in those patients where complete surgical resection pre-chemotherapy is not deemed feasible (e.g., patients with bulky upper abdominal disease). Where the intent is to give chemotherapy first, followed by surgery, there is no consensus on what constitutes an adequate sample for diagnosis prior to starting chemotherapy. A cytology sample can be adequate to confirm the diagnosis of malignancy, but it is not clear at present whether tumor subtype (e.g., HGSC versus LGSC or clear cell carcinoma) can be made based on a cytology sample.

Macroscopic Features

The ovarian tumors in patients with pelvic HGSC are typically solid and cystic with areas of hemorrhage and necrosis (Fig. 6.18). The tumor tissue is soft and friable. Bilaterality and surface involvement are noted in most cases. The fallopian tube may or may not show macroscopic evidence of involvement. In many cases the fallopian tube is completely obliterated by the adnexal mass such that normal tube or ovarian tissue cannot be identified. In some cases the tumor can be seen to involve the fimbriated end of the fallopian tube as a polypoid growth (Fig. 6.19), while in other instances there is more diffuse enlargement of the fallopian tube by intraluminal tumor. The omentum can be diffusely involved (so-called omental cake) or show multiple discrete nodules of tumor. The peritoneal surfaces often are studded with tumor "implants" of metastatic carcinoma.

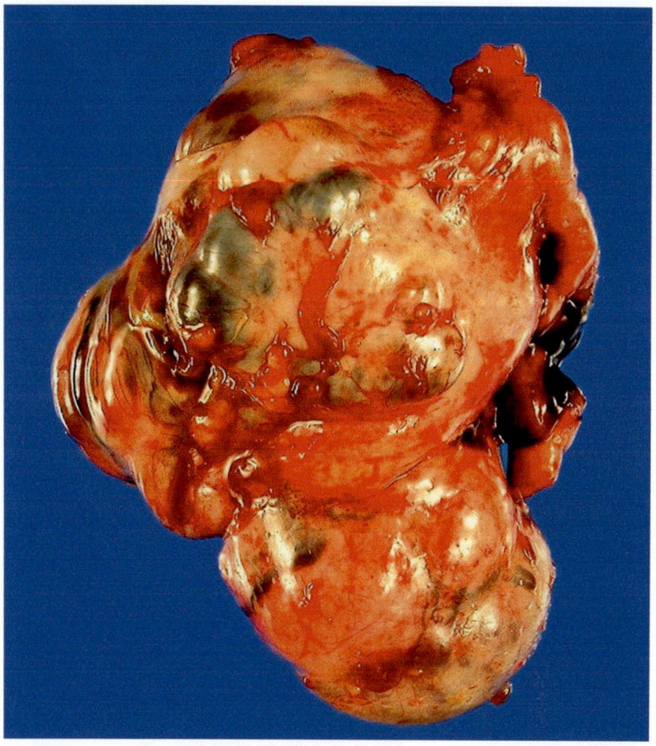

Fig. 6.18 Gross appearance of HGSC, with formation of an adnexal mass. This tumor does not show prominent surface involvement, which is commonly seen in HGSC. A distorted fallopian tube can be seen at the right side of the mass; the fallopian tube may be completely obliterated by carcinoma in HGSC

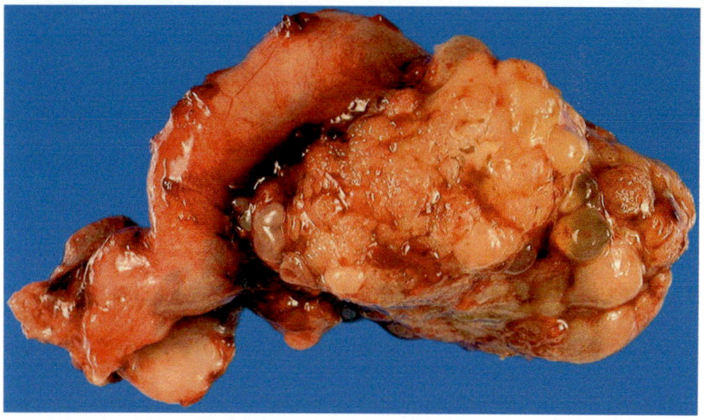

Fig. 6.19 An example of pelvic HGSC with the fimbriated end of the fallopian tube obliterated by the mass. The ovary in this case is uninvolved by tumor (*lower left*)

Microscopic Features

HGSC show a variety of architectural growth patterns. Two or more of these frequently coexist in the same tumor and, indeed, may coexist in the same tissue section. The most characteristic growth pattern is termed "papillary." In fact, papillae with well-formed lymphovascular cores are less common than tufting of highly stratified epithelium, with a fenestrated appearance and slit-like spaces (Fig. 6.20). Psammoma bodies may be present, but are rarely as numerous as in LGSC. Solid and glandular architectural patterns are also common. The solid pattern consists of sheets of cells without gland formation (Fig. 6.21). Focally, however, slit-like spaces may be seen (Fig. 6.22). Glandular differentiation is relatively common in HGSC, with well-formed glandular spaces lined by cells with high-grade nuclear features that are identical to those seen in the solid or papillary areas (Fig. 6.23). Rarely, the glandular pattern can be predominant. HGSC may show a microcystic pattern with small rounded spaces within sheets of tumor cells (Fig. 6.24). Finally, some HGSC bear a resemblance to high-grade transitional cell carcinoma of the urinary bladder (Fig. 6.25). There are no specific criteria for recognition of this transitional-like differentiation and, not surprisingly, it is

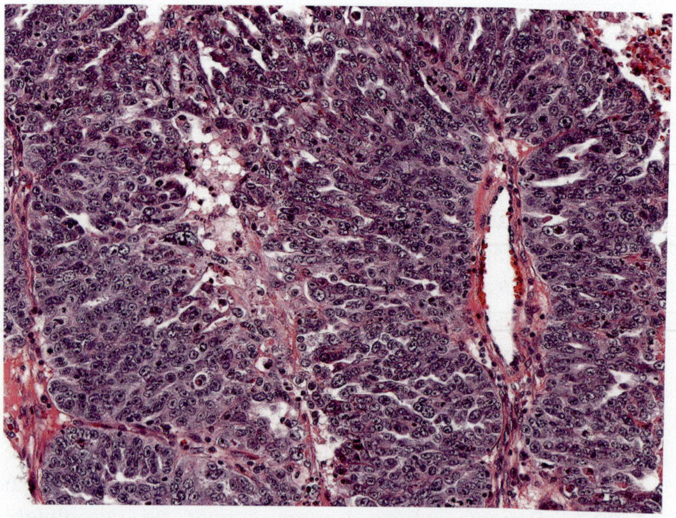

Fig. 6.22 Solid pattern of HGSC showing slit-like spaces

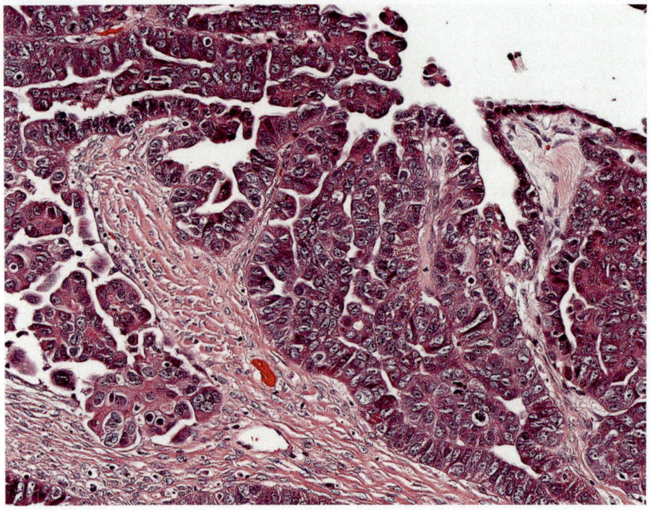

Fig. 6.20 Papillary pattern of HGSC

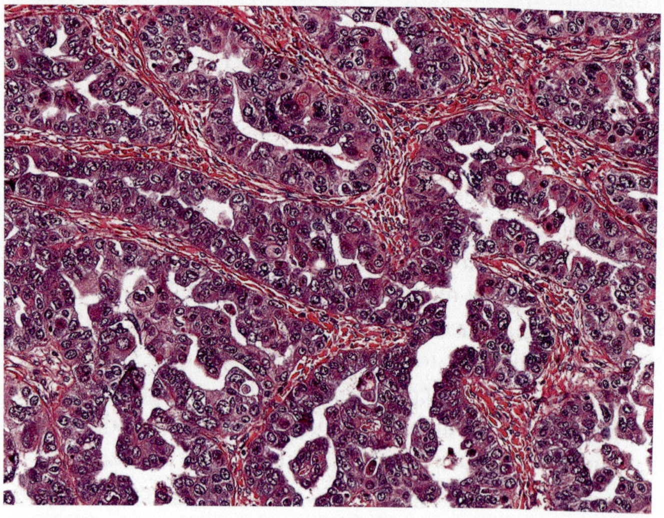

Fig. 6.23 Glandular pattern of HGSC. The cytological features are identical to those seen in other architectural patterns of HGSC, and higher grade than in usual endometrioid carcinoma of ovary

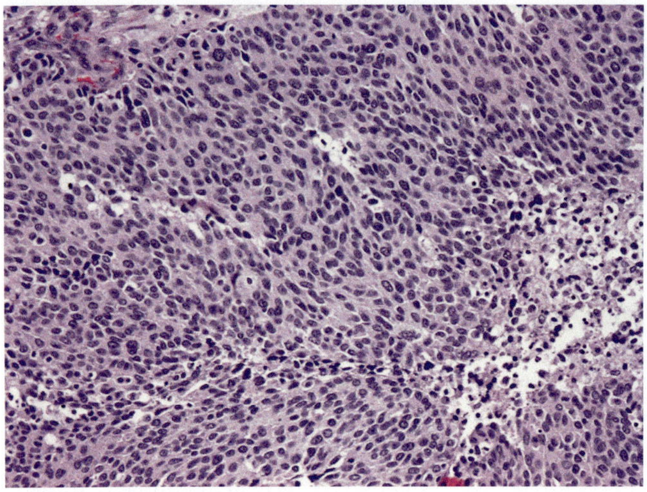

Fig. 6.21 Solid pattern of HGSC

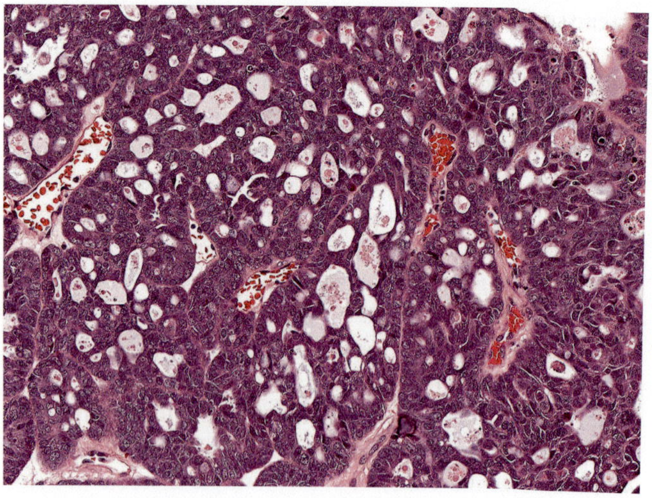

Fig. 6.24 Microcystic pattern of HGSC

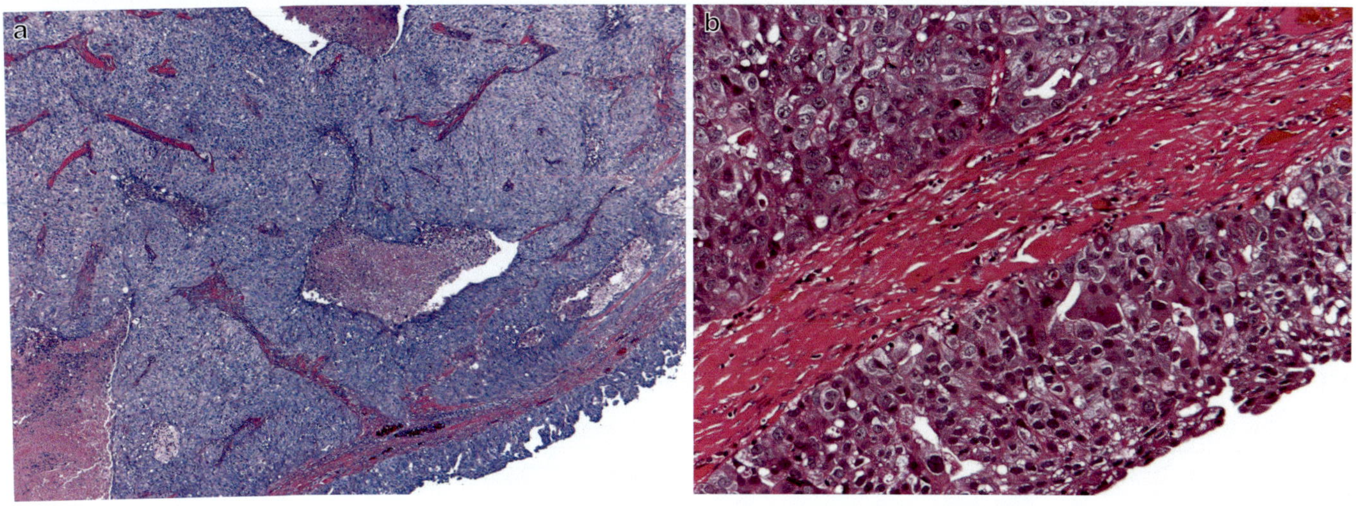

Fig. 6.25 Transitional-like architecture in HGSC at low power (**a**), with high-power examination showing typical HGSC cytological features (**b**)

the subject of considerable interobserver variability in recognition. Furthermore, so-called transitional cell carcinomas of the ovary have an immunophenotype that is indistinguishable from HGSC and different from urothelial carcinoma [26]. Transitional-like differentiation is therefore best regarded as a variant within the spectrum of HGSC. Fallopian tube mucosal involvement in high-grade serous carcinoma is relatively common, when searched for, even in cases of sporadic ovarian cancer. The characteristic mucosal lesion is tubal intraepithelial carcinoma (TIC), with the fallopian tube lined by cells indistinguishable from those found elsewhere within the tumor (Fig. 6.26). Fallopian tube mucosal involvement is usually unilateral, and these lesions are identical to TICs occasionally encountered in risk-reducing/prophylactic bilateral salpingo-oophorectomy specimens from patients with *BRCA1* or *BRCA2* mutations.

The cytological features of HGSC are distinctive, consisting of hyperchromatic nuclei with three-fold or greater variation in nuclear size (Fig. 6.27). A common finding is the presence of large malignant cells (tumor giant cells). Prominent nucleoli are common. Clear cell change as a focal finding in HGSC is relatively common (Fig. 6.28). This can be designated "HGSC with clear cell change" [27], but should not be diagnosed as mixed HGSC/clear cell carcinoma, a tumor type that is rarely if ever encountered in practice. The mitotic rate in HGSC is very high. The cut-off point for distinction between LGSC and HGSC is greater than 12 mitotic figures/10 high-power microscopic fields; however, the mitotic rates in HGSC are frequently more than double this figure. Abnormal mitotic figures are common.

After chemotherapy, a range of morphological features may be seen in the tumor cells [28], from increased numbers of large nuclei with smudgy chromatin, and decreased mitotic activity (Fig. 6.29), through to little or no microscopic alterations secondary to treatment, in the minority of cases that are platinum-resistant. In many cases, the tumor will have largely

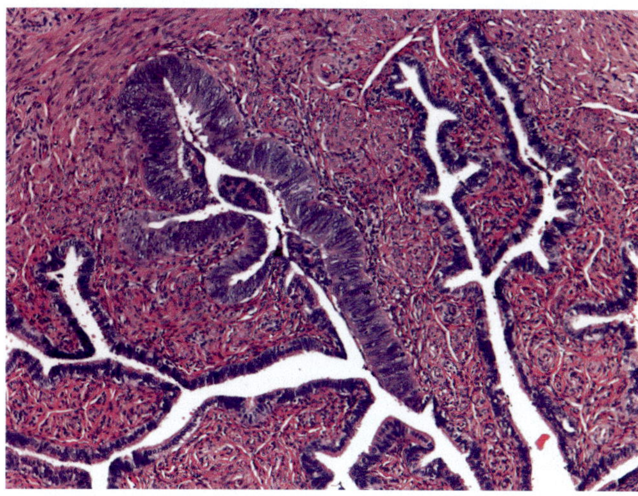

Fig. 6.26 Serous tubal intraepithelial tumor (STIC) in a patient with a germ-line *BRCA* mutation

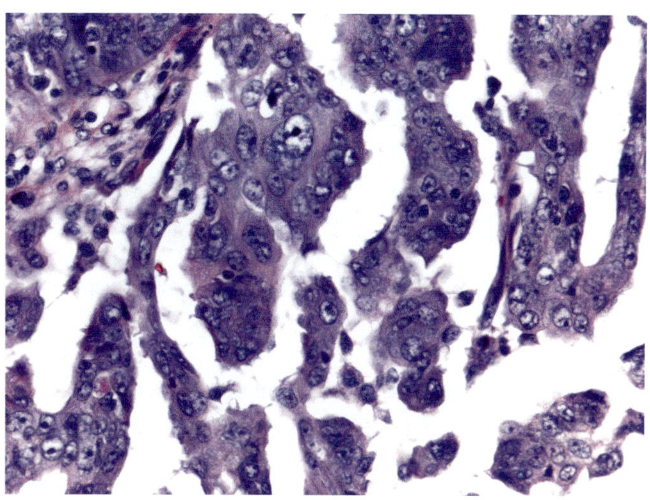

Fig. 6.27 Cytological features of HGSC, with greater than three-fold variation in nuclear size

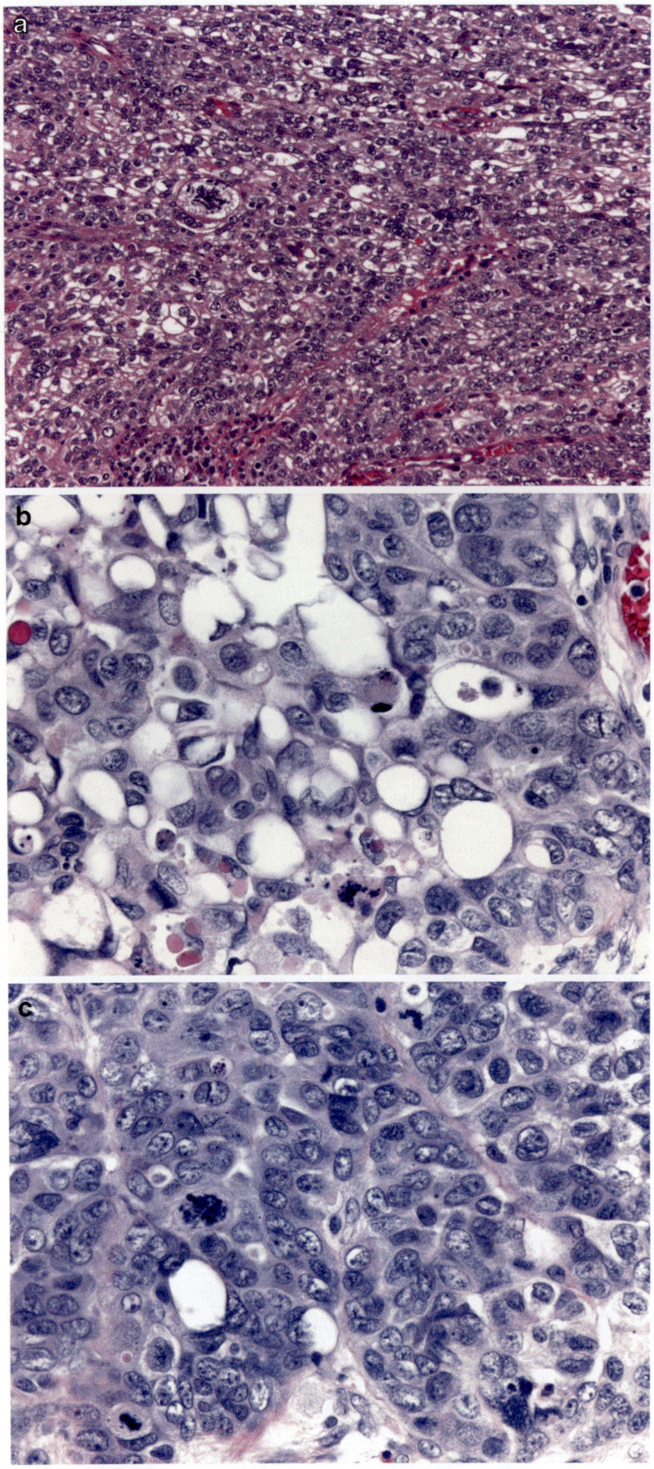

Fig. 6.28 Clear cell change in HGSC (**a** and **b**). In other areas, typical HGSC was present (**c**)

regressed; however, it is uncommon not to identify at least microscopic foci of residual tumor in the hysterectomy specimen after three or four cycles of chemotherapy. When treatment effects are marked, it can be difficult or impossible to comment on tumor cell type, but it is likely that

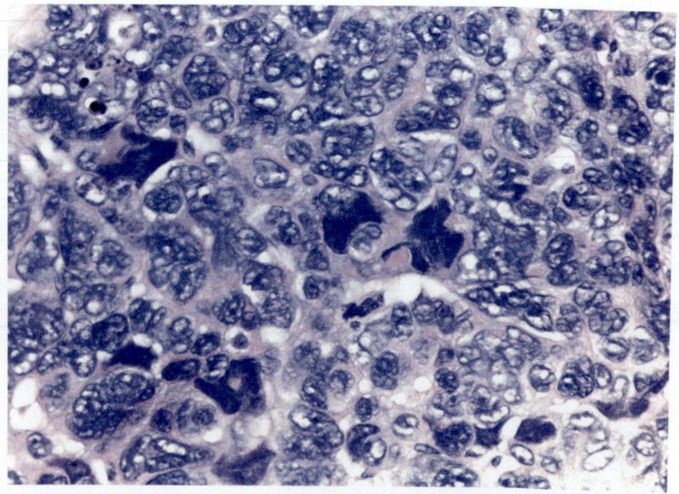

Fig. 6.29 HGSC after chemotherapy, showing large nuclei with smudgy chromatin

such tumors are HGSC, on a probabilistic basis, and immunostaining may be used to support this diagnosis. Sampling of omentum after chemotherapy typically shows areas of fibrosis and old hemorrhage. TIC lesions may persist after chemotherapy.

Genotype

The proportion of patients with pelvic HGSC who have a germline *BRCA1* or *BRCA2* mutation is quite variable within the literature, but there has been a trend toward a higher figure over the years, with improved methods for detection of mutations (which now include full gene sequencing and MLPA analysis to detect deletions). More recent studies have shown that greater than 10% of patients with pelvic HGSC have a germline *BRCA1* or *BRCA2* mutation [29, 30], a frequency that has led some centers to recommend referral of all patients with pelvic HGSC (irrespective of age) for genetic counseling and testing. In contrast, the likelihood of a *BRCA1* or *BRCA2* mutation in patients with ovarian tumors other than HGSC is sufficiently low that such testing is not indicated.

The importance of *BRCA1* and *BRCA2* germline mutations as a risk factor for development of high-grade serous carcinoma is widely appreciated, but somatic mutations in *BRCA1* and *BRCA2* also occur [29, 30]. Somatic mutations are less common than germline mutations. In patients with pelvic HGSC, as many as 20% or more of cases may be associated with either somatic or germline BRCA mutations. Epigenetic silencing of *BRCA1* can occur through promoter hypermethylation, with a dramatic decrease in expression of the gene and resulting decrease in intracellular protein [29]. More than half of high-grade serous

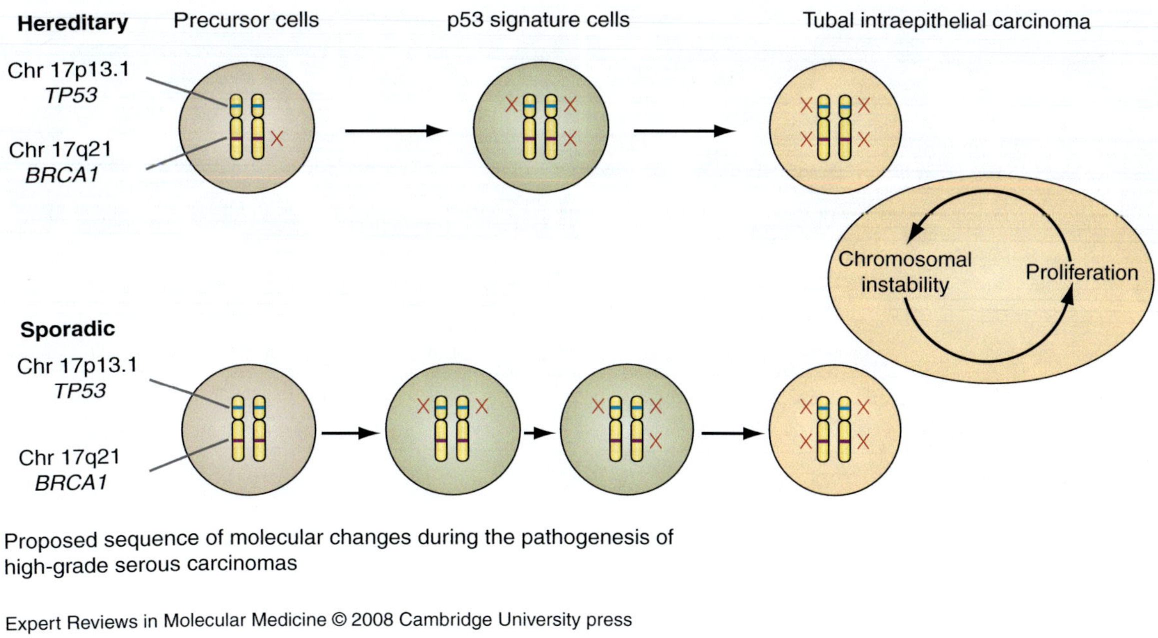

Fig. 6.30 A proposed sequence of molecular events in the pathogenesis of HGSC; hereditary ovarian carcinoma, with one copy of *BRCA1* mutated, is shown at the top, with sporadic HGSC at the bottom (Public Domain, reproduced from [34])

carcinomas will show either *BRCA1* or *2* mutation (germline or somatic), or *BRCA1* promoter hypermethylation with epigenetic silencing. These abnormalities are potentially important as treatments that target the defect in DNA repair that results from loss of *BRCA1* or *BRCA2* (PARP inhibitors) are under development and currently in clinical trials [30]. Loss of *BRCA1* or *BRCA2* function in a normal cell is lethal. It is thought that cells must first acquire a mutation or mutations that are permissive for loss of *BRCA*, as the first event in tumorigenesis. Loss of p53, for example, allows cells to survive subsequent loss of *BRCA1* or *BRCA2*. p53 loss of function, through mutation or deletion, is even more common than *BRCA* loss in HGSC, and can be found in greater than 90% of cases [31]. The putative precursor lesion of pelvic HGSC, the "p53 signature" in the fallopian tube epithelium, is a discrete microscopic area of the tubal epithelium where there is p53 loss of function (usually as a result of a missense mutation), but no significant cytological atypia and no increased proliferation [32]. These p53 signature lesions can be detected by immunohistochemistry for the mutant p53 protein that accumulates within the nuclei of the affected cells (p53 signatures, while of interest in studies on pathogenesis of HGSC, are not of clinical significance, as they can be seen in 20% of normal fallopian tubes, and this is NOT a diagnosis that should be used in routine practice). This combination of loss of function of p53 and either *BRCA1* or *BRCA2* is characteristic of pelvic HGSC (but is not seen in all cases). A result of loss of p53 and *BRCA* is chromosomal instability. A proposed sequence of genetic changes in the fallopian tube epithelium, leading to pelvic HGSC, is shown in Fig. 6.30 [33].

Abnormalities in the p16-cyclinD1-RB-E2F pathway are also common in HGSC, although their role in tumorigenesis is less well-understood. Increased expression of p16, which is thought to be a compensatory increase in expression of this anti-oncogene in response to pathway dysregulation, is a common feature of HGSC. As a result of chromosomal instability, HGSC are typically aneuploid, with very complex karyotypes.

Immunophenotype

Cells of pelvic HGSC are positive for WT1, estrogen receptor (ER), and PAX8 in most cases, and negative for HNF-1β (Table 6.3) [34]. They are also negative for the mesothelioma marker calretinin in most instances.

Table 6.3 Immunophenotype of high-grade serous carcinoma and low-grade serous carcinoma

Marker	High-grade serous	Low-grade serous
WT1	80%	70%
ER	76%	80%
HNF-1beta	6%	0
PAX8	80%	70%

From Kobel et al. [34]

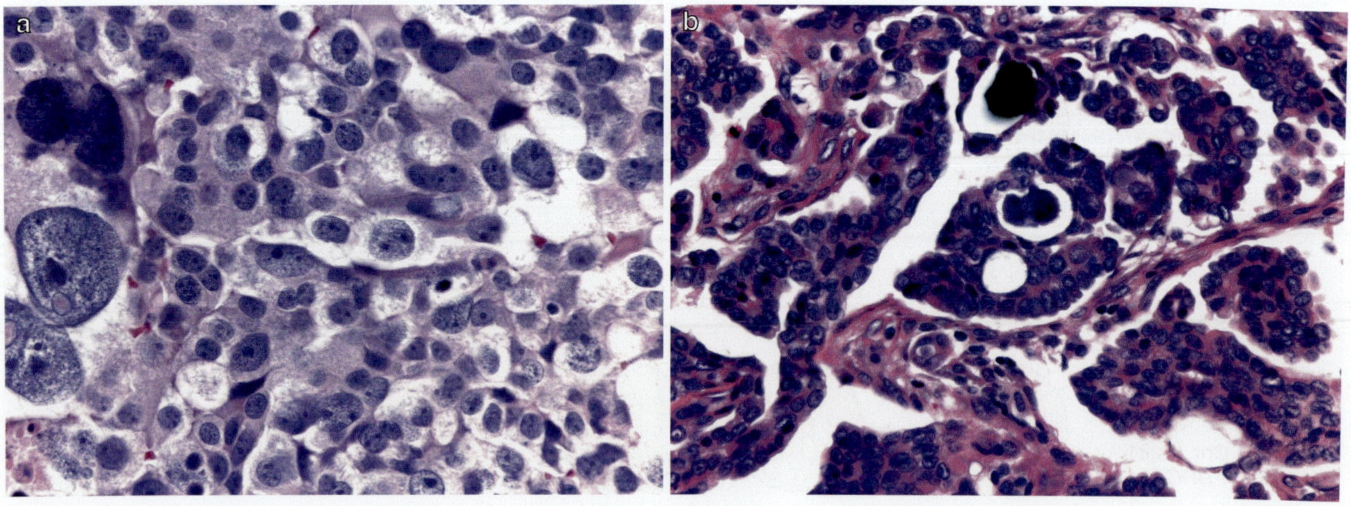

Fig. 6.31 Cytological features of HGSC (**a**) and LGSC (**b**), showing the greater variation in nuclear size and higher mitotic rate in the former

Differential Diagnosis

1. *HGSC versus LGSC*. The differential diagnosis between HGSC and LGSC is based on greater than three-fold variation in nuclear size in HGSC; as a secondary criterion, should the diagnosis be problematic after consideration of nuclear features, the presence of greater than 12 mitotic figures/10 high power microscopic fields is indicative of HGSC (Fig. 6.31) [4]. Rarely is there a problem in separating these two entities based on examination of several representative sections of tumor. On a small biopsy, however, this can be a problematic differential diagnosis. Strong, diffuse p53 positivity, indicative of a missense mutation in p53, supports a diagnosis of HGSC in such a case. In theory, Ki-67 staining could also help, but the diagnostic cut point between HGSC and LGSC has not been determined. In a very small sample, therefore, it may not be possible to make the distinction between LGSC and HGSC. Rarely, LGSC may progress to high-grade carcinoma; such tumors are not well-characterized.
2. *HGSC versus endometrioid carcinoma*. The distinction between HGSC and endometrioid carcinoma has historically been problematic. With recognition that WT1 immunostaining is a feature of HGSC but not endometrioid carcinomas [35], the distinction has become much more reliable. It turns out that many tumors diagnosed as high-grade endometrioid carcinomas in the past are best classified as HGSC [36]. The diagnosis of endometrioid carcinoma should be reserved for those tumors that are indistinguishable from their counterparts in the endometrium. Glandular architecture with high-grade cytological features is more in keeping with HGSC than endometrioid carcinoma, as high grade (grade 3) endometrioid carcinomas of the ovary are rare, accounting for less than 1% of ovarian surface epithelial carcinomas. The presence of squamous differentiation, on the other hand, is in favor of the diagnosis of endometrioid carcinoma. Similarly, a low-grade endometrioid component giving rise to a high-grade tumor is more in keeping with endometrioid carcinoma. Immunostains can be used in problematic cases; however, WT1 is only positive in approximately 80% of HGSC and occasionally is positive in endometrioid carcinomas, albeit in less than 5% of cases. Cases diagnosed as mixed HGSC/endometrioid carcinoma are rare in practice, although they do occur (just as they occur in the endometrium, where they represent progression of a low-grade endometrioid carcinoma). Most tumors so designated in the past are best considered HGSC.
3. *HGSC versus undifferentiated carcinoma*. HGSC frequently show solid or undifferentiated architecture. The diagnosis of undifferentiated carcinoma of the ovary should be reserved for those tumors which show no evidence of serous differentiation. If there is any convincing serous differentiation, a diagnosis of HGSC is preferred, even if these differentiated areas comprise a minority of the tumor. The diagnosis of mixed HGSC/undifferentiated carcinoma should not be used in practice. Because solid growth pattern is so common in high-grade serous carcinoma, all such tumors should simply be diagnosed as HGSC. This single modification of the WHO classification of ovarian surface epithelial carcinoma dramatically improves the interobserver reproducibility of diagnosis of ovarian carcinoma subtypes.
4. *HGSC versus clear cell carcinoma*. HGSC with clear cell change can be difficult to distinguish from clear cell carcinoma (Fig. 6.32). The presence of more typical HGSC patterns usually permits recognition of serous differentiation. In problematic cases an immuno panel of WT1, ER, and HNF-1beta can be used [27, 37]. WT1 and ER are positive in most HGSC, and HNF-1beta is positive in clear cell carcinoma; this panel is informative in all cases except where all three markers are negative. As noted

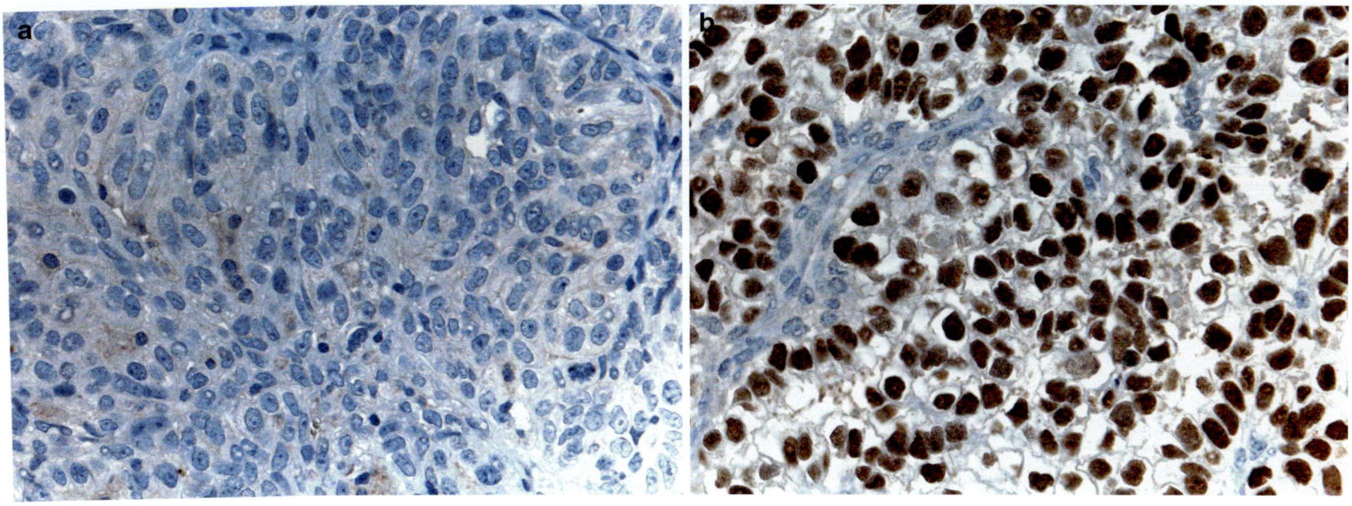

Fig. 6.32 The cells of HGSC are negative for HNF-1beta expression (**a**), while most clear cell carcinomas show strong expression (**b**)

previously, the diagnosis of mixed HGSC/clear cell carcinoma should be made rarely, if ever.

5. *Pelvic HGSC versus endometrial serous carcinoma.* Distinction between HGSC arising from fallopian tube, ovary, or peritoneum versus high-grade serous carcinoma arising from the endometrium is problematic. The histological features are often very similar. The distinction should be made based on examination of the endometrium. It is now well accepted that small primary endometrial carcinomas of serous type, even when confined to a polyp, without invasion of myometrium, can spread to the peritoneal cavity [38]. When there is endometrial involvement by high-grade serous carcinoma, a diagnosis of primary endometrial serous carcinoma should be favored except in those rare cases where the tumor shows a pattern of endometrial involvement typical of metastases, with no endometrial surface involvement, infiltration through the endometrial stroma between remnant benign endometrial glands, with prominent lymphovascular invasion in the endometrium and myometrium (Fig. 6.33). There will be some cases where uncertainty about the primary site of a gynecological HGSC will have to be acknowledged in a diagnostic report, especially if only a single site has been sampled; however, in most cases it should be possible to distinguish between pelvic HGSC and endometrial serous carcinoma when uterus, tubes, and ovaries are available for examination. At present, unfortunately, there are no reliable immunostains to aid in this distinction; although WT1 is more commonly expressed in pelvic HGSC than in endometrial serous carcinoma [39], there are too many exceptions for this to be a useful test for primary site in clinical practice.
6. *HGSC versus mesothelioma.* Pelvic HGSC can show a pattern of intraperitoneal involvement that raises the possibility of peritoneal mesothelioma. The cells of mesothelioma are typically more cuboidal and more uniform cytologically, without the stratification, pleomorphism, and high mitotic rate of HGSC (Fig. 6.34). Immunomarkers are usually not required for distinction between these two diagnostic possibilities; however, the combination of calretinin as a positive mesothelioma marker and epithelial markers such as Ber-EP4, B72.3, and MOC31 can be used.
7. *HGSC versus metastatic breast carcinoma.* A special circumstance that one encounters in practice is the patient with a history of breast cancer who now presents with advanced-stage tumors involving both ovaries, and the differential diagnosis is metastatic breast carcinoma versus pelvic HGSC. This is particularly challenging when the breast carcinoma, as is frequently the case in

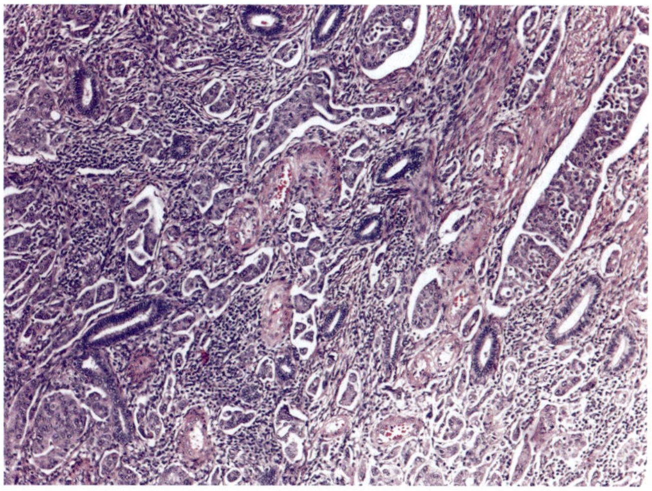

Fig. 6.33 A rare case of involvement of the endometrium by pelvic HGSC. Note the infiltration around normal endometrial glands, with preservation of endometrial architecture, and prominent lymphovascular invasion. Such cases do not show involvement of the endometrial surface epithelium, in contrast to serous carcinomas arising in the endometrium

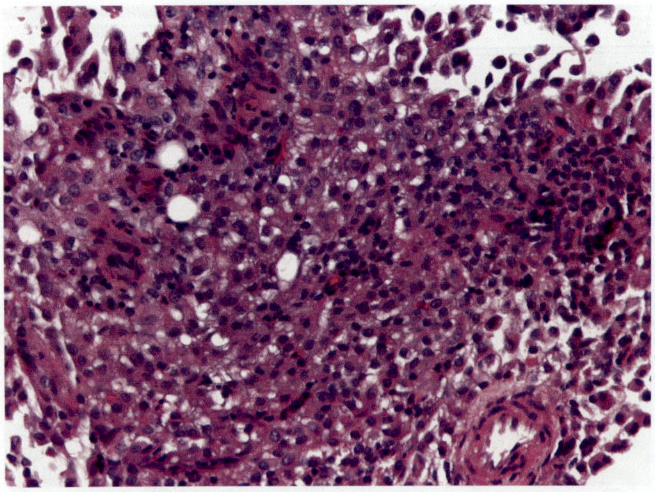

Fig. 6.34 Mesothelioma of the peritoneum; the cells are more uniform than those of HGSC (see Fig. 17.4)

patients with *BRCA1* mutations, is of basal-like type, as these tumors are typically high grade, and negative for HER2, ER, and progesterone receptor (PR). However, it should be noted that the statistical likelihood of a new primary adnexal HGSC far outweighs the probability of metastatic breast carcinoma. Papillary differentiation of serous carcinoma and the presence of psammoma bodies are helpful. The high cytological grade and brisk mitotic activity of basal-like breast carcinoma is indistinguishable from HGSC, and one frequently has to resort to immunostaining. GCDFP-15 and mammoglobin-beta are markers of breast carcinoma, while WT1 and PAX8 are markers of HGSC [40].

8. *HGSC versus metastatic carcinoma*. Endometrial and breast carcinoma metastases to the ovary have been discussed previously. In theory, any high-grade carcinoma metastatic to the ovary could mimic HGSC. As it happens, non-gynecological cancers that present with ovarian metastases are more likely to be from the gastrointestinal tract and do not resemble HGSC microscopically. In patients with a history of lung carcinoma or bladder carcinoma, it is important to be aware of the previously diagnosed tumor as otherwise confusion with primary HGSC could result if there is development of ovarian metastases. Occasional HGSCs express TTF-1 focally, so this marker should only be used as part of a panel that also includes ER, PR, PAX8, and WT1.

Intraoperative Frozen Section Diagnosis

Intraoperative frozen section diagnosis of ovarian masses only provides limited information, and definitive diagnosis of cell type may not be possible. Similarly, it may not be possible, based on frozen section, to distinguish between HGSC and LGSC, or even primary versus secondary carcinoma. The differential diagnoses for each of these possibilities are discussed above. The limitations of frozen section examination (limited sampling, artifactual distortion, inability to use immunostaining) can introduce interpretive difficulties that preclude definitive intraoperative diagnosis, but these will typically disappear on examination of permanent sections. Nonetheless, in most cases of HGSC it is possible to indicate to the surgeon that there is high-grade carcinoma and that it is consistent with a primary pelvic HGSC – a diagnosis that will typically prove sufficient to guide completion of the appropriate surgical procedure.

References

1. Singer G, Kurman RJ, Chang HW, et al. Diverse tumorigenic pathways in ovarian serous carcinoma. Am J Pathol. 2002;160:1223–8.
2. Singer G, Stöhr R, Cope L, et al. Patterns of p53 mutations separate ovarian serous borderline tumors and low- and high-grade carcinomas and provide support for a new model of ovarian carcinogenesis: a mutational analysis with immunohistochemical correlation. Am J Surg Pathol. 2005;2:218–24.
3. Vang R, Shih IeM, Kurman RJ. Ovarian low-grade and high-grade serous carcinoma: pathogenesis, clinicopathologic and molecular biologic features, and diagnostic problems. Adv Anat Pathol. 2009;5:267–82.
4. Malpica A, Deavers MT, Tornos C, et al. Interobserver and intraobserver variability of a two-tier system for gradinig ovarian serous carcinoma. Am J Surg Pathol. 2007;31:1168–74.
5. Szyfelbein WM, Young RH, Scully RE. Cystic struma ovarii: a frequently unrecognized tumor. A report of 20 cases. Am J Surg Pathol. 1994;18:785–8.
6. Gershenson DM, Sun CC, Lu KH, et al. Clinical behavior of stage II-IV low-grade serous carcinoma of the ovary. Obstet Gynecol. 2006;108:361–8.
7. Lim-Tan SK, Cajigas HE, Scully RE. Ovarian cystectomy for serous borderline tumors: a follow-up study of 35 cases. Obstet Gynecol. 1988;72:775–81.
8. Segal GH, Hart WR. Ovarian serous tumors of low malignant potential (serous borderline tumors). The relationship of exophytic surface tumor to peritoneal "implants". Am J Surg Pathol. 1992;16((6):577–83.
9. Gilks CB, Alkushi A, Yue J, et al. Advanced stage serous borderline tumors of the ovary: a clinicopathological study of 49 cases. Int J Gynecol Pathol. 2003;22:29–36.
10. Silva EG, Gershenson DM, Malpica A, Deavers M. The recurrence and the overall survival rates of ovarian serous borderline neoplasms with noninvasive implants is time dependent. Am J Surg Pathol. 2006;30:1367–71.
11. Shvartsman HS, Sun CC, Bodurka DC, et al. Comparison of the clinical behavior of newly diagnosed stages II-IV low-grade serous carcinoma of the ovary with that of serous ovarian tumors of low malignant potential that recur as low-grade serous carcinoma. Gynecol Oncol. 2007;105:625–9.
12. Longacre TA, McKenney JK, Tazelaar HD, et al. Ovarian serous tumors of low malignant potential (borderline tumors): outcome-based study of 276 patients with long-term (> or=5-year) follow-up. Am J Surg Pathol. 2005;29:707–23.
13. McKenney JK, Balzer BL, Longacre TA. Patterns of stromal microinvasion in ovarian serous tumors of low malignant potential

(borderline tumors): a reevaluation of the concept of stromal microinvasion. Am J Surg Pathol. 2006;30:1209–21.
14. McKenney JK, Balzer BL, Longacre TA. Lymph node involvement in ovarian serous tumors of low-malignant potential (borderline tumors): pathology, prognosis, and proposed classification. Am J Surg Pathol. 2006;30:614–24.
15. Eichhorn JH, Bell DA, Young RH, et al. Ovarian serous borderline tumors with micropapillary and cribriform patterns: a study of 40 cases and comparison with 44 cases without these patterns. Am J Surg Pathol. 1999;23:397–409.
16. Singer G, Oldt R, Cohen Y, et al. Mutations in BRAF and KRAS characterize the development of low-grade serous carcinoma. J Natl Cancer Inst. 2003;95:484–6.
17. Bell DA, Longacre TA, Prat J, et al. Serous borderline (low malignant potential, atypical proliferative) ovarian tumors: workshop perspectives. Hum Pathol. 2004;35:934–48.
18. Sangoi AR, Soslow RA, Teng NN, et al. Ovarian clear cell carcinoma with papillary features: a potential mimic of serous tumor of low malignant potential. Am J Surg Pathol. 2008;32:269–74.
19. Köbel M, Kalloger SE, Huntsman DG, et al. Differences in tumor type in low-stage versus high-stage ovarian carcinomas. Int J Gynecol Pathol. 2010;29:203–11.
20. Lee Y, Medeiros F, Kindelberger D, et al. Advances in the recognition of tubal intraepithelial carcinoma: applications to cancer screening and the pathogenesis of ovarian cancer. Adv Anat Pathol. 2006;1:1–7.
21. Carlson JW, Miron A, Jarboe EA, et al. Serous tubal intraepithelial carcinoma: its potential role in primary peritoneal serous carcinoma and serous cancer prevention. J Clin Oncol. 2008;26(25):4160–5.
22. Salvador S, Gilks B, Köbel M, Huntsman D, Rosen B, Miller D. The fallopian tube: primary site of most pelvic high-grade serous carcinomas. Int J Gynecol Cancer. 2009;19(1):58–64.
23. Köbel M, Kalloger SE, Santos JE, et al. Tumor type and substage predict survival in stage I and II ovarian carcinoma: insights and implications. Gynecol Oncol. 2010;116:50–6.
24. Vergote I, Trope CG, Amant F, Kristensen GB, Ehlen T, et al. Neoadjuvant chemotherapy or primary surgery in Stage IIIc-IV ovarian cancer. N Engl J Med. 2010;363:943–53.
25. Chi DS, Eisenhauer EL, Zivanovic O, et al. Improved progression-free and overall survival in advanced ovarian cancer as a result of a change in surgical paradigm. Gynecol Oncol. 2009;114:26–31.
26. Soslow RA, Rouse RV, Hendrickson MR, et al. Transitional cell neoplasms of the ovary and urinary bladder: a comparative immunohistochemical analysis. Int J Gynecol Pathol. 1996;15:257–65.
27. Han G, Gilks CB, Leung S, et al. Mixed ovarian epithelial carcinomas with clear cell and serous components are variants of high-grade serous carcinoma: an interobserver correlative and immunohistochemical study of 32 cases. Am J Surg Pathol. 2008;7:955–64.
28. Miller K, Price JH, Dobbs SP, et al. An immunhistochemical and morphological analysis of post-chemotherapy ovarian carcinoma. J Clin Pathol. 2008;61:652–7.
29. Hennessy BT, Timms KM, Carey MS, et al. omatic mutations in BRCA1 and BRCA2 could expand the number of patients that benefit from poly (ADP ribose) polymerase inhibitors in ovarian cancer. J Clin Oncol. 2010;28(22):3570–6. Epub ahead of print.
30. Fong PC, Boss DS, Yap TA, et al. Inhibition of poly(ADP-ribose) polymerase in tumors from BRCA mutation carriers. N Engl J Med. 2009;361:123–34.
31. Ahmed AA, Etemadmoghadam D, Temple J, et al. Driver mutations in TP53 are ubiquitous in high grade serous carcinoma of the ovary. J Pathol. 2010;221:49–56.
32. Jarboe E, Folkins A, Nucci MR, et al. Serous carcinogenesis in the fallopian tube: a descriptive classification. Int J Gynecol Pathol. 2008;27:1–9.
33. Köbel M, Huntsman D, Gilks CB. Critical molecular abnormalities in high-grade serous carcinoma of the ovary. Expert Rev Mol Med. 2008;10:e22.
34. Köbel M, Kalloger SE, Boyd N, et al. Ovarian carcinoma subtypes are different diseases: implications for biomarker studies. PLoS Med. 2008;5:e232.
35. Al-Hussaini M, Stockman A, Foster H, et al. WT-1 assists in distinguishing ovarian from uterine serous carcinoma and in distinguishing between serous and endometrioid ovarian carcinoma. Histopathology. 2005;46:468.
36. Gilks CB, Ionescu D, Kalloger S, et al. Tumor cell type can be reproducibly diagnosed and is of independent significance in patients with maximally debulked ovarian carcinoma. Hum Pathol. 2008;39:1239–51.
37. Köbel M, Kalloger SE, Carrick J, et al. A limited panel of immunomarkers can reliably distinguish between clear cell and high-grade serous carcinoma of the ovary. Am J Surg Pathol. 2009;33:14–21.
38. Silva EG, Jenkins R. Serous carcinoma in endometrial polyps. Mod Pathol. 1990;3(2):120–8.
39. Al-Hussaini M, Stockman A, Foster H, McCluggage WG. WT-1 assists in distinguishing ovarian from uterine serous carcinoma and in distinguishing between serous and endometrioid ovarian carcinoma. Histopathology. 2004;44:109–15.
40. Nonaka D, Chiriboga L, Soslow RA. Expression of pax8 as a useful marker in distinguishing ovarian carcinomas from mammary carcinomas. Am J Surg Pathol. 2008;32:1566–71.

Pathology of Endometrioid Tumors

7

Melissa P. Murray and Kay J. Park

Keywords
Ovary • Endometrioid • Adenocarcinoma • Borderline • Tumor

Introduction

Endometrioid adenocarcinoma of the ovary is significantly less prevalent than has been traditionally reported, representing approximately 10% of all ovarian carcinomas, yet it also comprise the majority (57%) of low-stage ovarian carcinomas [1]. This is due to the misclassification of other histologic subtypes, mainly high-grade serous carcinoma, as high-stage endometrioid adenocarcinoma [2]. However, endometrioid adenocarcinoma is still the second-most common ovarian carcinoma subtype in North America and Europe, with most cases being FIGO stage I or II [1].

Benign Endometrioid Tumors

Endometrioid Cystadenoma and Adenofibroma

Prevalence

Endometrioid cystadenomas and adenofibromas are uncommon and usually unilateral tumors with a mixed-solid and cystic-cut surface. Adenofibromas typically occur in association with borderline tumors, and generous sampling of a borderline tumor often reveals the adenofibromatous components.

Macroscopic Features

Grossly, adenofibromas are typically solid with small cystic spaces, resembling the surface of a sponge, while cystadenomas are simple cystic structures. The cysts are usually filled with a clear yellow fluid and can occasionally have small papillations lining the cyst wall.

Microscopic Features

Microscopically, endometrioid adenofibromas are composed of simple glands lined by cells that resemble proliferative, weakly proliferative, inactive endometrium, or even simple non-atypical hyperplasia, embedded in a fibromatous stroma (Fig. 7.1). Nuclear grade is low, and squamous/morular metaplasia is a common finding. Cystadenomas are simple cysts lined by a single layer of similar endometrioid cells without atypia or extensive proliferation.

Differential Diagnosis

Endometrioid cystadenomas and adenofibromas are not difficult to diagnose. It is important that large tumors are well-sampled so that any borderline areas are not missed. A good gross examination should reveal any suspicious areas for sampling.

Clinical Behavior

Endometrioid cystadenomas and adenofibromas are benign and successfully treated by oophorectomy.

K.J. Park (✉)
Department of Pathology, Memorial Sloan-Kettering Cancer Center, 1275 York Avenue, New York, NY 10065, USA

R.A. Soslow and C. Tornos (eds.), *Diagnostic Pathology of Ovarian Tumors*,
DOI: 10.1007/978-1-4419-9751-7_7, © Springer Science+Business Media, LLC 2011

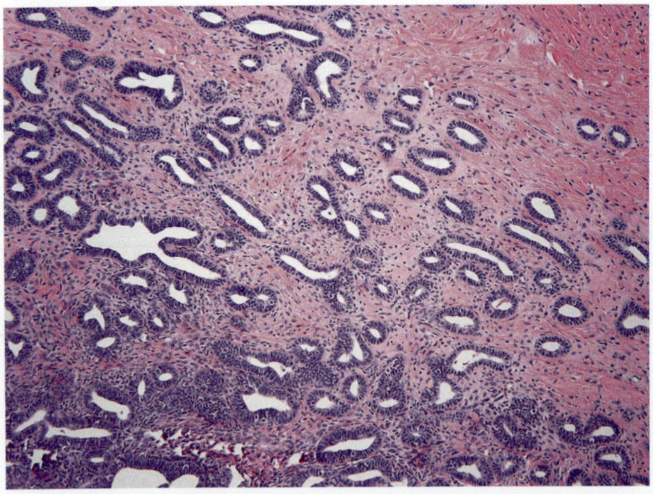

Fig. 7.1 Endometrioid adenofibroma. Variably sized glands in a fibromatous stroma [10×]

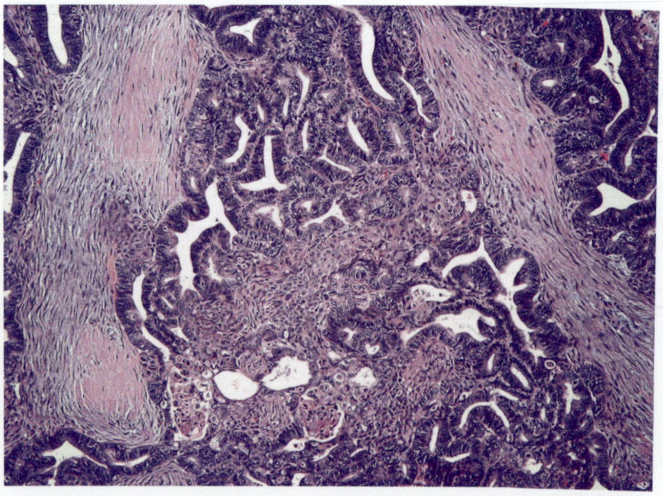

Fig. 7.2 Endometrioid borderline tumor. Complex glandular architecture set in a fibromatous background [10×]

Borderline Endometrioid Tumors

As with benign and malignant endometrioid tumors, endometrioid borderline tumors occur most commonly in women in the older reproductive and postmenopausal age groups [3]. Endometrioid borderline tumors can be classified as "tumors of low malignant potential," but they are benign if they are accurately diagnosed [4, 5]. This finding has led to a proposal that the "borderline" and "low malignant potential" nomenclature be replaced with "atypical proliferative tumor." The exact prevalence is not known because of variation in diagnostic criteria. However, Norris reported that borderline tumors comprised 2–19% of endometrioid tumors and 2–10% of all borderline tumors [6].

Macroscopic Features

Grossly, borderline tumors are similar on gross examination to benign endometrioid tumors. Borderline tumors are unilateral and predominantly solid with a smooth outer surface. The cut surface is firm and solid or variegated and may have small cysts with clear mucoid or hemorrhagic fluid.

Microscopic Features

Microscopically, most tumors have an adenofibromatous pattern with nodular architecture; however, they have more proliferative epithelium than adenofibromas. On low power, the appearance is often similar to that of complex atypical hyperplasia of the endometrium, with both architectural and cytologic atypia. Endometrioid borderline tumors typically show more architectural complexity, with higher gland-to-stroma ratios (as compared to adenofibromas) and closely packed cribriforming glands, and, by definition, have cytologic atypia, although it is usually mild (Fig. 7.2). Squamous morules are common; when they are extensive, it can be a diagnostic challenge to distinguish a borderline lesion from a well-differentiated endometrioid adenocarcinoma. Endometriosis is a common finding in endometrioid borderline tumors reported in approximately 35% of cases [4].

Criteria for distinguishing borderline tumor from adenocarcinoma with expansile invasion are the same as those that permit distinction of complex atypical hyperplasia from endometrioid adenocarcinoma of the endometrium [7, 8]; extensive gland fusion, large gland cribriforming, maze-like lumens, and extensive papillary architecture are considered evidence of invasion [4, 5, 9] (Table 7.1). Microinvasion in borderline tumors can also be encountered. The designation of endometrioid borderline tumor with microinvasion is applied when there are one or more small foci of destructive stromal invasion and each focus measures less than 5 mm in greatest linear dimension or less than 10 mm² in greatest area in any one slide [4]. Borderline tumors with intraepithelial carcinoma show the expected borderline architectural features, but contain cells with marked cytologic atypia with nuclear enlargement and

Table 7.1 Endometrioid adenocarcinoma versus borderline tumor

Adenocarcinoma with expansile invasion	Borderline tumor
Extensive gland fusion	Back-to-back glands without fusion
Large gland cribriforming and maze-like lumens	No cribriforming
Extensive papillary architecture	No extensive papillary architecture

pleomorphism with coarse chromatin or hyperchromasia and large prominent nucleoli (grade 3 nuclei).

Differential Diagnosis of Borderline Tumors

Benign Endometrioid Tumor

The degree of architectural and cytologic atypia of the epithelial component distinguishes between benign and borderline endometrioid tumors. A diagnosis of borderline tumor is favored when cribriform glandular or papillary architecture and epithelial cytologic atypia are present, and criteria for adenocarcinoma (discussed above) are not met.

Clinical Behavior and Treatment of Borderline Tumors

The prognosis of borderline endometrioid tumors is excellent, including those with microinvasion or intraepithelial carcinoma [4–6]. The majority of borderline tumors present as stage I disease [4, 5, 10]. Recurrences and metastases are extremely rare. Borderline tumors can be associated with endometrial polyps, and there is a reported coexistence of borderline endometrioid tumors and hyperplasia of the endometrium, and even endometrioid adenocarcinoma of the endometrium [4]. Synchronous atypical polypoid adenomyoma has also been diagnosed.

Bilateral salpingo-oophorectomy would be prudent when retention of fertility is no longer an issue; however, the infrequency of bilaterality and the very low rate of metastasis mean that a staging procedure is not indicated after confirmation of the diagnosis on permanent section. Lymphadenectomy is unnecessary. For women of childbearing age, unilateral salpingo-oophorectomy with close follow-up for early detection of any subsequent carcinoma is appropriate.

Malignant Endometrioid Tumors

Endometrioid Adenocarcinoma

Macroscopic Features

Malignant endometrioid tumors are variably cystic and solid, with generally smooth outer surfaces. They often have foci of residual endometriosis. Sixty percent of endometrioid adenocarcinomas are reportedly unilateral, but our institutional experience indicates that 91% of primary ovarian endometrioid adenocarcinomas are unilateral.

Microscopic Features

Ovarian endometrioid adenocarcinomas are similar to invasive endometrial adenocarcinoma of the uterus.

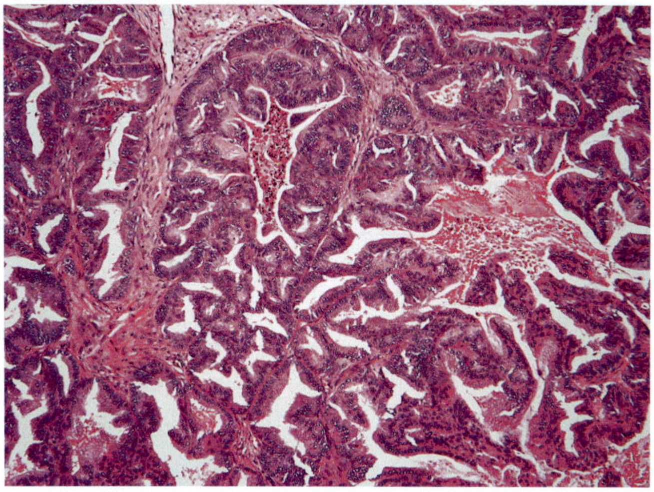

Fig. 7.3 Endometrioid adenocarcinoma of the ovary. Glands are lined by tall, stratified columnar cells [100×]

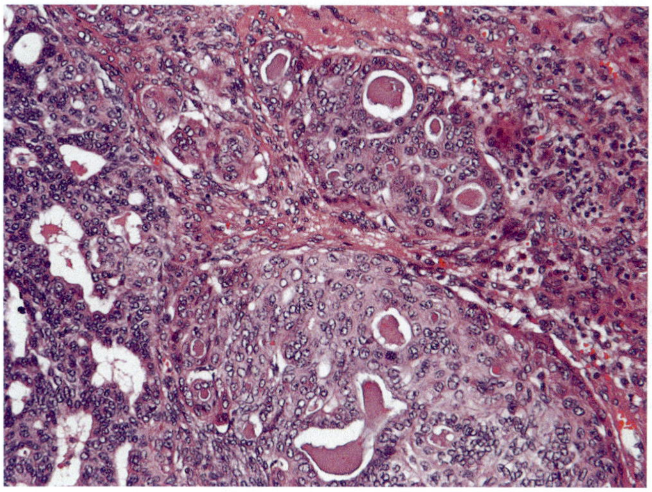

Fig. 7.4 Endometrioid adenocarcinoma of the ovary. Eosinophilic colloid-like material within the gland lumens [200×]

Well-differentiated adenocarcinomas, which account for the majority, predominantly have a glandular, cribriform, and/or villoglandular pattern with a back-to-back growth of round-to-elongated glands with smooth luminal contours in a fibrotic stroma. The glands are typically lined by stratified tall columnar cells with scant eosinophilic cytoplasm (Fig. 7.3). The cytoplasm may be focally mucinous, with mucin accumulating at the apical portion. Mucin or eosinophilic colloid-like material within the gland lumens is seen in some tumors (Fig. 7.4). In higher-grade tumors, solid sheet-like growth is present. In general, the cytologic features are concordant with the architectural features such that markedly atypical and highly proliferative cells are not encountered in simple tubules or papillae (Fig. 7.5).

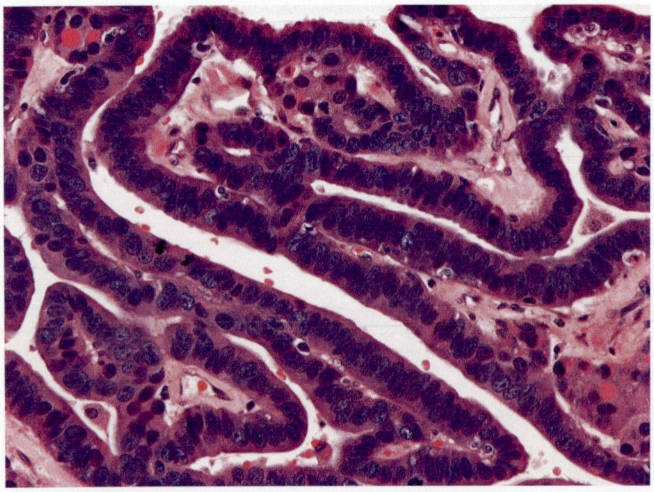

Fig. 7.5 Endometrioid adenocarcinoma of the ovary. Cytologic features of classic endometrioid adenocarcinoma [400×]

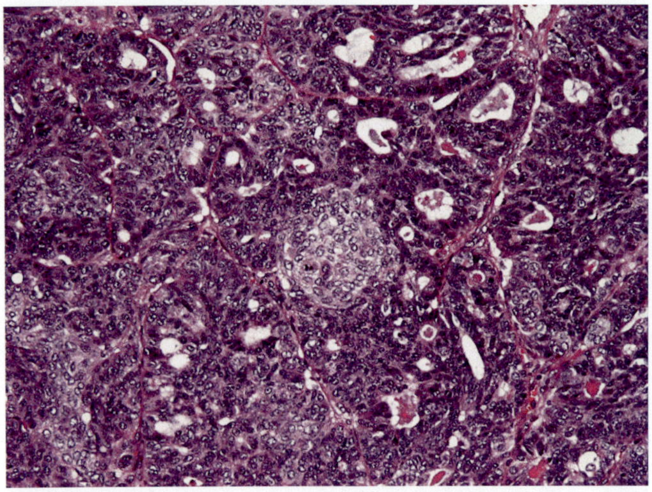

Fig. 7.6 Endometrioid adenocarcinoma of the ovary. Squamous morular metaplasia [200×]

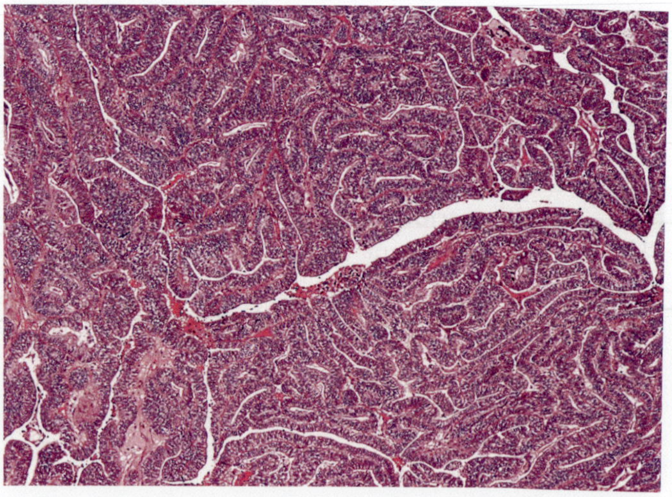

Fig. 7.7 Endometrioid adenocarcinoma of the ovary. Expansile pattern of invasion with confluent growth of epithelium [100×]

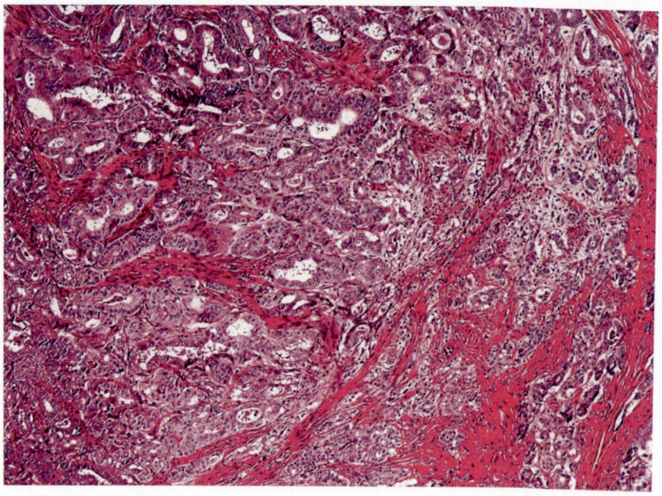

Fig. 7.8 Endometrioid adenocarcinoma of the ovary. Infiltrative pattern of invasion with small nests of irregularly shaped glands infiltrating the stroma [200×]

Squamous metaplasia in the form of morules or larger sheets is common (Fig. 7.6). The morules typically contain small, bland, immature-appearing squamous cells; however, nests or diffuse areas of cytologically malignant squamous epithelium may also be present. Endometriosis is often present [4]. These findings are often present in the context of an easily recognized endometrial-like background. It is not uncommon to see a spectrum of lesions from endometriosis to borderline tumor to adenocarcinoma.

Most endometrioid adenocarcinomas exhibit an expansile or confluent pattern of invasion, whereas infiltrative (destructive) invasion, the more easily recognizable form, is seen less often. In the expansile pattern of invasion, confluent growth of epithelium excludes stroma, which implies invasion. These patterns include large cribriform glands, extensive gland fusion, maze-like patterns, and extensive papillae (Fig. 7.7). Neither stromal reaction nor jagged infiltration as seen in infiltrative (destructive) invasion is present (Fig. 7.8). Since endometrioid adenocarcinoma should only be diagnosed when there is convincing evidence of invasion, recognition of the architectural features of expansile invasion is important to avoid underdiagnosis of an endometrioid adenocarcinoma as borderline tumor. Destructive stromal invasion resembles the typical pattern of myometrial invasive endometrial cancer, with irregularly shaped glands and small nests surrounded by reactive stroma.

Most well- to moderately differentiated ovarian endometrioid adenocarcinomas exhibit at least one histologic "confirmatory endometrioid feature" (Table 7.2; also see Table 1.4). In an unpublished review of our institutional experience, we

Table 7.2 Typical endometrioid features

Resemblance to endometrium and endometrioid neoplasms of the uterine corpus
• Tubules, cribriform structures, solid, sheet-like growth and papillae with smooth luminal contours, composed of columnar cells
• Cytologic features are concordant with the architectural features[a]
• Squamous, morular, hobnail, eosinophilic, and mucinous metaplasia, secretory change
• Associated endometriosis, ovarian endometrioid adenofibroma, synchronous endometrioid neoplasm of endometrium

[a]i.e., simple tubules or papillae are lined by cells with low nuclear grade, lacking extensive budding

studied 50 cases diagnosed as ovarian endometrioid adenocarcinoma from 1992 to 2008. Twenty-three cases showed well-established histologic features of endometrioid adenocarcinoma; 20 showed one or more of the following: squamous metaplasia; endometriosis; and endometrioid adenofibromatous tumor, including endometrioid borderline tumor. One had spindle and corded growth with osseous metaplasia, and two cases were grade 1 tumors with cribriform and villoglandular growth patterns. All endometrioid adenocarcinomas were International Federation of Gynecology and Obstetrics (FIGO) grade 1 or 2, although it should be noted that grade 3 tumors are occasionally seen in practice (Figs. 7.9 and 7.10). Nineteen endometrioid adenocarcinomas were stage I/II, two were stage III/IV, and two were unstaged. Only 2 of 22 were bilateral (91% were unilateral). Follow-up information was available for 19 patients: two died of disease in 16–44 months, one patient is alive with disease, and 15 patients have no evidence of disease. Conversely, 23 cases lacked confirmatory histologic features of endometrioid adenocarcinoma (Tables 1.4 and 7.2). These cases demonstrated solid sheet-like growth and papillary and micropapillary architecture with evident slit-like spaces (at least focally), marked nuclear atypia, and brisk mitotic activity that, in our opinion, represented high-grade serous carcinoma (Figs. 7.11 and 7.12). Nineteen of these cases were stage III/IV, and only four were stage I/II. Follow-up information was available for 22 patients: 12 died of disease in 19–124 months, 2 are alive with disease, and 8 have no evidence of disease. Well-accepted "confirmatory endometrioid features" for endometrioid adenocarcinoma were able to separate these tumors into distinctly different and clinically relevant groups [11]. Our practice is to avoid diagnosing endometrioid adenocarcinoma when confirmatory endometrioid features are lacking except when immunohistochemical or genotypic information is most consistent with endometrioid adenocarcinoma. It is acknowledged that some pathologists would make a diagnosis of "high-grade endometrioid adenocarcinoma" in the presence of a highly atypical gland-forming and cribriforming ovarian carcinoma, but we believe that objective data supporting that practice are lacking.

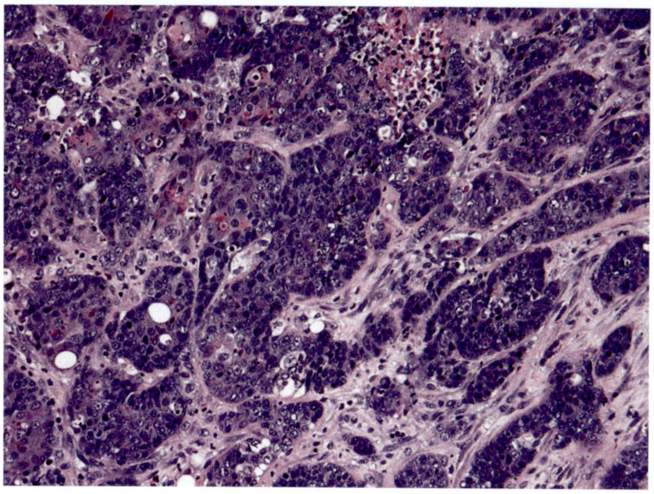

Fig. 7.9 Endometrioid adenocarcinoma of the ovary. High-grade endometrioid adenocarcinoma [200×]

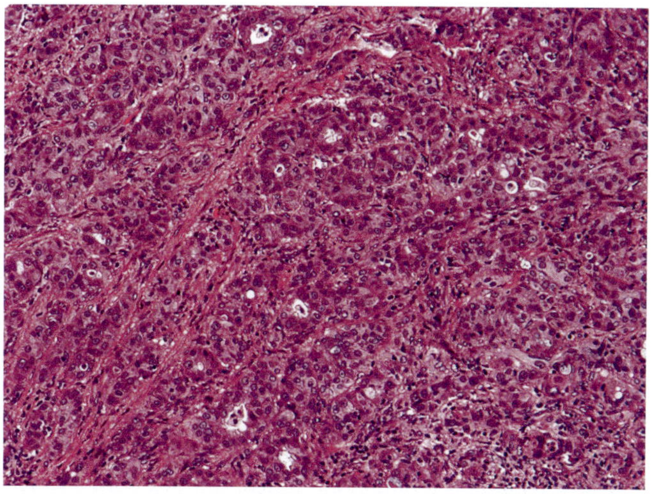

Fig. 7.10 Endometrioid adenocarcinoma of the ovary. High-grade nuclear features that may suggest a serous carcinoma or a metastasis [200×]

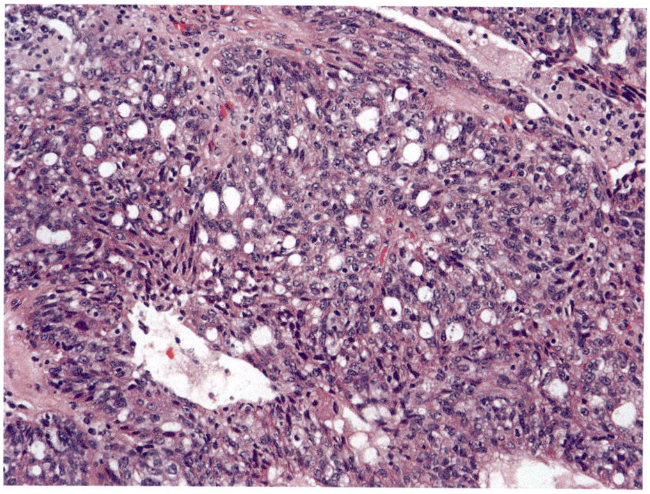

Fig. 7.11 Serous carcinoma, high grade. The tumor cells grow in sheets and form pseudoglandular spaces [200×]

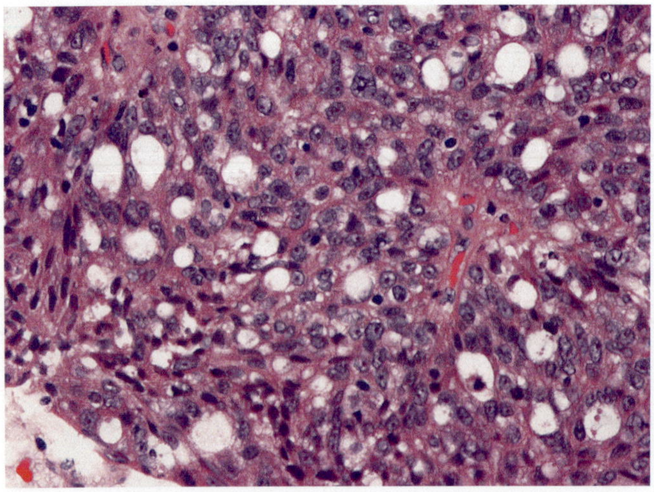

Fig. 7.12 Serous carcinoma, high grade. The tumor cells grow in sheets and form pseudoglandular spaces. The epithelial cells are pleomorphic with prominent nuclei [400×]

Variants of Endometrioid Adenocarcinoma

There are several less-common morphologic patterns that can be seen in both ovarian endometrioid adenocarcinomas and their uterine counterparts. In the ovary, these patterns typically coexist with the more common patterns of endometrioid adenocarcinoma.

Sex Cord-Like

Ovarian endometrioid adenocarcinomas can be histologically similar to sex cord stromal tumors and mimic adult-type granulosa cell tumors, Sertoli cell tumors, and Sertoli-Leydig cell tumors [13]. The microscopic appearance can vary within each tumor, but unlike true sex cord stromal tumors, the endometrioid variants will almost always have areas of more conventional endometrioid adenocarcinoma with larger, more irregular glands, squamous metaplasia, and luminal and intracytoplasmic mucin. Even papillary foci may be present, as well as an adenofibromatous background, which are additional helpful clues to the diagnosis. Areas of transition between the sex cord and endometrioid components can frequently be observed. Common microscopic appearances include small hollow tubules lined by cuboidal or low columnar epithelium separated by varying amounts of fibromatous stroma, which can sometimes be luteinized, resembling well-differentiated Sertoli-Leydig cell tumor (Fig. 7.13). There may even be eosinophilic luminal secretions. Anastomosing cords of solid tubules (Sertoli-like) and nests/islands of small blue cells with scant cytoplasm (granulosa-like) are other patterns that may be observed in these variants (Fig. 7.14). The cytologic features will resemble that of endometrioid adenocarcinoma displaying round-to-oval nuclei with a single prominent nucleolus. Pale, grooved nuclei of granulosa cell tumors and Crystalloids of Reinke of Leydig cell tumors are not present, although the eosinophilic secretions within some tumor islands can mimic Call-Exner bodies. Mitotic activity can range from less than one to greater than ten per ten high-power fields, and the cytologic atypia can also range from mild to severe.

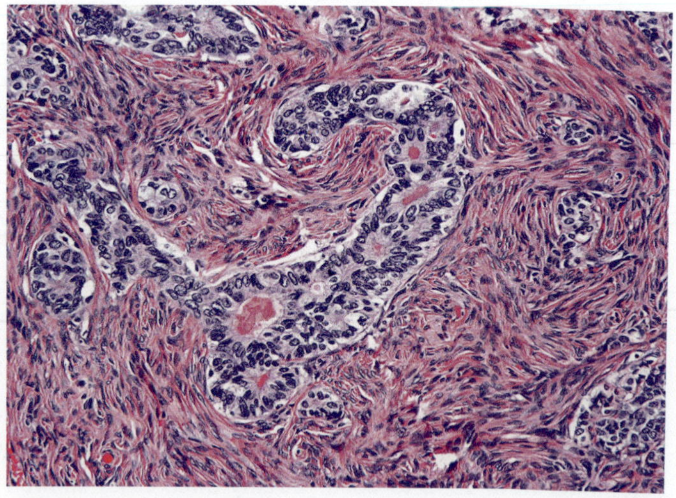

Fig. 7.13 Variant of endometrioid adenocarcinoma. Small sertoliform tubules mimicking a sex cord stromal tumor [200×]

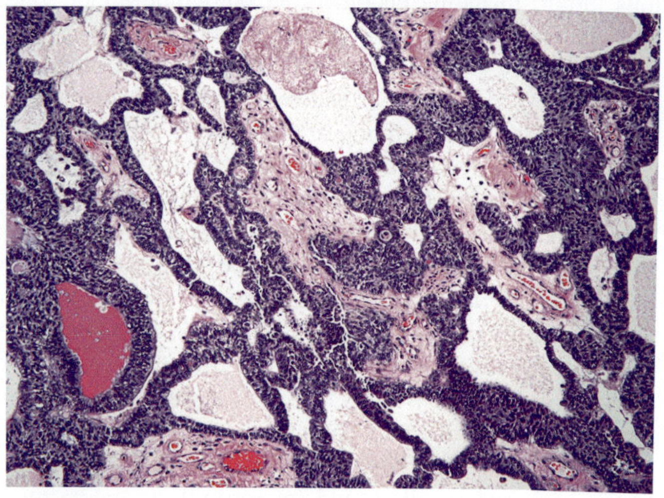

Fig. 7.14 Variant of endometrioid adenocarcinoma. A follicular pattern simulating a granulosa cell tumor [100×]

Spindled

Ovarian endometrioid adenocarcinomas can also have a prominent spindled appearance (Fig. 7.15), which can cause diagnostic confusion with other spindle cell neoplasms and mixed tumors such as Malignant mixed mesodermal tumor (MMMT)/carcinosarcoma. In the original description by Tornos et al., the tumors all had a spindle cell component comprising between 30% and 90% of the entire tumor, with most containing 50% or more spindle cells [14]. All cases

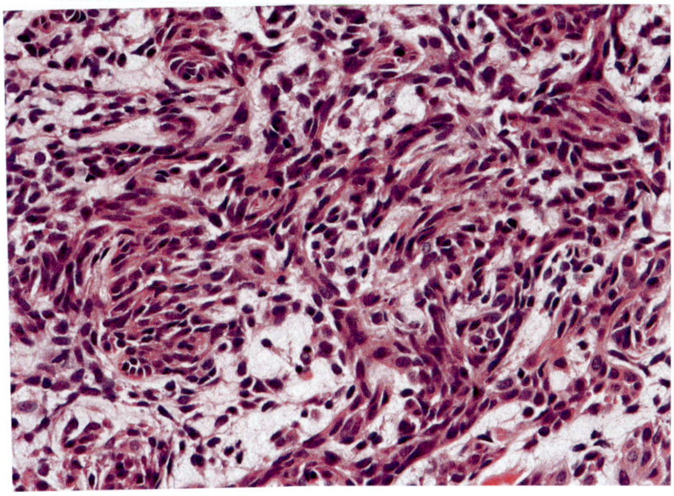

Fig. 7.15 Variant of endometrioid adenocarcinoma. Spindled cell growth pattern may mimic a MMMT/carcinosarcoma [200×]

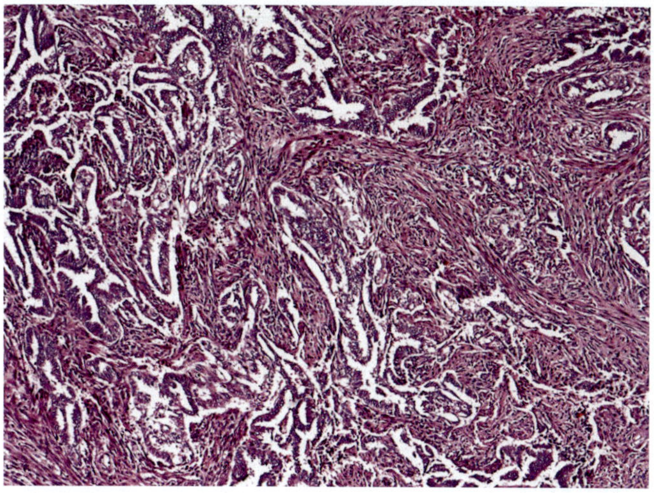

Fig. 7.16 Variant of endometrioid adenocarcinoma. Glands are admixed with spindle cells. Nuclear grade is the same in glandular and spindled cell components [200×]

also had some conventional endometrioid glands, either admixed with or separate from the spindle cells.

The spindle cells can be arranged in a well-circumscribed nested pattern, separated by relatively acellular stroma, imparting a lobular appearance. They can also be diffusely distributed, form whorls, interlacing fascicles, and interlacing palisades resembling schwannoma [14]. The spindle cells may even become fibroblast like, with collagen deposition blending with adjacent stroma. The endometrioid glands can be identified by low columnar or cuboidal cells that are tightly packed, sometimes forming tiny round tubular glands. The glands can be intimately admixed with the spindle cells, located centrally, located peripherally, or distributed throughout the spindle cells so that they are difficult to recognize.

The nuclear grade of the spindle cell component is usually the same as the glandular component, typically grade 1, and sometimes grade 2 (Fig. 7.16). The mitotic index of both spindle and glandular components tends to be similar, although a significantly higher mitotic rate can be seen in the glandular component. This is one of the main features that can help distinguish between an endometrioid adenocarcinoma with spindle cells and a carcinosarcoma. The spindle cells of an MMMT are high grade with marked pleomorphism and bizarre cells, with a high mitotic rate and abnormal mitotic figures. There can also be heterologous mesenchymal metaplasia within the spindle cell component, including osteoid and osteoclast-like giant cells (Fig. 7.17). This in and of itself, however, does not necessitate a diagnosis of MMMT. Similar changes have been described in endometrioid tumors of the endometrium [15]. Like usual types of endometrioid tumors, the spindle cell variant can and usually does have some of the other more typical patterns seen in endometrioid adenocarcinomas, including squamous metaplasia, an adenofibromatous background, intraluminal mucin, and endometriosis. Other unusual features, such as sex cord-like differentiation, can also be seen in combination with spindle cells.

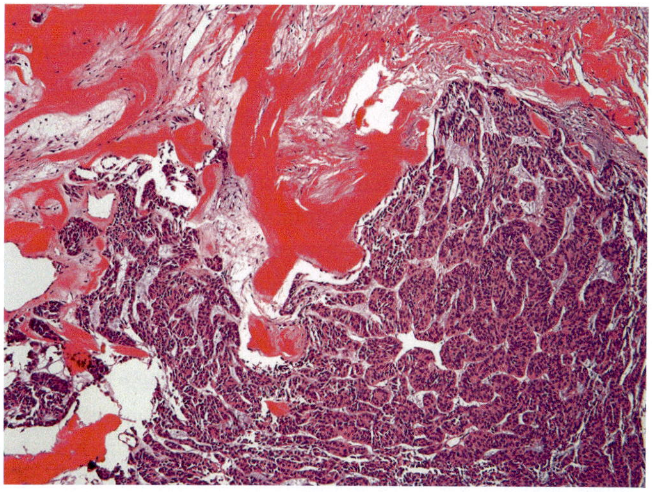

Fig. 7.17 Variant of endometrioid adenocarcinoma. Heterologous mesenchymal metaplasia (osteoid) [100×]

Immunohistochemistry can be helpful in identifying this variant as an adenocarcinoma. Mucin, cytokeratins, and EMA should all show at least focal reactivity in endometrioid adenocarcinomas with spindle cells in both the glandular and spindle components. They can be especially useful in distinguishing the glands when they are crowded by spindle cells or only focally developed. The EMA highlights the luminal side of glands, and cytokeratins usually have a strong, diffuse staining pattern in the glandular component. Unlike true sex-cord tumors, endometrioid adenocarcinomas with spindle cells are inhibin negative, other than the occasional luteinized stromal cells [16].

Grading of endometrioid adenocarcinoma with prominent spindle cells can be problematic since the spindle cells represent a non-morular solid growth pattern which, according to the FIGO grading system, could upgrade the tumor to grade 2 or 3, depending on the percentage of solid growth. However, the spindle cells are thought to represent a metaplastic process, much like the morular squamous metaplasia that is commonly seen in these tumors. The cytology of the spindle cells is similar to the glandular component and typically low grade, and the percentage of spindle cell component appears to have no clinical impact [14]. Therefore, the spindle component should be discounted in the estimation of solid growth when grading these tumors.

Clear Cells

In ovarian endometrioid adenocarcinoma, clear cells are well-described findings that are readily recognized when they have the secretory pattern or squamous metaplasia with clear cytoplasm. When present in small areas, these findings do not present a diagnostic challenge.

In endometrioid adenocarcinoma with secretory features, the glandular lining cells contain subnuclear (and sometimes supranuclear) glycogen vacuoles resembling the glandular cells of an early secretory endometrium (Fig. 7.18). Areas of typical endometrioid adenocarcinoma are usually present and facilitate the correct interpretation. Squamous metaplastic changes may have large glycogen vacuoles or balloon-like hydropic vacuoles that form clear cells (Table 7.3).

When there are extensive clear cell changes or the endometrioid tumor shows clear cells with papillary architecture, the lesion needs to be distinguished from a clear cell carcinoma. Silva and Young described 21 cases of endometrioid neoplasms with clear cells. All of the neoplasms had the typical architecture of endometrioid tumors, but they differed markedly in their cytoplasmic features. In the majority of the tumors, at least one-third of the cells had clear cytoplasm in which the clearing was not of typical secretory type [17].

Other Rare Variants

Rare examples of ciliated cell [18, 19] and oxyphilic [20] types of ovarian endometrioid adenocarcinoma have been described. In the ciliated cell endometrioid adenocarcinoma, a cribriform pattern is characteristic, and most of the glandular lining cells bear cilia. The oxyphilic variant has a prominent component of large polygonal tumor cells with abundant eosinophilic cytoplasm, and round central nuclei with prominent nucleoli.

Rare endometrioid adenocarcinomas with yolk sac tumor component have also been described [21]. These combine typical endometrioid adenocarcinoma with the microcystic and glandular pattern of yolk sac tumor, which is positive for alpha fetoprotein.

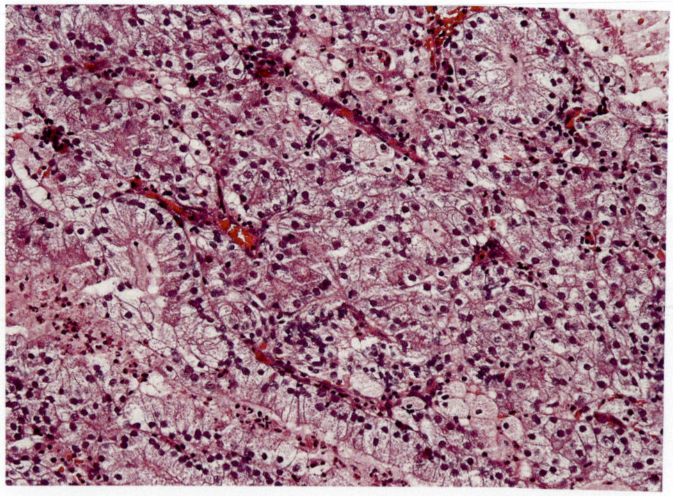

Fig. 7.18 Variant of endometrioid adenocarcinoma. Secretory pattern [200×]

Table 7.3 Endometrioid adenocarcinoma versus other surface epithelial tumors of the ovary

Features that favor endometrioid adenocarcinoma over high-grade serous carcinoma:
• Stratified tall columnar cells
• Uniformly atypical cells without marked pleomorphism
• Cytologic features concordant with architectural features
• Smooth luminal contours without slit-like spaces
• Absence of small complex papillae with cellular budding
• Background adenofibroma or borderline tumor
• Presence of endometriosis
• Squamous metaplasia
• P53 and p16 low level or patchy positive
• WT1 negative or weak
Features that favor clear cell carcinoma over endometrioid adenocarcinoma:
• Hobnail or flat cells with clear or eosinophilic cytoplasm
• Papillary architecture with single layer of epithelial cells lining papillae
• Dense eosinophilic hyaline deposition within stroma
• ER/PR negative, HNF1-β positive

ER estrogen receptor, *PR* progesterone receptor

Carcinomas of the ovary with undifferentiated components are uncommon tumors and, like their counterparts in the endometrium, are underrecognized. Proper recognition requires attention to histologic and cytologic features. One type of undifferentiated carcinoma that may be associated with endometrioid neoplasms is characterized by patternless, sheet-like growth of dyshesive cells often separated by delicate fibrovascular septa, without any evidence of gland formation, trabecular or nested growth patterns (Fig. 7.19). Tumor cells are uniform, small to medium size, and round to ovoid with vesicular nuclei and prominent eosinophilic nucleoli. The cells may vary from small and basophilic with scant cytoplasm to large with clear or vacuolated cytoplasm

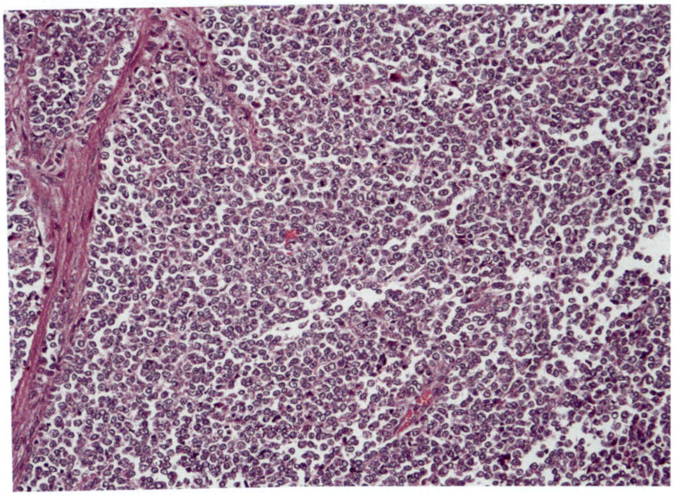

Fig. 7.19 Undifferentiated carcinoma. Sheets of dyshesive, ovoid, monotonous cells without evidence of gland formation [200×]

and there may be focal areas with more pleomorphic or rhabdoid cells (large cells with eosinophilic cytoplasm, eccentric nuclei, and prominent nucleoli). Undifferentiated carcinomas can also be associated with a differentiated component, most commonly well to moderately differentiated endometrioid adenocarcinoma. Other types of undifferentiated carcinomas are covered in Chap. 10.

The presence of an undifferentiated component in the background of a well-differentiated endometrioid adenocarcinoma (dedifferentiated carcinoma) [22] may prompt a diagnosis of a FIGO grade 2 or 3 tumor. Although endometrioid adenocarcinomas can have areas with solid growth, these areas frequently show glandular or trabecular architecture and the nuclear features in the glandular and solid areas are similar. A carcinosarcoma/MMMT is also in the differential given its biphasic appearance. However, the epithelial component of MMMT is usually of higher grade and the sarcomatous component shows pleomorphic spindle cells, not the ovoid epithelioid cells of undifferentiated carcinoma. Neuroendocrine carcinoma can be differentiated from undifferentiated carcinoma based on immunohistochemistry since undifferentiated carcinomas show no more than focal staining for synaptophysin and chromogranin [23].

Undifferentiated carcinomas often show only patchy, focal positive staining for keratins and EMA; however, they may lack any expression of epithelial markers. CK18 and EMA have been reported as the most useful stains to support epithelial differentiation and these tumors sometimes require staining of multiple blocks [24]. Focal staining for synaptophysin and chromogranin may be seen and has no impact on clinical outcome [23]. Immunohistochemical stains can be helpful in differentiating undifferentiated carcinoma from FIGO 2 or 3 endometrioid tumors since keratins and EMA are diffusely positive in endometrioid adenocarcinomas of all grades.

Table 7.4 FIGO grading of endometrioid ovarian adenocarcinoma

- Grade 1: ≤5% non-squamous, solid growth pattern
- Grade 2: 6–50% non-squamous, solid growth pattern
- Grade 3: >50% non-squamous, solid growth pattern

Table 7.5 Shimizu/Silverberg grading of ovarian carcinomas

Score	Predominant architectural pattern	Cytologic atypia[a]	MFs/10 HPF
1	Glandular	Slight	0–9
2	Papillary	Moderate	10–24
3	Solid	Marked	≥25

Total score: 3–5 = grade 1; 6 or 7 = grade 2; 8 or 9 = grade 3
MFs mitotic figures
[a]For assessment of cytologic atypia the most atypical areas occupying at least one-half of a low-power field (4× objective, 10× ocular) is assessed

Undifferentiated carcinomas in the endometrium and ovary have been shown to pursue an aggressive clinical course [24, 25]. The presence of even a small undifferentiated component in tumors with associated differentiated carcinoma appears to be associated with poor clinical outcomes [22].

Grading

The most commonly suggested grading criteria for ovarian endometrioid adenocarcinoma is the FIGO system (Gynecologic Oncology Group) that uses the same criteria as uterine endometrioid adenocarcinoma (Table 7.4). Only those cells that are of glandular type should be considered in the grading schema. Areas with squamous and/or spindled cell differentiation are not included in the assessment of the percentage of solid component. Bizarre nuclear atypia raises a grade 1 or grade 2 tumor by one grade. The Shimizu/Silverberg grading system can also be used. This grading system has been proposed for all cell types of ovarian carcinoma [11, 26]. Architecture, nuclear grade, and mitotic activity are independently assessed (Table 7.5). The Shimizu/Silverberg grading system has been shown to correlate with prognosis for all histologic types of ovarian carcinomas except for clear cell carcinoma. A recent study, however, suggests that distinguishing grade 1 from grade 2 endometrioid adenocarcinoma is not clinically relevant [27].

Immunophenotype

Immunophenotypically, endometrioid tumors express vimentin and epithelial markers, including cytokeratin and epithelial membrane antigen (EMA). Like other primary epithelial ovarian tumors, endometrioid tumors exhibit a CK7 positive, CK20 negative profile. The tumor cells are also typically positive for estrogen receptor (ER) and progesterone receptor (PR). PAX 8 is a transcription factor involved in the development of the Müllerian system and is expressed as positive nuclear staining in ovarian endometrioid adenocarcinomas, as well as other ovarian epithelial tumors except mucinous tumors [28].

Typically, there is positive staining for CA125 in endometrioid adenocarcinomas, which is of limited value since most epithelial ovarian cancers, as well as many other non-gynecologic cancers, have also been shown to express CA125. Some primary ovarian endometrioid adenocarcinomas are positive for nuclear β-catenin (particularly in areas showing squamous/morular metaplasia) since, like their uterine counterparts, they may be associated with β-catenin gene mutation [29–33]. Endometrioid adenocarcinomas with confirmatory endometrioid features usually show low-level, patchy expression of p16, whereas high-grade serous carcinomas and related tumors typically show diffuse, intense expression. High-grade serous carcinomas with a glandular pattern (that some pathologists interpret as high-grade endometrioid adenocarcinoma) will show diffuse and intense expression of this marker. Similarly, p53 is overexpressed (in greater than 75% of tumor cell nuclei) along with WT1 in high-grade serous carcinoma and those tumors diagnosed as high-grade endometrioid adenocarcinoma with glandular and cribriform features. Only low-level or patchy p53 expression with negative or weak WT1 is seen in typical endometrioid adenocarcinomas with confirmatory endometrioid features. Endometrioid adenocarcinomas are negative for inhibin, calretinin, and monoclonal CEA.

Genotype

Various mutations have been reported in ovarian endometrioid adenocarcinomas, including CTNNB-1 (β-catenin) [29–31, 34–36], PIK3CA [36], and PTEN [33, 37], as well as high levels of microsatellite instability [29, 38, 39]. A recent study describes *ARID1A* as a tumor-suppressor gene frequently disrupted in ovarian clear-cell and endometrioid adenocarcinomas [12].

Differential Diagnosis

Primary Versus Metastatic Ovarian Neoplasms

The ovary is a common site of metastases, many of which can mimic primary ovarian epithelial neoplasms, both clinically and histologically. Tumors that can mimic primary endometrioid adenocarcinoma of the ovary include metastatic colorectal carcinoma and endocervical adenocarcinoma. Colorectal carcinoma is one of the most common metastatic tumors to involve the ovary and can occur in a setting of known history or present as a primary ovarian mass. Features that suggest metastases include bilaterality, infiltrative growth pattern, surface tumor involvement, and lymphovascular invasion, especially in the hilar region [40]. Grossly, metastases can be solid, nodular, or cystic. Unfortunately, none of these features is specific, and it is not uncommon for primary endometrioid adenocarcinomas of the ovary to be bilateral, cystic, or both. Histologically, the presence of dirty, segmental, or "garland-like" necrosis is highly suggestive of colorectal carcinoma [41]. Features that favor a primary endometrioid adenocarcinoma include a background of endometriosis, adenofibroma or borderline tumor, and squamous metaplasia (Table 7.6). Cytologically, colorectal carcinoma typically has more pleomorphic cells with nuclear hyperchromasia and more numerous mitotic figures.

Table 7.6 Primary endometrioid adenocarcinoma versus metastatic adenocarcinoma resembling endometrioid adenocarcinoma

	Primary	Metastatic
Size	Large	Small
Laterality	Usually unilateral	Often bilateral
Surface involvement	No	Yes
Infiltrative growth	No	Yes
Nodularity	No	Yes
Endometriosis/adenofibroma	Yes	No
Squamous metaplasia	Yes	No

Metastatic endocervical adenocarcinoma to the ovary is a less common phenomenon, but can present synchronously or metachronously and resemble ovarian mucinous or endometrioid borderline tumor or adenocarcinoma. Cases of extensive cervical adenocarcinoma in situ without definitive evidence of stromal invasion have also been known to metastasize to the ovaries, possibly because the patterns of early invasion in endocervical carcinoma are difficult to detect on histology [42, 43]. The pattern of metastasis can be a helpful clue to the diagnosis, as in the case of all metastases. However, unilateral metastatic endocervical adenocarcinomas with expansile growth without obvious ovarian stromal invasion can be particularly challenging. Cytologically, endocervical adenocarcinoma nuclei are elongated, columnar, pseudostratified, and hyperchromatic, and the cells contain abundant apical mitoses and apoptotic bodies [44].

Immunohistochemical stains can be very useful in distinguishing primary endometrioid ovarian adenocarcinoma from metastatic colorectal carcinoma (Table 7.7). CK7 and CA125 are typically positive, while CK20 and CEA are negative in endometrioid tumors; the opposite is true for colorectal tumors [44–46]. ER and PR are also helpful, especially in well-differentiated tumors, since they are expressed in endometrioid but not in colorectal carcinomas. CDX-2, an intestinal-specific transcription factor, is also useful in differentiating colorectal (positive) from endometrioid (negative) adenocarcinoma, although endometrioid adenocarcinoma with extensive mucinous differentiation can express CDX-2 [47].

Endocervical adenocarcinoma will have similar staining patterns with endometrioid tumors for CK7 and CK20. Using p16 or some form of human papillomavirus (HPV) detection method has been shown to be able to differentiate between

Table 7.7 Immunohistochemical stains in primary and metastatic carcinoma

IHC stains	Endometrioid adenocarcinoma	Metastatic colorectal carcinoma	Metastatic endocervical adenocarcinoma
CK7	Positive	Negative	Positive
CK20	Negative	Positive	Negative
CDX-2	Negative[a]	Positive	Negative[a]
p16	Negative or focally positive	Negative	Diffusely positive
CEA	Negative	Positive	Positive
CA125	Positive	Negative	Positive
ER/PR	Positive	Negative	Negative or focally positive

IHC immunohistochemistry, *ER* estrogen receptor, *PR* progesterone receptor
[a]Endometrioid and endocervical adenocarcinomas with extensive mucinous differentiation may be positive for CDX-2

Table 7.8 Differential diagnosis for endometrioid adenocarcinoma

- High- and low-grade serous carcinoma
- Primary ovarian mucinous adenocarcinoma
- Metastatic mucinous adenocarcinoma (i.e., colorectal, endocervical)
- Metastatic endometrial endometrioid adenocarcinoma
- Granulosa cell tumor
- Sertoli and Sertoli-Leydig cell tumor
- Yolk sac tumor
- Wolffian tumor (FATWO)
- Ependymoma
- Carcinoid tumor
- Primitive neuroectodermal tumor
- Carcinosarcoma/MMMT
- Transitional cell carcinoma
- Squamous cell carcinoma

the two tumors since p16 is considered a surrogate marker for high-risk HPV and typical endometrioid tumors with confirmatory endometrioid features do not show the strong, diffuse pattern expected in endocervical adenocarcinomas [42, 43]. Serous carcinomas with a glandular pattern that some pathologists interpret as high-grade endometrioid adenocarcinoma will express p16 in a strong, diffuse pattern despite the absence of high-risk HPV.

Other less-common tumors involving the ovary that can morphologically mimic endometrioid adenocarcinoma include pulmonary adenocarcinoma, yolk sac tumor with endometrioid-like glandular pattern, ependymoma, and Sertoli/Sertoli-Leydig cell tumors with endometrioid-like glands (Table 7.8). Stains that may be used in these cases are TTF-1 (positive in lung adenocarcinoma), AFP (positive in yolk sac tumor), GFAP (positive in ependymoma), and inhibin and calretinin (positive in Sertoli/Sertoli-Leydig cell tumors) [48–51].

Synchronous Versus Metastatic Ovarian Tumors

The coexistence of carcinoma in the ovary and endometrium is uncommon but not rare, occurring in about 10% of ovarian carcinoma and slightly over 5% in endometrial carcinoma [52]. Of patients with endometrioid ovarian tumors, 10% had concurrent endometrial malignancy compared with only 0.2% of patients with serous carcinoma, and those with concurrent endometrial malignancy had better survival than those with only ovarian tumor [53]. Although endometrial endometrioid adenocarcinoma is predominantly a disease of postmenopausal women, 2–14% occur in women 40 years old or younger [54–57]. Rates of synchronous ovarian carcinoma in the setting of a clinically apparent endometrial cancer in young patients range from 11% to 23% [58, 59].

The criteria for determining whether the tumors represent two independent primaries or metastatic disease have been long established, although not all cases fit neatly into the categories and it is admittedly difficult to determine with certainty in some cases (Table 7.9). If the ovarian tumor shows a multinodular growth pattern with separate nodules involving both the cortex and the medulla, it favors being a metastasis from the endometrium [60] (Fig. 7.20). Without this feature, if the tumor meets two or more of the following criteria, then it also favors a metastasis from the endometrium [60].

- Small ovary(ies) (<5 cm)
- Bilateral ovarian involvement
- Deep myometrial invasion (>1/3 of myometrium)
- Vascular invasion
- Tubal involvement

The background lesions in the ovary should also be taken into consideration (i.e., the presence of endometriosis or borderline tumor favors the presence of an ovarian primary). The incidence of extrauterine and extraovarian spread of disease is higher in tumors that are deemed metastatic versus those that are thought to be independent synchronous primaries [60]. Also, the majority of metastatic tumors involving the endometrium and ovary are likely from an endometrial primary since there is a high incidence of associated endometrial hyperplasia providing the premalignant background for the development of endometrial cancer [60]. It has been estimated that ovarian tumor metastasis to the endometrium is less than one-third as frequent as metastases to the ovary [61], although in practice this diagnosis is usually difficult to establish with certainty and is probably more commonly encountered with ovarian and tubal serous carcinomas rather than ovarian endometrioid adenocarcinomas.

Surface involvement of the ovaries is usually suggestive of metastasis; however, in contrast to this usual situation, there are cases in which the ovarian endometrioid adenocarcinoma arises in a background of surface endometriosis and the tumor protrudes from the ovarian surface, mimicking implants [62].

Table 7.9 Synchronous versus metastatic endometrioid adenocarcinoma

	Favor synchronous ovarian and endometrial carcinoma	Favor metastatic endometrial carcinoma to ovary
Uterine features	• No or minimal myometrial invasion • No vascular invasion	• Deep myometrial invasion (>1/3 of myometrium) • Vascular invasion • Tubal involvement
Ovarian features	• Large ovary • Unilateral • Expansile growth • Background endometriosis or borderline tumor	• Small ovaries (<5 cm) • Bilateral ovarian involvement • Multinodular growth pattern • Separate nodules involving both ovarian cortex and medulla • Ovarian surface involvement

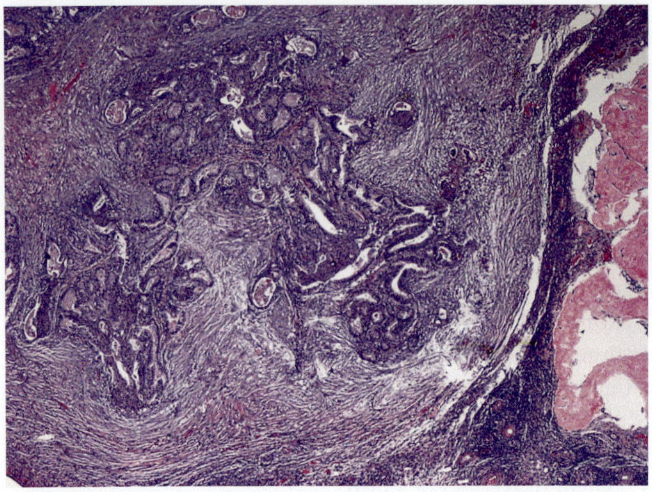

Fig. 7.20 Metastatic carcinoma to ovary. Nodular pattern of invasion [40×]

Serous Carcinoma

Serous carcinomas generally lack confirmatory endometrioid features. The finding of irregular slit-like spaces, with smaller, more complex papillae, cellular budding, severe cytologic atypia, and the presence of psammoma bodies support the diagnosis of high-grade serous carcinoma. Low-grade serous carcinoma with cytologically bland cribriform architecture may mimic endometrioid adenocarcinoma. Immunohistochemistry can be employed when there is uncertainty about the diagnosis. Endometrioid tumors are negative for WT 1 and lack overexpression of p53 (patchy staining of variable intensity is commonly present). Low-grade serous carcinomas express WT 1 and do not overexpress p53 whereas high-grade serous carcinomas express WT 1 and either overexpress p53 (diffuse, uniformly intense expression in more than 75% of tumor cell nuclei) or demonstrate a p53 null phenotype (no staining) (Table 7.3).

Mucinous Carcinoma

Mucinous carcinoma contains abundant luminal mucin and goblet cells with mucin-rich cytoplasm. Mucin should be demonstrable in at least 50% of tumor cells to qualify for mucinous carcinoma, and confirmatory endometrioid features should be lacking. In contrast to endometrioid adenocarcinoma, mucinous carcinoma is ER and PR negative, and diffusely CEA positive.

Sex Cord-Like Endometrioid Adenocarcinoma

There are clinical differences between patients with endometrioid adenocarcinoma of the ovary versus sex cord stromal tumors. The average age of patients with Sertoli-Leydig cell tumors is 24–38 years, with very few patients over 50 years of age, while the average age of patients with endometrioid adenocarcinomas is 60–70 years [13, 63–66]. There is some overlap, however, with adult-type granulosa cell tumor patients whose average age at presentation is between 50 and 55 years [67].

The presenting symptoms are also quite distinct in that patients with Sertoli-Leydig cell tumors typically present with symptoms related to androgen overproduction, as opposed to ovarian carcinoma patients who usually present with abdominal pain, fullness, a mass, or ascites [68]. However, patients with the sex cord variant of endometrioid adenocarcinoma can also present with androgenic symptoms due to the luteinized cells within the stroma of the ovary, which can further confound the diagnosis [16].

In endometrioid adenocarcinomas with sex cord-like features, findings such as tubular and cord-like growth or grooved nuclei may mimic a sex cord stromal tumor. The following features are helpful in distinguishing an endometrioid adenocarcinoma with sex cord-like features from a sex cord stromal tumor: foci of typical endometroid adenocarcinoma; squamous metaplasia; mucinous metaplasia; intraluminal mucin; adenofibromatous background; and associated endometriosis. When grooved nuclei are seen in an endometrioid adenocarcinoma with sex cord-like features, they are typically rounder and more hyperchromatic than those of a granulosa cell tumor.

Immunohistochemical stains should be helpful in distinguishing the various entities (Table 7.10). α-Inhibin and calretinin are positive in sex cord stromal tumors and are most useful in differentiating these from epithelial neoplasms,

Table 7.10 Variants of endometrioid adenocarcinoma of the ovary and differential diagnoses

Variant	Differential diagnosis				
Sex cord-like endometrioid ACA	Adult granulosa cell tumor	Sertoli cell tumor	Sertoli-Leydig cell tumor		
CK+ EMA+ CK7+ Inhibin – Calretinin –	[a]CK–/+ EMA– CK7– Inhibin + Calretinin +	CK–/+ EMA– CK7– Inhibin + Calretinin +	CK–/+ EMA– CK7– Inhibin + Calretinin +		
Endometrioid adenocarcinoma with spindle cells	Carcinosarcoma/MMMT	Granulosa cell tumor, diffuse type	Sertoli-Leydig cell tumor, intermediate or poorly differentiated	FATWO	ESS
CK+ EMA+(luminal) Mucin+	CK+/– EMA+/–	CK– EMA– CK7– Inhibin+ Calretinin+ Mucin–	CK–/+ EMA– CK7– Inhibin+/– Calretinin+/– Mucin–	EMA– Mucin–	CK–/+ CD10+
Endometrioid adenocarcinoma with clear cells	Clear cell carcinoma				
ER+ HNF1 β–	ER– HNF1 β +				

FATWO female adnexal tumor of probable Wolffian origin, *ESS* endometrioid stromal sarcoma
[a]Pankeratin can show punctate cytoplasmic staining in some sex cord stromal tumors

with calretinin being slightly more sensitive and inhibin being more specific [68–73]. The staining in granulosa cell tumors can vary from focal weak to diffuse strong [73, 74], although cases can also be entirely negative for both markers [75]. Epithelial markers CK7, and EMA are also very useful as they should be positive in carcinomas and negative in sex cord stromal tumors (although focal EMA positivity has been reported in juvenile granulosa cell tumors) [45, 46, 76–78]. A panel using a combination of these various markers will have the highest utility in correctly classifying these tumors. Pankeratin marker CK AE1/AE3 is less useful since granulosa cell tumors and other sex cord stromal neoplasms can show punctate cytoplasmic staining [79–81].

Spindle Variant

The differential diagnosis includes any tumor that can have a spindled appearance, including intermediate or poorly differentiated Sertoli-Leydig cell tumors, diffuse-type adult granulosa cell tumor, female adnexal tumor of probable Wolffian origin (FATWO), carcinosarcoma/MMMT, and endometrial stromal sarcoma (ESS) (Table 7.10). The histologic features that are usually seen in endometrioid adenocarcinomas, such as the presence of typical endometrioid glands, squamous metaplasia, intraluminal mucin, and background adenofibroma, as well as the differing immunohistochemical profiles of the various tumors, should be enough to distinguish most from true endometrioid adenocarcinoma with spindle cells. Endometrioid tumors are EMA and cytokeratin positive, and mucin positive, while this has only rarely been reported in Sertoli-Leydig or adult-type granulosa cell tumors, typically in a focal and weak pattern [82]. FATWOs are another histologically similar tumor which can also show low-power lobulation, tubular structures, and cystic formation. These tumors do not show squamous metaplasia, only exceptionally show intraluminal mucin, and are EMA negative [41].

Carcinosarcoma/MMMT is a potential pitfall for the obvious reasons of malignant epithelial and mesenchymal components that comprise the tumor. Unlike endometrioid adenocarcinomas, however, MMMT usually presents at high stage (III or IV), with high-grade carcinomatous and sarcomatous components, in contrast to the usual low-stage, low-grade presentation of endometrioid adenocarcinomas with spindle cells. The metastases of endometrioid adenocarcinomas with spindle cells are typically composed of the spindle component alone or in combination with the glands, in contrast to MMMTs where the metastases tend to be carcinomatous [14, 83]. Immunohistochemical stains for cytokeratins and EMA will not be helpful since they stain MMMT in both epithelial and mesenchymal components. Both entities can show heterologous bone and cartilage. Endometrioid stromal sarcoma (ESS) in the ovary is histologically identical to its uterine counterpart. The tumor is composed of small round or oval cells that resemble proliferative phase endometrial stroma and are CD10 positive, as opposed to the more spindled cells of endometrioid adenocarcinoma. There are also the typical spiral arteriole-like delicate vessels in ESS, with tumor cells arranged in concentric whorls around them, which is not seen in endometrioid adenocarcinomas. Focal glandular formation has been seen in ESS, but recognizing the typical endometrial stromal background in such cases should not lead to diagnostic difficulties with endometrioid adenocarcinoma (although distinction from adenosarcoma and endometriosis may be problematic).

Clear Cell Carcinoma

Clear cell carcinoma often is associated with endometriosis and may be admixed with endometrioid adenocarcinoma. Clear cell carcinomas have papillary, tubulocystic, and solid growth patterns. The tumor cell cytoplasm can be scant, abundant and clear, or eosinophilic. Cells are cuboidal with abundant cytoplasm, hobnail with scant cytoplasm, or flattened. Nuclei range from flat and almost inconspicuous, to wrinkled and koilocytic, to round and vesicular with a prominent nucleolus. The nuclei are rounded or angular, often eccentric, and hyperchromatic without prominent nuceloli. Endometrioid adenocarcinomas may appear similar to clear cell carcinomas when they are papillary, display secretory changes, and have squamous metaplasia with clear cytoplasm. However, endometrioid adenocarcinomas are generally composed of columnar cells with elongate nuclei and lack the typical architectural patterns, hobnail cells, and high-grade nuclei of clear cell carcinoma.

Hepatocyte nuclear factor-1 beta (HNF1β) has recently emerged as a sensitive marker for ovarian clear cell carcinoma, with reported positive rates ranging from 82% to almost 100%.

ER expression is typical of most endometrioid neoplasms; however, clear cell carcinomas have, at most, focal ER expression.

Clinical Behavior

The behavior of ovarian carcinomas is dependent primarily on stage; however, it has recently been shown that histologic subtype is also an important independent prognostic factor [72]. Stage I endometrioid adenocarcinomas have greater than 90% 10-year disease-specific survival as compared with approximately 70% for clear cell and mucinous tumors, and 40% for serous carcinomas [84]. The grade of endometrioid adenocarcinomas, whether stratified by stage or not, appears to have no prognostic significance when pseudoendometrioid, high-grade serous carcinoma are excluded from analysis [84]. Endometrioid adenocarcinomas are more likely to present at an early stage, with fewer patients presenting with ascites as compared to serous carcinomas, even at higher stages [53]. Treatment for endometrioid adenocarcinoma with high-risk factors (stage IC or greater, grade 2 or greater) is typically a combination of platinum- and taxane-based chemotherapy, which has been shown to have a better overall response rate than single-agent or other platinum combinations (77% versus 52% and 59%, respectively) [53]. It should be noted, however, that these recommendations are based upon several assumptions, many of which are likely *not* valid when endometrioid adenocarcinomas, specifically, are studied: grade 2 endometrioid adenocarcinoma is more aggressive than grade 1 endometrioid adenocarcinoma; and endometrioid adenocarcinoma present on the ovarian surface has the same prognostic significance as intraoperative capsular rupture. Important independent predictors of survival after platinum-based chemotherapy on multivariate analysis include debulking status and disease stage [53]. Patients who undergo optimal debulking with residual disease less than 2 cm postsurgery prior to chemotherapy have longer survival and longer time to progression than those who do not undergo debulking surgery.

References

1. Seidman JD, Horkayne-Szakaly I, Haiba M, Boice CR, Kurman RJ, Ronnett BM. The histologic type and stage distribution of ovarian carcinomas of surface epithelial origin. Int J Gynecol Pathol. 2004;23(1):41–4.
2. Leitao Jr MM, Boyd J, Hummer A, et al. Clinicopathologic analysis of early-stage sporadic ovarian carcinoma. Am J Surg Pathol. 2004;28(2):147–59.
3. Bell KA, Kurman RJ. A clinicopathologic analysis of atypical proliferative (borderline) tumors and well-differentiated endometrioid adenocarcinomas of the ovary. Am J Surg Pathol. 2000;24 (11):1465–79.
4. Costa MJ, DeRose PB, Roth LM, Brescia RJ, Zaloudek CJ, Cohen C. Immunohistochemical phenotype of ovarian granulosa cell tumors: absence of epithelial membrane antigen has diagnostic value. Hum Pathol. 1994;25(1):60–6.
5. Roth LM, Emerson RE, Ulbright TM. Ovarian endometrioid tumors of low malignant potential: a clinicopathologic study of 30 cases with comparison to well-differentiated endometrioid adenocarcinoma. Am J Surg Pathol. 2003;27(9):1253–9.
6. Norris HJ. Proliferative endometrioid tumors and endometrioid tumors of low malignant potential of the ovary. Int J Gynecol Pathol. 1993;12(2):134–40.
7. Longacre TA, Chung MH, Jensen DN, Hendrickson MR. Proposed criteria for the diagnosis of well-differentiated endometrial carcinoma. A diagnostic test for myoinvasion. Am J Surg Pathol. 1995;19(4):371–406.
8. Kurman RJ, Norris HJ. Evaluation of criteria for distinguishing atypical endometrial hyperplasia from well-differentiated Carcinoma. Cancer. 1982;49(12):2547–59.
9. Chen S, Leitao MM, Tornos C, Soslow RA. Invasion patterns in stage I endometrioid and mucinous ovarian carcinomas: a clinicopathologic analysis emphasizing favorable outcomes in carcinomas without destructive stromal invasion and the occasional malignant course of carcinomas with limited destructive stromal invasion. Mod Pathol. 2005;18(7):903–11.
10. Creasman WT, Park R, Norris H, Disaia PJ, Morrow CP, Hreshchyshyn MM. Stage I borderline ovarian tumors. Obstet Gynecol. 1982;59(1):93–6.
11. Shimizu Y, Kamoi S, Amada S, Akiyama F, Silverberg SG. Toward the development of a universal grading system for ovarian epithelial

carcinoma: testing of a proposed system in a series of 461 patients with uniform treatment and follow-up. Cancer. 1998;82(5):893–901.

12. Wiegand KC, Shah SP, Al-Agha OM, et al. *ARID1A* mutations in endometriosis-associated ovarian Carcinomas. N Engl J Med. 2010;363:1532–43.
13. Young RH, Prat J, Scully RE. Ovarian endometrioid adenocarcinomas resembling sex cord-stromal tumors. A clinicopathological analysis of 13 cases. Am J Surg Pathol. 1982;6(6):513–22.
14. Tornos C, Silva EG, Ordonez NG, Gershenson DM, Young RH, Scully RE. Endometrioid adenocarcinoma of the ovary with a prominent spindle-cell component, a source of diagnostic confusion. A report of 14 cases. Am J Surg Pathol. 1995;19(12):1343–53.
15. Murray SK, Clement PB, Young RH. Endometrioid adenocarcinomas of the uterine corpus with sex cord-like formations, hyalinization, and other unusual morphologic features: a report of 31 cases of a neoplasm that may be confused with carcinosarcoma and other uterine neoplasms. Am J Surg Pathol. 2005;29(2):157–66.
16. Ordi J, Schammel DP, Rasekh L, Tavassoli FA. Sertoliform endometrioid adenocarcinomas of the ovary: a clinicopathologic and immunohistochemical study of 13 cases. Mod Pathol. 1999;12(10):933–40.
17. Silva EG, Young RH. Endometrioid neoplasms with clear cells: a report of 21 cases in which the alteration is not of typical secretory type. Am J Surg Pathol. 2007;31(8):1203–8.
18. Eichhorn JH, Scully RE. Endometrioid ciliated-cell tumors of the ovary: a report of five cases. Int J Gynecol Pathol. 1996;15(3):248–56.
19. Hendrickson MR, Kempson RL. Ciliated carcinoma–a variant of endometrial adenocarcinoma: a report of 10 cases. Int J Gynecol Pathol. 1983;2(1):1–12.
20. Pitman MB, Young RH, Clement PB, Dickersin GR, Scully RE. Endometrioid adenocarcinoma of the ovary and endometrium, oxyphilic cell type: a report of nine cases. Int J Gynecol Pathol. 1994;13(4):290–301.
21. Nogales FF, Bergeron C, Carvia RE, Alvaro T, Fulwood HR. Ovarian endometrioid tumors with yolk sac tumor component, an unusual form of ovarian neoplasm. Analysis of six cases. Am J Surg Pathol. 1996;20(9):1056–66.
22. Silva EG, Deavers MT, Bodurka DC, Malpica A. Association of low-grade endometrioid adenocarcinoma of the uterus and ovary with undifferentiated carcinoma: a new type of dedifferentiated carcinoma? Int J Gynecol Pathol. 2006;25(1):52–8.
23. Taraif SH, Deavers MT, Malpica A, Silva EG. The significance of neuroendocrine expression in undifferentiated carcinoma of the endometrium. Int J Gynecol Pathol. 2009;28(2):142–7.
24. Tafe LJ, Garg K, Chew I, Tornos C, Soslow RA. Endometrial and ovarian carcinomas with undifferentiated components: clinically aggressive and frequently underrecognized neoplasms. Mod Pathol. 2010;23(6):781–9.
25. Altrabulsi B, Malpica A, Deavers MT, Bodurka DC, Broaddus R, Silva EG. Undifferentiated carcinoma of the endometrium. Am J Surg Pathol. 2005;29(10):1316–21.
26. Silverberg SG. Histopathologic grading of ovarian carcinoma: a review and proposal. Int J Gynecol Pathol. 2000;19(1):7–15.
27. Clarke BA, Gilks CB. Endometrial carcinoma: controversies in histopathological assessment of grade and tumour cell type. J Clin Pathol. 2010;63(5):410–5.
28. Nonaka D, Chiriboga L, Soslow RA. Expression of pax8 as a useful marker in distinguishing ovarian carcinomas from mammary carcinomas. Am J Surg Pathol. 2008;32(10):1566–71.
29. Moreno-Bueno G, Gamallo C, Perez-Gallego L, de Mora JC, Suarez A, Palacios J. Beta-Catenin expression pattern, beta-catenin gene mutations, and microsatellite instability in endometrioid ovarian carcinomas and synchronous endometrial carcinomas. Diagn Mol Pathol. 2001;10(2):116–22.
30. Irving JA, Catasus L, Gallardo A, et al. Synchronous endometrioid adenocarcinomas of the uterine corpus and ovary: alterations in the beta-catenin (CTNNB1) pathway are associated with independent primary tumors and favorable prognosis. Hum Pathol. 2005;36(6): 605–19.
31. Oliva E, Sarrio D, Brachtel EF, et al. High frequency of beta-catenin mutations in borderline endometrioid tumours of the ovary. J Pathol. 2006;208(5):708–13.
32. Shimizu M, Toki T, Takagi Y, Konishi I, Fujii S. Immunohistochemical detection of the Wilms' tumor gene (WT1) in epithelial ovarian tumors. Int J Gynecol Pathol. 2000;19(2):158–63.
33. Catasus L, Bussaglia E, Rodrguez I, et al. Molecular genetic alterations in endometrioid adenocarcinomas of the ovary: similar frequency of beta-catenin abnormalities but lower rate of microsatellite instability and PTEN alterations than in uterine endometrioid adenocarcinomas. Hum Pathol. 2004;35(11):1360–8.
34. Palacios J, Gamallo C. Mutations in the beta-catenin gene (CTNNB1) in endometrioid ovarian carcinomas. Cancer Res. 1998;58(7):1344–7.
35. Wu R, Hendrix-Lucas N, Kuick R, et al. Mouse model of human ovarian endometrioid adenocarcinoma based on somatic defects in the Wnt/beta-catenin and PI3K/Pten signaling pathways. Cancer Cell. 2007;11(4):321–33.
36. Wu R, Zhai Y, Fearon ER, Cho KR. Diverse mechanisms of beta-catenin deregulation in ovarian endometrioid adenocarcinomas. Cancer Res. 2001;61(22)):8247–55.
37. Sato N, Tsunoda H, Nishida M, et al. Loss of heterozygosity on 10q23.3 and mutation of the tumor suppressor gene PTEN in benign endometrial cyst of the ovary: possible sequence progression from benign endometrial cyst to endometrioid adenocarcinoma and clear cell carcinoma of the ovary. Cancer Res. 2000;60(24):7052–6.
38. Pal T, Permuth-Wey J, Kumar A, Sellers TA. Systematic review and meta-analysis of ovarian cancers: estimation of microsatellite-high frequency and characterization of mismatch repair deficient tumor histology. Clin Cancer Res. 2008;14(21):6847–54.
39. Zhai QJ, Rosen DG, Lu K, Liu J. Loss of DNA mismatch repair protein hMSH6 in ovarian cancer is histotype-specific. Int J Clin Exp Pathol. 2008;1(6):502–9.
40. Lee KR, Young RH. The distinction between primary and metastatic mucinous carcinomas of the ovary: gross and histologic findings in 50 cases. Am J Surg Pathol. 2003;27(3):281–92.
41. Young RH, Scully RE. Ovarian tumors of probable wolffian origin. A report of 11 cases. Am J Surg Pathol. 1983;7(2):125–35.
42. Vang R, Gown AM, Farinola M, et al. p16 expression in primary ovarian mucinous and endometrioid tumors and metastatic adenocarcinomas in the ovary: utility for identification of metastatic HPV-related endocervical adenocarcinomas. Am J Surg Pathol. 2007;31(5):653–63.
43. Ronnett BM, Yemelyanova AV, Vang R, et al. Endocervical adenocarcinomas with ovarian metastases: analysis of 29 cases with emphasis on minimally invasive cervical tumors and the ability of the metastases to simulate primary ovarian neoplasms. Am J Surg Pathol. 2008;32(12):1835–53.
44. Elishaev E, Gilks CB, Miller D, Srodon M, Kurman RJ, Ronnett BM. Synchronous and metachronous endocervical and ovarian neoplasms: evidence supporting interpretation of the ovarian neoplasms as metastatic endocervical adenocarcinomas simulating primary ovarian surface epithelial neoplasms. Am J Surg Pathol. 2005;29(3):281–94.
45. Berezowski K, Stastny JF, Kornstein MJ. Cytokeratins 7 and 20 and carcinoembryonic antigen in ovarian and colonic carcinoma. Mod Pathol. 1996;9(4):426–9.
46. Lagendijk JH, Mullink H, Van Diest PJ, Meijer GA, Meijer CJ. Tracing the origin of adenocarcinomas with unknown primary using immunohistochemistry: differential diagnosis between colonic and ovarian carcinomas as primary sites. Hum Pathol. 1998;29(5):491–7.
47. Groisman GM, Meir A, Sabo E. The value of Cdx2 immunostaining in differentiating primary ovarian carcinomas from colonic carcinomas metastatic to the ovaries. Int J Gynecol Pathol. 2004;23(1):52–7.
48. Irving JA, Young RH. Lung carcinoma metastatic to the ovary: a clinicopathologic study of 32 cases emphasizing their morphologic spectrum and problems in differential diagnosis. Am J Surg Pathol. 2005;29(8):997–1006.

49. Idowu MO, Rosenblum MK, Wei XJ, Edgar MA, Soslow RA. Ependymomas of the central nervous system and adult extra-axial ependymomas are morphologically and immunohistochemically distinct–a comparative study with assessment of ovarian carcinomas for expression of glial fibrillary acidic protein. Am J Surg Pathol. 2008;32(5):710–8.
50. McCluggage WG, Young RH. Ovarian sertoli-leydig cell tumors with pseudoendometrioid tubules (pseudoendometrioid sertoli-leydig cell tumors). Am J Surg Pathol. 2007;31(4):592–7.
51. Clement PB, Young RH, Scully RE. Endometrioid-like variant of ovarian yolk sac tumor. A clinicopathological analysis of eight cases. Am J Surg Pathol. 1987;11(10):767–78.
52. Zaino R, Whitney C, Brady MF, DeGeest K, Burger RA, Buller RE. Simultaneously detected endometrial and ovarian carcinomas–a prospective clinicopathologic study of 74 cases: a gynecologic oncology group study. Gynecol Oncol. 2001;83(2):355–62.
53. Storey DJ, Rush R, Stewart M, et al. Endometrioid epithelial ovarian cancer: 20 years of prospectively collected data from a single center. Cancer. 2008;112(10):2211–20.
54. Crissman JD et al. Endometrial carcinoma in women 40 years of age or younger. Obstet Gynecol. 1981;57(6):699–704.
55. Gallup DG, Stock RJ. Adenocarcinoma of the endometrium in women 40 years of age or younger. Obstet Gynecol. 1984;64(3):417–20.
56. Duska LR et al. Endometrial cancer in women 40 years old or younger. Gynecol Oncol. 2001;83(2):388–93.
57. Ota T et al. Clinicopathologic study of uterine endometrial carcinoma in young women aged 40 years and younger. Int J Gynecol Cancer. 2005;15(4):657–62.
58. Garg K, Shih K, Barakat R, Zhou Q, Iasonos A, Soslow RA. Endometrial carcinomas in women aged 40 years and younger: tumors associated with loss of DNA mismatch repair proteins comprise a distinct clinicopathologic subset. Am J Surg Pathol. 2009;33:1869–77.
59. Soliman PT, Slomovitz BM, Broaddus RR, et al. Synchronous primary cancers of the endometrium and ovary: a single institution review of 84 cases. Gynecol Oncol. 2004;94(2):456–62.
60. Ulbright TM, Roth LM. Metastatic and independent cancers of the endometrium and ovary: a clinicopathologic study of 34 cases. Hum Pathol. 1985;16(1):28–34.
61. Mazur MT, Hsueh S, Gersell DJ. Metastases to the female genital tract. Analysis of 325 cases. Cancer. 1984;53(9):1978–84.
62. Young RH. From Krukenberg to today: the ever present problems posed by metastatic tumors in the ovary. Part II Adv. Anat Pathol. 2007;14(3):149–77.
63. Roth LM, Liban E, Czernobilsky B. Ovarian endometrioid tumors mimicking Sertoli and Sertoli-Leydig cell tumors: Sertoliform variant of endometrioid adenocarcinoma. Cancer. 1982;50(7):1322–31.
64. Young RH, Scully RE. Ovarian Sertoli cell tumors: a report of 10 cases. Int J Gynecol Pathol. 1984;2(4):349–63.
65. Young RH, Scully RE. Well-differentiated ovarian Sertoli-Leydig cell tumors: a clinicopathological analysis of 23 cases. Int J Gynecol Pathol. 1984;3(3):277–90.
66. Zaloudek C, Norris HJ. Sertoli-Leydig tumors of the ovary. A clinicopathologic study of 64 intermediate and poorly differentiated neoplasms. Am J Surg Pathol. 1984;8:405–18.
67. Fujibayashi M, Aiba M, Iizuka E, Igarashi A, Muraoka M, Takagi K. Granulosa cell tumor-like variant of endometrioid adenocarcinoma of the ovary exhibiting nuclear clearing with biotin activity: a subtype showing close macroscopic, cytologic, and histologic similarity to adult granulosa cell tumor. Arch Pathol Lab Med. 2005;129(10):1288–94.
68. Cao QJ, Jones JG, Li M. Expression of calretinin in human ovary, testis, and ovarian sex cord-stromal tumors. Int J Gynecol Pathol. 2001;20(4):346–52.
69. Cathro HP, Stoler MH. The utility of calretinin, inhibin, and WT1 immunohistochemical staining in the differential diagnosis of ovarian tumors. Hum Pathol. 2005;36(2):195–201.
70. McCluggage WG, Maxwell P. Immunohistochemical staining for calretinin is useful in the diagnosis of ovarian sex cord-stromal tumours. Histopathology. 2001;38(5):403–8.
71. Deavers MT, Malpica A, Ordonez NG, Silva EG. Ovarian steroid cell tumors: an immunohistochemical study including a comparison of calretinin with inhibin. Int J Gynecol Pathol. 2003;22(2):162–7.
72. Movahedi-Lankarani S, Kurman RJ. Calretinin, a more sensitive but less specific marker than alpha-inhibin for ovarian sex cord-stromal neoplasms: an immunohistochemical study of 215 cases. Am J Surg Pathol. 2002;26(11):1477–83.
73. Shah VI, Freites ON, Maxwell P, McCluggage WG. Inhibin is more specific than calretinin as an immunohistochemical marker for differentiating sarcomatoid granulosa cell tumour of the ovary from other spindle cell neoplasms. J Clin Pathol. 2003;56(3):221–4.
74. Flemming P, Wellmann A, Maschek H, Lang H, Georgii A. Monoclonal antibodies against inhibin represent key markers of adult granulosa cell tumors of the ovary even in their metastases. A report of three cases with late metastasis, being previously misinterpreted as hemangiopericytoma. Am J Surg Pathol. 1995;19(8)):927–33.
75. McCluggage WG, Young RH. Immunohistochemistry as a diagnostic aid in the evaluation of ovarian tumors. Semin Diagn Pathol. 2005;22(1):3–32.
76. Ji H, Isacson C, Seidman JD, Kurman RJ, Ronnett BM. Cytokeratins 7 and 20, Dpc4, and MUC5AC in the distinction of metastatic mucinous carcinomas in the ovary from primary ovarian mucinous tumors: Dpc4 assists in identifying metastatic pancreatic carcinomas. Int J Gynecol Pathol. 2002;21(4):391–400.
77. McCluggage WG. Immunoreactivity of ovarian juvenile granulosa cell tumours with epithelial membrane antigen. Histopathology. 2005;46(2):235–6.
78. Park SY, Kim HS, Hong EK, Kim WH. Expression of cytokeratins 7 and 20 in primary carcinomas of the stomach and colorectum and their value in the differential diagnosis of metastatic carcinomas to the ovary. Hum Pathol. 2002;33(11):1078–85.
79. Costa MJ, DeRose PB, Roth LM, Brescia RJ, Zaloudek CJ, Cohen C. Immunohistochemical phenotype of ovarian granulosa cell tumors: absence of epithelial membrane antigen has diagnostic value. Hum Pathol. 1994;25:60–6.
80. Czernobilsky B, Moll R, Levy R, Franke WW. Co-expression of cytokeratin and vimentin filaments in mesothelial, granulosa and rete ovarii cells of the human ovary. Eur J Cell Biol. 1985;37:175–90.
81. Otis CN, Powell JL, Barbuto D, Carcangiu ML. Intermediate filamentous proteins in adult granulosa cell tumors. An immunohistochemical study of 25 cases. Am J Surg Pathol. 1992;16(10):962–8.
82. Aguirre P, Thor AD, Scully RE. Ovarian endometrioid adenocarcinomas resembling sex cord-stromal tumors. An immunohistochemical study. Int J Gynecol Pathol. 1989;8(4):364–73.
83. Dictor M. Malignant mixed mesodermal tumor of the ovary: a report of 22 cases. Obstet Gynecol. 1985;65(5):720–4.
84. Gilks CB, Ionescu DN, Kalloger SE, et al. Tumor cell type can be reproducibly diagnosed and is of independent prognostic significance in patients with maximally debulked ovarian carcinoma. Hum Pathol. 2008;39(8):1239–51.

Pathology of Clear Cell Tumors

8

Robert A. Soslow and Deborah F. DeLair

Keywords
Clear cell carcinoma • Adenofibroma • Borderline tumor • Endometriosis • HNF-1β

Introduction

Prevalence

Ovarian clear cell carcinoma is the second or third most common is the ovarian carcinoma diagnosed in North America, where it accounts for approximately 5–10% of all ovarian tumors [1]. It constitutes a larger percentage of ovarian cancers in East Asia, particularly Japan [2], but it is not certain whether this is due to environmental or genetic factors. Like endometrioid carcinomas, clear cell carcinomas are disproportionately represented among FIGO stages I and II ovarian carcinomas. Approximately 20% [3] of low-stage ovarian carcinomas are clear cell carcinomas, but unlike serous carcinomas, probably less than one-third of clear cell carcinoma patients have peritoneal or nodal metastasis at presentation [4–6].

There are data that suggest that pure clear cell carcinomas of the ovary have been accurately and reproducibly diagnosed [7, 8], but there are also reports indicating that distinction from high-grade serous carcinoma and endometrioid carcinoma remains a problem. Difficulties separating mixed epithelial carcinomas with clear cell components from serous carcinoma have been emphasized recently [7]. The theoretical rationale for strict ovarian carcinoma subclassification is provided in Chap. 5. These matters will likely become increasingly more relevant in practice as treatment protocols become tailored for different tumor types. Furthermore, ovarian clear cell carcinoma has been shown to be present in patients with Lynch Syndrome (hereditary non-polyposis colon cancer) and also harbors the same mismatch repair (MMR) abnormalities found in the syndrome [9, 10].

Morphology

Macroscopic Features

There are a few characteristic gross presentations of ovarian clear cell carcinoma. Almost all ovarian clear cell carcinomas are unilateral. Rare exceptions are the uncommonly encountered, large, solid, and cystic tumors with contralateral surface metastases and the bilateral carcinomas that arise in the setting of clear cell adenofibroma. Many ovarian clear cell carcinomas are associated with surface adhesions that result from longstanding endometriosis. Some gynecologists consider this evidence of FIGO stage II disease, whereas others do not. The most typical gross appearance is a large, cystic and solid tumor that may be accompanied by endometriosis (Fig. 8.1a). When an endometriotic cyst is the predominant finding, evidence of clear cell carcinoma may take the form of mural nodules or clustered, round papillae attached to the cyst lining, protruding into the lumen. The least common presentation is that of an adenofibromatous tumor with innumerable, small cysts separated by fibromatous stroma (Fig. 8.1b).

Microscopic Features

The typical clear cell carcinoma is composed of cells with clear cytoplasm (Fig. 8.2), thought to be due to cytoplasmic accumulation of glycogen, but not mucin. Despite this,

R.A. Soslow (✉)
Department of Pathology, Memorial Sloan-Kettering Cancer Center, 1275 York Avenue, New York, NY 10065, USA

R.A. Soslow and C. Tornos (eds.), *Diagnostic Pathology of Ovarian Tumors*,
DOI: 10.1007/978-1-4419-9751-7_8,

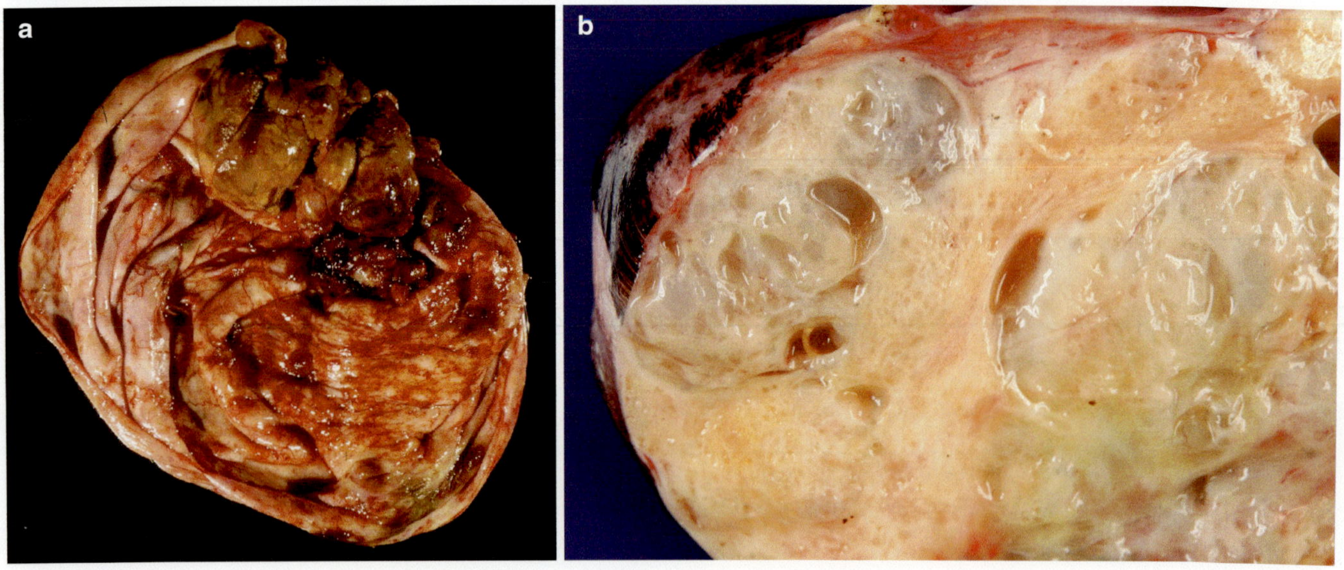

Fig. 8.1 Clear cell carcinoma, gross appearance. Panel (**a**) demonstrates the gross appearance of a large, cystic, and solid tumor arising in endometriosis. Panel (**b**) exhibits the gross appearance of an adenofibromatous clear cell carcinoma

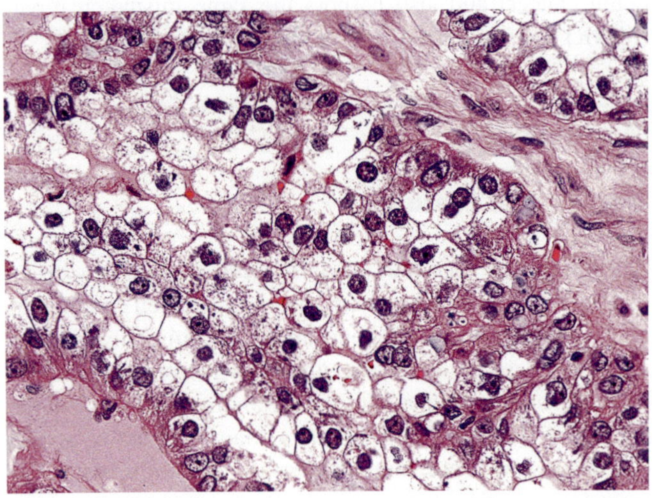

Fig. 8.2 Clear cell carcinoma displaying clear cytoplasm

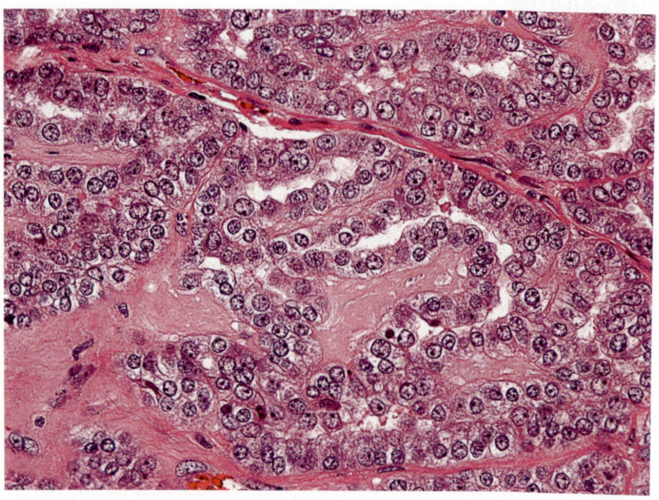

Fig. 8.3 Clear cell carcinoma with eosinophilic cytoplasm. This variant is usually described as "oxyphilic"

it should be emphasized that the presence of clear cytoplasm should not be used as the major diagnostic criterion since many other ovarian tumor types feature clear cells and rare ovarian clear cell carcinomas, referred to as "oxyphilic clear cell carcinomas," have eosinophilic cytoplasm (Fig. 8.3). Instead, ovarian clear cell carcinoma should be diagnosed based on the identification of cuboidal cell shape, characteristic nuclear features, and, probably most importantly, the presence of typical architectural patterns (Table 8.1) [11].

The predominant cell encountered in ovarian clear cell carcinoma is cuboidal in shape (Fig. 8.4); less frequently, the tumor cells are flattened (Fig. 8.5). Columnar cells are much more frequently observed in mimics of clear cell carcinoma such as serous and endometrioid carcinoma and serous borderline tumor. "Hobnail" cells are typically observed (Fig. 8.6), but these may be rare or absent in individual

Table 8.1 Morphologic summary

- Cuboidal or flattened or hobnail cells
- Round, mostly uniform nuclei, some with prominent nucleoli
- Papillary, tubulocystic and, occasionally, solid architectural patterns
- Low mitotic rate
- Clear cells

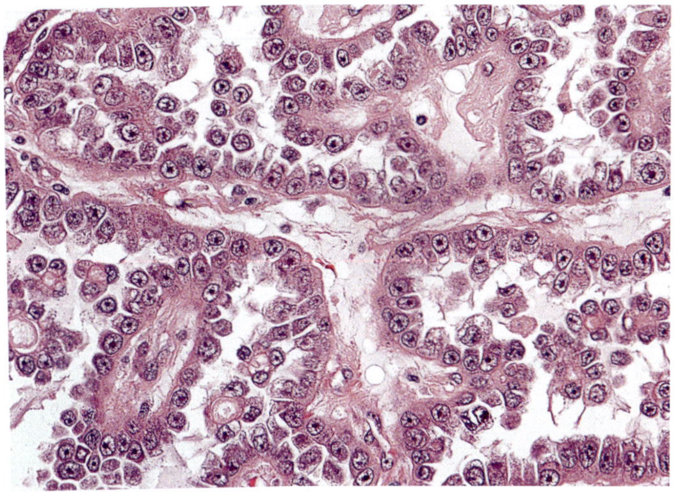

Fig. 8.4 Clear cell carcinoma. This clear cell carcinoma is composed of cuboidal cells with uniform nuclei that contain prominent nucleoli

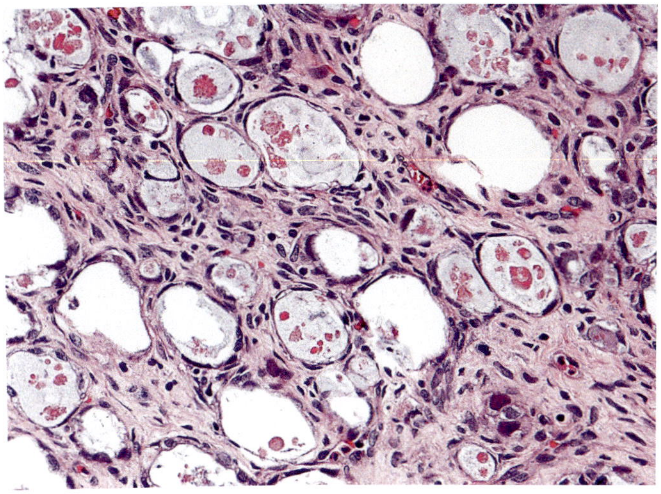

Fig. 8.5 Tubulocystic clear cell carcinoma with flattened tumor cells

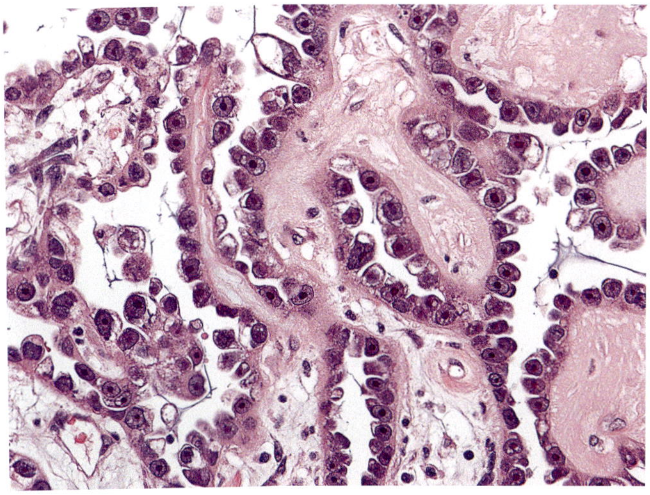

Fig. 8.6 Clear cell carcinoma showing "hobnail" cells

tumors and may be present in ovarian tumors other than clear cell carcinoma. The hobnail appearance is imparted by tumor cell nuclei that protrude beyond the luminal border of a gland or papilla. Signet ring cells with intraluminal mucin have been reported, but they are very rarely encountered. The nuclei of clear cell carcinomas are usually large and round, may have a prominent nucleolus, and tend to be of uniform size and appearance (Fig. 8.4). Striking variations in size and shape from one cell to the next occur, but bizarre cells tend to be scattered throughout the tumor or aggregated in small clusters (Fig. 8.7a, b). Rarely, recurrent and high stage tumors may be composed of cells with an anaplastic appearance. Occasional clear cell carcinomas with a solid architecture contain cells with koilocyte-like nuclei (Fig. 8.8). Hobnail cells may have oval, darkly stained nuclei. The average mitotic rate is approximately five mitotic figures per ten high power fields [7, 11]; bizarre and multipolar forms are unusual.

Ovarian clear cell carcinomas display a limited architectural inventory that includes papillary, tubulocystic, and solid architectural varieties. Most clear cell carcinomas exhibit mixtures of these patterns, with the papillary patterns being by far the most common [11]. Tumors displaying purely tubulocystic or solid patterns are unusual. The papillae of clear cell carcinomas differ from those of serous and endometrioid carcinomas – tumors that may display similar morphologic features. Clear cell carcinoma papillae are short and round and may have eosinophilic, hyalinized, or myxoid stroma (Fig. 8.9a, b). There are generally only one or two layers of cells lining the papillae. This contrasts with the prominent tufting seen in serous carcinomas. Tubulocystic tumors display infiltrating glands, most of which are cystically dilated. Some have a flattened monolayer of tumor cells at the glands' periphery (Fig. 8.5). Flattened tumor cells may give the misleading impression that the tumor is benign, so recognition of the typical tubulocystic architecture remains the best way of identifying this type of clear cell carcinoma. The stroma surrounding the glands can show edema (stromal "loosening") or desmoplasia. Solid clear cell carcinoma is almost always present in a background of a tumor with obvious papillary and/or tubulocystic architecture (Fig. 8.10). Many solid ovarian clear cell carcinomas have a pavement or cobblestone-like appearance resulting from a flat sheet of square to rectangular tumor cells with sharp cytoplasmic boundaries.

Other morphologic features found in clear cell carcinoma include hyaline globules (Fig. 8.11) (present in up to one-third of tumors) and, less commonly, psammoma bodies (Fig. 8.12), basophilic secretions, open tumor rings (Fig. 8.13), nuclear pseudoinclusions, targetoid bodies (Fig. 8.14), or an associated lymphoplasmacytic or lymphocytic infiltrate [11] (Fig. 8.15). Clear cell carcinomas are not currently graded.

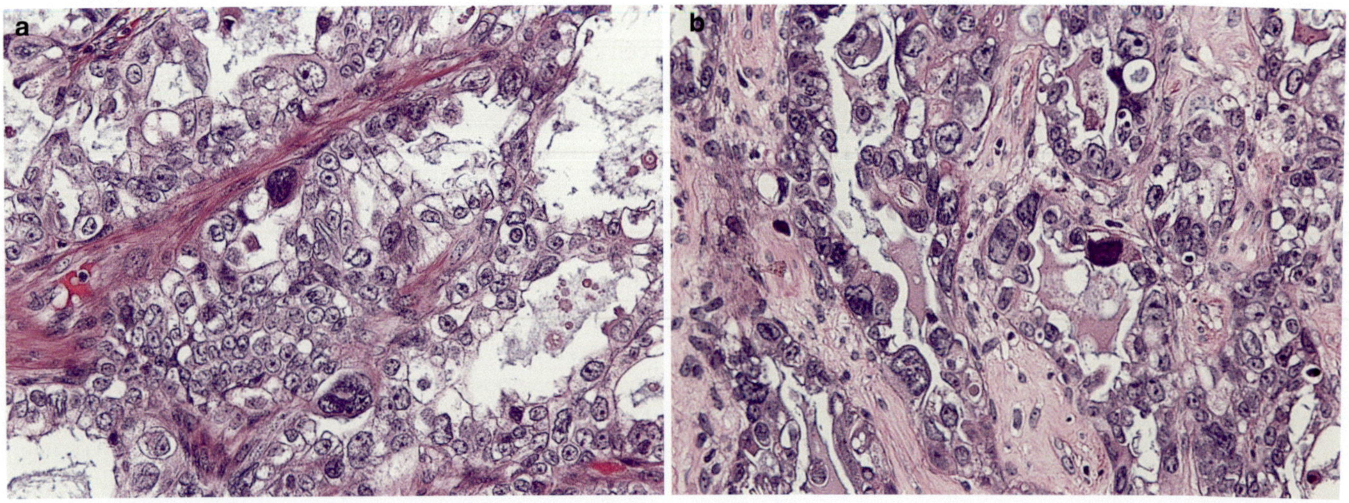

Fig. 8.7 (**a** and **b**). Clear cell carcinoma. Bizarre tumor cells are scattered throughout the tumor or aggregated in small clusters

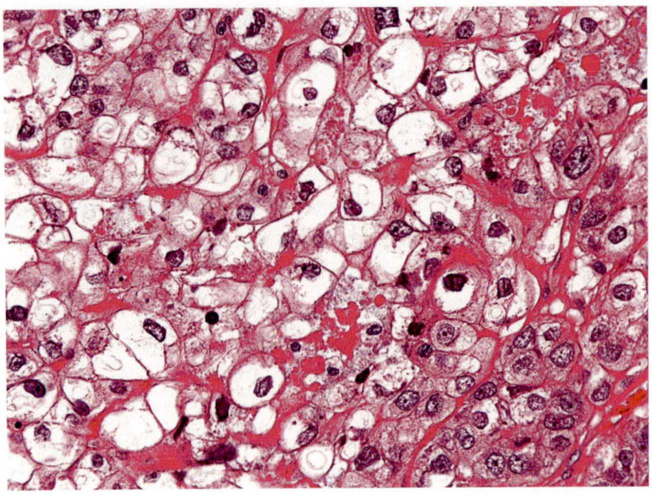

Fig. 8.8 Clear cell carcinoma. Solid variants often have cells with koilocyte-like nuclei

Mixed Epithelial Tumors Containing Clear Cells

A recent study of tumors diagnosed as mixed serous and clear cell carcinoma indicated only a low level of interobserver diagnostic agreement [7]. Using rigorous application of clear cell carcinoma diagnostic criteria, discussed above, and immunohistochemical staining, discussed subsequently, nearly all cases were reclassified as pure serous carcinomas without bona fide clear cell carcinoma components (Fig. 8.16). Therefore, tumors containing clear cells, but lacking typical clear cell carcinoma cytology or architecture, are likely not examples of clear cell carcinoma (Fig. 8.17). Mixed epithelial ovarian carcinoma with clear cell carcinoma components should be diagnosed sparingly, only after

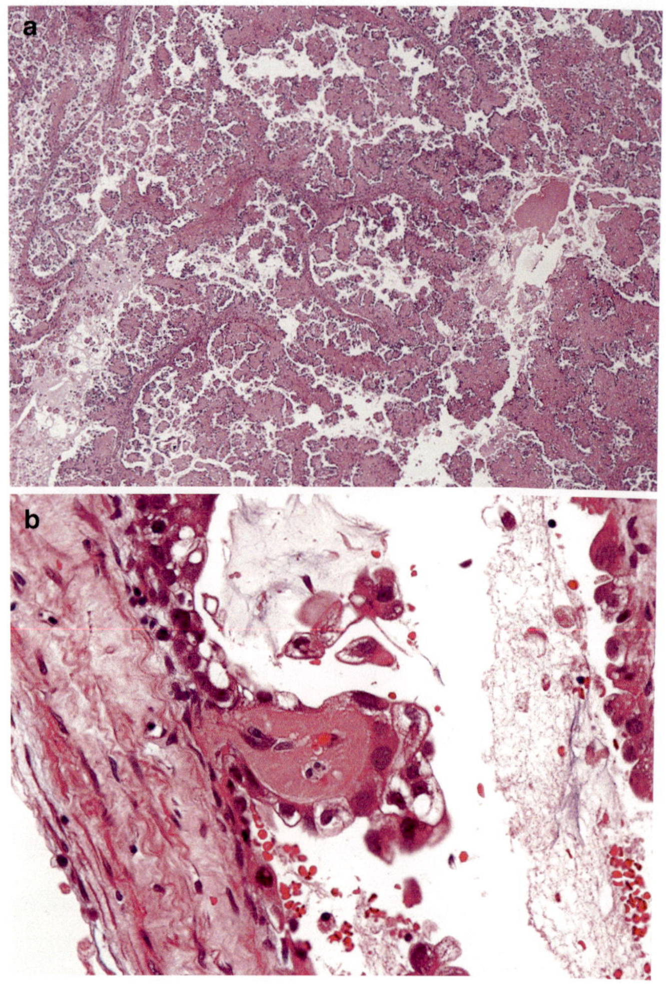

Fig. 8.9 (**a** and **b**) Clear cell carcinoma with papillary architecture. Papillae are short and round and may have eosinophilic, hyalinized or myxoid stroma

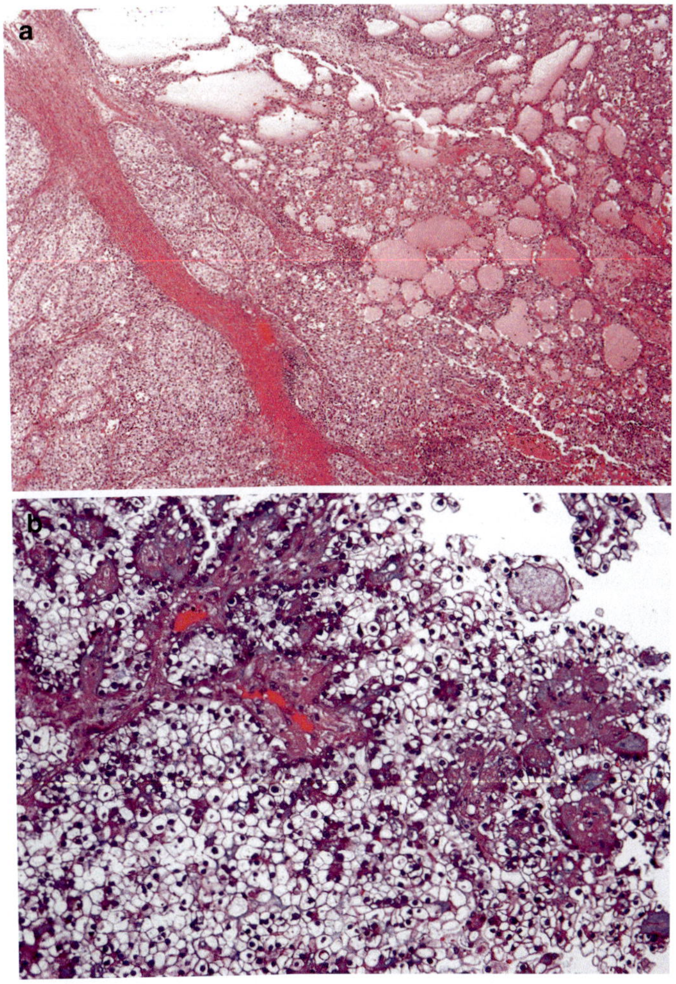

Fig. 8.10 Clear cell carcinoma with solid architecture. Solid clear cell carcinoma is almost always present in a background of a tumor with obvious tubulocystic (**a**) or papillary (**b**) architecture

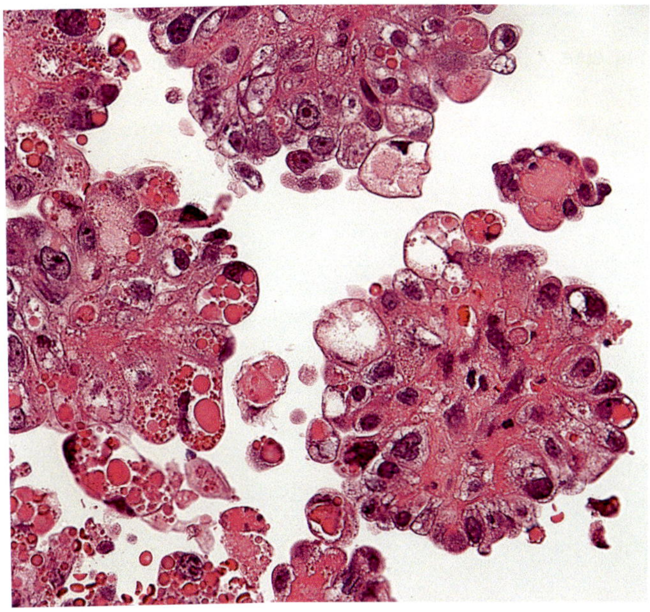

Fig. 8.11 Clear cell carcinoma with hyaline globules

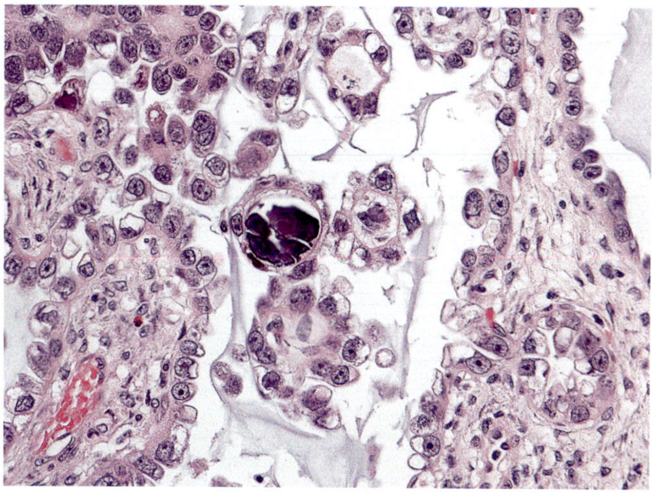

Fig. 8.12 Clear cell carcinoma with a psammoma body

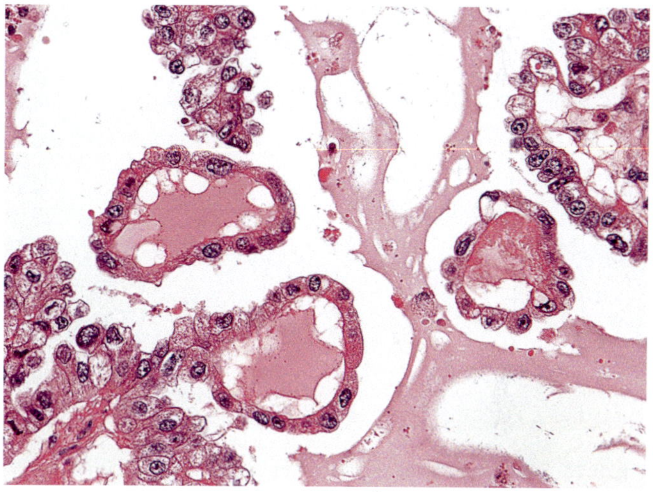

Fig. 8.13 Clear cell carcinoma showing open tumor rings

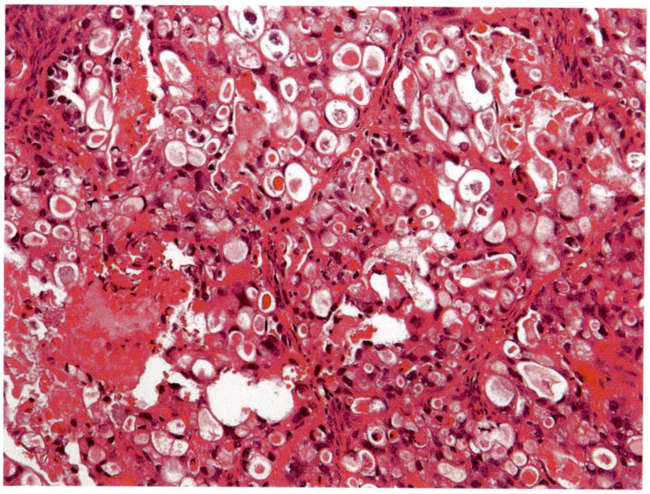

Fig. 8.14 Clear cell carcinoma showing targetoid bodies

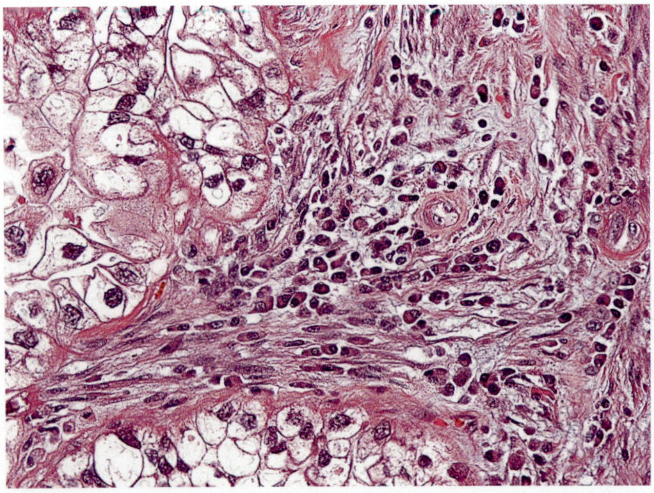

Fig. 8.15 Clear cell carcinoma with a lymphoplasmacytic infiltrate

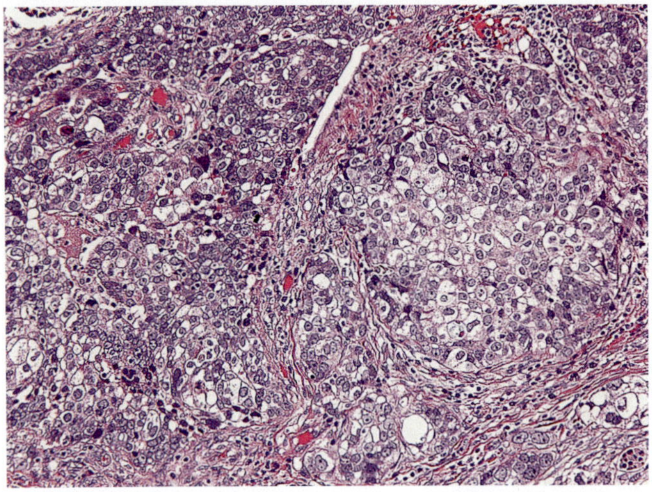

Fig. 8.16 Serous carcinoma containing clear cells. Pure clear cell carcinoma and clear cell carcinomatous components of mixed epithelial tumors do not display this degree of nuclear pleomorphism and mitotic activity

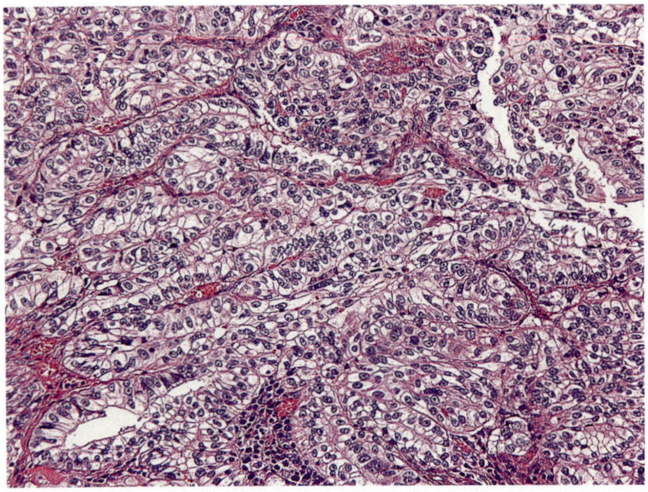

Fig. 8.17 Endometrioid adenocarcinoma containing clear cells. Columnar cells with low-grade nuclei and secretory changes are characteristic of endometrioid neoplasia; they are not seen in clear cell carcinoma

confirmation that an absolutely typical and distinctive clear cell carcinoma component is present.

Clear Cell Carcinoma Precursors

Endometriosis

At least half of clear cell carcinomas are associated with endometriosis, particularly atypical endometriosis [12–16] (Fig. 8.18). Some also arise in a background of endometriosis-associated tumors [17, 18], such as endometrioid borderline tumor and adenocarcinoma (Fig. 8.19), and seromucinous borderline tumor [19]. Associated endometrioid neoplasms may display obvious secretory change and/or mucinous differentiation. "Atypical endometriosis" describes a heterogeneous group of lesions that includes endometriosis with complex atypical hyperplasia and endometriosis with atypical hobnail metaplasia [20–22]. The latter lesion (Fig. 8.20)

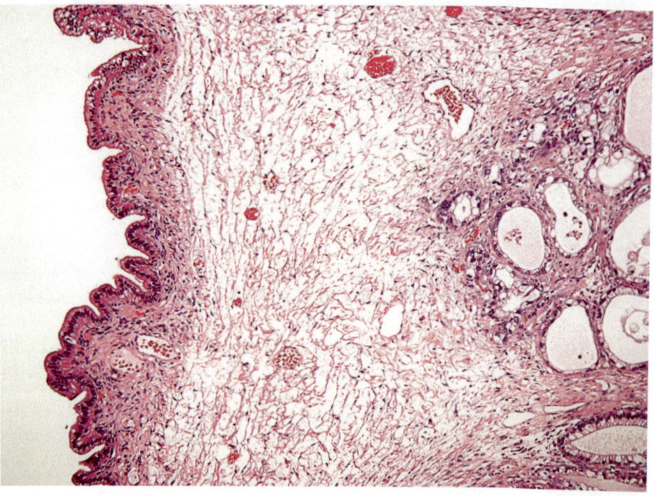

Fig. 8.18 Clear cell carcinoma (*right*), associated with endometriosis

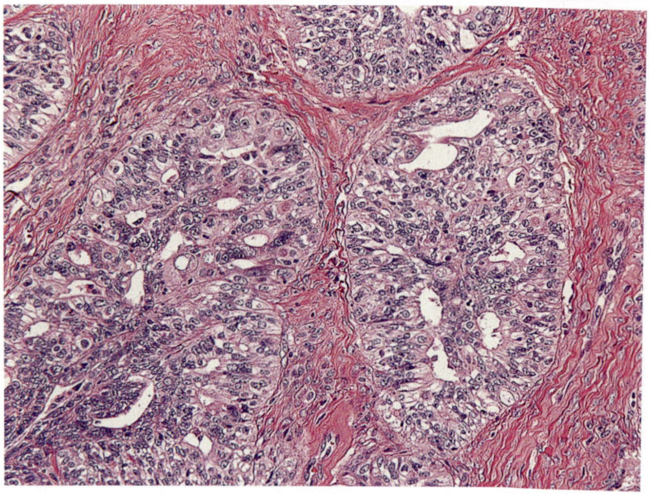

Fig. 8.19 Endometrioid neoplasia with clear cell features

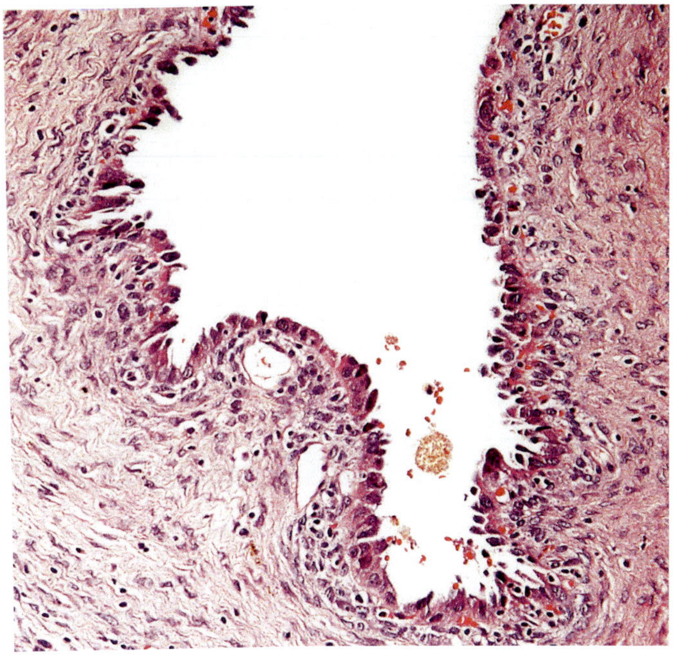

Fig. 8.20 Atypical endometriosis. The lesion demonstrates cytologically atypical hobnail cells

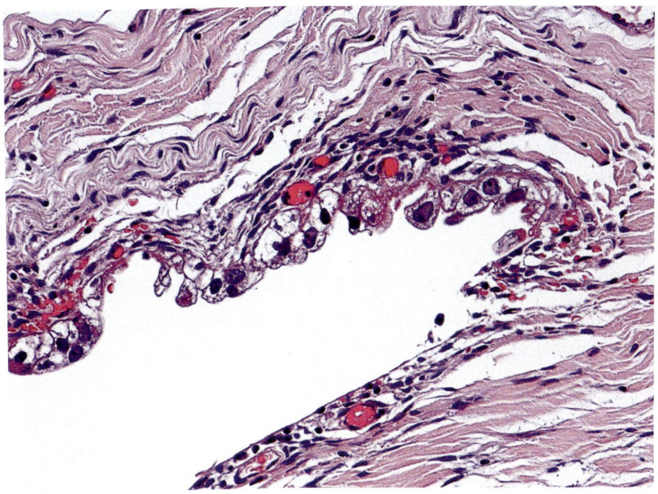

Fig. 8.21 Intraepithelial clear cell carcinoma. This focal lesion was identified incidentally in an endometrioma

is more frequently associated with clear cell carcinoma than the former. Endometriosis with atypical hobnail metaplasia can be distinguished from typical endometriosis by the presence of notable cytologic atypia; and the absence of papillary, tubulocystic, and solid growth patterns distinguishes this lesion from clear cell carcinoma. That being said, some examples of endometriosis with atypical hobnail metaplasia may represent intraepithelial clear cell carcinoma (Fig. 8.21). Atypical endometriotic lesions therefore should be sampled extensively with consideration for submission of all lesional tissue and liberal use of deeper recut sections.

Clear Cell Adenofibroma

Several recent publications [23–25] support the idea that clear cell carcinomas associated with clear cell adenofibromatous components are a subgroup of ovarian clear cell adenocarcinomas with distinct clinicopathologic characteristics. Approximately 15–20% of clear cell carcinomas have a predominant adenofibromatous component [23, 25]. "Clear cell adenofibroma" is a lesion composed of small, round glands with one layer of flat or low cuboidal clear cells lacking nuclear atypia, set in a fibromatous background (Fig. 8.22). These lesions usually demonstrate a gross appearance that sometimes resembles Swiss cheese. "Clear cell borderline tumor" has the same architecture and gross appearance, but the nuclei are cytologically atypical [26, 27]. Clear cell borderline tumor is distinguished from clear cell carcinoma by either the presence of papillary projections into the cysts

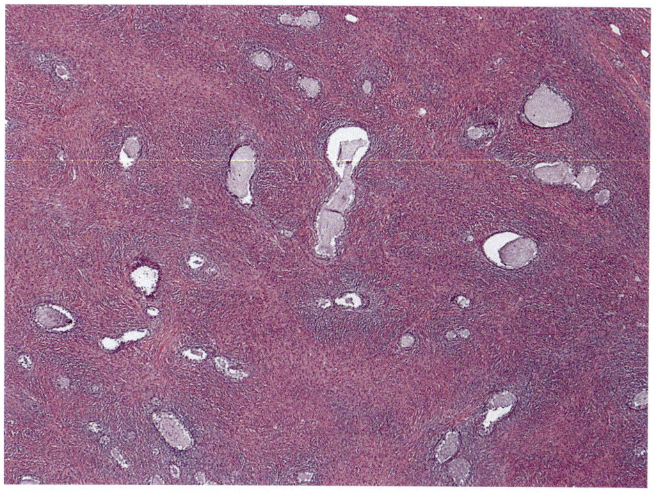

Fig. 8.22 Clear cell adenofibroma

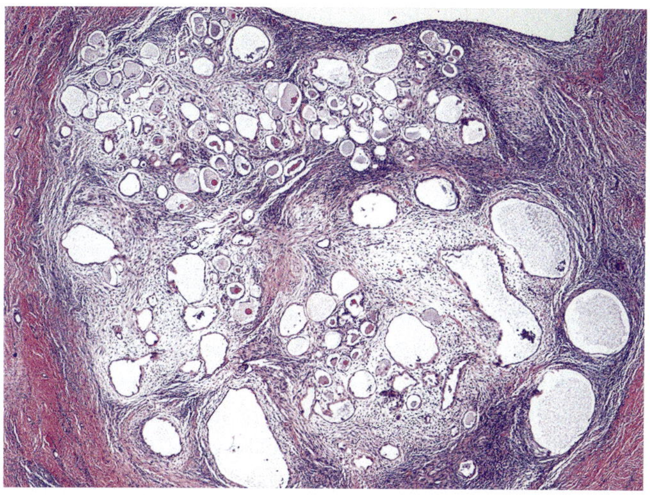

Fig. 8.23 Clear cell carcinoma arising from clear cell adenofibroma. Irregular clustering of variably sized glands surrounded by altered, loose stroma indicates the presence of invasive carcinoma

Table 8.2 Immunohistochemical summary

HNF-1β	Positive
ER	Negative
WT1	Negative
P53	Not overexpressed
CK 7	Positive

(Fig. 8.9b) or stromal invasion (Fig. 8.23). Borderline tumors retain densely collagenized or fibromatous stroma, whereas carcinomas demonstrate stromal edema, myxoid change, or desmoplasia in stroma surrounding the glands (Fig. 8.23). The tubules of adenofibroma and borderline tumor should be regularly distributed, and those of a carcinoma are usually described as haphazardly arranged and irregularly clustered. Extensive sampling of tumors that resemble clear cell borderline tumor is recommended as clear cell carcinoma can almost always be found on extensive sectioning. This also means that a diagnosis of clear cell adenofibroma or clear cell borderline tumor should not be rendered at the time of frozen section when only one or two slides are reviewed.

Immunophenotype

Clear cell carcinomas lack more than focal ER and WT1 expression (Table 8.2) [7, 28]. p53 expression can be encountered, but diffuse and strong overexpression similar to that seen in most high-grade serous carcinomas is not characteristic [7, 28]. Positive markers of clear cell differentiation have now been described. Recently reported examples include hypoxia-inducible factor 1 alpha (HIF-1α) [29], human kidney injury molecule-1 [30], glypican-3 [31], and hepatocyte nuclear factor-1 beta (HNF-1β). HNF-1β is a transcription factor related to glycogen metabolism that has recently emerged as a sensitive and specific marker for ovarian clear cell carcinoma. Positive staining for clear cell tumors of the ovary has ranged from 82% to 100% and only rare and focal positivity has been reported in other ovarian epithelial tumors such as endometrioid, serous, and mucinous carcinomas [28, 32–34] (Fig. 8.24).

Molecular Biology

Mutations in K-ras [35] and PTEN [36] have been reported, as well as several examples of microsatellite instability-high clear cell carcinoma [10, 37, 38]. Mutation of the oncogene PIK3CA has been demonstrated to occur frequently in ovarian clear cell carcinoma, occurring in up to 33% of cases. This mutation is believed to activate the PI3K/AKT pathway [39], which has been shown to play an important role in tumor oncogenesis in several types of human cancer as it promotes increased cell proliferation, invasion, and decreased apoptosis. Lynch Syndrome, a syndrome with defects in mismatch repair proteins due to germline mutations, increases the incidence of certain malignancies including ovarian and endometrial carcinoma. A recent series showed that ovarian clear cell carcinoma was overrepresented among ovarian carcinomas in young patients with MMR defects; 17% had clear cell histology [10]. Synchronous ovarian clear cell carcinomas have also been identified in two patients with endometrial carcinoma and MMR abnormalities [9, 40].

A recent study describes *ARID1A* as a tumor-suppressor gene frequently disrupted in ovarian clear-cell and endometrioid carcinomas [41].

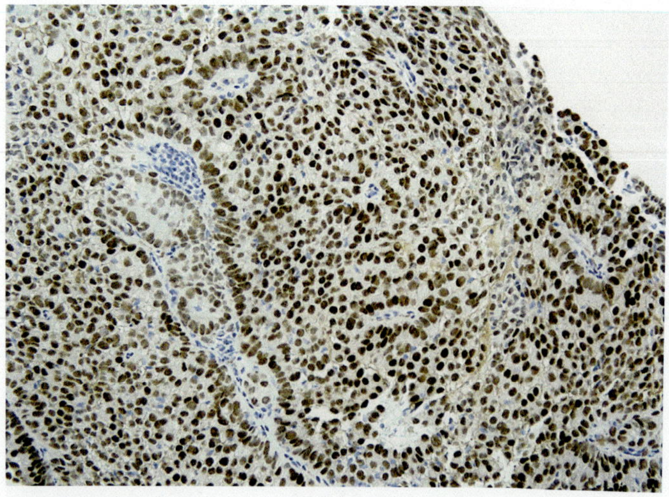

Fig. 8.24 Clear cell carcinoma with diffuse and strong nuclear expression of HNF-1β

Table 8.3 Differential diagnosis

Serous carcinoma and borderline tumor
Endometrioid carcinoma and borderline tumor
Yolk sac tumor
Peritoneal mesothelioma
Papillary cystadenoma of von Hippel-Lindau Disease
Metastatic carcinoma, including Krukenberg tumor
Dysgerminoma
Juvenile granulosa cell tumor
Arias Stella reaction
Tumors with oxyphilic and clear cells (see Table 8.8)

Differential Diagnosis (Table 8.3)

Serous Carcinoma (Fig. 8.16)

When confronted with a papillary ovarian carcinoma showing high nuclear grade, the choices are essentially restricted to serous and clear cell carcinomas if germ cell tumors and metastases are excluded. Rounded papillary cores with hyaline, surrounded by one or two layers of cuboidal cells with uniform, but atypical nuclei, favor clear cell carcinoma. Extensive multilayering and tufting of columnar cells with mitotic rates in excess of ten mitotic figures per ten high power fields as well as abnormal mitotic figures are findings much more frequently associated with serous carcinoma, even when tumor cells with cytoplasmic clearing are present. The presence of additional typical clear cell patterns such as tubulocystic architecture and adjacent endometriosis can also aid in an accurate diagnosis. WT1 expression and p53 overexpression, especially along with ER expression, would strongly favor a serous neoplasm over clear cell carcinoma; HNF-1β expression without WT1 or ER expression is characteristic of clear cell carcinoma.

Serous Borderline Tumor (Fig. 8.25, Table 8.4)

Occasional papillary clear cell carcinomas can resemble serous borderline tumor [41], especially at frozen section when only one or two slides are available for review. Clues to the correct diagnosis here include an adenofibromatous gross appearance, an associated endometriotic cyst, and the identification of even occasional cells with large, round nuclei containing nucleoli. Clinical features should also be contributory, as bilateral clear cell carcinomas and those that are disseminated at presentation are much less commonly encountered compared to serous neoplasms [7, 42]. The immunophenotypic guidelines discussed in reference to serous carcinoma above would apply to this differential diagnosis as well, although serous borderline does not overexpress p53.

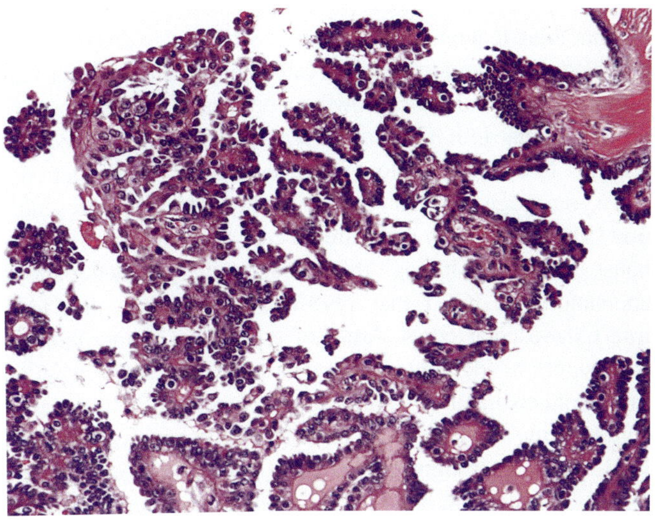

Fig. 8.25 Serous borderline tumor with hobnail cells. Distinction from clear cell carcinoma can sometimes be problematic

Table 8.4 Features favoring clear cell carcinoma over serous carcinoma

• Low stage, unilateral tumor
• Round papillae with minimal multilayering[a]
• Tubulocystic growth pattern[a]
• Cuboidal cell shape
• Low mitotic rate (less than ten mitotic figures per ten high power fields
• Associated endometriosis[a]
• HNF-1β positive[a]; WT1 and ER negative[a]; no p53 overexpression

[a]Features also useful in the distinction of serous borderline tumor

Endometrioid Carcinoma and Borderline Tumor (Figs. 8.17 and 8.19, Table 8.5)

Endometrioid carcinomas and borderline tumors may mimic clear cell carcinomas when they are papillary, show abundant cytoplasmic clearing, display secretory features, and when tumor cell cytoplasm is squamoid and contains abundant glycogen [43]. The traditional approach to this problem was to restrict a diagnosis of clear cell carcinoma in this setting to tumors with unequivocally high nuclear grade, and this is still absolutely relevant. However, rare endometrioid carcinomas display moderate or severe nuclear atypia. Typical endometrioid features such as columnar cell shape, the presence of squamous metaplasia showing either keratinization or unequivocal evidence of intercellular bridges, subnuclear cytoplasmic clearing, and cribriform architecture favor an endometrioid neoplasm. ER expression is typical of most endometrioid neoplasms as well. To diagnose a clear cell carcinoma in the context of a mixed epithelial ovarian carcinoma with an endometrioid component, look for typical clear cell architectural patterns, cuboidal cell shape, nuclear features that are distinct from the endometrioid component, and loss of ER expression.

Yolk Sac Tumor (Fig. 8.26)

Yolk sac tumors and clear cell carcinomas often show similar morphologic features including papillary and tubulocystic architecture, clear cells, and eosinophilic globules. Immunohistochemical features can also be misleading as a small number of clear cell carcinomas can show AFP positivity [44, 45] while this marker may be only patchily positive or negative in yolk sac tumors [46]. Table 8.6 lists

Table 8.5 Features favoring clear cell carcinoma over endometrioid carcinoma

• Round papillae
• Tubulocystic growth pattern
• No squamous or mucinous metaplasia; no cribriform growth
• Cuboidal cell shape
• HNF-1β positive; ER negative

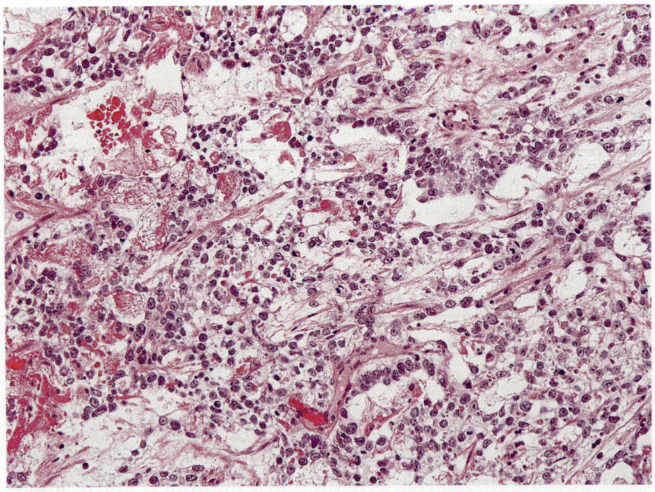

Fig. 8.26 Yolk sac tumor showing some similarity to clear cell carcinoma

Table 8.6 Features favoring clear cell carcinoma over yolk sac tumor

- Older patient
- No serum AFP elevation
- No associated germ cell tumor
- No primitive-appearing nuclei; low mitotic rate
- EMA and CK7 positive; SALL4 not diffuse and strong

features that will help differentiate these tumors. The typical age of patients with yolk sac tumor is at least four decades younger than clear cell carcinoma, though there can be some overlap in a small percentage of cases. There are also a number of robust morphologic features that differentiate these tumors. Yolk sac tumors, in general, have a wider range of architectural appearances compared to clear cell carcinoma, more primitive-appearing nuclei, a higher mitotic rate, elevated serum AFP levels, and can be associated with other germ cell tumors. Since these entities are treated differently and have disparate prognoses, many pathologists consider performing immunohistochemical stains when there is any doubt about the diagnosis. Clear cell carcinomas express EMA and CK7, while yolk sac tumor expresses AFP and SALL4 without EMA or CK7. SALL4 appears to be a very sensitive and specific marker when the differential diagnosis includes the entities discussed above, with diffuse and strong positivity in 100% of yolk sac tumors and only focal and weak labeling in 7% of clear cell carcinomas [45].

Dysgerminoma

As is the case with yolk sac tumors, the clinical features of patients with dysgerminomas and clear cell carcinomas may overlap, although in typical cases, the dysgerminoma patient is substantially younger. There may be certain morphologic similarities as well, including clear cytoplasm, prominent nucleoli, lymphocytic infiltrate, and a solid growth pattern. A thorough sampling of the tumor will almost always reveal an additional papillary or tubulocystic pattern in clear cell carcinoma [11] and may often reveal another germ cell component in dysgerminoma [47]. Additional morphologic features favoring a dysgerminoma include a granulomatous reaction and, rarely, syncytiotrophoblastic giant cells [48]. There are several immunohistochemical markers that may also aid in the diagnosis. Cytokeratins are usually strongly and diffusely positive in clear cell carcinoma and negative in dysgerminomas; however, they have been reported to be focally positive in dysgerminomas [49]. SALL4 shows the same sensitivity and specificity as that previously described in yolk sac tumors when the differential diagnosis includes dysgerminomas and clear cell carcinoma [45]. OCT4 also appears to be a sensitive and specific marker for dysgerminoma as it is strongly and diffusely positive in all studied cases and shows only focal immunoreactivity in clear cell carcinoma [50].

Juvenile Granulosa Cell Tumor (JGCT)

Morphologic features that may cause difficulty in this differential diagnosis include hobnail cells lining follicles that can resemble a tubulocystic growth pattern and the presence of pseudopapillae in JGCT. Unique features seen in JGCT include follicles lined by granulosa cells, estrogenic manifestations, and a young age, with 13 years being the mean age of presentation [51]. JGCTs may very rarely occur in women after menopause, however. Although the pseudopapillary growth pattern encountered in JGCT may resemble true papillae, they lack true fibrovascular cores. Areas of typical JGCT morphology can usually be identified in these cases [52]. A small panel of immunohistochemical stains can also facilitate the diagnosis. Cytokeratins and EMA may be only focally positive in JGCT, while they are strongly and diffusely positive in clear cell carcinoma. In addition, inhibin and calretinin are usually positive in JGCT while negative in clear cell carcinoma [53, 54].

Peritoneal Mesothelioma (Fig. 8.27, Table 8.7)

Papillary mesotheliomas are composed of flattened or cuboidal cells without extensive hierarchical branching, multilayering or tufting, just like clear cell carcinoma. Of course, typical clear cell carcinoma would present with a dominant ovarian mass, and the typical mesothelioma would present with diffuse or multifocal peritoneal disease. Exceptions occur and diagnosis can be difficult when confronted with

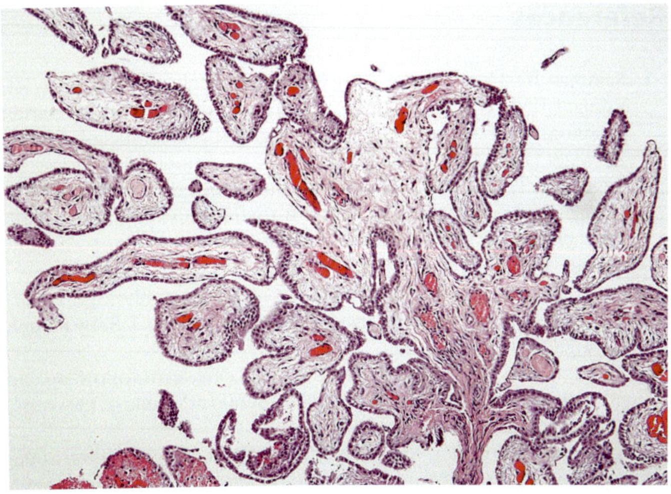

Fig. 8.27 Well-differentiated papillary mesothelioma. Papillae lined by a cellular monolayer are histologically similar to some clear cell carcinomas

Table 8.7 Features favoring clear cell carcinoma over peritoneal mesothelioma

- Unilateral ovarian tumor without peritoneal disease
- Prominent nucleoli
- B72.3 and BerEP4 positive; WT1, calretinin, and podoplanin negative

only a core biopsy or a frozen section. Both tumors have small, rounded papillae, but papillary mesotheliomas tend to have myxoid stroma in fibrovascular cores, whereas clear cell carcinomas frequently display hyalinized cores. Both clear cell carcinoma and mesothelioma can have tumor cells with uniform nuclei, but mesothelioma nuclei tend to be smaller and have fewer chromatin irregularities as compared to clear cell carcinoma. Although some mesotheliomas can have clear cells, most are composed of cells with eosinophilic cytoplasm. Some have circumferential accumulations of intermediate filaments recognizable on hematoxylin and eosin stains and, in some cases, fuzzy apical membrane projections can be appreciated. These tumors have different immunophenotypes, which can be exploited for diagnostic purposes. Mesotheliomas express WT1, calretinin, and podoplanin, markers lacking in clear cell carcinoma. EMA and CK7 expression would be expected in both tumor types, but only clear cell carcinoma would be expected to express at least one of the glycoprotein markers BerEP4, or B72.3.

Papillary Cystadenoma of Von Hippel-Lindau Disease

Papillary cystadenoma is an extremely rare tumor that may present in association with von Hippel-Lindau (VHL) disease. In females, VHL usually occurs in the broad ligament or mesosalpinx. Morphologically, it has several features that resemble clear cell carcinoma, including papillae lined by a single layer of cuboidal cells, clear cytoplasm, and hyalinized fibrous stroma. In addition, the immunophenotype is similar as both tumors are positive for CK7 and EMA [54, 55]. Typical characteristics of a papillary cystadenoma include a clinical history of other VHL tumors, a non-ovarian location, and absent mitotic activity, necrosis, nuclear atypia, and prominent nucleoli [55–57].

Metastatic Carcinoma, Including Krukenberg Tumor

Certain metastatic tumors to the ovary, including the typical Krukenberg tumor [58], can show morphologic similarities to ovarian clear cell carcinoma due to the presence of clear cytoplasm or signet ring cells. Features favoring metastasis include small size [59, 60], bilaterality, surface involvement (usually microscopic) without significant parenchymal involvement, a nodular growth pattern, lymphovascular invasion, and a desmoplastic or edematous reactive stroma [61]. Additional features favoring a primary clear cell carcinoma include a mixture of the typical clear cell patterns (solid, papillary, tubulocystic) and cell types (cuboidal, flattened, hobnail), as opposed to a metastatic tumor which characteristically shows a uniform architecture and cell type. Occasionally, immunohistochemistry may also contribute, especially when the differential diagnosis includes a carcinoma of lower gastrointestinal origin, typically strongly CK20 positive and CK7 negative.

Arias-Stella Change

Clear cell carcinoma and endometriosis with Arias-Stella change may appear very similar. The presence of clear cells, hobnail forms, nuclear atypia, HNF-1β expression, and even mitotic activity can be seen in both. In contrast to Arias-Stella change, clear cell carcinoma with associated endometriosis usually presents as a mass lesion within an endometriotic cyst, a microscopic, infiltrative tubulocystic tumor in the wall of the cyst, or a mixed papillary and tubulocystic neoplasm associated with clear cell adenofibroma. Arias-Stella change, however, is usually present in a background of non-neoplastic and metaplastic changes in ectopic endometrium, including endometrioid stroma. It lacks an invasive profile and does not, by itself, form a mass lesion. Other histologic manifestations of progestational effect (i.e., pseudodecidua) are frequently present.

Table 8.8 Tumors with oxyphilic and clear cells

- Clear cell carcinoma of ovary
- Metastatic clear cell carcinoma
- Endometrioid adenocarcinoma
- Hepatoid carcinoma [67, 68]
- Hepatoid yolk sac tumor [69]
- Metastatic hepatocellular carcinoma [70]
- Steroid cell tumor [71]
- Epithelioid smooth muscle tumor [72]
- Perivascular epithelioid cell tumor (PEComa) [73]
- Undifferentiated carcinomas with rhabdoid morphology [74]
- Gastrointestinal stromal tumors [75]
- Intermediate trophoblastic tumors (placental site trophoblastic tumor; epithelioid trophoblastic tumor) [76]
- Clear cell sarcoma
- Melanoma
- Alveolar soft part sarcoma
- Rhabdoid tumor
- Epithelioid sarcoma
- Epithelioid angiosarcoma

Tumors with oxyphilic and clear cells also enter into the differential diagnosis with clear cell carcinoma (Table 8.8)

Treatment and Clinical Outcomes

When serous carcinomas composed of clear cells are excluded, ovarian clear cell carcinoma is a tumor that is confined to the pelvis in about two-thirds of cases [4, 5]. Survival is stage-dependent. No prognostically relevant grading scheme has been devised. Paraneoplastic hypercalcemia and thromboembolic disease are known associations with ovarian clear cell carcinoma.

Five-year survivals for stage I clear cell carcinoma are 70–90% [3, 25, 62], intermediate between those expected for high grade serous carcinoma (which are significantly worse) and endometrioid carcinoma (which are significantly better) [8, 63]. Patients with low stage tumors that lack expression of IGF2BP3 (IMP3) have the best prognosis [61]. Adenofibromatous tumors have also been noted to be less clinically aggressive than others [23], although this has not been confirmed by other groups [25]. Platinum and taxane-based chemotherapy, similar to serous carcinoma, is standard therapy for these patients, although clear cell carcinoma is now appreciated to be relatively chemoresistant to these agents [64, 65]. Patients who present with extraovarian disease therefore have very disappointing survivals. Some investigators have reported positive outcomes using radiation therapy [66]. Targeted therapy against components of the PIK3CA pathway and the tyrosine kinase receptors VEGF-R and PDGF-R is currently under study.

References

1. Seidman JD, Horkayne-Szakaly I, Haiba M, Boice CR, Kurman RJ, Ronnett BM. The histologic type and stage distribution of ovarian carcinomas of surface epithelial origin. Int J Gynecol Pathol. 2004;23(1):41–4.
2. Sugiyama T, Kamura T, Kigawa J, et al. Clinical characteristics of clear cell carcinoma of the ovary: a distinct histologic type with poor prognosis and resistance to platinum-based chemotherapy. Cancer. 2000;88(11):2584–9.
3. Leitao Jr MM, Boyd J, Hummer A, et al. Clinicopathologic analysis of early-stage sporadic ovarian carcinoma. Am J Surg Pathol. 2004;28(2):147–59.
4. Mizuno M, Kikkawa F, Shibata K, et al. Long-term follow-up and prognostic factor analysis in clear cell adenocarcinoma of the ovary. J Surg Oncol. 2006;94(2):138–43.
5. Takano M, Kikuchi Y, Yaegashi N, et al. Clear cell carcinoma of the ovary: a retrospective multicentre experience of 254 patients with complete surgical staging. Br J Cancer. 2006;94(10):1369–74.
6. Behbakht K, Randall TC, Benjamin I, Morgan MA, King S, Rubin SC. Clinical characteristics of clear cell carcinoma of the ovary. Gynecol Oncol. 1998;70:255–8.
7. Han G, Gilks CB, Leung S, et al. Mixed ovarian epithelial carcinomas with clear cell and serous components are variants of high-grade serous carcinoma: an interobserver correlative and immunohistochemical study of 32 cases. Am J Surg Pathol. 2008;32(7):955–64.
8. Gilks CB, Ionescu DN, Kalloger SE, et al. Tumor cell type can be reproducibly diagnosed and is of independent prognostic significance in patients with maximally debulked ovarian carcinoma. Hum Pathol. 2008;39(8):1239–51.
9. Garg K, Leitao Jr MM, Kauff ND, et al. Selection of endometrial carcinomas for DNA mismatch repair protein immunohistochemistry using patient age and tumor morphology enhances detection of mismatch repair abnormalities. Am J Surg Pathol. 2009;33(6):925–33.
10. Jensen KC, Mariappan MR, Putcha GV, et al. Microsatellite instability and mismatch repair protein defects in ovarian epithelial neoplasms in patients 50 years of age and younger. Am J Surg Pathol. 2008;32(7):1029–37.
11. DeLair D, Soslow R, Gilks B, Macias A, Oliva E. The morphologic spectrum of immunohistochemically characterized clear cell carcinoma of the ovary: a study of 83 cases. Mod Pathol. 2009;22 Suppl 1:211A. abstract 961.
12. Fukunaga M, Nomura K, Ishikawa E, Ushigome S. Ovarian atypical endometriosis: its close association with malignant epithelial tumours. Histopathology. 1997;30(3):249–55.
13. Moll UM, Chumas JC, Chalas E, Mann WJ. Ovarian carcinoma arising in atypical endometriosis. Obstet Gynecol. 1990;75:537–9.
14. Mandai M, Yamaguchi K, Matsumura N, Baba T, Konishi I. Ovarian cancer in endometriosis: molecular biology, pathology, and clinical management. Int J Clin Oncol. 2009;14(5):383–91.
15. Kobayashi H. Ovarian cancer in endometriosis: epidemiology, natural history, and clinical diagnosis. Int J Clin Oncol. 2009;14(5):378–82.
16. LaGrenade A, Silverberg SG. Ovarian tumors associated with atypical endometriosis. Hum Pathol. 1988;19:1080–4.
17. Brescia RJ, Dubin N, Demopoulos RI. Endometrioid and clear cell carcinoma of the ovary. Factors affecting survival. Int J Gynecol Pathol. 1989;8:132–8.
18. Czernobilsky B, Silverman BB, Enterline HT. Clear cell carcinoma of the ovary. A clinicopathologic analysis of pure and mixed forms and comparison with endometrioid carcinoma. Cancer. 1970;25:762–72.
19. Asad H, Soslow RA. Clinicopathologic and immunohistochemical analysis of mixed epithelial ovarian tumors with a clear cell carcinoma component. Mod Pathol. 2008;21 Suppl 1:195A. abstract 894.

20. Czernobilsky B, Morris WJ. A histologic study of ovarian endometriosis with emphasis on hyperplastic and atypical changes. Obstet Gynecol. 1979;53:318–23.
21. Seidman JD. Prognostic importance of hyperplasia and atypia in endometriosis. Int J Gynecol Pathol. 1996;15(1):1–9.
22. Ballouk F, Ross JS, Wolf BC. Ovarian endometriotic cysts analysis cytologic atypia DNA ploidy patterns. Am J Clin Pathol. 1994;102:415–9.
23. Yamamoto S, Tsuda H, Yoshikawa T, et al. Clear cell adenocarcinoma associated with clear cell adenofibromatous components: a subgroup of ovarian clear cell adenocarcinoma with distinct clinicopathologic characteristics. Am J Surg Pathol. 2007;31(7):999–1006.
24. Yamamoto S, Tsuda H, Takano M, et al. Expression of platelet-derived growth factors and their receptors in ovarian clear-cell carcinoma and its putative precursors. Mod Pathol. 2008;21(2):115–24.
25. Veras E, Mao TL, Ayhan A, et al. Cystic and adenofibromatous clear cell carcinomas of the ovary: distinctive tumors that differ in their pathogenesis and behavior: a clinicopathologic analysis of 122 cases. Am J Surg Pathol. 2009;33(6):844–53.
26. Bell DA, Scully RE. Benign and borderline clear cell adenofibromas of the ovary. Cancer. 1985;56:2922–31.
27. Roth LM, Langley FA, Fox H, Wheeler JE, Czernobilsky B. Ovarian clear cell adenofibromatous tumors: benign, low malignant potential, and associated with invasive clear cell carcinoma. Cancer. 1984;53:1156–63.
28. Kobel M, Kalloger SE, Carrick J, et al. A limited panel of immunomarkers can reliably distinguish between clear cell and high-grade serous carcinoma of the ovary. Am J Surg Pathol. 2009;33(1): 14–21.
29. Lee S, Garner EI, Welch WR, Berkowitz RS, Mok SC. Over-expression of hypoxia-inducible factor 1 alpha in ovarian clear cell carcinoma. Gynecol Oncol. 2007;106(2):311–7.
30. Lin F, Zhang PL, Yang XJ, et al. Human kidney injury molecule-1 (hKIM-1): a useful immunohistochemical marker for diagnosing renal cell carcinoma and ovarian clear cell carcinoma. Am J Surg Pathol. 2007;31(3):371–81.
31. Stadlmann S, Gueth U, Baumhoer D, Moch H, Terracciano L, Singer G. Glypican-3 expression in primary and recurrent ovarian carcinomas. Int J Gynecol Pathol. 2007;26(3):341–4.
32. Yamamoto S, Tsuda H, Aida S, Shimazaki H, Tamai S, Matsubara O. Immunohistochemical detection of hepatocyte nuclear factor 1beta in ovarian and endometrial clear-cell adenocarcinomas and non-neoplastic endometrium. Hum Pathol. 2007;38(7):1074–80.
33. Kato N, Sasou S, Motoyama T. Expression of hepatocyte nuclear factor-1beta (HNF-1beta) in clear cell tumors and endometriosis of the ovary. Mod Pathol. 2006;19(1):83–9.
34. Tsuchiya A, Sakamoto M, Yasuda J, et al. Expression profiling in ovarian clear cell carcinoma: identification of hepatocyte nuclear factor-1 beta as a molecular marker and a possible molecular target for therapy of ovarian clear cell carcinoma. Am J Pathol. 2003;163(6):2503–12.
35. Amemiya S, Sekizawa A, Otsuka J, Tachikawa T, Saito H, Okai T. Malignant transformation of endometriosis and genetic alterations of K-ras and microsatellite instability. Int J Gynaecol Obstet. 2004;86(3):371–6.
36. Sato N, Tsunoda H, Nishida M, et al. Loss of heterozygosity on 10q23.3 and mutation of the tumor suppressor gene PTEN in benign endometrial cyst of the ovary: possible sequence progression from benign endometrial cyst to endometrioid carcinoma and clear cell carcinoma of the ovary. Cancer Res. 2000;60(24):7052–6.
37. Cai KQ, Albarracin C, Rosen D, et al. Microsatellite instability and alteration of the expression of hMLH1 and hMSH2 in ovarian clear cell carcinoma. Hum Pathol. 2004;35(5):552–9.
38. Gras E, Catasus L, Argueelles R, et al. Microsatellite instability, MLH-1 promoter hypermethylation, and frameshift mutations at coding mononucleotide repeat microsatellites in ovarian tumors. Cancer. 2001;92(11):2829–36.
39. Kuo KT, Mao TL, Jones S, et al. Frequent activating mutations of PIK3CA in ovarian clear cell carcinoma. Am J Pathol. 2009;174(5): 1597–601.
40. Garg K, Shih K, Barakat RR, Zhou Q, Iasonos A, Soslow RA. Endometrial carcinomas in women age 40 years and younger: tumors associated with loss of DNA mismatch repair proteins appear to be a distinct clinicopathologic subset 2009. Am J Surg Pathol. 2009;33(2):1869–77.
41. Wiegand KC, Shah SP, Al-Agha OM, et al. *ARID1A* mutations in endometriosis-associated ovarian carcinomas. N Engl J Med. 2010;363:1532–43.
42. Sangoi AR, Soslow RA, Teng NN, Longacre TA. Ovarian clear cell carcinoma with papillary features: a potential mimic of serous tumor of low malignant potential. Am J Surg Pathol. 2008;32(2):269–74.
43. Silva EG, Young RH. Endometrioid neoplasms with clear cells: a report of 21 cases in which the alteration is not of typical secretory type. Am J Surg Pathol. 2007;31(8):1203–8.
44. Esheba GE, Pate LL, Longacre TA. Oncofetal protein glypican-3 distinguishes yolk sac tumor from clear cell carcinoma of the ovary. Am J Surg Pathol. 2008;32(4):600–7.
45. Cao D, Guo S, Allan RW, Molberg KH, Peng Y. SALL4 is a novel sensitive and specific marker of ovarian primitive germ cell tumors and is particularly useful in distinguishing yolk sac tumor from clear cell carcinoma. Am J Surg Pathol. 2009;33(6):894–904.
46. Ramalingam P, Malpica A, Silva EG, Gershenson DM, Liu JL, Deavers MT. The use of cytokeratin 7 and EMA in differentiating ovarian yolk sac tumors from endometrioid and clear cell carcinomas. Am J Surg Pathol. 2004;28(11):1499–505.
47. Gordon A, Lipton D, Woodruff JD. Dysgerminoma: a review of 158 cases from the Emil Novak Ovarian Tumor Registry. Obstet Gynecol. 1981;58:497–504.
48. Zaloudek CJ, Tavassoli FA, Norris HJ. Dysgerminoma with syncytiotrophoblastic giant cells: a histologically and clinically distinctive subtype of dysgerminoma. Am J Surg Pathol. 1981;5:361–7.
49. Lifschitz-Mercer B, Walt H, Kushnir I, et al. Differentiation potential of ovarian dysgerminoma: an immunohistochemical study of 15 cases. Hum Pathol. 1995;26:62–6.
50. Cheng L, Thomas A, Roth LM, Zheng W, Michael H, Karim FW. OCT4: a novel biomarker for dysgerminoma of the ovary. Am J Surg Pathol. 2004;28(10):1341–6.
51. Young RH, Dickersin GR, Scully RE. Juvenile granulosa cell tumor of the ovary. A clinicopathological analysis of 125 case. Am J Surg Pathol. 1984;8:575–96.
52. Irving JA, Young RH. Granulosa cell tumors of the ovary with a pseudopapillary pattern: a study of 14 cases of an unusual morphologic variant emphasizing their distinction from transitional cell neoplasms and other papillary ovarian tumors. Am J Surg Pathol. 2008;32(4):581–6.
53. Movahedi-Lankarani S, Kurman RJ. Calretinin, a more sensitive but less specific marker than alpha-inhibin for ovarian sex cord-stromal neoplasms – An immunohistochemical study of 215 cases. Am J Surg Pathol. 2002;26(11):1477–83.
54. Deavers MT, Malpica A, Liu J, Broaddus R, Silva EG. Ovarian sex cord-stromal tumors: an immunohistochemical study including a comparison of calretinin and inhibin. Mod Pathol. 2003;16(6):584–90.
55. Gaffey MJ, Mills SE, Boyd JC. Aggressive papillary tumor of middle ear/temporal bone and adnexal papillary cystadenoma: manifestations of von Hippel-Lindau disease. Am J Surg Pathol. 1994;18: 1254–60.
56. Aydin H, Young RH, Ronnett BM, Epstein JI. Clear cell papillary cystadenoma of the epididymis and mesosalpinx: immunohistochemical differentiation from metastatic clear cell renal cell Carcinoma. Am J Surg Pathol. 2005;29(4):520–3.

57. Gersell DJ, King TC. Papillary cystadenoma of the mesosalpinx in von Hippel-Lindau disease. Am J Surg Pathol. 1988;12:145–9.
58. Kiyokawa T, Young RH, Scully RE. Krukenberg tumors of the ovary: a clinicopathologic analysis of 120 cases with emphasis on their variable pathologic manifestations. Am J Surg Pathol. 2006;30(3):277–99.
59. Seidman JD, Kurman RJ, Ronnett BM. Primary and metastatic mucinous adenocarcinomas in the ovaries: incidence in routine practice with a new approach to improve intraoperative diagnosis. Am J Surg Pathol. 2003;27(7):985–93.
60. Khunamornpong S, Suprasert P, Chiangmai WN, Siriaunkgul S. Metastatic tumors to the ovaries: a study of 170 cases in northern Thailand. Int J Gynecol Cancer. 2006;16 Suppl 1:132–8.
61. Lee KR, Young RH. The distinction between primary and metastatic mucinous carcinomas of the ovary: gross and histologic findings in 50 cases. Am J Surg Pathol. 2003;27(3):281–92.
62. Kobel M, Xu H, Bourne PA, et al. IGF2BP3 (IMP3) expression is a marker of unfavorable prognosis in ovarian carcinoma of clear cell subtype. Mod Pathol. 2009;22(3):469–75.
63. Kobel M, Kalloger SE, Santos JL, Huntsman DG, Gilks CB, Swenerton KD. Tumor type and substage predict survival in stage I and II ovarian carcinoma: insights and implications. Gynecol Oncol. 2010;116(1):50–6.
64. Itamochi H, Kigawa J, Terakawa N. Mechanisms of chemoresistance and poor prognosis in ovarian clear cell carcinoma. Cancer Sci. 2008;99(4):653–8.
65. Eltabbakh GH, Mount SL, Beatty B, Simmons-Arnold L, Cooper K. Clinical and molecular differences between clear cell and papillary serous ovarian carcinoma. J Surg Oncol. 2006;93(5):379–86.
66. Nagai Y, Inamine M, Hirakawa M, et al. Postoperative whole abdominal radiotherapy in clear cell adenocarcinoma of the ovary. Gynecol Oncol. 2007;107(3):469–73.
67. Ishikura H, Scully RE. Hepatoid carcinoma of the ovary newly described tumor. Cancer. 1987;60:2775–84.
68. Matsuta M, Ishikura H, Murakami K, Kagabu T, Nishiya I. Hepatoid carcinoma of the ovary: a case report. Int J Gynecol Pathol. 1991;10(3):302–10.
69. Prat J, Bhan AK, Dickersin GR, Robboy SJ, Scully RE. Hepatoid yolk sac tumor of the ovary (endodermal sinus tumor with hepatoid differentiation): a light microscopic, ultrastructural and immunohistochemical study of seven cases. Cancer. 1982;50:2355–68.
70. Young RH, Gersell DJ, Clement PB, Scully RE. Hepatocellular carcinoma metastatic to the ovary: a report of three cases discovered during life with discussion of the differential diagnosis of hepatoid tumors of the ovary. Hum Pathol. 1992;23:574–80.
71. Hayes MC, Scully RE. Ovarian steroid cell tumors (not otherwise specified). A clinicopathological analysis of 63 cases. Am J Surg Pathol. 1987;11:835–45.
72. Kurman RJ, Norris HJ. Mesenchymal tumors of the uterus. VI. Epithelioid smooth muscle tumors including leiomyoblastoma and clear cell leiomyoma: a clinical and pathological analysis of 26 cases. Cancer. 1976;37:1853–65.
73. Vang R, Kempson RL. Perivascular epithelioid cell tumor ('PEComa') of the uterus: a subset of HMB-45-positive epithelioid mesenchymal neoplasms with an uncertain relationship to pure smooth muscle tumors. Am J Surg Pathol. 2002;26(1):1–13.
74. Altrabulsi B, Malpica A, Deavers MT, Bodurka DC, Broaddus R, Silva EG. Undifferentiated carcinoma of the endometrium. Am J Surg Pathol. 2005;29(10):1316–21.
75. Irving JA, Lerwill MF, Young RH. Gastrointestinal stromal tumors metastatic to the ovary: a report of five cases. Am J Surg Pathol. 2005;29(7):920–6.
76. Shih IM, Kurman RJ. Epithelioid trophoblastic tumor – A neoplasm distinct from choriocarcinoma and placental site trophoblastic tumor simulating carcinoma. Am J Surg Pathol. 1998;22:1393–403.
77. Soslow R. Histologic Subtypes of ovarian carcinoma: an overview. Int J Gynecol Pathol. 2008;27:161–74.

Pathology of Mucinous Tumors

9

Jeffrey D. Seidman and Anna Yemelyanova

Keywords
Atypical proliferative mucinous tumor • Mucinous adenocarcinoma • Mucinous borderline tumor • Mucinous cystadenoma • Ovarian neoplasms • Pseudomyxoma peritonei

Mucinous cystadenomas are the most commonly encountered ovarian mucinous neoplasms, whereas atypical proliferative mucinous tumors (APMT, mucinous borderline tumors) and primary ovarian mucinous carcinomas are much less common [1]. Diagnosis of the mucinous cell type in ovarian epithelial tumors is based solely on hematoxylin and eosin-stained sections. Mucin stains do not play a role in classification as they can be positive in other cell types.

Both metastatic mucinous carcinomas and low-grade mucinous tumors of appendiceal origin associated with disseminated peritoneal adenomucinosis (DPAM, the low-grade form of pseudomyxoma peritonei or PMP) can simulate primary ovarian mucinous tumors [1–3]. Refined criteria for the diagnosis of ovarian mucinous tumors in recent years have led to more accurate distinction of these non-ovarian tumors from true primary ovarian mucinous tumors. It is now clear that true primary ovarian mucinous carcinomas are much less common than previously believed, comprising only 2–3% of ovarian carcinomas in North America [4–6] and nearly 6–7% in Asia. Worldwide, however, the reported figure remains over 12%, although there is a downward trend over the last three FIGO annual reports (2001–2006) from 13.8% to 12.4% [7]. It is important to recognize this when evaluating the literature up to and including the 1990s, which undoubtedly includes a significant proportion of misclassified cases [6].

J.D. Seidman (✉)
Department of Pathology, Washington Hospital Center,
110 Irving St. N.W., Washington, DC 20010, USA

Benign Mucinous Tumors

Clinical and Macroscopic Features

The mean patient age for mucinous cystadenoma, which comprises 13% of benign ovarian epithelial neoplasms, is about 50 years. Nearly all are of the intestinal type. In as much as they tend to be large, presenting manifestations often include a pelvic mass and abdominal fullness or distention. These are unilocular or multilocular tumors of variable size, ranging from a few centimeters to over 30 cm, with a mean of about 10 cm. They are typically unilateral (98%). The capsule is thick and white with a smooth outer surface. The cysts contain thick gelatinous or mucinous material. The contents may be brown in areas of hemorrhage or infarction. An associated Brenner tumor may appear as a solid yellowish nodule, and an associated mature cystic teratoma may contain hair or sebaceous material (Table 9.1).

Table 9.1 Uncommon features and associated lesions with mucinous cystadenomas and APMT (mucinous borderline tumors)

Teratoma	5%
Brenner tumor	18%
Mural nodules	1–2%
Adenofibromatous pattern	<4%
Pseudomyxoma ovarii	20%
Seromucinous cell type	10%
Endometriosis	10%

Note: The above percentages refer to cystadenomas; the proportions for APMT/mucinous borderline tumors are lower

R.A. Soslow and C. Tornos (eds.), *Diagnostic Pathology of Ovarian Tumors*,
DOI: 10.1007/978-1-4419-9751-7_9, © Springer Science+Business Media, LLC 2011

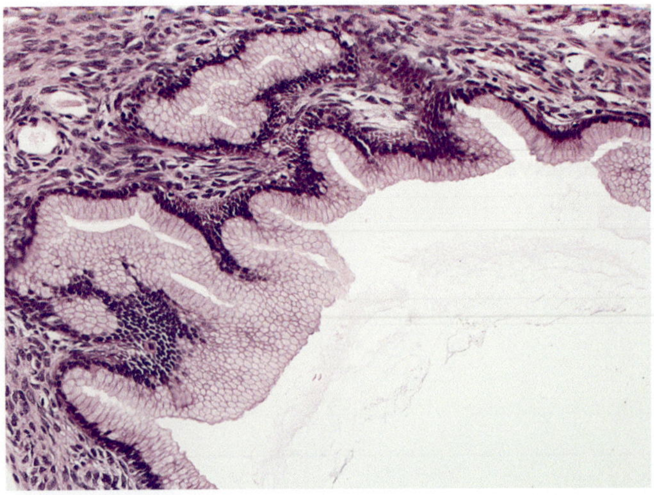

Fig. 9.1 Mucinous cystadenoma. Simple cyst lined by gastrointestinal-type epithelium with abundant mucinous cytoplasm and small bland basally placed nuclei

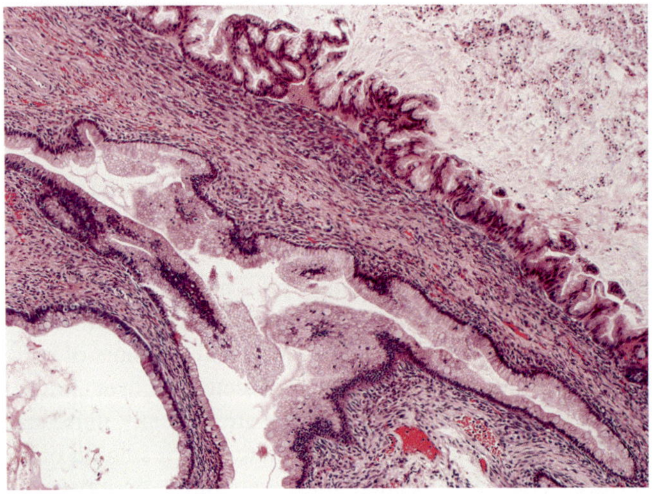

Fig. 9.2 Mucinous cystadenoma with focal epithelial proliferation. Multiloculated cyst with most of the epithelial lining consistent with mucinous cystadenoma. Foci of epithelial proliferation manifested by intraluminal tufting constitute less than 10% of the tumor

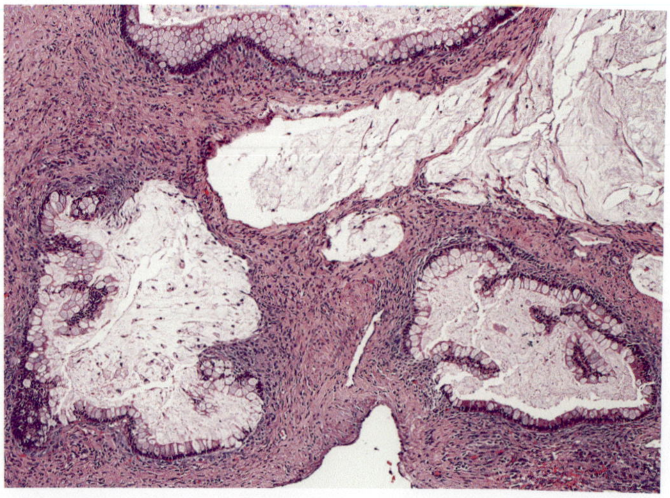

Fig. 9.3 Gland rupture associated with dissecting mucin (pseudomyxoma ovarii) can be seen in primary ovarian mucinous tumors

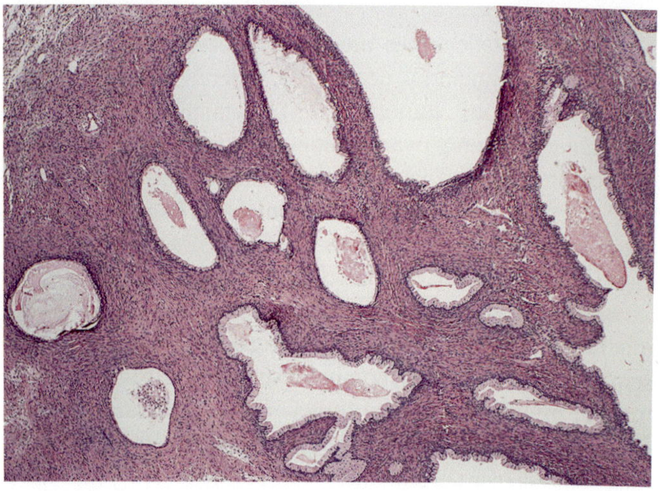

Fig. 9.4 Mucinous adenofibroma. Simple glands lined by bland gastrointestinal-type mucinous epithelium associated with prominent fibrous-stroma

Microscopic Features

The typical appearance is characterized by glands and cysts lined by simple non-stratified mucinous epithelium resembling gastric foveolar-type epithelium or intestinal epithelium containing goblet cells (Fig. 9.1). Epithelial stratification and tufting are generally absent; when such epithelial proliferation is present and resembles APMT, this feature must be limited to less than10% for the tumor to qualify as a cystadenoma. Tumors composed predominantly of cystadenoma with less than 10% of APMT are diagnosed as mucinous cystadenoma with focal proliferation or focal atypia (Fig. 9.2) [1, 8]. The majority of tumors contain calcification, which is usually spiculated and non-psammomatous. Muciphages, pseudoxanthoma cells, and luteinized stromal cells are each present in 40–50%. Dissection of acellular mucin in the tumor stroma, referred to as "pseudomyxoma ovarii," is seen in 10% (Fig. 9.3), and 10% also display multinucleated giant cells which may be associated with mucin granulomas [9]. Rarely, benign mucinous neoplasms are solid and adenofibromatous; these are referred to as mucinous adenofibroma (Fig. 9.4). Up to 18% of mucinous cystadenomas contain transitional cell nests, also referred to as a Brenner tumor component, often in a discrete nodule which is usually interpreted as a concurrent Brenner tumor [9]. When the Brenner component is prominent or intimately admixed with the mucinous component,

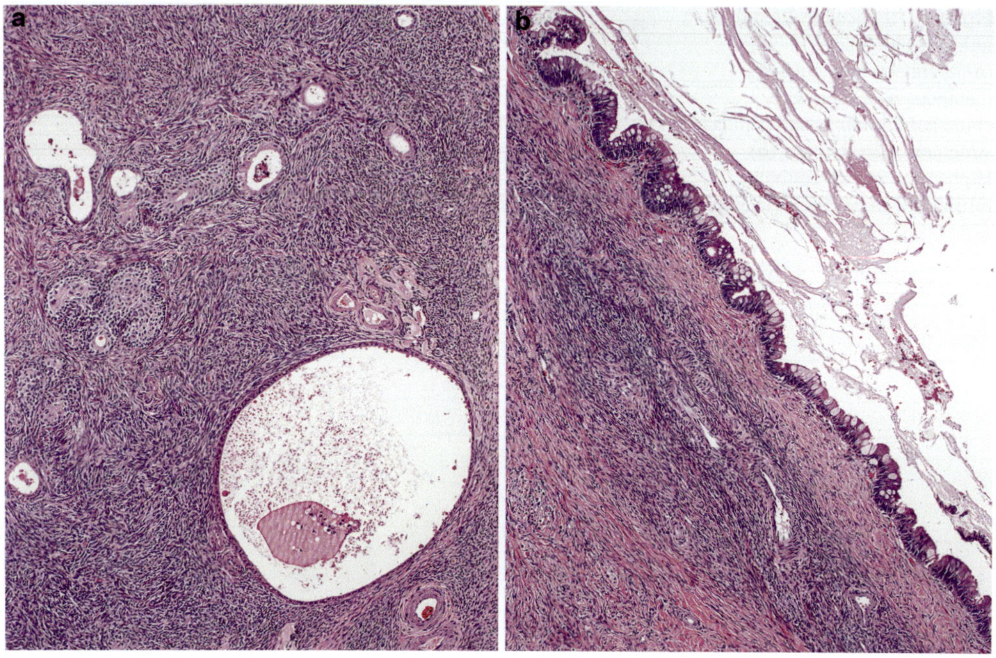

Fig. 9.5 Mixed mucinous cystadenoma/Brenner tumor. (**a**) Brenner tumor component represented by nests of transitional epithelium coexists with a few mucinous glands within ovarian-type stroma. (**b**) Adjacent cystic area of mucinous cystadenoma

the tumor can be classified as metaplastic Brenner tumor or mixed mucinous-Brenner tumor (Fig. 9.5) (see Chap. 10). Rarely, cystadenomas have endocervical-type mucinous epithelium. These are seromucinous cystadenomas (also referred to as Müllerian mucinous or endocervical-like mucinous cystadenomas). They are cystic tumors that are often unilocular and display a papillary architecture in contrast to the predominantly glandular pattern of the intestinal-type mucinous cystadenoma. The epithelium closely resembles endocervical mucinous epithelium with columnar cells having apical mucin and basal nuclei without atypia. Most seromucinous tumors fall into the APMT category [1, 8] (see below).

Clinical Outcomes

Mucinous cystadenomas are benign. If there is a recurrence, the possibility of an extraovarian primary tumor mimicking an ovarian mucinous cystadenoma should be considered.

Atypical Proliferative Mucinous Tumor (APMT) (Mucinous Borderline Tumor)

The terms "atypical proliferative mucinous tumor," "mucinous borderline tumor," and "mucinous tumor of low malignant potential" are synonymous, but as these tumors are not malignant, the term "mucinous tumor of low malignant potential" is not favored. There are two types of APMT: the gastrointestinal type and the seromucinous (endocervical-like or Müllerian) type [1, 8]. Both types are nearly always confined to the ovary. The seromucinous type APMT is grossly, microscopically, and immunophenotypically distinct from the gastrointestinal-type tumor. The modifier "seromucinous" reflects the morphologic and behavioral features that are shared with the serous tumors and that differ from the typical gastrointestinal-type mucinous tumor.

Macroscopic Features

APMT, Intestinal Type (Intestinal Mucinous Borderline Tumor)

This type comprises about 90% of APMT, and is typically large with a smooth white capsule. It is usually multilocular (80%), and nearly always unilateral (>98%) with a mean size of about 18–22 cm [10–14]. The locules of the tumor are usually filled with mucinous material and papillary excrescences may be present. Tumors with microinvasion and/or intraepithelial carcinoma cannot be distinguished grossly from APMT without these features.

APMT, Seromucinous Type (Seromucinous, Müllerian Mucinous, or Mixed Epithelial Borderline Tumor)

In comparison to the intestinal type of APMT, the seromucinous tumors are much less common, smaller (mean of 8 cm), more frequently bilateral (up to 40%), architecturally resemble APST (i.e., often unilocular and have prominent exophytic

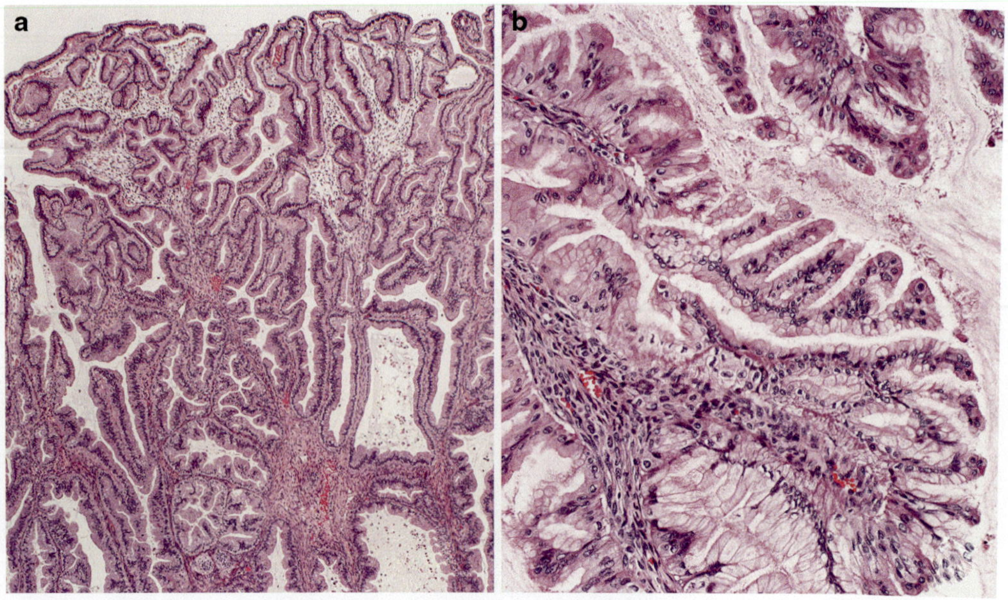

Fig. 9.6 Atypical proliferative mucinous tumor. (**a**) Architecturally complex mucinous neoplasm with epithelial infoldings and papillary structures at low power. (**b**) Higher power with epithelial tufting and some cytologic atypia

and intracystic papillary excrescences), and are much more often associated with endometriosis. The cyst contents are typically yellow and mucoid; those arising within an endometriotic cyst often contain chocolate-colored fluid.

Microscopic Features

APMT, Gastrointestinal Type (Intestinal Mucinous Borderline Tumor)

Microscopically, the glands and cysts are lined by stratified, proliferative gastrointestinal-type mucinous epithelium exhibiting tufting, gland crowding, and villoglandular or papillary intraglandular or cribriform growth [1, 8, 10–14]. Secondary cysts and gland outpouchings may produce a complex architecture. Pseudomyxoma ovarii is present in about 20% [10]. Mild or moderate nuclear atypia is present. If severe cytologic atypia is present, a diagnosis of APMT with intraepithelial carcinoma is warranted (see below). Stromal invasion is absent (Figs. 9.6–9.8).

APMT, Seromucinous Type (APSMT; Seromucinous, Müllerian Mucinous or Mixed Epithelial Borderline Tumor)

These tumors have a papillary architecture with hierarchical branching similar to APST (serous borderline tumor). Over one-third can be seen arising within an endometriotic cyst. They frequently display acute inflammation of the stroma, and have both endocervical mucinous and serous (ciliated) types of epithelium, often admixed with minor components (<10%) of other cell types [14–19]. The cytoplasm may display basophilic mucin or may be eosinophilic (Figs. 9.9 and 9.10).

APMT with Intraepithelial (Noninvasive) Carcinoma (Mucinous Borderline Tumor with Intraepithelial Carcinoma)

APMT that lack invasion but display marked nuclear atypia are considered intraepithelial carcinomas (Figs. 9.11 and 9.12), and are nearly always of the gastrointestinal type [1, 8, 11–14, 19, 20]. In the majority, the intraepithelial carcinoma is focal, usually comprising <5% of the tumor, but on occasion may constitute a much larger area, rarely up to 90% of the tumor [10]. Excessive epithelial stratification and other complex intraglandular growth patterns can be seen in typical APMT and, although seen commonly with intraepithelial carcinoma, should not be used to diagnose intraepithelial carcinoma in the absence of severe atypia [1, 8]. When considering a diagnosis of APMT with intraepithelial carcinoma, the confluent or expansile pattern of invasion should be sought (see below Sect. "Differential Diagnosis").

APMT with Microinvasion (Mucinous Borderline Tumor with Microinvasion)

Microinvasion is seen in about 10% of intestinal type APMT and up to 20% of seromucinous APMT [12, 15, 19, 20]. Microinvasion manifests as either small foci of stromal invasion characterized by single cells, glands or small clusters or nests of mucinous epithelial cells within the stroma, or as small foci of confluent glandular or cribriform growth within the stroma (Figs. 9.13–9.15) [1, 8, 12, 14, 19–21]. These

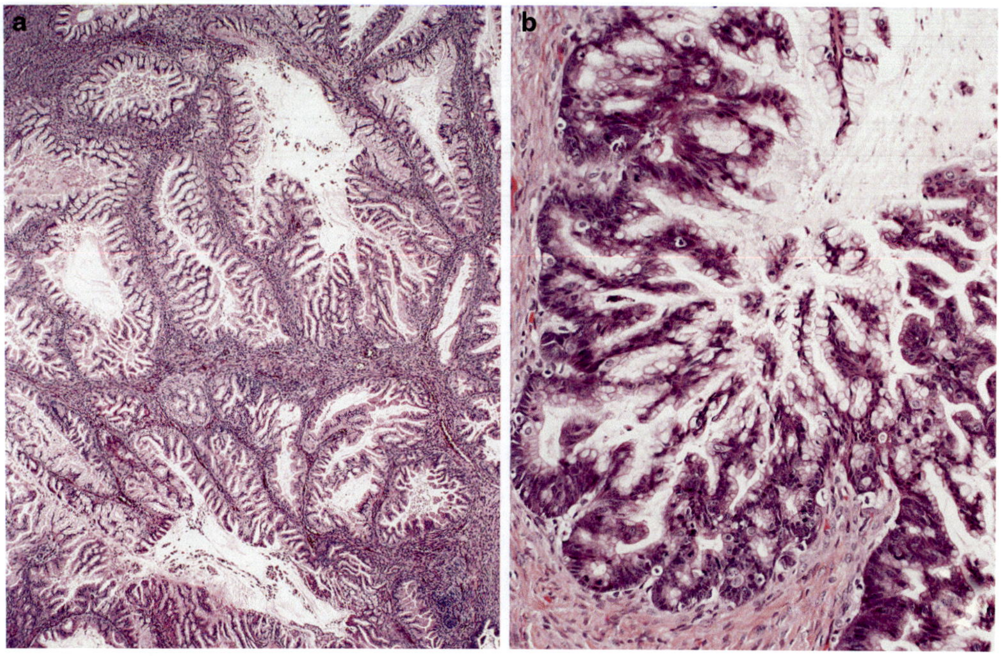

Fig. 9.7 Atypical proliferative mucinous tumor. (**a**) Large glandular spaces with prominent intraglandular epithelial tufting separated by fibrous stroma. (**b**) Tufted mucinous epithelium with epithelial stratification and cytologic atypia

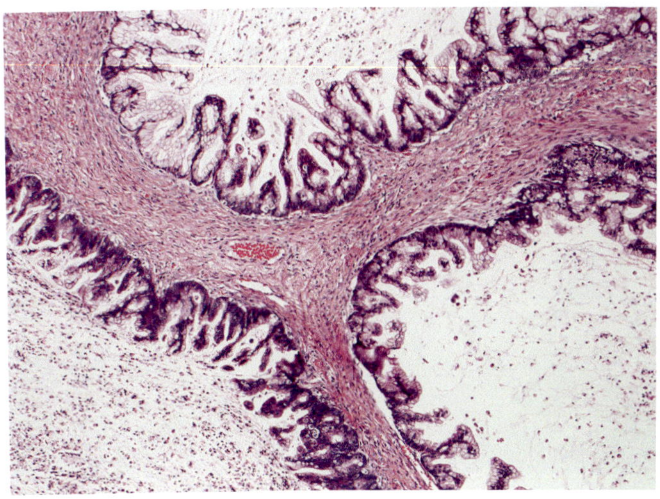

Fig. 9.8 Atypical proliferative mucinous tumor. Mucinous glands with epithelial tufting and stratification evident at medium power

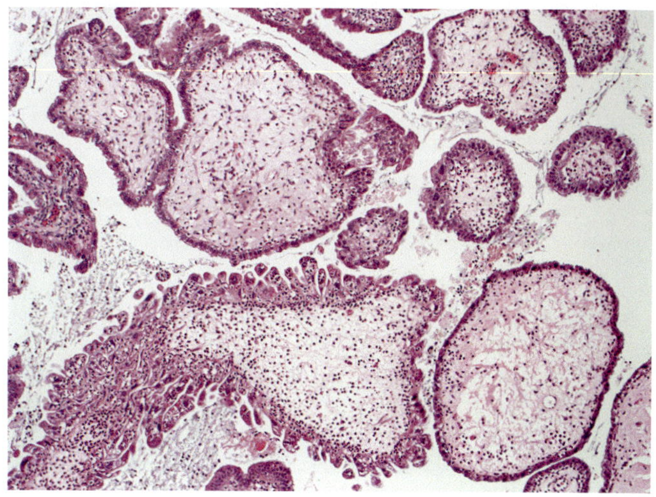

Fig. 9.9 Atypical proliferative seromucinous tumor. Papillary structures lined by eosinophilic epithelium with epithelial proliferation (*lower left*) and acute inflammation

foci are often surrounded by a space which contains extracellular mucin. The size criterion has varied; 5 mm is the preferred upper limit, with no requirement regarding the number of foci allowed [1, 8]. Some tumors with microinvasion also have intraepithelial carcinoma. The possibility of sclerosis and entrapment of glands due to infarction and reparative changes should be considered when diagnosing microinvasion. In addition, neither mucin granulomas [21] nor mucin extravasation in the stroma (pseudomyxoma ovarii) is considered evidence of invasion. However, if there is extensive pseudomyxoma ovarii, the possibility of an appendiceal adenoma with DPAM should be considered (see pseudomyxoma peritonei, later in this chapter).

Associated Peritoneal Lesions

Peritoneal mucinous epithelial lesions associated with intestinal type APMT, previously classified as "implants," are in the vast majority of cases examples of PMP which is of extraovarian origin (see below, and Chap. 11) [1, 3, 8]. True "implants" of APMT have not been described [1, 8]. Rarely, a mucinous cystadenoma or APMT is associated with a localized collection of mucin involving the pelvic

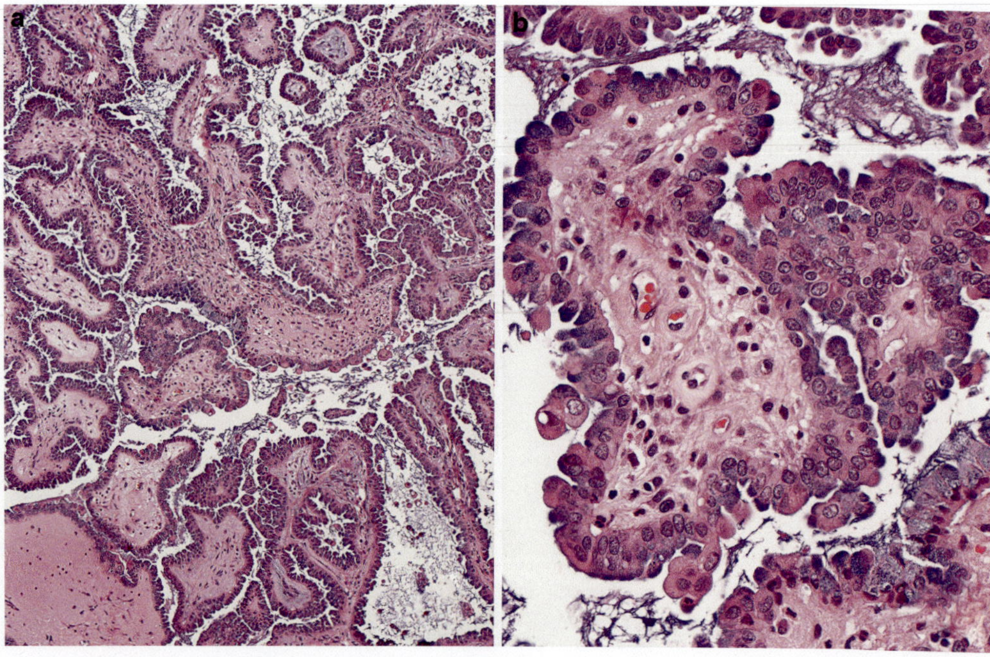

Fig. 9.10 Atypical proliferative seromucinous tumor. (**a**) Complex papillary architecture with detached small cell clusters and individual cells. (**b**) Papillae lined by atypical epithelium containing two cell types: cells with eosinophilic cytoplasm, and cells with basophilic cytoplasm containing mucin

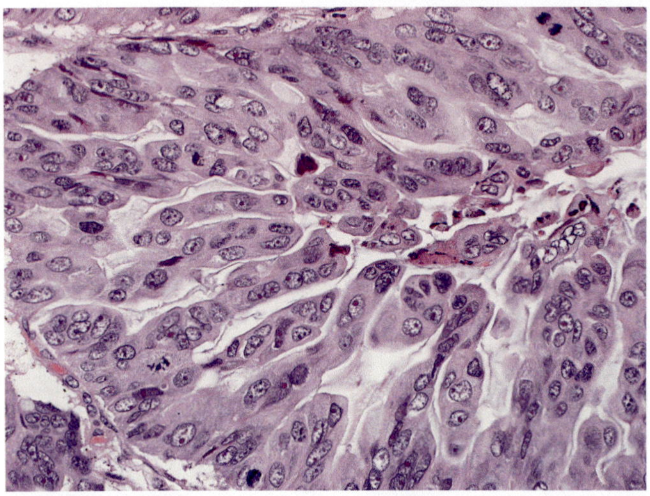

Fig. 9.11 Atypical proliferative mucinous tumor with intraepithelial carcinoma. Epithelial tufts lining a glandular structure composed of cytologically malignant cells with frequent mitotic figures

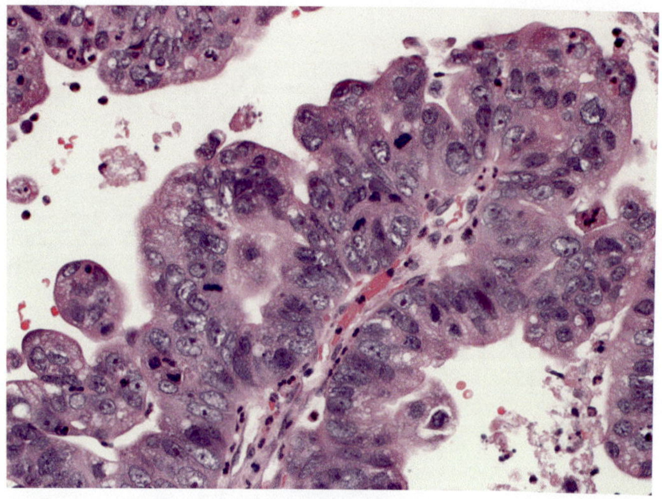

Fig. 9.12 Atypical proliferative mucinous tumor with intraepithelial carcinoma. Tumor papilla lined by markedly atypical cells with frequent mitotic figures

peritoneum, and may even contain a few strips of benign mucinous epithelium. This is probably due to tumor rupture and should not be classified as PMP or APMT with peritoneal implants. Rarely, the seromucinous type of APMT is associated with peritoneal implants resembling those associated with APST (see Chap. 6); more often, peritoneal endometriosis is seen.

Clinical Outcomes

APMTs (MBTs) are generally benign. Reported recurrences, most often in patients with pseudomyxoma peritonei, are now generally regarded as extraovarian primary tumors. Data on APMT/MBT with microinvasion and/or intraepithelial carcinoma are limited. Nearly all reported cases have had a benign behavior. Rare reports of advanced stage and/or fatalities may be due to unsampled foci of more extensive invasion.

Mucinous Tumors with Mural Nodules

Ovarian mucinous neoplasms of all types, but most commonly APMT, can occasionally contain mural nodules. This is the most poorly understood variant of ovarian mucinous neoplasm. True sarcomas, sarcoma-like mural nodules, and

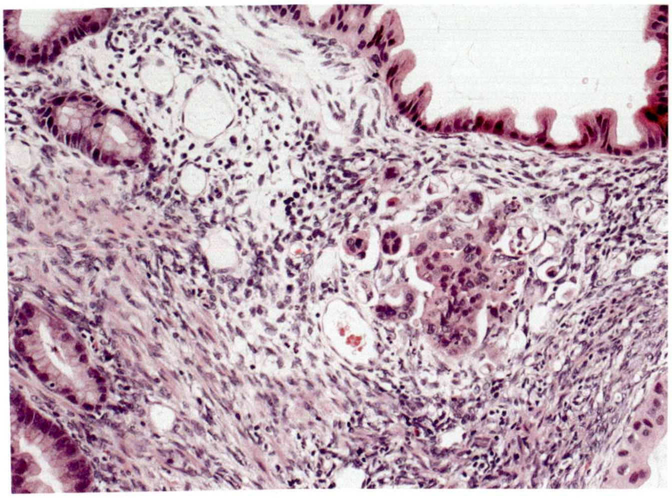

Fig. 9.13 Atypical proliferative mucinous tumor with microinvasion. Focus of single cells with eosinophilic cytoplasm and small cell clusters within the stroma

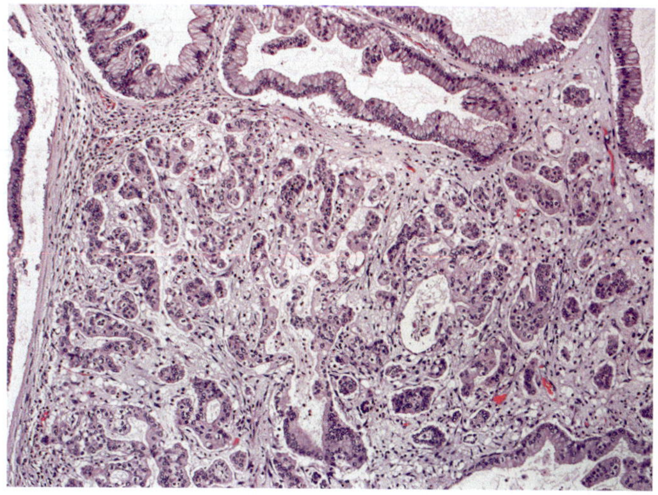

Fig. 9.15 Atypical proliferative mucinous tumor with microinvasion. Larger focus of invasive irregularly-shaped glands and single cells within the stroma, measuring less than 5 mm

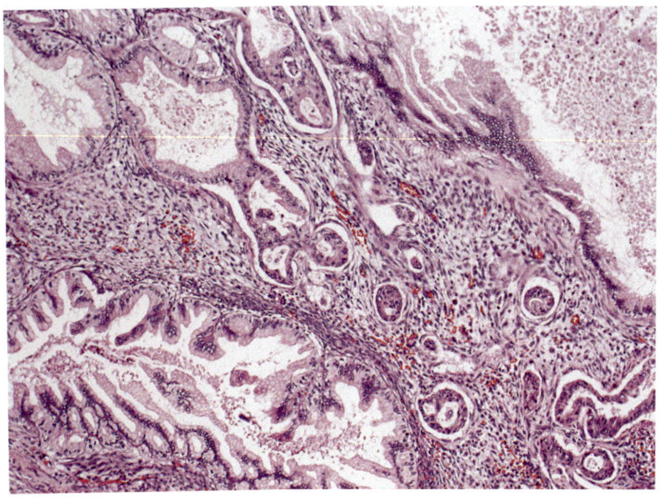

Fig. 9.14 Atypical proliferative mucinous tumor with microinvasion. Individual cells and small irregularly-shaped glands, often surrounded by cleft-like spaces within the stroma

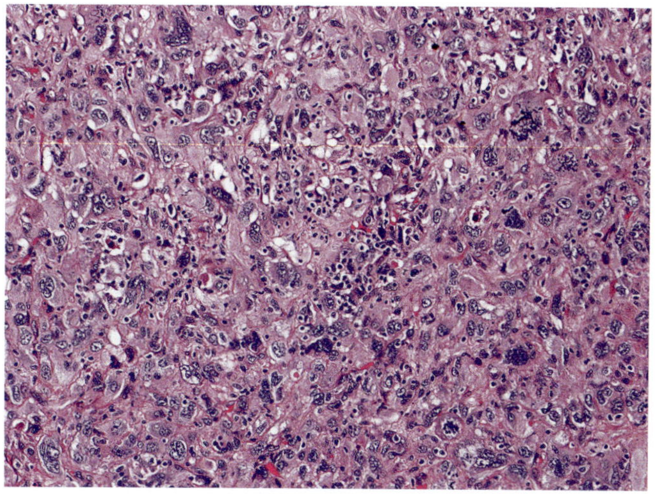

Fig. 9.16 "Sarcoma-like" mural nodule associated with an atypical proliferative mucinous tumor composed of a mixture of markedly pleomorphic spindled and epithelioid cells, epulis-like multinucleated giant cells, and inflammatory cells

foci of anaplastic carcinoma have been described [22, 23]. In general, the sarcomatous nodules and foci of anaplastic carcinoma are grossly characterized by large, poorly circumscribed firm nodules, multiple in almost half the cases, and display a malignant spindle cell population in the former, and evidence of carcinomatous differentiation in the latter. In contrast, sarcoma-like mural nodules are associated with mucinous carcinomas in about half of cases and may coexist with mural nodules of anaplastic carcinoma in 20% [22]. They are characterized by smaller size, sharp demarcation, and a heterogeneous cell population. Anaplastic carcinomatous nodules generally have spindled and rhabdoid cells with unequivocal cytologic features of a high-grade malignancy, and are associated with intraepithelial or invasive mucinous carcinoma in 72% of cases [23]. Sarcomatous nodules may resemble several types of high-grade sarcoma or undifferentiated sarcoma The sarcoma-like mural nodules can have various histologic appearances, including a pleomorphic and epulis-like type with multinucleated giant cells, a pleomorphic and spindle cell type, and a giant cell-histiocytic type [22] (Fig. 9.16). The sarcoma-like nodules can be solitary or multiple and mixtures of histologic types can be encountered, with the pleomorphic and epulis-like type being the most common. Hemorrhage, necrosis, and a mixed inflammatory cell component are commonly seen.

The classification of mucinous tumors with malignant mural nodules is not clear. Some would classify these as variants of mucinous carcinoma (for anaplastic carcinomatous nodules) or carcinosarcoma (for sarcomatous nodules), while others prefer descriptive diagnoses such as "high-grade sarcoma associated with a mucinous adenocarcinoma, see note," or "mural nodule of high-grade sarcoma associated with mucinous adenocarcinoma." Sarcoma-like mural nodules in noninvasive mucinous tumors are believed to be benign, but should be regarded with caution as their histologic appearance is quite worrisome, there is morphologic overlap with true sarcomatous nodules [23], and published data are limited to small numbers of cases with relatively short follow-up.

Invasive Mucinous Carcinomas

Macroscopic Features

Mucinous carcinomas of the intestinal type are relatively rare and comprise only 2–3% of ovarian carcinomas [2, 4–6]. Primary ovarian mucinous carcinomas most often present as large unilateral ovarian masses, usually without evidence of ovarian surface involvement or extraovarian disease, similar to APMT. About 75–80% are stage I. The gross findings are similar to APMT in that they are typically large, unilateral, multicystic mucus-containing tumors with smooth white capsules and have a mean size in the 18–22 cm range. They may contain solid areas and foci of necrosis and hemorrhage.

Microscopic Features

Mucinous carcinomas of intestinal type typically are architecturally well-differentiated and display a variety of glandular patterns, with most tumors having areas of mucinous cystadenoma and/or APMT adjacent to areas of carcinoma [10, 12–14, 20]. The standard definition of mucinous carcinoma requires the presence of destructive stromal invasion by malignant mucinous epithelium (Fig. 9.17). A second pattern of invasion, termed the "confluent glandular" or "expansile" pattern, is more difficult to recognize [10, 24–26]. In this pattern, the glandular epithelium is markedly crowded, with little intervening stroma, and interconnected in a confluent or labyrinthine pattern (Figs. 9.18 and 9.19). About half of primary ovarian mucinous carcinomas exhibit this pattern of invasion [10], and the presence of an infiltrative pattern of stromal invasion should raise concern for metastatic mucinous carcinoma, particularly if noninvasive areas are not seen. Mucinous carcinomas with the expansile pattern of invasion are usually stage I and, with rare exceptions [26], have an excellent prognosis, while those with destructive invasion are more likely to have extraovarian disease. For tumors composed predominantly of APMT, the foci of invasion or confluent growth should measure at least 5 mm (the upper limit for microinvasion) to qualify for the diagnosis of invasive carcinoma [1, 8].

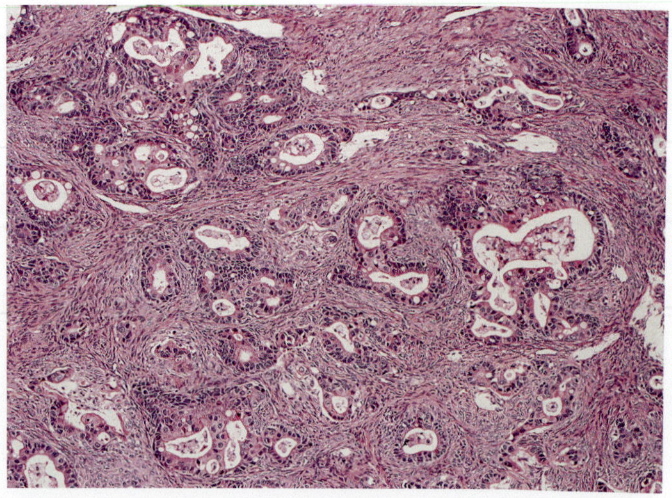

Fig. 9.17 Mucinous carcinoma with destructive stromal invasion. Irregularly-shaped malignant glands infiltrating ovarian stroma. This pattern of invasion is commonly present in metastases to the ovary and warrants immunohistochemical work-up

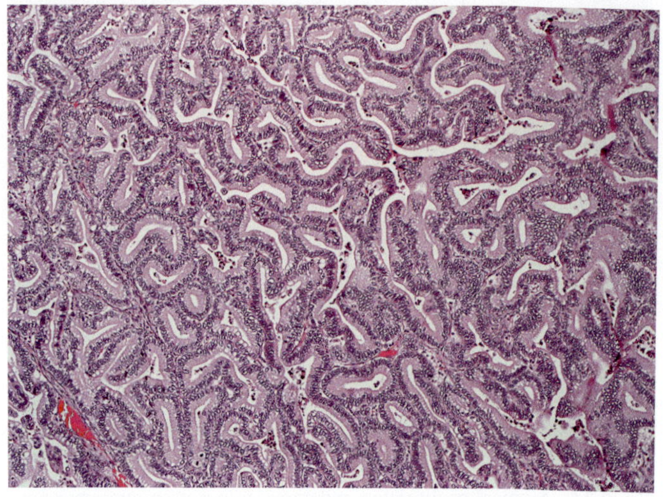

Fig. 9.18 Mucinous carcinoma with confluent glandular pattern of invasion. Back-to-back mucinous glands without intervening stroma are considered a manifestation of stromal invasion

Occasionally mucinous carcinomas arise within mature cystic teratomas [27, 28]. These carcinomas usually display morphologic and immunohistochemical features similar to mucinous adenocarcinoma of lower gastrointestinal tract origin (Fig. 9.20). Occasionally, mucinous cystadenomas resembling appendiceal adenomas can arise in teratomas, and rarely are associated with pseudomyxoma peritonei [28].

Mucinous carcinomas of the seromucinous (Müllerian or endocervical-like) type are quite uncommon [10]. They are

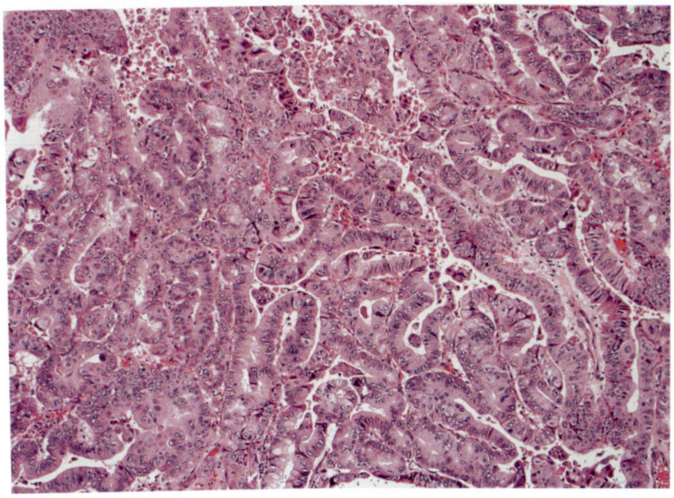

Fig. 9.19 Mucinous carcinoma with confluent-glandular pattern of invasion. Back-to-back glands and ribbons of atypical mucinous epithelium may have a "falling apart" appearance that should not preclude a diagnosis of glandular confluence

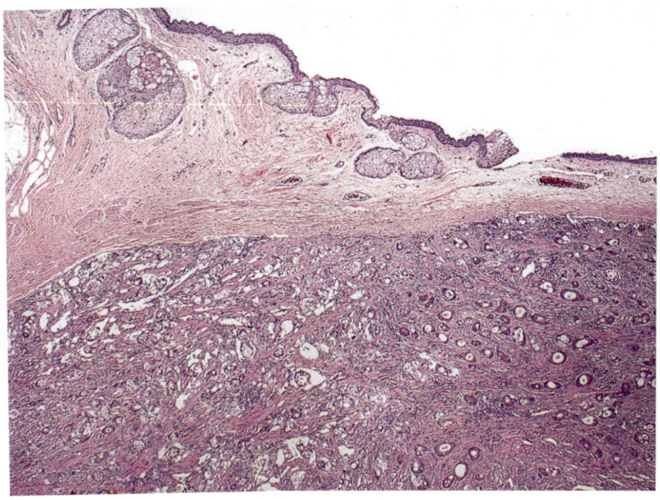

Fig. 9.20 Mucinous carcinoma arising in association with mature cystic teratoma. Benign teratomatous elements consist of skin and sebaceous glands with adjacent invasive mucinous carcinoma

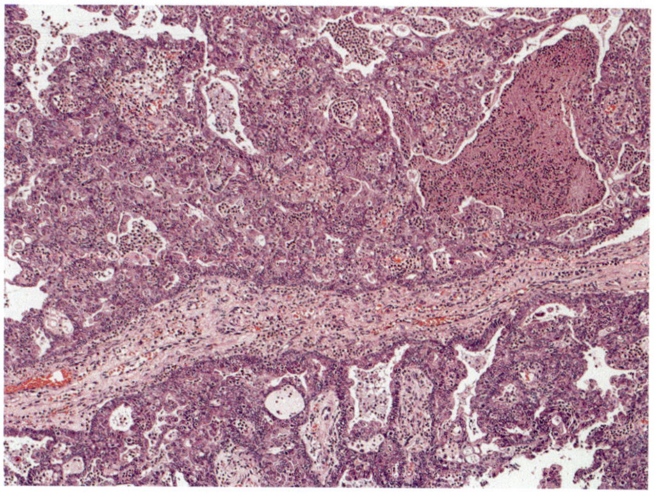

Fig. 9.21 Seromucinous carcinoma. Confluent seromucinous glands with extensive luminal debris and marked inflammation

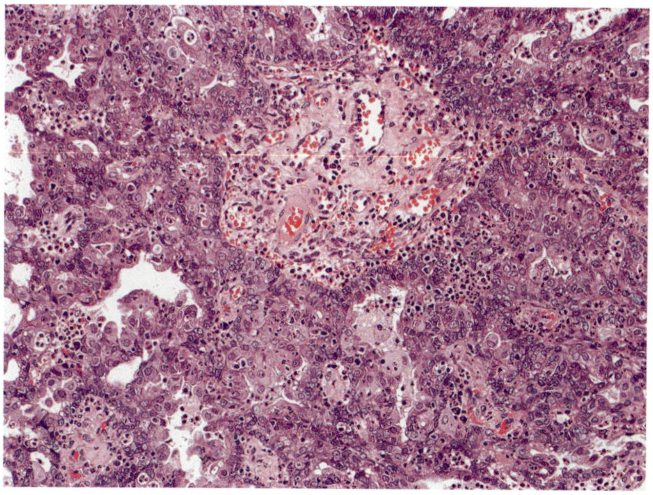

Fig. 9.22 Seromucinous carcinoma. Cribriform glands formed by atypical cells with evidence of serous and mucinous differentiation

clinically and grossly similar to the atypical proliferative tumors of this type. Microscopically, they also resemble APSMT but exhibit destructive stromal invasion or sufficient complex papillary or confluent glandular growth to be classified as carcinoma [15, 18, 19] (Figs. 9.21 and 9.22).

Clinical Outcomes

Like the other cell types of ovarian surface epithelial tumors, stage I mucinous carcinoma has an excellent prognosis with about 90% 5-year survival. Advanced stage mucinous carcinoma documented to be primary in the ovary is very rare, and limited data suggest the behavior may be somewhat worse than serous carcinoma of similar stage. Some studies have shown that the confluent/expansile pattern of invasion is associated with a better prognosis than the infiltrative pattern of invasion.

Immunophenotype

Ovarian mucinous tumors of gastrointestinal type, including cystadenomas, APMTs, and carcinomas, are characterized by diffuse expression of CK7 and variable expression of CK20, with the majority being positive for CK20 but usually with focal staining [29–34] (Fig. 9.23). They can express CDX2 (nuclear staining) [29, 30, 34], exhibit generally

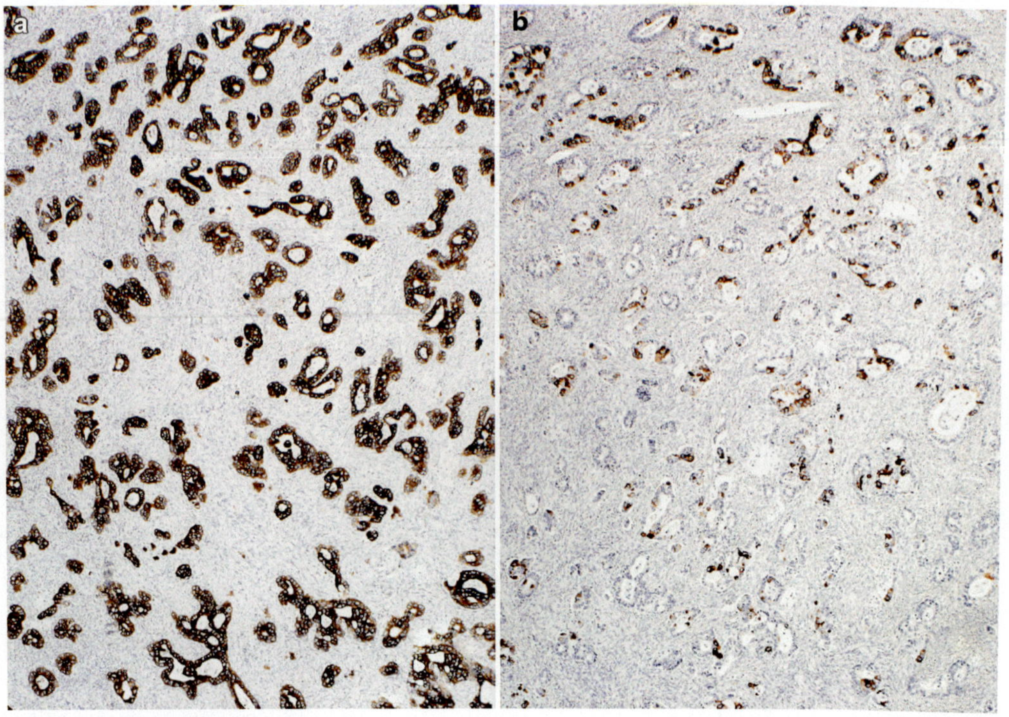

Fig. 9.23 Typical immunohistochemical profile of primary ovarian mucinous carcinoma. Diffuse expression of cytokeratin 7 (**a**) and focal expression of cytokeratin 20 (**b**)

Table 9.2 Key histologic features of primary ovarian mucinous tumors.

	Atypia and stratification	Marked cytologic atypia	Invasion
Cystadenoma	N	N	N
Cystadenoma with focal atypia	Y, <10%	N	N
APMT (MBT)	Y, >10%	N	N
APMT (MBT) with intraepithelial carcinoma	Y	Y	Usually N
APMT (MBT) with microinvasion	Y	+/–	Y, <5 mm
Mucinous carcinoma	Y	Usually Y	Y, >5 mm

APMT atypical proliferative mucinous tumor, *MBT* mucinous borderline tumor

patchy expression of p16 [35], are often positive for DPC4 [32] and are negative for ER, PR [36], WT1, and CA125 [29, 30]. Seromucinous APMTs express CK7, ER, PR, and CA125. CK20, WT1, and CDX2 are negative [29, 30, 33]. Sarcoma-like mural nodules are usually positive for histiocytic markers and vimentin [22]. Staining for cytokeratins is generally weak and focal in sarcoma-like nodules, and strongly positive in anaplastic carcinomatous nodules [23].

Molecular Biology

In primary ovarian mucinous carcinomas, *KRAS* mutation is a very common molecular alteration, occurring in over 75% of ovarian mucinous carcinomas [37–39]. Identical *KRAS* mutations have been detected in mucinous carcinomas and adjacent mucinous cystadenoma and borderline tumor, a molecular finding supporting this morphological continuum of tumor progression in ovarian mucinous neoplasia [40].

The pattern of gene expression of mucinous carcinomas is largely distinguishable from serous, endometrioid, and clear cell carcinomas based on comprehensive gene expression profiling. Ovarian mucinous tumors express several mucin genes (MUC2, MUC3, and MUC17) that are characteristic of mucinous carcinomas irrespective of their tissue origins. A recent report indicates *HER2neu* gene amplification and overexpression in almost 20% of mucinous carcinomas [41].

Differential Diagnosis

The key histologic features of primary ovarian mucinous neoplasms are summarized in Table 9.2. Assessment of the proportion of the tumor that displays atypia, the presence of foci of marked cytologic atypia, and the presence of invasion

Table 9.3 Distinction of primary and metastatic mucinous carcinomas in the ovary

Feature	Primary ovarian mucinous carcinomas	Metastatic mucinous carcinoma
Laterality	Unilateral (>95%)	Commonly bilateral (65–75%)
Size	Large (mean, 18–22 cm)	Typically smaller (mean, 10–12 cm, varies depending on primary site)
Gross features	Multicystic, may have solid areas, smooth capsule	Often nodular with surface involvement but can be multicystic with smooth capsule
Location of tumor	Entirely within stroma; often intracystic	Surface nodules, superficial cortical stroma
Microscopic features	Morphologically noninvasive areas with well-differentiated mucinous epithelium forming organized cysts with peripheral crypts; invasive areas with well differentiated mucinous epithelium most commonly with either confluent glandular/cribriform pattern or destructive infiltrative pattern	Typically infiltrative and haphazard pattern of mucinous glands, often with nodules throughout or in superficial cortex, but can have confluent glandular and "borderline-like" patterns; nuclear atypia is often greater than expected for the low-grade architecture
Extraovarian disease	Usually absent	Often present

and its extent, allow the discrimination of mucinous neoplasms into cystadenomas with or without focal atypia, APMT with or without microinvasion and/or intraepithelial carcinoma, and mucinous adenocarcinoma.

It is critical when evaluating a mucinous carcinoma involving the ovary to consider the possibility of an extraovarian primary [1, 2] (see also Chap. 11). Metastatic mucinous carcinomas usually exhibit at least two of the following features: bilaterality, small size (typically less than 10–12 cm), ovarian surface involvement, a nodular pattern of ovarian involvement, and a haphazard infiltrative pattern of stromal invasion (Table 9.3) [35]. Other tip-offs to a likely metastatic tumor include a degree of cytologic atypia out of proportion to the degree of epithelial proliferation (i.e., cytologic features of malignancy in an architecturally simple mucinous cyst), signet ring cells, pseudomyxoma ovarii, and mucinous carcinomatosis.

An algorithm has been proposed to assist in the distinction of primary versus metastatic mucinous carcinoma (Table 11.2), particularly at the time of intraoperative evaluation when only limited histologic material can be examined. As originally proposed, bilateral mucinous carcinomas and unilateral ones smaller than 10 cm are metastatic, and unilateral ones larger than 10 cm are primary [2]. This algorithm correctly classified 90% of mucinous carcinomas. Subsequent studies have confirmed and refined the algorithm, also with close to 90% accuracy, and suggest that 12 or 13 cm is a better cutoff than 10 cm [42]. Recent studies have shown that the algorithm performs as well for APMTs and endometrioid carcinomas [42, 43]. As such measurements are rarely exact and may be performed after the tumor has been incised and deflated, 10–15 cm might be considered a size range of lesser predictive value [43].

Immunohistochemistry can be of great value in the distinction of primary versus metastatic mucinous carcinoma. This is further discussed in Chap. 11.

Diagnostic pitfalls are highlighted in Table 9.4.

1. *Mucinous cystadenoma versus APMT (mucinous borderline tumor)*

   Focal cytologic atypia and epithelial stratification and proliferation resembling APMT may occur in mucinous cystadenoma. If these findings exceed 10% of the tumor, a diagnosis of APMT is warranted. If less than 10%, the term mucinous cystadenoma with focal atypia or focal proliferation may be used. Note: Both gross and microscopic extent should be evaluated for the 10% determination.
2. *APMT (mucinous borderline tumor) versus mucinous carcinoma*

   Destructive infiltrative growth defines mucinous carcinoma, and this must exceed 5 mm in diameter (the cutoff for microinvasion; see below) in at least one area. A less obvious form of invasion, the expansile or confluent pattern, is diagnosed when extensive back to back or complex papillary or cribriform growth occurs; this must also exceed 5 mm.
3. *APMT (mucinous borderline tumor) versus APMT with intraepithelial carcinoma (mucinous borderline tumor with intraepithelial carcinoma)*

   APMT displays cytologic atypia which may be focally marked. If there are unequivocal cytologic features of carcinoma, but invasion is not demonstrated, a diagnosis of APMT with intraepithelial carcinoma is warranted.
4. *APMT (mucinous borderline tumor) versus APMT with microinvasion (mucinous borderline tumor with microinvasion)*

   Microinvasion is characterized by single cells and cell clusters with an infiltrative pattern, which seem to bud off from the base of the atypical epithelium. Each focus should measure less than 5 mm; multiple foci are permitted. Dissection of extracellular mucin through the ovarian/tumor stroma (pseudomyxoma ovarii) does not constitute microinvasion regardless of the features of the cells floating in the mucin. An infiltrative pattern of atypical cells within a solid area of tumor stroma is required for a diagnosis of microinvasion. Single cells and small, often angulated, nests of atypical cells are usually present.

Table 9.4 Diagnostic pitfalls

1. Intracytoplasmic mucin in greater than 50% of tumor cells is needed to qualify for mucinous differentiation
2. Limited foci of atypia or proliferation should not lead to a diagnosis of APMT. If less than 10%, diagnose mucinous cystadenoma with focal atypia (or focal proliferation)
3. In the absence of atypia, back-to-back glands with preservation of scanty intervening stroma can be seen in mucinous cystadenoma
4. If microinvasion and/or intraepithelial carcinoma is present, sample the tumor generously and look carefully for more extensive invasion, particularly the confluent/expansile pattern
5. If mucinous carcinoma or peritoneal mucinous carcinomatosis is present, an extraovarian primary is more likely than a primary ovarian tumor
6. If pseudomyxoma peritonei is present, a primary in the appendix is most likely
7. Bilateral intestinal type mucinous cystadenomas are very rare; consider (a) a metastatic tumor which may display bland areas; (b) a seromucinous cystadenoma; (c) additional sampling
8. If it is a primary ovarian mucinous tumor and endometriosis and/or bilaterality is present, the seromucinous type is much more likely than the intestinal type
9. A very simple adenomatous architecture with marked cytologic atypia, even if the atypia is focal, is a clue to consider metastatic carcinoma
10. At the time of intraoperative consultation and/or prior to rendering a final diagnosis, speak directly to the surgeon and obtain as much clinical information as possible, particularly regarding any history of carcinoma, the appearance of the appendix, and any other abnormalities found intraoperatively
11. Never hesitate to examine additional sections, particularly if the tumor is large
12. If mucinous carcinoma comprises 7–8% or more of ovarian carcinomas in your practice, you may be overlooking some metastases or over-diagnosing intraepithelial carcinomas as invasive (exception: cases referred specifically for pathology consultation; do not count APMT with microinvasion as ovarian carcinoma)

5. *Primary mucinous carcinoma versus metastatic mucinous carcinoma*

   It is essential to have complete clinical information. Mucinous carcinomatosis is generally of extraovarian origin. Metastases to the ovaries are generally bilateral, small (<10–12 cm), have a nodular pattern of cortical involvement, and often display ovarian surface involvement. Primary ovarian mucinous carcinomas are generally large (>15 cm), unilateral, and display extensive areas of mucinous cystadenoma and APMT. Immunohistochemistry can be of great value and is further discussed in Chap. 11.
6. *Intestinal-type APMT (mucinous borderline tumor) versus seromucinous type APMT (seromucinous borderline tumor)*

   About 90% of APMT are intestinal type. These typically display either goblet cells or gastric type foveolar epithelium, lack significant acute inflammation, and only rarely are associated with endometriosis. The architecture is glandular, cystic, and multilocular; when papillary, a villoglandular pattern is common. The seromucinous type of APMT is most often unilocular, commonly associated with endometriosis, usually displays significant acute inflammation, and has a papillary architecture resembling APST.
7. *APMT versus low-grade appendiceal neoplasm with PMP*

   When involving the ovary, DPAM most often mimics APMT of the intestinal type. Characteristic features of PMP in the ovary include bilaterality, a right-sided predominance when unilateral, extensive pseuxomyxoma ovarii, and ovarian surface involvement. It is important to be aware of the clinical history and extraovarian findings, particularly in the appendix and remainder of the GI/pancreaticobiliary tract.

Metastatic Mucinous Carcinomas and Pseudomyxoma Peritonei

Metastatic Mucinous Carcinoma

Some metastatic mucinous carcinomas, especially those derived from the colorectum, pancreaticobiliary tract, appendix, and endocervix, can exhibit gross and microscopic features simulating a primary ovarian mucinous tumor. In particular, metastases can be large, unilateral (thus misclassified by the algorithm discussed earlier), and multicystic, and can display deceptive patterns of ovarian involvement. If these patterns are not recognized, these metastatic mucinous carcinomas can be misinterpreted as primary ovarian APMT with intraepithelial carcinoma or well-differentiated mucinous carcinoma. Occasionally some of these metastases display low-grade or well-differentiated areas adjacent to areas of invasion, simulating benign and APMT precursor lesions. When mucinous carcinomas in the ovaries are rigorously classified, metastatic mucinous carcinomas are much more commonly encountered than primary ovarian mucinous carcinoma by a factor of about 3:1 [2]. In general, the presence of a mucinous carcinoma in the ovary, particularly if extraovarian disease is present, should always prompt the pathologist and surgeon to consider the possibility of metastatic mucinous carcinoma. Immunohistochemical analysis can be useful for identifying some metastatic mucinous carcinoma that simulates primary ovarian mucinous tumors. Clinical evaluation is usually required to exclude an extraovarian source. Metastatic mucinous carcinomas are discussed in more detail in Chap. 11.

Mucinous Tumors Associated with Pseudomyxoma Peritonei (PMP)

PMP has historically referred to the presence of mucinous ascites or mucoid nodules adherent to peritoneal surfaces and has not had a consistently applied histopathological description [3]. Morphologic, immunohistochemical, and molecular genetic studies have provided strong evidence that nearly all cases of PMP are derived from low-grade mucinous adenomas, usually of the appendix, and that the ovarian involvement is secondary. On very rare occasion, PMP may arise in an intestinal component of a mature cystic teratoma [27, 28].

The terms "disseminated peritoneal adenomucinosis" (DPAM) and "involvement by low-grade appendiceal mucinous neoplasm" are now recommended as diagnostic terms for these low-grade mucinous tumors. The term PMP is restricted to use as a clinical descriptor for the syndrome and is maintained as a general term for historical continuity. These entities are further discussed in Chap. 11.

References

1. Ronnett BM, Kajdacsy-Balla A, Gilks CB, et al. Mucinous borderline ovarian tumors: points of general agreement and persistent controversies regarding nomenclature, diagnostic criteria, and behavior. Hum Pathol. 2004;35:949–60.
2. Seidman JD, Kurman RJ, Ronnett BM. Primary and metastatic mucinous adenocarcinomas in the ovaries: incidence in routine practice with a new approach to improve intraoperative diagnosis. Am J Surg Pathol. 2003;27:985–93.
3. Ronnett BM, Zahn CM, Kurman RJ, et al. Disseminated peritoneal adenomucinosis and peritoneal mucinous carcinomatosis: a clinicopathologic analysis of 109 cases with emphasis on distinguishing pathologic features, site of origin, prognosis, and relationship to "pseudomyxoma peritonei". Am J Surg Pathol. 1995;19:1390–408.
4. Seidman JD, Horkayne-Szakaly I, Haiba M, Boice CR, Kurman RJ, Ronnett BM. The histologic type and stage distribution of ovarian carcinomas of surface epithelial origin. Int J Gynecol Pathol. 2004;23:41–4.
5. Gilks CB, Kobel M, Kalloger SE, et al. Significant differences in tumor cell type in early versus advanced stage ovarian carcinoma. Mod Pathol. 2009;22 Suppl 1:215A.
6. Gilks CB. Subclassification of ovarian surface epithelial tumors based on correlation of histologic and molecular pathologic data. Int J Gynecol Pathol. 2004;23:200–5.
7. Heintz AP, Odicino F, Maisonneuve P, et al. Carcinoma of the ovary. In: FIGO 26th annual report on the results of treatment in gynecological cancer. Int J Gynaecol Obstet. 2006;95 suppl 1:S161–92.
8. Seidman JD, Soslow RA, Vang R, et al. Borderline ovarian tumors: diverse contemporary viewpoints on terminology and diagnostic criteria with illustrative images. Hum Pathol. 2004;35:918–33.
9. Seidman JD, Khedmati F. Exploring the histogenesis of ovarian mucinous and transitional cell (Brenner) tumors: a study of 120 tumors. Arch Pathol Lab Med. 2008;132:1753–60.
10. Lee KR, Scully RE. Mucinous tumors of the ovary: a clinicopathologic study of 196 borderline tumors (of intestinal type) and carcinomas, including an evaluation of 11 cases with 'pseudomyxoma peritonei'. Am J Surg Pathol. 2000;24:1447–64.
11. Nomura K, Aizawa S, Hano H. Ovarian mucinous borderline tumors of intestinal type without intraepithelial carcinoma: are they still tumors of low malignant potential? Pathol Int. 2004; 54:420–4.
12. Riopel MA, Ronnett BM, Kurman RJ. Evaluation of diagnostic criteria and behavior of ovarian intestinal- type mucinous tumors: atypical proliferative (borderline) tumors and intraepithelial, microinvasive, invasive, and metastatic carcinomas. Am J Surg Pathol. 1999;23:617–35.
13. Rodriguez IM, Prat J. Mucinous tumors of the ovary: a clinicopathologic analysis of 75 borderline tumors (of intestinal type) and carcinomas. Am J Surg Pathol. 2002;26:139–52.
14. Hart WR. Mucinous tumors of the ovary: a review. Int J Gynecol Pathol. 2005;24:4–25.
15. Shappell HW, Riopel MA, Smith-Sehdev AE, Ronnett BM, Kurman RJ. Diagnostic criteria and behavior of ovarian seromucinous (endocervical-type mucinous and mixed cell-type) tumors: atypical proliferative (borderline) tumors, intraepithelial, microinvasive, and invasive carcinoma. Am J Surg Pathol. 2002;26: 1529–41.
16. Rodriguez IM, Irving JA, Prat J. Endocervical-like mucinous borderline tumors of the ovary: a clinicopathologic analysis of 31 cases. Am J Surg Pathol. 2004;28:1311–8.
17. Nagai Y, Kishimoto T, Nikaido T, et al. Squamous predominance in mixed- epithelial papillary cystadenomas of borderline malignancy of mullerian type arising in endometriotic cysts: a study of four cases. Am J Surg Pathol. 2003;27:242–7.
18. Lee KR, Nucci MR. Ovarian mucinous and mixed epithelial carcinomas of mullerian (endocervical-like) type: a clinicopathologic analysis of four cases of an uncommon variant associated with endometriosis. Int J Gynecol Pathol. 2003;22:42–51.
19. Dube V, Roy M, Plante M, et al. Mucinous ovarian tumors of mullerian-type: an analysis of 17 cases including borderline tumors and intraepithelial, microinvasive, and invasive carcinomas. Int J Gynecol Pathol. 2005;24:138–46.
20. Nomura K, Aizawa S. Noninvasive, microinvasive, and invasive mucinous carcinomas of the ovary: a clinicopathologic analysis of 40 cases. Cancer. 2000;89:1541–6.
21. Kim KR, Lee HI, Lee SK, et al. Is stromal microinvasion in primary mucinous ovarian tumors with "mucin granuloma" true invasion? Am J Surg Pathol. 2007;31:546–54.
22. Bague S, Rodriguez IM, Prat J. Sarcoma-like mural nodules in mucinous cystic tumors of the ovary revisited: a clinicopathologic analysis of 10 additional cases. Am J Surg Pathol. 2002;26:1467–76.
23. Provenza C, Young RH, Prat J. Anaplastic carcinoma in mucinous ovarian tumors: a clinicopathologic study of 34 cases emphasizing the crucial impact of stage on prognosis, their histologic spectrum, and overlap with sarcomalike mural nodules. Am J Surg Pathol. 2008;32:383–9.
24. Leitao MM, Boyd J, Hummer A, et al. Clinicopathologic analysis of early-stage sporadic ovarian carcinoma. Am J Surg Pathol. 2004;28:147–59.
25. Chen S, Leitao MM, Tornos C, Soslow RA. Invasion patterns in stage I endometrioid and mucinous ovarian carcinomas: a clinicopathologic analysis emphasizing favorable outcomes in carcinomas without destructive stromal invasion and the occasional malignant course of carcinomas with limited destructive stromal invasion. Mod Pathol. 2005;18:903–11.
26. Ludwick C, Gilks CB, Miller D, et al. Aggressive behavior of stage I ovarian mucinous tumors lacking extensive infiltrative invasion: a

report of four cases and review of the literature. Int J Gynecol Pathol. 2005;24:205–17.
27. McKenney JK, Soslow RA, Longacre TA. Ovarian mature teratomas with mucinous epithelial neoplasms: morphologic heterogeneity and association with pseudomyxoma peritonei. Am J Surg Pathol. 2008;32:645–55.
28. Vang R, Gown AM, Zhao C, et al. Ovarian mucinous tumors associated with mature cystic teratomas: morphologic and immunohistochemical analysis identifies a subset of potential teratomatous origin that shares features of lower gastrointestinal tract mucinous tumors more commonly encountered as secondary tumors in the ovary. Am J Surg Pathol. 2007;31:854–69.
29. McCluggage WG. Immunohistochemical and functional biomarkers of value in female genital tract lesions. Int J Gynecol Pathol. 2006;25:101–20.
30. Baker PM, Oliva E. Immunohistochemistry as a tool in the differential diagnosis of ovarian tumors: an update. Int J Gynecol Pathol. 2005;24:39–55.
31. Vang R, Gown AM, Barry TS, et al. Cytokeratins 7 and 20 in primary and secondary mucinous tumors of the ovary: analysis of coordinate immunohistochemical expression profiles and staining distribution in 179 cases. Am J Surg Pathol. 2006;30:1130–9.
32. Ji H, Isacson C, Seidman JD, et al. Cytokeratins 7 and 20, Dpc4, and MUC5AC in the distinction of metastatic mucinous carcinomas in the ovary from primary ovarian mucinous tumors: Dpc4 assists in identifying metastatic pancreatic carcinomas. Int J Gynecol Pathol. 2002;21:391–400.
33. Vang R, Gown AM, Barry TS, et al. Ovarian atypical proliferative (borderline) mucinous tumors: gastrointestinal and seromucinous (endocervical-like) types are immunophenotypically distinctive. Int J Gynecol Pathol. 2006;25:83–9.
34. Vang R, Gown AM, Wu LS, et al. Immunohistochemical expression of CDX2 in primary ovarian mucinous tumors and metastatic mucinous carcinomas involving the ovary: comparison with CK20 and correlation with coordinate expression of CK7. Mod Pathol. 2006;19:1421–8.
35. Vang R, Gown AM, Farinola M, et al. p16 expression in primary ovarian mucinous and endometrioid tumors and metastatic adenocarcinomas in the ovary: utility for identification of metastatic HPV-related endocervical adenocarcinomas. Am J Surg Pathol. 2007;31:653–63.
36. Vang R, Gown AM, Barry TS, et al. Immunohistochemistry for estrogen and progesterone receptors in the distinction of primary and metastatic mucinous tumors in the ovary: an analysis of 124 cases. Mod Pathol. 2006;19:97–105.
37. Cuatrecasas M, Villanueva A, Matias-Guiu X, Prat J. K-ras mutations in mucinous ovarian tumors: a clinicopathologic and molecular study of 95 cases. Cancer. 1997;79:1581–6.
38. Gemignani ML, Schlaerth AC, Bogomolniy F, Barakat RR, Lin O, Soslow R, et al. Role of KRAS and BRAF gene mutations in mucinous ovarian carcinoma. Gynecol Oncol. 2003;90:378–81.
39. Mayr D, Hirschmann A, Lohrs U, Diebold J. KRAS and BRAF mutations in ovarian tumors: a comprehensive study of invasive carcinomas, borderline tumors and extraovarian implants. Gynecol Oncol. 2006;103:883–7.
40. Kurman RJ, Shih IeM. Pathogenesis of ovarian cancer: lessons from morphology and molecular biology and their clinical implications. Int J Gynecol Pathol. 2008;27:151–60.
41. McAlpine JN, Wiegand KC, Vang R, et al. HER2 overexpression and amplification is present in a subset of ovarian mucinous carcinomas and can be targeted with trastuzumab therapy. BMC Cancer. 2009;9:433.
42. Yemelyanova AV, Vang R, Judson K, et al. Distinction of primary and metastatic mucinous tumors involving the ovary: analysis of size and laterality data by primary site with reevaluation of an algorithm for tumor classification. Am J Surg Pathol. 2008;32:128–38.
43. Khunamornpong S, Suprasert P, Pojchamarnwiputh S, et al. Primary and metastatic mucinous adenocarcinomas of the ovary: evaluation of the diagnostic approach using tumor size and laterality. Gynecol Oncol. 2006;101:152–7.

Uncommon Epithelial Ovarian Tumors

10

Jeffrey D. Seidman and Anna Yemelyanova

Keywords

Atypical proliferative Brenner tumor • Malignant transitional cell tumor • Malignant Brenner tumor • Transitional cell carcinoma • Squamous tumor

Brenner (Transitional Cell) Tumors

Transitional cell tumors comprise 3% of ovarian epithelial tumors. Nearly all of these are benign Brenner tumors; atypical proliferative (borderline) and malignant forms are very uncommon. The transitional epithelial cell type, characterized by a uniform population of stratified cells with ovoid nuclei displaying nuclear grooves, is named because of its resemblance to urothelium. Evidence suggests that benign Brenner tumors have some true urothelial differentiation, while transitional cell carcinoma (unassociated with a benign Brenner component) is a variant of high-grade serous carcinoma. Peritoneal Walthard cell nests also display transitional cell differentiation.

Brenner (Benign Transitional Cell) Tumors

Clinical Features

The mean patient age is 50–56 years. As these tumors tend to be small, they are usually incidental findings and unrelated to clinical symptoms. Uncommon large examples may present with signs and symptoms of a pelvic mass.

J.D. Seidman (✉)
Department of Pathology, Washington Hospital Center,
110 Irving St. N.W., Washington, DC, 20010, USA

Macroscopic Features

Brenner tumors are common incidental findings and are often microscopic in size. Most tumors are 2 cm or smaller, and although a 1 cm minimum size is used for the other cell types of ovarian epithelial tumors, this criterion has not been applied to Brenner tumors, which frequently measure only a few millimeters. Multiple microscopic tumors are seen in 17% of patients. Occasionally, they are large and may exceed 10 cm. These tumors are well-circumscribed, firm, and rubbery, with a smooth or bosselated outer surface. The cut surfaces are typically solid and fibrous, usually white or yellow, and may be whorled or lobulated; occasionally, a cystic component is present. Small tumors are usually close to the hilus or ovarian surface.

Microscopic Features

The characteristic microscopic feature is sharply demarcated epithelial nests in a dense fibrous stroma (Fig. 10.1). The epithelial cells are uniform in size, with prominent cell borders and pale to eosinophilic cytoplasm. The nuclei are oval, often with small nucleoli, and longitudinal grooves are present (Fig. 10.2). Atypia and mitotic activity are generally absent [1]. The nests often become cystic and contain eosinophilic debris or mucin. In 30%, metaplastic mucinous epithelium lines the cystic nests (see Chap. 9, Fig. 9.5); when extensive, some investigators have referred to these as "metaplastic Brenner tumors" [2]. The metaplastic mucinous component may occasionally form the dominant part of the tumor and

R.A. Soslow and C. Tornos (eds.), *Diagnostic Pathology of Ovarian Tumors*,
DOI: 10.1007/978-1-4419-9751-7_10, © Springer Science+Business Media, LLC 2011

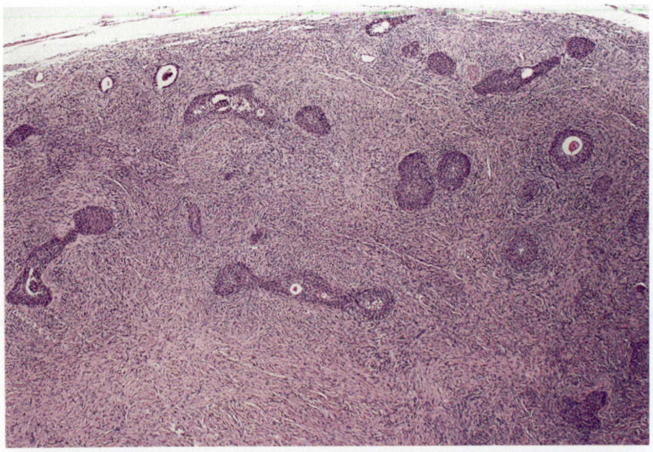

Fig. 10.1 Benign Brenner tumor. The neoplasm is composed of a fibromatous stromal component and a transitional epithelial component

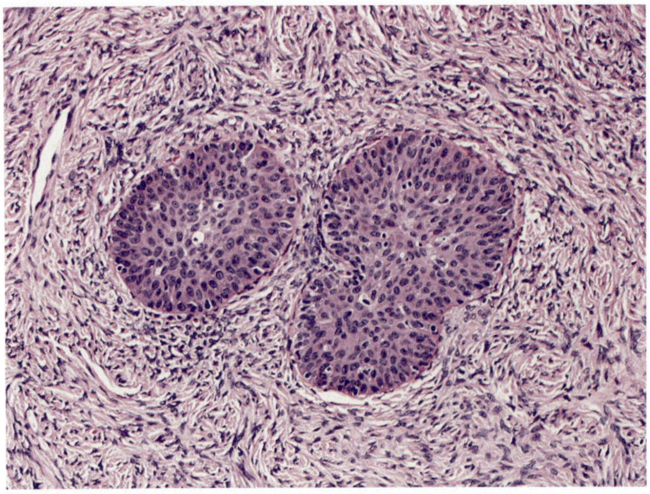

Fig. 10.2 Benign Brenner tumor. Sharply demarcated epithelial nests are composed of relatively uniform transitional type cells with elongated nuclei. Occasional longitudinal nuclear grooves are seen

accounts for the association of Brenner tumors with mucinous cystadenomas. It has been suggested that the majority of non-teratomatous mucinous tumors arise in Brenner tumors as an overgrowth of the mucinous component [3] (see Sect. "Mixed Brenner-mucinous Tumors"). The stroma varies from closely resembling ovarian cortical stroma to densely fibrous. Hyalinized areas are common, and dystrophic, spiculated calcification is present in about half of cases. Calcification is seen in 20% of tumors smaller than 1 cm and in over 80% of larger tumors. Rare examples of "epidermoid cyst" of the ovary may represent cystic variants of Brenner tumor with squamous metaplasia (see Sect. "Squamous Tumors"). Walthard cell nests are seen in 50% of patients with Brenner tumors, significantly more often than those with serous, clear cell, or endometrioid tumors, but similar to those with mucinous tumors [3].

Ancillary Studies

Immunophenotype

The epithelial nests stain strongly for epithelial markers including keratins and EMA. CK7 is typically positive, and CK20, negative. Calretinin, inhibin, and WT1 are negative. Uroplakin III, a urothelium-specific marker, is focally positive in a luminal pattern in a majority of cases. Thrombomodulin, a sensitive but less-specific marker of urothelium, shows strong membranous staining in 82%. P63, CEA, and CA 19-9 are positive in the majority of cases. The tumors are weakly reactive or negative for p16, cyclin D1, EGFR, and Ras protein, and are negative for p53 and Rb [4–10].

Molecular Biology

Available data are limited. A recent study found a mutation in *CTNNB1* in 1 of 4 Brenner tumors tested. No mutations were found in *p53, EGFR, B-raf, K-ras*, or *PIK3CA* [9].

Mixed Brenner-Mucinous Tumors

Tumors containing both Brenner and mucinous components are more common than previously appreciated. These are variants of Brenner tumor and can be classified as metaplastic Brenner tumor [2] or mixed Brenner-mucinous tumor (see Sect. "Mixed Epithelial Tumors"). One-fourth of benign ovarian epithelial tumors that have a mucinous component also contain a Brenner component. Conversely, 16% of tumors with a Brenner component contain a mucinous component. Tumors with both components are unilateral and most often have discrete Brenner and mucinous components, but in 30% the components are admixed. Most other clinical and pathologic features are similar to their pure counterparts except that the mixed tumors occur in an older age group with a mean age of 68 years [3].

Atypical Proliferative Brenner (Transitional Cell) Tumors

These rare neoplasms, also termed "proliferating" or "borderline" transitional cell or Brenner tumor, or transitional cell or Brenner tumor "of low malignant potential," are always unilateral and confined to the ovary.

Clinical Features

The mean patient age is 59. The clinical presentation is nonspecific. Those with larger tumors often present with signs and symptoms of a pelvic mass.

Macroscopic Features

These tumors are typically much larger than their benign counterparts, usually cystic, and measure 10–28 cm with a mean diameter of 18 cm. Friable papillary or polypoid masses project into the cyst lumens, and there is usually a benign Brenner component present with a more solid and fibrous cut surface.

Microscopic Features

The intracystic papillary component is composed of transitional-type epithelium resembling noninvasive low-grade (grade 1 of 3) papillary transitional cell neoplasms of the urinary tract (Figs. 10.3–10.5). Mucinous metaplasia may be present in the epithelium lining the papillae (Fig. 10.6). Underneath the papillae and within the wall, there may be small solid areas of transitional epithelium with little intervening stroma. Nearly all cases display areas of benign Brenner tumor, and occasionally, the proliferating component can be seen to be arising directly from the benign component. The cytologic features are similar to those in benign transitional cell tumors, but occasionally, significant atypia and mitotic activity are present. Roth and associates have separately classified "proliferative" and "low malignant potential" transitional cell tumors. The former are characterized by a resemblance to grade 1–2 urothelial papillary transitional cell carcinoma, whereas the latter display high-grade nuclear atypia resembling grade 3 urothelial papillary transitional cell carcinoma or in situ squamous cell

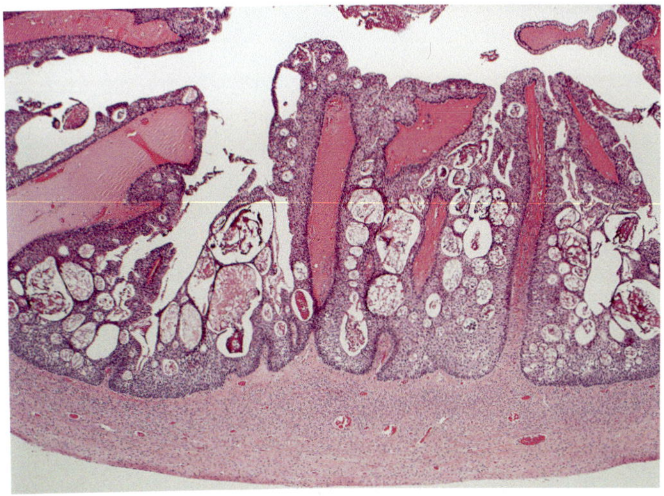

Fig. 10.3 Atypical proliferative Brenner tumor (borderline Brenner tumor). Cystic tumor with numerous broad papillae lined by transitional-type epithelium. Intraepithelial microcystic change is common

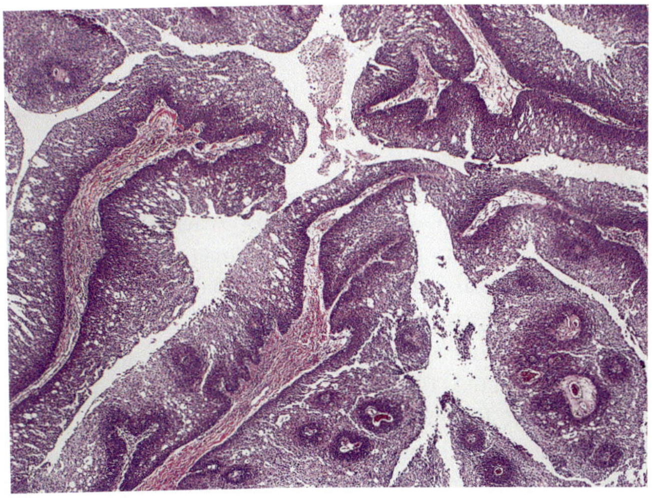

Fig. 10.4 Atypical proliferative Brenner tumor (borderline Brenner tumor). Broad papillary fronds are lined by transitional-like epithelium

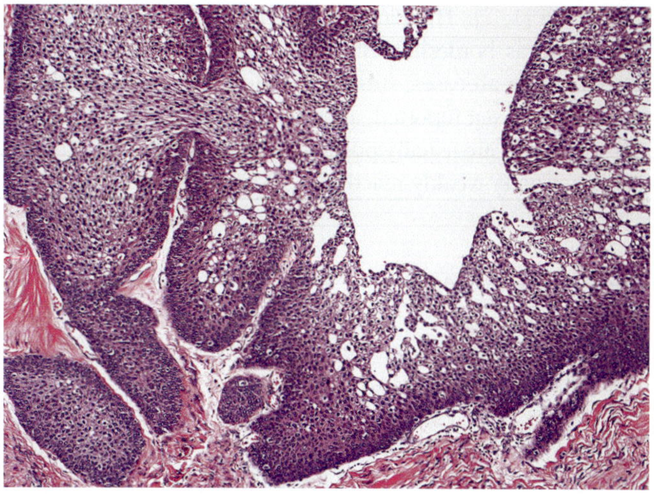

Fig. 10.5 Atypical proliferative Brenner tumor (borderline Brenner tumor). Neoplastic epithelium composed of cells with abundant cytoplasm and moderately atypical nuclei

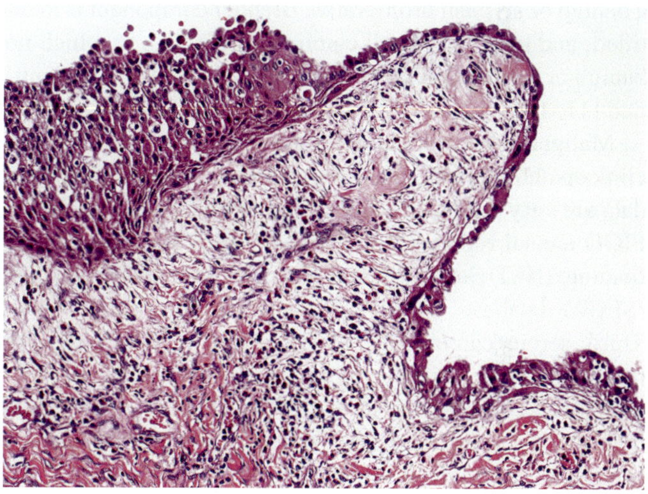

Fig. 10.6 Atypical proliferative Brenner tumor (borderline Brenner tumor). Areas of mucinous metaplasia are occasionally seen within the epithelium lining the papillae

carcinoma [1, 2]. The latter tumors have more recently been referred to as borderline Brenner tumors with intraepithelial carcinoma. In all cases, stromal invasion is absent; microinvasion has not been reported, although it can occur. EGFR, Ras, CEA, and p63 are usually positive [8]. Rb, cyclin D1, p16, and p53 are usually weakly reactive or negative [9].

Clinical Outcomes

The clinical course is uneventful. Among over 50 reported cases, there has been 1 local recurrence.

Malignant Transitional Cell Tumors (Malignant Brenner Tumor and Transitional Cell Carcinoma)

Two clinicopathologic types of malignant transitional cell tumor have been defined: malignant Brenner tumor, in which a benign or atypical proliferative Brenner component is identified; and transitional cell carcinomas (TCCs), in which no benign or atypical proliferative Brenner component is identified [11, 12].

Malignant transitional cell tumors are difficult to define clinicopathologically for several reasons. First, published data are very limited. Second, large databases, including the FIGO annual report and the United States National Cancer Institute (NCI) Surveillance, Epidemiology, and End Results (SEER) database, do not separately classify ovarian TCC. Third, serous carcinomas not infrequently have focal areas that display features resembling TCC. Nearly all tumors that have been classified as ovarian TCC are now regarded as high-grade serous carcinomas with transitional-like features and are best classified as serous rather than transitional [13–18] (Chaps. 4 and 6).

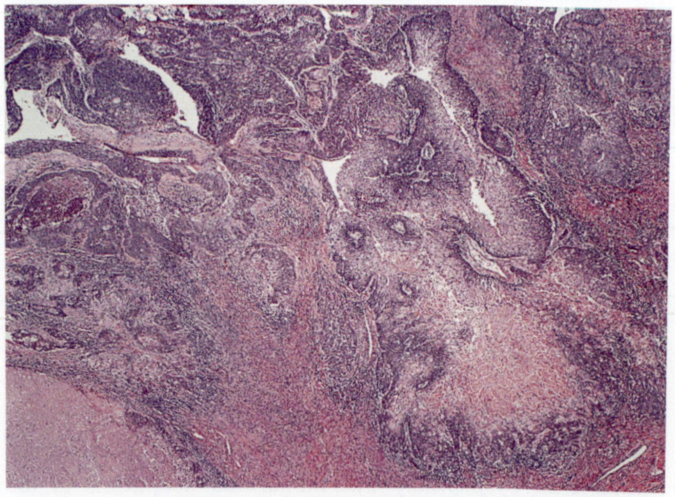

Fig. 10.7 Malignant Brenner tumor. Papillary neoplasm composed of transitional-type epithelium with stromal invasion. Benign Brenner tumor component is also present (not shown)

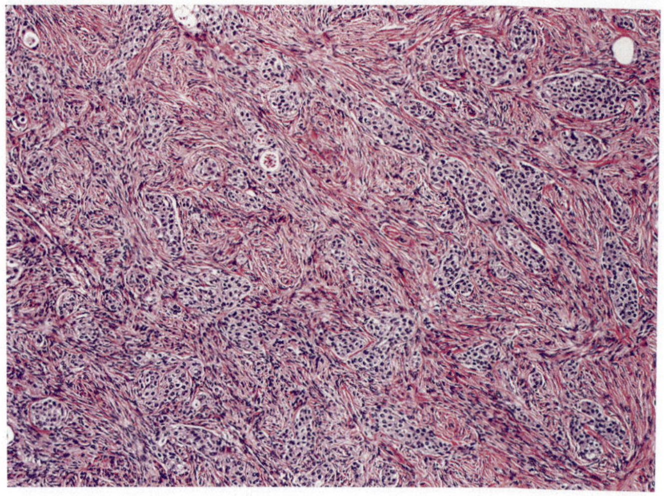

Fig. 10.8 Malignant Brenner tumor. Solid neoplasm composed of epithelial nests resembling benign Brenner tumor arranged in haphazard infiltrative pattern

Malignant Brenner Tumor

Malignant Brenner tumor is rare, comprising less than 0.5% of ovarian carcinomas. The stage distribution is as follows: stage I, 64%; stage II, 12%; stage III 18%; stage IV, 6%. TCC is more common than malignant Brenner tumor.

Clinical Features

The mean patient age is 63 years. The clinical presentation is nonspecific.

Macroscopic Features

The size of malignant Brenner tumors ranges up to 25 cm, with a mean diameter of 14 cm. In the malignant Brenner tumor, the benign Brenner component may be identifiable as a solid fibrous nodule within a cyst wall. Sometimes malignant Brenner tumor is completely solid.

Microscopic Features

Malignant Brenner tumor typically resembles invasive, high-grade papillary urothelial carcinoma (see below, Sects. "Transitional Cell Carcinoma", "Microscopic Features"). If these features are intimately associated with a benign Brenner component, a diagnosis of malignant Brenner tumor is indicated (Fig. 10.7). Another architectural pattern of malignant Brenner tumor is characterized by a solid tumor resembling benign Brenner tumor, but the epithelial nests are more angulated and have a disorderly, infiltrative growth pattern (Fig. 10.8). Stromal desmoplasia is occasionally present. In the solid type of malignant Brenner tumor with angulated and disorderly nests, mild or moderate atypia may be present (Fig. 10.8). Mitotic activity may be prominent. In addition, malignant Brenner tumors, like their benign counterparts, often have prominent stromal calcification, which is usually spiculated, in contrast to TCCs in which calcification is less common and more often psammomatous.

Immunophenotype

Published data on the immunohistochemistry of malignant Brenner tumors are very scant; however, the available information suggests similarities with urothelial neoplasms but not with transitional cell carcinomas of ovary. Based on a handful of case reports and small series, malignant Brenner tumors are negative for p16, p53, Rb [9], and GATA3 [19], and positive for cyclinD1, Ras, EGFR [9], S100P, p63 [19], CEA, CA19-9 [10], and CA125 [18, 20]. They may be positive for CK7 [20] and CK20 [18]. In one study, Liao and associates found that only one of six malignant Brenner tumors was positive for p63, and interestingly, in the remaining five cases, the noninvasive benign and proliferative Brenner component showed diffuse p63 positivity but lost expression in the invasive component [8].

Molecular Biology

Based on limited data, malignant Brenner tumors lack *p53* mutations [17].

Clinical Outcomes

For patients who present in stage I, the prognosis is excellent, with greater than 90% survival. Too few cases of advanced-stage malignant Brenner tumor have been reported to identify a survival difference with the other cell types of ovarian carcinoma.

Transitional Cell Carcinoma

Some investigators have found that approximately 10–20% of advanced-stage ovarian carcinomas contain a transitional cell component, which often predominates. In contrast, we found that only 3% of advanced-stage serous carcinoma were classifiable as TCC, although we do not diagnose TCC unless the features are absolutely classic [14]. The distinction depends upon the degree to which one considers a transitional-like pattern with high-grade nuclear features to represent a variant pattern of serous carcinoma, and, at present, most experts classify these as high-grade serous carcinomas. Molecular and immunohistochemical data support this interpretation [13, 15–18].

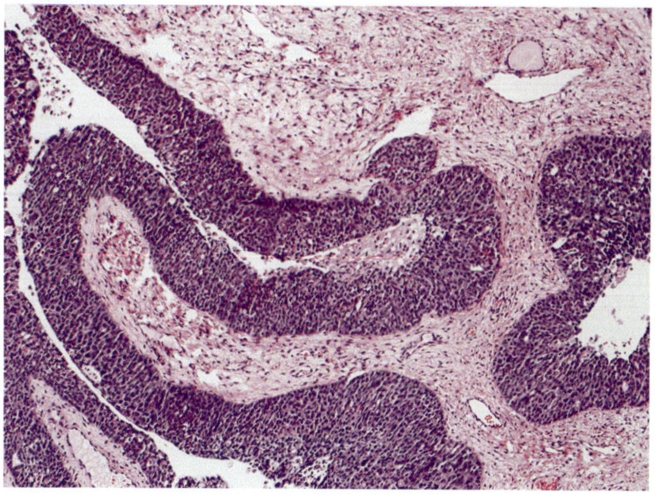

Fig. 10.9 Transitional cell carcinoma. Blunt elongated papillary fronds lined by transitional-type epithelium resembling urothelium

Clinical Features

The mean age at presentation is 56 years, slightly younger than the 60–63 years for typical high-grade serous carcinoma. These tumors present like high-grade serous carcinomas, typically with abdominal distention, ascites, gastrointestinal disturbances, and other evidence of peritoneal carcinomatosis.

Macroscopic Features

The gross appearance of TCC is similar to high-grade serous carcinoma – typically solid and cystic. The cysts may exhibit polypoid, friable mural nodules, and the cyst fluid may be serous, brown, bloody, or mucinous. Hemorrhage and necrosis may be prominent, and half have foci of gritty calcification.

Microscopic Features

The histologic patterns of TCC are usually mixed with serous carcinoma, and, as noted earlier, most experts now believe that TCC represents high-grade serous carcinoma with a transitional cell-like pattern. Approximately 10% of ovarian carcinomas containing the TCC pattern are said to be pure, although some series have a much lower proportion, which reflects variation in the belief that these should be classified as high-grade serous carcinomas [14]. More than 50% of the tumor should display the patterns of TCC for a diagnosis of TCC.

The characteristic microscopic feature of TCC is thick, blunt, and often elongated papillary folds with fibrovascular cores, lined by transitional-type epithelium resembling urothelium (Figs. 10.9 and 10.10). The papillae often appear to arise from a cyst wall with a similar lining of stratified and atypical transitional cells. A solid pattern is present in about half the cases. Microspaces, large cysts, and necrosis are seen in the majority (Fig. 10.9). Foci of squamous or glandular differentiation are seen in less than 20%. Stromal invasion is present and characterized by haphazard infiltrative

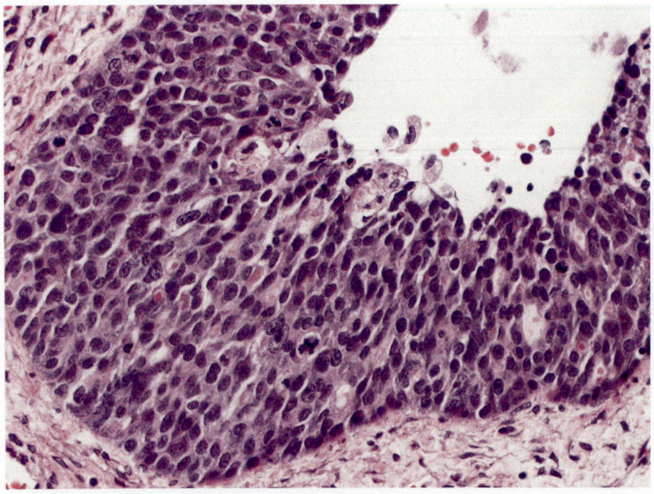

Fig. 10.10 Transitional cell carcinoma. Markedly atypical epithelium with frequent mitotic figures resembling high-grade transitional cell carcinoma of the urinary tract

growth of epithelium at the base of the papillae into the cyst wall, or extensive areas of solid epithelial proliferation with scant or no fibrovascular support. Slit-like fenestrations are found in half of TCCs, highlighting the resemblance to serous carcinoma. Cytologic atypia is usually prominent. High-grade pleomorphic nuclear features and bizarre giant cells occur in about one-third, further supporting that these tumors are transitional-like variants of serous carcinoma [11].

Immunophenotype

Ovarian TCCs are positive for CK7, as well as WT1, p16, p53, and CA125, just like high-grade serous carcinoma. They are generally negative for CK 20 in contrast to TCCs of the urinary tract, which are usually CK20 positive. EGFR, Ras, and cyclin D1 are usually weakly reactive or negative. A minority of TCCs are positive for CEA and CA 19-9. Several urothelial markers have been studied: uroplakin III (and other uroplakins) and thrombomodulin are positive in only a small minority; GATA3 is positive in one-third; and S100P in 17% [4, 6, 7, 9, 18, 19]. P63 was found to be positive in 6 of 12 cases (strongly in 2 cases) by Esheba and associates [19], but was always negative in 24 cases studied by Liao and associates [8]. In support of the classification of TCC as a variant of high-grade serous carcinoma, this immunophenotype is very similar to that of serous carcinoma (see Chap. 6).

Differential Diagnosis

The key features of transitional cell neoplasms are shown in Table 10.1.

1. *Benign Brenner tumor versus atypical proliferative (borderline) transitional cell tumor.* Brenner tumors are typically small and solid as compared to atypical proliferative tumors which are large and more cystic. An exuberant papillary epithelial proliferation reliably distinguishes atypical proliferative (borderline) transitional cell tumor from the Brenner tumor in which epithelial growth is confined to small nests.
2. *Atypical proliferative transitional cell tumor versus malignant Brenner tumor.* Destructive infiltrative growth, or stromal invasion, must be present to diagnose a malignant Brenner tumor.
3. *High-grade serous carcinoma versus transitional cell carcinoma.* The appearance of thick, blunt papillae of transitional cell differentiation contrasts with the papillae of typical high-grade serous carcinoma, which are generally thinner, but as noted earlier, TCC is generally considered a variant of high-grade serous carcinoma. On occasion, a predominantly solid serous carcinoma resembles TCC due to dropout of large areas of necrosis, which gives rise to an undulating or apparent macropapillary pattern. If any serous component is present, a diagnosis of serous carcinoma is preferred.
4. *Primary ovarian transitional cell carcinoma versus metastatic urothelial carcinoma.* This is rarely a problem because urinary tract TCCs that metastasize to the ovaries are generally deeply invasive and clinically evident. In addition to the morphologic features of metastatic carcinoma (nodularity, bilaterality, ovarian surface involvement; see also Chap. 11), urothelial carcinomas are CK7+ and CK20+ while ovarian TCCs are CK7+ and usually CK20−.

Table 10.1 Key features of transitional cell (Brenner) neoplasms

	Benign Brenner component	Disorderly Brenner nests	Cytologic atypia	Stromal invasion	Papillary growth	Extraovarian involvement
Benign Brenner tumor	Present	Often focally present	None	Absent	Absent	Absent
Atypical proliferative (borderline) Brenner tumor	Present	May be present	Present, variable degree	Absent	Usually present	Absent
Malignant Brenner tumor	Present	Often present	Variable, may be severe	Present	May be present	May be present (36%)
Transitional cell carcinoma	Absent	Absent	Severe	Present	Present	Usually present

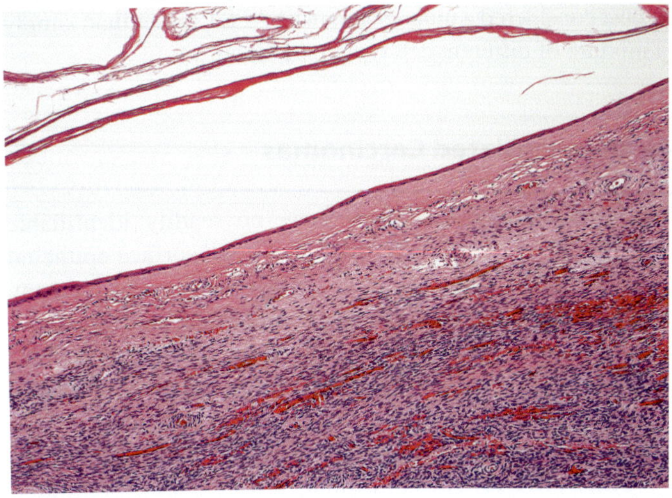

Fig. 10.11 Epidermoid cyst. Unilocular cyst lined by flattened squamous epithelium and contains keratinaceous debris

Clinical Outcomes

There are conflicting data on the behavior of TCC. Some studies have shown that these tumors are more chemosensitive and, accordingly, have a better survival as compared to serous carcinomas, while others have shown no survival difference [1, 21]. Two possible explanations for these discrepant findings have been suggested. Westfall et al. suggest that variation in diagnostic criteria for TCC has led to the inclusion of many serous carcinomas in some series which have concluded that the outcome is similar to serous carcinoma [21]. A prospective randomized trial of 302 ovarian carcinomas comparing TCC to non-TCC found a significantly better 5 year survival for TCC. In that study, TCC were significantly more likely to present with smaller (<1 cm) extraovarian tumor (i.e., more likely to be stage IIIA or IIIB as compared to IIIC) and, accordingly, were more likely to be optimally cytoreduced [22]. TCC may therefore represent a serous carcinoma variant with a similar prognosis after correction for substage and volume of residual disease. Additional work focusing on the biological characteristics of ovarian TCC is necessary.

Squamous Tumors

Benign Squamous Tumors (Epidermoid Cyst)

Epidermoid cysts comprise 1.4% of ovarian epithelial tumors. They are lined by squamous epithelium and lack teratomatous elements (Fig. 10.11). About half of these have either contralateral mature cystic teratomas, suggesting that they are insufficiently sampled or monodermal teratomas (and are therefore germ cell tumors, not surface epithelial tumors); foci of Brenner tumor, suggesting that they are Brenner tumors with squamous metaplasia; or features indicating a metaplastic change in endometriosis. Occasionally, other teratomatous elements are identified after extensive sampling of an apparent epidermoid cyst. Accordingly, pure epidermoid cyst is quite rare and comprises less than 1% of surface epithelial tumors [23, 24].

Squamous Cell Carcinoma

Pure invasive squamous cell carcinoma is very rare, comprising 0.5% of ovarian carcinomas. In a series of 18 cases, endometriosis was found in 7 cases, and no underlying lesion was found in 11 cases [25]. Invasive squamous cell carcinoma in the ovary is most commonly due to malignant transformation of a mature cystic teratoma; in such cases, the tumor is classified as a germ cell tumor with malignant transformation (see Chap. 13, Pathology of Germ Cell Tumors). Rarely, metastatic disease from the uterine cervix is the source of squamous cell carcinoma in the ovary, and very rarely, in situ squamous cell carcinoma of the cervix spreads intraepithelially into the endometrium and fallopian tubes, and may involve the ovaries. Rare reports of nonteratomatous in situ squamous cell carcinoma of the ovary not extending from the cervix probably reflect atypical proliferative transitional cell tumors (see Sect. "Atypical Proliferative Transitional Cell Tumors", above). Rarely, an endometrioid carcinoma with a prominent metaplastic squamous component will mimic a pure squamous cell carcinoma.

Differential Diagnosis

The differential diagnostic considerations when squamous differentiation is identified in the ovary are shown in Table 10.2.

Table 10.2 Differential diagnosis of squamous differentiation in the ovary

Benign	Epidermoid cyst Mature cystic teratoma Brenner (transitional cell) tumor with squamous metaplasia[a] Endometriosis with squamous metaplasia Endometrioid or seromucinous tumor with squamous differentiation[a]
Malignant	Metastatic squamous cell carcinoma (often of cervical origin) Endometrioid carcinoma with squamous differentiation[b] Squamous cell carcinoma arising in a mature cystic teratoma Squamous cell carcinoma extending intraepithelial from cervix Squamous cell carcinoma NOS

NOS not otherwise specified
[a]May be benign or atypically proliferative (borderline)
[b]May be primary or metastatic (typically from the endometrium)

Clinical Outcomes

Based on very few reported cases, when stratified by stage, the prognosis is not significantly different from that of the other cell types of ovarian carcinoma.

Mixed Epithelial Tumors

The presence of two or more epithelial cell types in an ovarian epithelial neoplasm, each comprising at least 10% of the tumor, warrants a designation of a mixed epithelial tumor. In many cases, minor components can be safely ignored unless the minor component is borderline or malignant in an otherwise benign tumor, or is of a higher grade than the predominant component.

Serous and endometrioid differentiation not infrequently appear to occur together. In particular, high-grade serous and endometrioid carcinomas often display overlapping features, and, at times, assignment to one or the other group is arbitrary and varies among observers. It has been suggested that there is no clinical or molecular basis for separating high-grade serous and endometrioid carcinomas [17]. Most experts currently believe that most of these tumors are preferably classified as serous, reserving endometrioid for those tumors that display easily recognizable and characteristic features of eutopic endometrioid adenocarcinomas [15, 16]. It has been shown that in true mixed serous and endometrioid tumors, the survival more closely approximates that expected for serous carcinoma and, therefore, if a serous component is present, it should be noted in the diagnosis, even if it is small [26].

Endometrioid and clear cell differentiation are mixed on occasion, as both frequently arise in endometriosis.

Mixed serous-clear cell carcinoma is overdiagnosed. It is not rare to find foci of cells with clear cytoplasm in otherwise ordinary high-grade serous carcinomas. This is considered part of the spectrum of serous differentiation [15, 27]. True clear cell carcinoma differentiation is characterized by unique histologic patterns, including hobnail, tubulocystic, and tubulopapillary, and frequently has a characteristically hyalinized stroma (see Chap. 8, Pathology of Clear Cell Tumors). A useful panel is ER, WT1, and p53, all of which are typically positive in serous carcinoma and negative in clear cell carcinoma [27].

A mucinous component is often observed in benign transitional cell neoplasms (see above, "Mixed Transitional Cell-Mucinous Tumors"). Among atypical proliferative (borderline) tumors, if "mixed type" is strictly defined, the most common mixed type is seromucinous (APSMT), also called mullerian mucinous or endocervical-like type atypical proliferative (borderline) mucinous tumor. Although serous (papillary) architectural features are mixed with mucinous cytologic features (and also may be mixed with other cell types), APSMT is best considered a unique variant of APST rather than simply a mixture of different cell types (see Chap. 9).

Undifferentiated Carcinomas

Undifferentiated carcinomas show no readily identifiable features of any of the cell types of ovarian surface epithelial neoplasms [28–30]. Therefore, when strictly defined, any element of glands, papillae, or psammoma bodies removes a tumor from this category. Accordingly, solid carcinomas with rare foci of glands or papillae are preferably classified as serous, although some investigators permit up to 5% of glandular or papillary growth in undifferentiated carcinoma.

There are five types of primary ovarian undifferentiated carcinomas: undifferentiated carcinoma not otherwise specified (NOS), non-small cell neuroendocrine carcinoma, the hypercalcemic type of small cell carcinoma, undifferentiated carcinoma associated with endometrioid adenocarcinoma, and the pulmonary type of small cell carcinoma (Chap. 14). Mucinous neoplasm with mural nodule of anaplastic carcinoma are usually categorized separately. Relative prevalence data are unavailable, but anecdotally, either the hypercalcemic type or the undifferentiated carcinoma associated with endometrioid adenocarcinoma is probably the most common. When strict pathologic criteria are employed for these diagnoses, all types are rare, together comprising less than 1% of invasive carcinomas. At the Washington Hospital Center, among over 400 consecutive ovarian and peritoneal carcinomas classified using uniform current criteria, classifying predominantly undifferentiated tumors with any foci of papillary or glandular growth as serous or endometrioid depending on the patterns present, there was a solitary case (0.2%) of undifferentiated small cell carcinoma, hypercalcemic type [14]. Significant worldwide variation in the criteria for the diagnosis of undifferentiated carcinoma is demonstrated by the 2006 FIGO annual report which found that 5.4% of ovarian carcinomas were classified as undifferentiated [32], a figure similar to that found in the NCI SEER database. In all cases when a diagnosis of undifferentiated carcinoma is considered, it is very important to exclude metastatic tumor, particularly from the lungs [33], and differentiated tumors, such as serous carcinoma, with a solid growth pattern. The distinguishing features of the five types are shown in Table 10.3.

Clinical Features

The mean age at presentation is slightly higher than the other cell types and is in the seventh decade, with the exception of small cell carcinoma, hypercalcemic type, which peaks in

Table 10.3 Undifferentiated carcinoma of the ovary

	Cell size	Follicle-like spaces	Immunohistochemistry	Associated lesion	Differential diagnosis
Undifferentiated carcinoma NOS	Large	Absent	CK7+, CK20–, EMA+, CAM5.2+, B72.3+, CA125+, may be WT1+	None	Solid serous carcinoma, metastatic carcinoma
Small cell carcinoma, hypercalcemic type [30]	Small; large cells may be present	Present	WT1+, TTF1–, calretinin+, p53+, CD10+, EMA+, AE1/3, CAM5.2+, NSE+, vimentin+, synaptophysin may be focally+, CD56 may be +; CD99–, inhibin–, desmin–, chromogranin– [28, 30, 34]	Mucinous carcinoma in 9%; benign mucinous epithelium in 3% [30]	Granulosa cell tumor (adult or juvenile type) Metastatic melanoma, other small, round, blue cell tumors, Desmoplastic small round cell tumor Poorly differentiated Sertoli-Leydig cell tumor
Small cell carcinoma, pulmonary type	Small to medium	Absent	CK20+ (perinuclear dot-like) [35], EMA+, NSE+, may be chromogranin+	Endometrioid carcinoma (in 44%), Brenner component may be present	Metastatic carcinoma (lung) [33]
Undifferentiated carcinoma associated with low grade endometrioid carcinoma [36, 37]	Variable	Absent	CK18 focally +, EMA+[36]	Endometrioid carcinoma	Metastatic endometrial carcinoma; lymphoma, plasmacytoma; granulosa cell tumor, endometrioid stromal sarcoma; rhabdoid tumor
Non-small cell carcinoma, neuroendocrine type [29]	Medium to large	Occasionally present	Synaptophysin usually +, chromogranin often+[29]	Mucinous or endometrioid carcinoma (54%); mucinous borderline tumor (17%); teratoma [29]	Metastatic neuroendocrine carcinoma (variety of sites)

NOS not otherwise specified

the third decade. Presenting manifestations are nonspecific and similar to typical serous carcinoma.

Macroscopic Features

All types of undifferentiated carcinoma are typically large and solid, with prominent hemorrhage and necrosis. The usual type of undifferentiated carcinoma NOS shares many clinical and pathologic features with high-grade serous carcinoma. About 80% present in stages III and IV.

Microscopic Features

Undifferentiated carcinoma NOS and non-small cell neuroendocrine carcinoma are composed of solid sheets of large pleomorphic cells with high-grade nuclear features, and usually with abundant cytoplasm that is often eosinophilic (Fig. 10.12). Undifferentiated carcinoma associated with endometrioid carcinoma has uniform, round monomorphic cells [36, 37]. The pulmonary type of undifferentiated small cell carcinoma is characterized by small- to medium-sized cells. Non-small cell neuroendocrine carcinoma is usually associated with another cell type of surface epithelial carcinoma, most often mucinous carcinoma. Similarly, the pulmonary type of small cell carcinoma is associated with endometrioid carcinoma in nearly half of cases, and the undifferentiated type associated with endometrioid carcinoma contains endometrioid differentiation in all cases. Classification of these tumors as high-grade mucinous and endometrioid carcinomas could be a source of variation in the frequency of undifferentiated carcinoma, and experts disagree on whether these tumors should be classified according to the better differentiated component. Classification by histogenesis would support designating these tumors based on the better differentiated component, while classification by behavior with an eye toward more aggressive treatment would support classification based on the poorly differentiated component.

The small cell type of undifferentiated carcinoma associated with hypercalcemia occurs (Chap. 16) at a mean age of 24 years. This is a distinctive neoplasm characterized by predominantly diffuse growth of closely packed cells with areas of nests, cords, and clusters, and most characteristically, follicle-like spaces containing eosinophilic fluid (Figs. 10.13, and 10.14). The cells are small and round, with scanty cytoplasm and hyperchromatic nuclei. Mitotic figures are frequent. Up to half of cases may contain a large cell component with abundant eosinophilic cytoplasm. Mucinous carcinoma is present in 9%, and benign mucinous epithelium in 3% [30].

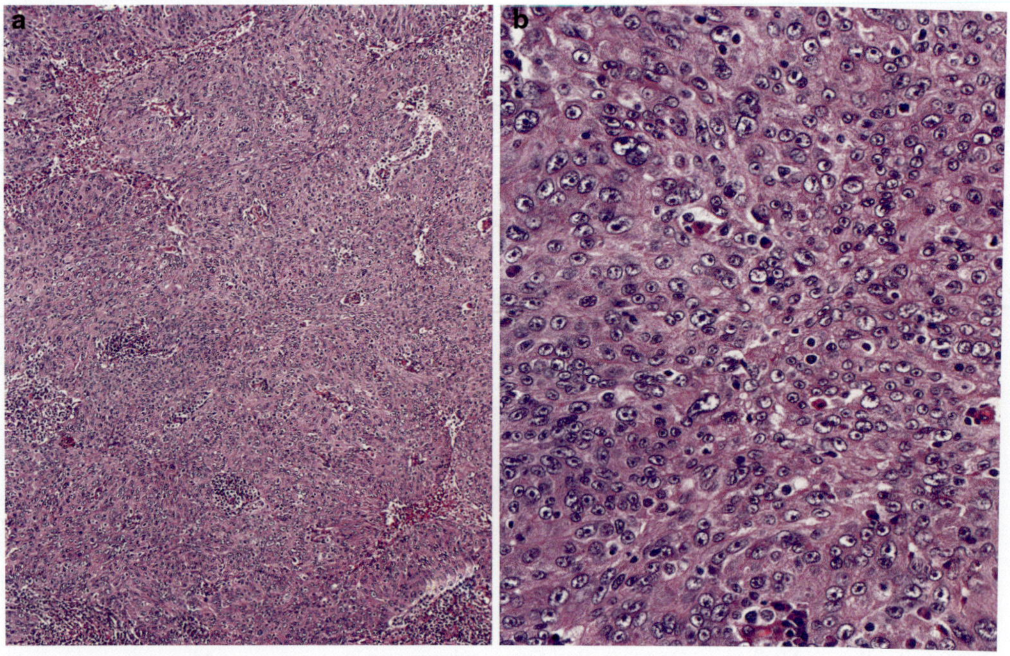

Fig. 10.12 Undifferentiated carcinoma. (**a**) The tumor is composed of solid sheets of cells without apparent differentiation. (**b**) Markedly atypical cells with large pleomorphic nuclei, abundant eosinophilic cytoplasm, and prominent nucleoli

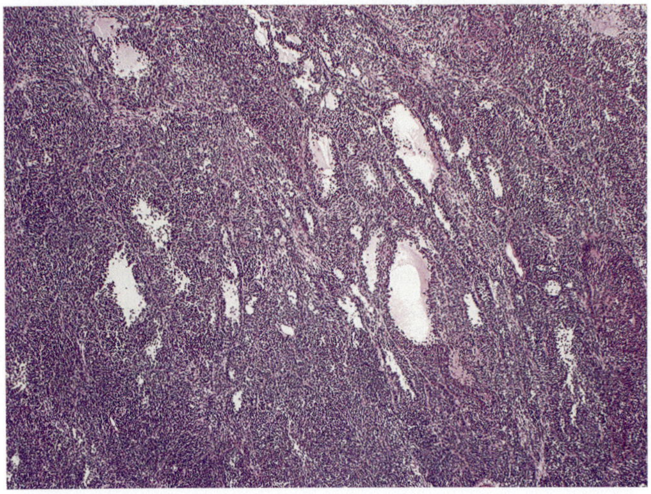

Fig. 10.13 Small cell carcinoma, hypercalcemic type. Diffusely cellular neoplasm composed of nests and cords of basophilic cells. Follicle-like spaces are characteristic

Ancillary Studies

Immunohistochemical findings are shown in Table 10.3. Data on the molecular features of these tumors are scant.

Clinical Outcomes

Although a few reports of small numbers of patients suggest a highly aggressive behavior, according to the 2006 FIGO annual report, when stratified by stage, the survival for undifferentiated carcinoma is not significantly different from that of the other cell types of ovarian carcinoma [32]. Whether these tumors respond well to platinum-paclitaxel therapy like serous carcinoma is not known.

Malignant Mixed Mesodermal Tumor (Carcinosarcoma)

Previously thought to be rare, malignant mixed mesodermal tumors (MMMTs) comprise about 7.5% of ovarian carcinomas in the United States when classified by uniform current criteria [14], 2.4% when based on SEER data, and 2.5% of ovarian carcinomas in Asia. The mean patient age is about 65 years. These tumors are typically large and solid, ranging from 15 to 20 cm in diameter, and often have hemorrhage and necrosis. Cystic areas may be present. The stage distribution is identical to that of serous carcinoma [32, 38, 39]. The morphology is similar to its uterine counterpart. The characteristic microscopic feature is an intimate admixture of malignant epithelial and stromal elements (Fig. 10.15). The malignant epithelial element is most commonly a high-grade serous or endometrioid carcinoma, but can resemble any of the surface epithelial cell types of ovarian tumors. The stromal component usually contains sheets of hyperchromatic rounded to spindled cells with marked nuclear atypia and a high mitotic index (Fig. 10.16). Heterologous elements, most commonly cartilage, osteoid, and rhabdomyoblasts (Fig. 10.17), are commonly found, and, as in the uterine counterpart, their frequency depends on the diligence with which they are sought. Occasionally, a tumor that

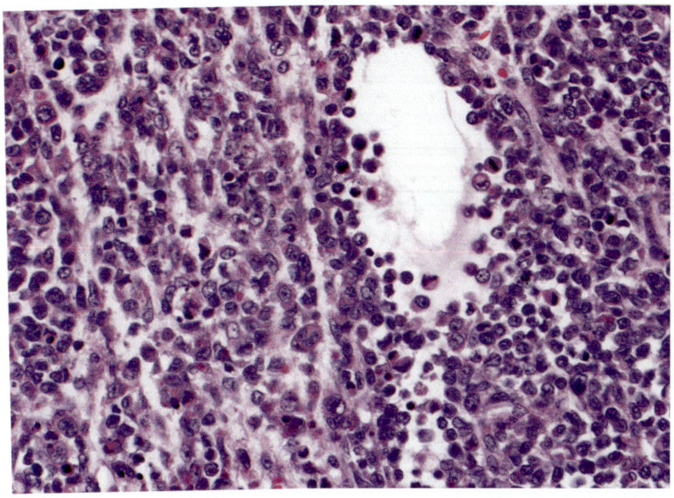

Fig. 10.14 Small cell carcinoma, hypercalcemic type. Atypical cells with hyperchromatic nuclei, scant cytoplasm, and indistinct cellular borders. Frequent mitotic figures are evident

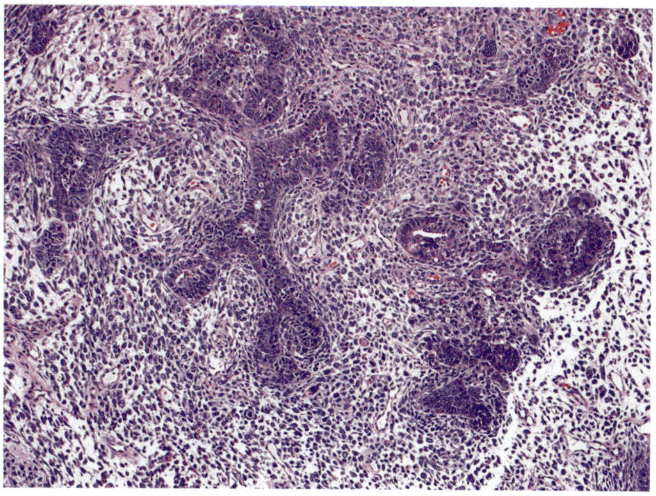

Fig. 10.15 Malignant mixed mesodermal tumor (carcinosarcoma). Biphasic neoplasm composed of malignant epithelial and stromal components that are intimately admixed

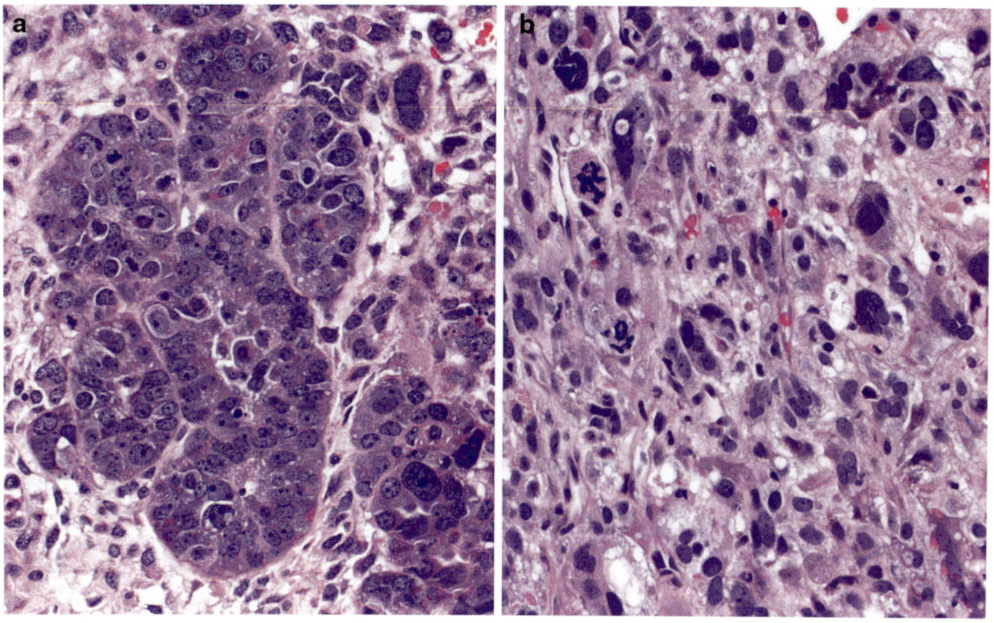

Fig. 10.16 Malignant mixed mesodermal tumor (carcinosarcoma). (**a**) Epithelial component represented by high-grade adenocarcinoma. (**b**) Malignant stromal component composed of markedly pleomorphic cells. Abnormal mitotic figures are seen

is otherwise a typical carcinoma has a small focus of malignant stroma. Although there are few published data on such tumors, they are classified as MMMTs.

Ancillary Studies

Immunohistochemical stains for epithelial markers are often positive in the sarcomatous component. Immunohistochemistry for skeletal muscle markers such as MyoD1 and myogenin can be used to help identify heterologous rhabdomyoblasts. Like endometrial carcinosarcomas, the majority have been demonstrated to be monoclonal. These observations suggest that these tumors should be classified as metaplastic carcinomas like their uterine counterparts.

Clinical Outcomes

In the older literature, MMMTs were considered aggressive, rapidly fatal tumors with a median survival of approximately 1 year, but more recent data suggest that with cisplatin and either a taxane or ifosfamide therapy, survival rates approaching those for serous carcinoma can be achieved. When stratified by stage,

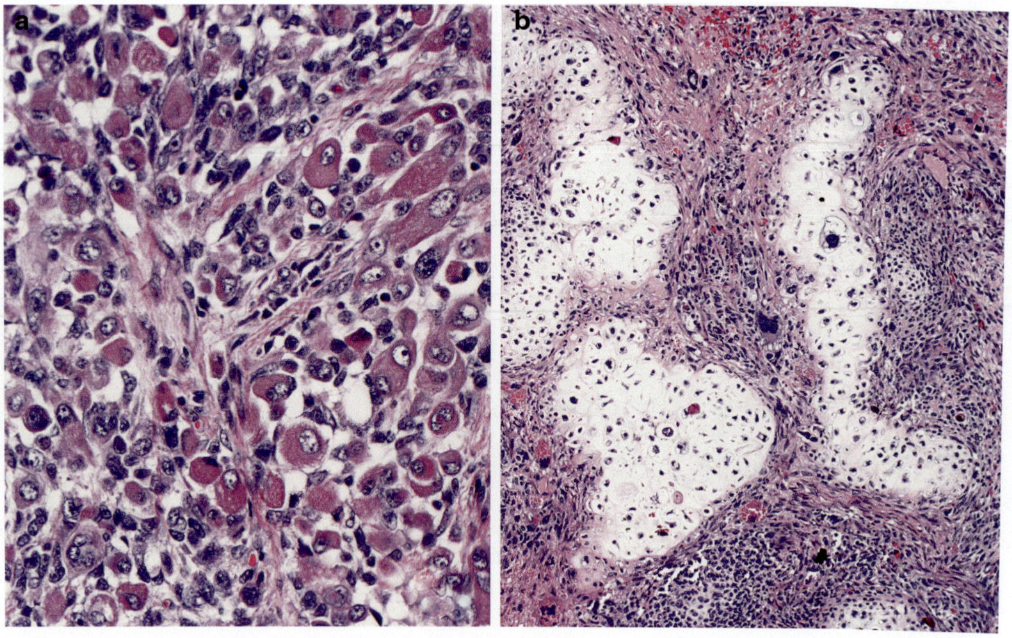

Fig. 10.17 Malignant mixed mesodermal tumor (carcinosarcoma). Heterologous elements are represented by foci of rhabdomyoblastic (**a**) and chondrosarcomatous (**b**) differentiation

Table 10.4 Diagnostic pitfalls

1. Although nuclear grooves are nearly always present in benign and atypical proliferative transitional cell tumors, they are nonspecific as they can be seen in all types of gonadal stromal tumors, particularly granulosa cell tumors.
2. Foci of somewhat disorderly and occasionally angulated nests are not uncommon in benign Brenner tumor. Unequivocal invasion is required to diagnose a malignant Brenner tumor.
3. Small Brenner tumors are common incidental findings. A carcinoma associated with a benign Brenner tumor should only be classified as a malignant Brenner tumor if the two are intimately associated or admixed, and only with an invasive component having the appropriate pattern.
4. A squamous-lined cyst is not diagnostic of a teratoma. At least one other teratomatous element must be identified to distinguish a teratoma from an epidermoid cyst. Further sampling can be of value.
5. If squamous cell carcinoma is present in the ovary, exclude metastatic tumor, particularly from the cervix and lung, before classifying as primary in the ovary.
6. If what appears to be squamous cell carcinoma in situ is seen, and invasion is not identified, consider atypical proliferative transitional cell tumor (borderline Brenner tumor) or direct extension of a squamous intraepithelial lesion from the cervix. Mature cystic teratoma with carcinoma in situ is also possible.
7. Diagnoses of mixed serous-clear cell carcinoma and mixed serous-endometrioid carcinoma should be used sparingly, if at all. The vast majority of ostensibly mixed cases ultimately prove to be pure high-grade serous carcinoma after careful morphologic evaluation, which may be supplemented by immunohistochemistry.
8. Serous carcinomas may be predominantly solid. If there are any glands, characteristic slit-like spaces, papillae, or psammoma bodies, a diagnosis of undifferentiated carcinoma is not warranted.
9. When strictly defined, all types of undifferentiated carcinoma of the ovary are rare and in total should comprise less than 1% of all ovarian carcinomas. Always consider the possibility of metastatic carcinoma or a solid high-grade serous carcinoma.

however, some investigators have found a prognosis that does appear to be worse than that for serous carcinoma [40]. A summary of diagnostic pitfalls is provided in Table 10.4.

References

1. Roth LM, Gersell DJ, Ulbright TM. Ovarian Brenner tumors and transitional cell carcinoma: recent developments. Int J Gynecol Pathol. 1993;12:128–33.
2. Roth LM, Dallenbach-Hellweg G, Czernobilsky B. Ovarian Brenner tumors: I. Metaplastic, proliferating, and of low malignant potential. Cancer. 1985;56:582–91.
3. Seidman JD, Khedmati F. Exploring the histogenesis of ovarian mucinous and transitional cell (Brenner) tumors: a study of 120 tumors. Arch Pathol Lab Med. 2008;132:1753–60.
4. Baker PM, Oliva E. Immunohistochemistry as a tool in the differential diagnosis of ovarian tumors: an update. Int J Gynecol Pathol. 2005;24:39–55.
5. Cathro HP, Stoler MH. The utility of calretinin, inhibin and WT1 immunohistochemical staining in the differential diagnosis of ovarian tumors. Hum Pathol. 2005;36:195–201.

6. Logani S, Oliva E, Amin MB, Folpe AL, Cohen C, Young RH. Immunoprofile of ovarian tumors with putative transitional cell (urothelial) differentiation using novel urothelial markers: histogenetic and diagnostic implications. Am J Surg Pathol. 2003;27: 1434–41.
7. McCluggage WG. Immunohistochemical and functional biomarkers of value in female genital tract lesions. Int J Gynecol Pathol. 2006;25:101–20.
8. Liao XY, Xue WC, Shen DH, Ngan HYS, Siu MK, Cheung AN. p63 expression in ovarian tumours: a marker for Brenner tumours but not transitional cell carcinomas. Histopathology. 2007;51:477–83.
9. Cuatrecasas M, Catasus L, Palacios J, Prat J. Transitional cell tumors of the ovary: a comparative clinicopathologic, immunohistochemical, and molecular genetic analysis of Brenner tumors and transitional cell carcinomas. Am J Surg Pathol. 2009;33:556–67.
10. Soslow RA, Rouse RV, Hendrickson MR, Silva EG, Longacre TA. Transitional cell neoplasms of the ovary and urinary bladder: a comparative immunohistochemical analysis. Int J Gynecol Pathol. 1996;15:257–65.
11. Eichorn JH, Young RH. Transitional cell carcinoma of the ovary: a morphologic study of 100 cases with emphasis on differential diagnosis. Am J Surg Pathol. 2004;28:453–63.
12. Roth LM, Czernobilsky B. Ovarian Brenner tumors: II. Malignant. Cancer. 1985;56:592–601.
13. McCluggage WG. My approach to and thoughts on the typing of ovarian carcinomas. J Clin Pathol. 2008;61:152–63.
14. Seidman JD, Horkayne-Szakaly I, Haiba M, Boice CR, Kurman RJ, Ronnett BM. The histologic type and stage distribution of ovarian carcinomas of surface epithelial origin. Int J Gynecol Pathol. 2004;23:41–4.
15. Soslow RA. Histologic subtypes of ovarian carcinoma: an overview. Int J Gynecol Pathol. 2008;27:161–74.
16. Gilks CB. Subclassification of ovarian surface epithelial tumors based on correlation of histologic and molecular pathologic data. Int J Gynecol Pathol. 2004;23:200–5.
17. Gilks CB, Prat J. Ovarian carcinoma pathology and genetics: recent advances. Hum Pathol. 2009;40:1213–23.
18. Riedel I, Czernobilsky B, Lifschitz-Mercer B, Roth LM, Wu X, Sun T, et al. Brenner tumors but not transitional cell carcinomas of the ovary show urothelial differentiation: immunohistochemical staining of urothelial markers, including cytokeratins and uroplakins. Virchows Arch. 2001;438:181–91.
19. Esheba GE, Longacre TA, Atkins KA, Higgins JP. Expression of the urothelial differentiation markers GATA3 and Placental S100 (S100P) in female genital tract transitional cell proliferations. Am J Surg Pathol. 2009;33:347–53.
20. Ceausu M, Terzea D, Georgescu A, et al. Transitional cell tumors of the ovary: a compact group with a heterogeneous histological and immunophenotypical pattern. Rom J Morphol Embryol. 2008;49: 513–6.
21. Westfall D, Roma AA, Silva EG. High grade serous carcinoma of the ovary. Ann Diagn Pathol. 2009;13:285–90.
22. Kommoss F, Kommoss S, Schmidt D, Trunk MJ, Pfisterer J, du Bois A, et al. Survival benefit patients advanced stage transitional cell carcinomas vs. other subtypes ovarian carcinoma after chemotherapy platinum paclitaxel. Gynecol Oncol. 2005;97:195–9.
23. Khedmati F, Chirolas C, Seidman JD. Ovarian and paraovarian squamous-lined cysts (epidermoid cysts): a clinicopathologic study of 18 cases with comparison to mature cystic teratomas. Int J Gynecol Pathol. 2009;28:193–6.
24. Fan LD, Zang HY, Zhang XS. Ovarian epidermoid cyst: report of eight cases. Int J Gynecol Pathol. 1996;15:69–71.
25. Pins MR, Young RH, Daly WJ, et al. Primary squamous cell carcinoma of the ovary: a report of 37 cases. Am J Surg Pathol. 1996;20:823–33.
26. Tornos C, Silva EG, Khorana SM, Burke TW. High stage endometrioid carcinoma of the ovary: prognostic significance of pure versus mixed histologic types. Am J Surg Pathol. 1994;18:687–93.
27. Han G, Gilks CB, Leung S, et al. Mixed ovarian epithelial carcinomas with clear cell and serous components are variants of high-grade serous carcinoma: an interobserver correlative and immunohistochemical study of 32 cases. Am J Surg Pathol. 2008; 32:955–64.
28. McCluggage WG. Ovarian neoplasms composed of small round cells: a review. Adv Anat Pathol. 2004;11:288–96.
29. Veras E, Deavers MT, Silva EG, Malpica A. Ovarian nonsmall cell neuroendocrine carcinoma: a clinicopathologic and immunohistochemical study of 11 cases. Am J Surg Pathol. 2007;31:774–82.
30. Young RH, Oliva E, Scully RE. Small cell carcinoma of the ovary, hypercalcemic type: a clinicopathological analysis of 150 cases. Am J Surg Pathol. 1994;18:1102–16.
31. Kobel M, Kalloger SE, Huntsman DG, et al. Differences in tumor type in low-stage versus high-stage ovarian carcinomas. Int J Gynecol Pathol. 2010;29:203–11.
32. Heintz APM, Odicino F, Maisonneuve P, et al. Carcinoma of the ovary: FIGO 26th annual report on the results of treatment in gynecological cancer. Int J Gynaecol Obstet. 2006;95 Suppl 1:S161–92.
33. Irving JA, Young RH. Lung carcinoma metastatic to the ovary: a clinicopathologic study of 32 cases emphasizing their morphologic spectrum and problems in differential diagnosis. Am J Surg Pathol. 2005;29:997–1006.
34. McCluggage WG, Oliva E, Connolly LE, McBride HA, Young RH. An immunohistochemical analysis of ovarian small cell carcinoma of hypercalcemic type. Int J Gynecol Pathol. 2004;23:330–6.
35. Rund CR, Fischer EG. Perinuclear dot-like cytokeratin 20 staining in small cell neuroendocrine carcinoma of the ovary (pulmonary type). Appl Immunohistochem Mol Morphol. 2006;14:244–8.
36. Tafe LJ, Garg K, Chew I, Tornos C, Soslow RA. Endometrial and ovarian carcinomas with undifferentiated components: clinically aggressive and frequently underrecognized neoplasms. Mod Pathol. 2010;23(6):781–9.
37. Silva EG, Deavers MT, Bodurka DC, Malpica A. Association of low grade endometrioid carcinoma of the uterus and ovary with undifferentiated carcinonoma: a new type of dedifferentiated carcinoma? Int J Gynecol Pathol. 2006;25:52–8.
38. Silasi DA, Illuzzi JL, Kelly MG, Rutherford TJ, Mor G, Azodi M, et al. Carcinosarcoma of the ovary. Int J Gynecol Cancer. 2008;18:22–9.
39. Leiser AL, Chi DS, Ishill NM, Tew WP. Carcinosarcoma of the ovary treated with platinum and taxane: the memorial Sloan-Kettering Cancer Center experience. Gynecol Oncol. 2007;105: 657–61.
40. Mano MS, Rosa DD, Azambuja E, et al. Current management of ovarian carcinosarcoma. Int J Gynecol Cancer. 2007;17:316–24.

Metastatic Tumors

11

Anna Yemelyanova and Jeffrey D. Seidman

Keywords
Breast carcinoma • Colon carcinoma • Endometrioid adenocarcinoma • Gastric carcinoma • Immunohistochemistry • Metastatic carcinoma • Mucinous adenocarcinoma • Ovarian neoplasms • Pancreatic carcinoma • Signet-ring cell adenocarcinoma

Introduction

Historically, metastatic tumors to the ovaries have been considered to comprise about 5–14% of cancers involving the ovaries. In the 1990s, however, it became more widely appreciated that metastatic carcinoma in the ovary can closely mimic primary ovarian carcinoma, both grossly and microscopically. This is particularly true of mucinous carcinomas. While mucinous carcinoma had previously been considered one of the more common types of ovarian cancer, it is now clear that the majority of mucinous carcinomas involving the ovaries, about 75–85% [1–4], are metastatic from elsewhere. Furthermore, it is now agreed that pseudomyxoma peritonei (PMP), a low-grade mucinous neoplasm historically considered to be of gastrointestinal (GI) or ovarian origin, is virtually always primary in the GI tract, usually the appendix [5–9]. The older literature on ovarian mucinous carcinoma is now considered unreliable as studies prior to and even including the 1990s undoubtedly include an unknown but significant proportion of misclassified extraovarian tumors. To a lesser extent, this may also be true of endometrioid adenocarcinomas, as one of the most common patterns of metastatic colorectal carcinoma closely resembles endometrioid adenocarcinoma.

It cannot be overemphasized that one should maintain a high index of suspicion to accurately diagnose ovarian metastases. Careful attention to the history, operative findings, gross features, and histology are essential, particularly if there is even a single unusual finding. We find it a safe and useful practice to consider a mucinous carcinoma in the ovary to be metastatic until proved otherwise. Assigning the primary site to a mucinous carcinoma in the ovary without considering clinical information is analogous to diagnosing a bone tumor without knowledge of the radiological findings, or a cytopathological specimen without knowing the anatomic site. Knowledge of the history of prior malignancies, even if remote, and particularly of GI, pancreaticobiliary tract, or breast origin, is essential. Immunohistochemical studies should be readily considered in evaluating all mucinous carcinomas, as well as all apparently noninvasive mucinous neoplasms with even slightly unusual features for a primary ovarian tumor.

Morphology

Macroscopic Features

It is critical to know whether an ovarian tumor is bilateral and whether the patient has obvious peritoneal carcinomatosis.

J.D. Seidman (✉)
Department of Pathology, Washington Hospital Center, 110 Irving St. N.W., Washington, DC, 20010, USA

R.A. Soslow and C. Tornos (eds.), *Diagnostic Pathology of Ovarian Tumors*,
DOI: 10.1007/978-1-4419-9751-7_11, © Springer Science+Business Media, LLC 2011

As a general rule, metastatic carcinomas involving the ovaries are small and bilateral, but there are many exceptions. Furthermore, serous carcinomas are often small and bilateral. As non-serous tumors are confined to the ovaries in the majority of cases, the presence of peritoneal carcinomatosis of non-serous histology is an important clue when considering the possibility of secondary ovarian involvement. Peritoneal carcinomatosis is typically manifested as a firm, white, nodular omental cake with a chalky cut surface, often with calcification and soft yellowish areas of necrosis. Mucinous carcinomatosis, reflected grossly by gelatinous, mucoid nodules, or mucinous ascites (pseudomyxoma peritonei; see later in this chapter) is very rarely primary in the ovary [1, 2, 5–9].

Both primary and metastatic tumors to the ovaries are commonly cystic and, consequently, a multilocular cystic or solid and cystic tumor is not necessarily primary, even though this is the typical appearance of many primary ovarian carcinomas [1, 2, 4, 10–13]. Careful examination of the ovarian surface may reveal exophytic papillary growth; this finding is more suggestive of a primary ovarian carcinoma, usually of serous type. Nodularity is suggestive of metastatic tumor. Small bilateral solid tumor is another common appearance of metastatic carcinoma, particularly for signet-ring cell carcinomas and breast carcinoma, especially lobular carcinoma [14]. Metastatic breast carcinoma tends to be small and solid, often with grossly normal or only slightly enlarged ovaries, as the median tumor size is 1–2 cm. Undoubtedly, many of these size calculations were made by direct measurement from the slides, as the tumors are often not grossly evident [14]. Metastatic colonic carcinoma to the ovary, when unilateral, is equally likely to be ipsilateral or contralateral to the primary colonic tumor [15].

Several gross features can suggest a primary ovarian carcinoma. A cyst containing chocolate-colored fluid suggests endometriosis, which can be the origin of an ovarian carcinoma, usually of clear cell or endometrioid type. A mucinous neoplasm may arise in a Brenner tumor, which can be identified as a small, solid, yellowish nodule, often with calcification, in a larger cystic tumor [16]. Origin in a mature cystic teratoma rarely occurs. In such cases, hair and sebaceous material may be seen in conjunction with a markedly thickened or nodular wall, or solid areas containing the focus of malignant transformation. Primary ovarian mucinous carcinoma is typically large, in the range of 18–20 cm, and therefore a small mucinous carcinoma should raise the suspicion of metastasis [1, 2, 13, 16].

Microscopic Features

When metastatic carcinoma forms a tumorous mass, low-power examination of the ovary typically reveals a nodular architecture [1–4, 8, 10–13]. Varying-sized nodules of malignant glands compress the ovarian stroma around them and create a characteristic pattern of metastatic carcinoma (Figs. 11.1 and 11.2). Metastatic nodules are typically small, in the 2–4 mm range. Another common low-power appearance is a diffuse, haphazard infiltrative pattern (Figs. 11.3–11.5). It is important to examine the entire tumor at low magnification, as metastasis to the ovary can have noninvasive and sometimes cytologically bland-appearing areas in addition to foci of haphazard infiltrative growth (Figs. 11.6–11.8). It is also important to examine the ovarian surface at low magnification. In a relatively normal-sized ovary, metastases may appear as small plaques or nodules that are adherent or appear stuck onto

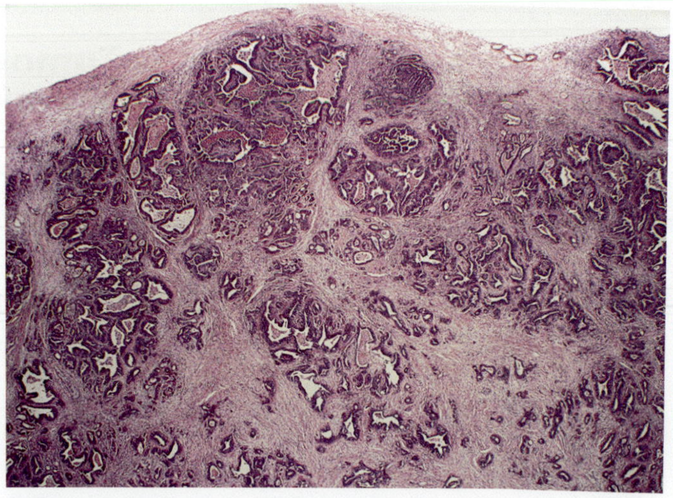

Fig. 11.1 Metastatic colonic adenocarcinoma. The ovary is replaced by tumor composed of infiltrating glands of varying sizes and shapes arranged in variably sized nodules

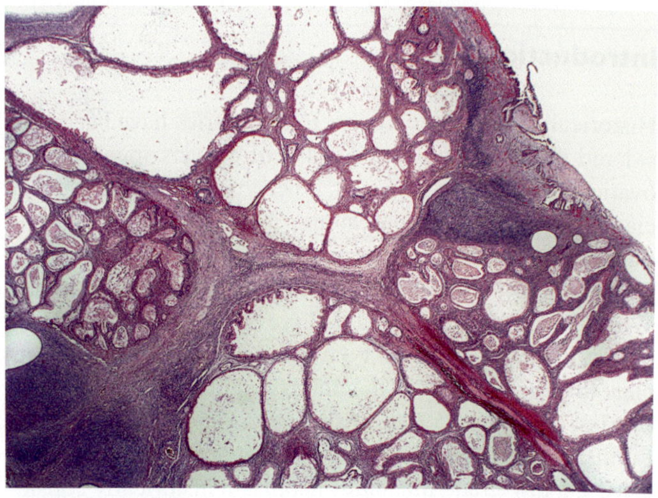

Fig. 11.2 Metastatic pancreatic adenocarcinoma. A distinctly nodular architecture and wedge-shaped cortical lesions are typical patterns of metastatic mucinous carcinoma

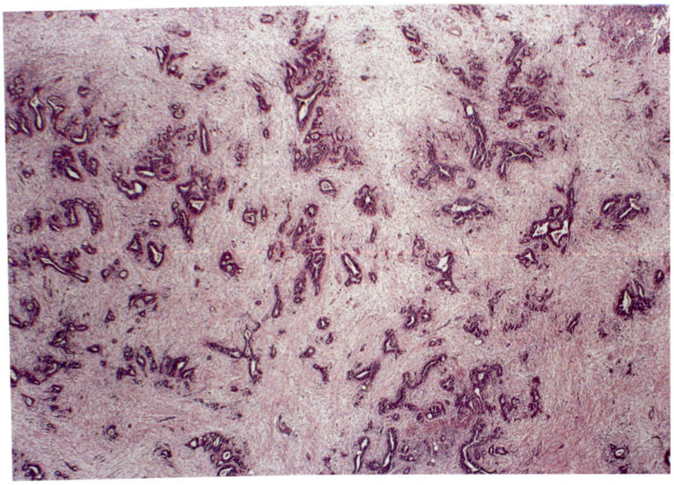

Fig. 11.3 Metastatic colonic adenocarcinoma. Small angulated glands arranged in a haphazard infiltrative pattern

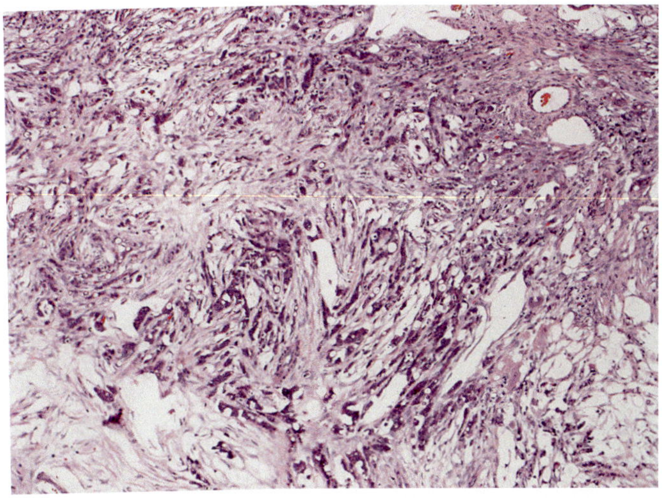

Fig. 11.4 Metastatic gastric adenocarcinoma. Poorly differentiated carcinoma with cords of malignant epithelial cells infiltrating the stroma interspersed with foci of dissecting mucin

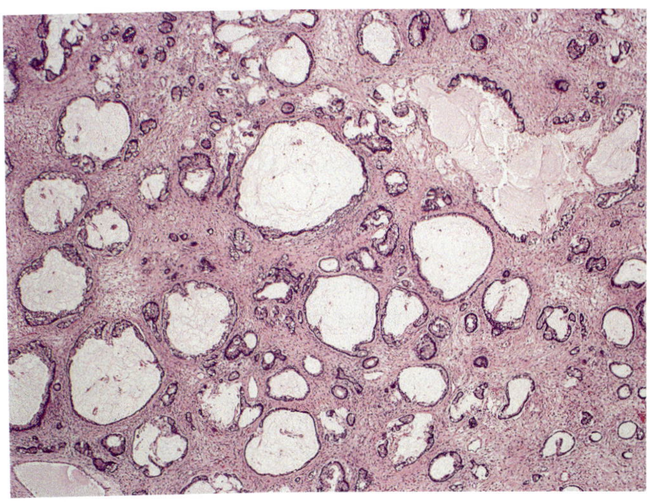

Fig. 11.5 Metastatic colonic adenocarcinoma. Simple mucinous glands of different sizes infiltrating the ovarian stroma with foci of gland rupture and mucin dissection

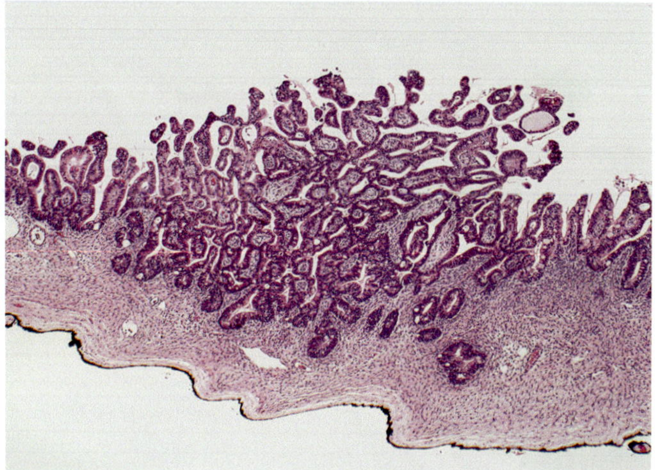

Fig. 11.6 Metastatic pancreatic adenocarcinoma. Thin-walled cyst with papillary excrescences on the inner surface mimics an atypical proliferative (borderline) mucinous tumor

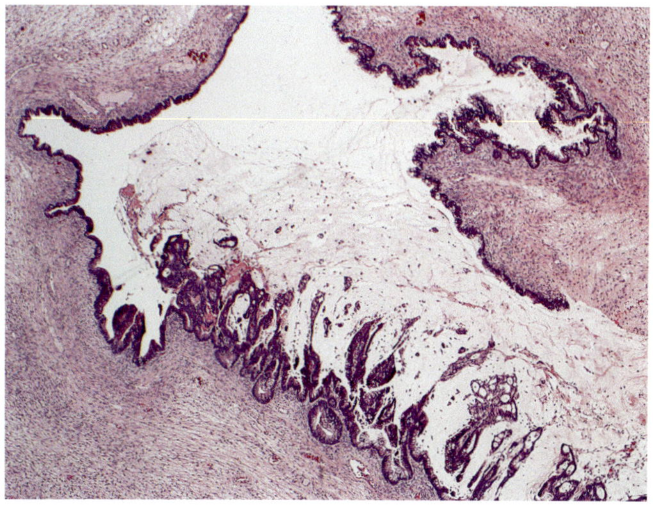

Fig. 11.7 Metastatic colonic adenocarcinoma. Simple cyst with foci of epithelial tufting may suggest the diagnosis of mucinous cystadenoma with focal epithelial proliferation

the ovarian surface (Fig. 11.9). This appearance is rare in primary ovarian carcinomas, although occasionally, a large primary ovarian carcinoma can display such an appearance in the contralateral, normal-sized ovary. The characteristic features of metastatic carcinoma in the ovaries are summarized in Table 11.1.

At medium and high magnification, the cell type should be determined. Mucinous differentiation should immediately raise the suspicion of metastatic carcinoma (Fig. 11.10). The presence of marked nuclear atypia disproportionate to the simplicity of the architecture and low degree of epithelial proliferation is also incriminating for a metastasis (Fig. 11.11). An endometrioid pattern is often seen in metastases, particularly from the colon [10–13, 15].

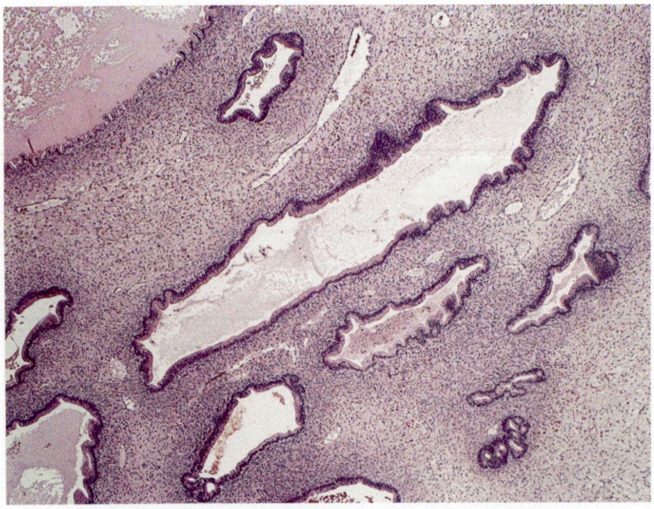

Fig. 11.8 Metastatic pancreatic adenocarcinoma. Simple mucinous glands with rather bland mucinous epithelium may suggest mucinous cystadenoma; other areas with more obviously infiltrative patterns are usually present

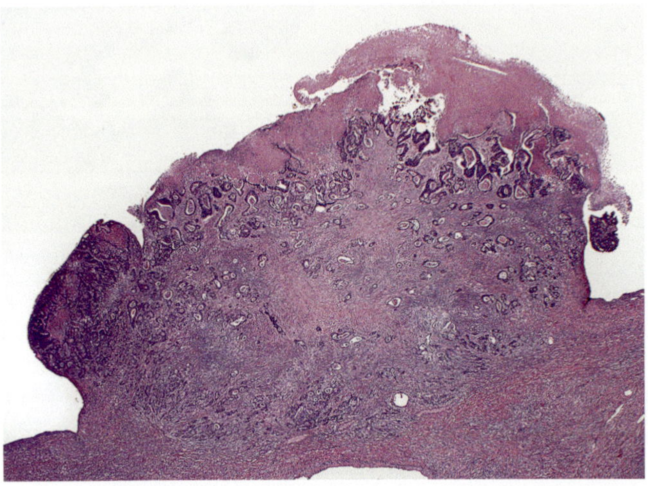

Fig. 11.9 Metastatic adenocarcinoma from the small bowel. An isolated tumor nodule appears stuck on the ovarian surface of a normal-sized ovary

A colonic mucinous carcinoma with mucin depletion will often closely resemble an endometrioid carcinoma (Fig. 11.12). The presence of endometriosis, particularly if the tumor involves the lining of an endometriotic cyst, suggests an ovarian primary. Squamous differentiation is commonly seen in endometrioid adenocarcinomas, but not in intestinal carcinomas and only rarely in breast carcinomas. A typical serous morphology with high-grade cytologic features and papillae, and often with psammoma bodies, can be confidently considered of gynecological origin. Although likely to be ovarian, primary peritoneal, tubal, and endometrial primary sites should be considered in light of other findings, particularly the anatomic distribution of tumor (see Chaps. 6 and 17, and later in this chapter).

Table 11.1 Key features that suggest an extraovarian primary neoplasm

Clinical/gross
• Small size, especially <8–10 cm (non-serous tumors only) • Bilaterality (non-serous) • Nodularity • History of carcinoma, particularly of gastrointestinal origin
Microscopic
• Nodularity • Ovarian surface involvement (for non-serous tumors) • Extensive infiltrative stromal invasion, especially single-cell infiltration • Absence of a noninvasive component, i.e., background of atypical proliferative mucinous tumor (ovarian intestinal mucinous borderline tumor), when mucinous carcinoma is a consideration • Extensive pseudomyxoma ovarii • Signet-ring cells • A degree of cytologic atypia out of proportion to the degree of epithelial proliferation • Abundant extracellular mucin, especially if on ovarian surface • Lymphatic or vascular invasion, especially if extensive • Carcinoma within an otherwise architecturally normal ovary or within a corpus luteum or corpus albicans
Extraovarian features
• Peritoneal mucinous carcinomatosis • Pseudomyxoma peritonei • Invasion of ovarian hilus and broad ligament lymphatics • Parenchymal liver metastasis • Extensive involvement of mesenteric lymph nodes

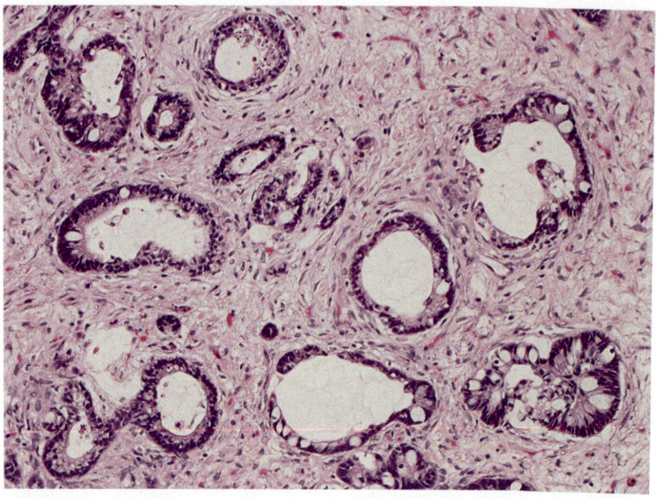

Fig. 11.10 Metastatic colonic adenocarcinoma. Small mucinous glands with goblet-cell differentiation and focal gland rupture should raise the suspicion for metastasis

Similarly, typical clear cell carcinoma is nearly always of gynecological origin, but can rarely be mimicked by metastases such as renal cell carcinoma [17]. Rarely are neuroendocrine features present, and it is useful to remember that although usually metastatic, carcinoid tumors may rarely arise in a mature cystic teratoma or a Sertoli-Leydig cell tumor (see Chaps. 13 and 14). Luteinized stromal cells are often prominent between nodules of metastatic carcinoma. However, these are also often seen in primary ovarian tumors and are therefore not discriminatory.

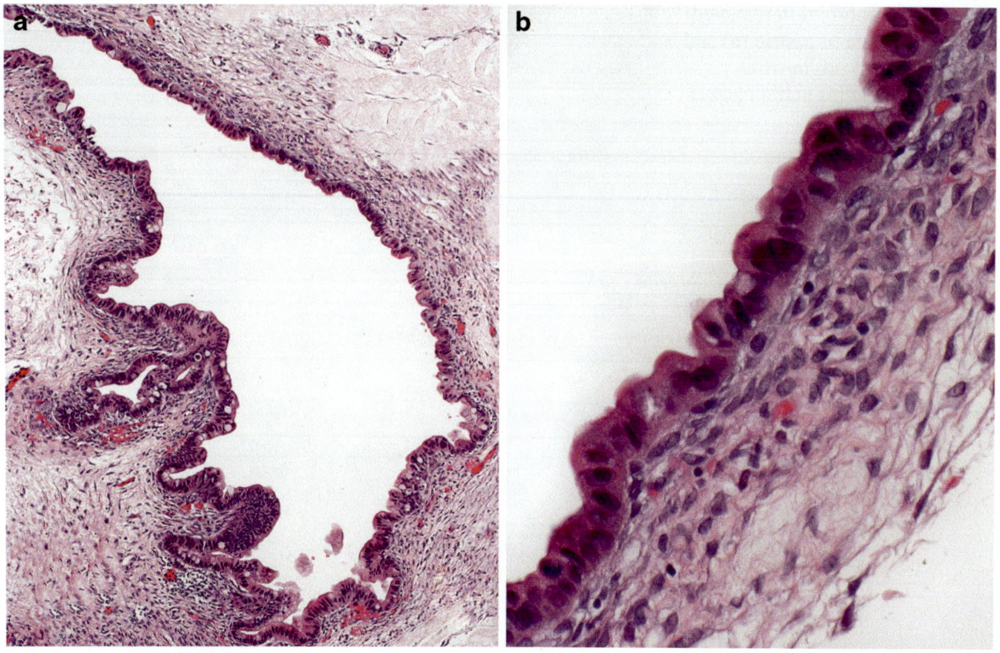

Fig. 11.11 Metastatic pancreatic adenocarcinoma. (**a**) Large simple gland suggesting mucinous cystadenoma from low power. (**b**) Foci of marked nuclear atypia with nuclear enlargement, hyperchromasia, and prominent nucleoli, disproportionate to the simplicity of the architecture

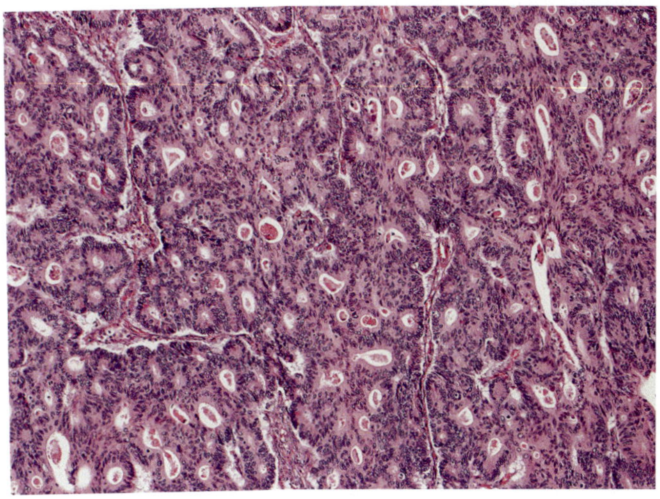

Fig. 11.12 Metastatic colonic adenocarcinoma. Complex glandular architecture with a strikingly endometrioid appearance may suggest a primary endometrioid adenocarcinoma. A higher degree of nuclear atypia and brisk mitotic activity are more typical of metastasis rather than a primary

The most common non-gynecological sources of metastatic carcinoma in the ovary are the breast and GI tract. A signet-ring cell pattern is common in metastatic breast and gastric carcinomas (Fig. 11.13), and can also be seen in metastases from other sites in the GI tract. The most common pattern of metastatic colorectal carcinoma to the ovaries mimics endometrioid carcinoma [10–13, 15]. This pattern is characterized by garlands with central areas of dirty necrosis (Figs. 11.14 and 11.15), although this pattern can occasionally be seen in primary ovarian endometrioid carcinoma. The typical appearance of lobular or signet-ring cell breast carcinoma metastatic to the ovary is manifested by malignant cells in a single-file pattern (Fig. 11.16) [14].

Metastases from the pancreas are usually large and cystic, and often have foci with minimal atypia resembling mucinous cystadenoma [18]. Metastases from the biliary tract are uncommon and usually mucinous [19]. Metastases from the liver are quite rare, and for hepatoid-appearing tumors, the differential diagnosis should include these rare primary ovarian neoplasms: hepatoid yolk sac tumor; hepatoid carcinoma; oxyphilic clear cell carcinoma; and lipid cell (steroid cell) tumor [20]. An ovarian tumor that resembles urothelial carcinoma is much more likely to be a primary ovarian serous or transitional cell carcinoma rather than a metastasis from the urinary tract (see Chap. 10). An undifferentiated carcinoma, either small cell or large cell, can rarely be primary, but may also be metastatic, most commonly from the lung [21] (see Chap. 10).

Metastases from endocervical adenocarcinomas are often large and unilateral, and microscopically display a "borderline" -like pattern mimicking an ovarian atypical proliferative mucinous or endometrioid tumor (mucinous or endometrioid borderline tumor) [22] (Fig. 11.17). Interestingly, these can be indolent with prolonged survival. Occasionally, appendiceal metastases have features of goblet-cell carcinoid or mixed carcinoid-adenocarcinoma [23] (Fig. 11.18).

Ancillary Studies

Immunohistochemistry

A useful panel in the distinction of primary and metastatic adenocarcinoma is CK7 and CK20. Serous, mucinous, and endometrioid ovarian carcinomas are typically diffusely

Fig. 11.13 Poorly differentiated adenocarcinoma of gastric (**a**) and appendiceal (**b**) origin with signet-ring features at high power

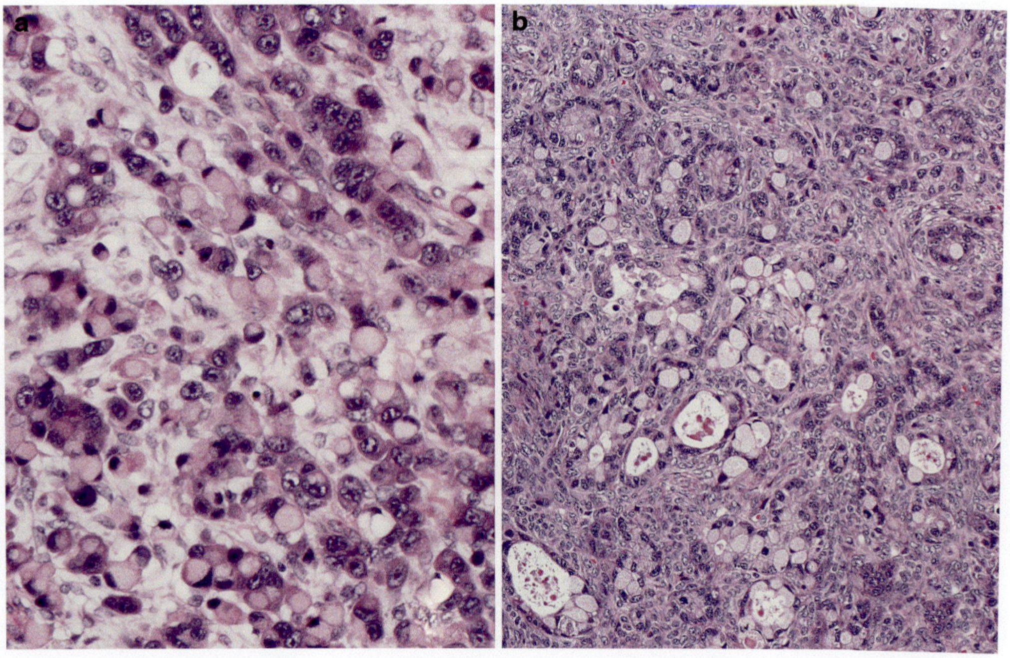

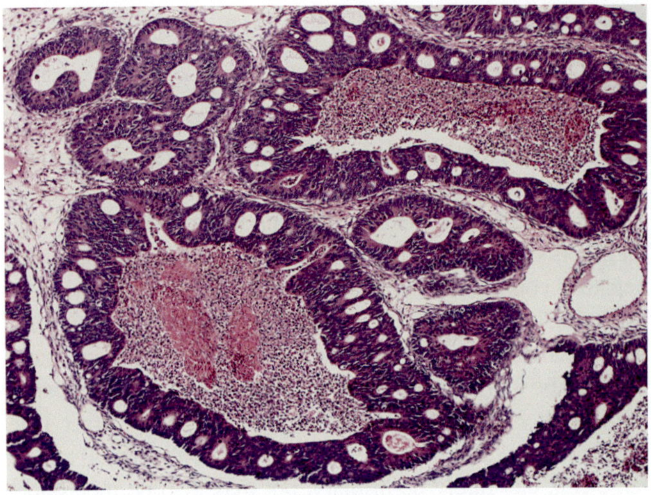

Fig. 11.14 Metastatic colonic adenocarcinoma. Cribriform glands lining larger glandular spaces filled with necrotic debris are referred to as "garlands" and are typical of colorectal adenocarcinoma

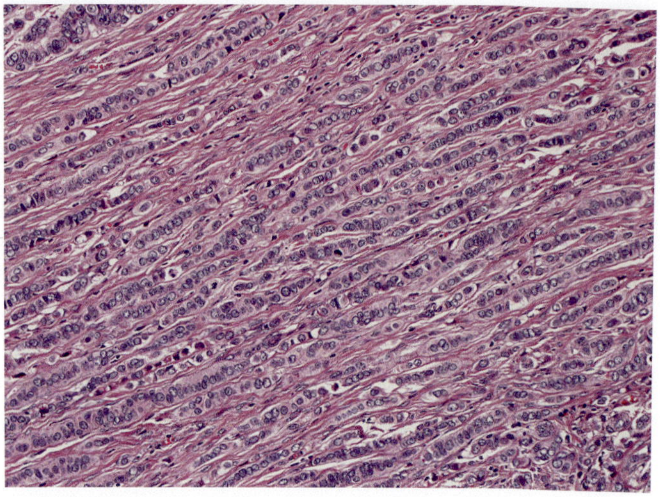

Fig. 11.16 Metastatic breast carcinoma. Small, monotonous epithelial cells arranged in a single-file pattern are characteristic of lobular breast carcinoma

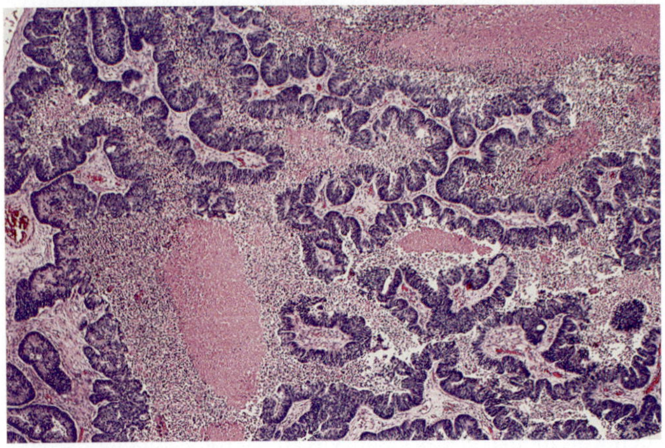

Fig. 11.15 Metastatic colonic adenocarcinoma. Garlands containing "dirty" necrosis, typical of colorectal adenocarcinoma

positive for CK7. CK20 may also be positive, but less often, and usually focally (see Chap. 9, Fig. 9.23) [18, 24–26]. CDX2 is often positive in mucinous carcinomas, with a nuclear staining pattern. Colorectal carcinomas are typically CDX2 positive, while endometrioid carcinomas tend to be CDX2 negative [12, 24, 25, 27]. Lower GI tract primary carcinomas are typically CK7 negative and CK20 positive, but gastric and pancreaticobiliary adenocarcinomas often have a CK7+/CK20– profile. In addition, right-sided and high-grade colonic carcinomas are often CK7+ [24]. Lung and breast carcinoma are typically CK7+ and CK20–. A negative DPC4 can provide evidence of a pancreatic primary when tumor cell nuclei are negative for this marker (50% of pancreatic ductal carcinomas lose DPC4 expression). Expression is retained in essentially

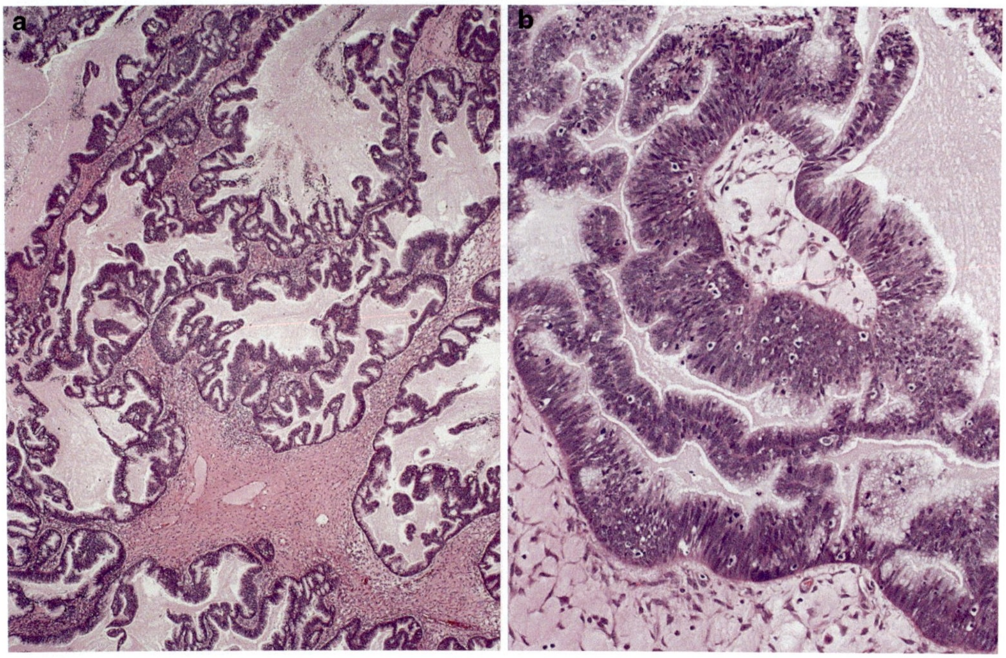

Fig. 11.17 Metastatic endocervical adenocarcinoma. (**a**) Low power with architectural pattern mimicking atypical proliferative mucinous tumor. (**b**) Higher power displays frequent mitotic figures in the apical portion of the cytoplasm and apoptotic bodies that should raise suspicion for endocervical origin

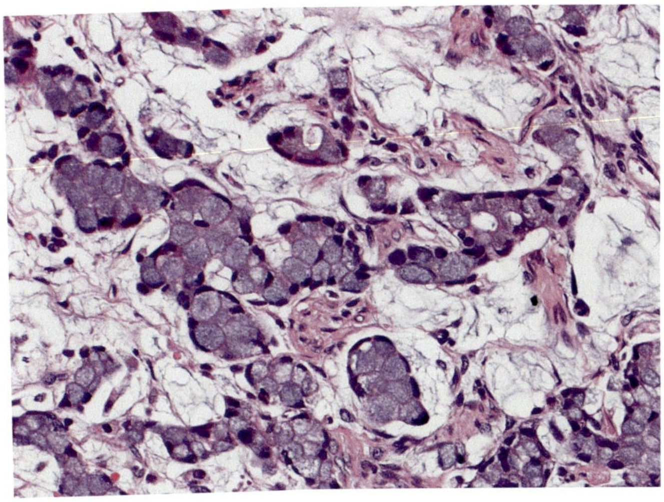

Fig. 11.18 Metastatic appendiceal adenocarcinoma with goblet-cell carcinoid-like features. Uniform cells with abundant basophilic cytoplasm and peripherally displaced nuclei arranged in small nests and cords

all ovarian mucinous carcinomas and most colorectal carcinomas [18, 24, 25]. Estrogen receptor (ER) and progesterone receptor (PR) are often positive in endometrioid carcinomas, may be positive in serous carcinoma, and are usually negative in mucinous carcinoma and non-breast metastatic carcinomas [28]. Colon carcinoma is typically positive for carcinoembryonic antigen, while 35% of ovarian mucinous carcinomas are negative [24]. High-grade serous carcinomas of all sites are typically p53 positive. WT1 is typically positive in ovarian serous carcinoma and negative in endometrial serous carcinoma [29–33]. Breast carcinomas are often gross cystic disease fluid protein and mammaglobin positive and negative for WT1, Pax8, and CA125 [33–35]. CA125 tends to be negative in primary ovarian mucinous tumors. Pax8 is positive in about 90% of ovarian serous, endometrioid, and clear cell carcinoma, but usually negative in mucinous carcinoma [35]. Mammaglobin is negative in serous carcinomas [34]. If there is a suspicion of an endocervical adenocarcinoma, p16 stain is positive in endocervical metastases, and in situ polymerase chain reaction for human papillomavirus can be diagnostic [22, 36, 37]. TTF1, a widely used marker of lung and thyroid tumors, can occasionally be positive in serous and endometrioid carcinomas [38].

Differential Diagnosis

When the gross and microscopic features have been evaluated, an algorithm can be used to estimate the probability of an ovarian versus an extraovarian primary site. For mucinous carcinoma, if all bilateral tumors and unilateral tumors smaller than 10 cm are tentatively classified as metastatic, and larger unilateral tumors tentatively classified as primary, this rule will correctly classify close to 90% of tumors. As initially proposed, this algorithm applied only to mucinous carcinomas [2], but evidence suggests it can be used for low-grade endometrioid-appearing carcinomas as well [39]. More recent data suggest that 12 cm or 13 cm may be a better cut-off than 10 cm [13] (Table 11.2). The algorithm is of particular value at the time of intraoperative consultation when histologic sampling is very limited, and can be used to estimate probabilities to guide surgical management. It is essential to obtain clinical information, particularly regarding any history of carcinoma, even if remote, as well as the findings at exploratory laparotomy. Diagnostic pitfalls and important features of ovarian metastases by site of origin are summarized in Tables 11.3 and 11.4, respectively.

Table 11.2 Performance of the algorithm using original (10 cm) and modified (12 and 13 cm) size criteria

Tumor type	*N*	10 cm cutoff % correctly classified	12 cm cutoff % correctly classified	13 cm cutoff % correctly classified
Primary ovarian mucinous tumors	52	100	100	98
Metastatic mucinous tumors (all)	142	77	80	82
Colorectal carcinoma	46	74	76	80
Disseminated peritoneal adenomucinosis from appendix	28	79	79	79
Metastatic appendiceal carcinoma	20	95	100	100
Pancreaticobiliary carcinomas	20	95	95	100
Small intestinal carcinomas	3	33	33	33
Gastric carcinomas	5	80	100	100
Endocervical carcinomas	20	55	60	70
Total	194	84	86	87

Modified from Yemelyanova et al. [13]
The algorithm designates unilateral ovarian tumors larger than the cutoff size as primary, and all smaller and bilateral tumors as metastatic

Table 11.3 Diagnostic pitfalls

1. Large unilateral metastases
   - Colorectal and endocervical carcinomas are the most likely to produce large unilateral metastases
2. Endometrioid pattern in metastatic colorectal carcinoma
   - The classic garlands of metastatic colon carcinoma simulate endometrioid carcinoma. Endometriosis and/or squamous differentiation can provide evidence of an ovarian primary
3. CK7+ and CK20− in certain metastases, especially pancreas, breast, and stomach
   - Take all clinical and pathologic features into account; do not rely solely on immunostains
4. An apparent transition from a noninvasive component (benign or atypical proliferative) to invasive carcinoma in occasional metastases
   - This occurs most often in mucinous carcinomas; a transition from benign to malignant, or the presence of noninvasive-appearing areas, is supportive but not diagnostic of an ovarian primary
5. Typical histology of clear cell carcinoma or endometrioid carcinoma in combination with features suggestive of metastasis such as bilaterality, nodularity, and/or surface involvement
   - Consider metastatic carcinoma from the endometrium or endocervix
6. Rare signet ring-like cells in primary ovarian carcinomas
   - May rarely be present in clear cell carcinoma or serous carcinoma
7. Lack of clinical or operative information suggesting an extraovarian primary site
   - Especially with mucinous carcinoma, always keep the possibility of metastasis in mind, even if the patient is relatively young

Pseudomyxoma Peritonei: Disseminated Peritoneal Adenomucinosis (DPAM) and Mucinous Carcinomatosis

Pseudomyxoma peritonei (PMP) is an old term that is used for historical continuity. This entity is now divided into two groups.

DPAM refers to the classical gross appearance of pseudomyxoma peritonei as described in the old literature: an abdominal cavity filled with gelatinous material that adheres to peritoneal surfaces. "Disseminated low-grade mucinous appendiceal neoplasm" is an alternative term that describes the same entity. Neither the term "mucinous borderline tumor" nor the term "well-differentiated mucinous adenocarcinoma" is favored. Nearly all cases are associated with a GI primary site, usually the appendix (Fig. 11.19). Typically, both ovaries are involved, and there is a right-sided predominance for unilateral ovarian involvement. Microscopically, there are large hypocellular pools of extracellular mucin with rare, scattered strips of mucinous epithelium and mucinous glands displaying mild or, at most, moderate cytological atypia [5–9]. Dissection of hypocellular or acellular pools of extracellular mucin through the ovarian stroma (pseudomyxoma ovarii) is a prominent feature (Figs. 11.20–11.22). Although peritoneal involvement is extensive, there is no destructive invasion of subperitoneal tissue or parenchymal organs. Lymph node and distant (extra-abdominal) metastases are exceedingly rare, and such a finding suggests an unsampled or unrecognized component of high-grade mucinous carcinoma. The distinguishing features of DPAM and ovarian atypical proliferative mucinous tumors, the tumor with which these are most often confused, are shown in Table 11.5. Rare DPAM cases result from appendix-like tumors that arise in ovarian teratomas [40, 41].

Mucinous Carcinomatosis

Typical high-grade mucinous adenocarcinoma characterizes this group. A haphazard infiltrative architectural pattern is characteristic. Cytologic atypia is usually high grade. A signet-ring cell component may be present, and invasion of subperitoneal tissue and parenchymal organs such as the liver are frequent. In the ovary, the typical features are described earlier in this chapter. Lymph node metastases or distant metastases may be present. These are rarely, if ever, primary in the ovary.

Table 11.4 Important features of common sources of metastatic adenocarcinoma

Primary site	Gross features	Microscopic features	Ancillary studies
Colon	Often large, multiloculated cystic. May be unilateral	Garlands Dirty necrosis Endometrioid-like features	CK7 < CK20, CDX2+
Breast	Small, normal size, firm	Signet-ring cell features or single-file pattern	Often GCDFP-15+, mammaglobin +, ER/PR may be+; WT-1–, Pax8–, CA125–
Appendix (DPAM)	Mucus coating the surface	Pseudomyxoma ovarii Bland epithelium	CK7–, CK20+
Appendix (carcinoma)	Often small	Signet-ring cells; often have goblet-cell carcinoid-like features	CK7–, CK20+
Pancreas	Small and bilateral; peritoneal carcinomatosis	Areas of overtly invasive carcinoma and benign-appearing areas	Often CK7+, CD20–, DPC4–
Stomach	Peritoneal carcinomatosis	Signet-ring cell features	CK7/20 variable; CK7+, CK20–; or CK7 > CK20
Endocervix	Often large and unilateral	Often "borderline"-like pattern; frequent mitotic figures and apoptotic bodies	P16 +, HPV+

ER estrogen receptor, *PR* progesterone receptor, *DPAM* disseminated peritoneal adenomucinosis, *HPV* human papillomavirus

Ovarian Serous Carcinoma Versus Fallopian Tube or Peritoneal Serous Carcinoma

Primary peritoneal serous carcinoma may account for 25% or more of patients with serous carcinomatosis (excluding those of endometrial origin) [42, 43], although newer definitions for primary fallopian tubal carcinoma will likely diminish that number significantly (Chap. 17). The Gynecologic Oncology Group criteria for separating ovarian from primary peritoneal carcinoma are shown in Table 11.6 [44]. Typically, ovaries in women with peritoneal serous carcinomas are normal sized or only slightly enlarged, and display predominantly surface involvement (Fig. 11.23). It has recently been suggested that a significant proportion of apparent primary peritoneal serous carcinomas are of tubal fimbrial origin [45, 46]. At present, although a malignant intraepithelial serous proliferation in the fimbriae may be considered evidence of a tubal primary site, it remains to be proved that this appearance does not simply reflect one mode of tumor spread from the ovary or peritoneum. The staging, treatment, and prognosis of ovarian, peritoneal, and tubal serous carcinomas are, for practical purposes, identical. Nonetheless, accurate assignment of the primary site is very important for studies of pathogenesis, epidemiology, and other investigative research. See Chaps. 6 and 17 for further discussion.

Synchronous Endometrioid Adenocarcinomas Involving the Endometrium and Ovary

Approximately 14% of women with endometrioid ovarian carcinoma also have endometrioid adenocarcinoma of the uterine corpus [47–50], and as many as 25% of endometrial carcinoma patients younger than 40 years have synchronous ovarian endometrioid carcinoma. Endometrial hyperplasia is also commonly present. As both tumors are often well-differentiated, they resemble each other, and excluding the possibility that the ovarian tumor is metastatic can be difficult. Usually, this can be determined based on evaluation of all of the pathologic features. If the uterine endometrial tumor is low grade with no or only inner-half myometrial invasion, its metastatic potential is extremely low, and the ovarian tumor can be regarded as independent. If the endometrial tumor is high grade and/or deeply myoinvasive, the features of the ovarian tumors become important. Bilaterality and a multinodular pattern, as well as other patterns characteristic of metastatic disease as discussed earlier, indicate metastatic tumor. Close association of the ovarian tumors with either an underlying endometrioid adenofibroma or endometriosis can support an independent origin of the ovarian tumor. Molecular studies can be of value in difficult cases, and there still remains a group for which it can be impossible to make a definitive determination of site of origin. See Chap. 7 for further discussion.

Ovarian Serous Carcinoma Versus Endometrial Serous Carcinoma with Ovarian Metastasis

In occasional cases of high-grade serous carcinomatosis, both the ovaries and endometrium display features suggesting a primary. This is a relatively uncommon situation that has not been extensively studied. WT1 stain has been used, as it is usually positive in ovarian and peritoneal serous carcinomas and negative in endometrial serous carcinomas [29–33, 51–53]. It is possible that some WT1 positive endometrial serous carcinomas represent drop metastases from the fallopian tube or ovary [53].

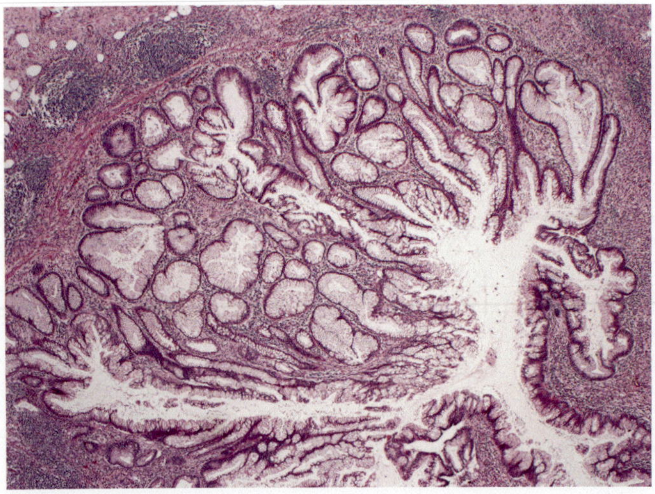

Fig. 11.19 Appendix with mucinous adenoma, the source of disseminated peritoneal adenomucinosis (DPAM)

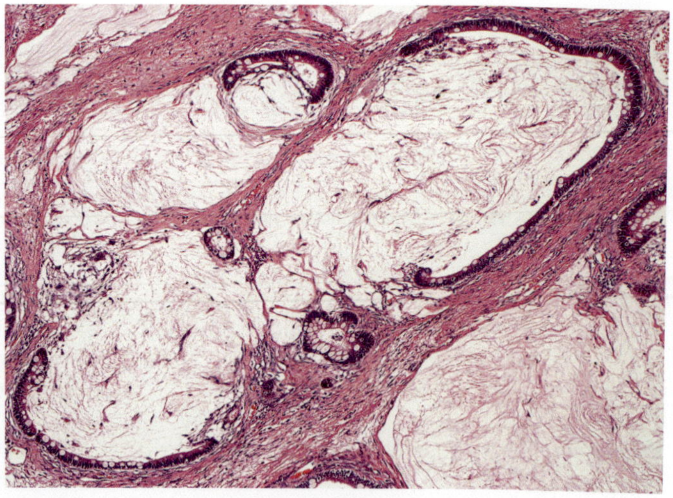

Fig. 11.21 Secondary ovarian involvement by low-grade mucinous neoplasm of appendiceal origin. Ruptured mucinous glands lined by bland epithelium and dissecting mucin

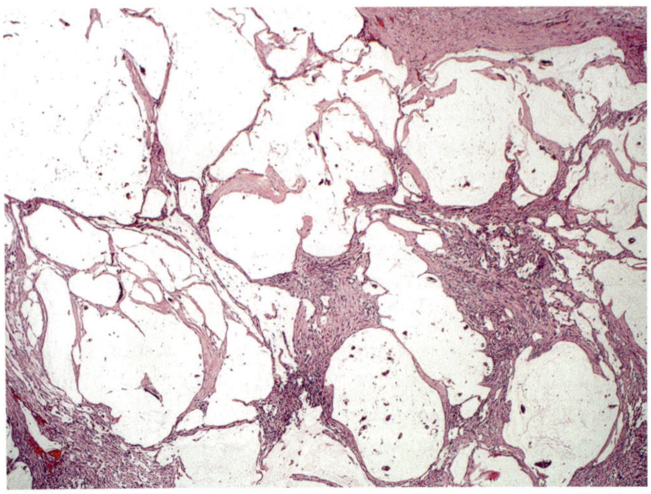

Fig. 11.20 Ovary with extensive pools of dissecting mucin (pseudomyxoma ovarii) and rare strips of minimally atypical mucinous epithelium, typical of secondary involvement by a low-grade mucinous neoplasm of appendiceal origin

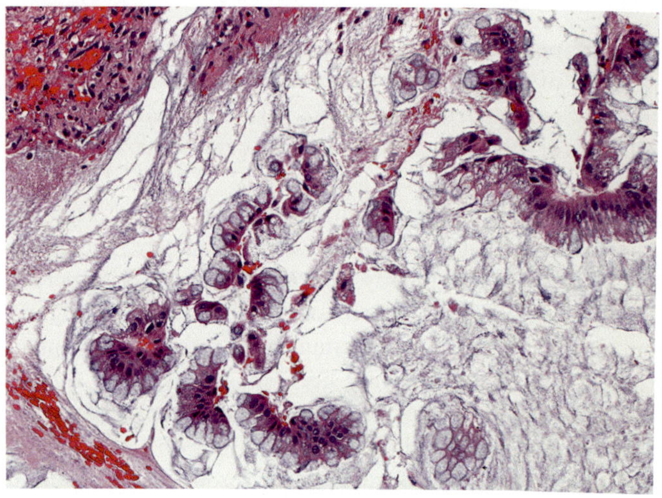

Fig. 11.22 Secondary ovarian involvement by low-grade mucinous neoplasm of appendiceal origin. Strips of bland mucinous epithelium floating in pools of mucin

Table 11.5 Distinction of ovarian mucinous tumors associated with disseminated peritoneal adenomucinosis from primary ovarian atypical proliferative mucinous tumors

Feature	Ovarian mucinous tumors in pseudomyxoma peritonei	Primary ovarian atypical proliferative mucinous tumors (intestinal mucinous borderline tumors)
Size	Variable (7–16 cm)	Large (18–22 cm)
Laterality	Bilateral (80%)	Unilateral (>95%)
Location of tumor	Surface, superficial cortex, stroma	Within stroma (rarely on surface when ruptured)
Pseudomyxoma ovarii	Often prominent	Usually absent or limited
Quantity of mucinous epithelium	Usually scant; haphazard when more abundant	Abundant; organized cysts with peripheral crypts
Associated appendiceal or other gastrointestinal mucinous tumor	Virtually all cases	None; rarely arises in the intestinal component of a dermoid cyst

Modified from Seidman et al. [2]

Table 11.6 Gynecologic Oncology Group (GOG) criteria for designation of ovarian versus primary peritoneal carcinoma [44]

GOG criteria for designation of ovarian versus primary peritoneal carcinoma [44]
1. Normal-sized ovaries or enlargement by a benign process
2. Extraovarian involvement greater than ovarian involvement
3. Ovaries have either: no tumor; tumor confined to the ovarian surface; or no tumor in ovarian stroma larger than 5 × 5 mm
4. Serous histology

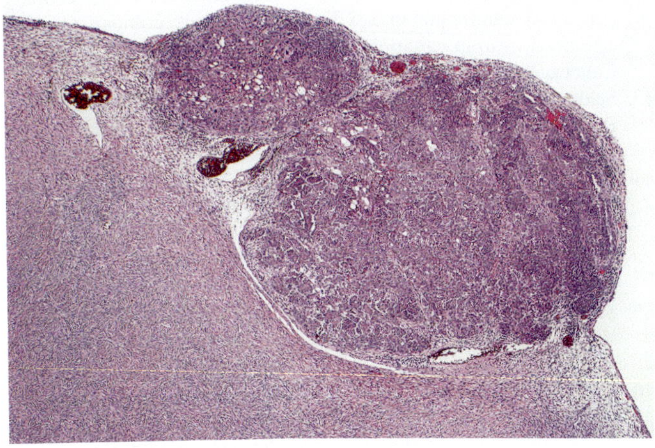

Fig. 11.23 Small surface nodule of high-grade serous carcinoma consistent with a peritoneal primary or a metastasis from a fallopian tube primary

Rare Primary Sites

On rare occasions, the epithelial type of malignant mesothelioma will mimic an ovarian primary tumor, but it should be remembered that diffuse malignant peritoneal mesothelioma is exceedingly rare in women [51, 54–56]. See Chap. 18 for further discussion. Adrenal and thyroid carcinomas rarely metastasize to the ovaries. Other tumors that may rarely create differential diagnostic problems when metastatic to the ovary include melanoma, lymphoma, and sarcomas. See Chap. 15 for further discussion.

References

1. Ronnett BM, Kajdacsy-Balla A, Gilks CB, et al. Mucinous borderline ovarian tumors: points of general agreement and persistent controversies regarding nomenclature, diagnostic criteria, and behavior. Hum Pathol. 2004;35:949–60.
2. Seidman JD, Kurman RJ, Ronnett BM. Primary and metastatic mucinous adenocarcinomas in the ovaries: incidence in routine practice with a new approach to improve intraoperative diagnosis. Am J Surg Pathol. 2003;27:985–93.
3. Khunamornpong S, Suprasert P, Pojchamarnwiputh S, et al. Primary and metastatic mucinous adenocarcinomas of the ovary: evaluation of the diagnostic approach using tumor size and laterality. Gynecol Oncol. 2006;101:152–7.
4. Riopel MA, Ronnett BM, Kurman RJ. Evaluation of diagnostic criteria and behavior of ovarian intestinal- type mucinous tumors: atypical proliferative (borderline) tumors and intraepithelial, microinvasive, invasive, and metastatic carcinomas. Am J Surg Pathol. 1999;23:617–35.
5. Ronnett BM, Yan H, Kurman RJ, et al. Patients with pseudomyxoma peritonei associated with disseminated peritoneal adenomucinosis have a significantly more favorable prognosis than patients with peritoneal mucinous carcinomatosis. Cancer. 2001;92: 85–91.
6. Ronnett BM, Kurman RJ, Zahn CM, et al. Pseudomyxoma peritonei in women: a clinicopathologic analysis of 30 cases with emphasis on site of origin, prognosis, and relationship to ovarian mucinous tumors of low malignant potential. Hum Pathol. 1995;26:509–24.
7. Ronnett BM, Zahn CM, Kurman RJ, et al. Disseminated peritoneal adenomucinosis and peritoneal mucinous carcinomatosis. A clinicopathologic analysis of 109 cases with emphasis on distinguishing pathologic features, site of origin, prognosis, and relationship to "pseudomyxoma peritonei". Am J Surg Pathol. 1995;19:1390–408.
8. Seidman JD, Cho KR, Ronnett BM, Kurman RJ. Surface epithelial tumors of the ovary. In: Kurman RJ, Ellenson LH, Ronnett BM, eds. Blaustein's Pathology of the Female Genital Tract, 6th edition. New York: Springer 2011, pages 679-784.
9. Szych C, Staebler A, Connolly DC, et al. Molecular genetic evidence supporting the clonality and appendiceal origin of pseudomyxoma peritonei in women. Am J Pathol. 1999;154:1849–55.
10. Lee KR, Young RH. The distinction between primary and metastatic mucinous carcinomas of the ovary: gross and histologic findings in 50 cases. Am J Surg Pathol. 2003;27:281–92.
11. Lewis MR, Deavers MT, Silva EG, Malpica A. Ovarian involvement by metastatic colorectal adenocarcinoma: still a diagnostic challenge. Am J Surg Pathol. 2006;30:177–84.
12. Judson K, McCormick C, Vang R, et al. Women with undiagnosed colorectal adenocarcinoma presenting with ovarian metastases: clinicopathologic features and comparison with women having known colorectal adenocarcinomas and ovarian involvement. Int J Gynecol Pathol. 2008;27:182–90.
13. Yemelyanova AV, Vang R, Judson K, et al. Distinction of primary and metastatic mucinous tumors involving the ovary: analysis of size and laterality data by primary site with reevaluation of an algorithm for tumor classification. Am J Surg Pathol. 2008;32:128–38.
14. Gagnon Y, Tetu B. Ovarian metastases of breast carcinoma: a clinicopathologic study of 59 cases. Cancer. 1989;64:892–8.
15. Lash RH, Hart WR. Intestinal adenocarcinomas metastatic to the ovaries: a clinicopathologic evaluation of 22 cases. Am J Surg Pathol. 1987;11:114–21.
16. Seidman JD, Khedmati F. Exploring the histogenesis of ovarian mucinous and transitional cell (Brenner) tumors: a study of 120 tumors. Arch Pathol Lab Med. 2008;132:1753–60.
17. Young RH, Hart WR. Renal cell carcinoma metastatic to the ovary: a report of three cases emphasizing possible confusion with ovarian clear cell adenocarcinoma. Int J Gynecol Pathol. 1992;11:96–104.
18. Ji H, Isacson C, Seidman JD, et al. Cytokeratins 7 and 20, Dpc4, and MUC5AC in the distinction of metastatic mucinous carcinomas in the ovary from primary ovarian mucinous tumors: Dpc4 assists in identifying metastatic pancreatic carcinomas. Int J Gynecol Pathol. 2002;21:391–400.
19. Khunamornpong S, Lerwill MF, Siriaunkgul S, et al. Carcinoma of extrahepatic bile ducts and gallbladder metastatic to the ovary: a report of 16 cases. Int J Gynecol Pathol. 2008;27:366–79.
20. Young RH, Gersell DJ, Clement PB, Scully RE. Hepatocellular carcinoma metastatic to the ovary: a report of three cases discovered during life with discussion of the differential diagnosis of hepatoid tumors of the ovary. Hum Pathol. 1992;23:574–80.

21. Young RH, Scully RE. Ovarian metastases from cancer of the lung: problems in interpretation- a report of seven cases. Gynecol Oncol. 1985;21:337–50.
22. Ronnett BM, Yemelyanova A, Vang R, et al. Endocervical adenocarcinomas with ovarian metastases: analysis of 29 cases with emphasis on minimally invasive cervical tumors and the ability of the metastases to simulate primary ovarian neoplasms. Am J Surg Pathol. 2008;32:1835–53.
23. Hristov AC, Young RH, Vang R, Yemelyanova AV, Seidman JD, Ronnett BM. Ovarian metastases of appendiceal tumors with goblet cell carcinoidlike and signet ring cell patterns: a report of 30 cases. Am J Surg Pathol. 2007;31:1502–11.
24. Baker PM, Oliva E. Immunohistochemistry as a tool in the differential diagnosis of ovarian tumors: an update. Int J Gynecol Pathol. 2005;24:39–55.
25. McCluggage WG. Immunohistochemical and functional biomarkers of value in female genital tract lesions. Int J Gynecol Pathol. 2006;25:101–20.
26. Vang R, Gown AM, Barry TS, et al. Cytokeratins 7 and 20 in primary and secondary mucinous tumors of the ovary: analysis of coordinate immunohistochemical expression profiles and staining distribution in 179 cases. Am J Surg Pathol. 2006;30:1130–9.
27. Vang R, Gown AM, Wu LS, et al. Immunohistochemical expression of CDX2 in primary ovarian mucinous tumors and metastatic mucinous carcinomas involving the ovary: comparison with CK20 and correlation with coordinate expression of CK7. Mod Pathol. 2006;19:1421–8.
28. Vang R, Gown AM, Barry TS, et al. Immunohistochemistry for estrogen and progesterone receptors in the distinction of primary and metastatic mucinous tumors in the ovary: an analysis of 124 cases. Mod Pathol. 2006;19:97–105.
29. Cathro HP, Stoler MH. The utility of calretinin, inhibin and WT1 immunohistochemical staining in the differential diagnosis of ovarian tumors. Hum Pathol. 2005;36:195–201.
30. Euscher ED, Malpica A, Deavers MT, Silva EG. Differential expression of WT-1 in serous carcinomas in the peritoneum with or without associated serous carcinoma in endometrial polyps. Am J Surg Pathol. 2005;29:1074–8.
31. Hashi A, Yuminamochi T, Murata S-I, Iwamoto H, Honda T, Hoshi K. Wilms tumor gene reactivity in primary serous carcinomas of the fallopian tube, ovary, endometrium, and peritoneum. Int J Gynecol Pathol. 2003;22:374–7.
32. Highlander B, Repasky E, Shrikant P, et al. Expression of Wilm's tumor gene (WT1) in epithelial ovarian cancer. Gynecol Oncol. 2006;101:12–7.
33. Tornos C, Soslow R, Chen S, Akram M, Hummer A, Abu-Rustum N, et al. Expression of WT1, CA 125 and GCDFP-15 as useful markers in the differential diagnosis of primary ovarian carcinomas versus metastatic breast cancer to the ovary. Am J Surg Pathol. 2005;29:1482–9.
34. Kanner WA, Galgano MY, Stoler MH, Mills SE, Atkins KA. Distinguishing breast carcinoma from Mullerian serous carcinoma with mammaglobin and mesothelin. Int J Gynecol Pathol. 2008;27:491–5.
35. Nonaka D, Chiriboga L, Soslow RA. Expression of pax8 as a useful marker in distinguishing ovarian carcinomas from mammary carcinomas. Am J Surg Pathol. 2008;32:1566–71.
36. Vang R, Gown AM, Farinola M, et al. p16 expression in primary ovarian mucinous and endometrioid tumors and metastatic adenocarcinomas in the ovary: utility for identification of metastatic HPV-related endocervical adenocarcinomas. Am J Surg Pathol. 2007;31:653–63.
37. Elishaev E, Gilks CB, Miller D, et al. Synchronous and metachronous endocervical and ovarian neoplasms: evidence supporting interpretation of the ovarian neoplasms as metastatic endocervical adenocarcinomas simulating primary ovarian surface epithelial neoplasms. Am J Surg Pathol. 2005;29:281–94.
38. Kubba LA, McCluggage WG, Liu J, Malpica A, Euscher ED, Silva EG, et al. Thyroid transcription factor-1 expression in ovarian epithelial neoplasms. Mod Pathol. 2008;21:485–90.
39. Stewart CJ, Brennan BA, Hammond IG, Leung YC, McCartney AJ. Accuracy of frozen section in distinguishing primary ovarian neoplasia from tumors metastatic to the ovary. Int J Gynecol Pathol. 2005;24:356–62.
40. McKenney JK, Soslow RA, Longacre TA. Ovarian mature teratomas with mucinous epithelial neoplasms: morphologic heterogeneity and association with pseudomyxoma peritonei. Am J Surg Pathol. 2008;32(5):645–55.
41. Vang R, Gown AM, Zhao C, et al. Ovarian mucinous tumors associated with mature cystic teratomas: morphologic and immunohistochemical analysis identifies a subset of potential teratomatous origin that shares features of lower gastrointestinal tract mucinous tumors more commonly encountered as secondary tumors in the ovary. Am J Surg Pathol. 2007;31(6):854–69.
42. Jaaback KS, Ludeman L, Clayton NL, Hirschowitz L. Primary peritoneal carcinoma in a UK cancer center: comparison with advanced ovarian carcinoma over a 5-year period. Int J Gynecol Cancer. 2006;16 Suppl 1:123–8.
43. Seidman JD, Horkayne-Szakaly I, Haiba M, Boice CR, Kurman RJ, Ronnett BM. The histologic type and stage distribution of ovarian carcinomas of surface epithelial origin. Int J Gynecol Pathol. 2004;23:41–4.
44. Bloss JD, Brady MF, Liao SY, Rocereto T, Partridge EE, Clarke-Pearson DL. Extraovarian peritoneal serous papillary carcinoma: a phase II trial of cisplatin and cyclophosphamide with comparison to a cohort with papillary serous ovarian carcinoma – a Gynecologic Oncology Group Study. Gynecol Oncol. 2003;89:148–54.
45. Salvador S, Gilks B, Kobel M, et al. The fallopian tube: primary site of most pelvic high grade serous carcinomas. Int J Gynecol Cancer. 2009;19:58–64.
46. Seidman JD, Zhao P, Yemelyanova A. Primary peritoneal high-grade serous carcinoma is very likely metastatic from serous tubal intraepithelial carcinoma: assessing the new paradigm of ovarian and pelvic serous carcinogenesis and its implications for screening for ovarian cancer. Gynecol Oncol. 2011;120:470-473.
47. Fuji H, Matsumoto T, Yoshida M, et al. Genetics of synchronous uterine and ovarian endometrioid carcinoma: combined analyses of loss of heterozygosity, PTEN mutation, and microsatellite instability. Hum Pathol. 2002;33:421–8.
48. Matias-Guiu X, Lagarda H, Catasus L, Bussaglia E, Gallardo A, Gras E, et al. Clonality analysis in synchronous or metachronous tumors of the female genital tract. Int J Gynecol Pathol. 2002;21:205–11.
49. Zaino R, Whitney C, Brady MF, DeGeest K, Burger RA, Buller RE. Simultaneously detected endometrial and ovarian carcinomas – a prospective clinicopathologic study of 74 cases: a Gynecologic Oncology Group study. Gynecol Oncol. 2001;83:355–62.
50. Soliman PT, Slomovitz BM, Broaddus RR, et al. Synchronous primary cancers of the endometrium and ovary: a single institution review of 84 cases. Gynecol Oncol. 2004;94:456–62.
51. Comin CE, Saieva C, Messerini L. h-caldesmon, calretinin, estrogen receptor, and Ber-EP4: a useful combination of immunohistochemical markers for differentiating epithelioid peritoneal mesothelioma from serous papillary carcinoma of the ovary. Am J Surg Pathol. 2007;31:1139–48.
52. McCluggage WG. WT1 is of value in ascertaining the site of origin of serous carcinomas within the female genital tract. Int J Gynecol Pathol. 2004;23:97–9.
53. Jarboe EA, Miron A, Carlson JW, et al. Coexisting intraepithelial serous carcinomas of the endometrium and fallopian tube: frequency and potential significance. Int J Gynecol Pathol. 2009;28(4):308–15.
54. Ordonez NG. Value of estrogen and progesterone receptor immunostaining in distinguishing between peritoneal mesotheliomas and serous carcinomas. Hum Pathol. 2005;36:1163–7.
55. Ordonez NG. What are the current best immunohistochemical markers for the diagnosis of epithelioid mesothelioma? A review and update. Hum Pathol. 2007;38:1–16.
56. Che M, Tornos C, Deavers MT, et al. Ovarian mixed-epithelial carcinomas with a microcystic pattern and signet ring cells. Int J Gynecol Pathol. 2001;20:323–8.

Germ Cell and Sex Cord Stromal Tumors: Clinical Introduction

12

Ginger J. Gardner and Jason A. Konner

Keywords
Germ cell tumor • Dysgerminoma • Yolk sac tumor • Teratoma • Granulosa cell tumor • Sertoli-Leydig cell tumor

Introduction

Non-epithelial ovarian tumors, which include germ cell and sex cord stromal tumors, comprise approximately 10% of all ovarian cancers [1]. In contrast to ovarian carcinomas, non-epithelial ovarian malignancies more frequently present in younger patients such as those in the reproductive age group, and the prognosis is frequently good. Unlike ovarian carcinomas, there is no defined genetic predisposition, such as *BRCA* status, for the most common tumors within this subgroup. The tumors themselves may demonstrate biologic activity with the production of excess endogenous hormones. Clinical differences between epithelial and non-epithelial ovarian tumors are summarized in Table 12.1. This chapter will discuss clinical issues pertaining to the presentation and management of germ cell and sex cord stromal tumors.

Germ Cell Tumors

Germ cell tumors of the ovary include a number of histologically different tumor types, all derived from the primitive germ cell of the embryonic gonad. While most germ cell tumors arise in the gonad, they can occasionally arise at locations along the line of migration of the primitive germ cells such as in the retroperitoneum or mediastinum.

Malignant ovarian germ cell tumors can be subdivided into the most common type, the dysgerminoma, and the less frequent non-dysgerminomatous tumors which include the yolk sac tumor (endodermal sinus tumor), immature teratoma, mixed germ cell tumors, embryonal carcinoma, non-gestational choriocarcinoma, and polyembryoma. Dysgerminomas are the most common malignant ovarian germ cell tumor, comprising 50% of cases, followed by immature teratomas and yolk sac tumors. Mixed tumors arise approximately 10% of the time, and embryonal carcinoma, polyembryoma, and non-gestational choriocarcinoma are very rare.

Clinical Presentation of Germ Cell Tumors

Ovarian germ cell tumors primarily occur in girls and young women. Approximately 70% of ovarian tumors that occur in girls under age 20 are of germ cell origin, 30% of which are malignant. Germ cell tumors are seen up to age 30, but thereafter the incidence is quite rare [1, 2].

Germ cell tumors present with abdominal pain and a palpable pelvic or abdominal mass in 85% of patients. Acute symptoms – due to ovarian torsion, hemorrhage, or rupture – occur in 10% of patients. Acute abdominal symptoms in this age group can be confused with appendicitis [3]. In some cases, menstrual irregularity and abdominal distension can occur and thereby masquerade as early pregnancy. A small proportion of patients exhibit isosexual precocity from tumors that produce HCG.

Diagnostic testing includes a complete history and physical examination, pelvic ultrasound, complete blood count, liver function tests, and serum tumor markers. Serum tumor markers include: LDH, AFP, HCG. Table 12.2 highlights the serum tumor marker expression associated with each subtype of ovarian germ cell tumor [4]. Often inhibin is also drawn as a serum tumor marker because its expression cor-

G.J. Gardner (✉)
Department of Surgery, Gynecology Service,
Memorial Sloan-Kettering Cancer Center,
1275 York Avenue, New York, NY 10065, USA

R.A. Soslow and C. Tornos (eds.), *Diagnostic Pathology of Ovarian Tumors*,
DOI: 10.1007/978-1-4419-9751-7_12, © Springer Science+Business Media, LLC 2011

Table 12.1 Clinical characteristics of malignant ovarian tumors [1, 3, 21, 64]

	Epithelial	Germ cell	Sex cord stromal
Incidence (%)	80–90	5–10	5–10
Age range (years)	30–79+	<35	Variable
Peak incidence	65	20	Variable
BRCA-associated	Yes	No	No
Hormonally productive	No	+/–	Yes
Stage distribution			
I/II	30	70	80+
III/IV	70	30	<20
5-year survival (%)			
Stage I/II	70–95	95+	75–90
Stage III/IV	20–50	80+	75–90

Table 12.2 Serum tumor markers associated with ovarian germ cell tumors

Tumor type	LDH	AFP	HCG
Dysgerminoma	+	–	+/–
Immature teratoma	+/–	+/–	–
Yolk sac tumor	+/–	+	–
Embryonal tumor	+/–	+	+
Choriocarcinoma	+/–	–	+
Mixed tumor	+	+	+

Modified from: Morrow and Curtin [4]

relates with granulosa cell tumors, a sex cord stromal tumor which can also occur in young patients. The hallmark serum tumor markers for ovarian germ cell tumors are LDH in dysgerminoma, AFP in yolk sac tumor, and HCG in choriocarcinoma. These serum tumor markers are not only helpful from a diagnostic standpoint, but if a malignant germ cell tumor is confirmed on final pathology, subsequent measurement of these serum tumor markers can be useful to monitor response to therapy and for post-treatment surveillance.

Based on the above diagnostic tests, a CT scan may be warranted. Chest imaging is important because germ cell tumors can metastasize to the chest. In premenarchal girls, a preoperative karyotype should be obtained because of the propensity of these tumors to arise in dysgenetic gonads.

Surgical exploration to achieve a histologic diagnosis is recommended for ovarian masses >2 cm in premenarchal girls, or in post-menarchal girls with either an ovarian cyst >8 cm or a complex ovarian lesion of any size. Simple ovarian cysts ≤8 cm in size in post-menarchal girls with negative tumor markers can undergo a trial of hormonal suppression for 2 months and reserve surgical exploration for only those cases that demonstrate interval enlargement.

Non-epithelial ovarian cancers can progress via intraperitoneal spread or through the lymphatics, similar to epithelial ovarian cancers; the staging maneuvers take this into account. The following section provides a synopsis of clinical issues related to each type of germ cell tumor, and is followed by a description of the general principles guiding the management of these diseases.

Clinical Features of Specific Germ Cell Tumors

Dysgerminoma

Dysgerminoma is the most common malignant ovarian germ cell tumor, accounting for up to 50% of cases. These tumors can cause an elevated LDH level. Approximately 5% of dysgerminomas contain syncytiotrophoblastic cells, which accounts for the potential of these tumors to secrete HCG as well [1].

Dysgerminoma occurs unilaterally in the right ovary (50%), in the left ovary (33–35%), or bilaterally (15–17%) [5, 6]. This rate of bilaterality is higher than for other malignant germ cell tumors of the ovary for which bilaterality is rare. The majority of cases have disease confined to the ovary at the time of diagnosis; however, these tumors have a greater propensity for lymphatic spread than for intraperitoneal metastases. For clinical stage IA disease, up to 38% of cases can harbor microscopic para-aortic lymph node metastasis, which underscores the importance of surgical staging [7, 8].

Approximately 5% of dysgerminomas are identified in patients of female phenotype but with abnormal gonadal development [1, 9]. Patients with pure gonadal dysgenesis (46XY, bilateral streak gonads), mixed gonadal dysgenesis (45X/46XY, unilateral streak gonad, contralateral testis), and androgen insensitivity syndrome (46XY, testicular feminization) frequently develop gonadoblastomas. Gonadoblastomas are benign ovarian tumors composed of germ cells and sex cord stroma which undergo germinomatous overgrowth and develop into an ovarian dysgerminoma over time [10]. If the gonads are left in situ in patients with gonadal dysgenesis, up to 50% will develop an ovarian malignancy, especially a dysgerminoma, and the risk increases with age. Patients are therefore recommended to undergo a prophylactic bilateral gonadectomy [11, 12]. In pre-menarchal girls that present with a pelvic mass, the karyotype should be determined preoperatively in order to appropriately plan for either a bilateral resection or a fertility sparing approach.

For surgical stage IA dysgerminoma, no adjuvant treatment is recommended. This approach results in a 5-year disease-free survival of greater than 95% [13, 14]. More advanced stage disease is treated with bleomycin, etoposide, and cisplatin (BEP) for three to four cycles or until serum tumor markers normalize. Dysgerminomas are extraordinarily chemosensitive and radiosensitive. Radiation treatment was historically given to patients with dysgerminoma in the primary treatment setting; however, due to high response rates with combination chemotherapy and the detrimental effects of radiation treatment on subsequent fertil-

Table 12.3 Grading of immature teratomas

Grade	Amount of immature neuroepithelium present on a single slide
0	None. All mature elements
1	Rare foci. ≤1 low power field
2	2–3 low power fields
3	≥4 low power fields

Adapted from O'Connor and Norris [15]

ity, radiation is now primarily reserved for the salvage setting.

Teratoma

Teratomas are derived from all three primitive cell layers of the embryo: the ectoderm, endoderm, and mesoderm. Mature teratomas are benign and demonstrate well-differentiated tissue which can include fully developed sebaceous glands, hair, and teeth. Benign mature cystic teratoma is the most common benign germ cell tumor in the young patient, and it occurs bilaterally in 5–10% of cases.

Unlike the mature teratoma, immature teratomas are malignant. Approximately 50% occur between the ages of 10–20 years, and rarely occur in the postmenopausal age group. These tumors rarely produce steroid hormones. They are primarily unilateral, and up to 30% have extra-ovarian spread at diagnosis.

The most common immature element, and therefore the one that is evaluated for grading, is the primitive neuroepithelium (Table 12.3). Some authors have proposed that immature teratoma grade be grouped into two categories: those with a slight degree of immaturity (Grade 1), which are not treated with combination chemotherapy, and those with a more marked degree of immaturity (Grade 2 and 3), which are treated [15]. As such, treatment for surgically staged grade 1 (low grade) immature teratoma confined to one ovary is unilateral salpingo-oophorectomy, without adjuvant treatment. All other patients receive adjuvant combination chemotherapy with bleomycin, etoposide, and cisplatin (BEP).

Among mature teratomas, malignant transformation is a rare event. For monodermal teratomas, the malignant cell type is directly related to the predominant mature cell type present. For example, struma ovarii can develop a focus of papillary thyroid carcinoma (<5%). In biphasic and triphasic mature teratomas, squamous cell carcinoma is the most common type of malignant transformation (0.5–2%) and typically occurs in postmenopausal patients [16, 17]. Treatment is based on the histologic cell type of the tumor.

Yolk Sac Tumor

Yolk sac tumor (endodermal sinus tumor) is the third most frequent malignant germ cell tumor of the ovary, and represents 10–15% of all such tumors. Most of these tumors secrete AFP and have unilateral ovarian involvement. Some authors have shown a good correlation between the AFP level and the extent of disease as identified intraoperatively [14, 18]. As with other germ cell tumors, treatment consists of surgical staging and debulking followed by BEP chemotherapy for three to four cycles or until normalization of serum tumor markers.

Other Germ Cell Tumors

Embryonal carcinoma is extremely rare. The primary lesion tends to be large, and approximately two-thirds are confined to the ovary at presentation. These tumors frequently secrete AFP and HCG [18, 19]. Polyembromas are also rare. The tumor demonstrates structures of the three somatic layers: endoderm, mesoderm, and ectoderm. The tumor tends to occur in very young pre-menarchal girls with signs of pseudopuberty with elevated AFP and HCG levels. Mixed germ cell tumors of the ovary contain two or more elements of the lesions described previously. The most frequent combination is a dysgerminoma and yolk sac tumor. Mixed tumors have the potential to secrete LDH, AFP, and HCG dependent on the tumor components present [18, 20].

Treatment for all of these germ cell tumors consists of surgical staging and debulking followed by BEP chemotherapy.

Choriocarcinoma of the Ovary

There are very few reported cases of pure non-gestational choriocarcinoma of the ovary. This tumor occurs in females <20 years old, and patients may present with isosexual precocity. Histologically, this tumor has a similar appearance as gestational choriocarcinoma metastatic to the ovary, and both secrete HCG [21]. Non-gestational choriocarcinoma of the ovary however, has a worse prognosis than metastatic gestational choriocarcinoma, most likely due to the difference in chemosensitivity of these two entities. Unlike other malignant germ cell tumors, chemotherapy with a methotrexate-based regimen as given for gestational trophoblastic disease (EMACO, etoposide, methotrexate, actinomycin D, cyclophosphamide, and vincristine) is frequently used, although BEP as given for other germ cell tumors can also be considered. Given the rarity of this tumor, data on long-term outcomes are lacking.

Surgery for Malignant Ovarian Germ Cell Tumors

The cornerstone of surgery in the management of malignant ovarian germ cell tumors is surgical resection and staging, similar to epithelial ovarian cancer. Non-epithelial ovarian cancers, however, principally occur in young reproductive age females, a group in which fertility-sparing surgery is often desirable. In 1977, Kurman and Norris observed no worsening of prognosis with fertility-sparing surgery in 182 patients who had tumors grossly confined to the ovary [22].

Table 12.4 Surgical staging for malignant germ cell tumors of the ovary

- Unilateral salpingo-oophorectomy**
- Peritoneal washings
- Omentectomy
- Pelvic and para-aortic lymphadenectomy
- Peritoneal biopsies
- Thorough exploration of the peritoneal structures with sampling any areas that appear suspicious for extra-ovarian disease

**The uterus and contralateral ovary are retained for future fertility

Table 12.5 Surgery for malignant germ cell tumors of the ovary

- Unilateral salpingo-oophorectomy and surgical staging
- Cytoreductive (debulking) surgery for advanced disease
- Second-look surgery is an option for incompletely resected immature teratoma
- If dysgenic gonads are present, perform a bilateral salpingo-oophorectomy to prevent development of a dysgerminoma

Over subsequent years, fertility-sparing surgery with unilateral salpingo-oophorectomy has become standard practice, even in the presence of metastatic disease. In addition to unilateral salpingo-oophorectomy, the surgical staging for malignant ovarian germ cell tumors includes: peritoneal washings, omentectomy, pelvic and para-aortic lymphadenectomy, and peritoneal biopsies including the posterior cul-de-sac, bladder reflection, bilateral pelvic sidewalls, bilateral paracolic spaces, and diaphragmatic surfaces (Table 12.4).

For those patients that have completed salpingo-oophorectomy and for whom the diagnosis of malignant germ cell tumor was not available until the post-operative final pathology, the question of whether to then recommend a second surgery in order to complete surgical staging arises. Recent intergroup analysis from the Pediatric Oncology Group and the Children's Study Group found that deviations from surgical staging guidelines did not adversely affect survival [23]. This has led some authors to consider a CT scan, and if the ovarian mass has been satisfactorily managed and no abnormalities are seen on CT, to proceed directly to adjuvant chemotherapy [3]. Exceptions to this include an unstaged dysgerminoma or an unstaged grade 1 immature teratoma. In these cases, the identification of extra-ovarian disease by surgical staging would change the management from observation to adjuvant combination chemotherapy and therefore surgical staging must be completed.

If metastatic disease is encountered during surgical resection for a newly diagnosed germ cell tumor, the same principles of cytoreductive surgery that have been applied to surgically manage advanced epithelial ovarian cancer are recommended with resection of as much tumor as is technically feasible and safe. In a Gynecologic Oncology Group (GOG) study, only 15/54 (28%) completely resected patients failed chemotherapy with vincristine, dactinomycin, and cyclophosphamide (VAC) compared to 15/22 (68%) of patients with incompletely resected disease [24]. Similar results were seen in a subsequent GOG trial of cisplatin, vinblastine, and bleomycin chemotherapy (PVB) in which completely resected disease patients failed chemotherapy 55% of the time compared to a rate of 82% for patients that had bulky residual disease [25]. Similarly, there appears to be some value to secondary cytoreductive surgery for disease recurrence [26–28]. Recently, Lai et al. demonstrated a favorable overall survival for patients with minimal residual disease following salvage surgery [29].

Second-look surgery to assess for residual disease upon completion of adjuvant chemotherapy is generally not recommended. In a report by the GOG, 117 patients with malignant germ cell tumors underwent second-look surgery following platinum-based chemotherapy. This study concluded that patients with completely resected disease at primary surgery, or incompletely resected disease without teratomatous elements, rarely benefit from this procedure; however, the authors suggested that the procedure could be considered for patients with incompletely resected tumor that contains teratoma. One limitation of this study, however, is that it included patients with clinically evident tumor before undergoing second-look surgery [30].

The role for surgery in the diagnosis, staging, debulking, and resection of recurrent disease for malignant germ cell tumors largely parallels the surgical approach for ovarian carcinoma. However, the age of the patient population and the generally good prognosis of malignant germ cell tumors support a fertility-sparing approach (Table 12.5).

Chemotherapy for Malignant Ovarian Germ Cell Tumors

In 1946, Gunna Teilum, a Danish pathologist, was the first to describe the homology between the male and female counterparts of germ cell tumors [31]. The parallels between testicular seminoma and ovarian dysgerminoma (ovarian seminoma) have provided the basis for utilizing similar treatment algorithms for both tumor types, although these superficial similarities belie some genetic differences. Because the incidence of malignant germ cell tumors of the testis is ten times greater than the incidence of malignant ovarian germ cell tumors, effective treatment regimens identified in testicular seminoma patients have frequently been extrapolated for use in ovarian dysgerminomas, often with good response.

For stage IA dysgerminomas and stage IA, grade I immature teratomas, no adjuvant therapy is recommended. All other patients with malignant ovarian germ cell tumors receive combination chemotherapy, usually with bleomycin, etoposide, and cisplatin (BEP) for three to four cycles or until serum tumor markers normalize [3, 29].

BEP is the standard adjuvant treatment for malignant ovarian germ cell tumors, yielding a 95% cure rate for

Table 12.6 Chemotherapy for malignant ovarian germ cell tumors

Bleomycin (B) 30 units/week Etoposide (E) 100 mg/m² day 1–5 Cisplatin (P) 20 mg/m² day 1–5	Given every 3 weeks, × 3 cycles
Recommended for all malignant ovarian germ cell tumors as adjuvant treatment except the following:* Dysgerminoma, stage IA Immature teratoma, stage IA, grade 1	

*Thorough surgical staging must be completed in order to assure early stage disease

completely resected stage I–III disease [32, 33] (Table 12.6). Historically, the vincristine, actinomycin D, and cyclophosphamide (VAC) regimen was used for postoperative adjuvant treatment in the early 1980s. Subsequently, the drugs found to be curative in metastatic testicular cancer, vinblastine, bleomycin, and cisplatin (VBP), were used [25]. The current regimen of choice is bleomycin, etoposide, and cisplatin (BEP). In randomized trials on patients with advanced testicular cancer, BEP is equivalent to VBP as adjuvant therapy and avoids vinblastine-related neuromuscular toxicity [33]. In two independent multicenter studies, BEP was significantly better than EP for patients with testicular germ cell tumors [34, 35].

The exquisite radiosensitivity of dysgerminoma is unique to this tumor, specifically in comparison to other types of malignant ovarian germ cell tumor. Given the detrimental effects of radiation on subsequent fertility, and the chemosensitivity of dysgerminoma to platinum-based chemotherapy, dysgerminomas are now primarily treated with BEP, and radiation is reserved for the salvage setting.

For platinum-refractory persistent disease, or recurrent disease, the prognosis is much worse, and salvage rates are less than 50%. Secondary cytoreductive surgery for recurrent disease can be considered. There is no standard chemotherapeutic regimen for persistent or recurrent disease.

Long-Term Treatment Effects

Given the young age at diagnosis and high likelihood of cure, the long-term treatment effects in survivors of malignant ovarian germ cell tumors have been described. Gershenson reported 40 patients with malignant ovarian germ cell tumors that were successfully treated with fertility-sparing surgery followed by combination chemotherapy, mostly non-platinum-based [36]. Eighty-three percent resumed regular menstrual function. Of the 16 patients who attempted pregnancy after chemotherapy, 11 patients delivered 22 healthy infants, none with major birth defects. A subsequent report by Brewer reported 26 patients with dysgerminoma treated with fertility-sparing surgery followed by platinum-based chemotherapy [37]. Ninety-three percent returned to their prechemotherapy menstrual pattern, and multiple reports have documented post-treatment live births [38–42].

In addition to menstrual and reproductive function, a recent GOG report evaluated sexual functioning and ability to form close relationships (dyadic adjustment) among 137 malignant ovarian germ cell survivors compared to 137 age-matched controls [43]. Among survivors initially treated with fertility-sparing surgery followed by combination chemotherapy, 87.3% reported still having menstrual function. Twenty-four survivors delivered 37 offspring after completing their cancer treatment. Survivors demonstrated greater concern about reproductive function and less sexual pleasure, but better dyadic consensus, satisfaction, and cohesion. As a group, survivors of malignant ovarian germ cell tumors are very likely to retain menstrual function and fertility over the ensuing years in the setting of primary fertility-sparing surgery.

BEP also poses a risk of long-term medical complications, including pulmonary toxicity, secondary malignancies, and permanent neuropathy or hearing loss. Pulmonary fibrosis from bleomycin is dose-related and can be debilitating. The risk increases cumulatively with dose; however, it is relatively uncommon with a total exposure under 400 units. Several factors, such as advanced age, exposure to oxygen or radiation, and renal dysfunction, frequently due to cisplatin, can increase an individual's risk. Radiography or pulmonary function testing are sometimes used for screening during therapy.

In men treated with BEP, there is a prolonged, elevated (1.5–2-fold) risk of developing secondary solid tumors, and this risk appears to persist for more than a decade. Lung, bladder, and gastrointestinal cancers constitute the majority of these cancers. Patients who receive etoposide are at risk of developing myeloid leukemia, which can develop 2–3 years after therapy. They are typically characterized by an 11q23 translocation and can be refractory to therapy.

Sex Cord–Stromal Tumors

Tumors in this category derive from the supporting tissue within the ovary including the stroma, and the sex cords supporting ovarian germ cell development. The third and most recent update of the World Health Organization classification of ovarian sex cord–stromal cell tumors is listed in Table 12.7 [44]. Granulosa cell tumors are the most frequent malignant ovarian sex cord stromal tumor. These tumors have an indolent growth rate and overall favorable prognosis with long-term survival ranging from 75% to 90% for all stages [45]. Surgery alone is recommended at the time of initial diagnosis for most cases. Patients with metastatic granulosa cell tumor or high grade Sertoli-Leydig cell tumors may require adjuvant therapy.

Table 12.7 Classification of sex cord stromal ovarian tumors granulosa–stromal tumors

Granulosa cell tumor Adult type Juvenile type	
Tumors in the thecoma–fibroma group Thecoma Fibroma–fibrosarcoma Sclerosing stromal tumor	
Sertoli-Leydig cell tumors	
Sertoli cell tumor Stromal-Leydig cell tumor Sertoli-Leydig (androblastomas) Well differentiated Intermediate differentiation Poorly differentiated Retiform	
Sex cord–stromal cell tumors of mixed or unclassified cell type Gynandroblastoma Sex cord tumor with annular tubules Steroid cell tumors	

Modified from Tavassoli and Devilee [44]

Granulosa–Stromal Tumors

Granulosa–stromal tumors include granulosa cell tumors, thecomas, and fibromas. Thecomas and fibromas are primarily benign. Granulosa cell tumors, however, account for approximately 70% of all malignant ovarian sex cord stromal tumors. Most of these tumors occur in the perimenopausal age group, but they can occur at any age. Five percent of granulosa cell tumors occur in prepubertal girls, and compose most of the juvenile subtype of this tumor.

Granulosa cell tumors are mostly unilateral (>95%). The most common signs and symptoms of granulosa cell tumors are nonspecific and common to most ovarian malignancies. Granulosa cell tumors, however, frequently secrete estrogen. In prepubertal girls with juvenile granulosa cell tumor, isosexual precocious pseudo-puberty can occur. In the reproductive and post-menopausal patients, abnormal vaginal bleeding is frequently seen. The increased systemic levels of estrogen secreted by granulosa cell tumors can cause a secondary endometrial hyperplasia in 25–50% of cases, and endometrial cancer in at least 5% of cases [46–48]. Any patient that undergoes fertility-sparing surgery for granulosa cell tumor requires endometrial curettings to evaluate for hyperplasia or carcinoma. Granulosa cell tumors tend to be very hemorrhagic tumors, and can therefore present as emergent operative cases with hemoperitoneum.

Granulosa cell tumors are usually stage I at diagnosis, but the adult-type granulosa cell tumor may recur 5–30 years later. Common sites for recurrence are the upper abdomen and pelvis. Occasionally, hematogenous spread to the liver, lungs, and brain can occur many years after initial diagnosis. Recurrence rate correlates with initial stage of disease. Inhibin, a polypeptide hormone that is produced from granulosa cells, is frequently secreted by granulosa cell tumors and is a useful marker for the disease surveillance. Recently, müllerian inhibitory substance (MIS) has shown promising activity as a tumor marker as well [49].

Table 12.8 Surgery for sex cord stromal tumors

- Total abdominal hysterectomy/ bilateral salpingo-oophorectomy, if no further fertility desired or postmenopausal
- Unilateral salpingo-oophorectomy only, if further fertility desired
- If uterus is retained, a uterine curettage is performed in estrogen-producing tumors to evaluate for concurrent endometrial cancer
- Staging is recommended for
  - Granulosa cell tumors
  - Sertoli-Leydig tumors, intermediate and high grade
  - Steroid tumors not otherwise specified
- For recurrent disease, surgical resection is recommended when feasible

Surgery is the most effective treatment for granulosa cell tumors (Table 12.8). In young patients, fertility-sparing surgery with unilateral salpingo-oophorectomy, staging, and debulking is performed. Endometrial curettings should be completed at the time of fertility-sparing surgery due to the associated risk of endometrial hyperplasia and carcinoma. In peri- or post-menopausal women, a hysterectomy and contralateral salpingo-oophorectomy should be performed.

Surgical staging is recommended for all patients with newly diagnosed granulosa cell tumor; it should include pelvic washings as well as pelvic and para-aortic lymphadenectomy similar to the staging maneuvers for an epithelial ovarian cancer. Surgical stage is considered to be the most important prognostic factor. Reported 5-year survival rates for stage I adult granulosa cell tumor (AGCT) range from 75% to 95%, and are in contrast to stage III/IV disease with 5-year survival of 22–50% [46]. Patients with AGCT that do not complete full surgical staging have a 15% incidence of nodal metastasis at initial recurrence, likely due to occult nodal disease not identified at initial diagnosis [50]. Large tumor size, tumor rupture, and high mitotic rate have also been associated with worse prognosis in some series [51, 52]. For those AGCT patients who recur, the median time to recurrence is 6 years, and the median survival after recurrence is 5.6 years, consistent with the indolent growth pattern of these tumors.

In addition to surgical staging, surgery remains the mainstay of initial management for patients with granulosa cell tumors in order to achieve cytoreductive surgery to a complete gross resection. The presence of residual disease has been shown to be an important predictor of survival [53]. For completely resected disease, there are no strict guidelines regarding adjuvant treatment. Some authors have suggested platinum-based chemotherapy for advanced stage patients or in early stage patients with intraoperative tumor rupture or high mitotic activity; however, supporting data are limited by the low incidence of this disease [54].

In the recurrent setting, these tumors are again primarily managed by surgical resection. Secondary surgical debulking is recommended whenever technically feasible. Following secondary debulking or in the presence of unresectable disease, targeted radiation or combination chemotherapy can be considered. There is no generally accepted standard regimen. Cisplatin, vinblastine, bleomycin (PVB) has demonstrated activity in two separate Italian studies and in the European Organization for Research and Treatment of Cancer (EORTC) with response rates for metastatic disease 57–82% [54–56]. Bleomycin, etoposide, and cisplatin (BEP) have also demonstrated activity in advanced and recurrent disease [57, 58]. More recently, there is promising antitumor activity reported with paclitaxel. Taxane-based therapy has demonstrated activity and lower toxicity. Further investigation with taxane-based treatment is underway (GOG 187) [59], as taxane-based therapy appears to have lower toxicity, and may serve as an alternate chemotherapeutic strategy for recurrent disease [60]. For unresectable metastatic disease, there is a rationale for the use of hormonal agents such as progestins, antiestrogens, or aromatase inhibitors. Data are limited, but hormonal treatment is considered a low toxicity option.

Sertoli-Leydig Tumors

Sertoli-Leydig tumors occur primarily in women <40 years old. These tumors are rare and account for less than 0.2% of ovarian cancers. Sertoli-Leydig cell tumors are mostly low-grade (of well or intermediate differentiation) and unilateral (98%), although a more aggressive high-grade subtype can occur.

Sertoli-Leydig cell tumors produce androgens, and virilization is present in 70–85% of patients. Signs of virilization include oligomenorrhea, amenorrhea, breast atrophy, acne, hirsutism, clitoralmegaly, deepening of the voice, and receding hairline. Elevated testosterone and androstenedione can occur, with normal or slightly elevated dehydroepiandrosterone sulfate. Less commonly, Sertoli-Leydig cell tumors can be associated with manifestations of estrogenization including isosexual precocity [61].

Surgical resection with unilateral salpingo-oophorectomy is recommended. In older patients in whom reproductive potential is not an issue, hysterectomy and bilateral salpingo-oophorectomy is appropriate. Surgical staging is performed for high-grade lesions. Adjuvant treatment is similar to that for granulosa cell tumors, and is generally recommended in the setting of high grade or advanced stage disease. For low-grade tumors, the 5-year survival is 70–90%. Most deaths are associated with poorly differentiated and heterologous lesions.

Sex Cord - Stromal Tumors: Mixed and Unclassified Types

Ovarian sex cord tumor with annular tubules (SCTAT) typically occurs between ages 30–35. This tumor can be associated with Peutz–Jeghers Syndrome, an autosomal dominant syndrome characterized by buccal pigmentation, intestinal polyposis, and a 5% incidence of gonadal stromal tumors. In addition, carriers have an increased risk for adenoma malignum, a well-differentiated mucinous adenocarcinoma of the cervix. For SCTATs associated with Peutz–Jeghers syndrome, surgical staging is not required [62]. In contrast, SCTATs unassociated with Peutz-Jeghers Syndrome may pursue a clinically malignant course.

Gynandroblastoma is very rare tumor composed of admixed well-differentiated Sertoli cell and granulosa cell components. The lesions can be clinically estrogenic, androgenic, or nonfunctional. Almost all cases reported have been stage I, and the prognosis is favorable [63].

Conclusion

In summary, non-epithelial ovarian cancer represents a diverse group of malignant cell types. Appropriate tumor classification drives the surgical approach and selection of adjuvant therapy. Fertility-sparing surgery for malignant ovarian germ cell tumors and sex cord stromal tumors provides excellent long-term oncologic outcomes. Platinum-based chemotherapy is frequently effective in these tumors. Over the last half century, much has been accomplished in understanding the classification, clinical outcome, and treatment options for these rare tumors. With this foundation, future investigation to understand the molecular basis of these tumors and improve on the therapeutic outcomes for these patients is possible.

References

1. Scully RE, Young RH, Clement RB. Tumors of the ovary, maldeveloped gonads, fallopian tube, and broad ligament, Atlas of tumorpathology, fascicle 23, vol. 3. Washington, DC: Armed Forces Institute of Pathology; 1998. p. 169–498.
2. Chen LM, Berek JS. Ovarian and fallopian tubes. In: Haskell CM, editor. Cancer treatment. 5th ed. Philadelphia: WB Saunders; 2000. p. 55.
3. Gershenson DM. Management of ovarian germ cell tumors. J Clin Oncol. 2007;25(20):2938–43.
4. Morrow CP, Curtin JP. Synopsis of gynecologic oncology. 5th ed. London: Churchill Livingstone, 1998; 11:295.
5. Mueller CW, Topkins P, Lapp WA. Dysgerminoma of the ovary. An analysis of 427 cases. Am J Obstet Gynecol. 1950;60:153.
6. Santesson L. Clinical and pathological survey of ovarian tumours treated at the Radiumhemmet – dysgerminoma. Acta Radiol. 1947;28:643.
7. DePalo G, Lattuada A, Kenda R, et al. Germ cell tumors of the ovary: the experience of the National Cancer Institute of Milan I. Dysgerminoma. Int J Radiat Oncol Biol Phys. 1987;13:853.
8. Buskirk SJ, Schray MF, Podratz KC, et al. Ovarian dysgerminoma: a retrospective analysis of results of treatment, sites of treatment failure, and radiosensitivity. Mayo Clin Proc. 1987;62(12):1149–57.

9. Obata NJ, Nakashima N, Kawai, et al. Gonadoblastoma with dysgerminoma in one ovary and gonadoblastoma with dysgerminoma and yolk sac tumor in the contralateral ovary in a girl with 46XX karyotype. Gynecol Oncol. 1995;58:124–8.
10. Scully RE. Gonadoblastoma: a review of 74 cases. Cancer. 1970;25:1340–56.
11. Bremer GL, Land JA, Tiebosch A, et al. Five different histologic subtypes of germ cell malignancies in an XY female. Gynecol Oncol. 1993;5:225–31.
12. Krasna IH, Lee MI, Smilow P, Sciorra, et al. Risk of malignancy in bilateral streak gonads: the role of the Y chromosome. J Pediatr Surg. 1992;27:1376–80.
13. Thomas GM, Dembo AJ, Hacker NF, DePetrillo AD. Current therapy for dysgerminoma of the ovary. Obstet Gynecol. 1987;70:268–75.
14. Talerman A. Germ cell tumors of the ovary. Curr Opin Obstet Gynecol. 1997;9:44–7.
15. O'Connor DM, Norris HJ. The influence of grade on the outcome of stage I ovarian immature (malignant) teratomas and the reproducibility of grading. Int J Gynecol Pathol. 1994;13(4):283–9.
16. Hackethal A, Brueggmann D, Bohlmann MK, et al. Squamous-cell carcinoma in mature cystic teratoma of the ovary: systematic review and analysis of published data. Lancet Oncol. 2008;9:1173–80.
17. Dos Santos L, Mok E, Iasonos A, et al. Squamous cell carcinoma arising in mature cystic teratoma of the ovary: a case series and review of the literature. Gynecol Oncol. 2007;105:321–4.
18. Kawai M, Kano T, Kikkawa F, et al. Seven tumor markers in benign and malignant germ cell tumors of the ovary. Gynecol Oncol. 1992;45:248–53.
19. Ueda G, Abe Y, Yoshida M, Fujiwara T. Embryonal carcinoma of the ovary: a six-year survival. Gynecol Oncol. 1990;31:287–92.
20. Gershenson DM, Del Junco G, Copeland LJ, Rutledge FN. Mixed germ cell tumors of the ovary. Obstet Gynecol. 1984;64:200–6.
21. Oliva E, Andrada E, Pezzica E, Prat J. Ovarian carcinomas with choriocarcinomatous differentiation. Cancer. 1993;72:2441–6.
22. Kurman RJ, Norris HJ. Malignant germ cell tumors of the ovary. Hum Pathol. 1977;8:551–64.
23. Billmire D, Vinocur C, Rescoria F, et al. Outcome and staging evaluation in malignant germ cell tumors of the ovary in children and adolescents: an intergroup study. J Pediatr Surg. 2004;39:424–9.
24. Slayton RE, Park RC, Silverberg SG, et al. Vincristine, dactinomycin, and cyclophosphamide in the treatment of malignant germ cell tumors of the ovary: a Gynecologic Oncology Group study. Cancer. 1985;56:243–8.
25. Willaims SD, Blessing JA, Moore DH, et al. Cisplatin, vinblastine, and bleomycin in advanced and recurrent ovarian germ-cell tumors. Ann Intern Med. 1989;111:22–7.
26. Messing MJ, Gershenson DM, Morris M, et al. Primary treatment failure in patients with malignant ovarian germ cell neoplasms. Int J Gynecol Cancer. 1992;2:295–300.
27. Munkarah A, Gershenson DM, Levenback C, et al. Salvage surgery for chemorefractory ovarian germ cell tumors. Gynecol Oncol. 1994;55:217–23.
28. Rezk Y, Sheinfeld J, Chi DS. Prolonged survival following salvage surgery for chemorefractory ovarian immature teratoma: a case report and review of the literature. Gynecol Oncol. 2005;96(3):883–7.
29. Lai CH, Chang TC, Hsueh S, et al. Outcome and prognostic factors in ovarian germ cell malignancies. Gynecol Oncol. 2005;96:784–91.
30. Williams SD, Blessing JA, DiSaia PJ, et al. Second-look laparotomy in ovarian germ cell tumors: the Gynecologic Oncology Group experience. Gynecol Oncol. 1994;52:287–91.
31. Teilum G. Gonocytoma: homologous ovarian and testicular tumors. Acta Pathol Microbiol Scand. 1946;23:242.
32. Gershenson DM, Morris M, Cangir A, et al. Treatment of malignant germ cell tumors of the ovary with bleomycin, etoposide, and cisplatin. J Clin Oncol. 1990;8:715–20.
33. Williams S, Blessing JA, Liao SY, et al. Adjuvant therapy of ovarian germ cell tumors with cisplatin, etoposide, and bleomycin: a trial of the Gynecologic Oncology Group. J Clin Oncol. 1994;12:701–6.
34. Loehrer PJ, John SD, Elson P, Eihorn LH, Trump D. Importance of bleomycin in favorable-prognosis disseminated germ cell tumors: an Eastern Cooperative Oncology Group trial. J Clin Oncol. 1995;13:470–6.
35. De Wit R, Stoler G, Kaye SB, et al. Importance of bleomycin in combination chemotherapy for good-prognosis testicular nonseminoma: a randomized study of the European Organization for Research and Treatment of Cancer Genitourinary Tract Cancer Cooperative Group. J Clin Oncol. 1997;15:1837–43.
36. Gershenson DM. Menstrual and reproductive function after treatment with combination chemotherapy for malignant ovarian germ cell tumors. J Clin Oncol. 1988;6:270–5.
37. Brewer M, Gershenson DM, Herzog CE, et al. Outcome and reproductive function after chemotherapy for ovarian dysgerminoma. J Clin Oncol. 1999;17:1670–5.
38. Low JH, Perrin LC, Crandon J, et al. Conservative surgery to preserve ovarian function in patients with malignant ovarian germ cell tumors: a review of 74 cases. Cancer. 2000;89:391–8.
39. Zanetta G, Bonazzi C, Cantu MG, et al. Survival and reproductive function after treatment of malignant germ cell ovarian tumors. J Clin Oncol. 2001;19:1015–20.
40. Tangir J, Zelterman D, Ma W, et al. Reproductive function after conservative surgery and chemotherapy for malignant germ cell tumors of the ovary. Obstet Gynecol. 2003;101:251–7.
41. Kang H, Kim T, Kim WY, et al. Outcome and reproductive function after cumulative high-dose combination chemotherapy with bleomycin, etoposide and cisplatin (BEP) for patients with ovarian endodermal sinus tumor. Gynecol Oncol. 2008;111:106–10.
42. De La Motte Rouge T, Pautier P, Duvillard P, et al. Survival and reproductive function of 52 women treated with surgery and bleomycin, etoposide, cisplatin (BEP) chemotherapy for ovarian yolk sac tumor. Ann Oncol. 2008;19:1435–41.
43. Gershenson DM, Miller AM, Champion VL, et al. Reproductive and sexual function after platinum-based chemotherapy in long-term ovarian germ cell tumor survivors: a Gynecologic Oncology Group Study. J Clin Oncol. 2007;25(19):2792–7.
44. Tavassoli FA, Devilee P, editors. Pathology and genetics of tumours of the breast and female genital tract. Lyon: IARC; 2003.
45. Colombo N, Parma G, Zanagnolo V, Insinga A. Management of ovarian stromal cell tumors. J Clin Oncol. 2007;25(20):2944–51.
46. Schumer ST, Cannistra SA. Granulosa cell tumor of the ovary. J Clin Oncol. 2003;21:1180–9.
47. Gusberg SB, Kardon P. Proliferative endometrial response to theca-granulosa cell tumors. Am J Obstet Gynecol. 1971;111:633–43.
48. Stenwig JT, Hazekamp JT, Beecham JB. Granulosa cell tumors of the ovary. A clinicopathological study of 118 cases with long-term follow-up. Gynecol Oncol. 1979;7:136–52.
49. Rey RA, Lhomme C, Marcillac I, et al. Anti-müllerian hormone as a serum marker of granulosa cell tumors of the ovary: comparative study with serum alpha inhibin and estradiol. Am J Obstet Gynecol. 1996;174:958–65.
50. Abu-Rustum N, Restivo A, Ivy J, et al. Retroperitoneal nodal metastasis in primary and recurrent granulosa cell tumors of the ovary. Gynecol Oncol. 2006;103:31–4.
51. Miller BE, Barron BA, Wan JY, et al. Prognostic factors in adult granulosa cell tumor of the ovary. Cancer. 1997;79:1951–5.
52. Fujimoto T, Sakuragi N, Okuyama K, et al. Histopathological prognostic factors of adult granulosa cell tumors of the ovary. Acta Obstet Gynecol Scand. 2001;80:1069–74.
53. Holland DR, Le Riche J, Swenerton KD, et al. Flow cytometric assessment of DNA ploidy is a useful prognostic factor for patients with granulosa cell ovarian tumors. Int J Gynecol Cancer. 1991;1:227–32.

54. Colombo N, Sessa C, Landoni F, et al. Cisplatin, vinblastine, and bleomycin combination chemotherapy in metastatic granulosa cell tumor of the ovary. Obstet Gynecol. 1986;67:265–8.
55. Zambetti M, Escobedo A, Pilotti S, et al. Cisplatin, vinblastine, bleomycin combination chemotherapy in advanced or recurrent granulosa cell tumors of the ovary. Gynecol Oncol. 1990;36:317–20.
56. Pecorelli S, Wagenaar HC, Vergote IB, et al. Cisplatin, vinblastine, and bleomycin combination chemotherapy in recurrent or advanced granulosa cell tumor of the ovary: an EORTC Gynecologic Cancer Cooperative Group study. Eur J Cancer. 1999;35:1331–7.
57. Gershenson DM, Morris M, Burke TW, et al. Treatment of poor-prognosis sex cord-stromal tumors of the ovary with the combination of bleomycin, etoposide and cisplatin. Obstet Gynecol. 1996;87:527–31.
58. Homesley HD, Bundy BN, Hurteau JA, Roth LM. Bleomycin, etoposide, and cisplatin combination therapy of ovarian granulosa cell tumors and other stromal malignancies: a Gynecologic Oncology Group study. Gynecol Oncol. 1999;72:131–7.
59. Homesley HD, Bundy BN, Hurteau JA, et al. Bleomycin, etoposide and cisplatin combination therapy of ovarian granulosa cell tumors and other stromal malignancies: a Gynecologic Oncology Group study. Gynecol Oncol. 1999;72:131–7.
60. Brown J, Shvartsman HS, Deavers MT, et al. The activity of taxanes compared with bleomycin, etoposide, and cisplatin in the treatment of sex cord-stromal ovarian tumors. Gynecol Oncol. 2005;97(2):489–96.
61. Roth LM, Anderson MC, govan AD, et al. Sertoli-Leydig cell tumors: a clinicopathologic study of 34 cases. Cancer. 1981;48:187–97.
62. Young RH, Welch WR, Dickersin GR, Scully RE. Ovarian sex cord tumor with annular tubules: review of 74 cases including 27 with Peutz-Jeghers syndrome and four with adenoma malignum of the cervix. Cancer. 1982;50:1384–402.
63. Roth LM. Recent advances in the pathology and classification of ovarian sex cord-stromal tumors. Int J Gynecol Pathol. 2006;25:199–215.
64. Wong LC, Ngan HYS, Ma HK. Primary treatment with vincristine, dactinomycin, and cyclophosphamide in non-dysgerminomatous germ cell tumour of the ovary. Gynecol Oncol. 1989;34:155–8.

13 Pathology of Germ Cell Tumors

Charles J. Zaloudek

Keywords
Dysgerminoma • Embryonal carcinoma • Neuroectodermal Tumors • PNET • Struma Ovarii • Teratoma • Yolk sac tumor

The neoplasms in this group are almost all derived from germ cells, although rare examples in older women appear to originate from somatic cells in ovarian carcinomas. Dysgerminoma and embryonal carcinoma are tumors of undifferentiated germ cells, but a majority of germ cell tumors show differentiation along embryonic (teratoma) or extraembryonic (choriocarcinoma, yolk sac tumor) lines (Table 13.1) [1]. Benign cystic teratomas are among the most common ovarian tumors, but all other types of germ cell tumors are rare.

Table 13.1 Germ cell tumors of the ovary

Dysgerminoma
Embryonal carcinoma
Yolk sac tumor
Polyembryoma
Non-gestational choriocarcinoma
Mature teratoma
Benign or malignant neoplasm arising in a mature teratoma
Immature teratoma
Monodermal teratoma
Neuroectodermal tumor
Struma ovarii
Carcinoid tumor
Gonadoblastoma

Dysgerminoma

Dysgerminoma is one of the two most common malignant germ cell tumors of the ovary [2, 3], but it is still an uncommon neoplasm accounting for only 1–2% of all malignant ovarian tumors.

Clinical Features

Dysgerminoma occurs mainly in children and young women [4–6]. The average age is 22 years, and 90% of patients are less than 30 years. Occasional dysgerminomas are detected during pregnancy [7]. The most common clinical presentation is with abdominal distention, an abdominal mass, or abdominal pain. Rare patients have hypercalcemia [8, 9]. Serum lactic dehydrogenase (LDH) is often elevated, and can serve as a tumor marker. Occasional patients with dysgerminoma (3–5%) have a modest elevation of human chorionic gonadotropin, which is secreted by syncytiotrophoblastic giant cells (STGC) in the tumor [10]. High levels of human chorionic gonadotropin or elevated levels of alpha-fetoprotein (AFP) suggest the presence of other germ cell elements, such as choriocarcinoma, yolk sac tumor, or immature teratoma. Dysgerminoma is the most common malignant gonadal tumor in patients with gonadal dysgenesis.

Dysgerminoma tends to be limited to the ovaries (stage I) at diagnosis and it is usually unilateral. It is unique among

C.J. Zaloudek (✉)
Department of Pathology, University of California San Francisco, San Francisco, CA, USA

R.A. Soslow and C. Tornos (eds.), *Diagnostic Pathology of Ovarian Tumors*,
DOI: 10.1007/978-1-4419-9751-7_13, © Springer Science+Business Media, LLC 2011

the malignant germ cell tumors as it is the only type in which there is a significant risk of bilateral disease; the contralateral ovary contains dysgerminoma in 5–15% of cases. Involvement of the contralateral ovary is visible or palpable at surgery in half of the bilateral cases, but it is microscopic in the rest, which is why some oncologists biopsy the contralateral ovary even if it appears normal. Dysgerminoma spreads via lymphatics to the para-aortic lymph nodes, and subsequently to the mediastinal lymph nodes, and transperitoneally to the pelvic and abdominal peritoneum [11].

Unilateral encapsulated dysgerminoma (FIGO stage Ia) is typically treated by unilateral salpingo-oophorectomy [12]. After surgery, patients are either managed by close surveillance or they are given adjuvant chemotherapy; the current trend appears to be against routine adjuvant chemotherapy due to its potential adverse effect on fertility [13, 14]. Relapses occur in approximately 20% of patients, and are generally successfully managed by chemotherapy [13]. If a dysgerminoma is detected in a dysgenetic gonad the appropriate treatment is bilateral gonadectomy, since the gonads are nonfunctional and are at increased risk for the development of neoplasia. The treatment of advanced disease (stage > IA) varies according to the clinical situation. If fertility is not an issue, the standard treatment is total abdominal hysterectomy, bilateral salpingo-oophorectomy, limited debulking, and postoperative chemotherapy or radiotherapy. The uterus and the contralateral ovary may be conserved in advanced cases where preservation of fertility is important, provided that they are not involved by tumor [15]. Chemotherapy with platinum-based regimens is highly effective and has fewer complications than radiation therapy. In particular, it is less likely to cause ovarian failure and infertility. Current therapy is so successful that the overall survival rate for optimally treated patients with dysgerminoma is greater than 90% [3, 12]. Recurrences are generally detected within the first few years after primary treatment.

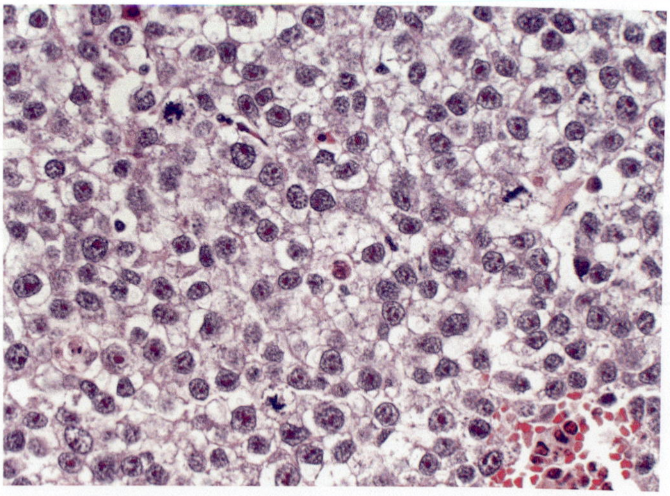

Fig. 13.1 Dysgerminoma. Dysgerminoma cells are polygonal with relatively uniform round vesicular nuclei, conspicuous nucleoli, and abundant clear cytoplasm. Mitotic figures are readily identified

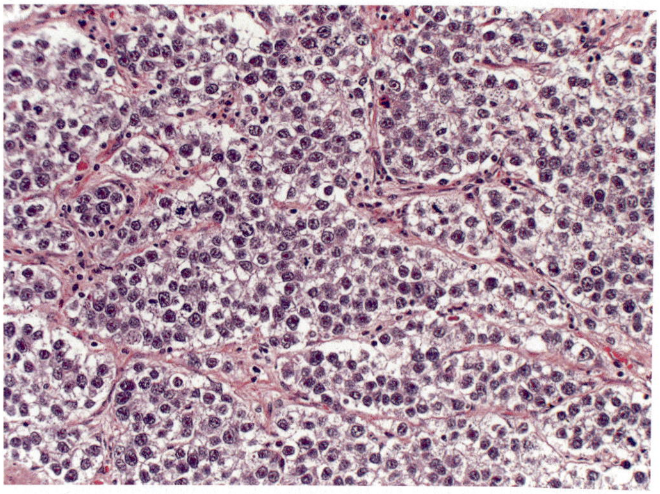

Fig. 13.2 Dysgerminoma. The tumor cells grow in nests separated by fibrous septa

Macroscopic Features

Dysgerminoma tends to be large (usually >10 cm diameter) and solid with a smooth outer surface. The cut surface is fleshy and gray, tan, or white. Foci of hemorrhage, necrosis, or cystic degeneration are often present.

Microscopic Features

Microscopically, dysgerminoma consists of polygonal cells with abundant granular eosinophilic or clear cytoplasm and distinct cell membranes (Fig. 13.1). The nuclei are round, medium sized, and relatively uniform with vesicular chromatin and prominent nucleoli. Mitotic figures are usually numerous. The cells grow in sheets, nests, or trabeculae that are separated by fibrous septa (Fig. 13.2). The septa vary from thin wisps of connective tissue to thick fibrous bands. Lymphocytes are usually noted within the septa, and are also present among the tumor cells. Most of the tumor infiltrating lymphocytes are T-cells, with B-cells largely confined to the germinal centers that are present in some tumors [16, 17]. Epithelioid cells and multinucleated Langhans-type giant cells are present in many dysgerminomas and often coalesce to form sarcoid-like granulomas (Fig. 13.3).

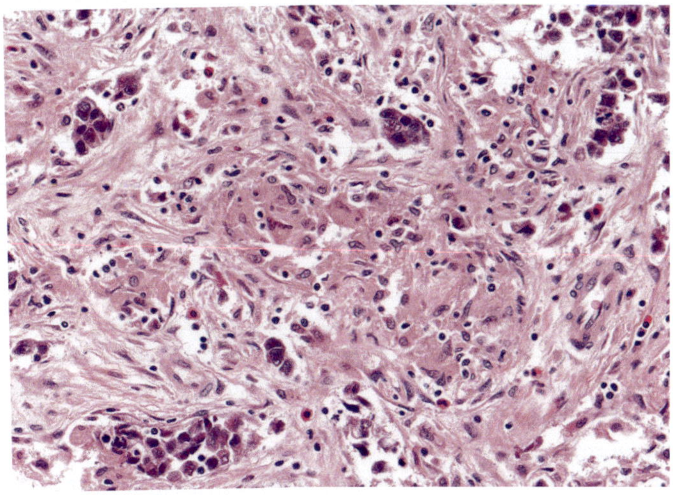

Fig. 13.3 Dysgerminoma. Small clusters of tumor cells are separated by epithelioid granulomas and fibrous tissue

Table 13.2 Variant patterns in dysgerminoma: unusual histologic features that can cause diagnostic difficulties

Extensive necrosis
Extensive fibrosis
Extensive granulomatous reaction
Tubular or microcystic pattern
Anaplastic pattern
Poor fixation artifact
Syncytiotrophoblastic cells

The diagnosis of dysgerminoma is straightforward in typical cases, but there are a number of variants that can cause diagnostic problems (Table 13.2), particularly in small biopsies or frozen sections. Occasionally, necrosis is so extensive that it obscures the underlying tumor morphology. The nature of some tumors is masked by focal or extensive fibrosis secondary to infarction or some other insult. Granulomatous inflammation is so extensive in some dysgerminomas that it is difficult to see the tumor cells among the epithelioid cells. In some dysgerminomas there is edema and loss of intercellular cohesion resulting in formation of gland-like spaces or microcysts, a pattern that has been called "tubular" dysgerminoma. When the degree of nuclear atypia is greater than usual and there are numerous mitotic figures (more than 30 mitotic figures per 10 high power fields) dysgerminoma can mimic embryonal carcinoma. Such tumors were designated "anaplastic" dysgerminomas in the past, although now they are simply viewed as a variant histologic pattern of dysgerminoma. Poor fixation can result in clumping of the nuclear chromatin and dense basophilic cytoplasm, also raising the question of embryonal carcinoma. About 3% of dysgerminomas contain syncytiotrophoblastic giant cells; patients with such tumors can have positive pregnancy tests and elevated levels of beta-human chorionic gonadotropin. No other malignant germ cell elements are present, so a dysgerminoma with syncytiotrophoblastic giant cells is classified as a variant of dysgerminoma, not a mixed germ cell tumor. Although these dysgerminoma variants can be difficult to diagnose, there is no evidence that their clinical behavior is any different than that of typical dysgerminomas. The correct diagnosis can usually be established by study of additional slides that reveal more typical histologic features or by performing immunohistochemical stains.

Table 13.3 Most useful immunohistochemical markers for dysgerminoma

D2-40	Membrane staining
CD117	Membrane staining
PLAP	Membrane and cytoplasmic staining
OCT4	Nuclear staining
NANOG	Nuclear staining
SALL4	Nuclear staining

Immunohistochemistry

A variety of immunohistochemical stains can be used to confirm a diagnosis of dysgerminoma (Table 13.3). These fall into two main groups: antibodies against cytoplasmic and membranous antigens and antibodies against nuclear antigens. The first group includes placental alkaline phosphatase (PLAP), CD117 (c-kit), and D2-40 [18, 19]. All are excellent markers for dysgerminoma. Staining for PLAP and limited staining for D2-40 can be seen in other types of germ cell tumors. Cytoplasmic staining is often present but only membrane staining is considered to be indicative of dysgerminoma (Fig. 13.4). In general, staining is strongest for D2-40, while CD117 is least likely to stain necrotic background debris. The second group of immunostains, those for nuclear antigens, includes OCT-4, NANOG, and SALL4 [18, 20–22]. These are nuclear transcription factors that are present in primitive germ cell tumors. Diffuse strong nuclear staining is characteristic of dysgerminoma (Fig. 13.5) but these stains are not specific. All three are positive in embryonal carcinoma and SALL4 also stains yolk sac tumor. Immunostains for cytokeratin frequently show limited cytoplasmic dot or rim-like staining in dysgerminoma but the type of diffuse strong cytoplasmic or membranous staining that is seen in embryonal carcinoma and yolk sac tumor does not occur [23]. Dysgerminoma is generally negative for epithelial

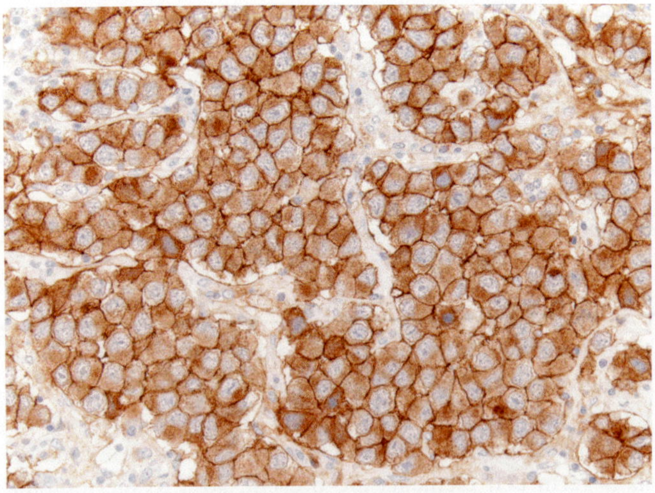

Fig. 13.4 Dysgerminoma. An immunostain for CD117 (c-kit) shows strong membrane staining in all of the tumor cells

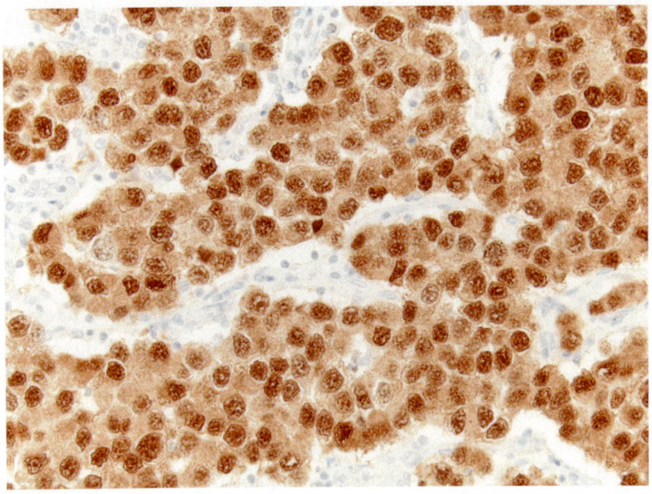

Fig. 13.5 Dysgerminoma. There is strong staining of tumor cell nuclei with an immunostain for OCT4

membrane antigen (EMA) [23]. Given the lack of specificity of many of the stains for dysgerminoma, a panel of stains, usually including CD117 and/or D2-40, OCT4 and/or SALL4, and cytokeratin must be used to confirm the diagnosis. Staining for human chorionic gonadotropin is generally negative, but positive cytoplasmic staining is seen in the syncytiotrophoblastic giant cells that are found in dysgerminomas from patients with elevated serum beta-human chorionic gonadotropin. Immunostains for AFP are negative.

Differential Diagnosis

A number of tumors enter the differential diagnosis of dysgerminoma. Several of these are malignant germ cell tumors, but some non-germ cell tumors can also be confused with dysgerminoma (Table 13.4). Dysgerminoma is the most common element in mixed germ cell tumors of the ovary, so it is important to search carefully to exclude other germ cell elements before making a diagnosis of a pure dysgerminoma.

Table 13.4 Dysgerminoma differential diagnosis

Mixed germ cell tumor
Embryonal carcinoma
Yolk sac tumor
Clear cell carcinoma
Lymphoma
Melanoma

Embryonal carcinoma rarely occurs in pure form in the ovary. It is mainly seen as a minor component in mixed germ cell tumors. When the cells grow in solid sheets they can resemble dysgerminoma, particularly if the cytoplasm is pale or clear. Embryonal carcinoma cells have larger more pleomorphic nuclei than dysgerminoma and the chromatin is coarser. The cytoplasm tends to be amphophilic or basophilic rather than clear or eosinophilic. Neither granulomas nor a lymphoid stroma are present in embryonal carcinoma. Careful evaluation often reveals poorly formed glands in embryonal carcinoma. Both embryonal carcinoma and dysgerminoma show positive staining for the nuclear transcription factors OCT4, NANOG, and SALL4, so these do not differentiate the two. Dysgerminoma is positive for CD117 and D2-40, while embryonal carcinoma is positive for CD30 and cytokeratin.

Yolk sac tumor grows in a confusing variety of patterns, but most do not resemble dysgerminoma. A solid pattern of growth in which the cells have clear cytoplasm can, however, resemble dysgerminoma. The tumor cell nuclei tend to be smaller and darker in yolk sac tumor, and the overall cell size is also smaller. Adequate study generally reveals other more characteristic patterns of yolk sac tumor. Immunostains are helpful in this differential diagnosis. Both dysgerminoma and yolk sac tumor are positive for SALL4, but yolk sac tumor does not stain for OCT4 or NANOG, nor is there staining for CD117 or D2-40. Staining for PLAP is less extensive in yolk sac tumor than in dysgerminoma. Yolk sac tumor shows diffuse strong staining for cytokeratin, and most examples are positive for AFP and glypican-3, which are negative in dysgerminoma.

Clear cell carcinoma occurs in older patients than dysgerminoma, and it is often associated with endometriosis. The solid pattern of clear cell carcinoma can mimic dysgerminoma, but the tumor cell nuclei are larger. Tubulocystic and papillary growth patterns can usually be identified in clear cell carcinoma, in addition to the solid pattern. Immunohistochemical stains for cytokeratin and epithelial membrane antigen are strongly positive in clear cell carcinoma, but are either negative or only focally and weakly positive in dysgerminoma. Clear cell carcinoma is negative for most of the characteristic stains for dysgerminoma, PLAP being the only one that might show staining in a carcinoma.

Lymphoma can resemble dysgerminoma because it is a monotonous proliferation of cells with medium-sized nuclei and the cell cytoplasm can be clear or pale. Lymphoma often surrounds residual ovarian structures, rather than displacing them, as occurs in dysgerminoma. Dysgerminoma cell nuclei tend to be vesicular, while the nuclear chromatin is often coarsely granular in lymphoma and lymphoma cells usually have less cytoplasm. Most lymphomas of the ovary are B-cell lymphomas, and accordingly stain for leukocyte common antigen and various markers of B-cell differentiation, all of which are negative in dysgerminoma. Lymphomas do not stain for the characteristic dysgerminoma markers.

Melanoma of the ovary is most often metastatic, but it can also be primary, arising in a teratoma. The histologic appearance is notoriously variable, and melanoma can mimic a wide variety of neoplasms, including when the cells have an epithelioid appearance and pale cytoplasm, dysgerminoma. Melanoma often exhibits a variety of growth patterns and the tumor cell nuclei tend to be pleomorphic while dysgerminoma is monotonous, so examination of an adequate number of sections generally differentiates melanoma from dysgerminoma. Melanoma cells sometimes contain visible brown melanin pigment. Immunostains for melanoma markers such as S100, HMB45, and Melan A are almost invariably positive in melanoma, but negative in dysgerminoma. Dysgerminoma markers are likewise negative in melanoma, although it should be noted that cytoplasmic, but not membranous, staining for CD117 occurs in melanoma.

Embryonal Carcinoma

Embryonal carcinoma is a rare ovarian neoplasm that is morphologically identical to embryonal carcinoma of the testis.

Clinical Features

Embryonal carcinoma occurs mainly in children and young women less than 30 years old [24] with occasional reports of cases in older women [25]. Only one large series has been reported. In 1976 Kurman and Norris reported 14 ovarian tumors in which embryonal carcinoma was the predominant element. These were found in a series of 85 germ cell tumors on file at the Armed Forces Institute of Pathology (AFIP) that also included 71 yolk sac tumors [24]. A survey of our files and the literature reveals that embryonal carcinoma is even more uncommon in general practice than Kurman and Norris found in the AFIP files. The average patient is about 15 years old and the most common clinical presentation is with pelvic or abdominal pain or a palpable abdominal mass. Precocious pseudopuberty can occur in premenarchal patients while older patients often have menstrual abnormalities. Postpubertal patients are sometimes initially thought to be pregnant since many have a positive pregnancy test or an elevated serum β-HCG.

Since most patients are young, unilateral salpingo-oophorectomy is the most appropriate surgical treatment when the tumor is confined to the ovary. Treatment is individualized in patients with more advanced disease. Unilateral salpingo-oophorectomy and limited debulking surgery are performed if the contralateral ovary and uterus are uninvolved while patients with more advanced disease are generally treated by total abdominal hysterectomy, bilateral salpingo-oophorectomy, and debulking.

Prior to the availability of effective combination chemotherapy embryonal carcinoma was often rapidly fatal, and survival, even in stage I, was less than 50% [24]. At present, patients with completely resected embryonal carcinoma are treated with the same postoperative cisplatin-based adjuvant chemotherapy that is used for other nondysgerminomatous germ cell tumors with nearly complete success. Many patients with residual or recurrent tumors can be cured with combination chemotherapy.

Serum assays for β-HCG and AFP are useful in patient follow-up. If serum levels of either or both markers remain elevated after treatment, or if they increase during follow-up, the patient is likely to have recurrent or metastatic tumor and requires further evaluation and, usually additional therapy.

Macroscopic Pathology

Embryonal carcinomas tend to be large solid tumors with an average diameter of around 15 cm. The cut surface is soft and fleshy and small or large cysts filled with mucoid secretions can be present. The tumor tissue is tan or gray and there are generally prominent areas of hemorrhage and necrosis.

Microscopic Pathology

The microscopic appearance is similar to that of embryonal carcinoma of the testis [24, 26]. The tumor cells are large and polygonal with vesicular nuclei that contain coarse basophilic chromatin and one or two prominent nucleoli (Fig. 13.6). The cell membranes are well defined and the cytoplasm is abundant and usually amphophilic or clear. Mitotic figures and apoptotic bodies are numerous. The tumor cells grow in nests or disorderly sheets surrounded by a loose myxoid, fibrous, or cellular spindle cell stroma

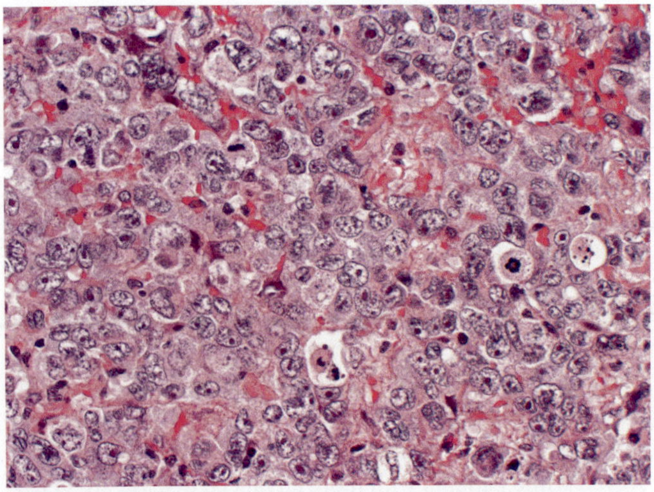

Fig. 13.6 Embryonal carcinoma. Sheet-like growth of poorly differentiated tumor cells with atypical vesicular nuclei that contain conspicuous nucleoli. Mitotic figures are readily identified. The cytoplasm is moderate in amount and amphophilic. No glands or other patterns of differentiation are present

containing numerous blood vessels. While some tumors consist entirely of sheets of undifferentiated tumor cells, in most there are at least focal areas in which clefts, glands, or papillae are lined by cuboidal to columnar tumor cells. The cells lining glands often have a disorderly arrangement. Syncytiotrophoblastic giant cells (STGC) are present in most embryonal carcinomas, either among the tumor cells or, more commonly, in loose connective tissue around the stromal blood vessels. The STGC have abundant eosinophilic or amphophilic cytoplasm and multiple large atypical nuclei. The nuclei have coarse chromatin and most contain conspicuous nucleoli. These cells are the source of the human chorionic gonadotropin that causes most of the endocrine symptoms in patients with embryonal carcinoma. Cases of pure embryonal carcinoma do occur in the ovary, but embryonal carcinoma is more often present as a component of a mixed germ cell tumor. Whenever embryonal carcinoma is identified in an ovarian tumor a careful search for other germ cell tumor types is mandatory.

Immunohistochemistry

Embryonal carcinoma is so rare that it is always desirable to confirm the diagnosis with immunohistochemistry (Table 13.5). Embryonal carcinoma shows evidence of epithelial differentiation. Immunostains performed with wide spectrum cytokeratin antibodies such as AE1/AE3 are positive, with a distinctive membranous pattern of staining (Fig. 13.7). Staining for epithelial membrane antigen, however, is negative in embryonal carcinoma. The staining pattern to be expected with cytokeratin subsets often employed in ovarian pathology, particularly cytokeratin 7 (CK7), is unclear, with some authors reporting positive staining for CK7 [27] and others absence of staining for CK7 [28]. CD30 is a positive marker for embryonal carcinoma, with a membranous pattern of staining, while CD117 and D2-40, which are diffusely positive in dysgerminoma, are negative or only focally positive in embryonal carcinoma. Stains for a variety of stem cell markers are positive in embryonal carcinoma, reflecting its primitive nature. Embryonal carcinoma shows strong positive nuclear staining for OCT4, NANOG, and SALL4 (Fig. 13.8) [18, 20]. Dysgerminoma also stains for all three of these markers and yolk sac tumor stains for SALL4. Embryonal carcinoma is apparently unique among the primitive germ cell tumors in that it shows positive nuclear staining for SOX2, although staining is not observed in every case [18, 26]. The syncytiotrophoblastic giant cells present in many embryonal carcinomas show intense homogeneous cytoplasmic staining for cytokeratin and human chorionic gonadotropin. Kurman and Norris reported cytoplasmic staining for AFP in embryonal carcinoma cells in 70% of cases [24], but this seems high based on experience with testicular tumors and no studies using modern immunohistochemistry have been performed to confirm their data.

Differential Diagnosis

Embryonal carcinoma must be differentiated from dysgerminoma, yolk sac tumor, and choriocarcinoma, all of which have some overlapping histologic features with it, and from high-grade surface epithelial carcinomas such as serous carcinoma. The immunostains that are most useful in the differential diagnosis of embryonal carcinoma are listed in Table 13.5.

Dysgerminoma consists of polygonal cells with clear, eosinophilic, or even basophilic cytoplasm and when the nuclei are more atypical than usual there is some resemblance to embryonal carcinoma. Embryonal carcinoma cells tend to be larger than dysgerminoma cells, the nuclei are more pleomorphic and contain coarser nuclear chromatin, and the mitotic rate is higher. Immunohistochemical stains are helpful in the differential diagnosis. Embryonal carcinoma shows diffuse membranous staining for cytokeratins such as AE1/AE3 and for CD30, while there is little or no staining for these markers in dysgerminoma. Dysgerminoma is positive for D2-40 and CD117 and these markers are negative in embryonal carcinoma. The nuclear markers OCT4, NANOG, and SALL4 are positive in both dysgerminoma and embryonal carcinoma, but SOX2 is positive only in embryonal carcinoma.

Yolk sac tumor can resemble embryonal carcinoma when the tumor cells grow in sheets or when there are glands lined by columnar cells. Generally, the cells in yolk sac tumor are significantly smaller than those in embryonal carcinoma, and

Table 13.5 Immunohistochemical features of embryonal carcinoma and their use in differential diagnosis

Stain	Embryonal carcinoma	Yolk sac tumor	Dysgerminoma	Serous carcinoma
Cytokeratin	+, membranes	+, cytoplasm	–	+
Cytokeratin 7	Unknown	–	–	+
EMA	–	–	–	+
CD30	+, membranes	–	–	–
CD117	–	–	+	–
D2-40	–	–	+	–
OCT4	+, nucleus	–	+	–
NANOG	+, nucleus	–	+	–
SALL4	+, nucleus	+	+	–
SOX2	+, nucleus	–	–	–

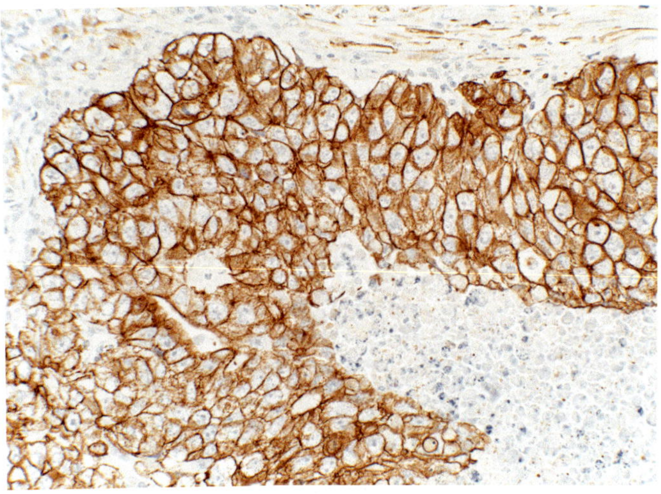

Fig. 13.7 Embryonal carcinoma. An immunostain for cytokeratin AE1/AE3 reveals a distinctive strong membranous pattern of staining. A similar pattern of staining is observed with antibodies against CD30

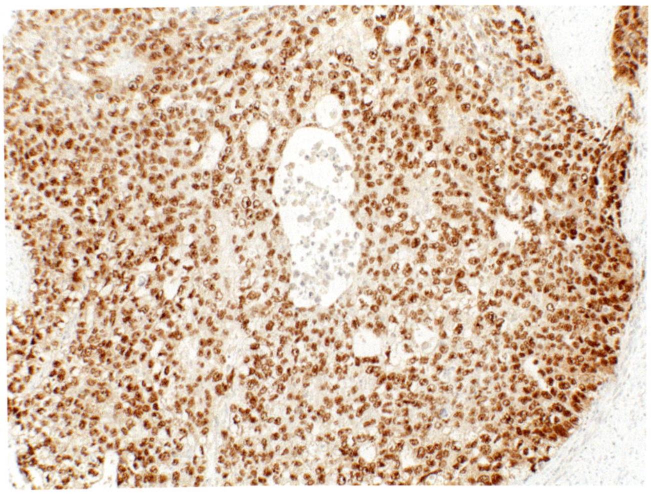

Fig. 13.8 Embryonal carcinoma shows strong positive nuclear staining for OCT4. A similar pattern of staining is observed with antibodies against NANOG and SALL4

they have smaller, more uniform nuclei. Adequate sampling generally reveals other patterns of yolk sac tumor, permitting an accurate diagnosis. Immunohistochemical studies can be helpful. Staining for keratin is predominantly cytoplasmic in yolk sac tumor, while it is membranous in embryonal carcinoma. Embryonal carcinoma stains for CD30, while yolk sac tumor does not. Yolk sac tumor is more likely to stain for AFP, and staining is more widespread. Yolk sac tumor shows positive staining for glypican-3, which is negative in embryonal carcinoma [27]. Both yolk sac tumor and embryonal carcinoma show positive staining for SALL4, but only embryonal carcinoma stains for OCT4 and NANOG, and yolk sac tumor does not stain for SOX2 [18, 26].

Many embryonal carcinomas contain syncytiotrophoblastic giant cells, which can cause concern that *choriocarcinoma* is present. In embryonal carcinoma only occasional STGC are present, generally scattered among embryonal carcinoma cells or in the stroma adjacent to blood vessels. In choriocarcinoma, on the other hand, there is a characteristic admixture of STGC and mononuclear trophoblastic cells, including cytotrophoblastic cells and intermediate trophoblastic cells. This intermixture of cell types is required for the diagnosis of choriocarcinoma and it is not present in embryonal carcinomas that contain STGC.

Finally, some *surface epithelial carcinomas*, such as high-grade serous carcinoma, transitional cell carcinoma, and undifferentiated carcinoma can contain sheets of high-grade tumor cells and poorly formed glands lined by large tumor cells with markedly atypical pleomorphic nuclei. Ovarian carcinomas generally occur in patients who are outside the usual age range for embryonal carcinoma, and they are frequently bilateral, which would be unusual for embryonal carcinoma. Serum levels of CA125 may be markedly elevated, while human chorionic gonadotropin and alpha-fetoprotein levels are within the normal range. Microscopically, ovarian carcinomas often contain distinctive areas of papillary or glandular growth and psammoma bodies are often present. Simple immunohistochemical stains can differentiate surface epithelial carcinomas from embryonal carcinoma, as carcinomas show strong staining for both cytokeratin and epithelial membrane antigen, while the latter is negative in embryonal carcinoma. Surface epithelial carcinomas lack staining for such embryonal carcinoma markers as CD30, OCT4, NANOG, and SALL4.

Yolk Sac Tumor

Yolk sac tumor, formerly known as endodermal sinus tumor, is a primitive malignant germ cell tumor that shows various patterns of endodermal differentiation, frequently toward yolk sac structures. In studies of pathology consultation cases, yolk sac tumor appeared to be the second most common malignant germ cell tumor of the ovary [29] but studies of clinical populations reveal it to be the third most common, after immature teratoma and dysgerminoma [2].

Clinical Features

Yolk sac tumor mainly occurs in children and young women, although rare examples have been reported in women over 50 years of age. The median age is 19 years.

The most common presenting complaints are abdominal pain, abdominal enlargement or distention, or a palpable abdominal mass. Occasional patients present with acute abdominal symptoms caused by rupture or torsion of the tumor. Most patients have high serum levels of α-fetoprotein (AFP) and many have increased levels of CA-125 as well [30].

Yolk sac tumor appears to be confined to one ovary at diagnosis in about 50% of patients; bilateral tumors are virtually never present in women with early stage tumors. When yolk sac tumor spreads beyond the ovary it metastasizes to the peritoneum and omentum, the para-aortic lymph nodes, and the liver. Metastases are within the pelvis (stage II) in about 10% of cases, while in the remaining 40% of cases there is more widespread metastasis (stage III and IV).

The standard surgical treatment for yolk sac tumor is by unilateral salpingo-oophorectomy with limited debulking of extraovarian tumor [31]. The contralateral ovary and uterus do not need to be removed in patients with advanced disease if they are uninvolved. In the past, the clinical course was characterized by the rapid development of metastases and high mortality. The prognosis was poor even in patients with stage I tumors, indicating that subclinical metastases are present at diagnosis in most patients with yolk sac tumors. Radiotherapy is not effective treatment for yolk sac tumor. The 3-year survival was only 13% in a study of patients treated prior to 1975 [32]. The development of effective combination chemotherapy has completely changed the outlook for patients with yolk sac tumor. With the introduction of VAC (vincristine, actinomycin-D, and cyclophosphamide) combination chemotherapy, survival rates greatly increased to about 80% for patients with stage I tumors and to about 50% for those with advanced disease [33]. Standard chemotherapy is currently with platinum containing regimens such as BEP (bleomycin, etoposide, and cisplatin), and these provide even better results than those obtained with VAC [34–36]. The serum AFP level is useful for monitoring the response to treatment and for detecting recurrences.

Table 13.6 Yolk sac tumor histologic patterns

Microcystic
Macrocystic/cystic
Endodermal sinus
Papillary
Glandular/alveolar
Solid
Polyvesicular vitelline
Hepatoid
Myxoid
Parietal

Gross Pathology

Yolk sac tumors are typically large unilateral neoplasms with smooth exterior surfaces. The average diameter is 16 cm. The cut surfaces are tan, white, or gray, and predominantly solid. Variably sized cysts are often present, as are areas of hemorrhage and necrosis.

Microscopic Pathology

Yolk sac tumors can be difficult to diagnose because they exhibit a wide variety of histologic patterns and multiple patterns are typically present in an individual tumor. A good strategy for dealing with these tumors is to search for the easiest patterns to recognize, the microcystic and endodermal sinus patterns. Recognition of one of these patterns provides a clue to the correct diagnosis, which can be confirmed by identification of other patterns and the demonstration of characteristic immunophenotypic features (Table 13.6).

A mixture of growth patterns is present in most yolk sac tumors. The two most common and distinctive patterns are the microcystic pattern and the endodermal sinus pattern [32]. The *microcystic* pattern consists of a reticular network of small spaces lined by a single layer of flattened or cuboidal cells or aggregates of signet ring-like cells with large cytoplasmic vacuoles that compress the tumor cell nuclei against the peripheries of the cells (Fig. 13.9). The tumor cells have

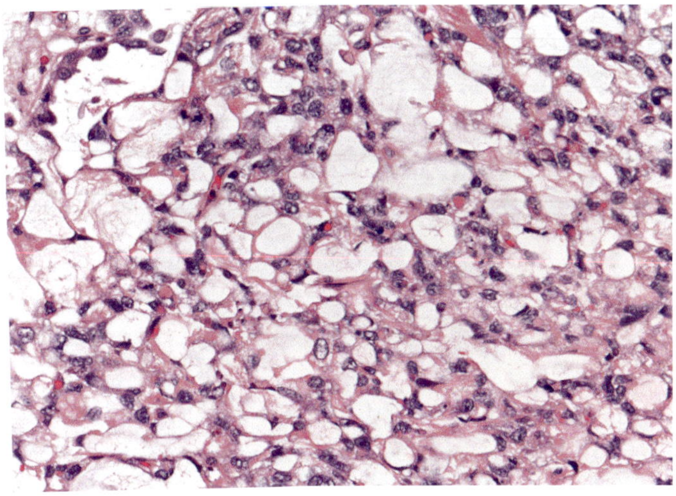

Fig. 13.9 Yolk sac tumor. The microcystic pattern in which small spaces are surrounded by flattened cells with primitive atypical nuclei is one of the most distinctive patterns of growth in yolk sac tumors

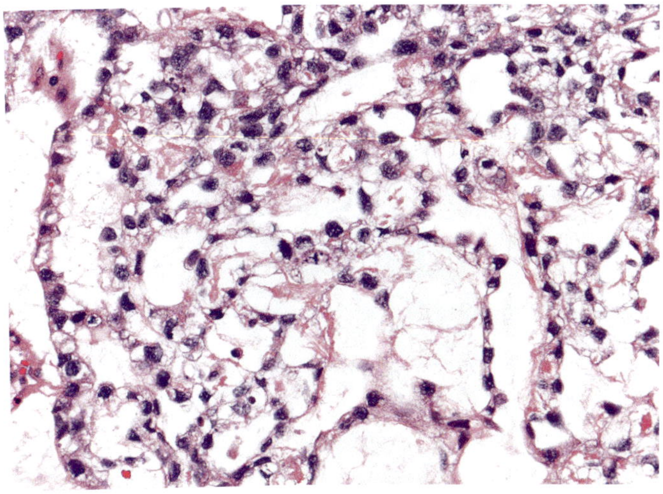

Fig. 13.10 Yolk sac tumor. Labyrinthine spaces are lined ("festooned") by cuboidal tumor cells with clear cytoplasm and hyperchromatic nuclei

clear or amphophilic cytoplasm and small to medium sized primitive atypical hyperchromatic nuclei. Sometimes the cystic spaces are larger, resulting in a *macrocystic* appearance. The *endodermal sinus* pattern is also known as the festoon pattern. It consists of labyrinthine anastomosing spaces lined by columnar cells with clear or amphophilic cytoplasm and round or fusiform hyperchromatic nuclei (Fig. 13.10). Schiller-Duval bodies are vaguely glomeruloid structures formed when fibrovascular papillae covered by cuboidal to columnar tumor cells project into cystic spaces (Fig. 13.11). Schiller-Duval bodies are a characteristic feature of yolk sac tumor and are present in about two thirds of cases. They are most often found in association with the endodermal sinus pattern. Yolk sac tumors frequently contain variably sized cysts lined by flat, cuboidal, or columnar tumor cells. Papillae

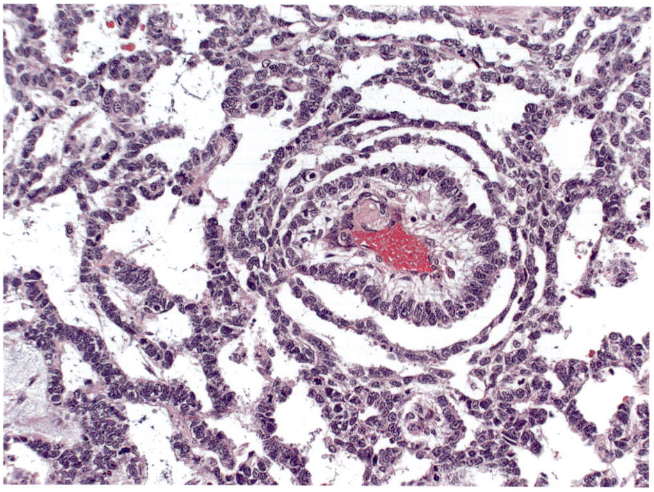

Fig. 13.11 Yolk sac tumor. A Schiller-Duval body, present in the center, is a vaguely glomeruloid structure with a central fibrovascular core and a lining of cuboidal to columnar tumor cells. The Schiller-Duval body is present in a background of anastomosing spaces lined by cuboidal to columnar tumor cells, characteristic of the "festoon" or endodermal sinus pattern

with fibrovascular cores or elongated micropapillary tufts of tumor cells growing into cystic spaces characterize the *papillary* pattern of growth. A variety of *alveolar or glandular* patterns are seen in yolk sac tumors, surrounded by a myxoid or spindle cell stroma. The glands are lined by one or more layers of cuboidal or columnar cells that sometimes form tufts. Some areas of glandular growth have distinctive appearances. The *endometrioid-like* variant of yolk sac tumor consists of tubular glands lined by a single layer of columnar cells having clear supra- or subnuclear cytoplasmic vacuoles, resulting in a resemblance to secretory endometrial glands [37]. While the term "endometrioid variant of yolk sac tumor" seems well established, some pathologists find the glands to be more reminiscent of immature endodermal glands (enteroblastic variant of yolk sac tumor) or fetal lung. The endometrioid-like variant of yolk sac tumor could potentially be mistaken for endometrioid carcinoma. Yolk sac tumors with an *intestinal* pattern of glandular differentiation can mimic a primary or metastatic mucinous tumor. This variant is characterized by aggregates of primitive endodermal glands lined by columnar or low columnar cells growing in loose or cellular stroma. In the *solid* pattern medium-sized undifferentiated cells with a moderate amount of amphophilic or clear cytoplasm grow in nests or sheets. The tumor cells have medium sized hyperchromatic nuclei with inconspicuous nucleoli. The *polyvesicular vitelline* pattern is one in which small cysts resembling yolk sac vesicles are surrounded by immature mesenchyme. The cysts are lined by cuboidal, columnar, or mucinous epithelial cells [38]. Small foci of hepatocytic cells are frequently observed in yolk sac tumors [39, 40], but when large cells resembling hepatocytes grow in sheets or trabeculae throughout most or all of the tumor it is appropriate to make a diagnosis of the *hepatoid* variant of yolk sac tumor. The hepatoid cells have vesicular nuclei with prominent nucleoli and abundant granular eosinophilic cytoplasm [41, 42], and they may exhibit immunohistochemical staining for markers of hepatocytic differentiation such as anti-hepatocyte antibody (hepatocyte paraffin 1) and AFP [43]. Occasional yolk sac tumors are composed predominantly of hepatoid cells but generally the hepatoid pattern is a focal microscopic finding [40]. Some yolk sac tumors have abundant myxoid stroma and relatively scant epithelial elements. When myxoid stroma is prominent the tumor is designated as a *myxoid* or myxomatous variant of yolk sac tumor. Small amounts of eosinophilic hyaline material are commonly present in the stroma of a yolk sac tumor, but when large amounts of this material, which resemble basement membrane material electron microscopically and immunohistochemically, are present the pattern is designated as the *parietal* pattern of yolk sac tumor [40].

Eosinophilic, PAS-positive, diastase-resistant hyaline globules are a characteristic finding in yolk sac tumors, and are most often found in parts of the tumor growing in a microcystic or endodermal sinus pattern. Extracellular eosinophilic hyaline material is frequently present in yolk sac tumors, particularly in areas where the tumor cells grow in a microcystic or solid pattern. The hyaline material is PAS-positive and is immunoreactive for laminin and type IV collagen, similar to basement membrane material [40]. Spindle or stellate cells that stain with both cytokeratin and vimentin are a prominent feature in 25% of yolk sac tumors [44]. Such stromal cells can differentiate into mesenchymal elements like cartilage, striated muscle, and bone, which are occasionally seen in yolk sac tumors. Small, bland, enteric glands lined by columnar and goblet cells are present in 50% of yolk sac tumors. When prominent and especially when present in clusters, luteinized stromal cells, which are present in or adjacent to about 15% of yolk sac tumors, can be mistaken for hepatoid cells. Stromal cells exhibit positive immunohistochemical staining for inhibin, calretinin, and steroidogenic factor-1, and do not stain for yolk sac tumor markers.

Rare neoplasms in which yolk sac tumor is admixed with a mucinous carcinoid have been described [45]. Occasionally, a yolk sac tumor occurs in an older woman in association with an endometrioid or mucinous epithelial tumor; such tumors likely have a somatic rather than a germ cell origin [46–48].

Immunohistochemistry

Yolk sac tumor is cytokeratin positive but epithelial membrane antigen negative [49]. With cytokeratin AE1/AE3 the tumor cells show cytoplasmic staining, as opposed to the membrane pattern of staining that is characteristic of embryonal carcinoma. Most cases of yolk sac tumor are cytokeratin 7 negative, although focal weak staining is occasionally noted [49]. There are a number of fairly specific positive stains for yolk sac tumor. The first to be described and the most characteristic is positive cytoplasmic staining for AFP (Fig. 13.12). There is staining of tumor cell cytoplasm for AFP in more than 75% of yolk sac tumors, but staining can be weak and focal [20, 49]. Glypican-3, initially used as a tumor marker for hepatocellular carcinoma, stains the tumor cell cytoplasm in nearly all yolk sac tumors (Fig. 13.13) [27, 50, 232]. It is worth noting that some cases of clear cell carcinoma show positive staining for Glypican-3, but it tends to be more limited than the staining seen in yolk sac tumor [50]. Staining for the stem cell factor SALL4 is almost invariably

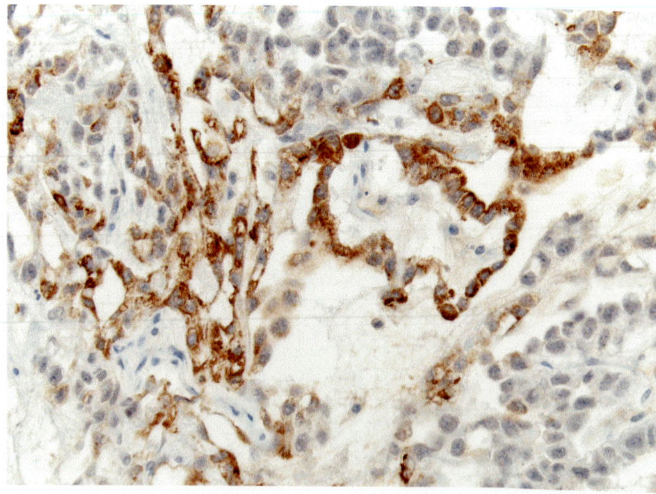

Fig. 13.12 Yolk sac tumor. Positive cytoplasmic staining for alpha-fetoprotein is characteristic of yolk sac tumor. Staining for alpha-fetoprotein typically varies in intensity and not all tumor cells stain

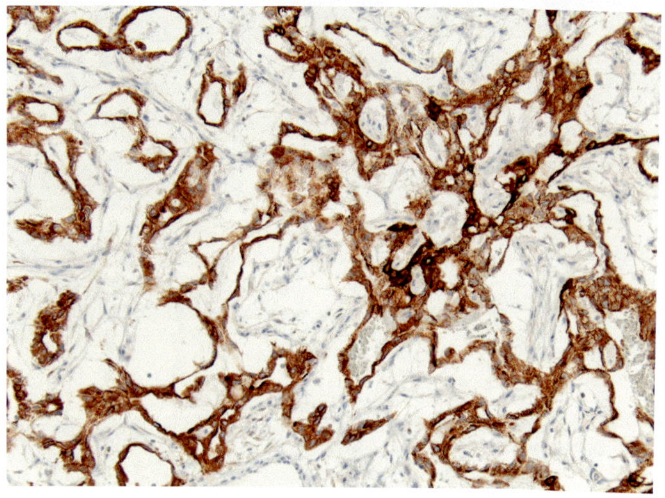

Fig. 13.13 Yolk sac tumor. Glypican-3 is a useful tumor marker for yolk sac tumor, and shows strong cytoplasmic staining in most examples, as shown here. In general, staining for glypican-3 is stronger and more widely distributed than staining for alpha-fetoprotein

positive in yolk sac tumor, with diffuse strong nuclear staining in most tumor cells [20]. SALL4 is also positive in dysgerminoma and embryonal carcinoma. Positive nuclear staining for OCT4 and NANOG is present in the latter two tumors, but there is no staining in yolk sac tumor. Immunohistochemical stains for α1-antitrypsin and placental alkaline phosphatase are positive in yolk sac tumor cells and the extracellular hyaline material is laminin-positive. Yolk sac tumor does not stain for CD30 or CD117. Immunostains for human chorionic gonadotropin are negative except in those rare tumors that contain syncytiotrophoblastic giant cells. Foci of hepatoid differentiation in yolk sac tumor stain with anti-hepatocyte antibody [43].

Differential Diagnosis

The differential diagnosis of yolk sac tumor includes juvenile granulosa cell tumor, dysgerminoma, embryonal carcinoma, clear cell carcinoma, and hepatoid carcinoma (Table 13.7).

Juvenile granulosa cell tumor is in the differential diagnosis partly because it, like yolk sac tumor, occurs in young patients [51, 52]. However, juvenile granulosa cell tumors secrete estrogens, often resulting in such symptoms as precocious pseudopuberty or menstrual abnormalities and serum AFP levels are not elevated. Microscopically, the solid and macrofollicular patterns of juvenile granulosa cell tumor can simulate the solid, macrocystic, and polyvesicular vitelline patterns of yolk sac tumor, and when the stroma of a juvenile granulosa cell is edematous or myxoid the appearance can mimic a microcystic or myxoid pattern of yolk sac tumor growth. Yolk sac tumor exhibits many other patterns of growth, as described above, that are not seen in juvenile granulosa cell tumor and the tumor cells have a primitive or embryonal appearance unlike the appearance of the cells in a juvenile granulosa cell tumor. The luteinized tumor cells of a juvenile granulosa cell tumor have abundant eosinophilic cytoplasm, in contrast to the modest amount of pale cytoplasm in most yolk sac tumor cells. However, the hepatoid cells and luteinized background stromal cells that occur in yolk sac tumor could be mistaken for luteinized tumor cells. Immunohistochemically, yolk sac tumor shows strong positive staining for cytokeratin and the tumor cells stain for AFP, glypican-3, and SALL4 (Table 13.8). These stains are negative in juvenile granulosa cell tumor, which also usually does not stain for keratin. Juvenile granulosa cell tumor generally shows positive staining for sex cord-stromal tumor markers such as inhibin, calretinin, and steroidogenic factor-1, which do not stain yolk sac tumor cells although they may stain the luteinized cells that are sometimes present in the stroma of a yolk sac tumor.

When yolk sac tumor grows in a solid pattern and the tumor cells have clear cytoplasm the histologic appearance can mimic *dysgerminoma*. Dysgerminoma cells are larger and have medium to large vesicular nuclei with clear chromatin and prominent nucleoli. In solid variants of yolk sac tumor, both the cell and the nuclear size are smaller than in dysgerminoma, and the nuclei are hyperchromatic with inconspicuous nucleoli. These tumors can be differentiated using immunohistochemistry (Table 13.8) as dysgerminoma is keratin negative, shows membrane staining for CD117 and D2-40, and shows nuclear staining for OCT-4 and NANOG as well as SALL4 [18, 20]. Yolk sac tumor is negative for all of these stains except SALL4, and shows positive staining for keratin. It also shows staining for AFP and glypican-3, which are negative in dysgerminoma.

Embryonal carcinoma and yolk sac tumor have overlapping morphologies and can be difficult to distinguish from

Table 13.7 Yolk sac tumor differential diagnosis

Juvenile granulosa cell tumor
Dysgerminoma
Embryonal carcinoma
Clear cell carcinoma
Hepatoid carcinoma

Table 13.8 Immunohistochemical features of yolk sac tumor and their use in differential diagnosis

Stain	Yolk sac tumor	Juvenile granulosa cell tumor	Dysgerminoma	Embryonal carcinoma	Clear cell carcinoma
Cytokeratin	+	−	−	+	+
Cytokeratin 7	−				+
EMA	−	−	−	−	+
AFP	+	−	−	−	−
Gylpican-3	+	−	−	−	−
Sall4	+	−	+	+	−
OCT4	−	−	+	+	−
CD117	−	−	+	−	−
CD30	−	−	−	+	−
Inhibin	−	+	−	−	−
SF-1	−	+	−	−	−

EMA epithelial membrane antigen, *AFP* alpha-fetoprotein, *SF-1* steroidogenic factor-1

each other unless specific patterns of yolk sac tumor are identified. In general, the tumor cells are larger in embryonal carcinoma, and the nuclei tend to be much larger and more pleomorphic than those in yolk sac tumor, with coarser nuclear chromatin and more conspicuous nucleoli. Syncytiotrophoblastic giant cells are more often present in embryonal carcinoma, but they are occasionally noted in yolk sac tumors as well. Immunohistochemistry generally resolves this differential diagnosis, as embryonal carcinoma shows positive nuclear staining for OCT4, NANOG, and Sox-2, which are all negative in yolk sac tumor [18, 26]. Staining for cytokeratin AE1/AE3 is membranous in embryonal carcinoma, but cytoplasmic in yolk sac tumor, and while embryonal carcinoma shows membrane staining for CD30, that stain is negative in yolk sac tumor. Yolk sac tumor is positive for glypican-3, which is negative in embryonal carcinoma [27], and for AFP. While early reports suggested that embryonal carcinoma frequently shows staining for AFP [24], I have not noted this using contemporary immunohistochemical techniques, and strong staining for AFP favors a diagnosis of yolk sac tumor.

Several types of epithelial carcinomas of the ovary have overlapping morphology with yolk sac tumor. *Clear cell carcinoma* and yolk sac tumor were at one time thought to be variants of the same tumor type, the so-called mesonephroma, although they are now known to be completely different tumor types. Clear cell carcinoma tends to occur in an older patient population and is associated with endometriosis in the ovary or elsewhere in the pelvis. Grossly, it can have a cystic or adenofibromatous appearance. Microscopically, the tubulocystic, papillary, and solid patterns of growth can mimic yolk sac tumor patterns, but the tumor cells tend to have larger more vesicular nuclei with prominent nucleoli. The cytoplasm can be clear or eosinophilic and hobnail cells are often present. Mitotic activity tends to be lower in clear cell carcinoma than in yolk sac tumor. Further compounding the difficulty of the differential diagnosis, hyaline bodies and stromal hyaline material are present in clear cell carcinoma as well as in yolk sac tumor. Clear cell carcinoma shows positive staining for CK7 and EMA, both of which tend to be negative in yolk sac tumor. Clear cell carcinoma tends to be negative for characteristic markers of yolk sac tumor, such as AFP, glypican-3, and Sall4, but some show at least focal positive staining for these markers [20, 50], so an appropriate panel of stains needs to be performed. One cannot rely on one or two stains to make the differential diagnosis between clear cell carcinoma and yolk sac tumor. Knowledge of the patient age and the clinical circumstances of the case are also helpful.

Hepatoid carcinoma is a rare type of ovarian carcinoma that occurs mainly in elderly women. The tumor cells are polygonal with eosinophilic cytoplasm and they grow in sheets and trabeculae, resulting in a similarity to hepatocellular carcinoma [53, 54]. Hepatoid carcinoma needs to be differentiated from the hepatoid variant of yolk sac tumor. The clinical setting in an elderly patient is helpful. Hepatoid carcinoma is often mixed with other patterns of epithelial carcinoma, such as endometrioid carcinoma or serous carcinoma [55], while hepatoid yolk sac tumor tends to be mixed with other patterns of yolk sac tumor. Immunohistochemistry may help with the differential diagnosis, although both hepatoid carcinoma and the hepatoid variant of yolk sac tumor show positive staining for anti-hepatocyte antibody and AFP [43]. Positive staining for the yolk sac tumor marker SALL4, which would not be expected in hepatoid carcinoma, would help confirm a hepatoid ovarian neoplasm to be a variant of yolk sac tumor. As discussed above, it is always worth remembering that there are rare tumors, generally occurring in older women, in which yolk sac tumor is admixed with some other type of ovarian tumor, such as an endometrioid carcinoma or a carcinoid tumor.

Polyembryoma

Polyembryoma is a rare type of malignant germ cell tumor that is invariably found as a component of a mixed germ cell tumor [56].

Clinical Features

Only 15 cases have been reported. The patient age ranges from 3 to 44 years of age, with an average of 24.5 years. The clinical presentation is variable and depends partly on what other elements are present in the tumor along with polyembryoma. Typical presentations are with abdominal pain, an abdominal mass, or abnormal vaginal bleeding. The serum AFP is usually elevated when yolk sac tumor is present, while the presence of choriocarcinoma or syncytiotrophoblastic giant cells generally results in elevation of the serum human chorionic gonadotropin [57]. This in turn can result in endocrinologic symptoms such as amenorrhea, abnormal bleeding or, if the patient is a child, precocious pseudopuberty. Most reported patients have had stage I tumors. The treatment is the same as for other malignant germ cell tumors, by salpingo-oophorectomy and limited debulking of extraovarian tumor. Patients treated before effective chemotherapy was available often had recurrences and died of tumor, but all patients treated since modern combination chemotherapy became available in the 1970s have been cured.

Pathology

Grossly, the tumors tend to be large solid neoplasms that contain small cysts. The cut surfaces tend to be tan or brown,

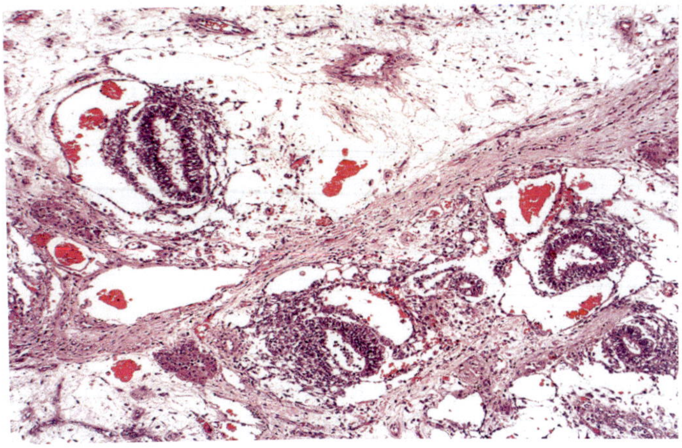

Fig. 13.14 Polyembryoma. Scattered embryoid bodies are present in a fibromyxoid background. The embryoid bodies are variably well formed, and tend to resemble 14–20 day embryos

often with areas of hemorrhage or necrosis. Microscopically, polyembryoma is composed of numerous embryoid bodies surrounded by fibromyxoid stroma (Fig. 13.14). Embryoid bodies resemble 14–20 day embryos. In one tumor, there were numerous less mature blastocyst-like bodies as well as occasional more typical embryoid bodies [58]. Well-formed embryoid bodies have a central embryonic disc composed of tall columnar cells with stratified hyperchromatic nuclei. These cells show positive staining with embryonal carcinoma markers such as OCT4 and CD30 [26]. On one side of the disc is an amnionic cavity lined by flat to cuboidal cells that are immunoreactive for calretinin, and on the other side is a yolk sac lined by AFP positive cells [56, 58]. While well-formed embryoid bodies are easy to recognize, embryoid bodies are often distorted or incomplete and difficult to recognize.

Choriocarcinoma

Ovarian choriocarcinoma of germ cell origin is rare; only about 40 cases had been reported as of 2006 [59]. In one review of malignant ovarian germ cell tumors, less than 1% were choriocarcinomas [29]. When it occurs in the ovary, choriocarcinoma is almost always present as a component of a mixed germ cell tumor.

Clinical Features

Choriocarcinoma of the ovary occurs in children and young women. The average age of reported patients is about 14 years [60, 61], but one case occurred in a postmenopausal woman [62]. The clinical presentation is with abdominal pain and abnormal vaginal bleeding. A third of cases occur in prepubertal girls, and about a third of them have precocious pseudopuberty. Patients invariably have a positive pregnancy test or an elevated serum β-HCG [63]. An ectopic pregnancy is often suspected clinically [64].

Choriocarcinoma of the ovary is unilateral and is treated by salpingo-oophorectomy. More extensive surgery is required only if the contralateral ovary or uterus is involved. Combination chemotherapy, currently usually with the BEP regimen (bleomycin, etoposide, and cisplatin), is administered after surgery either as adjuvant therapy or as treatment for advanced stage disease. The prognosis is favorable for patients who are treated with chemotherapy, but the survival rate is only about 30% for patients who do not receive chemotherapy [60].

The chemotherapy regimen and prognosis are different for patients who have gestational choriocarcinoma, so it is important to determine whether ovarian choriocarcinoma is of gestational or germ cell origin. In premenarchal patients the choriocarcinoma is certain to be of germ cell origin. In postpubertal patients, gestational choriocarcinoma, either metastatic from the uterus or fallopian tube, or primary in the ovary, and choriocarcinoma of germ cell origin are equally likely to occur. They are difficult to differentiate as there are no morphologic differences between gestational choriocarcinoma and choriocarcinoma of germ cell origin. The clinical history may help in the differential diagnosis. The finding of a corpus luteum of pregnancy in the ovary favors gestational choriocarcinoma. The presence of other germ cell elements in the tumor (i.e., a mixed germ cell tumor) indicates that the tumor is of germ cell origin, although in a pure ovarian choriocarcinoma of germ cell origin other elements are not present. Choriocarcinoma can be proven to be of gestational origin if a paternal component can be identified by DNA analysis [59, 61, 65, 66].

Pathology

Choriocarcinoma is a unilateral solid or solid and cystic tumor that ranges from 4 to 25 cm in diameter. The cut surface is soft and purple-red with hemorrhagic and necrotic areas.

Microscopically, there is usually extensive hemorrhage and necrosis with viable tumor cells being located mainly at the periphery of necrotic zones. Mononuclear trophoblastic cells and multinucleated syncytiotrophoblastic cells grow in a characteristic plexiform pattern with syncytiotrophoblastic cells rimming nests and sheets of mononuclear cells (Fig. 13.15). The mononuclear component consists of a mixture of cytotrophoblastic and intermediate trophoblastic cells. The mononuclear trophoblastic cells have abundant clear or lightly stained cytoplasm and well-defined cell borders. Their nuclei are irregular and vesicular and some

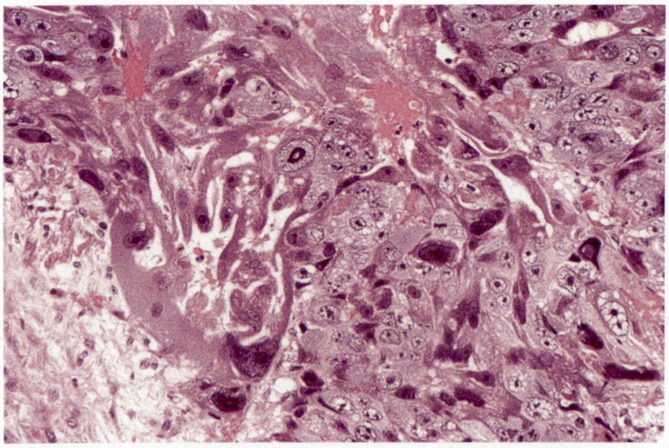

Fig. 13.15 Choriocarcinoma rarely occurs as a pure tumor in the ovary; it is usually a component of a mixed germ cell tumor. It consists of syncytiotrophoblastic cells, in which multiple atypical nuclei are variably distributed in abundant purple cytoplasm, and mononuclear cells which are polygonal with vesicular nuclei with conspicuous nucleoli and moderate amounts of amphophilic cytoplasm

contain macronucleoli. Intermediate trophoblastic cells are larger than cytotrophoblastic cells. Syncytiotrophoblastic cells have multiple hyperchromatic nuclei surrounded by abundant vacuolated basophilic or amphophilic cytoplasm. All trophoblastic cells show positive cytoplasmic staining for cytokeratin [67], inhibin [68], and CD10 [69]. Staining for keratin is dense and homogeneous in the cytoplasm of syncytiotrophoblastic cells, which also show strong positive cytoplasmic staining for β-HCG.

While occasional cases of pure choriocarcinoma occur in the ovary, choriocarcinoma is more often seen as a constituent of a mixed germ cell tumor. Choriocarcinoma is a component of 10–20% of mixed germ cell tumors [70, 71]. Very rarely, choriocarcinoma is mixed with a poorly differentiated epithelial carcinoma of the ovary [72]; presumably the choriocarcinoma in such tumors arises from the epithelial component and is not of germ cell origin. Nevertheless, the histologic appearance and immunohistochemistry of the choriocarcinoma is identical to that of germ cell and gestational choriocarcinoma, with the characteristic plexiform arrangement of mononuclear and syncytiotrophoblastic cells. The correct diagnosis is established by identifying the somatic carcinoma component of the tumor.

Teratoma

The benign cystic teratoma, which clinicians commonly call a "dermoid" or "dermoid cyst," accounts for about a quarter of all ovarian tumors and is the most common ovarian neoplasm. Other types of teratoma are uncommon. Benign teratomas are thought to be derived from postmeiotic germ cells and have a normal diploid 46XX karyotype.

Mature (Benign) Teratoma

These are cystic or, occasionally, solid tumors that contain various mature tissues derived from one or more of the embryonic germ layers: the ectoderm, mesoderm, and endoderm. It has been proposed that a mature teratoma should be defined as one composed of tissues with a similar maturity to what is seen in a developing fetus of 8 or more weeks of fertilization age [73].

Clinical Features

Benign teratomas can occur at any age. Most occur in women between 20 and 50 years of age with the peak incidence being between 20 and 29 years. Only about 20% of benign teratomas are detected in postmenopausal women [74, 75]. Some patients are asymptomatic; their tumors are discovered during a routine examination or during evaluation or treatment for some other condition. If the tumor grows to a large enough size it can cause symptoms such as pelvic pressure or pain. Serious complications include torsion, which occurs in 3–10% of cases, and rupture, which occurs in about 1% [75]. Torsion causes acute abdominal symptoms. Rupture can cause acute peritonitis or, rarely, chronic granulomatous peritonitis. Rare patients develop an unusual form of sudden onset paraneoplastic encephalitis that causes severe psychiatric and/or neurologic symptoms including coma in some cases [76–79]. The encephalitis is a form of autoimmune disease caused by antibodies to *N*-methyl-D-aspartate receptors (anti-NMDAR). Most patients are women, and many of them have ovarian teratomas. The teratoma can be either mature or immature, but it invariably contains neural tissue, which likely serves as the antigen that initiates development of the anti-NMDAR antibodies. Treatment includes immunotherapy and removal of the teratoma if one is present; full recovery is more likely in patients with a teratoma than in those without one.

Mature teratoma is a benign neoplasm that can be treated conservatively by cystectomy or salpingo-oophorectomy. Cystectomy is a common form of treatment in young patients and it can be performed laparoscopically or by laparotomy. Ovarian teratomas are bilateral in 10–15% of patients. Peritoneal implants composed of mature neural tissue occur in rare patients with mature teratoma of the ovary [80, 81]. Mature neural implants do not adversely affect survival.

Macroscopic Pathology

Benign teratomas are nearly always cystic tumors. The size varies from only a few cm to large tumors 15 cm or more in diameter; the average diameter is 7–8 cm. The cysts are unilocular or, less often, multilocular and typically a solid protuberance called the dermal papilla or Rokitansky tubercle projects into the lumen. The cysts contain hair, grumous material, or oily or serous liquid. Hard areas corresponding

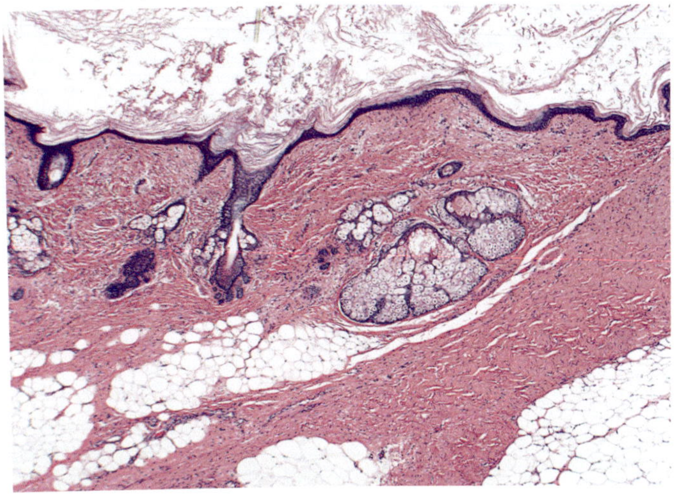

Fig. 13.16 Benign cystic teratoma. Tumors such as this, in which the cyst is lined by stratified squamous epithelium with adnexal structures such as hair shafts and sebaceous glands, are frequently called "dermoid cysts." The lumen contains keratinous debris, hair, and sebaceous material

to cartilage or bone may be present, and fully formed teeth are occasionally identified on gross examination. Foci of soft gray-tan tissue resembling brain, often present in the lining of the cysts, and glistening green or brown thyroid tissue are commonly seen. Extensive dense solid regions are unusual in a benign cystic teratoma and as these raise the question of an immature teratoma they should be carefully sampled for histologic study. Rare mature (benign) teratomas are completely solid. These can only be differentiated from an immature teratoma by microscopic evaluation [82].

Microscopic Pathology

Benign teratomas contain a variable mixture of ectodermal, mesodermal, and endodermal tissues. Tissue from at least two germ layers is present in two thirds of mature teratomas and tissue from all three germ layers is present in about a third of them [83]. The most common tissues in benign teratomas are ectodermal structures such as skin, hair follicles, and sebaceous and sweat glands (Fig. 13.16). When they predominate, as is often the case, the tumor is typically called a "dermoid cyst." Other ectodermal elements that are often seen are brain, including glia and neurons (Fig. 13.17), choroid plexus, peripheral nerve, and developing tooth structures. The most common endodermal tissues are digestive tract tissues, including esophageal squamous mucosa, gastric mucosa, and intestinal mucosa with or without endocrine cells, respiratory mucosa, and thyroid tissue. Ciliated respiratory mucosa is sometimes present in the lining of the large main cyst and at other times it lines small cysts in the dermal papilla. The most common mesodermal tissues are fat, smooth or striated muscle, bone, cartilage

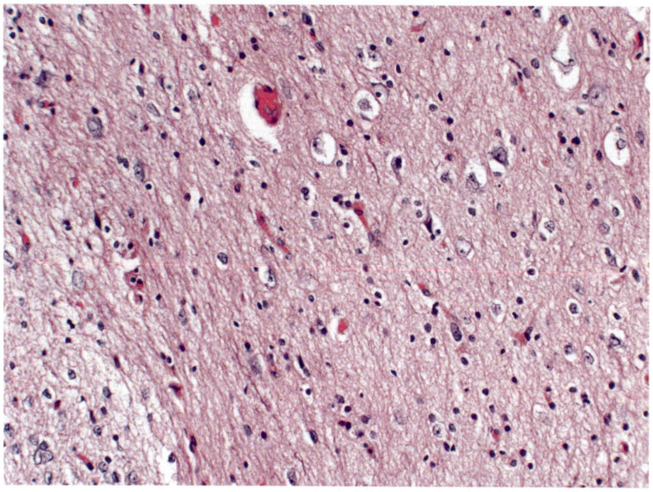

Fig. 13.17 Neural tissue, including glia and neurons, as seen here, is frequently present in benign cystic teratomas

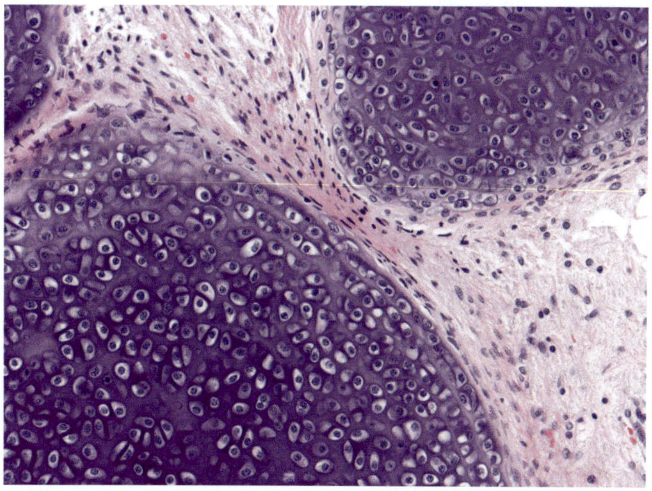

Fig. 13.18 Benign cartilage in a teratoma. Small chondrocytes lie in lacunae a hyaline matrix. The nodules of cartilage are surrounded by mature fibrous tissue of low cellularity

(Fig. 13.18), and a loose connective tissue framework that surrounds the other elements. Male tissues such as prostate are occasionally present [84, 85]. While ectodermal tissues are common in the cyst wall, endodermal and mesodermal tissues are most likely to be found in sections from the dermal papilla. If an individual element greatly dominates, (i.e., >50%) the tumor is classified as a monodermal teratoma (e.g., struma ovarii).

The epithelium lining the large or small cysts is commonly eroded resulting in escape of the cyst contents into the surrounding tissues. As a result, one sees cystic spaces lined by granulation tissue, histiocytes, and multinucleated foreign body giant cells, and pericystic zones of fibrosis, fat necrosis, and a foreign body reaction. This type of a reaction is a helpful clue to the presence of a teratoma.

Benign peritoneal or omental implants of neural tissue are seen mainly in patients with immature teratomas, and they are discussed in detail in the section on immature teratoma. They also rarely occur in patients with mature teratomas, in whom they have a similar gross and microscopic appearance and distribution. They consist predominantly of glial cells and fibers, and are accordingly often referred to as glial implants or, if they are numerous, as gliomatosis. Neurons and neurofilaments are also invariably present in the implants, however, and it might be more correct to refer to them as neural implants.

Secondary Neoplasms in Mature (Benign Cystic) Teratomas

Benign or malignant tumors can develop from any of the tissues present in a mature cystic teratoma, but this is a rare occurrence. A malignant tumor, usually squamous cell carcinoma, develops in only 1–3% of mature cystic teratomas [86, 87].

Clinical Features

Secondary malignancies in mature teratomas are most common in postmenopausal women, but they can occur in patients of any age. The average patient age is 55 years [88]. The clinical presentation is usually with signs or symptoms caused by an adnexal mass, such as abdominal or pelvic pain, abdominal distention, or a palpable mass. Occasional patients, most of whom have squamous cell carcinoma, have hypercalcemia [89]. Tumor markers such as CA125 are frequently elevated, and squamous cell carcinoma antigen is typically elevated when the secondary malignancy is a squamous cell carcinoma. About 50% of women with secondary malignancies have tumors that are confined to the ovary (stage I), and those women have a favorable prognosis [90]. Women with more advanced tumors have an unfavorable prognosis [91]. Most patients with advanced tumors do not respond to chemotherapy or radiotherapy and die within a year or two of diagnosis.

Pathology

Grossly, a secondary neoplasm may present as a nodule or papillary growth projecting into the cyst lumen or there may be thickening or an area of firmness in the cyst wall. In some cases the secondary neoplasm is only detected microscopically. Secondary neoplasms are usually unilateral, but the contralateral ovary may contain a benign cystic teratoma.

A variety of benign skin and noncutaneous tumors have been reported to arise in benign cystic teratomas. These include nevi, sebaceous adenomas, and other cutaneous adnexal neoplasms, benign salivary gland type tumors, meningiomas, glomus tumors, and hemangiomatous vascular proliferations [92]. Malignant tumors can be either in situ or invasive. Squamous cell carcinoma in situ [93, 94] and Paget's disease [95] are types of in situ malignant tumors that have been reported in benign cystic teratomas. Squamous cell carcinoma is by far the most common type of invasive carcinoma in benign cystic teratomas, accounting for about 80% of all malignancies [88]. Grossly, the tumor typically forms a nodule in the wall of the cyst. Features such as necrosis, destruction of tissue, and growth onto the exterior of the ovary suggest malignancy. Microscopically, most are moderately to poorly differentiated squamous cell carcinomas (Fig. 13.19), sometimes with associated squamous cell carcinoma in situ [93]. Many other types of malignancies have been reported in benign cystic teratomas. These include other types of skin cancer such as basal cell carcinoma or sebaceous carcinoma [96], melanoma [97–99], adenocarcinoma [100], and sarcomas [101], and other rare tumor types.

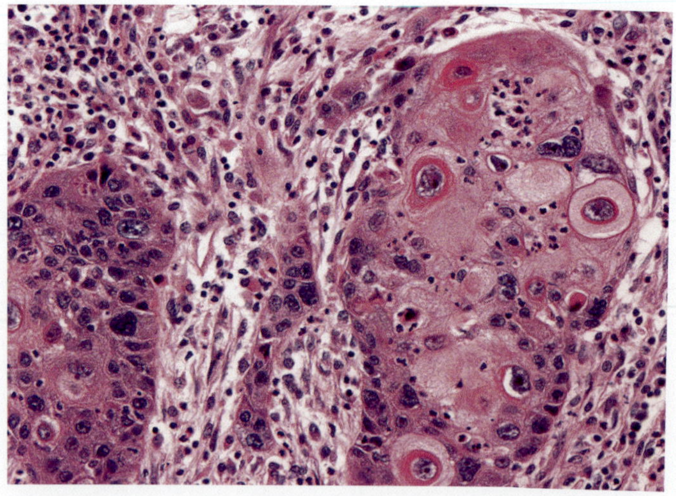

Fig. 13.19 Moderately differentiated squamous cell carcinoma that arose in a benign cystic teratoma. Nests of malignant squamous cells invade an inflamed fibrous stroma. The tumor cells have variable but sometimes abundant pink cytoplasm and several cells have dense eosinophilic keratinized cytoplasm

Ovarian mucinous tumors are occasionally associated with teratomas, raising the possibility that some mucinous tumors are derived from teratomas. All types of mucinous tumors are associated with teratomas. Two large series contained 41 mucinous cystadenomas, 44 borderline mucinous tumors including 4 with intraepithelial carcinoma, and 11 mucinous carcinomas [102, 103]. Any combination of staining for cytokeratin 7 (CK7) and cytokeratin 20 (CK20) can be seen. About half of cystadenomas are CK7−/CK20+, but they can also be CK7+/CK20− or CK7+/CK20+. Borderline tumors tend to have either a CK7+/CK20+ or CK7−/CK20+ immunophenotype, and carcinomas are usually CK7−/CK20+. Tumors with a lower gastrointestinal immunophenotype (CK7−/CK20+) may originate from intestinal epithelium in the teratoma. The others could arise from nonintestinal mucinous epithelium in the teratoma, or could represent collision tumors, with the mucinous tumor being of surface epithelial, rather than germ cell origin. Pseudomyxoma ovarii is often conspicuous, especially in women with borderline tumors or carcinomas, and some women have pseudomyxoma peritonei, presumably of ovarian origin in this unusual setting [104–107]. The pseudomyxoma peritonei may have a more favorable prognosis than the more common form of pseudomyxoma peritonei caused by an appendiceal or colonic tumor [102, 103]. It is always necessary to rule out the possibility that the ovarian mucinous tumor could be a metastasis from an appendiceal or colonic tumor, particularly in patients whose tumors have a CK7–/CK20+ immunophenotype and in those with pseudomyxoma peritonei. If there is no evidence of a potential extraovarian primary, and sections of the ovary reveal evidence of a teratoma as well as a mucinous neoplasm, it is reasonable to conclude that the mucinous tumor is of primary ovarian origin, presumably in the teratoma, despite an unusual immunophenotype or clinical setting.

Immature Teratoma

Immature teratoma is the most common malignant germ cell tumor of the ovary, representing 20–35% of such tumors at major cancer centers [108, 109] and in the US population [2].

The grade of teratomas has been linked to stages of fetal development, with immature teratomas containing at least some of the types of tissues seen prior to a fertilization age of 8 weeks, and mature teratomas consisting exclusively of tissues similar to those seen at a fertilization age of 8 weeks or more [73]. Further, it has been proposed that high-grade immature teratoma is characterized by the presence of immature neuroepithelial structures, and low-grade immature teratoma by the presence of somite organogenesis [73].

Genetic studies of a limited number of cases have revealed that an isochromosome 12p, which is typical of primitive germ cell tumors, is often present in the immature teratoma component of a mixed germ cell tumor [110], compatible with its derivation from a less differentiated component of the tumor, while no i12p is typically identified in pure immature teratomas, suggesting a different pathogenesis for such tumors [110, 111]. On the other hand, immature teratoma is typically diploid, whether in a pure tumor or in the immature teratoma component of a mixed germ cell tumor, while other germ cell elements such as yolk sac tumor are aneuploid [112, 113], and an i12p has been detected in a pure immature teratoma [114]. The histogenesis of these tumors remains to be fully elucidated.

Clinical Features

Immature teratoma occurs predominantly in children and young women. Patients are rarely younger than 7 or older than 40; the average patient age is about 20. These tumors only exceptionally occur in older or postmenopausal women [115]. The clinical presentation is generally with nonspecific signs or symptoms of an ovarian mass such as pelvic or abdominal pain, abdominal distention, or a palpable abdominal mass. Occasional patients present with acute abdominal symptoms caused by infarction or rupture of the tumor. Rare patients have a paraneoplastic syndrome associated with severe neurologic and/or psychiatric findings; this is discussed more fully in the section on mature teratoma [116, 117]. Alpha-fetoprotein is elevated in the serum in many patients with pure immature teratoma [30, 118]. CA 125 is also frequently elevated, albeit not to the levels seen in patients with serous tumors of the ovary [30]. Most (50–80%) patients have localized tumors (stage I) at diagnosis. Immature teratoma is almost invariably unilateral, although it can spread to the contralateral ovary in patients with advanced disease. There is a benign cystic teratoma in the contralateral ovary in 10–15% of cases. Immature teratoma spreads mainly to the pelvic and abdominal peritoneum and the omentum.

Patients with an immature teratoma are generally young and are accordingly usually treated with fertility preserving therapy. Surgery for patients with tumors confined to the ovary (stage IA) is generally a unilateral salpingo-oophorectomy. Occasional patients have been successfully treated by cystectomy, in some cases followed by adjuvant chemotherapy [119]. Advanced tumors in young patients are treated by unilateral salpingo-oophorectomy and excision of extraovarian tumor [109, 120]. In older patients where preservation of fertility is not an issue or in cases where the contralateral ovary or uterus is involved, the treatment is hysterectomy and bilateral salpingo-oophorectomy.

Patients with stage IA grade 1 immature teratoma have an excellent prognosis and are usually treated by surgery alone [109, 121]. Adult patients with localized grade 3 immature teratomas and those who have advanced disease require postoperative chemotherapy. Whether or not adult patients with stage IA grade 2 tumors should have chemotherapy is controversial. In children, the prognosis appears to be more favorable and independent of tumor grade [122]. Children accordingly generally do not have chemotherapy unless they have metastases containing other malignant germ cell elements such as yolk sac tumor [118, 123]. Cisplatin-containing regimens such as BEP (cisplatin, etoposide, and bleomycin) are highly effective as adjuvant chemotherapy for patients with no residual tumor after surgery, with survival rates of 90–100% [2, 124]. The prognosis is less favorable when there is residual gross tumor after primary surgery and in cases of recurrent immature teratoma.

The microscopic appearance of extraovarian tumor deposits is important in determining the need for additional therapy. Sometimes the extraovarian tumor deposits, which are usually in the omentum, on the peritoneum, or in lymph nodes, consist entirely of mature tissues. Such deposits are graded as grade 0. Grade 0 tumor deposits do not adversely affect the prognosis, and do not require chemotherapy. Masses composed of mature teratoma are sometimes detected after chemotherapy in patients with incompletely resected immature teratoma. Such masses are generally resected to prevent local complications and to avoid development of a "growing teratoma syndrome" [125, 126]. However, we have seen small grade 0 tumor deposits that have persisted for more than 40 years after resection of an ovarian teratoma and caused no signs or symptoms. Rare examples of a malignant tumor arising in long-standing incompletely resected low-grade teratoma implants have been reported [127–129].

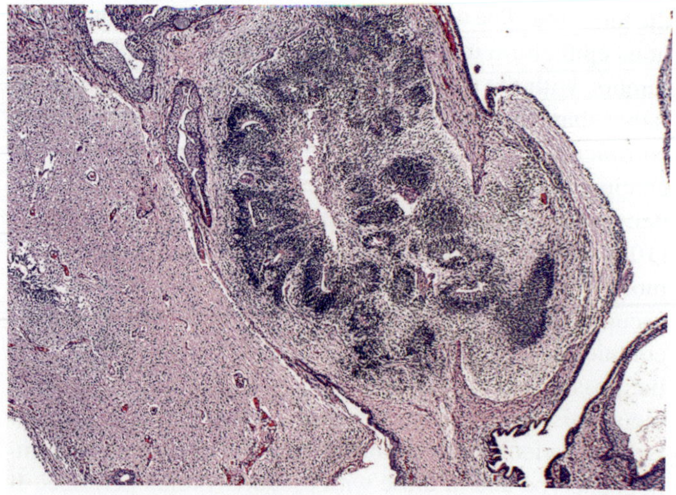

Fig. 13.20 Immature teratoma. Immature neuroepithelium (*on the right*) is the easiest marker of immaturity to recognize and to quantitate for purposes of grading the tumor

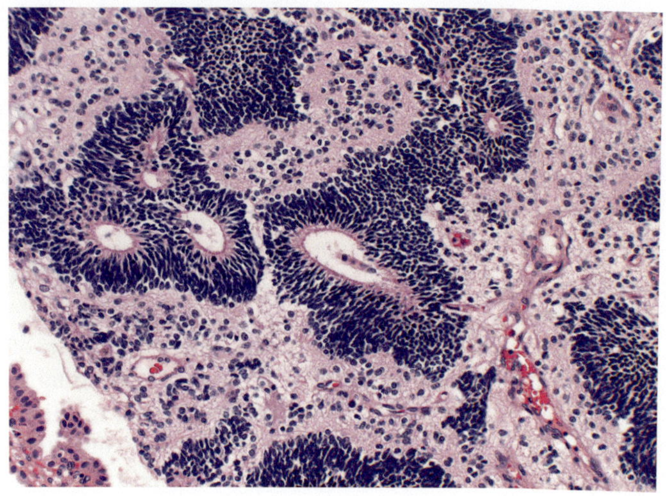

Fig. 13.21 Immature neuroepithelium in an immature teratoma. The tubules are lined by mitotically active stratified cells with hyperchromatic nuclei and modest amounts of eosinophilic cytoplasm

Macroscopic Pathology

Immature teratoma is predominantly solid and usually unilateral. The size varies but the average diameter is 18 cm. Solid areas are white, tan, gray, or brown. The tissue can be soft or firm, and if cartilage or bone is present, hard and gritty. Scattered small cysts are typically present and in about a quarter of immature teratomas large cysts that contain keratinous debris or hair are present, suggestive of a benign cystic teratoma [130].

Microscopic Pathology

Tissues derived from any of the three germ cell layers can be present, although immature ectodermal and mesodermal tissue is most common. A mixture of mature and immature tissues is present in most tumors, with the former usually predominating. In most immature teratomas, immature neuroectodermal elements are the dominant immature element, and they are also the easiest immature tissue to recognize and quantitate for purposes of grading (Fig. 13.20) [131, 132]. Immature neuroectodermal tissues include neuroepithelial tubules or trabeculae lined by mitotically active columnar cells with stratified hyperchromatic nuclei (Fig. 13.21), sheets and nests of neuroblasts (Fig. 13.22), sometimes containing neuropil and Homer Wright rosettes, hypercellular mitotically active immature glia, and primitive retinal tissue with melanin pigmentation [133, 134]. Prominent but benign vascular proliferations are occasionally associated with the neural elements in an immature teratoma [92].

Immature mesodermal tissue is hypercellular and consists of small spindle cells with hyperchromatic nuclei. Mitotic figures are usually present. Immature cartilage is often present in immature teratomas (Fig. 13.23). Two main features distinguish it from the more common "fetal" or "mature" cartilage that is often seen in mature teratomas. First, the chondroid cells within the lacunae have medium-sized round nuclei with open chromatin, as opposed to the small nuclei with dark chromatin seen in fetal cartilage. Second, foci of immature cartilage are surrounded by primitive small round mesenchymal cells rather than by the fibroblasts that typically surround mature cartilage (Fig. 13.18).

Endodermal tissues are usually less extensive than ectodermal or mesodermal tissues in an immature teratoma, and are only infrequently the predominant type of immature tissue present. Immature endodermal tissues that can be seen include primitive glands lined by columnar cells with subnuclear and supranuclear vacuoles resulting in an "enteroblastic" appearance, glands lined by partially differentiated

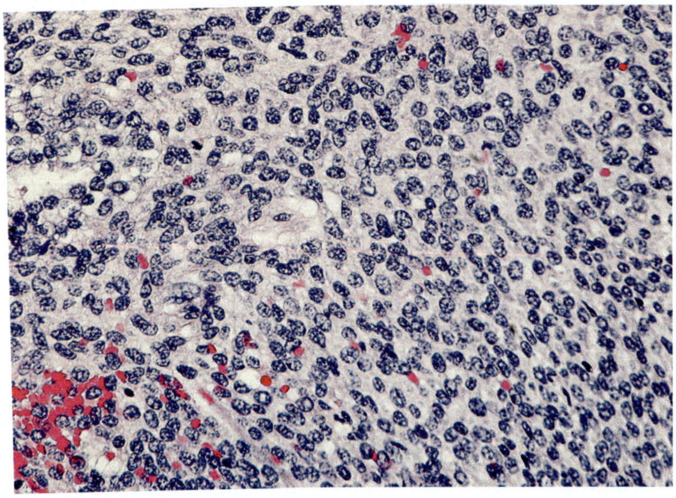

Fig. 13.22 Neuroblastic cells are another form of immature neural tissue that can be present in an immature teratoma. If large amounts of this type of tissue are present a diagnosis of primitive neuroectodermal tumor would be appropriate

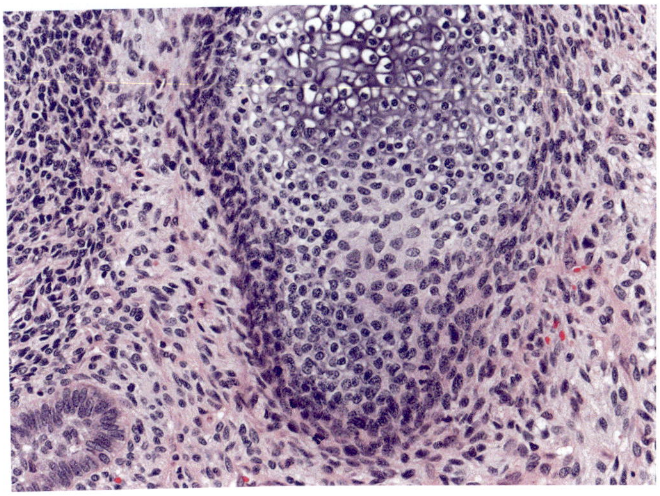

Fig. 13.23 Immature cartilage in an immature teratoma. The cartilage cells are plump with round nuclei and the stroma surrounding the cartilage is cellular and immature. Mitotic activity can be present in the stroma that surrounds the cartilage, as is the case here

stratified columnar intestinal epithelium with goblet cells, and islands of fetal liver tissue [39, 135]. Pediatric pathologists have proposed that foci of immature liver and glands with subnuclear vacuoles similar to immature endoderm or fetal lung are well-differentiated forms of yolk sac tumor [122], although gynecologic pathologists tend to view limited amounts of these elements as components of the teratoma [135]. Immature renal (metanephrogenic) tissue is another rare type of immature tissue that can be seen in teratomas [136].

Immature teratomas are graded using a system that rates them from grade 1 for a neoplasm composed almost entirely of mature tissues with only limited immaturity to grade 3 for a neoplasm containing abundant immature tissue (Table 13.9). Theoretically, any of the immature elements might be quantitated for grading purposes, but immature neuroepithelium is the easiest to assess [132]. An alternative two-grade system in which grade 1 tumors are designated as low grade and grades 2 and 3 as high grade has been proposed but it is not widely used [137]. Occasionally, a mature teratoma contains only one or a few microscopic foci of immature tissue (Fig. 13.24). A study of 360 ovarian teratomas that contained immature tissues included 350 immature teratomas and only ten mature cystic teratomas with microfoci of immature tissue [130]. Follow-up of 11 months to 7 years was available in nine of the latter ten cases. No patient had a recurrence, suggesting that a diagnosis of immature teratoma is not appropriate when only a miniscule amount of immature tissue is present in an otherwise mature ovarian teratoma.

Table 13.9 Histologic grading of immature teratoma of the ovary

Grade	Amount of immature tissue	Amount of neuroepithelium
1	+	Rare; not >1 LPF/slide
2	++	Common; but not >3 LPF/slide
3	+++	Prominent; ≥4 LPF/slide

LPF low power field

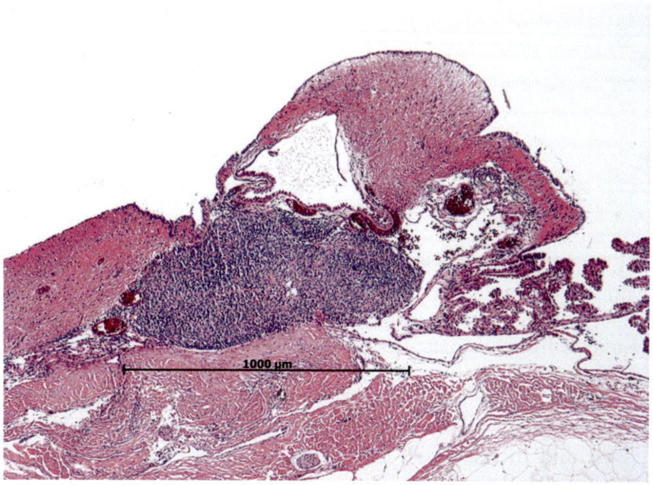

Fig. 13.24 Microfoci of immature tissue do not warrant a diagnosis of immature teratoma. This teratoma consisted entirely of mature tissues except for a 1 mm focus of immature neural tissue. The diagnosis was "benign cystic teratoma with a microscopic focus of immature tissue"

Microscopic foci of yolk sac tumor are occasionally detected in an immature teratoma. As long as there are only a few of them (≤3) and they are less than 3 mm in diameter they do not appear to adversely impact the prognosis and do not warrant a diagnosis of mixed germ cell tumor [137].

Extraovarian tumor deposits in the omentum, on the peritoneum, and in lymph nodes are not uncommonly detected in patients with immature teratomas of the ovary, and they occasionally also occur in patients with a mature teratoma. While immature tissue is occasionally present in an implant, the

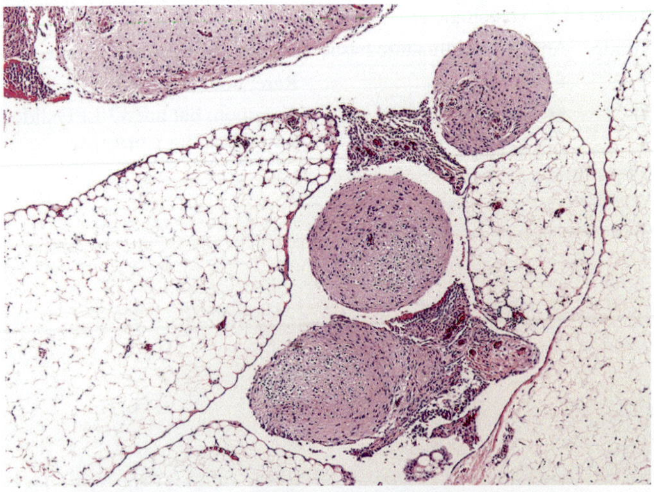

Fig. 13.25 Nodules of mature neural tissue in the omentum in a patient with an immature teratoma. Mature neural tissue, usually mainly glia, can be present in the omentum, on the peritoneum, and in lymph nodes. Occasionally, other mature tissues are mixed with the neural tissue

implants generally consist predominantly of mature neural tissue and the deposits are accordingly often designated as "gliomatosis" [138]. However, the deposits are not composed purely of glial cells as neurons and neurofilaments are also invariably present (Fig. 13.25) and other benign mesenchymal and epithelial elements are sometimes seen as well. Florid vascular proliferation of the type that occasionally occurs in the ovary in immature teratomas can also be seen in the implants [139]. When only mature tissues are present in an implant it can be designated as grade 0. Occasional implants with an admixture of mature teratomatous elements and endometriosis have been reported [140, 141]. In the past, the implants were thought to represent a form of extraovarian spread from the teratoma. Several recent molecular pathology studies have however shown genetic differences between grade 0 tumor deposits and the primary ovarian tumors, but a similar genetic pattern to normal tissue from the patient [142, 143]. In another study, multiple omental and peritoneal implants were analyzed and found to have mutually exclusive genetic differences [144]. These findings suggest to some that the extraovarian tumor deposits develop by a form of metaplasia in the omentum, the peritoneum, in a lymph node, or elsewhere and do not represent metastases from the ovarian tumor, a conclusion that is not universally accepted [145].

Immunohistochemistry

Immunohistochemistry does not play a major role in the diagnosis of immature teratoma. In most cases, the diagnosis is readily established based on the appearance of the tumor on hematoxylin and eosin stained slides. It has been proposed that expression of glial cell line-derived neurotropic factor receptor alpha-1 (GFRalpha-1) might be useful for the identification of immature neuroepithelium [146], but this antibody is not widely available, and is usually not needed. Primitive neuroepithelium stains for CD99 and bcl-2, and immature cartilage stains for CD34 and bcl-2 [73]. Staining for glial fibrillary acidic protein (GFAP) can help identify glial differentiation. Primitive neuroepithelial cells do not stain with GFAP, but they sometimes stain for neurofilaments or neuron specific enolase (NSE) [134, 147]. Intestinal and respiratory epithelium contains argyrophilic cells that stain for synaptophysin and chromogranin and with antibodies to a variety of neurohormonal peptides.

Some of the tissues in immature teratomas can show staining with antibodies generally thought of as markers of more primitive germ cell tumors; this can complicate identification of primitive elements in a mixed germ cell tumor. For example, AFP, generally used as a marker for yolk sac tumor, shows positive staining in immature liver and immature endodermal glands [73, 135, 148]. SALL4, a marker for embryonal carcinoma, dysgerminoma, and yolk sac tumor, SOX2, a marker for embryonal carcinoma, and glypican-3, a marker for yolk sac tumor, all show occasional staining in immature teratomas. Staining can be seen in immature neuroepithelium, blastomatous stroma, and immature glands [18, 20, 149].

Differential Diagnosis

It can be difficult to decide whether a teratoma is immature or mature. The question of whether a teratoma that has only a miniscule amount of immature tissue should be classified as an immature teratoma is discussed above; our approach is to classify such tumors as mature teratomas with microscopic foci of immature tissue [130]. Sometimes it is not clear whether a component of a teratoma should be viewed as immature or mature. One approach is to view any tissue type that could be seen in fetal development after a fertilization age of 8 weeks as "fetal" type tissue indicative of a mature teratoma [73]. A few specific findings seem especially likely to cause diagnostic problems. Ependymal tubules are occasionally seen in mature teratomas and can be mistaken for immature neuroepithelial tubules. However, they are lined by a single layer of cells, show no mitotic activity, and show minimal staining for proliferation markers such as MIB-1. Cerebellar type tissue is sometimes detected in mature teratomas. A fetal external granular layer can be present in this tissue and should not be mistaken for the primitive neuroectodermal cells of an immature teratoma. Glial tissue in mature teratomas can be surprisingly cellular but unless it is hypercellular and mitotically active it should be viewed as compatible with a mature teratoma [133]. The fetal type cartilage seen in mature teratomas consists of hyaline matrix with chondrocytes in lacunae. The chondroid foci are surrounded by fibrous stroma. Embryonal cartilage, which is indicative of an immature teratoma, consists of chondroid matrix containing plump chondroblasts with pale round nuclei. This type of cartilage is surrounded by cellular immature stroma.

Immature teratoma is a frequent constituent of *mixed germ cell tumors*, so whenever it is identified a careful search for other malignant germ cell elements is necessary. Generally, this requires evaluation of one block per cm of tumor diameter. Microfoci of yolk sac tumor are present in 5–10% of immature teratomas [137]. As long as they are small (<3 mm) and there are only a few of them, they do not warrant designating the tumor as a mixed germ cell tumor, although their presence should certainly be noted in the pathology report. Larger foci of yolk sac tumor are indicative of a mixed germ cell tumor.

Mixed mesodermal tumor or *carcinosarcoma* is a high-grade neoplasm that can be mistaken for an immature teratoma because, like a teratoma, it contains a mixture of epithelial and mesenchymal tissues [147]. However, it almost invariably arises in older women, outside of the age range of immature teratoma. It is entirely of mesodermal derivation and is composed of a mixture of carcinoma and sarcoma. Neither ectodermal nor endodermal elements are present. The epithelial component can be any type of ovarian carcinoma but endometrioid, serous, squamous cell, and undifferentiated carcinoma patterns are most common. Likewise, the mesenchymal component can be any type of homologous or heterologous sarcoma with a nonspecific fibrous sarcoma and rhabdomyosarcoma being the most common types of homologous and heterologous sarcomatous elements.

Teratoma Variants

In some teratomas there is marked prominence or complete overgrowth of one element. That element determines the clinical presentation or clinical behavior of the tumor. The most common monodermal teratoma variants are neuroectodermal tumors, struma ovarii, and carcinoid tumors.

Neuroectodermal Tumors

Rare germ cell tumors contain proliferations of neuroectodermal cells that resemble various tumors of the central nervous system. By analogy with similar tumors of the testis, the criterion for diagnosis of a neuroectodermal tumor of the ovary is the presence of a neuroectodermal component that is at least 1 cm in diameter [150]. In practice, it is usually not necessary to measure the neuroectodermal component, as it generally comprises most or all of a large ovarian neoplasm. Additional teratomatous tissues may be present along with the neuroectodermal elements. There are two main types, primitive neuroectodermal tumors (PNET) and differentiated neuroectodermal tumors growing in patterns that resemble glioblastoma multiforme or various lower grade gliomas [151].

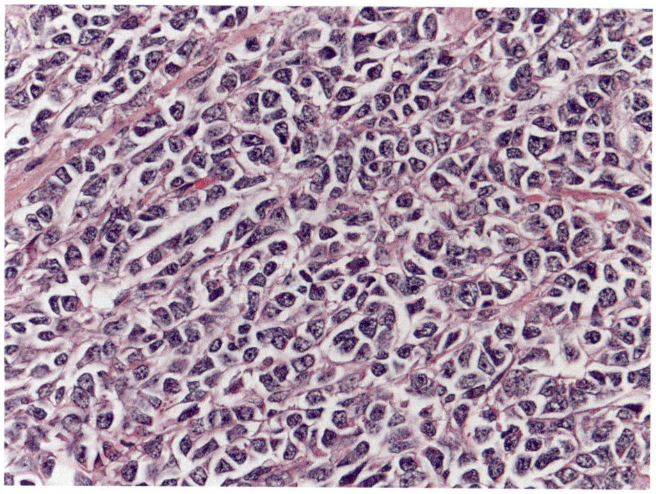

Fig. 13.26 Primitive neuroectodermal tumor. This central type primitive neuroectodermal tumor is composed of small round cells with hyperchromatic nuclei and scanty cytoplasm growing in patternless sheets

Primitive neuroectodermal tumors (PNET) are the most common neuroectodermal tumors of the ovary. Most occur in teenagers or young women 10–30 years of age. These are aggressive neoplasms that are generally treated with chemotherapy following surgical excision. Most ovarian PNET resemble primitive tumors of the brain and are referred to as "central" types of PNET. Rare PNET have a distinctive chromosomal translocation, usually t(11;22), and thus are "peripheral" types of PNET of similar lineage as Ewing sarcoma and peripheral PNET of bone or soft tissue origin [152]. Immunostains for CD99 generally show conspicuous membrane staining in pPNET, and their nature can be confirmed by cytogenetics or by molecular testing to demonstrate an EWS/FLI-1 fusion. If a pPNET is diagnosed, the possibility that it could be metastatic must be excluded [153]. PNET of both central and peripheral types generally consist predominantly of nests and sheets of small cells with hyperchromatic nuclei and scanty, sometimes fibrillar, cytoplasm (Fig. 13.26). Mitotic figures are numerous. Rosettes with central lumens, neuroblastic rosettes, neuropil, or foci of glial differentiation are present in some PNET. Some tumors are composed entirely of sheets of primitive cells while others have variable amounts of tumor that resemble medulloepithelioma, ependymoblastoma, or neuroblastoma. The differential diagnosis includes other small cell tumors that can occur in the ovary, including small cell carcinoma of hypercalcemic type, small cell carcinoma of neuroendocrine type, desmoplastic small round cell tumor, lymphoma, and rhabdomyosarcoma [154].

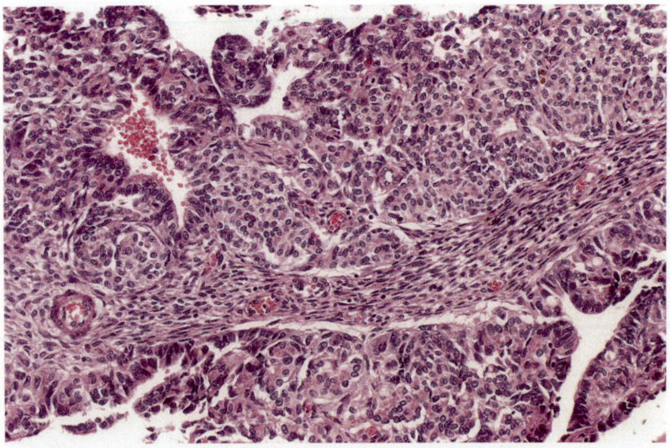

Fig. 13.27 Ependymoma. The partly papillary, partly solid growth pattern can be confused with various other types of ovarian tumors. The tumor cells tend to be fairly monotonous with round to oval dark nuclei and modest amounts of eosinophilic cytoplasm

Differentiated types of neuroectodermal tumor show extensive glial differentiation. These tumors occur over a wide age range with anaplastic tumors occurring over a similar age range as PNET and tumors resembling low-grade gliomas occurring in somewhat older women, usually in their 30s or 40s. Some tumors are anaplastic and resemble a glioblastoma multiforme [151]. These are composed of pleomorphic glial cells with variably sized hyperchromatic nuclei, often with prominent nucleoli, and variable amounts of eosinophilic cytoplasm. Multinucleated cells and numerous mitotic figures are present. The background can be fibrillary. The tumor cells generally show positive staining for glial fibrillary acidic protein (GFAP). Other differentiated neuroectodermal tumors resemble various types of low-grade gliomas. Tumors with features of an ependymoma are most common [151, 155, 156]. These are composed of round or columnar cells with eosinophilic or clear cytoplasm. In some tumors, the cytoplasm has a fibrillar appearance, and in others there are cytoplasmic vacuoles. The tumor cells grow in sheets or line cysts, papillae or glands, or form cribriform nests (Fig. 13.27). Perivascular pseudorosettes are almost always present and some tumors contain true ependymal rosettes with central lumens. Some of the growth patterns mimic various epithelial neoplasms, especially serous carcinoma, and it is not uncommon for ependymomas to be misdiagnosed as epithelial tumors. Immunohistochemistry helps to establish the diagnosis, as ependymomas always show positive staining for GFAP but unlike central ependymomas, they only rarely show diffuse strong membrane staining for CD99. Of note, positive staining for high molecular weight keratin, cytokeratin 8, cytokeratin 7, WT-1 and estrogen, and progesterone receptors is the rule, and unless the tumor is stained for GFAP the immunophenotype may further contribute to misdiagnosis as an epithelial tumor [156, 157].

Tumors resembling other types of low-grade gliomas occur, but they are quite rare. A few tumors resembling oligodendrogliomas have been reported, and there is one report of a tumor resembling a pilocytic astrocytoma [158].

Tumors resembling low-grade or anaplastic astrocytoma have not been specifically reported, but we have seen immature teratomas containing highly cellular mitotically active glial elements; it is likely that such tumors have simply been diagnosed as immature teratomas.

Primitive neuroectodermal tumors and anaplastic tumors resembling glioblastoma multiforme are often admixed with other teratomatous elements but usually none are found in an ependymoma. These tumors are generally classified as monodermal teratomas although they might also be viewed as variants of immature teratoma.

Primitive neuroectodermal tumors and malignant glial tumors have an unfavorable prognosis if they have spread beyond the ovary, but ependymoma appears to behave in an indolent fashion even when metastases are present [151, 156]. Nevertheless, occasional tumor-related deaths have been reported [159].

Struma Ovarii

Struma ovarii is a teratoma that is composed predominantly or exclusively of thyroid tissue. More than 50% of a teratoma should consist of thyroid tissue before it is designated as a struma ovarii. Although thyroid tissue is a common finding in mature and immature teratomas, struma ovarii only accounts for 1–3% of benign ovarian teratomas [160].

Clinical Features

Struma ovarii occurs mainly in women older than 40 years. The median age of women with malignant struma ovarii is 41 years [161]. Usually, struma ovarii, whether benign or malignant, causes nonspecific symptoms such as abdominal or pelvic pain, abdominal distension, or abnormal vaginal bleeding or it is an incidental finding in a patient operated on for some other reason. A few patients have ascites or ascites and hydrothorax (pseudo-Meigs' syndrome). Presentation with symptoms of hyperthyroidism is uncommon; hyperthyroidism only occurs in about 5% of patients with struma ovarii [162, 163]. Struma ovarii is generally benign, so cystectomy or unilateral salpingo-oophorectomy is adequate treatment.

Tumors that are diagnosed as malignant struma ovarii on histologic grounds are infrequently clinically malignant and do not necessarily require radical treatment [162]. When a diagnosis of malignant struma ovarii is made based on histologic criteria alone and there is no evidence of extraovarian spread, treatment can be conservative. It generally consists of complete excision, usually by at least salpingo-oophorectomy, measurement of serum thyroglobulin to exclude occult metastatic disease and ultrasound evaluation of the thyroid to rule out the possibility that the ovarian neoplasm is a metastasis from a primary thyroid carcinoma [164–166]. About two thirds of patients with biologically malignant strumas have extraovarian tumor deposits at the time of diagnosis [161]. Recurrences are relatively uncommon in patients with tumors that appear confined to the ovary at diagnosis (stage Ia) but even patients with tumors apparently limited to the ovary require long-term follow-up, typically including physical examinations, serial thyroglobulin determinations, and abdominal imaging. In the largest series, 10 of 27 patients with biologically malignant strumas had stage Ia tumors at diagnosis; recurrences were detected 10 months to 29 years after initial treatment [163]. Clinically malignant thyroid type cancers arising in ovarian teratomas spread within the abdomen to the peritoneum or omentum or to distant sites such as lymph nodes, liver, bone, or lung. Tumors with extraovarian spread are generally treated by hysterectomy and bilateral salpingo-oophorectomy. This is often followed by thyroidectomy and radioactive iodine treatment [164, 167]. The prognosis appears to be favorable for patients with malignant struma ovarii, even when extraovarian tumor spread is present. In the largest series, 10-year survival rate for patients with histologically malignant struma ovarii was 81% and, for patients with extraovarian tumor spread at diagnosis, the median survival time was 9 years to death or a status of alive with tumor [161]. Among patients with biologically malignant strumas, those with papillary carcinoma (72% dead of tumor) appear to have a worse prognosis than those with follicular carcinoma (23% dead of tumor) [163]. Patients with large cellular tumors of the ovary, particularly those with papillary or anaplastic tumors and those with such adverse findings as adhesions and ascites, require long-term follow-up even if there is no evidence of extraovarian tumor at diagnosis.

Gross Pathology

Struma ovarii ranges in size from a microscopic neoplasm to a large solid or cystic mass more than 10 cm in diameter. The average size is 5–10 cm. The cut surface is red, green, or tan with a characteristic glairy, meaty appearance. Small cysts are common. Occasional tumors are largely or totally cystic with one or multiple locules [168]. These can be difficult to diagnose, because the lining may consist predominantly of flat epithelium; readily identifiable thyroid tissue with follicles and colloid may be present only focally. Gross pathologic findings that might correlate with clinically malignant behavior have been identified, although none are conclusive indicators of malignancy [161]. Tumors with a strumal component that is smaller than 4 cm virtually never metastasize, while above that size increasing diameter of the strumal component parallels an increased risk of recurrence. The average diameter of clinically malignant strumas is 13–14 cm [163]. Other gross features associated with an increased risk of recurrence are adhesions and ascites.

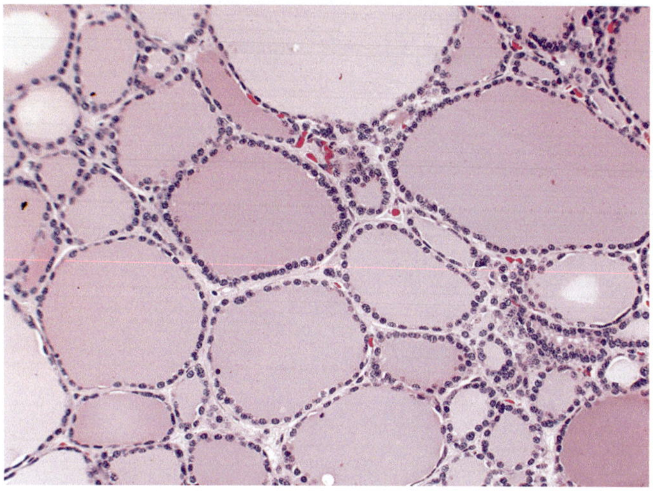

Fig. 13.28 Struma ovarii. The tumor is composed of thyroid-type follicles filled with eosinophilic colloid and lined by cuboidal follicular epithelial cells with uniform bland nuclei and scanty cytoplasm

Microscopic Pathology

Struma ovarii is composed of follicles lined by cuboidal or columnar epithelial cells with uniform round nuclei and moderate amounts of eosinophilic cytoplasm (Fig. 13.28). The follicles vary in size. They are filled with eosinophilic colloid, and it may be possible to demonstrate refractile oxalate crystals in the colloid; they are best seen with polarized light. Degenerative changes, including fibrosis, calcification, and hemosiderin-laden macrophages may be present. Some tumors consist exclusively of thyroid tissue, while in others residual mature cystic teratoma is present. Other teratomatous elements can be found in many cases if they are carefully searched for. Luteinized cells are commonly present in the ovarian stroma around struma ovarii [169].

Highly cystic variants of struma ovarii may be difficult to recognize because the cysts can be predominantly lined by a compressed layer of flattened epithelial cells, with only occasional areas showing easily recognized follicular growth and colloid [168]. Other problematic tumors include those composed of clear, oxyphilic, or signet-ring cells, those with cord-like arrangements of tumor cells, and hypercellular tumors in which the cells grow in microfollicular, trabecular, or solid patterns similar to those seen in a follicular adenoma of the thyroid [162, 169]. When the histologic appearance is similar to that of a thyroid follicular adenoma, with expansile noninvasive growth of crowded small follicles, solid sheets of thyroid type cells, or trabecular formations, the tumor is designated as "proliferative struma ovarii" (Fig. 13.29) [162].

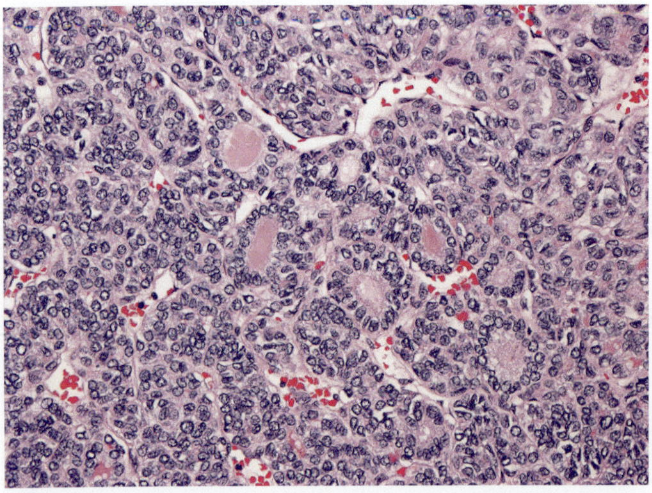

Fig. 13.29 Cellular struma ovarii. The tumor consists of crowded small follicles and areas of solid growth, but the periphery of the tumor was circumscribed and there was no evidence of vascular invasion. Elements of a benign cystic teratoma were present in other sections

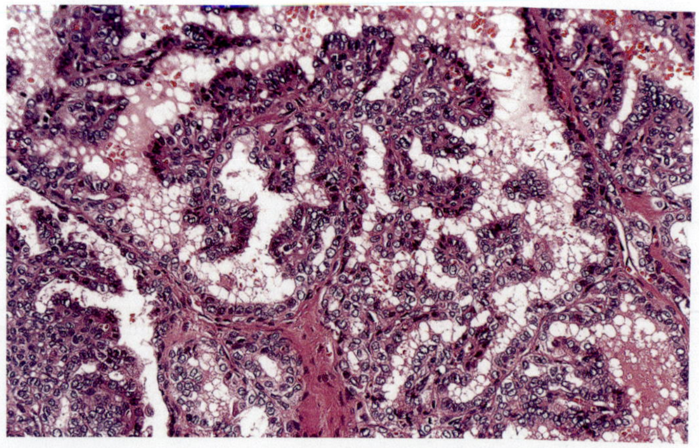

Fig. 13.30 Papillary thyroid carcinoma arising in struma ovarii. This tumor grows in a papillary pattern, with tumor cells lining papillae having distinct fibrovascular cores. The tumor cells have enlarged pale nuclei and moderate amounts of eosinophilic cytoplasm

Various types of thyroid cancer arise in struma ovarii, including papillary carcinoma and its follicular variant, follicular carcinoma, highly differentiated follicular carcinoma, poorly differentiated carcinoma, and very rare examples of other types [160, 170]. Malignant strumas are almost always entirely or predominantly (>50%) thyroid type neoplasms [163]. Rare examples arise in a strumal carcinoid [161, 171]. As a general rule, any tumor with a predominance of thyroid tissue and a size >5 cm should be viewed as having uncertain malignant potential [163].

Papillary carcinoma is the most common malignancy in struma ovarii and has the same histologic appearance as it does in the thyroid [161, 162, 164, 172]. The classical papillary variant contains follicles but well-developed papillae with fibrovascular cores lined by tumor cells are present at least focally (Fig. 13.30). In the follicular variant of papillary carcinoma, the growth is exclusively follicular. Regardless of the growth pattern, papillary carcinoma is composed of cells that have large, crowded, grooved nuclei with small nucleoli. Other characteristic findings are clearing of some tumor cell nuclei and the presence of intranuclear cytoplasmic inclusions. The cells have moderate or abundant eosinophilic cytoplasm. The colloid within the follicles often has a dense hypereosinophilic appearance. Some authors report that papillary carcinomas with areas of papillary growth are most common [161, 162], while others have found the follicular variant of papillary carcinoma to be more common than the classic papillary variant [164].

Follicular carcinoma is often difficult to differentiate from "proliferative" or "adenomatous" struma ovarii, but it is generally thought to be uncommon. In one large series, only 2 of 13 carcinomas were follicular carcinomas [162], while in another series there were no follicular carcinomas among ten malignant strumas [164]. Follicular carcinoma is a hypercellular neoplasm in which the tumor cells tend to grow in a microfollicular, trabecular, or solid pattern. One reason that it is difficult to differentiate follicular carcinoma from proliferative struma ovarii is the lack of a capsule to assess for transcapsular invasion. Features indicative of follicular carcinoma arising in struma ovarii include an invasive growth pattern, particularly with invasion onto the surface of the ovary, and vascular invasion. Proposed criteria for vascular invasion include conformation of intravascular tumor to the contour of the vessel lumen, adherence to the endothelium, and involvement of multiple blood vessels [161]. Unfortunately, criteria for identification of follicular carcinoma are less accurate than those for identification of papillary carcinoma. In the series with the most cases of malignant struma ovarii and the longest follow-up, 10 of 27 biologically malignant tumors recurred at some time after the initial treatment [161]. Only 2 of the 10 met histologic criteria for the diagnosis of carcinoma; both were papillary carcinomas. Eight had initially been interpreted as "adenomas" (i.e., proliferative struma ovarii) and their true nature was not recognized until metastases developed. Follicular carcinoma arising in struma ovarii is thus more common than is widely realized, but many cases are likely to be unrecognized unless they recur.

The term "strumosis" was used in the past to describe a condition in which histologically bland peritoneal or omental implants of thyroid tissue are found in a patient with what appears to be a benign struma ovarii. This term is inappropriate and should not be used, as based on the presence of metastatic spread of tumor to the peritoneum or omentum, progressive growth within the abdomen in some cases, and even distant metastatic spread, the peritoneal or omental tumor deposits represent metastases from a highly differentiated

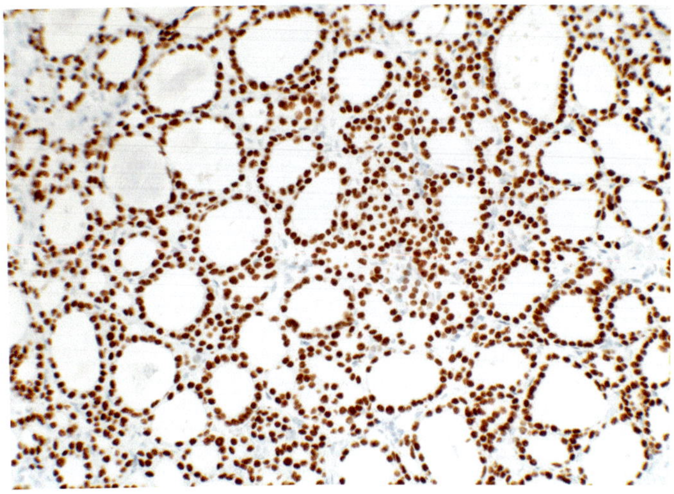

Fig. 13.31 Struma ovarii. An immunostain for thyroid transcription factor-1 (TTF-1) shows strong staining of the tumor cell nuclei

follicular carcinoma arising in struma ovarii [161, 173]. The ovarian tumor in these cases is so well differentiated that it resembles normal thyroid tissue, and, if extraovarian spread is not present at the time the ovarian tumor is removed a diagnosis of highly differentiated follicular carcinoma cannot be established until metastases are detected. Histologically bland, extremely well-differentiated variants of papillary carcinoma also occur, and these too may be difficult or impossible to recognize until the detection of metastases reveals the true nature of the ovarian tumor [164].

Immunohistochemistry and Molecular Pathology

Thyroglobulin is one of the main proteins produced by thyroid follicular cells. In struma ovarii, there is positive staining for thyroglobulin in the cytoplasm of the cells lining the neoplastic follicles and in the colloid within the follicles. A second useful immunostain is one for a nuclear antigen, thyroid transcription factor-1 (TTF-1). There is strong positive staining of tumor cell nuclei in struma ovarii (Fig. 13.31). In most cases the thyroidal nature of the tumor is obvious in routine hematoxylin and eosin stained slides and immunohistochemical studies are not necessary. Strumas with extensive cyst formation, those with a paucity of colloid, and those with unusual histologic patterns are more difficult to diagnose. In these, immunohistochemical stains that prove the thyroidal nature of the tumor cells can be very helpful in arriving at the correct diagnosis [174]. Immunohistochemistry may be of value in the diagnosis of papillary carcinoma, in that such tumors tend to show positive staining for CK19 and HBME1. In one series, 10 of 10 follicular variants of papillary carcinoma were positive for CK19 and 8 of 10 were positive for HBME1 [175]. Two of the ten carcinomas were small and arose in a benign struma ovarii, and while the carcinoma areas were positive for CK19 and HBME1 the benign areas showed no staining for either marker. Three benign strumas likewise showed no staining.

There are many similarities between the immunophenotype of struma ovarii and that of epithelial ovarian cancers, and this can result in diagnostic problems if the pathologist is not aware of the overlap in immunophenotypes. For example, immunostains for cytokeratin 7 and PAX8 are commonly positive in both ovarian epithelial tumors and in benign and malignant strumal ovarii.

Molecular testing is of potential value in the diagnosis of malignant struma ovarii, particularly the papillary carcinoma type. Genetic abnormalities appear to be uncommon in benign struma ovarii. None of 14 reported cases has had a BRAF or RAS mutation and only one had a RET/PTC rearrangement [176]. All cases of malignant struma ovarii that have been studied have been classic or follicular variants of papillary carcinoma. A BRAF mutation was detected in 29% of 17 malignant strumas in three reports and 41% had a RET/PTC rearrangement [175, 177, 178]. Of patients with a BRAF mutation, those with the classic type of papillary carcinoma have had only BRAF^{V600E} mutations, while those with follicular variants of papillary carcinoma had different BRAF mutations. RET/PTC rearrangement has been reported only in patients with the follicular variant of papillary carcinoma. One additional case of a follicular variant of papillary carcinoma had an NRAS mutation and a PAX8/PPARγ rearrangement [179]. BRAF mutation is the only one of these abnormalities that is specific for malignancy, so testing for a BRAF mutation may be of value in cases of struma ovarii where a diagnosis of papillary carcinoma is unclear. However, a mutation is likely to be found in only a minority of cases, and only in those where a papillary growth pattern is evident on light microscopy. It is unclear whether the presence of a BRAF mutation has any prognostic significance in malignant struma ovarii. RET/PTC rearrangements and RAS mutations are not specific for malignancy, but the presence of one of these abnormalities might provide support for a diagnosis of papillary carcinoma if other features favoring such a diagnosis were also present.

Carcinoid Tumor

Carcinoid tumors are uncommon ovarian neoplasms that are classified as monodermal teratomas as many are associated with other teratomatous elements [180]. A pure ovarian carcinoid has most likely overgrown the teratoma in which it originated, but other possible sites of origin include from neuroendocrine cells in a mucinous tumor or the mucinous component of a heterologous Sertoli-Leydig cell tumor [181–183], or from non-endocrine ovarian cells by neometaplasia. Also, when there is no associated teratoma the possibility that the carcinoid is metastatic from the gastrointestinal tract or elsewhere must be considered [184, 185].

Clinical Features

Ovarian carcinoids mainly occur in peri- or postmenopausal women who present with nonspecific symptoms such as pelvic or abdominal pain, abdominal enlargement, menstrual irregularities, or abnormal vaginal bleeding. A quarter to a third of women with ovarian carcinoid tumors have the carcinoid syndrome [186]. The venous drainage of the ovaries is into the inferior vena cava or the renal vein, bypassing the liver, where the vasoactive substances secreted by carcinoid tumors are normally inactivated. Women with ovarian carcinoids can therefore develop the carcinoid syndrome without having liver metastases. Most have large insular carcinoid tumors; the carcinoid syndrome rarely occurs in patients with other types of carcinoids [187]. Patients with the carcinoid syndrome typically have facial flushing, diarrhea, bronchospasm, and hypertension. Some develop carcinoid heart disease and have symptoms of right heart valvular disease or right heart failure, including heart murmurs or edema. The valve disease may be sufficiently severe as to require valve replacement [188]. A few patients have presented with severe chronic constipation caused by secretion of the intestinal hormone peptide YY by a trabecular or strumal carcinoid [189–191]. Rare patients with a strumal carcinoid develop hyperthyroidism [192].

Carcinoid tumors tend to be unilateral and confined to the ovary at the time of diagnosis. A benign cystic teratoma is present in the contralateral ovary in 10–15% of cases. The standard treatment is hysterectomy and bilateral salpingo-oophorectomy, but a unilateral salpingo-oophorectomy is sufficient in a young woman. The symptoms generally subside quickly once the tumor has been removed. Women with carcinoid heart disease, however, can have progressive cardiac disease despite the complete removal of the carcinoid tumor [187, 193]. The prognosis is favorable in most cases and metastases, and tumor-related deaths are rare. Progressive disease occurs mainly in women with extraovarian tumor spread at diagnosis. Both the strumal and the carcinoid components of strumal carcinoid appear capable of giving rise to metastases [194].

The ovary is a common site for metastases from carcinoids of the gastrointestinal tract, usually the small intestine or the appendix [185]. Metastatic carcinoids can cause identical symptoms as primary ovarian carcinoids, including the carcinoid syndrome. Findings that suggest that a carcinoid might be metastatic include bilateral ovarian involvement, multiple tumor nodules in one or both ovaries, and the presence of peritoneal metastases.

Macroscopic Pathology

Carcinoid tumors are unilateral solid tumors, sometimes containing small cysts. They vary considerably in size, ranging from small tumors detected only microscopically to large masses up to 28 cm in diameter. Patients with the carcinoid syndrome tend to have larger tumors, with an average diameter of about 9 cm, compared with an average diameter of only about 3 cm in patients who do not have the carcinoid syndrome [187]. Carcinoids are typically firm homogeneous tan or yellow tumors. Some are found in the wall of a benign cystic teratoma and rare examples are associated with a mucinous tumor [183].

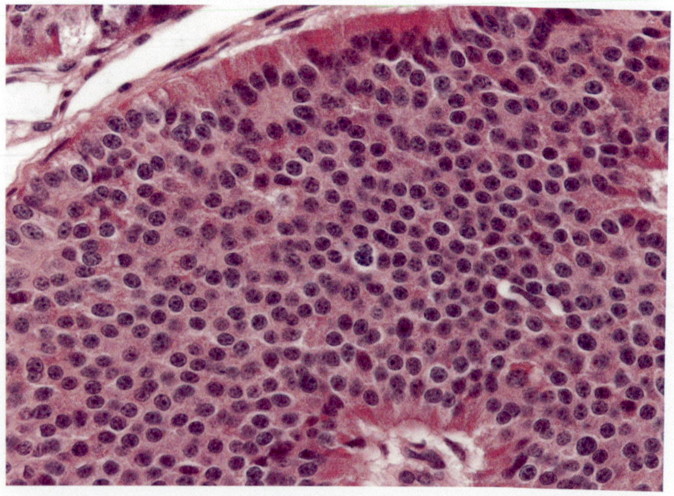

Fig. 13.32 Carcinoid tumor. The tumor cells have uniform round nuclei with coarse chromatin and modest amounts of eosinophilic cytoplasm. At the periphery of the nests the tumor cell nuclei are similar to those seen elsewhere, but the cells are columnar with more abundant cytoplasm that contains red granules

Microscopic Pathology

Ovarian carcinoids are composed of round, cuboidal, or columnar neuroendocrine cells that have relatively uniform round or oval nuclei with coarse chromatin and small nucleoli. Mitotic figures are generally infrequent. The cytoplasm ranges from moderate to abundant and can be eosinophilic or clear, or, occasionally, oxyphilic. The cytoplasm typically appears granular in hematoxylin and eosin stained sections and brightly stained basal eosinophilic cytoplasmic granules are sometimes visible (Fig. 13.32). The granules usually stain black with argentaffin or argyrophil stains. Luteinized stromal cells are distributed around the periphery of some carcinoids. Benign teratomatous elements are often detected adjacent to an ovarian carcinoid. Their identification is important evidence that the carcinoid is primary in the ovary.

Four histologic types of carcinoid tumors have been defined: insular, trabecular, strumal, and mucinous [195]. Many tumors contain a mixture of types. *Insular* carcinoids are the most common ones [187]. In this type, the carcinoid cells grow in nests or islands that coalesce or are surrounded by fibrous stroma (Fig. 13.33). Small tubules are sometimes present, particularly at the periphery of the tumor cell nests. In some tumors the tubules contain eosinophilic secretions

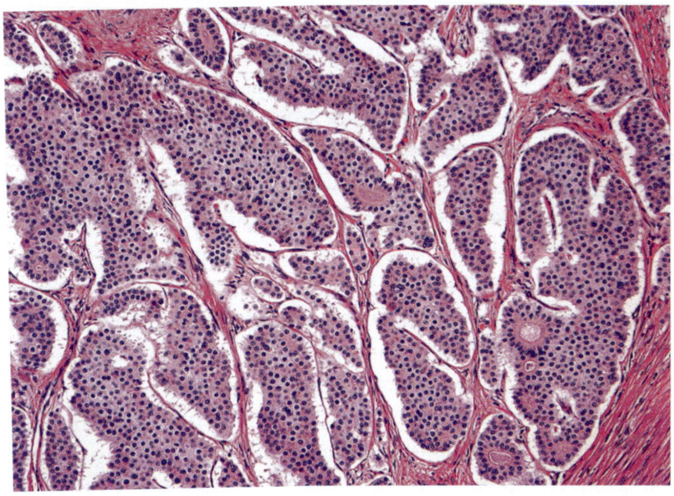

Fig. 13.33 Insular carcinoid tumor. The tumor cells grow in rounded nests that are surrounded by dense fibrous stroma. Most nests are composed of sheets of tumor cells, but in a few, seen on the right, the tumor cells line tubules

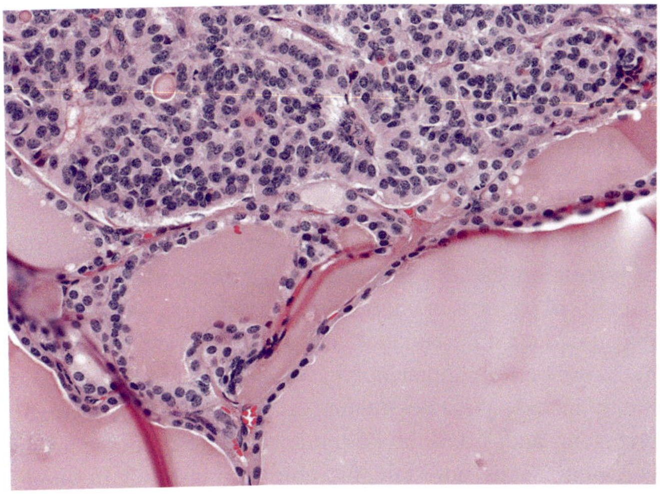

Fig. 13.34 Strumal carcinoid tumor. This biphasic tumor consists of thyroid follicles filled with colloid at the bottom and carcinoid tubules lined by low columnar cells with hyperchromatic nuclei at the top

and in other tumors psammoma bodies are present. The appearance of insular carcinoids simulates that of midgut carcinoids. *Trabecular* carcinoids are composed of tall columnar cells with central oval nuclei and granular eosinophilic cytoplasm [182, 196]. The cells grow in undulating cords, ribbons, or trabecula, a pattern similar to that of foregut and hindgut carcinoids. *Strumal* carcinoid is a mixed tumor that contains both carcinoid and strumal (thyroid) elements [192, 197]. The carcinoid component grows in a trabecular or mixed trabecular-insular pattern. In the strumal component thyroid-type follicles are filled with colloid and lined by columnar follicular cells. There are calcium oxalate crystals in the colloid in about half the cases. Although tumors with discrete areas of carcinoid and struma meet the definition of a strumal carcinoid, in most strumal carcinoids the two components abut each other and generally merge (Fig. 13.34). In areas where the carcinoid and strumal elements merge the carcinoid cells grow between and into the follicles where they undermine and replace the follicular cells. Some follicles are lined by thyroid cells, some by a mixture of thyroid and carcinoid cells, and some by carcinoid cells. Mucinous epithelium is a fairly common finding in strumal carcinoids. Amyloid is a rare finding in the stroma of a strumal carcinoid. While the carcinoid component is generally viewed as potentially malignant, the strumal component is usually histologically benign. Rare strumal carcinoids, however, give rise to a malignant thyroid type neoplasm [194]. *Mucinous* carcinoids, which resemble goblet cell carcinoids of the appendix, are the least common ovarian carcinoids [198]. In this type of carcinoid, round or tubular glands lined by a mixture of columnar or cuboidal neuroendocrine cells and goblet cells float in pools of mucin surrounded by fibrous stroma. Variants of mucinous carcinoid contain tumor cells with atypical nuclei, crowded glands, and foci of cribriform or microcystic growth. Some mucinous carcinoids contain areas of carcinoma; these are the ones that are most likely to behave aggressively [198].

Immunohistochemistry

Carcinoid tumors almost invariably show positive cytoplasmic staining for chromogranin and synaptophysin (Fig. 13.35). Staining for these neuroendocrine markers is now the preferred means for confirming the diagnosis of a carcinoid tumor. Strumal carcinoid has a distinctive immunophenotype, as it shows staining for both neuroendocrine and thyroid markers. The carcinoid elements stain for chromogranin and synaptophysin, while the thyroid component shows staining for thyroglobulin in the follicular cell cytoplasm and the colloid and there is strong follicular cell nuclear staining for thyroid transcription factor-1 (TTF-1) [174, 197, 199]. Ovarian carcinoids are cytokeratin positive, but they do not stain for cytokeratin 7, a feature that helps differentiate them from ovarian epithelial tumors such as trabecular or tubular variants of endometrioid carcinoma [200].

Trabecular and strumal carcinoids typically show cytoplasmic staining for prostate specific acid phosphatase (PSAP), as do hindgut carcinoids, but they do not stain for prostate specific antigen (PSA) [201]. A variety of peptide hormones can be detected in carcinoid cells by immunohistochemistry, including serotonin, gastrin, pancreatic polypeptide (PP), vasoactive intestinal polypeptide (VIP), insulin, glucagon, substance P, adrenocorticotrophin (ACTH), and somatostatin [202]. Trabecular and strumal carcinoids from patients with chronic constipation typically stain for protein YY.

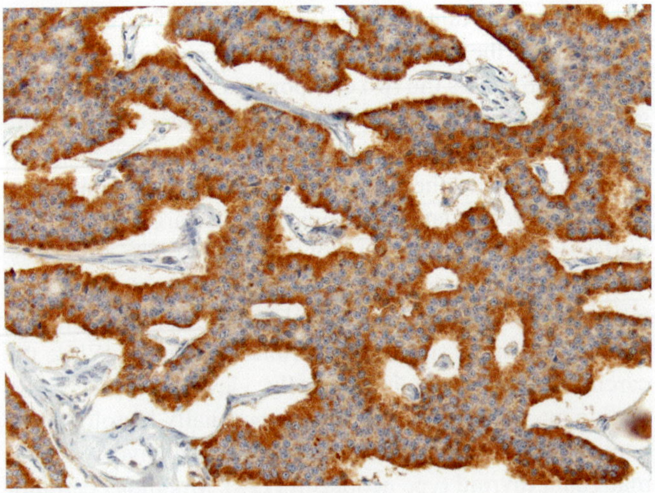

Fig. 13.35 Positive staining for synaptophysin and chromogranin, the latter manifested here as dense brown granular staining of the basal cytoplasm, is characteristic of carcinoid tumors

Differential Diagnosis

The possibility that an ovarian carcinoid tumor is *metastatic* should always be considered. Most metastatic carcinoid tumors in the ovary are from a primary tumor of the small or large intestine or the appendix, particularly the distal ileum [184], but other primary sites, such as the lung, are also possible. Metastatic carcinoid tumors are usually multinodular and bilateral and are not associated with a teratoma [184, 185]. Synchronous metastases are typically also present in the liver or on the peritoneum. Microscopically, an insular pattern of growth is most common. Pulmonary carcinoid tumors metastatic to the ovary are likely to express thyroid transcription factor-1 (TTF-1) while primary carcinoids other than strumal carcinoids do not express TTF-1 [200]. Staining for CDX-2, on the other hand cannot be used to differentiate between a metastatic intestinal carcinoid and a primary ovarian carcinoid as most insular ovarian carcinoids are CDX-2 positive [200]. Primary mucinous carcinoid tumors occur in the ovary, as discussed above, but metastasis from a mucinous or goblet cell carcinoid-like tumor of the appendix or intestine is equally likely [203]. It is common for both the appendiceal primary and the ovarian metastases from such appendiceal tumors to exhibit a varied appearance that includes not only features of goblet cell carcinoid but also tubuloglandular adenocarcinoma and signet ring cell adenocarcinoma, as well as pools of mucin. Staining for neuroendocrine markers is variable, with positive cells being present in only a quarter to a third of cases [203]. Positive staining for CDX-2 and CK20 is typically present in primary ovarian mucinous carcinoids as well as in metastatic appendiceal tumors, so distinction between primary and metastatic tumors depends mainly on standard pathologic criteria such as tumor size (metastases are generally large) and whether the tumor shows features of a metastasis such as multinodularity and bilateral ovarian involvement. Metastatic tumors from the appendix may have goblet cell carcinoid like features, but since they also typically exhibit features of adenocarcinoma and have a highly unfavorable prognosis it is best to view them as metastatic carcinomas rather than as metastatic goblet cell or mucinous carcinoids. Most have a significant component of signet ring cells and therefore qualify as Krukenberg tumors [203].

Granulosa cell tumors grow in patterns similar to those seen in carcinoid tumors, such as insular and trabecular patterns, and they are also composed of small cells with darkly stained nuclei. However, carcinoid tumor cells have coarsely granular nuclei and the cytoplasm may contain eosinophilic neuroendocrine granules. Granulosa cells have finely granular grooved nuclei and the cells lack granules. Carcinoid tumors show positive staining for cytokeratin, chromogranin, and synaptophysin while granulosa cells are usually cytokeratin negative. They never stain for chromogranin or synaptophysin, but are usually positive for inhibin, calretinin, and steroidogenic factor-1, which are negative in carcinoid tumors.

Brenner tumors can occasionally mimic a carcinoid tumor as the tumor cells form variably sized nests surrounded by fibrous stroma. However, the tumor cell nuclei often have grooves and they have finely granular chromatin. Brenner tumors are cytokeratin 7 positive but they do not stain for neuroendocrine markers.

Endometrioid carcinomas occasionally have a trabecular or tubular appearance that might suggest a carcinoid tumor, but typical patterns of endometrioid carcinoma, often including squamous differentiation, are usually present somewhere in the tumor. Immunophenotypic differences from a carcinoid tumor include positive staining for cytokeratin 7, estrogen receptors, and, often, progesterone receptors. Endometrioid carcinoma does not stain for chromogranin but it should be noted that occasional tumors show some staining for synaptophysin.

Finally, small foci of carcinoid tumor are rarely present in ovarian tumors such as Sertoli-Leydig cell tumors and carcinosarcomas. The carcinoid foci in such tumors are viewed as a form of heterologous differentiation [181].

Malignant Mixed Germ Cell Tumor

A majority of ovarian malignant germ cell tumors consist exclusively of one type of germ cell tumor; that is, they are pure tumors. Malignant mixed germ cell tumors contain a mixture of two or more germ cell elements. Mixed tumors are not uncommon and comprise 5–20% of all malignant germ cell tumors [70, 71]. In a clinical study of 72 patients with malignant germ cell tumors treated at one gynecologic oncology center, 21% of patients had mixed germ cell tumors

[204]. Mixed germ cell tumors accounted for about 9% of malignant germ cell tumors identified in a review of the SEER database [2]. A germ cell tumor that contains one malignant element mixed with benign teratomatous elements does not fall into this category.

Clinical Features

Mixed germ cell tumors occur over the same age range as pure germ cell tumors. Most occur in children and young women, and they are uncommon in women more than 40 years old. The average patient age is about 16 years. Most patients present with abdominal pain or swelling or a palpable abdominal mass. A few have acute abdominal symptoms caused by torsion or rupture of the tumor. About a third of premenarchal children with mixed germ cell tumors have precocious pseudopuberty. Amenorrhea or abnormal vaginal bleeding is a common complaint in older children and adults. Serum tumor marker levels depend on which germ cell elements are present. The serum β-HCG is elevated in about 50% of patients and the serum AFP is likewise elevated in about 50% of patients. The tumor is confined to the ovary at diagnosis (stage I) in about two thirds of the patients [70, 71].

Given the young age of most patients, the usual treatment is by unilateral salpingo-oophorectomy, with debulking of extraovarian tumor if any is present. If the contralateral ovary or the uterus is involved surgery may need to be more extensive and may include total abdominal hysterectomy and bilateral salpingo-oophorectomy. In the past, more than 50% of stage I tumors treated by surgery alone recurred, so treatment has typically included adjuvant combination chemotherapy as well as surgery, except for patients with stage IA tumors that contain only dysgerminoma and grade I immature teratoma. Most patients treated in this way are cured. Close observation without chemotherapy can also be considered [14]. Patients with more advanced tumors had a lower survival rate in the past, but improved survival is now being reported due to the availability of more effective chemotherapy [2]. In early reports the prognosis appeared to depend on the size of the tumor, the types and amounts of germ cell elements present, and the stage [2]. Modern chemotherapy has essentially eliminated survival differences among the various tumor types and contemporary reports indicate that stage is the only significant determinant of prognosis [70, 204].

Macroscopic Pathology

Mixed germ cell tumors are large, with an average diameter of 15 cm. Their gross appearance depends on which germ cell elements are present. Areas of dysgerminoma are fleshy and gray, tan, or white. Yolk sac tumor is tan or white with small or large cysts and foci of hemorrhage and necrosis. Immature teratoma is white or tan with cysts and, often, hard cartilaginous or bony foci. Choriocarcinoma is hemorrhagic and necrotic. Mixed germ cell tumors are generally unilateral, but if dysgerminoma is one of the components, involvement of the contralateral ovary is possible.

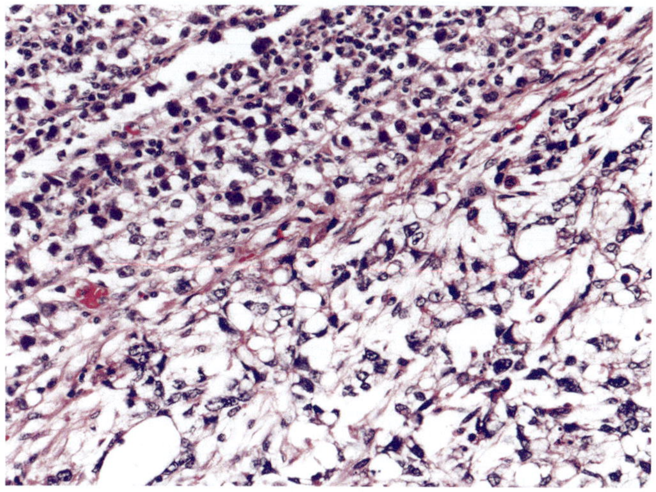

Fig. 13.36 Malignant mixed germ cell tumor. This tumor contains two malignant germ cell elements, dysgerminoma at the top and yolk sac tumor at the bottom

Microscopic Pathology

Two malignant elements are present in 80% of mixed germ cell tumors, three in 15% and four or more in the final 5%. The elements can be admixed or they can grow in separate but contiguous foci (Fig. 13.36). Dysgerminoma is the most common tumor type seen in mixed germ cell tumors, followed by yolk sac tumor and immature teratoma. Embryonal carcinoma, choriocarcinoma, and polyembryoma are less common, but when they occur in the ovary it is usually as a component of a mixed germ cell tumor.

Gonadoblastoma

It is in a sense a misnomer to call gonadoblastoma an ovarian tumor, since it mainly occurs in the abnormal gonads of individuals with a Y chromosome. However, most patients are phenotypic females, and gonadoblastoma has traditionally been grouped with the ovarian tumors. It can be viewed as a form of carcinoma in situ in which there is an admixture of germ cells (the carcinoma in situ) and sex cord cells [205]. Invasive malignant germ cell tumors, usually a dysgerminoma, can arise in gonadoblastoma [206].

Clinical Features

Gonadoblastoma has been diagnosed in patients of all ages but it is most often detected in the young. The average age at diagnosis is 18 years and 80% of gonadoblastomas are detected before the age of 20 years. Gonadoblastoma has been diagnosed in infants and young children less than 2 years old and dysgerminoma arising in gonadoblastoma has been detected by 15 years of age [207]. Some gonadoblastomas are incidental findings or are discovered when adnexal calcifications are seen on abdominal or pelvic x-rays. Most are found when a patient is evaluated for primary or secondary amenorrhea or for an abnormally formed genital tract. Most patients are phenotypic females, but gonadoblastoma also occurs in males, usually pseudohermaphrodites. Phenotypic females with gonadoblastoma can have a normal or a short vagina. The uterus and cervix are small in three quarters of patients and about a third have small or rudimentary fallopian tubes. More than half of female patients with gonadoblastoma are at least mildly virilized [206, 208].

Most patients have some form of gonadal dysgenesis [209]. The most common karyotypes are 46XY and 45X/46XY. A Y chromosome or a Y chromosome fragment is almost always present although rare patients with children and a 46XX karyotype have been reported. In some patients Y chromosome material cannot be detected by routine karyotyping but it is identified using more sensitive molecular techniques [210]. Some patients with Turner syndrome can be shown to harbor Y chromosome material and they, like patients with gonadal dysgenesis, are at risk for gonadoblastoma [211, 212]. In situ hybridization studies performed on gonadoblastomas in patients with mosaic karyotypes show that the gonadoblastoma is derived from cells that have a Y chromosome. There is a region on the short arm of the Y chromosome near the centromere, the GBY region (gonadoblastoma locus on the Y chromosome) [213] that contains a gene that is associated with susceptibility to gonadoblastoma in patients with dysgenetic gonads [214]. The most likely candidate gene is testis specific protein Y-encoded (TSPY) [214].

Gonadoblastoma is a form of carcinoma in situ and is clinically benign unless a malignant germ cell tumor develops in it. The treatment is bilateral gonadectomy since the gonads are abnormal and are not capable of normal reproductive function. They are removed to prevent virilization and to prevent development of an invasive malignant germ cell tumor. A gonadoblastoma or a malignant germ cell tumor develops in the abnormal gonads of patients with a Y-chromosome in more than a third of cases [215]. About 90% of the malignant tumors that arise in gonadoblastoma are germinomas (dysgerminomas) and more aggressive tumors such as yolk sac tumor and embryonal carcinoma account for the remaining 10%.

Macroscopic Pathology

Gonadoblastoma occurs almost exclusively in abnormal gonads, mainly those with features of pure or mixed gonadal dysgenesis [206]. Some gonadoblastomas are not grossly visible and are detected only when the abnormal gonads are examined microscopically. Those that are grossly visible are solid tan or white nodules. Gonadoblastoma can measure up to 2–3 cm in diameter, but most are smaller. Gritty areas of calcification are often detected when the tumor is sectioned. More than 40% of gonadoblastomas are bilateral. If a malignant germ cell tumor has developed in a gonadoblastoma it is usually grossly visible as a solid white or tan nodule; with time, it usually overgrows and obscures the underlying gonadoblastoma.

Microscopic Pathology

Microscopically, nests of tumor cells grow in dense fibrous stroma. Calcifications are typically present and visible at low magnification. Gonadoblastoma contains a mixture of germ cells and sex cord cells (Fig. 13.37) [216]. Two types of germ cells have been described in gonadoblastoma: immature germ cells with coarse chromatin, prominent nucleoli, and scanty cytoplasm, and more mature germ cells with smaller, denser nuclei, and abundant clear cytoplasm [217]. In practice, it tends to be difficult to differentiate among different populations of germ cells. The overall appearance of the germ cells is similar to dysgerminoma cells. Sex cord cells surround individual germ cells, balls of hyaline material, or palisade at the periphery of the tumor cell nests. The sex cord cells are difficult to classify. They have small dark nuclei and variable amounts of amphophilic cytoplasm. They have features in common with both granulosa cells and Sertoli cells; positive staining for cytokeratin suggests a closer relationship to Sertoli cells. In addition to growing around the germ cells, they often surround balls of hyaline eosinophilic basement membrane like material, which are often a conspicuous finding in gonadoblastoma. Involution occurs in some gonadoblastomas, in the form of a decrease in the number of germ cells and an increase in the amount of hyaline material in the nests. If the patient is postpubertal there are frequently luteinized or Leydig-like cells in the stroma surrounding the tumor cell nests.

A malignant germ cell tumor, usually a germinoma, can develop in gonadoblastoma. Initially there is proliferation of germ cells in gonadoblastoma nests such that the germ cells form clusters and eventually coalesce into sheets that replace the nodules. Eventually, the germ cells invade the surrounding stroma and grow in an expansile pattern forming a nodule of germinoma. Other types of malignant germ cell tumor, such as yolk sac tumor or embryonal carcinoma, also occur in gonadoblastoma, but much less often than germinoma.

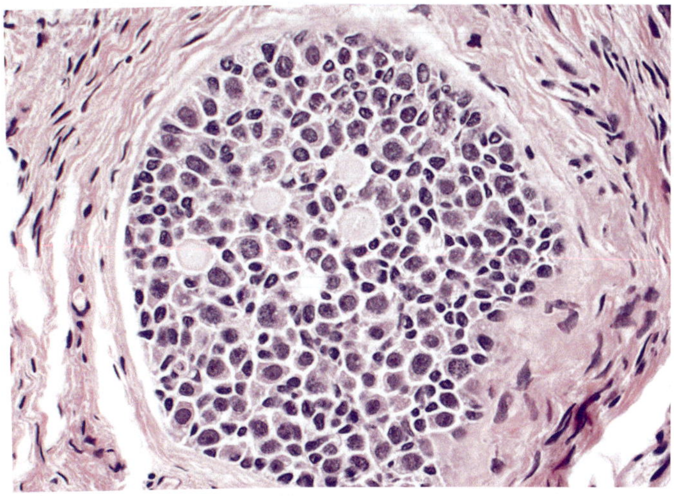

Fig. 13.37 Gonadoblastoma. The tumor cell nests are surrounded by dense fibrous stroma. The nests contain two types of cells, larger germ cells with vesicular nuclei, nucleoli and moderate amounts of pale cytoplasm, and smaller sex cord cells, which have dark nuclei and scanty cytoplasm. Rounded balls or cylinders of eosinophilic hyaline material, as seen here, are a typical finding in the nests

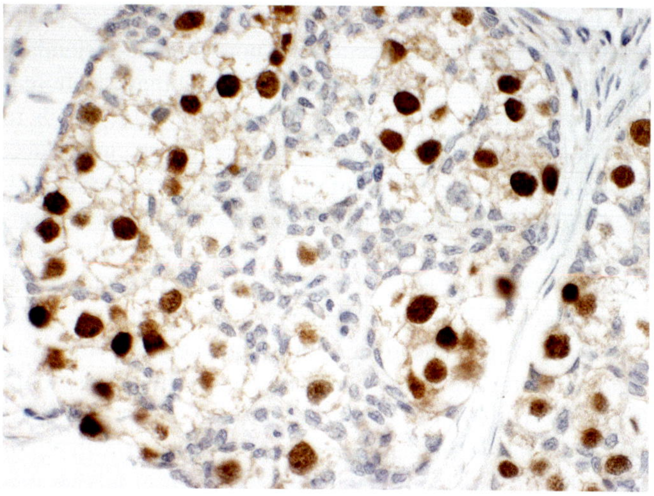

Fig. 13.38 Gonadoblastoma. Most germ cell nuclei show positive staining for OCT4, while the sex cord cell nuclei are negative. The staining pattern for NANOG and SALL4 is similar

Immunohistochemistry and Molecular Pathology

The germ cells in gonadoblastoma are immunophenotypically similar to dysgerminoma cells. They show positive membrane and cytoplasmic staining for placental alkaline phosphatase, CD117 (c-kit), and D2-40 (podoplanin), and they exhibit nuclear staining for OCT-4, NANOG, and SALL4 (Fig. 13.38) [20, 22]. Cytoplasmic and membrane staining for TSPY is present in the germ cells [218]. Reportedly, the immature germ cells show positive nuclear staining for OCT-4, the mature ones show cytoplasmic and membranous staining for TSPY, and occasional cells show positive staining for both [217]. In practice, positive staining for TSPY is generally present in gonadoblastoma [21].

The sex cord cells show positive cytoplasmic staining for inhibin, vimentin, and cytokeratin, and nuclear staining for WT-1 [219]. The hyaline material that forms balls or cylinders in the tumor cell nests is PAS positive and stains with anti-laminin antibodies. The luteinized cells in the stroma around the gonadoblastoma nests stain strongly for inhibin and calretinin.

FISH testing may reveal the presence of a Y chromosome in gonadoblastoma [21]. A complete Y chromosome may not be present in every case, so the success of FISH testing may depend on the portion of the Y chromosome that is tested for. In one case, a 55-year-old 46XX woman with two successful pregnancies was found to have a unilateral gonadoblastoma that gave rise to a small dysgerminoma [220]. FISH testing was performed using two Y chromosome probes, one centromeric and the other to the SRY region, revealing no Y signals but two X signals [221]. FISH testing using probes for the TSPY1 region and the X chromosome centromere revealed the presence of TSPY1 in all conventional gonadoblastomas arising in intersex individuals; all showed a signal for TSPY1 and one for X [222]. An unusual case in which a gonadoblastoma was associated with a dysgerminoma and a juvenile granulosa cell tumor in the ovary of a fertile 21-year-old woman with a 46XX karyotype revealed two X signals but no TSPY1 signal [222]. FISH studies that show an absence of Y chromosome material suggest that an alternative pathway of tumorigenesis not involving TSPY and the Y chromosome must be present in such individuals. RT-PCR testing of a gonadoblastoma revealed the presence of TSPY transcripts, but no transcripts of any of 13 other Y chromosome genes, providing evidence of the importance of TSPY in the development of gonadoblastoma [223]. FISH studies of dysgerminoma frequently reveal an isochromosome i12p or 12p overrepresentation. 12p abnormalities have not been identified in gonadoblastoma, but few cases have been studied [224].

Differential Diagnosis

A second type of combined tumor, the *mixed germ cell-sex cord-stromal tumor*, has been described [205, 225], and must be differentiated from gonadoblastoma. The mixed germ cell-sex cord-stromal tumor is a unilateral neoplasm that occurs in the ovaries of phenotypically normal females with a 46XX karyotype. Most are detected in young children but they occur in teenagers and adult women as well. These tumors can recur or metastasize, and about 10% of them, usually tumors occurring in postpubertal patients, are accompanied by a malignant germ cell tumor, usually a dysgerminoma.

The treatment differs from that of gonadoblastoma, since the tumor is unilateral and occurs in a normal gonad; standard therapy in a young patient is unilateral salpingo-oophorectomy. The mixed germ cell-sex cord-stromal tumor tends to be a large solid tumor with a gray, white, or tan cut surface. Although both germ cells and sex cord cells are present they are randomly intermixed and do not grow in the nested pattern typical of gonadoblastoma. The hyaline balls surrounded by tumor cells that are so characteristic of gonadoblastoma are not present. The germ cells resemble germinoma cells. The sex cord cells have dark nuclei and scanty cytoplasm. Mitotic figures are present in both components. Three main histologic patterns have been described: sheets of tumor cells, solid tubules, and anastomosing trabecular growth [205]. Heterologous mucinous epithelium and retiform tubules can be present in these tumors [226, 227]. Immunohistochemistry is not helpful in differentiating these tumors from gonadoblastoma, as the immunophenotypes are similar. The germ cells in the mixed germ cell-sex cord-stromal tumor stain for standard germ cell markers such as placental alkaline phosphatase and CD117 [228].

Microscopic gonadoblastoma-like lesions occur in fetal and infant ovaries in the absence of genetic abnormalities [229, 230]. They are often found in the walls of follicular cysts and appear to be the result of abnormal follicle formation. They do not express germ cell tumor markers such as placental alkaline phosphatase [229]. They are found in normal ovaries, they involute as the child grows and they appear to be unrelated to gonadoblastoma.

The nested pattern with sex cord cells surrounding balls and cylinders of hyaline material seen in gonadoblastoma is somewhat similar to the appearance of a *sex cord tumor with annular tubules* (SCTAT) [231]. However, the SCTAT arises in a different clinical setting. Patients have a normal 46XX karyotype and normal ovaries, and about a third of SCTAT develop in women with the Peutz-Jeghers syndrome. Syndromic SCTAT are small and bilateral, and hence more likely to be confused with gonadoblastoma. Non-syndromic tumors tend to be large and unilateral with extensive growth of sex cord cells and deposition of hyaline material. The main differentiating feature from gonadoblastoma is the absence of germ cells in the SCTAT.

References

1. Roth LM, Talerman A. Recent advances in the pathology and classification of ovarian germ cell tumors. Int J Gynecol Pathol. 2006;25(4):305–20.
2. Smith HO, Berwick M, Verschraegen CF, et al. Incidence and survival rates for female malignant germ cell tumors. Obstet Gynecol. 2006;107(5):1075–85.
3. Lee KH, Lee IH, Kim BG, et al. Clinicopathologic characteristics of malignant germ cell tumors in the ovaries of Korean women: a Korean gynecologic oncology group study. Int J Gynecol Cancer. 2009;19(1):84–7.
4. Asadourian LA, Taylor HB. Dysgerminoma. An analysis of 105 cases. Obstet Gynecol. 1969;33:370–9.
5. Bjorkholm E, Lundell M, Gyftodimos A, Silfversward C. Dysgerminoma. The radiumhemmet series 1927–1984. Cancer. 1990;65:38–44.
6. Gordon A, Lipton D, Woodruff JD. Dysgerminoma: a review of 158 cases from the Emil Novak Ovarian Tumor Registry. Obstet Gynecol. 1981;58:497–504.
7. Buller RE, Darrow V, Manetta A, Porto M, DiSaia PJ. Conservative surgical management of dysgerminoma concomitant with pregnancy. Obstet Gynecol. 1992;79(5 pt 2):887–90.
8. Masahito H, Hara F, Tomishige H, et al. 1,25-dihydroxyvitamin D-mediated hypercalcemia in ovarian dysgerminoma. Pediatr Hematol Oncol. 2008;25(1):73–8.
9. Inoue H, Kikuchi Y, Hirata J, Wada S, Seki K, Nagata I. Dysgerminoma of the ovary with hypercalcemia associated with elevated parathyroid hormone-related protein. Jpn J Clin Oncol. 1995;25:113–7.
10. Zaloudek CJ, Tavassoli FA, Norris HJ. Dysgerminoma with syncytiotrophoblastic giant cells: a histologically and clinically distinctive subtype of dysgerminoma. Am J Surg Pathol. 1981;5:361–7.
11. Kumar S, Shah JP, Bryant CS, et al. The prevalence and prognostic impact of lymph node metastasis in malignant germ cell tumors of the ovary. Gynecol Oncol. 2008;110(2):125–32.
12. Chan JK, Tewari KS, Waller S, et al. The influence of conservative surgical practices for malignant ovarian germ cell tumors. J Surg Oncol. 2008;98(2):111–6.
13. Patterson DM, Murugaesu N, Holden L, Seckl MJ, Rustin GJ. A review of the close surveillance policy for stage I female germ cell tumors of the ovary and other sites. Int J Gynecol Cancer. 2008;18(1):43–50.
14. Gershenson DM. Management of ovarian germ cell tumors. J Clin Oncol. 2007;25(20):2938–43.
15. Pectasides D, Pectasides E, Kassanos D. Germ cell tumors of the ovary. Cancer Treat Rev. 2008;34(5):427–41.
16. Dietl J, Horny HP, Ruck P, Kaiserling E. Dysgerminoma of the ovary: an immunohistochemical study of tumor-infiltrating lymphoreticular cells and tumor cells. Cancer. 1993;71:2562–8.
17. Stewart CJR, Farquharson MA, Foulis AK. Characterization of the inflammatory infiltrate in ovarian dysgerminoma: an immunocytochemical study. Histopathology. 1992;20:491–7.
18. Chang MC, Vargas SO, Hornick JL, Hirsch MS, Crum CP, Nucci MR. Embryonic stem cell transcription factors and D2-40 (Podoplanin) as diagnostic immunohistochemical markers in ovarian germ cell tumors. Int J Gynecol Pathol. 2009;28(4):347–55. 310.1097/PGP.
19. Sever M, Jones TD, Roth LM, et al. Expression of CD117 (c-kit) receptor in dysgerminoma of the ovary: diagnostic and therapeutic implications. Mod Pathol. 2005;18(11):1411–6.
20. Cao D, Guo S, Allan RW, Molberg KH, Peng Y. SALL4 is a novel sensitive and specific marker of ovarian primitive germ cell tumors and is particularly useful in distinguishing yolk sac tumor from clear cell carcinoma. Am J Surg Pathol. 2009;33(6):894–904.
21. Hoei-Hansen CE, Kraggerud SM, Abeler VM, Kaern J, Rajpert-De Meyts E, Lothe RA. Ovarian dysgerminomas are characterised by frequent KIT mutations and abundant expression of pluripotency markers. Mol Cancer. 2007;6:12.
22. Cheng L, Thomas A, Roth LM, Zheng W, Michael H, Karim FW. OCT4: a novel biomarker for dysgerminoma of the ovary. Am J Surg Pathol. 2004;28(10):1341–6.
23. Cossu-Rocca P, Jones TD, Roth LM, et al. Cytokeratin and CD30 expression in dysgerminoma. Hum Pathol. 2006;37(8):1015–21.
24. Kurman RJ, Norris HJ. Embryonal carcinoma of the ovary: a clinicopathologic entity distinct from endodermal sinus tumor

resembling embryonal carcinoma of the adult testis. Cancer. 1976;38:2420–33.
25. Kammerer-Doak D, Baurick K, Black W, Barbo DM, Smith HO. Endodermal sinus tumor and embryonal carcinoma of the ovary in a 53-year-old woman. Gynecol Oncol. 1996;63:133–7.
26. Cheng L, Zhang S, Talerman A, Roth LM. Morphologic, immunohistochemical, and fluorescence in situ hybridization study of ovarian embryonal carcinoma with comparison to solid variant of yolk sac tumor and immature teratoma. Hum Pathol. 2010;41(5):716–23.
27. Esheba GE, Pate LL, Longacre TA. Oncofetal protein glypican-3 distinguishes yolk sac tumor from clear cell carcinoma of the ovary. Am J Surg Pathol. 2008;32(4):600–7.
28. Damjanov I, Osborn M, Miettinen M. Keratin 7 is a marker for a subset of trophoblastic cells in human germ cell tumors. Arch Pathol Lab Med. 1990;114(1):81–3.
29. Kurman RJ, Norris HJ. Malignant germ cell tumors of the ovary. Hum Pathol. 1977;8:551–64.
30. Kawai M, Kano T, Kikkawa F, et al. Seven tumor markers in benign and malignant germ cell tumors of the ovary. Gynecol Oncol. 1992;45:248–53.
31. Dallenbach P, Bonnefoi H, Pelte MF, Vlastos G. Yolk sac tumours of the ovary: an update. Eur J Surg Oncol. 2006;32(10):1063–75.
32. Kurman RJ, Norris HJ. Endodermal sinus tumor of the ovary. A clinical and pathologic analysis of 71 cases. Cancer. 1976;38: 2404–19.
33. Gershenson DM, del Junco G, Herson J, Rutledge FN. Endodermal sinus tumor of the ovary: the M.D. Anderson experience. Obstet Gynecol. 1983;61:194–202.
34. de La Motte Rouge T, Pautier P, Duvillard P, et al. Survival and reproductive function of 52 women treated with surgery and bleomycin, etoposide, cisplatin (BEP) chemotherapy for ovarian yolk sac tumor. Ann Oncol. 2008;19(8):1435–41.
35. Kang H, Kim TJ, Kim WY, et al. Outcome and reproductive function after cumulative high-dose combination chemotherapy with bleomycin, etoposide and cisplatin (BEP) for patients with ovarian endodermal sinus tumor. Gynecol Oncol. 2008;111(1):106–10.
36. Shah JP, Kumar S, Bryant CS, et al. A population-based analysis of 788 cases of yolk sac tumors: a comparison of males and females. Int J Cancer. 2008;123(11):2671–5.
37. Clement PB, Young RH, Scully RE. Endometrioid-like variant of ovarian yolk sac tumor. A clinicopathological analysis of eight cases. Am J Surg Pathol. 1987;11:767–78.
38. Nogales FF, Matilla A, Nogales-Ortiz F, Galera-Davidson H. Yolk sac tumors with pure and mixed polyvesicular vitelline patterns. Hum Pathol. 1978;9:553–66.
39. Nakashima N, Fukatsu T, Nagasaka T, Sobue M, Takeuchi J. The frequency and histology of hepatic tissue in germ cell tumors. Am J Surg Pathol. 1987;11:682–92.
40. Ulbright TM, Roth LM, Brodhecker CA. Yolk sac differentiation in germ cell tumors. A morphologic study of 50 cases with emphasis on hepatic, enteric, and parietal yolk sac features. Am J Surg Pathol. 1986;10:151–64.
41. Prat J, Bhan AK, Dickersin GR, Robboy SJ, Scully RE. Hepatoid yolk sac tumor of the ovary (endodermal sinus tumor with hepatoid differentiation): a light microscopic, ultrastructural and immunohistochemical study of seven cases. Cancer. 1982;50:2355–68.
42. Devouassoux-Shisheboran M, Schammel DP, Tavassoli FA. Ovarian hepatoid yolk sac tumours: morphological, immunohistochemical and ultrastructural features. Histopathology. 1999;34(5):462–9.
43. Pitman MB, Triratanachat S, Young RH, Oliva E. Hepatocyte paraffin 1 antibody does not distinguish primary ovarian tumors with hepatoid differentiation from metastatic hepatocellular carcinoma. Int J Gynecol Pathol. 2004;23(1):58–64.
44. Michael H, Ulbright TM, Brodhecker CA. The pluripotential nature of the mesenchyme-like component of yolk sac tumor. Arch Pathol Lab Med. 1989;113:1115–9.
45. Nogales FF, Buritica C, Regauer S, Gonzalez T. Mucinous carcinoid as an unusual manifestation of endodermal differentiation in ovarian yolk sac tumors. Am J Surg Pathol. 2005;29(9):1247–51.
46. Rutgers JL, Young RH, Scully RE. Ovarian yolk sac tumor arising from an endometrioid carcinoma. Hum Pathol. 1987;18:1296–9.
47. Nogales FF, Bergeron C, Carvia RE, Alvaro T, Fulwood HR. Ovarian endometrioid tumors with yolk sac tumor component, an unusual form of ovarian neoplasm - analysis of six cases. Am J Surg Pathol. 1996;20(9):1056–66.
48. Lopez JM, Malpica A, Deavers MT, Ayala AG. Ovarian yolk sac tumor associated with endometrioid carcinoma and mucinous cystadenoma of the ovary. Ann Diagn Pathol. 2003;7(5):300–5.
49. Ramalingam P, Malpica A, Silva EG, Gershenson DM, Liu JL, Deavers MT. The use of cytokeratin 7 and EMA in differentiating ovarian yolk sac tumors from endometrioid and clear cell carcinomas. Am J Surg Pathol. 2004;28(11):1499–505.
50. Maeda D, Ota S, Takazawa Y, et al. Glypican-3 expression in clear cell adenocarcinoma of the ovary. Mod Pathol. 2009;22(6):824–32.
51. Young RH, Dickersin GR, Scully RE. Juvenile granulosa cell tumor of the ovary. A clinicopathological analysis of 125 cases. Am J Surg Pathol. 1984;8:575–96.
52. Zaloudek CJ, Norris HJ. Granulosa tumors of the ovary in children: a clinical and pathologic study of 32 cases. Am J Surg Pathol. 1982;6:503–12.
53. Ishikura H, Scully RE. Hepatoid carcinoma of the ovary. A newly described tumor. Cancer. 1987;60:2775–84.
54. Yigit S, Uyaroglu MA, Kus Z, Ekinci N, Oztekin O, Yigit S, et al. Hepatoid carcinoma of the ovary: immunohistochemical finding of one case and literature review. Int J Gynecol Cancer. 2006;16(3): 1439–41.
55. Tochigi N, Kishimoto T, Supriatna Y, Nagai Y, Nikaido T, Ishikura H. Hepatoid carcinoma of the ovary: a report of three cases admixed with a common surface epithelial carcinoma. Int J Gynecol Pathol. 2003;22(3):266–71.
56. Jondle DM, Shahin MS, Sorosky J, Benda JA. Ovarian mixed germ cell tumor with predominance of polyembryoma: a case report with literature review. Int J Gynecol Pathol. 2002;21(1):78–81.
57. Takemori M, Nishimura R, Yamasaki M, Kawabe Y, Hasegawa K. Ovarian mixed germ cell tumor composed of polyembryoma and immature teratoma. Gynecol Oncol. 1998;69:260–3.
58. King ME, Hubbell MJ, Talerman A. Mixed germ cell tumor of the ovary with a prominent polyembryoma component. Int J Gynecol Pathol. 1991;10:88–95.
59. Koo HL, Choi J, Kim KR, Kim JH. Pure non-gestational choriocarcinoma of the ovary diagnosed by DNA polymorphism analysis. Pathol Int. 2006;56(10):613–6.
60. Goswami D, Sharma K, Zutshi V, Tempe A, Nigam S. Nongestational pure ovarian choriocarcinoma with contralateral teratoma. Gynecol Oncol. 2001;80(2):262–6.
61. Tsujioka H, Hamada H, Miyakawa T, Hachisuga T, Kawarabayashi T. A pure nongestational choriocarcinoma of the ovary diagnosed with DNA polymorphism analysis. Gynecol Oncol. 2003;89(3): 540–2.
62. Park SH, Park A, Kim JY, Kwon JH, Koh SB. A case of non-gestational choriocarcinoma arising in the ovary of a postmenopausal woman. J Gynecol Oncol. 2009;20(3):192–4.
63. Axe SR, Klein VR, Woodruff JD. Choriocarcinoma of the ovary. Obstet Gynecol. 1985;66:111–4.
64. Balat O, Kutlar I, Ozkur A, Bakir K, Aksoy F, Ugur MG. Primary pure ovarian choriocarcinoma mimicking ectopic pregnancy: a report of fulminant progression. Tumori. 2004;90(1):136–8.
65. Fisher RA, Newlands ES, Jeffreys AJ, et al. Gestational and non-gestational trophoblastic tumors distinguished by DNA analysis. Cancer. 1992;69:839–45.
66. Yamamoto E, Ino K, Yamamoto T, et al. A pure nongestational choriocarcinoma of the ovary diagnosed with short tandem repeat

analysis: case report and review of the literature. Int J Gynecol Cancer. 2007;17(1):254–8.
67. Daya D, Sabet L. The use of cytokeratin as a sensitive and reliable marker for trophoblastic tissue. Am J Clin Pathol. 1991;95(2): 137–41.
68. Shih IM, Kurman RJ. Immunohistochemical localization of inhibin-alpha in the placenta and gestational trophoblastic lesions. Int J Gynecol Pathol. 1999;18(2):144–50.
69. Ordi J, Romagosa C, Tavassoli FA, et al. CD10 expression in epithelial tissues and tumors of the gynecologic tract: a useful marker in the diagnosis of mesonephric, trophoblastic, and clear cell tumors. Am J Surg Pathol. 2003;27(2):178–86.
70. Gershenson DM, del Junco G, Copeland LJ, Rutledge FN. Mixed germ cell tumors of the ovary. Obstet Gynecol. 1984;64:200–7.
71. Kurman RJ, Norris HJ. Malignant mixed germ cell tumors of the ovary. A clinical and pathologic analysis of 30 cases. Obstet Gynecol. 1976;48:579–89.
72. Oliva E, Andrada E, Pezzica E, Prat J. Ovarian carcinomas with choriocarcinomatous differentiation. Cancer. 1993;72:2441–6.
73. Cho NH, Kim YT, Lee JH, et al. Diagnostic challenge of fetal ontogeny and its application on the ovarian teratomas. Int J Gynecol Pathol. 2005;24(2):173–82.
74. Koonings PP, Campbell K, Mishell Jr DR, Grimes DA. Relative frequency of primary ovarian neoplasms: a 10-year review. Obstet Gynecol. 1989;74:921–6.
75. Comerci Jr JT, Licciardi F, Bergh PA, Gregori C, Breen JL. Mature cystic teratoma: a clinicopathologic evaluation of 517 cases and review of the literature. Obstet Gynecol. 1994;84:22–8.
76. Dalmau J, Gleichman AJ, Hughes EG, et al. Anti-NMDA-receptor encephalitis: case series and analysis of the effects of antibodies. Lancet Neurol. 2008;7(12):1091–8.
77. Dalmau J, Tuzun E, Wu HY, et al. Paraneoplastic anti-*N*-methyl-D-aspartate receptor encephalitis associated with ovarian teratoma. Ann Neurol. 2007;61(1):25–36.
78. Florance NR, Davis RL, Lam C, et al. Anti-*N*-methyl-D-aspartate receptor (NMDAR) encephalitis in children and adolescents. Ann Neurol. 2009;66(1):11–8.
79. Sabin TD, Jednacz JA, Staats PN. Case records of the Massachusetts General Hospital. Case 26–2008. A 26-year-old woman with headache and behavioral changes. N Engl J Med. 2008;359(8):842–53.
80. Robboy SJ, Scully RE. Ovarian teratoma with glial implants on the peritoneum. An analysis of 12 cases. Hum Pathol. 1970;1:643–53.
81. Nielsen SN, Scheithauer BW, Gaffey TA. Gliomatosis peritonei. Cancer. 1985;56(10):2499–503.
82. Steeper TA, Mukai K. Solid ovarian teratomas: an immunocytochemical study of thirteen cases with clinicopathologic correlation. Pathol Annu. 1984;19:81–92.
83. Russell P, Painter DM. The pathological assessment of ovarian neoplasms. V. The germ cell tumours. Pathology. 1982;14:47–72.
84. Nogales FF, Vergara E, Medina MT. Prostate in ovarian mature cystic teratoma. Histopathology. 1995;26:373–5.
85. Halabi M, Oliva E, Mazal PR, Breitenecker G, Young RH. Prostatic tissue in mature cystic teratomas of the ovary: a report of four cases, including one with features of prostatic adenocarcinoma, and cytogenetic studies. Int J Gynecol Pathol. 2002;21(3): 261–7.
86. Stamp GW, McConnell EM. Malignancy arising in cystic ovarian teratomas. A report of 24 cases. Br J Obstet Gynaecol. 1983;90: 671–5.
87. Hirakawa T, Tsuneyoshi M, Enjoji M. Squamous cell carcinoma arising in mature cystic teratoma of the ovary. Clinicopathologic and topographic analysis. Am J Surg Pathol. 1989;13:397–405.
88. Hackethal A, Brueggmann D, Bohlmann MK, Franke FE, Tinneberg HR, Munstedt K. Squamous-cell carcinoma in mature cystic teratoma of the ovary: systematic review and analysis of published data. Lancet Oncol. 2008;9(12):1173–80.
89. Ribeiro G, Hughesdon P, Wiltshaw E. Squamous carcinoma arising in dermoid cysts and associated with hypercalcemia: a clinicopathologic study of six cases. Gynecol Oncol. 1988;29:222–30.
90. Tseng CJ, Chou HH, Huang KG, et al. Squamous cell carcinoma arising in mature cystic teratoma of the ovary. Gynecol Oncol. 1996;63:364–70.
91. Kikkawa F, Ishikawa H, Tamakoshi K, Nawa A, Suganuma N, Tomoda Y. Squamous cell carcinoma arising from mature cystic teratoma of the ovary: a clinicopathologic analysis. Obstet Gynecol. 1997;89(6):1017–22.
92. Baker PM, Rosai J, Young RH. Ovarian teratomas with florid benign vascular proliferation: a distinctive finding associated with the neural component of teratomas that may be confused with a vascular neoplasm. Int J Gynecol Pathol. 2002;21(1):16–21.
93. Pins MR, Young RH, Daly WJ, Scully RE. Primary squamous cell carcinoma of the ovary. Report of 37 cases. Am J Surg Pathol. 1996;20:823–33.
94. Tobon H, Surti U, Naus GJ, Hoffner L, Hemphill RW. Squamous cell carcinoma in situ arising in an ovarian mature cystic teratoma. Report of one case with histopathologic, cytogenetic, and flow cytometric DNA content analysis. Arch Pathol Lab Med. 1991;115:172–4.
95. Monteagudo C, Torres JV, Llombart-Bosch A. Extramammary Paget's disease arising in a mature cystic teratoma of the ovary. Histopathology. 1999;35(6):582–4.
96. Vartanian RK, McRae B, Hessler RB. Sebaceous carcinoma arising in a mature cystic teratoma of the ovary. Int J Gynecol Pathol. 2002;21(4):418–21.
97. Davis GL. Malignant melanoma arising in mature ovarian cystic teratoma (dermoid cyst). Report of two cases and literature analysis. Int J Gynecol Pathol. 1996;15:356–62.
98. Gupta D, Deavers MT, Silva EG, Malpica A. Malignant melanoma involving the ovary: a clinicopathologic and immunohistochemical study of 23 cases. Am J Surg Pathol. 2004;28(6):771–80.
99. McCluggage WG, Bissonnette JP, Young RH. Primary malignant melanoma of the ovary: a report of 9 definite or probable cases with emphasis on their morphologic diversity and mimicry of other primary and secondary ovarian neoplasms. Int J Gynecol Pathol. 2006;25(4):321–9.
100. Kushima M. Adenocarcinoma arising from mature cystic teratoma of the ovary. Pathol Int. 2004;54(2):139–43.
101. Contreras AL, Malpica A. Angiosarcoma arising in mature cystic teratoma of the ovary: a case report and review of the literature. Int J Gynecol Pathol. 2009;28(5):453–7.
102. Vang R, Gown AM, Zhao C, et al. Ovarian mucinous tumors associated with mature cystic teratomas: morphologic and immunohistochemical analysis identifies a subset of potential teratomatous origin that shares features of lower gastrointestinal tract mucinous tumors more commonly encountered as secondary tumors in the ovary. Am J Surg Pathol. 2007;31(6):854–69.
103. McKenney JK, Soslow RA, Longacre TA. Ovarian mature teratomas with mucinous epithelial neoplasms: morphologic heterogeneity and association with pseudomyxoma peritonei. Am J Surg Pathol. 2008;32(5):645–55.
104. Ronnett BM, Seidman JD. Mucinous tumors arising in ovarian mature cystic teratomas: relationship to the clinical syndrome of pseudomyxoma peritonei. Am J Surg Pathol. 2003;27(5):650–7.
105. Hwang JH, So K-A, Modi G, et al. Borderline-like mucinous tumor arising in mature cystic teratoma of the ovary associated with pseudomyxoma peritonei. Int J Gynecol Pathol. 2009;28(4): 376–80.
106. Marquette S, Amant F, Vergote I, Moerman P. Pseudomyxoma peritonei associated with a mucinous ovarian tumor arising from a mature cystic teratoma. A case report. Int J Gynecol Pathol. 2006;25(4):340–3.
107. Stewart CJ, Tsukamoto T, Cooke B, Leung YC, Hammond IG. Ovarian mucinous tumour arising in mature cystic teratoma and

associated with pseudomyxoma peritonei: report of two cases and comparison with ovarian involvement by low-grade appendiceal mucinous tumour. Pathology. 2006;38(6):534–8.
108. Gershenson DM. Malignant germ-cell tumors of the ovary. Clin Obstet Gynecol. 1985;28(4):824–38.
109. Bonazzi C, Peccatori F, Colombo N, Lucchini V, Cantó MG, Mangioni C. Pure ovarian immature teratoma, a unique and curable disease: 10 years' experience of 32 prospectively treated patients. Obstet Gynecol. 1994;84:598–604.
110. Poulos C, Cheng L, Zhang S, Gersell DJ, Ulbright TM. Analysis of ovarian teratomas for isochromosome 12p: evidence supporting a dual histogenetic pathway for teratomatous elements. Mod Pathol. 2006;19(6):766–71.
111. Kraggerud SM, Szymanska J, Abeler VM, et al. DNA copy number changes in malignant ovarian germ cell tumors. Cancer Res. 2000;60(11):3025–30.
112. Baker BA, Frickey L, Yu IT, Hawkins EP, Cushing B, Perlman EJ. DNA content of ovarian immature teratomas and malignant germ cell tumors. Gynecol Oncol. 1998;71:14–8.
113. Kildal W, Kaern J, Kraggerud SM, et al. Evaluation of genomic changes in a large series of malignant ovarian germ cell tumors–relation to clinicopathologic variables. Cancer Genet Cytogenet. 2004;155(1):25–32.
114. Rodriguez E, Melamed J, Reuter V, Chaganti RS. Chromosome 12 abnormalities in malignant ovarian germ cell tumors. Cancer Genet Cytogenet. 1995;82:62–6.
115. Doss BJ, Jacques SM, Qureshi F, et al. Immature teratomas of the genital tract in older women. Gynecol Oncol. 1999;73(3):433–8.
116. van Altena AM, Wijnberg GJ, Kolwijck E, De Hullu JA, Massuger LF. A patient with bilateral immature ovarian teratoma presenting with paraneoplastic encephalitis. Gynecol Oncol. 2008;108(2):445–8.
117. de Lins e Horta H, de Castro AF, Fonseca RP, Fernandes Jr AS, Lima VS, Neuenschwander LC. Limbic encephalitis associated with immature teratoma. Eur J Gynaecol Oncol. 2009;30(3):329–31.
118. Marina NM, Cushing B, Giller R, et al. Complete surgical excision is effective treatment for children with immature teratomas with or without malignant elements: a Pediatric Oncology Group/Children's Cancer Group Intergroup Study. J Clin Oncol. 1999;17(7):2137–43.
119. Beiner ME, Gotlieb WH, Korach Y, et al. Cystectomy for immature teratoma of the ovary. Gynecol Oncol. 2004;93(2):381–4.
120. Low JJH, Perrin LC, Crandon AJ, Hacker NF. Conservative surgery to preserve ovarian function in patients with malignant ovarian germ cell tumors – a review of 74 cases. Cancer. 2000;89(2):391–8.
121. Lu KH, Gershenson DM. Update on the management of ovarian germ cell tumors. J Reprod Med. 2005;50(6):417–25.
122. Heifetz SA, Cushing B, Giller R, et al. Immature teratomas in children: pathologic considerations – a report from the combined Pediatric Oncology Group Children's Cancer Group. Am J Surg Pathol. 1998;22(9):1115–24.
123. Mann JR, Gray ES, Thornton C, et al. Mature and immature extracranial teratomas in children: the UK Children's Cancer Study Group Experience. J Clin Oncol. 2008;26(21):3590–7.
124. Li H, Hong W, Zhang R, Wu L, Liu L, Zhang W. Retrospective analysis of 67 consecutive cases of pure ovarian immature teratoma. Chin Med J (Engl). 2002;115(10):1496–500.
125. Hariprasad R, Kumar L, Janga D, Kumar S, Vijayaraghavan M. Growing teratoma syndrome of ovary. Int J Clin Oncol. 2008;13(1):83–7.
126. Zagame L, Pautier P, Duvillard P, Castaigne D, Patte C, Lhomme C. Growing teratoma syndrome after ovarian germ cell tumors. Obstet Gynecol. 2006;108(3):509–14.
127. Mengshol SC, DeMars LR, Schned AR. Gliomatosis peritonei and teratomatous implant with carcinomatous transformation presenting 54 years following oophorectomy for dermoid cyst. Gynecol Oncol. 2004;92(1):353–6.
128. Dadmanesh F, Miller DM, Swenerton KD, Clement PB. Gliomatosis peritonei with malignant transformation. Mod Pathol. 1997;10:597–601.
129. Shefren G, Collin J, Soriero O. Gliomatosis peritonei with malignant transformation: a case report and review of the literature. Am J Obstet Gynecol. 1991;164:1617–20; discussion 1620.
130. Yanai-Inbar I, Scully RE. Relation of ovarian dermoid cysts and immature teratomas: an analysis of 350 cases of immature teratoma and 10 cases of dermoid cyst with microscopic foci of immature tissue. Int J Gynecol Pathol. 1987;6:203–12.
131. Nogales FF, Favera BE, Major FJ, Silverberg SG. Immature teratoma of the ovary with a neural component ("solid" teratoma). A clinicopathologic study of 20 cases. Hum Pathol. 1976;7:625–42.
132. Norris HJ, Zirkin HJ, Benson WL. Immature (malignant) teratoma of the ovary. A clinical and pathologic study of 58 cases. Cancer. 1976;37:2359–72.
133. Ulbright TM. Germ cell tumors of the gonads: a selective review emphasizing problems in differential diagnosis, newly appreciated, and controversial issues. Mod Pathol. 2005;18 Suppl 2:S61–79.
134. Vance RP, Geisinger KR, Randall MB, Marshall RB. Immature neural elements in immature teratomas. An immunohistochemical and ultrastructural study. Am J Clin Pathol. 1988;90:397–411.
135. Nogales FF, Avila IR, Concha A, Del Moral E. Immature endodermal teratoma of the ovary: embryologic correlations and immunohistochemistry. Hum Pathol. 1993;24:364–70.
136. Nogales FF, Ortega I, Rivera F, Armas JR. Metanephrogenic tissue in immature ovarian teratoma. Am J Surg Pathol. 1980;4:297–9.
137. O'Connor DM, Norris HJ. The influence of grade on the outcome of stage I ovarian immature (malignant) teratomas and the reproducibility of grading. Int J Gynecol Pathol. 1994;13:283–9.
138. Harms D, Janig U, Gîbel U. Gliomatosis peritonei in childhood and adolescence. Clinicopathological study of 13 cases including immunohistochemical findings. Pathol Res Pract. 1989;184:422–30.
139. Nogales FF, Aguilar D. Florid vascular proliferation in grade 0 glial implants from ovarian immature teratoma. Int J Gynecol Pathol. 2002;21(3):305–7.
140. Calder CJ, Light AM, Rollason TP. Immature ovarian teratoma with mature peritoneal metastatic deposits showing glial, epithelial, and endometrioid differentiation: a case report and review of the literature. Int J Gynecol Pathol. 1994;13:279–82.
141. Dworak O, Knopfle G, Varchmin-Schultheiss K, Meyer G. Gliomatosis peritonei with endometriosis externa. Gynecol Oncol. 1988;29:263–6.
142. Ferguson AW, Katabuchi H, Ronnett BM, Cho KR. Glial implants in gliomatosis peritonei arise from normal tissue, not from the associated teratoma. Am J Pathol. 2001;159(1):51–5.
143. Kwan MY, Kalle W, Lau GT, Chan JK. Is gliomatosis peritonei derived from the associated ovarian teratoma? Hum Pathol. 2004;35(6):685–8.
144. Best DH, Butz GM, Moller K, Coleman WB, Thomas DB. Molecular analysis of an immature ovarian teratoma with gliomatosis peritonei and recurrence suggests genetic independence of multiple tumors. Int J Oncol. 2004;25(1):17–25.
145. Roth LM. Recent advances in the pathology and classification of ovarian sex cord-stromal tumors. Int J Gynecol Pathol. 2006;25(3):199–215.
146. Bing Z, Pasha TL, Lal P, Tomaszewski JE. Expression of glial cell line-derived neurotropic factor receptor alpha-1 in immature teratomas. Am J Clin Pathol. 2008;130(6):892–6.
147. Calame JJ, Schaberg A. Solid teratomas and mixed Mullerian tumors of the ovary: a clinical, histological, and immunocytochemical comparative study. Gynecol Oncol. 1989;33:212–21.

148. Perrone T, Steeper TA, Dehner LP. Alpha-fetoprotein localization in pure ovarian teratoma. An immunohistochemical study of 12 cases. Am J Clin Pathol. 1987;88:713–7.
149. Zynger DL, Everton MJ, Dimov ND, Chou PM, Yang XJ. Expression of glypican 3 in ovarian and extragonadal germ cell tumors. Am J Clin Pathol. 2008;130(2):224–30.
150. Michael H, Hull MT, Ulbright TM, Foster RS, Miller KD. Primitive neuroectodermal tumors arising in testicular germ cell neoplasms. Am J Surg Pathol. 1997;21:896–904.
151. Kleinman GM, Young RH, Scully RE. Primary neuroectodermal tumors of the ovary: a report of 25 cases. Am J Surg Pathol. 1993;17:764–78.
152. Kawauchi S, Fukuda T, Miyamoto S, et al. Peripheral primitive neuroectodermal tumor of the ovary confirmed by CD99 immunostaining, karyotypic analysis, and RT-PCR for EWS/FLI-1 chimeric mRNA. Am J Surg Pathol. 1998;22:1417–22.
153. Young RH, Kozakewich HPW, Scully RE. Metastatic ovarian tumors in children: a report of 14 cases and review of the literature. Int J Gynecol Pathol. 1993;12:8–19.
154. McCluggage WG. Ovarian neoplasms composed of small round cells: a review. Adv Anat Pathol. 2004;11(6):288–96.
155. Guerrieri C, Jarlsfelt I. Ependymoma of the ovary: a case report with immunohistochemical, ultrastructural, and DNA cytometric findings, as well as histogenetic considerations. Am J Surg Pathol. 1993;17:623–32.
156. Idowu MO, Rosenblum MK, Wei XJ, Edgar MA, Soslow RA. Ependymomas of the central nervous system and adult extra-axial ependymomas are morphologically and immunohistochemically distinct – a comparative study with assessment of ovarian carcinomas for expression of glial fibrillary acidic protein. Am J Surg Pathol. 2008;32(5):710–8.
157. Carr KA, Roberts JA, Frank TS. Progesterone receptors in bilateral ovarian ependymoma presenting in pregnancy. Hum Pathol. 1992;23:962–5.
158. Opris I, Ducrotoy V, Bossut J, Lamy A, Sabourin J-C. Oligodendroglioma arising in an ovarian mature cystic teratoma. Int J Gynecol Pathol. 2009;28(4):367–71.
159. Hirahara F, Yamanaka M, Miyagia E, et al. Pure ovarian ependymoma: report of a case treated with surgery, chemotherapy, irradiation and hyperthermotherapy. Eur J Obstet Gynecol Reprod Biol. 1997;75(2):221–3.
160. Roth LM, Talerman A. The enigma of struma ovarii. Pathology. 2007;39(1):139–46.
161. Robboy SJ, Shaco-Levy R, Peng RY, et al. Malignant struma ovarii: an analysis of 88 cases, including 27 with extraovarian spread. Int J Gynecol Pathol. 2009;28(5):405–22.
162. Devaney K, Snyder R, Norris HJ, Tavassoli FA. Proliferative and histologically malignant struma ovarii: a clinicopathologic study of 54 cases. Int J Gynecol Pathol. 1993;12:333–43.
163. Shaco-Levy R, Bean SM, Bentley RC, Robboy SJ. Natural history of biologically malignant struma ovarii: analysis of 27 cases with extraovarian spread. Int J Gynecol Pathol. 2010;29(3):212–27. 210.1097/PGP.1090b1013e3181bfb1133.
164. Garg K, Soslow RA, Rivera M, Tuttle MR, Ghossein RA. Histologically bland "extremely well differentiated" thyroid carcinomas arising in struma ovarii can recur and metastasize. Int J Gynecol Pathol. 2009;28(3):222–30.
165. Logani S, Baloch ZW, Snyder PJ, Weinstein R, Livolsi VA. Cystic ovarian metastasis from papillary thyroid carcinoma: a case report. Thyroid. 2001;11(11):1073–5.
166. Young RH, Jackson A, Wells M. Ovarian metastasis from thyroid carcinoma 12 years after partial thyroidectomy mimicking struma ovarii: report of a case. Int J Gynecol Pathol. 1994;13: 181–5.
167. DeSimone CP, Lele SM, Modesitt SC. Malignant struma ovarii: a case report and analysis of cases reported in the literature with focus on survival and I131 therapy. Gynecol Oncol. 2003;89(3): 543–8.
168. Szyfelbein WM, Young RH, Scully RE. Cystic struma ovarii: A frequently unrecognized tumor: A report of 20 cases. Am J Surg Pathol. 1994;18:785–8.
169. Szyfelbein WM, Young RH, Scully RE. Struma ovarii simulating ovarian tumors of other types. A report of 30 cases. Am J Surg Pathol. 1995;19:21–9.
170. Zhang X, Axiotis C. Thyroid-type carcinoma of struma ovarii. Arch Pathol Lab Med. 2010;134(5):786–91.
171. Pelosi G, Sonzogni A, Rosai J. Thyroid-type papillary microcarcinoma in ovarian strumal carcinoid. Int J Surg Pathol. 2008;16(4): 435–7.
172. Roth LM, Miller III AW, Talerman A. Typical thyroid-type carcinoma arising in struma ovarii: a report of 4 cases and review of the literature. Int J Gynecol Pathol. 2008;27(4):496–506.
173. Roth LM, Karseladze AI. Highly differentiated follicular carcinoma arising from struma ovarii: a report of 3 cases, a review of the literature, and a reassessment of so-called peritoneal strumosis. Int J Gynecol Pathol. 2008;27(2):213–22.
174. Hamazaki S, Okino T, Tsukayama C, Okada S. Expression of thyroid transcription factor-1 in strumal carcinoid and struma ovarii: an immunohistochemical study. Pathol Int. 2002;52(7):458–62.
175. Boutross-Tadross O, Saleh R, Asa SL. Follicular variant papillary thyroid carcinoma arising in struma ovarii. Endocr Pathol. 2007;18(3):182–6; Fall.
176. Elisei R, Romei C, Castagna MG, et al. RET/PTC3 rearrangement and thyroid differentiation gene analysis in a struma ovarii fortuitously revealed by elevated serum thyroglobulin concentration. Thyroid. 2005;15(12):1355–61.
177. Flavin R, Smyth P, Crotty P, et al. BRAF T1799A mutation occurring in a case of malignant struma ovarii. Int J Surg Pathol. 2007;15(2):116–20.
178. Schmidt J, Derr V, Heinrich MC, et al. BRAF in papillary thyroid carcinoma of ovary (struma ovarii). Am J Surg Pathol. 2007;31(9): 1337–43.
179. Celestino R, Magalhaes J, Castro P, et al. A follicular variant of papillary thyroid carcinoma in struma ovarii. Case report with unique molecular alterations. Histopathology. 2009;55(4):482–7.
180. Soga J, Osaka M, Yakuwa Y. Carcinoids of the ovary: an analysis of 329 reported cases. J Exp Clin Cancer Res. 2000;19(3): 271–80.
181. Young RH, Prat J, Scully RE. Ovarian Sertoli-Leydig cell tumors with heterologous elements. I. Gastrointestinal epithelium and carcinoid: A clinicopathologic analysis of 36 cases. Cancer. 1982;50:2448–56.
182. Robboy SJ, Scully RE, Norris HJ. Primary trabecular carcinoid of the ovary. Obstet Gynecol. 1977;49:202–7.
183. Robboy SJ. Insular carcinoid of ovary associated with malignant mucinous tumors. Cancer. 1984;54:2273–6.
184. Strosberg J, Nasir A, Cragun J, Gardner N, Kvols L. Metastatic carcinoid tumor to the ovary: a clinicopathologic analysis of seventeen cases. Gynecol Oncol. 2007;106(1):65–8.
185. Robboy SJ, Scully RE, Norris HJ. Carcinoid metastatic to the ovary. A clinicopathologic analysis of 35 cases. Cancer. 1974;33:798–811.
186. Davis KP, Hartmann LK, Keeney GL, Shapiro H. Primary ovarian carcinoid tumors. Gynecol Oncol. 1996;61:259–65.
187. Robboy SJ, Norris HJ, Scully RE. Insular carcinoid primary in the ovary. A clinicopathologic analysis of 48 cases. Cancer. 1975; 36:404–18.
188. Chaowalit N, Connolly HM, Schaff HV, Webb MJ, Pellikka PA. Carcinoid heart disease associated with primary ovarian carcinoid tumor. Am J Cardiol. 2004;93(10):1314–5.
189. Yaegashi N, Tsuiki A, Shimizu T, et al. Ovarian carcinoid with severe constipation due to peptide YY production. Gynecol Oncol. 1995;56:302–6.
190. Shigeta H, Taga M, Kurogi K, Kitamura H, Motoyama T, Gorai I. Ovarian strumal carcinoid with severe constipation: Immunohistochemical and mRNA analyses of peptide YY. Hum Pathol. 1999;30(2):242–6.

191. Matsuda K, Maehama T, Kanazawa K. Strumal carcinoid tumor of the ovary: a case exhibiting severe constipation associated with PYY. Gynecol Oncol. 2002;87(1):143–5.
192. Robboy SJ, Scully RE. Strumal carcinoid of the ovary. An analysis of 50 cases of a distinctive tumor composed of thyroid tissue and carcinoid. Cancer. 1980;46:2019–34.
193. Wilkowske MA, Hartmann LC, Mullany CJ, Behrenbeck T, Kvols LK. Progressive carcinoid heart disease after resection of primary ovarian carcinoid. Cancer. 1994;73:1889–91.
194. Armes JE, Ostor AG. A case of malignant strumal carcinoid. Gynecol Oncol. 1993;51:419–23.
195. Eichhorn JH, Young RH. Neuroendocrine tumors of the genital tract. Am J Clin Pathol. 2001;115(Suppl):S94–112.
196. Talerman A, Evans MI. Primary trabecular carcinoid tumor of the ovary. Cancer. 1982;50:1403–7.
197. Snyder RR, Tavassoli FA. Ovarian strumal carcinoid: immunohistochemical, ultrastructural, and clinicopathologic analysis. Int J Gynecol Pathol. 1986;5:187–201.
198. Baker PM, Oliva E, Young RH, Talerman A, Scully RE. Ovarian mucinous carcinoids including some with a carcinomatous component - a report of 17 cases. Am J Surg Pathol. 2001;25(5): 557–68.
199. Stagno PA, Petras RE, Hart WR. Strumal carcinoids of the ovary. An immunohistologic and ultrastructural study. Arch Pathol Lab Med. 1987;111:440–6.
200. Rabban JT, Lerwill MF, McCluggage WG, Grenert JP, Zaloudek CJ. Primary ovarian carcinoid tumors may express CDX-2: a potential pitfall in distinction from metastatic intestinal carcinoid tumors involving the ovary. Int J Gynecol Pathol. 2009;28(1): 41–8.
201. Sidhu J, Sanchez RL. Prostatic acid phosphatase in strumal carcinoids of the ovary: an immunohistochemical study. Cancer. 1993;72:1673–8.
202. Sporrong B, Falkmer S, Robboy SJ, et al. Neurohormonal peptides in ovarian carcinoids: an immunohistochemical study of 81 primary carcinoids and of intraovarian metastases from six midgut carcinoids. Cancer. 1982;49:68–74.
203. Hristov AC, Young RH, Vang R, Yemelyanova AV, Seidman JD, Ronnett BM. Ovarian metastases of appendiceal tumors with goblet cell carcinoidlike and signet ring cell patterns: a report of 30 cases. Am J Surg Pathol. 2007;31(10):1502–11.
204. Tewari K, Cappuccini F, DiSaia PJ, Berman ML, Manetta A, Kohler MF. Malignant germ cell tumors of the ovary. Obstet Gynecol. 2000;95(1):128–33.
205. Talerman A, Roth LM. Recent advances in the pathology and classification of gonadal neoplasms composed of germ cells and sex cord derivatives. Int J Gynecol Pathol. 2007;26(3):313–21.
206. Scully RE. Gonadoblastoma: a review of 74 cases. Cancer. 1970;25:1340–56.
207. Pena-Alonso R, Nieto K, Alvarez R, et al. Distribution of Y-chromosome-bearing cells in gonadoblastoma and dysgenetic testis in 45,X/46,XY infants. Mod Pathol. 2005;18(3):439–45.
208. Hoepffner W, Horn LC, Simon E, et al. Gonadoblastomas in 5 patients with 46,XY gonadal dysgenesis. Exp Clin Endocrinol Diabetes. 2005;113(4):231–5.
209. Kim KR, Kwon Y, Joung JY, Kim KS, Ayala AG, Ro JY. True hermaphroditism and mixed gonadal dysgenesis in young children: a clinicopathologic study of 10 cases. Mod Pathol. 2002;15(10):1013–9.
210. Mancilla EE, Poggi H, Repetto G, et al. Y chromosome sequences in Turner's syndrome: association with virilization and gonadoblastoma. J Pediatr Endocrinol Metab. 2003;16(8):1157–63.
211. Bianco B, Lipay Mn, Guedes A, Oliveira K, Verreschi ITN. SRY gene increases the risk of developing gonadoblastoma and/or nontumoral gonadal lesions in turner syndrome. Int J Gynecol Pathol. 2009;28(2):197–202. 110.
212. Brant WO, Rajimwale A, Lovell MA, et al. Gonadoblastoma and Turner syndrome. J Urol. 2006;175(5):1858–60.
213. Salo P, Kaariainen H, Petrovic V, Peltomaki P, Page DC, de la Chapelle A. Molecular mapping of the putative gonadoblastoma locus on the Y chromosome. Genes Chromosom Cancer. 1995;14: 210–4.
214. Lau YF, Li Y, Kido T. Gonadoblastoma locus and the TSPY gene on the human Y chromosome. Birth Defects Res C Embryo Today. 2009;87(1):114–22.
215. Cools M, Stoop H, Kersemaekers AM, et al. Gonadoblastoma arising in undifferentiated gonadal tissue within dysgenetic gonads. J Clin Endocrinol Metab. 2006;91(6):2404–13.
216. Woodcock AS, Govan AD, Gowing NF, Langley FA, Anderson MC. A report of the histological features in 12 cases of gonadoblastoma. Tumori. 1979;65:181–9.
217. Kersemaekers AM, Honecker F, Stoop H, et al. Identification of germ cells at risk for neoplastic transformation in gonadoblastoma: an immunohistochemical study for OCT3/4 and TSPY. Hum Pathol. 2005;36(5):512–21.
218. Li Y, Vilain E, Conte F, Rajpert-De Meyts E, Lau YF. Testis-specific protein Y-encoded gene is expressed in early and late stages of gonadoblastoma and testicular carcinoma in situ. Urol Oncol. 2007;25(2):141–6.
219. Hussong J, Crussi FG, Chou PM. Gonadoblastoma: immunohistochemical localization of Mullerian-inhibiting substance, inhibin, WT-1, and p53. Mod Pathol. 1997;10:1101–5.
220. Cooper C, Cooper M, Carter J, Russell P. Gonadoblastoma progressing to dysgerminoma in a 55-year-old woman with normal karyotype. Pathology. 2007;39(2):284–5.
221. Susman M, Cooper C, Cooper M, Carter J, Russell P. Gonadoblastoma progressing to dysgerminoma in a 55-year-old woman with normal karyotype: follow-up studies. Pathology. 2008;40(6):645–6.
222. Hertel JD, Huettner PC, Dehner LP, Pfeifer JD. The chromosome Y-linked testis-specific protein locus TSPY1 is characteristically present in gonadoblastoma. Hum Pathol. 2010;41(11): 1544–9.
223. Su MT, Lee IW, Kuo PL. Presence of TSPY transcript and absence of transcripts of other Y chromosomal genes in a case of microscopic gonadoblastoma. Gynecol Oncol. 2006;103(1): 357–60.
224. Cossu-Rocca P, Zhang S, Roth LM, et al. Chromosome 12p abnormalities in dysgerminoma of the ovary: a FISH analysis. Mod Pathol. 2006;19(4):611–5.
225. Talerman A. A distinctive gonadal neoplasm related to gonadoblastoma. Cancer. 1972;30:1219–24.
226. Zuntova A, Motlik K, Horejsi J, Eckschlager T. Mixed germ cell-sex cord stromal tumor with heterologous structures. Int J Gynecol Pathol. 1992;11:227–33.
227. Tokuoka S, Aoki Y, Hayashi Y, Yokoyama T, Ishii T. A mixed germ cell-sex cord-stromal tumor of the ovary with retiform tubular structure: a case report. Int J Gynecol Pathol. 1985;4(2):161–70.
228. Michal M, Vanecek T, Sima R, et al. Mixed germ cell sex cord-stromal tumors of the testis and ovary. Morphological, immunohistochemical, and molecular genetic study of seven cases. Virchows Arch. 2006;448(5):612–22.
229. Manivel JC, Dehner LP, Burke B. Ovarian tumorlike structures, biovular follicles, and binucleated oocytes in children: their frequency and possible pathologic significance. Pediatr Pathol. 1988;8:283–92.
230. Safneck JR, de Sa DJ. Structures mimicking sex cord-stromal tumours and gonadoblastomas in the ovaries of normal infants and children. Histopathology. 1986;10:909–20.
231. Young RH, Welch WR, Dickersin GR, Scully RE. Ovarian sex cord tumor with annular tubules. Review of 74 cases including 27 with Peutz-Jeghers syndrome and four with adenoma malignum of the cervix. Cancer. 1982;50:1384–402.
232. Kandil DH, Cooper K. Glypican-3: a novel diagnostic marker for hepatocellular carcinoma and more. Adv Anat Pathol. 2009; 16(2):125–9.

Sex Cord Stromal Tumors of the Ovary

14

Gkeok Stzuan Diana Lim and Esther Oliva

Keywords

Stromal hyperplasia • Hyperthecosis • Leydig cell hyperplasia • Massive edema/fibromatosis • Luteinized thecoma with sclerosing peritonitis • Pregnancy-related lesions • Fibroma • Fibrosarcoma • Thecoma • Sclerosing stromal tumor • Signet ring stromal tumor • Microcystic stromal tumor • Adult granulosa cell tumor • Juvenile granulosa cell tumor • Sertoli-Leydig cell tumor • Sertoli cell tumor • Stromal luteoma • Steroid cell tumor NOS • Leydig cell tumor

Stromal Hyperplasia and Hyperthecosis

Prevalence and Clinical Presentation

Stromal hyperplasia is a nonneoplastic condition characterized by the proliferation of ovarian stromal cells. When luteinized cells are also present, the term stromal hyperthecosis is applied. Both conditions typically affect postmenopausal women in their sixth or seventh decades. Patients may be asymptomatic but women with stromal hyperthecosis can present with clinical features of androgen excess [1], including virilization or less commonly, estrogenic manifestations such as abnormal vaginal bleeding, which may be secondary to endometrial hyperplasia or carcinoma [2]. Classical stromal hyperthecosis is associated with obesity, virilization, hypertension, and decreased glucose tolerance. A small group of patients have the HAIR-AN syndrome (insulin resistance, diabetes, acanthosis-nigricans, and hyperandrogenism) [3, 4]. Some ovarian neoplasms may be associated with stromal hyperthecosis, with stromal luteoma being one of the most common [5].

E. Oliva (✉)
Department of Pathology, Massachusetts General Hospital, 55 Fruit St., Boston, MA 02114, USA and
Harvard Medical School, Boston, MA, USA

Macroscopic and Microscopic Features

Typically, there is bilateral ovarian involvement. Both ovaries may be of normal size or enlarged (up to 7.0 cm), with a yellow to white cut surface. In stromal hyperplasia, low-power microscopic examination reveals a "blue" nodular or diffuse thickening of the ovarian cortex that may also involve the medulla in some instances. These areas are composed of a monotonous proliferation of stromal cells that have scant cytoplasm, ovoid nuclei, and inconspicuous nucleoli. Stromal hyperthecosis is characterized by single or aggregates of luteinized cells with round nuclei, small nucleoli, and abundant eosinophilic or pale vacuolated cytoplasm present in a background of stromal hyperplasia. When these aggregates are relatively large (but <0.5 cm in size), the term nodular hyperthecosis may be used [2] (Fig. 14.1).

Lesions with lipid laden cells considered in the differential diagnosis may include Leydig cell hyperplasia, stromal luteoma, luteinized thecoma, and steroid cell tumor. Leydig cell hyperplasia may be seen in association with stromal hyperplasia and hyperthecosis; however, the lesional cells are typically located within the ovarian hilum, often in close association with nerves, and may contain cytoplasmic Reinke crystals. Stromal luteoma is composed of lutein cells growing in sheets or nests, often arising in a background of stromal hyperthecosis. As both lesions are located in the cortex and composed of luteinized cells, the most helpful feature in

R.A. Soslow and C. Tornos (eds.), *Diagnostic Pathology of Ovarian Tumors*,
DOI: 10.1007/978-1-4419-9751-7_14, © Springer Science+Business Media, LLC 2011

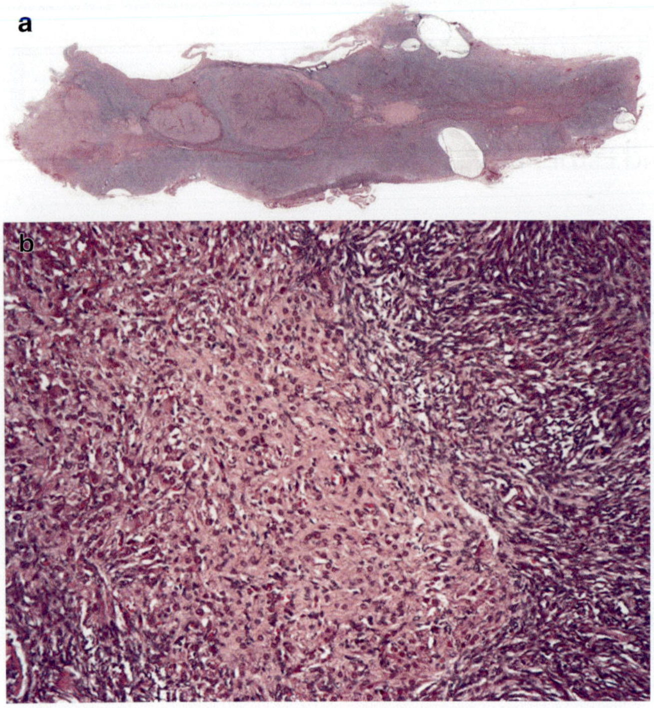

Fig. 14.1 Nodular hyperthecosis. Well-demarcated nodules (<5 mm) of luteinized cells are present in the ovarian cortex (**a**). The luteinized cells have abundant eosinophilic cytoplasm with indistinct borders and round nuclei with tiny nucleoli (**b**)

establishing the diagnosis of stromal luteoma is the presence of a diffuse growth of luteinized cells aggregating ≥0.5 cm in size. Although luteinized thecoma and steroid cell tumor may also have cells with lipid-rich cytoplasm, they typically form larger unilateral masses and the overall histologic features differ from that of stromal hyperthecosis.

Treatment

Patients with androgenic manifestations may respond to treatment with gonadotropin-releasing hormone agonists [6] or oral contraceptives, and if these fail, surgery (bilateral oophorectomy) may also be an option.

Leydig Cell Hyperplasia

Prevalence and Clinical Presentation

Leydig cell hyperplasia is characterized by a diffuse or nodular growth of Leydig cells within the ovarian hilum (hilus cell hyperplasia) or much less commonly, within the ovarian stroma (stromal Leydig cell hyperplasia). Leydig cell hyperplasia typically occurs in postmenopausal women (secondary to increased LH levels) but may also occur during pregnancy or after human chorionic gonadotropin administration. It may be seen in association with stromal hyperplasia, stromal hyperthecosis, Leydig cell tumor, and rarely, it can be found adjacent to other ovarian tumors [7–9]. Patients can present with androgenic manifestations if Leydig cell hyperplasia is prominent [10].

Macroscopic and Microscopic Features

Leydig cell hyperplasia is typically an incidental microscopic finding in oophorectomy specimens but rarely may be seen on cut section as minute yellow nodules within the ovarian hilum. At low-power magnification, small nests, clusters, irregular nodules (most common), or even diffuse sheets of cells are seen within the ovarian hilum frequently in close association with nerves (Fig. 14.2a). The cells are polygonal with round nuclei, often visible and centrally located nucleoli, and abundant eosinophilic cytoplasm. Some degree of nuclear pleomorphism, hyperchromasia, as well as rare mitotic figures may be found. Some cells contain eosinophilic Reinke crystals (Fig. 14.2b) and lipofuscin pigment. Leydig cells are immunoreactive for inhibin, calretinin, and relaxin-like factor [11], and they may stain for CD10 [12].

Differential Diagnosis

Stromal Leydig cell hyperplasia may be confused with stromal hyperthecosis. The finding of Reinke crystals establishes the diagnosis of Leydig cell hyperplasia in this location. Prominent Leydig cell hyperplasia may be confused with a Leydig cell tumor. However, the latter is more commonly associated with androgenic manifestations and forms a discrete mass [13].

Treatment

No specific therapy is required unless it is associated with androgenic manifestations, which typically resolves following oophorectomy.

Massive Edema and Fibromatosis

Prevalence and Clinical Presentation

These tumor-like lesions typically occur in young women in their first three decades of life (mean, 21 years) and are not infrequently bilateral [14, 15]. Patients may present

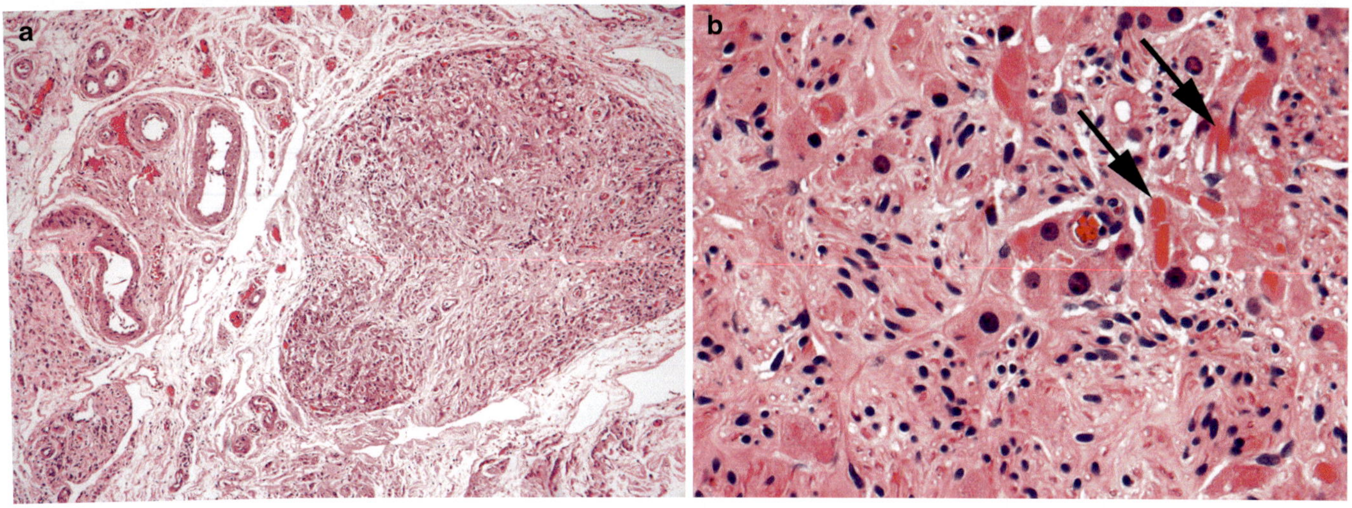

Fig. 14.2 Hilus cell hyperplasia. Hilus cells form nodular aggregates within the ovarian hilum (**a**). Some of the Leydig cells contain Reinke crystals (*arrows*) and are seen surrounding small nerves (**b**)

with abdominal pain, menstrual irregularity, or other hormonal manifestations [16, 17]. Although its exact etiology is uncertain, massive edema is postulated to be a consequence of venous and/or lymphatic obstruction secondary to intermittent ovarian torsion eventually resulting in fibromatosis [14, 18]. However, some authors postulate that ovarian fibromatosis may be the initial lesion, which as a consequence of repetitive torsion of the ovarian pedicle develops into massive edema [15]. The latter has also been reported in association with ovarian tumors, likely developing secondary to intermittent torsion of an enlarged ovary [19–21].

Macroscopic and Microscopic Features

Unilateral ovarian enlargement is seen in 80–90% of cases with the right ovary being more commonly affected. The involved ovaries range from 5.5 to 35 (mean, 11.5) cm in maximum dimension (massive edema often associated with larger size) and display a pale smooth and semitransparent capsule with a homogeneous white to gray cut surface that exudes watery fluid (massive edema) (Fig. 14.3) or a smooth lobulated outer surface with a yellow to white and firm cut surface (fibromatosis). Entrapped cystic follicles may be seen in both lesions.

Microscopic examination reveals either diffuse edema of the ovarian stroma (massive edema) (Fig. 14.4) or a diffuse hypocellular fibromatous proliferation (fibromatosis) separating and often entrapping preexisting follicles (Fig. 14.5). The latter is composed of spindle cells with banal cytologic features admixed with variable amounts of collagen. Alternating edematous and collagenous areas may be encountered in the same ovary. Luteinized stromal cells are present in 40% of lesions and extravasated red blood cells may also be discerned.

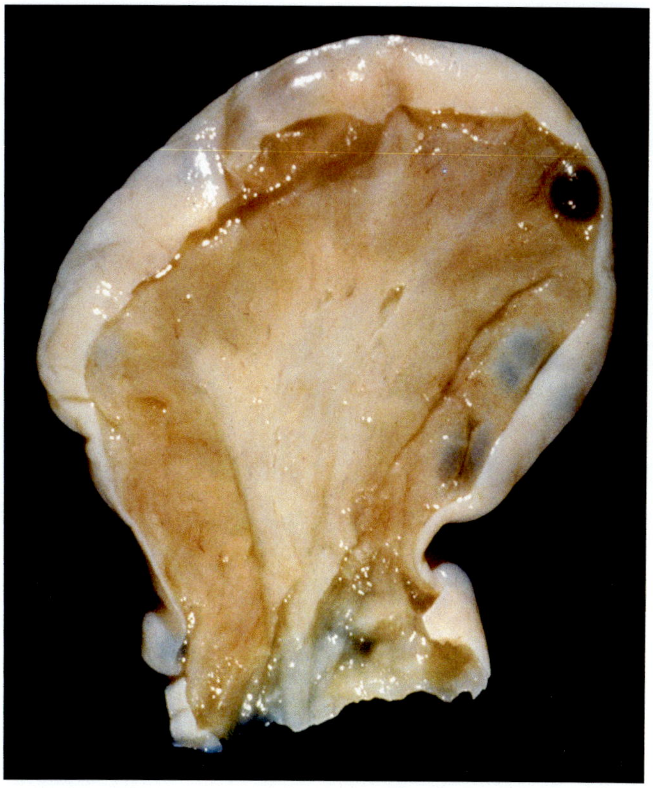

Fig. 14.3 Massive edema. The ovary is enlarged and has a homogeneous pale and shiny cut surface. Notice preservation of the preexistent structures in the ovarian cortex

Differential Diagnosis

These tumor-like lesions should be distinguished from ovarian fibroma with prominent edema, sclerosing stromal tumor, and myxoma. However, these tumors, although

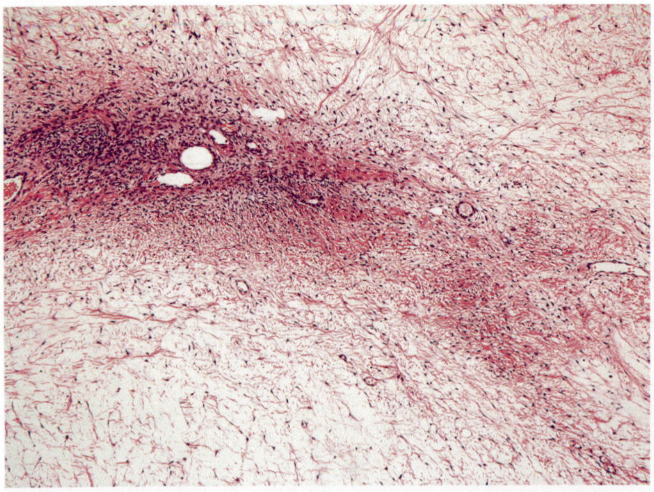

Fig. 14.4 Massive edema. There is striking edema of the ovarian stroma entrapping one small follicle

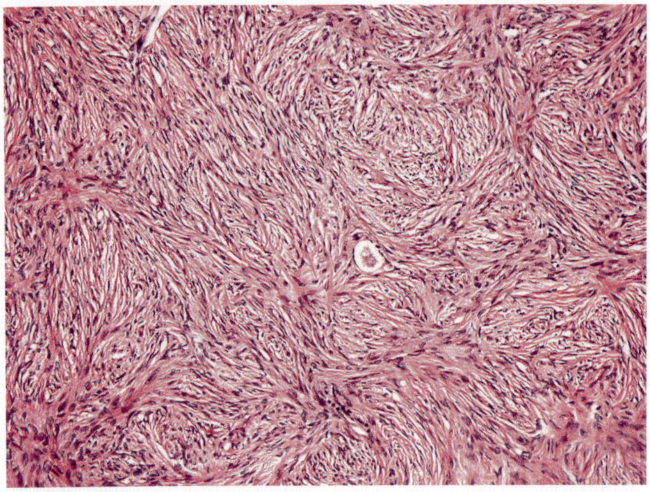

Fig. 14.5 Fibromatosis. Short fascicles of benign appearing spindle cells are set in a fibromatous stroma and surround a primordial follicle

typically non-encapsulated, form well-circumscribed masses and do not entrap preexisting follicular structures. Luteinized thecoma with sclerosing peritonitis may enter in the differential diagnosis as it often shows bilateral ovarian involvement, extensive edematous areas, luteinized cells, merges with the nonneoplastic ovarian stroma and may entrap ovarian follicles. The presence of hypercellular areas associated with brisk mitotic activity seen in this lesion are lacking in massive edema/fibromatosis. Finally, Krukenberg tumors may also enter in the differential diagnosis as they can be associated with prominent edema. However, they often show a nodular growth with hypercellular areas that on higher magnification reveal signet ring cells.

Treatment and Clinical Outcome

Massive edema/fibromatosis is a benign condition and conservative treatment should be considered in cases where fertility preservation is an issue. In rare instances, massive ovarian edema may result from lymphatic or venous obstruction by metastatic carcinoma [19, 22], and prognosis in these cases would be dependent on the underlying malignancy.

Luteinized Thecoma with Sclerosing Peritonitis

Prevalence and Clinical Presentation

Luteinized thecoma with sclerosing peritonitis is a rare and enigmatic entity that occurs most commonly in women in their second and third decades. Patients typically present with symptoms related to the associated peritoneal disease (sclerosing peritonitis) including abdominal distension, ascites, abdominal pain, or bowel obstruction. Rarely, sclerosing peritonitis may occur following the diagnosis and treatment of the ovarian lesions. Association with intake of antiepileptic medication has been reported but factors leading to sclerosing peritonitis remain unexplained [23, 24].

Macroscopic and Microscopic Features

This pseudoneoplastic proliferation typically involves both ovaries, which range in size from 2 to 31 cm in maximum dimension. A polypoid or cerebriform smooth and frequently at least focally hemorrhagic appearance of the ovarian surface is characteristic (Fig. 14.6). Not infrequently, diffuse thickening of the cortex with sparing of the medulla is seen on cut section (Fig. 14.7). These areas show a tan–pink or gray and usually homogeneous appearance, although areas of edema, cysts formation, and hemorrhage may be appreciated. The sclerosing peritonitis can involve the peritoneum, omentum, and intestinal serosa, and may be seen as white areas of thickening and/or induration, often associated with adhesions (Fig. 14.8).

Microscopically, there is a dense and often nodular proliferation of spindle cells that expands the ovarian cortex and may involve the medulla if the lesions are large (>10 cm). Areas of edema with secondary microcyst formation are often seen interspersed with these cellular foci (Fig. 14.9a). The spindle cells are small with scanty eosinophilic cytoplasm, oval nuclei, small nucleoli, and variable, but frequently brisk mitotic activity [>50 mitoses/10 high power fields (HPF)]. Admixed with the spindle cells are nests and clusters of small luteinized cells with clear or eosinophilic cytoplasm, round and often hyperchromatic nuclei, and tiny nucleoli (Fig. 14.10). Extravasated red blood cells may be

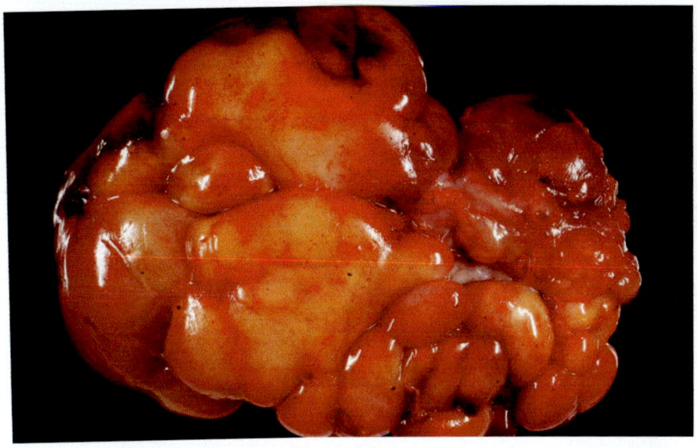

Fig. 14.6 Luteinized thecoma with sclerosing peritonitis. The ovary is enlarged with a striking cerebriform surface associated with areas of recent hemorrhage (courtesy of Dr F. Alameda, Barcelona, Spain)

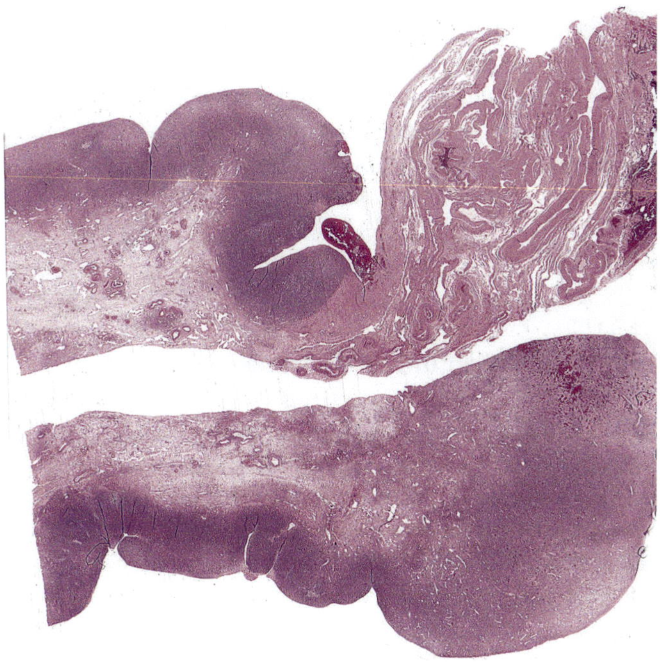

Fig. 14.7 Luteinized thecoma with sclerosing peritonitis. On low-power, there is diffuse "blue" thickening of the ovarian cortex with sparing of the medulla. Note the nodular configuration of the ovarian cortex

encountered. Preexisting ovarian follicles may be seen entrapped within this proliferation (Fig. 14.9b).

The peritoneal lesions are composed of a proliferation of bland fibroblastic cells separated by myxomatous, edematous, or collagenous stroma (Fig. 14.11), often associated with a chronic inflammatory infiltrate and a reactive mesothelial proliferation. In the omentum, involvement of the septa separating the fat lobules can impart a prominent lobulated architecture.

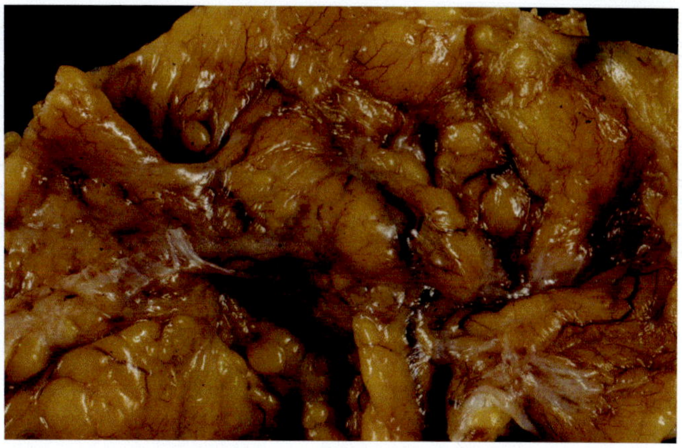

Fig. 14.8 Luteinized thecoma with sclerosing peritonitis. The omentum shows irregular white plaques associated with adhesions

Immunohistochemical Profile

The spindle cells may show positivity for smooth muscle actin, desmin, AE1/3, calretinin, CD56, and CD117 but are negative for inhibin. The luteinized cells stain diffusely for inhibin, calretinin, and CD56. The peritoneal lesions are strongly AE1/3 positive and may exhibit focal weak or moderate positivity for ER or PR [24]. There is no immunoreactivity for TGF-β [24].

Differential Diagnosis

This pseudoneoplastic lesion, if small, may sometimes be confused with stromal hyperthecosis because the latter is also often bilateral, may be multinodular, has luteinized cells within the dense spindle ovarian stroma, and may entrap normal ovarian structures. Stromal hyperthecosis, however, rarely causes ovarian enlargement, is not associated with edema, has spindle cells that do not exhibit the brisk mitotic activity seen in luteinized thecomas with sclerosing peritonitis, contains larger luteinized cells, and is not associated with peritoneal disease. If there is significant ovarian enlargement, entities in the differential diagnosis include cellular fibroma, luteinized thecoma, massive edema/fibromatosis, or rarely endometrioid stromal sarcoma. Cellular fibroma is typically composed of fascicles of spindle cells and tends to be well circumscribed from the surrounding ovarian parenchyma with no entrapment of preexisting ovarian structures. Although it may be mitotically active and can contain scattered luteinized cells, the mitotic activity is much less pronounced than in luteinized thecoma with sclerosing peritonitis. In contrast to the latter, typical luteinized thecoma either has a fibromatous background with admixed nests of large luteinized cells or when extensively luteinized, contains polygonal cells with abundant cytoplasm and is associated with a low mitotic rate.

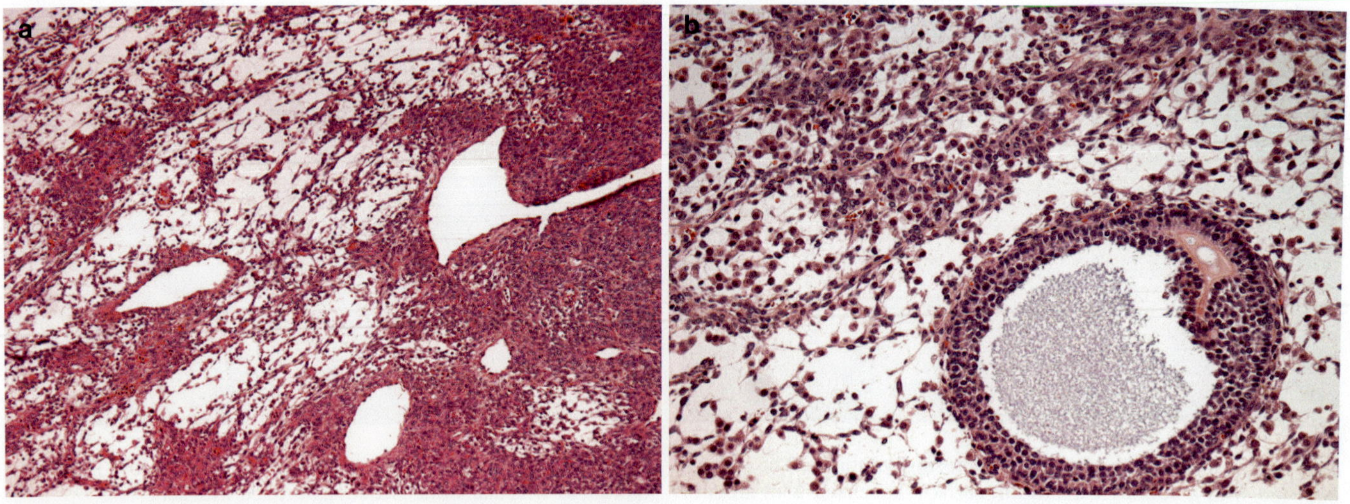

Fig. 14.9 Luteinized thecoma with sclerosing peritonitis. Densely cellular areas alternate with edematous hypocellular areas displaying a microcystic appearance (**a**). An entrapped follicle is seen (**b**)

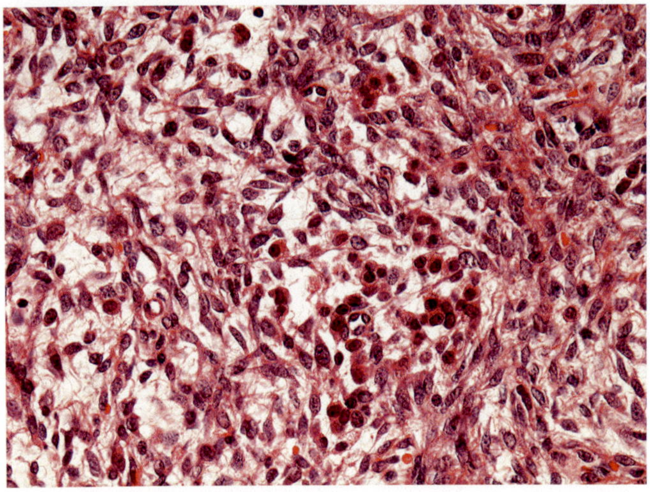

Fig. 14.10 Luteinized thecoma with sclerosing peritonitis. Small luteinized cells either single or in clusters are interspersed between the spindle cells

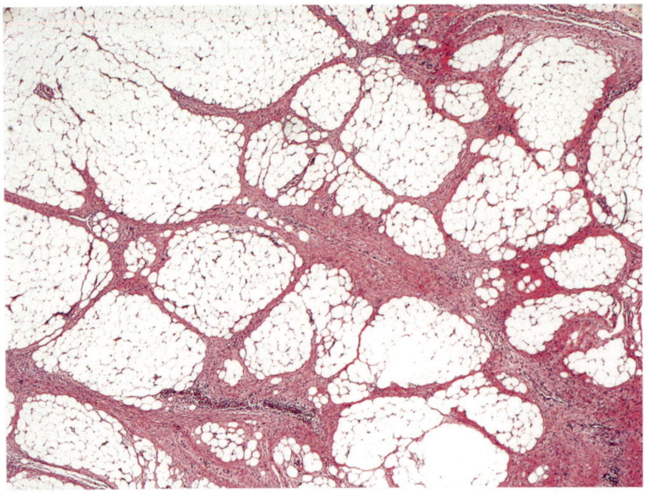

Fig. 14.11 Luteinized thecoma with sclerosing peritonitis. Omental fat lobules are separated by dense collagenous bands

Furthermore, all these tumors are typically unilateral (with the exception of bilateral fibromas in Gorlin's syndrome). Massive edema/fibromatosis may be bilateral and cause significant ovarian enlargement. In addition, they may contain luteinized cells and typically entrap preexistent follicles, but unlike luteinized thecomas with sclerosing peritonitis, they are usually hypocellular, exhibit marked areas of edema or collagen deposition, and lack mitotic activity.

Treatment and Clinical Outcome

Patients often undergo surgical excision of the ovarian masses and omentectomy. Adhesiolysis and bowel resection may be performed to relieve symptoms of bowel obstruction. The use of corticosteroids and hormonal therapy has also been shown to be of benefit in some patients [23, 25, 26]. Adverse outcome is typically secondary to complications from the peritoneal pathology.

Artifactual Displacement of Granulosa Cells

The artifactual displacement of granulosa cells from follicles into tissue or vascular spaces of the ovary (Fig. 14.12) can occur secondary to surgical trauma, tissue sectioning, or ovulation [27, 28]. The cells may have mildly atypical nuclei and scanty to abundant cytoplasm. An example of displacement of granulosa cells into the lumen and lamina propria of the fallopian tube has also been reported [29]. The presence of these cells within vascular channels may raise the differential diagnosis of a metastatic carcinoma or small cell carcinoma, pulmonary type, especially when prominent crushed artifact is present. Granulosa cells present on the ovarian surface post follicular rupture may also mimic mesothelial cells. The uniform nature of the displaced granulosa cells, their

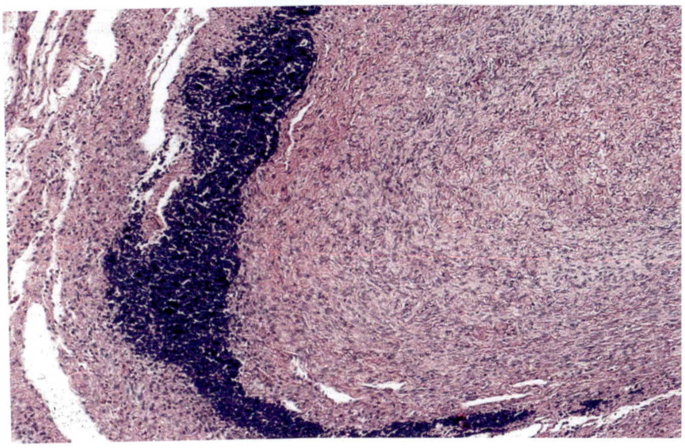

Fig. 14.12 Artifactual displacement of granulosa cells. Cells with prominent crushed artifact are present in a cleft within the ovarian stroma and potentially may simulate a malignancy

similarity to those present within adjacent follicles as well as immunoreactivity for inhibin and calretinin but not EMA, help to facilitate the diagnosis [27].

Large Solitary Luteinized Follicle Cyst of Pregnancy

Prevalence and Clinical Presentation

The large solitary luteinized follicle cyst is a rare nonneoplastic luteinized cyst that occurs in pregnant women [30]. It is often an incidental finding during delivery but some patients may present with a palpable pelvic mass on physical examination. Rarely, the cysts may only be evident during the postpartum period [30].

Macroscopic and Microscopic Features

The cyst is thin-walled and contains either clear, serosanguinous, or mucoid fluid, and can range in size from 8 to 55 (mean, 25) cm. It is lined by a single to multiple layers of luteinized granulosa and theca cells that may be morphologically indistinguishable from each other [31]. The luteinized cells have abundant eosinophilic or pale vacuolated cytoplasm and typically exhibit focal marked nuclear hyperchromatism and pleomorphism but mitotic activity is absent. No Call–Exner bodies are identified [30].

Differential Diagnosis

The differential diagnosis of a large solitary luteinized follicle cyst of pregnancy includes a conventional follicular cyst and more importantly, cystic granulosa cell tumors (both adult and juvenile types). Conventional follicular cysts are usually smaller in size and are often lined by distinct, granulosa and theca cells which may be luteinized. These cells may exhibit some degree of cytologic atypia but this is not as marked as that seen in a large solitary luteinized follicle cyst of pregnancy. Cystic adult granulosa cell tumors are lined by relatively uniform "blue" cells with nuclear grooves. The intracystic cellular proliferation tends to be more exuberant and Call–Exner bodies may be discerned in some tumors [32, 33]. Like large solitary luteinized follicle cyst, a cystic juvenile granulosa cell tumor is lined by cells with abundant eosinophilic cytoplasm that occasionally may show bizarre cytologic features; however, mitotic activity and a myxoid background are usually present.

Treatment

The cysts may require surgical excision because of their large size.

Hyperreactio Luteinalis

Prevalence and Clinical Presentation

Hyperreactio luteinalis is a nonneoplastic disorder characterized by bilateral ovarian enlargement secondary to the presence of multiple luteinized follicle cysts [34]. It is often found in women with high human chorionic gonadotropin (hCG) levels, such as those with multiple gestations [35] or gestational trophoblastic disease [36] but may also accompany a normal singleton pregnancy [37]. Ovarian hyperstimulation syndrome is an iatrogenic form of hyperreactio luteinalis, which occurs post induction of ovulation, either after administration of follicular stimulating hormone (FSH) followed by hCG or post clomiphene therapy [38]. Patients with preexisting polycystic ovaries appear to be more prone to developing this condition [39]. Rarely, the syndrome may occur spontaneously without exogenous stimulation [40]. Some of these cases have been associated with mutations of the transmembrane domain of the FSH receptor, which results in reduced ligand specificity, thus allowing for receptor stimulation by hCG and thyroid stimulating hormone [41, 42].

Patients are usually asymptomatic, but may present with a pelvic mass or abdominal pain when hemorrhage into the cysts occurs. Rarely, ovarian torsion or rupture with intraabdominal bleeding may result [43]. Androgenic manifestations have been reported in a small number of patients [44]. Ascites and hydrothorax may occur [45].

Macroscopic and Microscopic Features

There is variable but sometimes massive enlargement of both ovaries that have a bosselated tan to pink surface. On cut section, multiple thin-walled cysts of varying sizes, centered in the ovarian cortex and containing pale to hemorrhagic fluid,

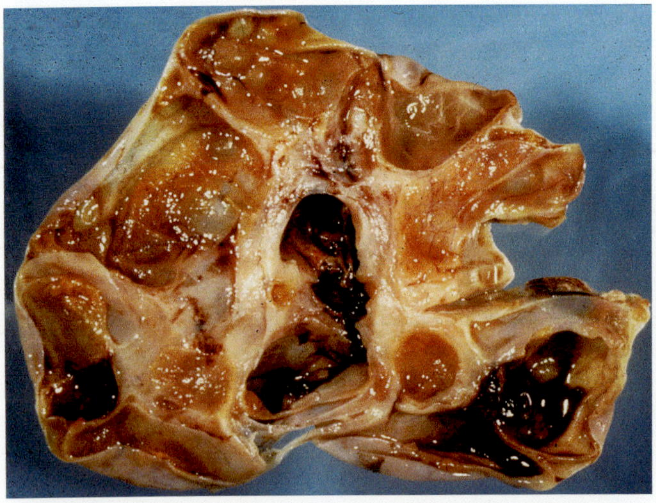

Fig. 14.13 Hyperreactio luteinalis. Cut section of the ovary shows multiple thin-walled cysts of varying sizes, some of them containing hemorrhagic fluid

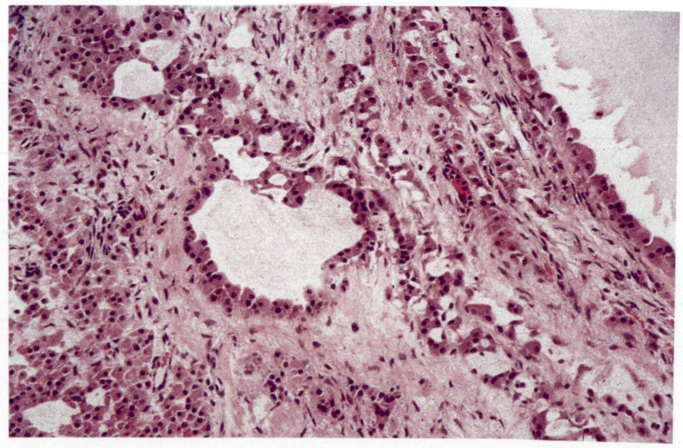

Fig. 14.14 Hyperreactio luteinalis. Luteinized granulosa and theca cells are lining cysts and are also present as clusters, cords, and single cells in the edematous ovarian stroma

are seen (Fig. 14.13). The cysts are lined by hyperplastic, luteinized theca interna and granulosa cells (Fig. 14.14). Typically, the theca cell layer is at least twice the thickness of the granulosa cell layer, which may show some degree of cytologic atypia. Often, there is also prominent edema of the theca layer and adjacent ovarian stroma.

Differential Diagnosis

Clinically, bilateral involvement of the ovaries with multiple hemorrhagic cysts may raise concern for malignancy but the gross and microscopic appearances are rather characteristic of this pregnancy-related lesion.

Treatment and Clinical Outcome

Hyperreactio luteinalis and ovarian hyperstimulation syndrome are self-limiting conditions that usually regress within weeks to months following cessation of excess hCG stimulation or ovulation induction, respectively [34]. Rarely, hyperreactio luteinalis may recur in subsequent pregnancies [46]. Conservative treatment is advocated but surgical intervention may be required in cases with secondary complications like torsion or rupture.

Ectopic Decidua

Prevalence and Clinical Presentation

Decidual metaplasia of ovarian stromal cells occurs in response to high estrogen and progesterone levels. Ectopic ovarian decidua is commonly seen in pregnancy and may be identified from the 9th week of gestation till term [47, 48]. Less commonly, it may be seen in association with trophoblastic disease, progestin treatment, an adjacent metastatic tumor, or a hormone-producing lesion of the ovary or adrenal gland. Occasionally, it may occur without a predisposing cause [49]. Besides the ovaries, ectopic decidua may be found on the serosa of other pelvic and abdominal organs [50–52], omentum [53], and within pelvic lymph nodes [54]. A case of pulmonary ectopic deciduosis has also been reported [55]. Patients are usually asymptomatic although rare lesions have been associated with intraperitoneal hemorrhage [56].

Macroscopic and Microscopic Features

Ectopic decidua is commonly an incidental microscopic finding but may be seen as small gray white or red nodules. The decidual cells are present within the superficial cortex or paraovarian adhesions. They may be seen as small clusters or large cellular sheets, often in association with a prominent vascular network of capillaries and a mild chronic inflammatory infiltrate. They are morphologically similar to decidualized endometrial stromal cells displaying a polygonal shape, an enlarged vesicular nuclei, occasional small nucleoli, and abundant eosinophilic or amphophilic cytoplasm. Cytoplasmic vacuoles, nuclear pleomorphism, and hemorrhagic necrosis may be encountered [34]. In postpartum patients, secondary hyalinization may occur.

Differential Diagnosis

Decidual cells exhibiting cellular atypia, especially when associated with signet ring-like cells, may mimic metastatic carcinoma. However, the uniform appearance of most of the cells and lack of mitotic activity should aid in establishing a benign diagnosis.

Treatment and Clinical Outcome

No treatment is required as the lesions typically regress when the underlying cause is removed.

Granulosa Cell Proliferations of Pregnancy

These are often incidental findings in the ovaries of pregnant women [57] but have also been reported in nonpregnant patients [58]. In the former group, these granulosa cell proliferations of pregnancy are commonly multiple and contained within atretic follicles. The granulosa cells have grooved nuclei and scanty cytoplasm and can be arranged in solid, insular, microfollicular, and trabecular patterns, surrounded by luteinized theca interna cells (Fig. 14.15). Rarely, the granulosa cells may contain moderate amount of finely vacuolated cytoplasm or show features of luteinization with round non-grooved nuclei [57].

The differential diagnoses to consider include either granulosa cell or Sertoli cell tumors. However, their occurrence during pregnancy, multifocality, microscopic dimension, and confinement to atretic follicles is more in keeping with a reactive process that occurs as a result of physiological response to the hormonal milieu of pregnancy [57].

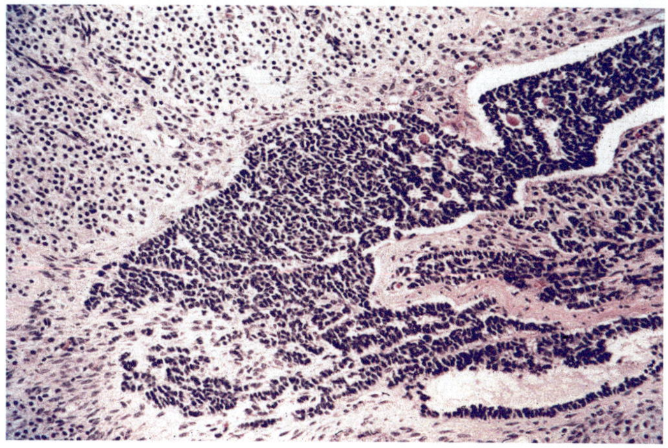

Fig. 14.15 Granulosa cell proliferation of pregnancy. There are microscopic aggregates of granulosa cells with insular and trabecular growths and Call–Exner bodies, reminiscent of an adult granulosa cell tumor. The granulosa cells are surrounded by luteinized thecal interna cells

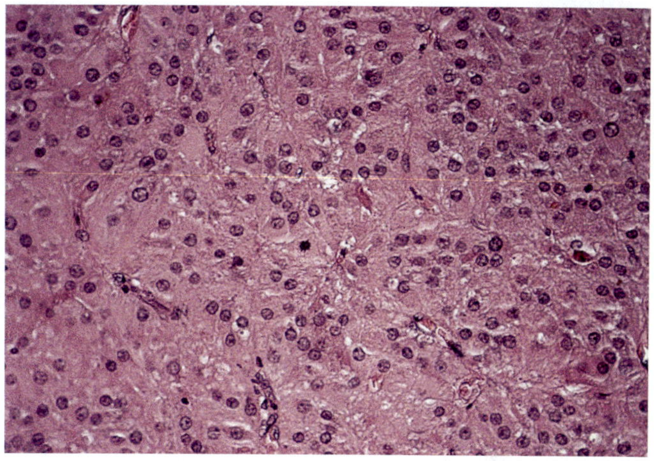

Fig. 14.16 Pregnancy luteoma. A diffuse growth of cells with abundant eosinophilic cytoplasm, central round nuclei, and occasional mitoses is seen

Pregnancy Luteoma

Prevalence and Clinical Presentation

Pregnancy luteoma is a tumor-like lesion that usually occurs in women during the later half of pregnancy. It is often an incidental finding during the time of cesarean section, but one quarter of women may present with androgenic manifestations [57, 59], which may also be evident in the fetus (if female) [59].

Macroscopic and Microscopic Features

Pregnancy luteomas are multifocal and bilateral in 30% of cases. The ovaries have a multinodular appearance, and on cut section they show well-demarcated, red to brown, solid, fleshy nodules that vary greatly in size (mean, 6 cm). The nodules are composed of luteinized cells that grow in diffuse sheets but may also form follicle-like spaces filled with pale or dense pink fluid. The cells have abundant eosinophilic cytoplasm, central round nuclei, and prominent nucleoli, and may exhibit some degree of nuclear pleomorphism (Fig. 14.16). Mitotic activity is often seen (mean, 2–3/10 HPF). Regressed lesions are characterized by degenerated luteinized cells with pyknotic nuclei associated with fibrosis and a chronic inflammatory infiltrate [34]. The luteinized cells show immunoreactivity for sex cord stromal markers including inhibin and calretinin [60].

Differential Diagnosis

Pregnancy luteoma should be differentiated from tumors composed of luteinized cells, including thecoma, Leydig cell tumor, and steroid cell tumor, NOS. Thecomas may occur during pregnancy, exhibit a nodular growth, and show cells with abundant cytoplasm. In contrast to pregnancy luteoma, they are typically solitary tumors usually composed of coalescent nodules with a yellow cut surface.

The neoplastic cells are associated with hyaline plaques and individual cells are invested by reticulin fibrils. Leydig cell and steroid cell tumors are also typically unilateral and they have an orange or yellow to tan cut surface. Leydig cell tumors are centered in the hilum and are frequently associated with Leydig cell hyperplasia. Reinke crystals as well as lipofuscin pigment may be found [61]. Steroid cell tumors are composed of cells that histologically resemble cells of the adrenal cortex, often containing multiple small cytoplasmic vacuoles.

Treatment and Clinical Outcome

These lesions should be treated conservatively as they typically regress within weeks of delivery [34].

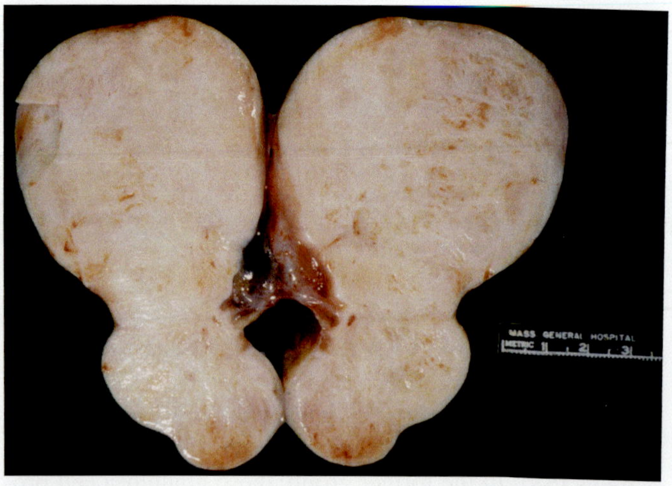

Fig. 14.17 Fibroma. The tumor is well circumscribed with a solid, trabeculated, homogeneous white cut surface

Fibroma

Prevalence and Clinical Presentation

Fibromas are the most common stromal neoplasms accounting for approximately 4% of all ovarian tumors. They occur most frequently in middle age women (mean, 48 years), but they can be seen in any age group [62]. Most patients present with an incidental or symptomatic ovarian or pelvic mass. Rarely, they may be associated with steroid hormone production or raised serum CA-125 levels [63, 64]. Approximately 15% of patients with fibromas measuring >10 cm present with ascites [65] and a minority of them also have Meigs' syndrome (ascites and pleural effusion) [64, 66], which usually resolve after tumor removal. Rarely, fibromas may also be associated with the nevoid basal cell carcinoma syndrome (Gorlin syndrome) and these tend to occur in younger patients [67, 68].

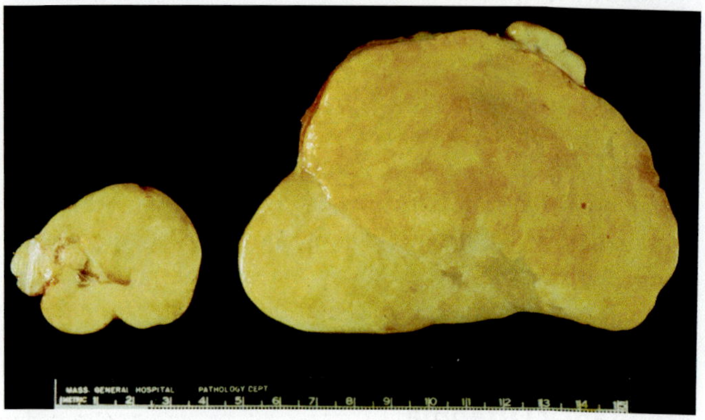

Fig. 14.18 Cellular fibroma. In contrast to typical fibroma, this tumor has a tan to yellow cut surface

Macroscopic and Microscopic Features

The vast majority of fibromas are unilateral but they may be bilateral (8%) [62], especially in patients with Gorlin syndrome [68]. They have a mean diameter of 6.0 cm [62], although cellular variants tend to be larger and are more frequently associated with ovarian surface and extraovarian adhesions and/or "implants" as well as rupture [63]. Fibromas have a smooth or lobulated surface except those associated with the Gorlin Syndrome, which often shows a multinodular growth as well as prominent calcification [69]. Some tumors may be pedunculated. Most neoplasms are solid, with a firm to rubbery, white to pale yellow cut surface (Fig. 14.17), but cellular fibromas tend to be softer and display a yellow to tan cut surface (Fig. 14.18). Areas of edema vand cyst formation may be present. Less commonly, hemorrhage and necrosis may be encountered, especially in cellular neoplasms.

Microscopically, fibromas typically show intersecting fascicles or a storiform arrangement of bland spindle cells with ovoid to elongated nuclei with "pointy ends," smooth nuclear contours, indistinct to occasionally small nucleoli, and scanty cytoplasm (Fig. 14.19). Less commonly, a herringbone pattern or Verocay-like areas may be found. Tumor cellularity may vary depending on the amount of hyalinization and/or edema within the neoplasm. The term cellular fibroma is applied to tumors with dense cellularity, minimal cytologic atypia, and low mitotic activity (≤3 mitoses/10 HPF) (Fig. 14.20). Cellular fibromas exhibiting ≥4 mitoses/10HPF (up to 19 in one series) [63] are diagnosed as mitotically active cellular fibromas. Lack of cytologic atypia is the most important criterion used to establish this diagnosis. Foci of hemorrhage and/or necrosis are

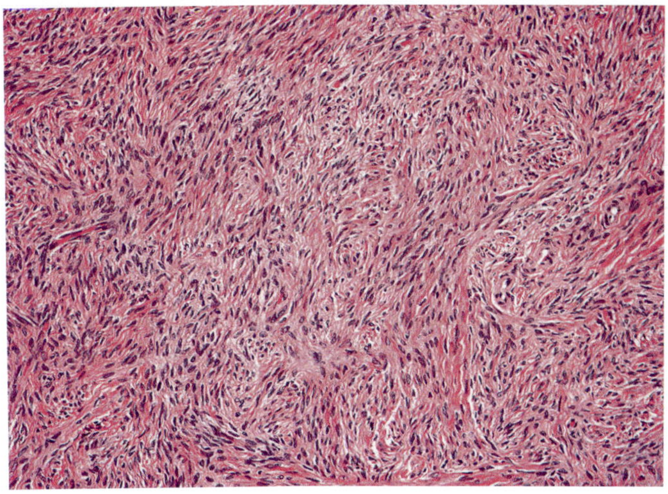

Fig. 14.19 Fibroma. The tumor is composed of short intersecting fascicles of bland spindle cells with ovoid to elongated nuclei and scanty cytoplasm

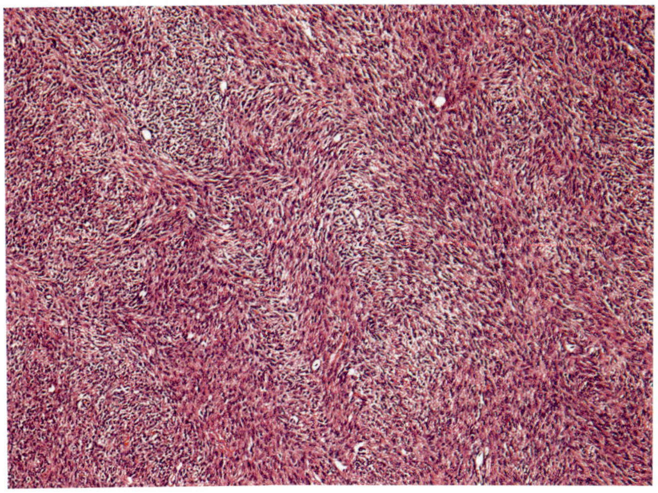

Fig. 14.20 Cellular fibroma. The tumor shows increased cellularity but no significant cytologic atypia

more commonly encountered in these fibroma subtypes. The latter may simulate tumor cell necrosis seen in malignant smooth muscle tumors, showing a sharp demarcation of the nonviable from the viable tumor with a characteristic perivascular growth. Viable tumor adjacent to these areas of necrosis can occasionally exhibit reactive cytologic atypia [63]. Fibromas can display sex cord-like elements [70] and when these account for <10% of the tumor, the term "fibroma with minor sex cord-like elements" should be used. These elements include nests or tubules of cells resembling granulosa, Sertoli, or undifferentiated sex cord cells [70]. Foci of dystrophic calcification (Fig. 14.21), scattered foamy histiocytes, cells with bizarre nuclei, intracytoplasmic hyalinized globules as well as scattered luteinized cells have also been described [63].

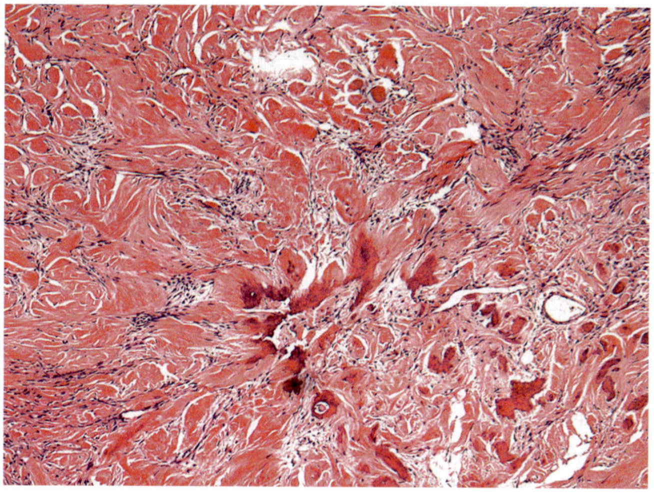

Fig. 14.21 Fibroma. The tumor is hypocellular due to prominent hyalinized stroma associated with foci of dystrophic calcification

Immunohistochemical Profile

Fibromas stain diffusely for vimentin [71] and can also show immunoreactivity for WT-1 [72], CD56 [72], smooth muscle actin [71, 72], desmin [73], and CD34 [71, 72]. Focal weak staining for inhibin and calretinin has also been reported [74–76] but they are typically negative for CD10 [12] and CD99 [77].

Genotype

Some typical and cellular fibromas have been reported to show loss of heterozygosity at 9q22.3 (PTCH gene) or 19p13.3 (STK11 gene). As mutations of the PTCH gene are associated with Gorlin syndrome, these findings suggest that sporadic fibromas and those occurring in association with Gorlin syndrome may arise through similar genetic pathways [78].

Differential Diagnosis

Stromal hyperplasia, fibromatosis, and thecoma should be considered in the differential diagnosis of conventional fibroma while fibrosarcoma, metastatic endometrial and primary endometrioid stromal sarcomas, smooth muscle tumors, as well as gastrointestinal stromal tumor (GIST) should be included in the differential diagnosis of cellular/mitotically active fibroma. Unlike fibroma, stromal hyperplasia is characterized by a diffuse proliferation of stromal cells within the ovarian cortex and medulla typically involving both ovaries, without the formation of a well-defined mass. Ovarian fibromatosis may closely mimic a fibroma but it is typically bilateral, lacks well-defined boundaries, and is associated with a dense collagenous background and entrapped native ovarian structures [15]. Even though thecomas and fibromas can show overlapping morphology, thecomas are frequently associated with hormonal manifestations, they show a nodular growth, and contain cells with more abundant lipid-rich amphophilic cytoplasm. In addition, in contrast to fibromas, they display extensive immunoreactivity for inhibin [71, 79]. Luteinized thecomas are more frequently associated with androgenic hormonal manifestations. They have a yellow cut surface due to abundant lipid content and either contain groups of luteinized cells in an otherwise fibromatous background or may be extensively luteinized [80]. Ovarian fibrosarcoma, although very rare, shares with mitotically active cellular fibromas the findings of dense cellularity and brisk mitotic activity. However, they typically exhibit striking cytologic atypia and may show atypical mitotic figures, features, which are lacking in cellular mitotically active fibromas [63]. Metastatic endometrial or primary endometrioid stromal tumors may closely mimic a fibroma. However, the typical morphology of small "blue" cells growing in sheets that are associated with the characteristic network of small arterioles as well as CD10 immunoreactivity will aid in establishing the correct diagnosis [12, 15, 81]. Smooth muscle tumors may superficially resemble cellular fibromas because of the spindled nature of the cells, similar cellularity and mitotic rate, and the presence of background collagen. However, they are composed of longer fascicles of spindle cells, which have more abundant eosinophilic cytoplasm and "cigar-shaped" nuclei, and the cells stain diffusely for smooth muscle markers [82]. Finally, ovarian involvement by GIST should be included in the differential diagnosis of a fibroma. A prior history of GIST, bilateral ovarian involvement, and focal epithelioid morphology should suggest a diagnosis of GIST, which can be confirmed with CD117 and CD34 immunostains [83].

Treatment and Clinical Outcome

Conventional fibromas are benign neoplasms that only require surgical excision. Cellular fibromas and mitotically active cellular fibromas are also usually associated with an indolent behavior; however, long-term follow-up may be warranted, especially if rupture or adhesions are present as these tumors may recur locally [63, 84].

Fibrosarcoma

Prevalence and Clinical Presentation

Fibrosarcomas are exceedingly rare. They usually occur in postmenopausal women (mean, 58 years) [84] and on occasion, may be associated with the Gorlin syndrome [85] or Maffucci's syndrome [86].

Macroscopic and Microscopic Features

They typically present as large unilateral pelvic masses that on gross examination, frequently show areas of hemorrhage and/or necrosis. These are densely cellular neoplasms composed of spindle cells arranged in a herringbone or storiform pattern. The cells have scant cytoplasm and elongated tapering nuclei showing diffuse moderate to severe cytologic atypia, the latter being the most important diagnostic feature separating these tumors from mitotically active cellular fibromas [63]. Mitotic count is typically high and abnormal mitotic figures are common. Fibrosarcomas also have a higher MIB-1 labeling index than cellular fibromas [87].

Immunohistochemical Profile and Cytogenetics

These tumors typically show immunoreactivity for vimentin and may also stain for CD117 [88]. Focal weak expression of CD10 has been described [12]. Some tumors have been reported to show a gain in trisomy 8 cells [87].

Treatment and Clinical Outcome

Patients with ovarian fibrosarcomas generally follow an aggressive clinical course with high recurrence and mortality rates [84, 89, 90]. Adjuvant chemotherapy or radiotherapy has been reported to prolong survival in some patients [84, 91]. However, it appears that some of the reported cases with long-term survival may, in fact, correspond to mitotically active cellular fibromas, based on the more recent diagnostic criteria [63].

Thecoma

Prevalence and Clinical Presentation

Thecomas occur most commonly in postmenopausal women (mean, 63 years) and are rare in prepubertal females [92]. They are frequently associated with steroid hormone production and the majority of these patients have clinical manifestations of estrogen excess [92, 93]. Abnormal vaginal bleeding is the most common clinical presentation, which may occur secondarily to concurrent endometrial pathology [93]. Like conventional thecomas, luteinized thecomas also occur mostly in older women, although approximately one-third are found in women below 30 years of age [80]. In contrast to conventional thecomas, these neoplasms are associated with a higher frequency of androgenic manifestations [80]. A subset of luteinized thecomas may be associated with sclerosing peritonitis [24, 94], but currently, these are favored to be nonneoplastic in nature.

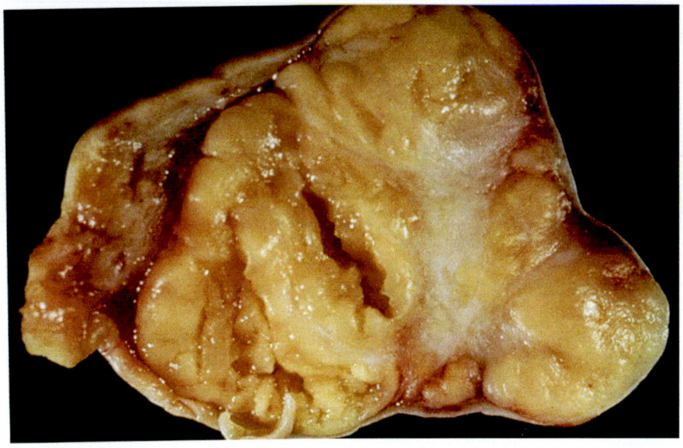

Fig. 14.22 Thecoma. The tumor has a solid, lobulated tan to yellow cut surface

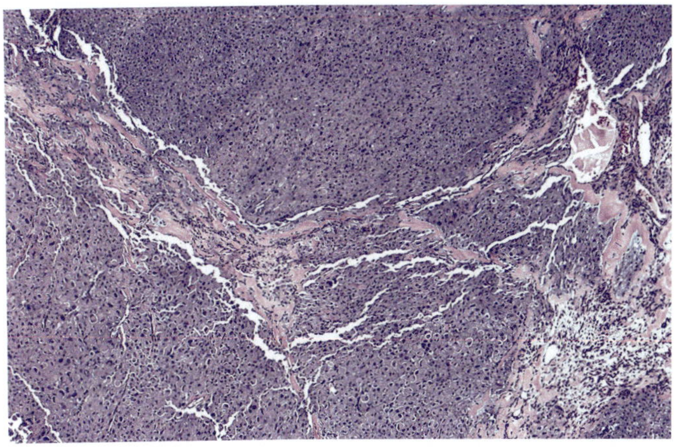

Fig. 14.23 Thecoma. The tumor is composed of confluent nodules of cells separated by hyaline plaques

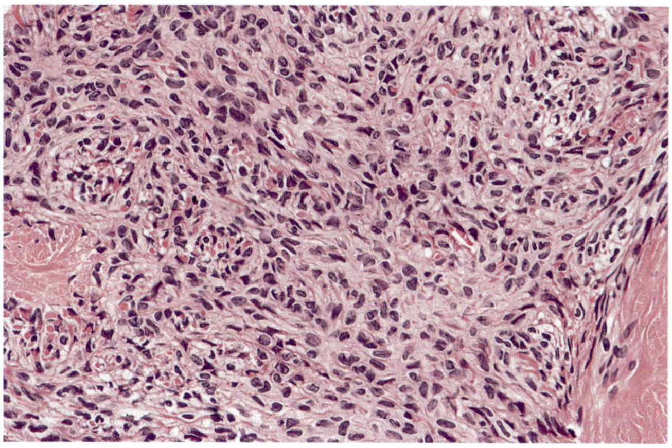

Fig. 14.24 Thecoma. The cells have irregular ovoid-shaped nuclei, inconspicuous nucleoli, and pale or vacuolated cytoplasm

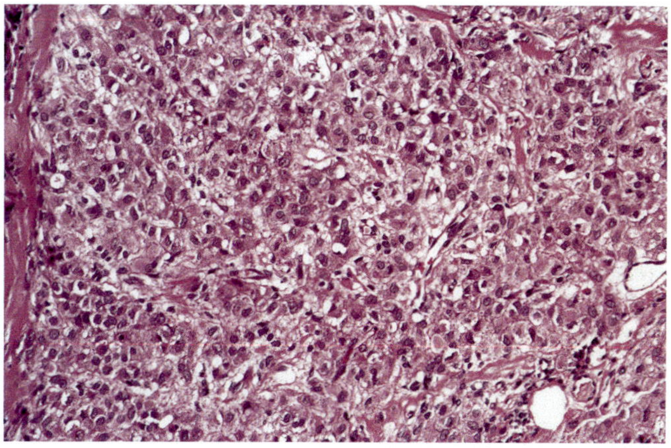

Fig. 14.25 Luteinized thecoma. The cells have round nuclei, central nucleoli, and abundant eosinophilic cytoplasm

Macroscopic and Microscopic Features

Most thecomas are unilateral, measuring 5–10 cm in largest dimension. The tumors are typically solid with a yellow to tan cut surface (Fig. 14.22). Occasionally, cystic change, hemorrhage, necrosis, and calcifications may be seen. They are composed of sheets or confluent nodules (Fig. 14.23), of ovoid to spindle cells with round to ovoid shape nuclei, inconspicuous nucleoli, ample pale or vacuolated, lipid-rich cytoplasm, and indistinct cellular borders (Fig. 14.24) [92, 95, 96]. There is usually minimal cytologic atypia and mitotic activity. Luteinized thecomas are characterized by single or groups of luteinized cells typically lying within a fibromatous and less commonly, thecomatous background. The luteinized cells have round nuclei, central prominent nucleoli, and abundant eosinophilic cytoplasm (Fig. 14.25) [80]. Conventional thecomas also frequently display hyaline plaques while stromal calcification may be prominent especially in tumors occurring in young women [97]. Rarely, bizarre nuclei [98] and osseous metaplasia [99] may be found. When Reinke crystals are identified within the luteinlike cells, these neoplasms are termed "stromal-Leydig cell tumor" [80, 100]. Uncommonly, thecomas may contain sex cord components in the form of granulosa cells, undifferentiated sex cord-type cells, or Sertoli cells. When these elements comprise <10% of the tumor, the term "thecoma with minor sex cord elements" should be applied [70].

Immunohistochemical Profile

A pericellular reticulin network (Fig. 14.26) is demonstrated on reticulin stain and intracytoplasmic lipid is highlighted on oil red O stain. Thecomas show immunoreactivity for inhibin, calretinin, CD10, and vimentin [12, 75, 76, 79], and they do not stain for cytokeratin or EMA [101].

Electron Microscopy

Ultrastructurally, thecoma cells show abundant intracytoplasmic lipid, smooth endoplasmic reticulum, and mitochondria.

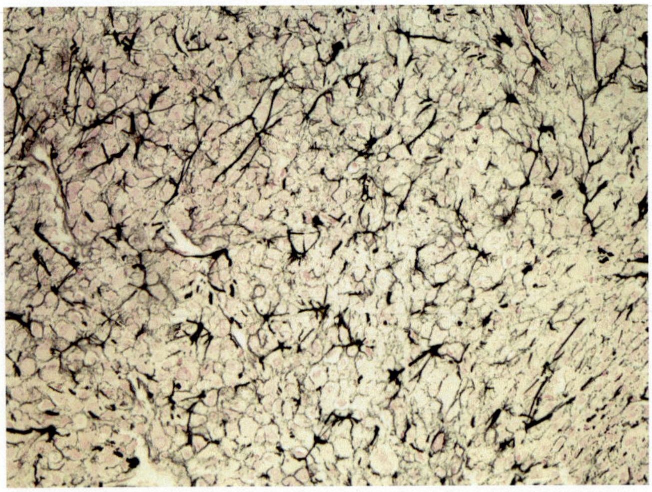

Fig. 14.26 Thecoma. Reticulin fibers surround individual cells

They are separated by collagen and lack desmosomes and basal lamina [102].

Genotype

A missense somatic mutation in the FOXL2 gene affecting codon 134 has been reported in some thecomas of the ovary [103, 104].

Differential Diagnosis

The distinction between thecoma and fibroma can be challenging as both tumors may show similar morphological features, thus the term "fibrothecoma" is used by some pathologists. However, in general, conventional thecomas are more commonly associated with steroid hormone production, they contain cells that have abundant pale or vacuolated cytoplasm and show diffuse staining for inhibin [71, 76]. Another entity in the differential diagnosis is adult granulosa cell tumor with a diffuse growth pattern as this may have an associated prominent thecomatous component. The presence of cells with grooved nuclei and the absence of a pericellular reticulin network in the granulosa areas will support the diagnosis of a granulosa cell tumor. Thecomas occurring in pregnant women may show prominent luteinization, causing confusion with a pregnancy luteoma. However, the latter typically shows bilateral and multifocal ovarian involvement, is not associated with a fibromatous or thecomatous background, and the cells do not have significant lipid content [34]. Luteinized thecomas may also resemble steroid cell tumors; however, the cytologic features of the cells differ and steroid cells typically contain abundant lipid vacuoles and show better defined cellular borders [105]. Finally, luteinized thecomas should also be separated from stromal hyperthecosis. Even though nests of luteinized cells are also seen in stromal hyperthecosis, they are present in a background of small stromal cells and are not associated with a fibromatous or thecomatous stroma, and there is typically bilateral ovarian involvement [2].

Treatment and Clinical Outcome

The vast majority of conventional and luteinized thecomas with rare exceptions [80, 106, 107] are benign neoplasms that may be adequately treated by unilateral salpingo-oophorectomy. Endometrial sampling is also advocated as a number of tumors may be associated with endometrial hyperplasia or carcinoma [93, 108].

Sclerosing Stromal Tumor

Prevalence and Clinical Presentation

This subtype of stromal tumor is a rare benign neoplasm comprising 2–6% of ovarian stromal tumors and <1% of all ovarian neoplasms. Eighty percent occur in women <30 years old with the most common clinical presentation being menstrual irregularity [109]. Much less frequently, patients may present with evidence of estrogen or androgen production [110, 111], raised CA-125 serum levels [112], or Meigs syndrome [113].

Macroscopic and Microscopic Features

The vast majority of tumors are unilateral, well circumscribed, with a solid or solid and cystic and white to yellow cut surface (Fig. 14.27). Rarely, they may present as unilocular cysts [114]. On microscopic examination, they are typically composed of cellular pseudolobules separated by hypocellular areas containing edematous or collagenous stroma (Fig. 14.28). The cellular areas contain a mixture of spindle cells resembling fibroblasts as well as rounded lutein cells (Fig. 14.29) [115]. The lutein cells have ovoid to round nuclei, prominent nucleoli, and abundant eosinophilic to clear vacuolated cytoplasm. Occasional signet ring-like cells may also be seen. Cytologic atypia and mitoses are minimal except when tumors occur in pregnant women. In this instance, the cells may have a higher nuclear to cytoplasmic ratio and display brisk mitotic activity. Another characteristic feature is the presence of angulated and often dilated thin-walled vessels, reminiscent of those seen in hemangiopericytoma. Calcification may be identified in some tumors [71, 116].

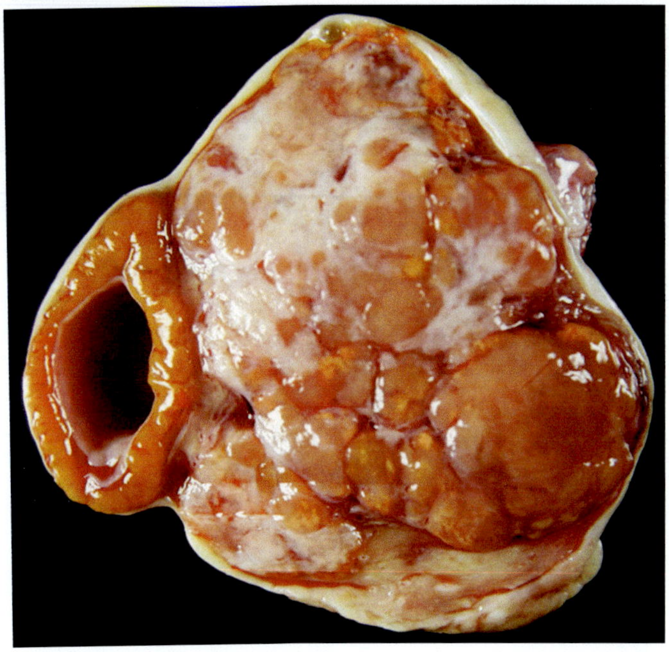

Fig. 14.27 Sclerosing stromal tumor. The tumor is well circumscribed with multiple solid and tan nodules separated by white septa. The tumor is juxtaposed to a corpus luteum cyst

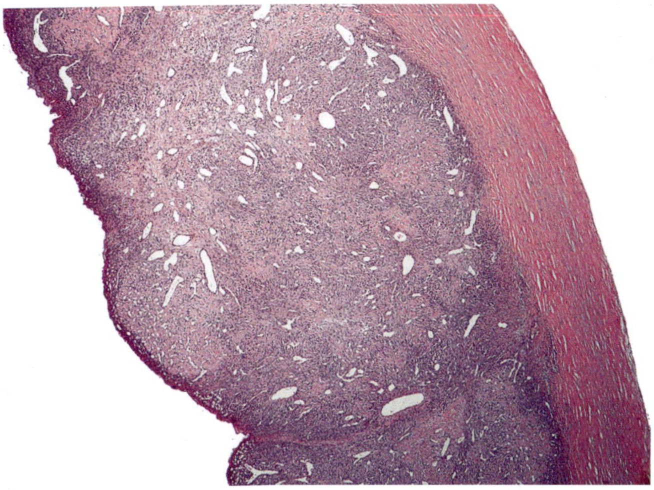

Fig. 14.28 Sclerosing stromal tumor. A striking pseudolobular architecture is seen. The lobules have hypo- and hypercellular areas with a prominent hemangiopericytoma-like vasculature

Immunohistochemical Profile

The tumor cells, especially the lutein cells, stain for calretinin [117] and may also express inhibin [75], CD10 [12], ER and PR, smooth muscle actin, desmin, CD99, and vimentin [115, 118–120]. Epithelial markers (cytokeratin and EMA) are negative.

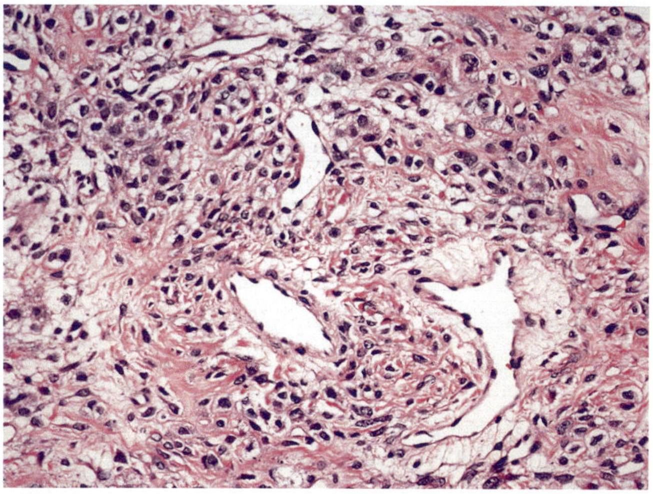

Fig. 14.29 Sclerosing stromal tumor. An intimate admixture of lutein and spindle cells is seen around blood vessels

Electron Microscopy

Ultrastructural studies have shown a mixture of lipid-laden epithelioid-type cells (lutein cells), fibroblast-like cells, undifferentiated mesenchymal as well as intermediated cells with varying degrees of differentiation [115].

Differential Diagnosis

Sclerosing stromal tumors may be confused with fibromas and more often with thecomas. Pure fibromas, however, typically occur in older women, lack lobulation and prominent vascularity as well as lutein-type cells [62]. Thecomas may be confused with sclerosing stromal tumors because of its lobular architecture, the presence of luteinized cells, and a similar vasculature, although the latter if present is typically focal. In contrast, thecomas contain hyaline plaques, and the majority are composed of a uniform population of luteinized cells (except for luteinized thecoma where spindle cells are the predominant component) [80]. Sclerosing stromal tumors containing prominent signet ring-like cells can mimic a Krukenberg tumor. However, these signet ring-like cells do not contain mucin or express epithelial markers but may instead contain cytoplasmic lipid [121]. The presence of "staghorn" type vessels in a sclerosing stromal tumor may raise the differential of a hemangiopericytoma [122]. The latter is a diagnosis of exclusion.

Treatment

Surgical excision is the treatment of choice for these benign tumors.

Signet Ring Stromal Tumor

Prevalence and Clinical Presentation

Signet ring stromal tumor is a very rare benign stromal neoplasm occurring in adult patients who may present with symptoms of abdominal pain or weight loss.

Macroscopic and Microscopic Features

Grossly, a unilateral and solid mass, with or without a cystic component, showing a gray, tan, or yellow cut surface is seen. Microscopically, it is composed of signet ring-like cells set in a fibromatous background [123]. The signet ring cells have eccentrically placed nuclei and cytoplasmic vacuoles that can contain pale eosinophilic to purplish material but no detectable mucin, glycogen, or lipid (Fig. 14.30) [124]. Ultrastructurally, the vacuoles are thought to be secondary to cytoplasmic edema, mitochondrial swelling, or pseudoinclusions of edematous extracellular matrix [124, 125]. Intra- and extracellular hyaline globules may occasionally be found [124].

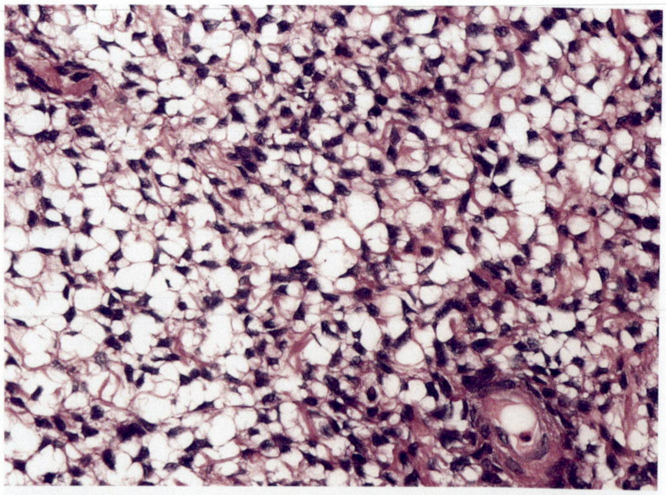

Fig. 14.30 Signet ring stromal tumor. The neoplastic cells grow in sheets and have eccentrically displaced nuclei and "empty" intracytoplasmic vacuoles

Immunohistochemical Profile

Immunoreactivity for vimentin, smooth muscle actin, inhibin, calretinin, CD56, and cytokeratin has been reported but there is typically no expression of EMA or S100 [124, 125].

Differential Diagnosis

It is important to distinguish this neoplasm from metastatic signet ring cell carcinoma (Krukenberg tumor) [121]. The latter typically shows epithelial structures in addition to mucin-containing signet ring cells [126]. Other differential diagnosis include goblet cell carcinoid tumor which has nests of cells either containing mucin filled cytoplasmic vacuoles or neuroendocrine granules, free floating in pools of extracellular mucin, and sclerosing stromal tumor, which typically has a lobulated architecture, contains two cell types (spindle and lutein), and a prominent staghorn-like vasculature. It is important to be aware that other stromal tumors as well as sex cord stromal tumors may have a minor component of signet ring-like cells of the type seen in the signet-ring stromal tumor.

Treatment

Signet ring stromal tumor is a benign tumor which is adequately treated by unilateral salpingo-oophorectomy.

Microcystic Stromal Tumor

Prevalence and Clinical Presentation

Microcystic stromal tumor is a recently described entity [127] that occurs in adult women ranging in age from 26 to 63 (mean, 45) years. The clinical presentation is nonspecific.

Macroscopic and Microscopic Features

Tumors are typically unilateral, averaging 8.7 cm in maximum diameter. They have well-defined borders with a solid, solid–cystic, or predominantly cystic cut surface. Microscopically, the tumors show a triad of microcysts, solid cellular areas, as well as fibrous stroma. The microcysts are of varying caliber and may contain light basophilic or clear contents. The cellular areas are composed of cells with bland, round to spindle shaped nuclei, small indistinct nucleoli, and finely granular, pale eosinophilic cytoplasm. Cells with intracytoplasmic vacuoles or bizarre nuclei and psammomatous calcifications may be identified. Mitotic rate is typically low (0–2/10HPF). The cells are frequently separated by fibrous bands and hyalinized plaques.

Immunohistochemical Profile

Diffuse staining for vimentin and CD10 is characteristic of these tumors and one quarter also express cytokeratin but are negative for EMA. Rare neoplasms also show focal immunoreactivity for inhibin and calretinin [127].

Differential Diagnosis

These tumors should be distinguished from conventional thecoma, luteinized thecoma associated with sclerosing peritonitis, steroid cell tumor, sclerosing stromal tumor, signet ring stromal tumor, yolk sac tumor with microcystic or reticular patterns, ovarian tumor of probable Wolffian origin, and struma ovarii. Unlike microcystic stromal tumors, thecoma occurs in older women who more frequently present with estrogenic manifestations. It is composed of polygonal rather than rounded cells with ample pale cytoplasm and ill-defined cellular borders that show diffuse staining for inhibin and calretinin but only focal CD10 expression [12, 128]. Luteinized thecoma with sclerosing peritonitis often exhibits a microcystic growth pattern, but this lesion is typically bilateral and is associated with sclerosing peritonitis. In addition, the lesional cells often display prominent mitotic activity and show diffuse staining for inhibin, calretinin, and CD56 [24]. Oxyphilic steroid cell tumor may resemble a microcystic stromal tumor, but it typically has a more yellow cut surface due to high lipid content and cells with central prominent nucleoli. Other findings including cytoplasmic lipochrome pigment, degenerative pseudovascular spaces, and immunoreactivity for inhibin and/or calretinin serve to separate this tumor from microcystic stromal tumor [105, 129]. Like microcystic stromal tumor, sclerosing stromal tumor and signet ring stromal tumor are unilateral neoplasms with a solid or solid–cystic gross appearance. Cellular islands but not microcysts are seen in sclerosing stromal tumor. The latter also typically contains staghorn-like vessels and a mixed population of spindle and luteinized cells [115] while signet-ring stromal tumor is characterized by an admixture of signet ring-like cells and fibroblasts [123]. Yolk sac tumor may show areas with a "microcystic" appearance. However, the presence of cells with "primitive" cytologic features associated with brisk mitotic activity as well as the presence of other typical growth patterns will help to distinguish this tumor from microcystic stromal tumor [130]. The female adnexal tumor of probable wolffian origin may be centered in the ovary and shares with microcystic stromal tumor, a variable microcystic and solid component as well as low-grade cytologic features. However, a pseudotubular pattern is also frequently discerned and the tumor lacks a stromal component as well as cells with intracytoplasmic vacuoles. Immunohistochemistry is not helpful in this differential diagnosis [75, 131, 132]. Struma ovarii with a microfollicular architecture may simulate the microcysts seen in the microcystic stromal tumor but other characteristic growth patterns are usually present and it is often juxtaposed to a mature cystic teratoma [133]. The use of thyroglobulin or TTF-1 immunostains would also be helpful.

Treatment and Clinical Outcome

This tumor appears to behave in a benign fashion with no reported recurrences to date; thus, excision is the treatment of choice.

Granulosa Cell Tumors

Adult Granulosa Cell Tumor

Prevalence and Clinical Presentation

Adult granulosa cell tumor (AGCT) is the most frequent neoplasm among all sex cord stromal tumors accounting for 1–2% of all ovarian neoplasms and 95% of all granulosa cell tumors. They can occur at any age but they are more commonly seen in perimenopausal or early postmenopausal patients with a peak age incidence between 50 and 55 years [134, 135]. Patients typically present with menometrorrhagia or postmenopausal bleeding as AGCT is the ovarian tumor most often associated with estrogenic manifestations. Approximately 5% of AGCT are associated with endometrial hyperplasia or adenocarcinoma, which occur secondary to prolonged estrogenic stimulation [130]. The adenocarcinoma tends to be well differentiated, low stage, and associated with good prognosis [134]. In prepubertal patients, isosexual pseudoprecocity is often the presenting complain. Rarely, AGCT can be associated with androgenic manifestations, especially when tumors are cystic [136]. In addition to endocrine manifestations, patients may also present with abdominal pain and distension secondary to a large pelvic mass and in about 10% of cases, hemoperitoneum occurs as a complication of tumor rupture.

Macroscopic and Microscopic Features

The great majority (90%) of AGCTs are unilateral with an average diameter of 12 cm, although size may vary from microscopic to 30 cm. They typically have a solid or solid and cystic cut surface (Fig. 14.31). Rarely, tumors may be completely cystic (Fig. 14.32) [32]. The solid component ranges from white to tan to yellow depending on the amount of associated stroma and the cysts may contain blood. At low-power magnification, the tumor is well circumscribed although not encapsulated. It displays a wide spectrum of histologic patterns and a combination of growth patterns is commonly appreciated in the same tumor. Diffuse sheets of tumor cells are most frequent, with trabecular (Fig. 14.33a), insular (Fig. 14.33b), tubular (solid or hollow) (Fig. 14.33c), corded (Fig. 14.33d), microfollicular, macrofollicular, watered silk (moiré-silk), and gyriform (zig-zag arrangement of the tumor cells) (Fig. 14.33e) patterns being less common but characteristic of this neoplasm. The tubular pattern overlaps with that seen in well-differentiated Sertoli cell tumors but it usually represents a minor component of the tumor. Call–Exner bodies,

Fig. 14.31 Adult granulosa cell tumor. The tumor is solid and cystic with the solid areas having tan cut surface

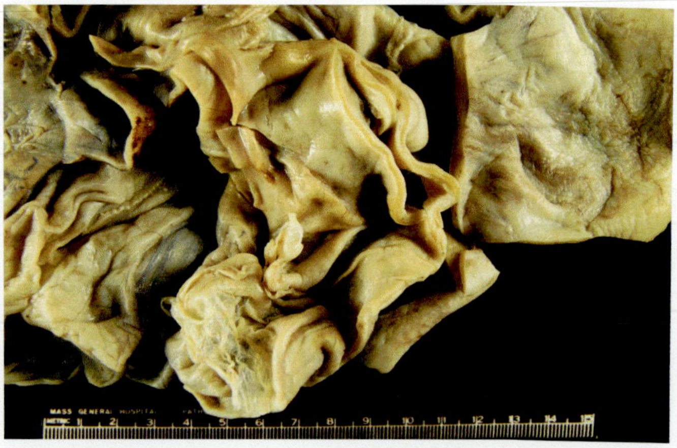

Fig. 14.32 Adult granulosa cell tumor. Variable-sized cysts have a smooth inner lining. Note a thin yellow rim present in the walls of some cysts

a characteristic feature of the microfollicular pattern, are microcystic spaces containing eosinophilic basal lamina material (Fig. 14.34) and may be found in 30–60% of AGCTs [135]. The granulosa cells typically have scant cytoplasm, uniform, ovoid and angular nuclei, often with a longitudinal nuclear groove ("coffee bean") (Fig. 14.35), but infrequently, granulosa cells may display moderate to abundant eosinophilic cytoplasm, round to ovoid nuclei with prominent nucleoli, and lack grooves (luteinized cells) (Fig. 14.36) [137]. Transition between the two components is often seen. Mitotic rate is variable but generally not brisk (<5/10HPF). Approximately 2% of AGCTs contain cells with large, bizarre hyperchromatic nuclei (Fig. 14.37) [98]. Rarely, GCTs with hepatic differentiation [138] or sarcomatous transformation [139] have been reported. The neoplastic granulosa cells are often associated with a variable fibromatous or fibrothecomatous component (Fig. 14.38).

Immunohistochemical Profile

Inhibin and calretinin are sensitive markers of AGCTs [75, 76, 79, 140–142], which often also stain for vimentin, WT-1, and CD99 [143, 144]. These tumors show variable immunoreactivity for CD56, AE1/3, CAM5.2, smooth muscle actin, desmin, S100, ER, and PR but CD34, CK7, and EMA are typically negative [145, 146]. Areas of hepatic differentiation are AFP, CEA, EMA, and CAM5.2 positive but vimentin and inhibin negative [147].

Genotype

A missense somatic mutation in the FOXL2 gene affecting codon 134 has recently been reported in adult and juvenile granulosa cell tumors [103, 148].

Differential Diagnosis

It is important to separate AGCTs from thecomas as the latter are typically not associated with recurrences, but this distinction may be especially difficult in tumors with a diffuse pattern. In most instances, thecomas tend to display at least a vague nodular growth of cells that contain more visible cytoplasm and classically have abundant intercellular reticulin fibrils whereas AGCT only shows reticulin fibers surrounding nests of granulosa cells, a feature that can be highlighted with a reticulin stain (Fig. 14.39). Endometrioid carcinomas may resemble AGCTs especially if they are composed of small glandular structures, simulating a microfollicular pattern [149]. In addition, they may grow in cords, trabeculae or have a predominantly spindled morphology, fibromatous stroma, scattered luteinized stromal cells surrounding epithelial elements, and grooved nuclei. However, typical areas of endometrioid carcinoma are almost always present, often with squamous and/or mucinous metaplasia, or even an adenofibromatous background. In contrast to AGCTs, endometrioid carcinomas with sex cord-like features are typically inhibin negative, but EMA and CK7 positive [145, 150]. At low power magnification, AGCT may have a pseudopapillary growth mimicking transitional cell carcinoma. At high power, even though both neoplasms have cells with grooved nuclei [151], transitional cell carcinoma is typically a high-grade malignant tumor with cytologic atypia, necrosis, and brisk mitotic activity [152]. Insular and trabecular carcinoid tumors may show prominent insular or trabecular growth as seen in some AGCT [153, 154]. However, tumor cells (especially those at the periphery of the islands) have abundant eosinophilic granular cytoplasm, nuclei with "salt and pepper chromatin," and are typically chromogranin positive and inhibin/calretinin negative. In addition, ovarian carcinoids often occur in association with a mature cystic teratoma.

Low-grade endometrioid stromal sarcoma (ESS) (primary or metastatic) may show sex cord-like differentiation set in a fibromatous background, simulating an AGCT. However, the

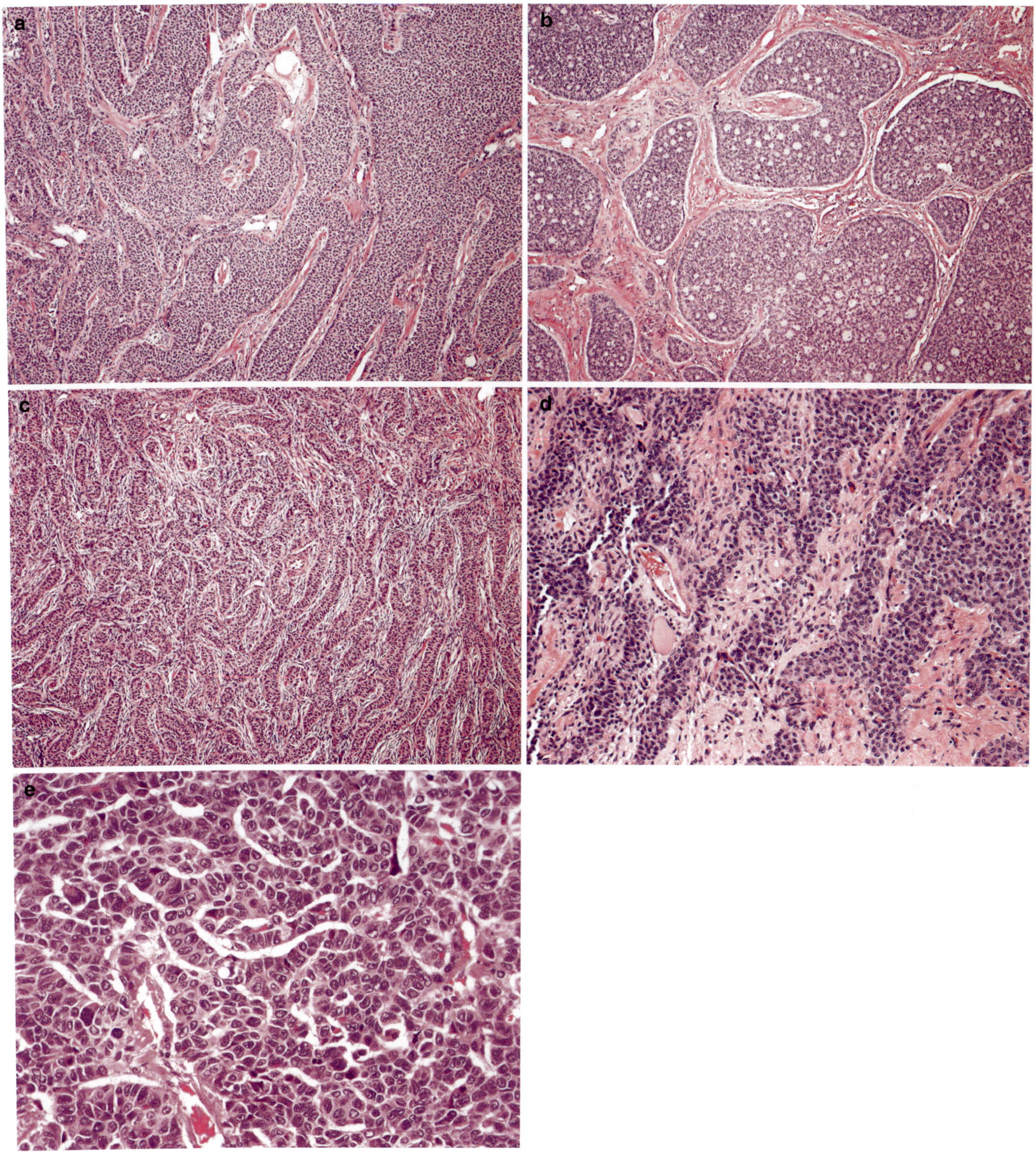

Fig. 14.33 Adult granulosa cell tumor. Trabecular (**a**), insular (**b**), tubular (**c**), corded (**d**), and gyriform (**e**) patterns

finding of typical areas of endometrial stromal neoplasia with a characteristic vascular network of small arterioles should be helpful in this differential diagnosis [155, 156]. Furthermore, primary ovarian ESS frequently shows the characteristic tongue-like pattern of growth at the hilum and extraovarian spread at the time of diagnosis. Inhibin and calretinin may also be helpful as these two markers are typically expressed in AGCT but not in conventional areas of ESS. CD10 is of limited utility as AGCT may also stain for this marker [12].

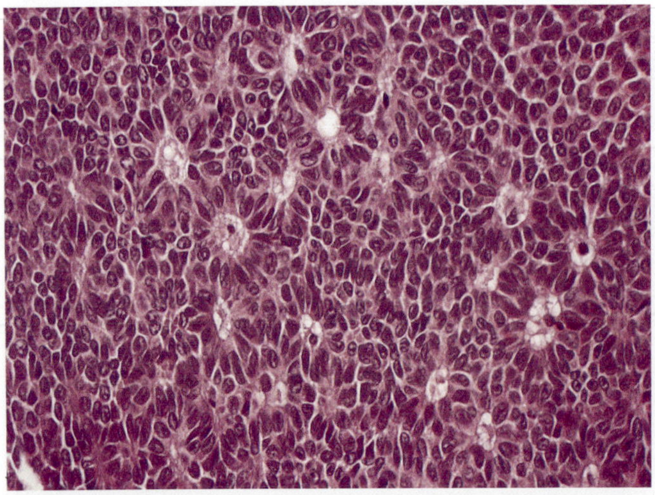

Fig. 14.34 Adult granulosa cell tumor. Call–Exner bodies are prominent in this neoplasm

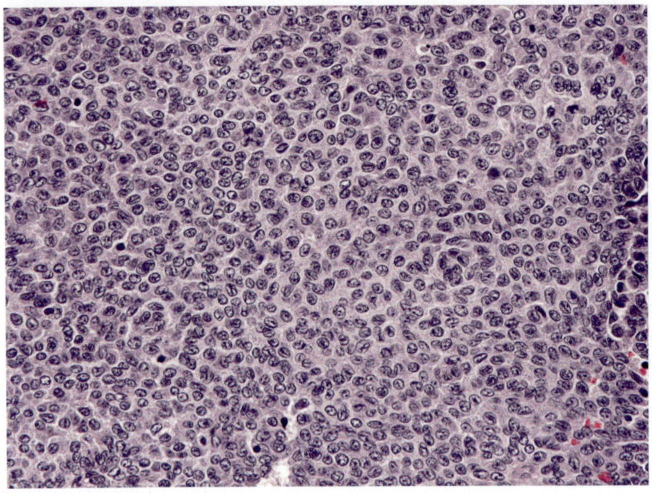

Fig. 14.35 Adult granulosa cell tumor. The cells typically have ovoid nuclei containing longitudinal grooves

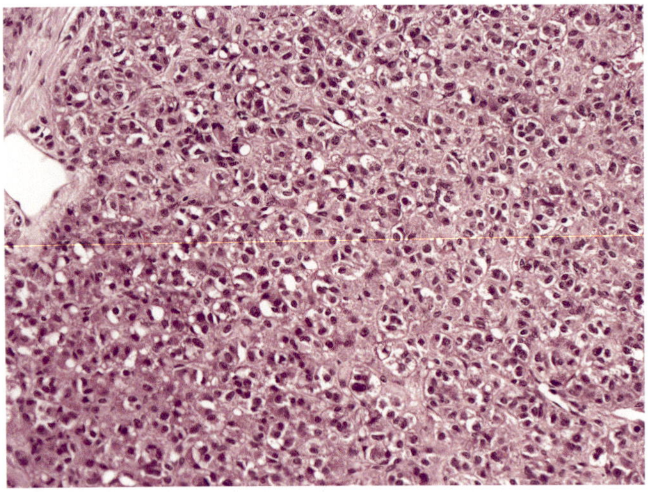

Fig. 14.36 Luteinized adult granulosa cell tumor. The neoplastic cells have moderate to abundant eosinophilic cytoplasm, round to ovoid nuclei with prominent nucleoli, and lack grooves

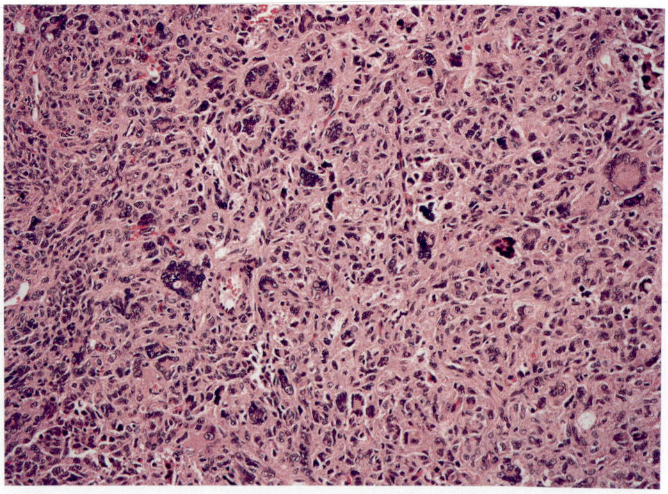

Fig. 14.37 Adult granulosa cell tumor. Cells with enlarged hyperchromatic bizarre nuclei are multifocally distributed within the tumor

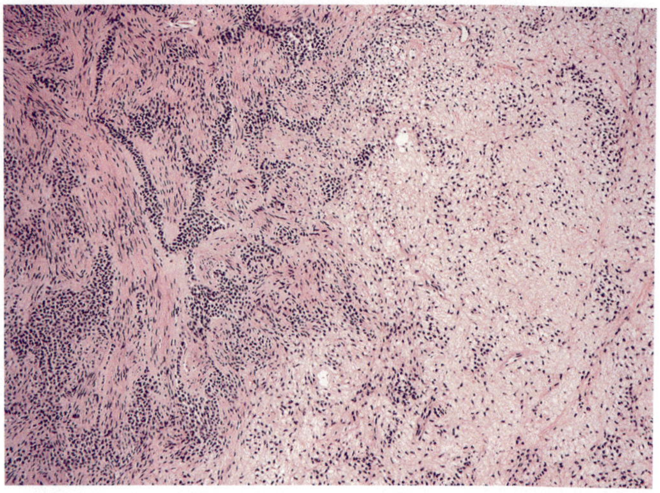

Fig. 14.38 Adult granulosa cell tumor. A fibromatous and edematous background may be striking, obscuring the typical growth of the tumor cells (*left*)

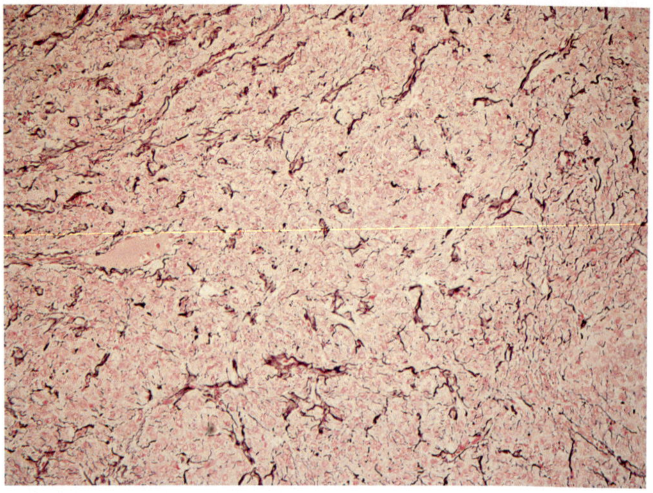

Fig. 14.39 Adult granulosa cell tumor. Reticulin fibers surround groups of granulosa cells

Another important entity in the differential diagnosis is small cell carcinoma of the hypercalcemic type (SCCHT). The dense cellularity (imparting a low power "blue" appearance), scant cytoplasm of the cells, and sometimes brisk mitotic activity as well as the presence of follicles in AGCT may lead to the misdiagnosis of SCCHT. However, the presence of grooved nuclei, lack of cytologic atypia, fibrothecomatous background, positivity for inhibin, and usual stage I at presentation help to separate AGCT from the highly aggressive SCCHT [157, 158]. Undifferentiated carcinoma may mimic AGCT with diffuse growth pattern, but the presence of nuclear hyperchromatism, cytologic atypia, brisk mitotic activity, and extra-ovarian tumor extension would favor the former. In addition, the malignant epithelial cells often stain for EMA but not inhibin. Metastatic or primary melanoma may be primarily composed of small cells that can form follicle-like spaces and exhibit nuclear grooves. The finding of extraovarian disease or a prior history, focal nevoid growth pattern, nuclear pseudoinclusions, melanin pigment, immunoreactivity for HMB45 and S100 will aid in establishing this diagnosis [159–161]. Metastatic breast carcinoma, especially the lobular variant, may have a diffuse growth pattern, mimicking AGCT. However, there is typically bilateral ovarian involvement and the carcinoma cells lack grooved nuclei but may contain intracytoplasmic vacuoles with mucin [162]. The tumor cells stain for GCDFP-15 and EMA but not inhibin [163].

Treatment and Clinical Outcome

All AGCTs have a malignant potential and may develop extra-ovarian extension or pelvic/lower abdominal recurrences. Distant metastases are rare but have been reported even 30 years following the initial surgery [164, 165]. Tumor stage is the most important prognostic indicator. Patients with stage I tumors have a much better 10-year survival rate than those with advanced stage disease (86% versus 49% in one series) [166]; however, the 25-year survival rate is significantly lower (40–60%) [167]. Most patients with AGCT have stage I disease. In these patients, younger age [168], tumor size (>5 cm) [135], mitotic activity (>5 mitoses/10HPF) [169, 170], lymphovascular invasion [171], tumor rupture [166], and aneuploidy [172] have been associated with worse prognosis in some studies, but not supported by others.

Total hysterectomy with bilateral salpingo-oophorectomy is the optimal treatment in peri- or postmenopausal women. In younger patients wishing to preserve fertility, unilateral salpingo-oophorectomy may be considered if there is no extra-ovarian spread or involvement of the contralateral ovary. Patients with tumor recurrences may benefit from secondary cytoreductive surgery, chemotherapy or hormonal therapy (including aromatase inhibitors), and/or radiotherapy [173, 174]. Because of the propensity of AGCTs to recur years after the initial diagnoses [130, 165], prolonged follow-up with serum inhibin to monitor for recurrences is advocated [175].

Juvenile Granulosa Cell Tumor

Prevalence and Clinical Presentation

Juvenile granulosa cell tumor (JGCT) accounts for 5–15% of all granulosa cell tumors and occurs almost exclusively (97%) within the first three decades of life [176]. Prepubertal patients commonly present with isosexual pseudoprecocity secondary to estrogen production [177–179] but rarely may exhibit features of virilization [180, 181]. In postpubertal patients, abdominal pain, distension, pelvic mass, and menstrual irregularities or amenorrhea are common presentations [176]. Less than 10% of patients (in particular, pregnant women) may present with an acute abdomen secondary to tumor rupture and intra-abdominal hemorrhage [182]. Association with Maffuci syndrome (enchondromatosis and hemangiomas) [176] and Ollier disease (enchondromatosis) [176, 183] has been reported.

Macroscopic and Microscopic Features

The vast majority of JGCTs are unilateral with only 2% being bilateral [176]. Tumors have a mean diameter of 12.5 cm but may range in size from 3 to 32 cm. Most neoplasms have a solid and cystic cut surface but some may be uniformly solid or cystic (comprising of unilocular or multilocular thin-walled cysts) (Fig. 14.40) [182, 184]. The solid areas have a tan yellow or gray color with occasional areas of necrosis and/or hemorrhage and the cysts usually contain serous or hemorrhagic contents [176].

Microscopic examination shows a nodular or diffuse proliferation of granulosa cells embedded in a myxoid or edematous stroma. A characteristic feature is the presence of follicle-like spaces of varying shapes and sizes (in contrast to AGCT) that contain eosinophilic or basophilic mucicarmine positive secretions (Fig. 14.41) [176]. The neoplastic granulosa cells have round, non-grooved, hyperchromatic nuclei, and abundant eosinophilic or vacuolated cytoplasm (Fig. 14.42). Occasional hobnail-type cells are present [176]. A variable degree of cytologic atypia including bizarre forms is often seen and can be striking (Fig. 14.43) [98]. Mitotic rate is also variable (ranging from <1 to 32/10 HPF) but generally brisk and atypical mitotic figures [185] as well as variable numbers of apoptotic cells can be identified. Call–Exner bodies are rarely seen. Other uncommon findings include a pseudopapillary pattern (Fig. 14.44), minor admixed foci of AGCT, hyaline globules, hyaline bands [98, 176]. In addition, theca cells that have a more spindled appearance and hyperchromatic nuclei may be seen surrounding or intermixed with the granulosa cells. Both the granulosa and theca cells may be luteinized [177]. During gestation, prominent edema, increased numbers of luteinized cells, and a

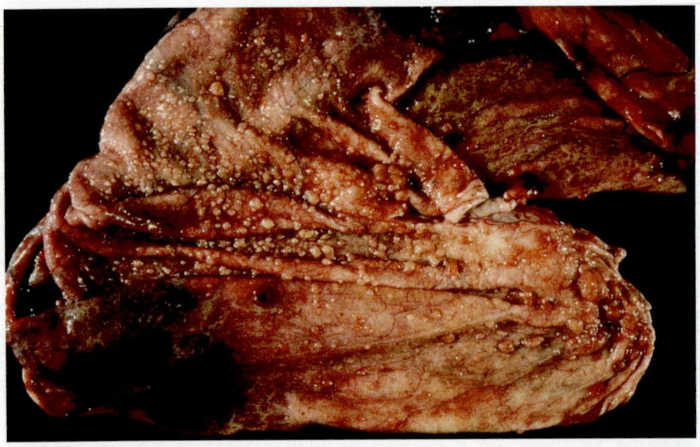

Fig. 14.40 Juvenile granulosa cell tumor. A unilocular cyst shows multiple small nodular excrescences

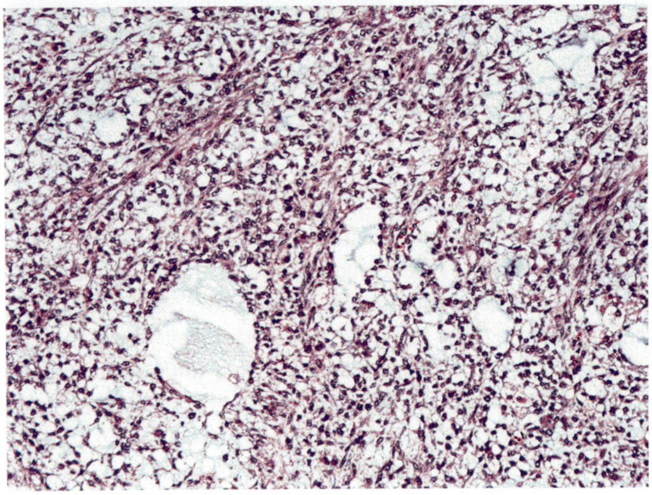

Fig. 14.41 Juvenile granulosa cell tumor. The tumor has a striking myxoid stroma and follicles of different sites are seen

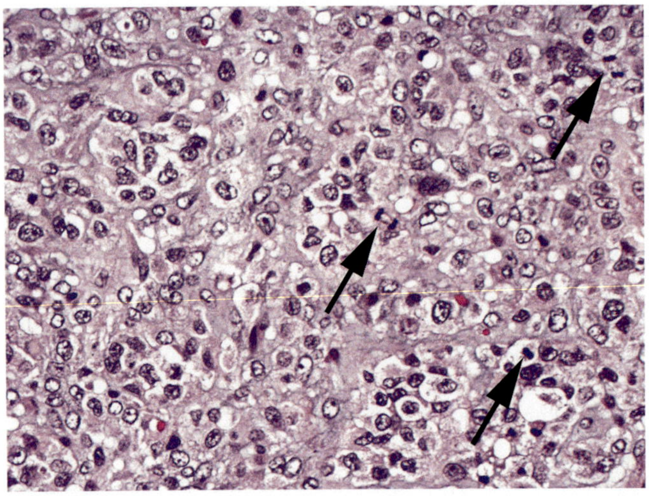

Fig. 14.42 Juvenile granulosa cell tumor. Granulosa cells with round, non-grooved nuclei and abundant eosinophilic or vacuolated cytoplasm are embedded in a myxoid stroma. Note brisk mitotic activity (*arrows*)

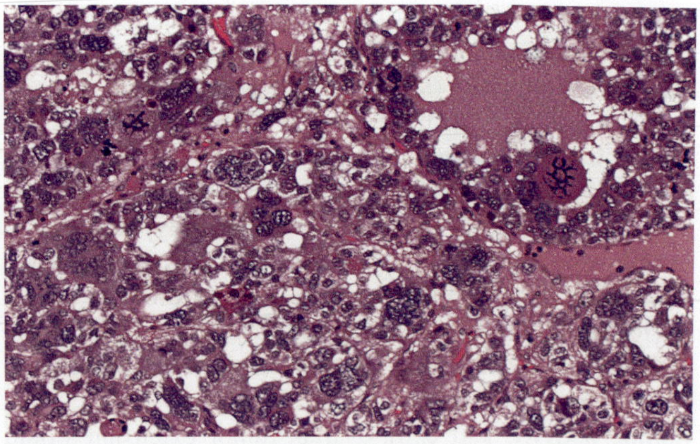

Fig. 14.43 Juvenile granulosa cell tumor. The cells show marked cytologic atypia as well as atypical mitoses

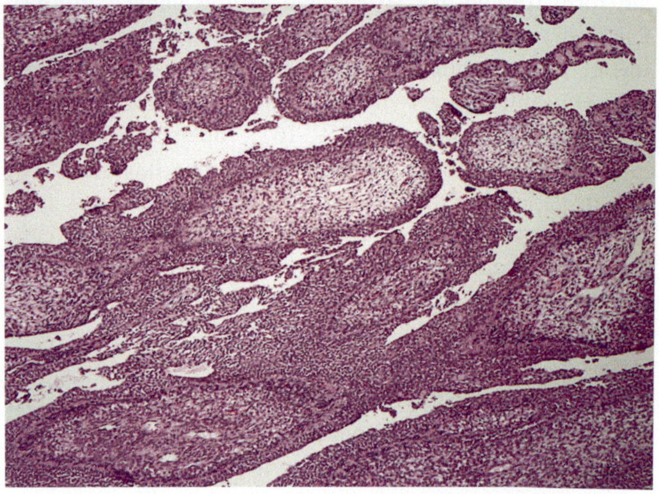

Fig. 14.44 Juvenile granulosa cell tumor. A striking pseudopapillary architecture is seen simulating at low power a transitional cell carcinoma

haphazard arrangement of the tumor cells may be appreciated [98, 176].

Immunohistochemical Profile

Reticulin stain shows fibrils surrounding groups of granulosa cells but investing individual theca cells. The tumor cells stain for inhibin [75, 142, 144, 186] and calretinin [76, 79, 185] and often express vimentin [186], WT-1 [79], CD56 [187], CD99 [144, 185], CD10 [12], SMA [188], S100, cytokeratin [186], and, rarely, EMA [79].

Electron Microscopy

The granulosa cells are interconnected by desmosomes as well as poorly formed intercellular junctions and are surrounded by basal lamina. The cells contain numerous rough endoplasmic reticulum, free ribosomes, and microfilaments as well as variable numbers of lipid droplets [176, 189–191].

Genotype

FOXL2 mutation, more commonly associated with AGCTs, has been reported in some JGCTs, especially those with a more aggressive clinical course [104, 192, 193].

Differential Diagnosis

JGCT has a broad differential diagnosis. Among the sex cord stromal category of neoplasms, JGCT is most commonly confused with AGCT. Both JGCT and AGCT often contain follicle-like spaces, but in JGCT, these tend to be more irregular in shape and variable in size with luminal basophilic mucicarminophilic contents [135, 194]. In addition, the cells of JGCT are typically set in a myxoid background, have more abundant cytoplasm with hyperchromatic, often nongrooved nuclei, and Call–Exner bodies are lacking. It is important to note that some JGCT may contain a small component of AGCT.

JGCT may mimic thecoma because both contain luteinized cells and often show a nodular growth. However, thecomas are uncommon in women below 30 years of age and lack the follicle-like spaces as well as conspicuous mitotic activity of JGCT. In difficult cases, a reticulin stain would be helpful as it highlights fibrils surrounding groups of granulosa cells in JGCT whereas a prominent pericellular fibrillary network is a feature of thecoma [130]. A JGCT with diffuse luteinized areas may superficially resemble a steroid cell tumor [105] but other more diagnostic foci would usually be present within the same tumor.

Like JGCT, small cell carcinoma of the hypercalcemic type also occurs in young patients, and follicle-like spaces as well as brisk mitotic activity are features of both tumors [158]. Furthermore, rare JGCTs have been reported to be associated with raised serum calcium levels [195, 196]. In contrast, small cell carcinoma of the hypercalcemic type is not associated with estrogenic manifestations and often shows extra-ovarian spread at the time of presentation. The tumor cells have scant cytoplasm and usually grow in sheets, cords, or nests. They exhibit a higher mitotic rate and stain for EMA but are negative for inhibin [157]. Follicle-like spaces are less common, and although large cells may be present, they should not be confused with theca cells as they have more abundant cytoplasm, sometimes with a rhabdoid appearance, and the nuclei are vesicular with very prominent nucleoli.

Some JGCTs may have hobnail-type cells or cells with abundant clear cytoplasm lining follicle-like spaces simulating clear cell carcinoma with a tubulocystic growth pattern. In contrast to JGCT, clear cell carcinoma occurs in an older cohort of women, is frequently associated with endometriosis or an adenofibromatous background, and stains for EMA and CK7 but not for inhibin and calretinin [197, 198]. JGCTs displaying high-grade cytologic atypia may raise the differential diagnosis with an undifferentiated carcinoma [199] while some JGCTs with pseudopapillary growth may resemble transitional cell carcinomas [151, 152]. Both types of carcinomas typically occur in older women, show other epithelial components either in the form of a low-grade carcinoma in dedifferentiated carcinoma or a serous carcinoma in transitional cell carcinoma, and are inhibin negative.

Yolk sac tumors are considered in the differential diagnosis of JGCTs as they also occur in young patients, and both tumors may have diffuse and microcystic architecture, myxomatous stroma, and brisk mitotic activity. However, the cytologic features of a yolk sac tumor are generally more primitive and the presence of Schiller Duval bodies as well as other germ cell components are helpful in establishing this diagnosis [20]. Rarely, primary or metastatic melanoma involving the ovaries may contain follicle-like spaces, mimicking JGCTs [159, 161]. A previous clinical history, bilateral ovarian involvement (in metastatic tumors), presence of melanin pigment, and immunoreactivity for melanoma markers help to distinguish these tumors from JGCTs. It should be noted, however, that focal immunoreactivity for inhibin has been reported in some melanomas involving the ovary [159].

Treatment and Clinical Outcome

The prognosis of patients with JGCT is stage dependent. The survival rate for patients with stage I tumors is approximately 97% whereas those with advanced stage tumors usually succumb to the disease [176, 201]. The presence of cytologic atypia or high mitotic rate does not affect prognosis in stage I tumors [186]. Unilateral salpingo-oophorectomy with staging should be considered in young patients with unilateral disease wishing to preserve fertility [200], whereas total hysterectomy with bilateral salpingo-oophorectomy is the treatment of choice in patients with advanced stage disease [202]. Adjuvant chemotherapy is also a consideration in this latter group as well as in patients with recurrent tumor [202, 203]. Unlike AGCT, recurrences of JGCT usually occur earlier (within 3 years of diagnosis) [185]. Serum inhibin levels can be used to monitor for tumor recurrences [204].

Sertoli-Leydig Cell Tumor

Prevalence and Clinical Presentation

Sertoli-Leydig cell tumors (SLCTs) are rare neoplasms constituting <0.5% of all ovarian tumors [205–207]. They can occur in patients of all age groups but the average age of presentation is 25 years with well-differentiated tumors occurring in slightly older (mean, 35 years) [208] and retiform variants in younger women (mean, 15 years) [206, 209]. Up to one-third of patients have androgenic manifestations, this being an infrequent feature of patients with

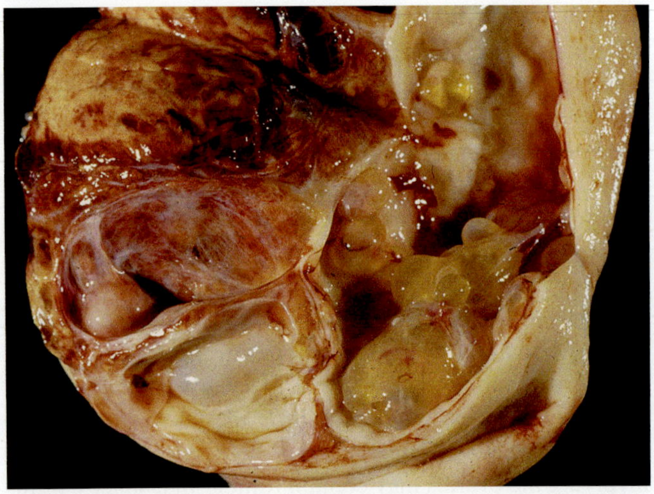

Fig. 14.45 Sertoli-Leydig cell tumor, intermediate differentiation. The tumor shows alternating solid and cystic areas. The solid component has a tan surface with focal hemorrhage, while the cysts contain clear fluid

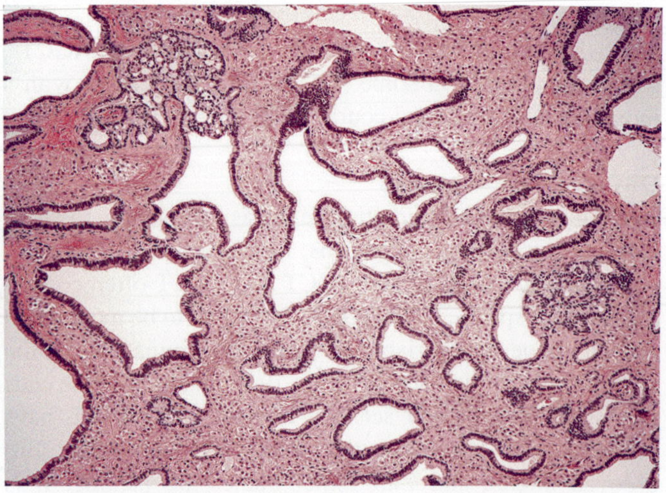

Fig. 14.46 Well-differentiated Sertoli-Leydig cell tumor. Tubules of varying sizes lined by Sertoli cells are separated by fibromatous stroma containing Leydig cells

tumors with a retiform component or containing heterologous elements [206]. Less commonly, neoplasms may be associated with hyperestrogenism [206], raised serum CA-125 [210], or raised alpha-fetoprotein levels [176, 211, 212]. Most patients, however, do not have hormonal manifestations but present with abdominal distension or discomfort.

Macroscopic and Microscopic Features

The vast majority (98%) of SLCTs are unilateral. Most tumors range in size from 5 to 20 cm, although they may vary from microscopic to very large, with the latter being a more common feature of poorly differentiated tumors [213]. On gross examination, SLCTs typically have solid, lobulated, and yellow cut surfaces (Fig. 14.45), but some tumors, especially when retiform or containing heterologous elements, may be cystic [207, 214, 215]. The cysts in retiform tumors can contain edematous or gelatinous papillary or polypoid projections, simulating a hydatidiform mole or a serous borderline tumor [209, 216]. Areas of hemorrhage and necrosis may be present and are more common in poorly differentiated tumors.

Histologically, SLCTs exhibit a wide range of morphologies and are classified into five main categories: (1) well-differentiated, (2) intermediate-differentiated, (3) poorly differentiated, (4) retiform, and (5) tumors with heterologous elements. Well-differentiated SLCTs are well circumscribed and are composed of lobules of hollow or solid tubules separated by fibromatous stroma (Fig. 14.46). The tubules (some of which may contain luminal eosinophilic secretions) are lined by cuboidal to columnar Sertoli cells with round to oval basally placed nuclei, small nucleoli, and abundant eosinophilic or pale vacuolated cytoplasm. Some

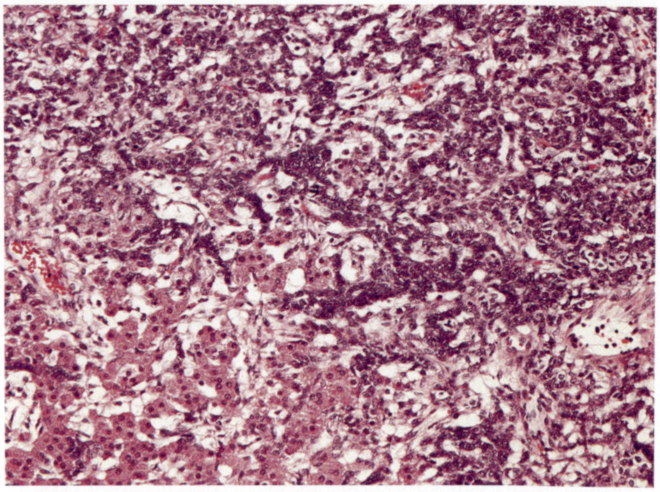

Fig. 14.47 Intermediate-differentiated Sertoli-Leydig cell tumor. Cellular lobules containing Sertoli cells are separated by edematous stroma mostly associated with Leydig cells

tubules may be composed of a mixture of Sertoli and Leydig cells although the latter tend to be more conspicuous in the fibromatous stroma. The Leydig cells, which are seen singly or in groups, have variable intracytoplasmic lipid contents and some may also contain lipochrome pigment or Reinke crystals [206]. Cytologic atypia and mitotic activity are rare. Retiform or heterologous elements are typically absent but a tumor with extensive ossification and calcification has been reported [217].

Tumors of intermediate differentiation also have a striking lobular architecture with cellular areas separated by edematous or much less frequently fibromatous stroma (Fig. 14.47). The cellular areas are composed of Sertoli cells arranged in nests, cords, columns, solid or poorly formed tubules or diffuse sheets. Infrequently, cysts containing

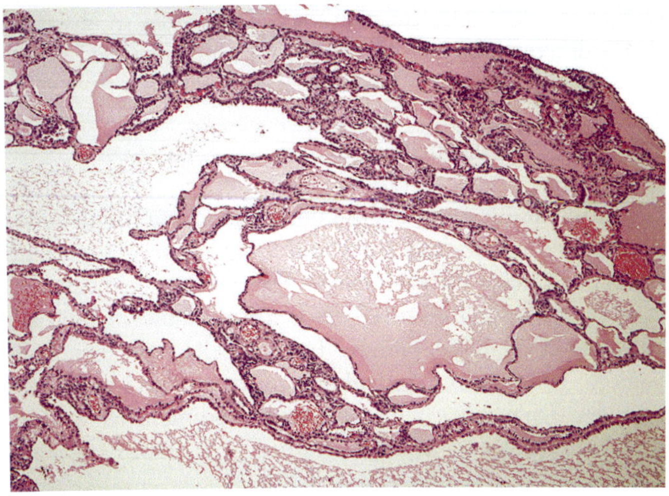

Fig. 14.48 Sertoli-Leydig cell tumor. Multiple cysts containing pale eosinophilic secretions simulate a struma ovarii

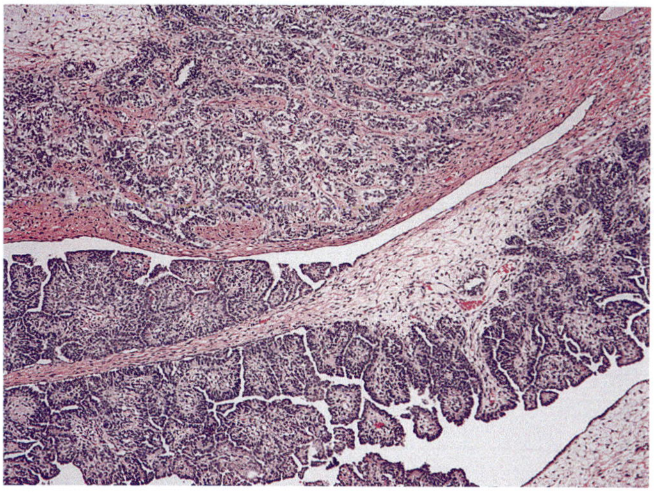

Fig. 14.50 Retiform Sertoli-Leydig cell tumor. Prominent papillary and retiform architecture is seen

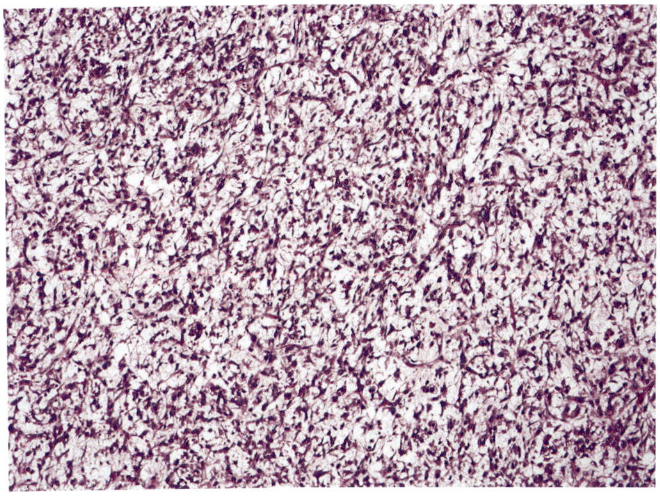

Fig. 14.49 Sertoli-Leydig cell tumor. Prominent stromal edema distorts the typical architecture, as is typically seen in pregnancy and postpartum

eosinophilic secretions as well as follicle-like spaces may be present (Fig. 14.48). The Sertoli cells appear immature with high nucleocytoplasmic ratio, small round to oval nuclei, and inconspicuous nucleoli. Leydig cells with ample eosinophilic cytoplasm can be found in between or surrounding the nodules of Sertoli cells but are more frequently encountered within the edematous stroma [207]. Both cell types may rarely display bizarre nuclear features [98]. Prominent stromal edema with distortion of the overall architecture (Fig. 14.49) and large aggregates of Leydig cells are seen in tumors encountered during pregnancy [208].

Poorly differentiated SLCTs are characterized by a diffuse or spindle (sarcomatoid) proliferation of poorly differentiated Sertoli cells showing hyperchromatic nuclei and scant cytoplasm often associated with brisk mitotic activity (>10/10HPF). Leydig cells are typically inconspicuous [206]. The histologic appearance may be indistinguishable from that of a pure sarcoma if well or intermediate differentiated areas are not seen.

A retiform component is found in 15% of SLCTs and is the predominant pattern in half of these neoplasms [209, 216, 218]. Microscopically, there is a network of irregularly branching narrow slit-like tubules and cysts with papillary or polypoid projections (Fig. 14.50) [216]. The cysts and tubules may contain eosinophilic secretions and are lined by cuboidal cells with scant cytoplasm and round to oval nuclei, often exhibiting varying degrees of cytologic atypia. Papillary and polypoid formations may be: (1) small and round or blunt with hyalinized cores (most common), (2) large and bulbous with edematous cores, or (3) delicate and branching and lined by stratified cells reminiscent of the papillary projections seen in serous borderline tumors. The surrounding stroma can by edematous, hyalinized, or densely cellular and immature [209].

Heterologous elements are found in 20% of SLCTs, most commonly in association with tumors of intermediate differentiation but can also be found in poorly differentiated or retiform subtypes [213, 215, 219, 220]. Glandular structures lined by gastrointestinal-type mucinous epithelium are the most common heterologous elements (Figs. 14.51 and 14.52) [215, 221]. The mucinous epithelium usually has a benign morphology but may exhibit cytologic features of a borderline or malignant tumor [207, 222]. Foci of insular (Fig. 14.53) or goblet cell carcinoid may also be identified [215, 223]. Much less commonly, especially in tumors with a sarcomatoid growth, heterologous stromal elements may be encountered, usually in the form of immature cartilage [215, 223, 224] or skeletal muscle [213, 219]. Rare tumors with hepatocytic differentiation [225, 226] or foci of neuroblastoma [213] have also been reported.

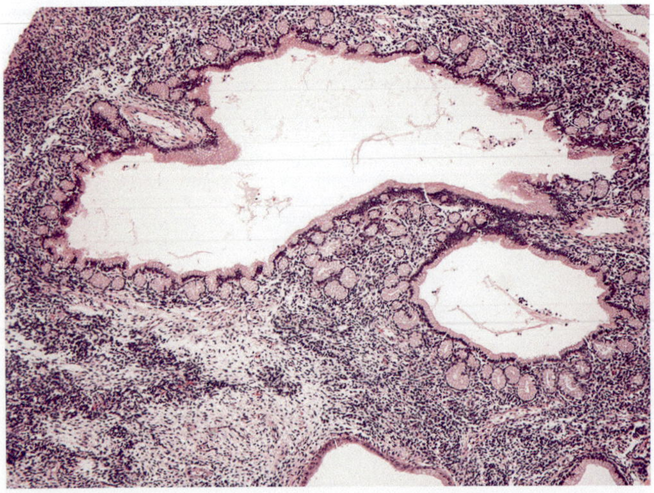

Fig. 14.51 Sertoli-Leydig cell tumor with heterologous elements. Glands lined by benign gastrointestinal-type epithelium are present within a Sertoli-Leydig cell tumor of intermediate differentiation

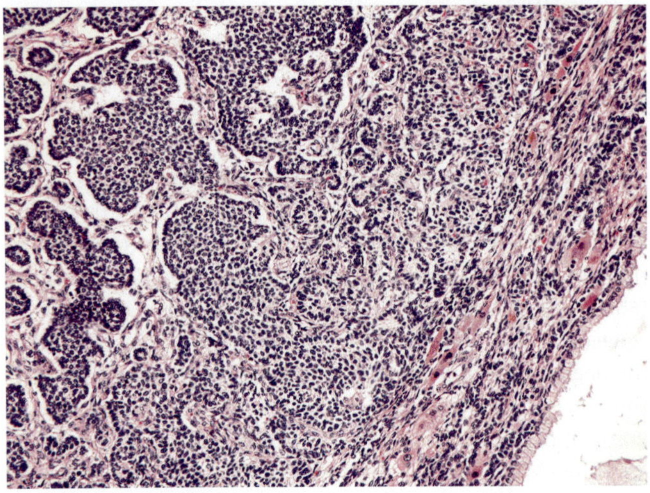

Fig. 14.52 Sertoli-Leydig cell tumor with heterologous elements. A focus of insular carcinoid (*left*) is seen adjacent to Sertoli tubules and mucinous epithelium

Although grading of SLCT is based on the degree of tubular differentiation of the Sertoli cell component and the quantity of the primitive gonadal stroma, no specific guidelines have been reported with regards to assignment of grade in these tumors; whether it should be based on the presence of a less differentiated component and on the amount of this component.

Immunohistochemical Profile

Sertoli-Leydig cell tumors commonly express inhibin [77, 142, 144], calretinin (often stronger in Leydig cells) [227], CD99 [77, 79, 144, 228], Melan-A [229], CD56, CD10 [12], WT-1 [79, 228], cytokeratin [79], and vimentin. Immunophenotypic expression of sex cord differentiation may be minimal or absent in poorly differentiated tumors. EMA is typically negative [79].

Electron Microscopy

Ultrastructurally, Sertoli cells contain abundant rough endoplasmic reticulum and lipid and may show cytoplasmic Charcot-Böttcher microfilaments. The cells are connected by tight junctions and desmosomes to form tubules, which are surrounded by basal lamina. Leydig cells have abundant smooth endoplasmic reticulum, mitochondria, and lipid droplets. Lysosomes with lipochrome pigment as well as Reinke crystals may also be identified [220, 230, 231].

Differential Diagnosis

SLCTs and Sertoli cell tumors (SCT) show widely overlapping histologic features, but finding of Leydig cells establishes the diagnosis of SLCT. Some endometrioid carcinomas may display sex cord-like features, forming cords and tubules reminiscent of those seen in SLCTs [231, 232]. Furthermore, the stroma is frequently fibromatous and may contain luteinized stromal cells. On the other hand, SLCT may contain pseudoendometrioid tubules [233]. Finding more conventional areas of endometrioid carcinoma with foci of squamous and/or mucinous metaplasia or associated endometriosis and immunoreactivity for EMA but negative inhibin staining will support this diagnosis [145, 150]. Female adnexal tumor of probable Wolffian origin (FATWO) may be centered within the ovary and form pseudotubular structures, mimicking a SLCT [234]. However, it typically lacks Leydig cells, often displays other growth patterns, and contains small oval or spindle-shaped cells, which are not characteristic of SLCT. Trabecular carcinoid may be confused with SLCT, especially of intermediate differentiation. However, the trabeculae seen in carcinoids are typically longer and thicker and are composed of cells whose nuclei are oriented perpendicular to the main axis of the trabeculae. Furthermore, the stroma is less cellular and frequently extensively fibromatous in contrast to the edematous stroma seen in intermediate differentiated SLCT [154]. The findings of a characteristic insular growth, cells with neuroendocrine granules and a teratomatous component would also support a diagnosis of carcinoid [215]. Krukenberg tumor may mimic a SLCT clinically, as patients may also present with androgenic manifestations secondary to stromal luteinization. Histologically, both tumors may show alternating hypercellular and hypocellular (edematous) areas as well as tubular formation. However, Krukenberg tumors are typically bilateral and are often composed, at least focally, of signet ring cells with cytoplasmic mucin [235].

The differential diagnosis of SLCT with retiform component includes serous neoplasms (mostly borderline tumors) and yolk sac tumor. Resemblance to a serous borderline tumor may be imparted by the presence of papillary structures within cyst lumens. Rarely, a complex branching architecture associated with cellular stratification and atypia may suggest a serous adenocarcinoma. Finding typical areas of intermediate differentiation associated with Leydig cells as well as the absence of hierarchical branching, budding of neoplastic cells and ciliated cells would support a diagnosis of SLCT. Like retiform SLCT, yolk sac tumors occur in young patients and may exhibit a papillary growth with edematous hypocellular stroma. Both tumors can also be associated with raised serum alpha-fetoprotein. However, up to one-third of women with SLCTs present with symptoms of virilization, which is not a feature of yolk sac tumors. The presence of other characteristic histologic patterns as well as immunoreactivity for alpha-fetoprotein in yolk sac tumors are helpful to distinguish them from SLCT (although SLCT may occasionally show focal staining for alpha-fetoprotein) [212].

SLCTs with heterologous elements (especially when poorly differentiated), may show morphologic resemblance to a carcinosarcoma, but the latter typically occurs in postmenopausal women and contains malignant and often high-grade epithelial as well as mesenchymal components. Heterologous SLCTs are often confused with teratomas [219], but other common constituents of the latter, such as squamous and respiratory epithelium, skin appendages, and neuroectodermal tissue are not found in SLCTs. When SLCTs contain prominent amounts of mucinous epithelium, they may be confused with pure mucinous cystic tumors on gross and microscopic examination. Although a history of virilization is suggestive of SLCT, occasional mucinous tumors containing luteinized stromal cells may be associated with androgenic manifestations. The diagnosis of SLCT lies in the finding of a Sertoli-Leydig cell component, which is almost always of intermediate differentiation, between the glands and cysts or at the periphery of the tumor. Heterologous SLCTs with mesenchymal elements may be confused with sarcoma when recognizable Sertoli and Leydig cells are scarce. Before a diagnosis of pure ovarian sarcoma is established, particularly in a young woman, a heterologous SLCT should be excluded by thorough sampling.

Treatment and Clinical Outcome

The majority of SLCTs are benign with malignant behavior reported in <15% of patients. Important prognostic factors include stage, degree of differentiation [207], tumor rupture [206, 207], and the presence of retiform component [209] or heterologous elements [207, 213]. Poorly differentiated tumors as well as tumors with retiform or heterologous elements are usually associated with more aggressive behavior [210]. As the majority of patients have stage 1a tumors at presentation (approximately 80%), salpingo-oophorectomy is the treatment of choice [206]. More aggressive surgery and chemotherapy are indicated in advanced stage tumors. Adjuvant therapy may also be considered in stage I tumors that are poorly differentiated or contain mesenchymal heterologous elements and for neoplasms of intermediated differentiation that have ruptured. In contrast to AGCT, which usually recur many years after treatment, SCLTs generally recur earlier and recurrences are often confined to the pelvis and abdomen [207].

Sertoli Cell Tumor

Prevalence and Clinical Presentation

Sertoli cell tumors are very rare, accounting for 4% of Sertoli stromal cell tumors. They typically occur in women of reproductive age (mean, 30 years) [236, 237] but have also been reported in children. Patients can present with abdominal pain, swelling or menstrual irregularities, although not infrequently, they are asymptomatic. Most Sertoli cell tumors are non-functioning; however 30% of patients may exhibit hyperestrogenic manifestations, and lipid-rich variants have been reported to cause isosexual pseudoprecocity in children [238, 239]. Less commonly, these tumors may cause androgenic manifestations. Association with progesterone [240], aldosterone, or renin [241, 242] production, and with the Peutz-Jeghers syndrome has been documented [239, 243].

Macroscopic and Microscopic Features

Sertoli cell tumors are unilateral and average 9.0 cm in diameter, with the majority ranging from 4 to 12 cm in size [236]. They have a lobulated, solid, yellow to brown cut surface, but areas of cystic degeneration can be seen. Sertoli cell tumors are classically composed of solid (Fig. 14.53a) to hollow (Fig. 14.53b) tubules that may be rounded, elongated, or branched and are commonly arranged in lobules separated by delicate fibromatous septa or rarely prominent hyalinized stroma (sclerosing variant). Other less commonly encountered growth patterns include: cords, trabeculae, diffuse, pseudopapillary, retiform, nested or alveolar, and spindled. The tubules are lined by columnar to cuboidal cells with moderate, pale vacuolated to eosinophilic cytoplasm. If vacuolated cells or cells with prominent eosinophilic cytoplasm are present in abundance, the terms "lipid-rich" (Fig. 14.54) or "oxyphilic variant" (Fig. 14.55) (associated with the Peutz-Jeghers syndrome)

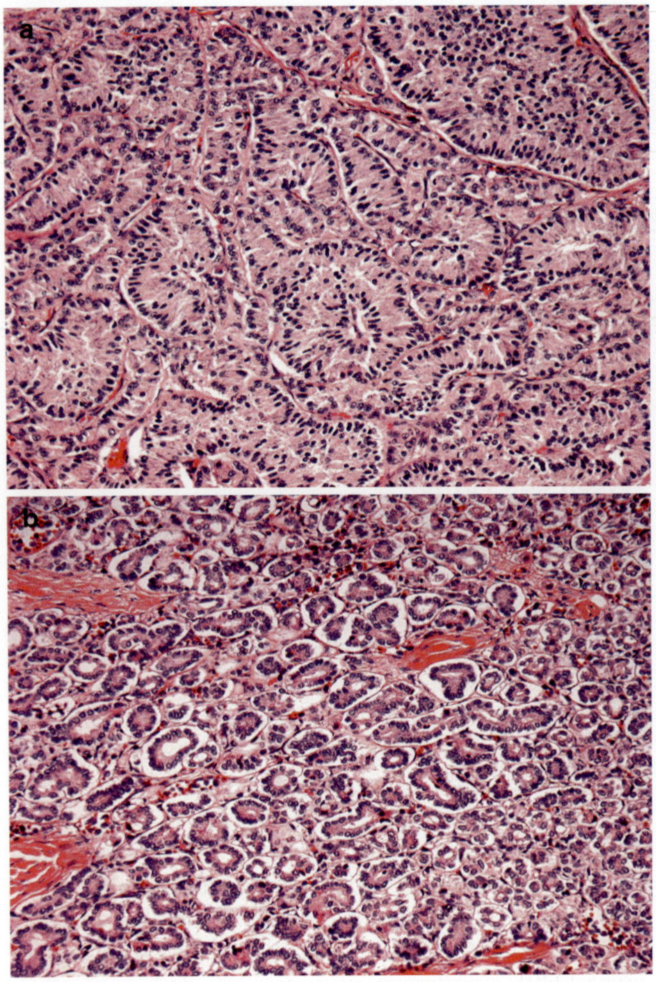

Fig. 14.53 Sertoli cell tumor. The cells form solid (**a**) and hollow tubules (**b**)

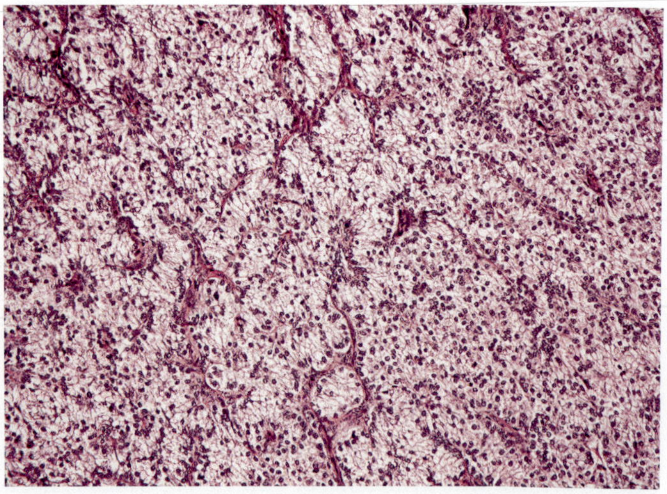

Fig. 14.54 Sertoli cell tumor, lipid-rich variant. The tumor is composed of cells with abundant pale vacuolated cytoplasm and small round nuclei

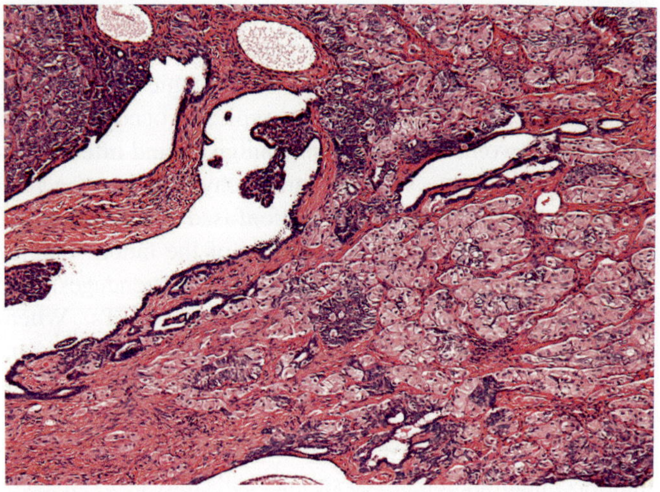

Fig. 14.55 Sertoli cell tumor. The cells form retiform structures (*upper*) and some have abundant eosinophilic cytoplasm (oxyphilic variant) (*bottom*)

are applied. Tubular lumens are usually empty but may contain eosinophilic secretions. These tumors usually exhibit minimal cytologic atypia and mitotic activity; however, cells with bizarre nuclei may be encountered [98]. Calcifications, necrosis, and hemorrhage may be seen. Very infrequently, a minor component of granulosa cell tumor or sex cord tumor with annular tumors as well as rare Leydig cells can also be identified [236].

Immunohistochemical Profile

The tumor cells stain for inhibin, calretinin, CD10, CD99, CD56, WT-1, vimentin, and keratin (AE1/3 and CAM5.2) [75, 142, 244–247]. CK7 is positive in approximately 10% of tumors [150] while focal EMA expression has been reported in rare tumors [146, 248]. Occasional SMA, NSE, chromogranin, and S100 expression have also been documented [236].

Electron Microscopy

By electron microscopy, basal lamina surrounds the tubules. The tumor cells have luminal microvilli, abundant rough endoplasmic reticulum and lipid, and they are connected by tight junctions and desmosomes [238, 249]. Some cells also contain Charcot-Böttcher microfilaments.

Differential Diagnosis

Tumors considered in the differential of Sertoli cell tumors include endometrioid carcinoma, female adnexal tumor of probable Wolffian origin (FATWO), carcinoid tumor, well-differentiated SLCT, and Krukenberg tumor.

Some endometrioid carcinomas may contain (a) cords and tubules in a fibromatous stroma reminiscent of those seen in Sertoli cell tumor (sertoliform endometrioid carcinoma) [231, 232, 250], (b) cells with prominent eosinophilic cytoplasm [251] mimicking the oxyphilic variant of Sertoli cell tumor, or (c) spindle cells [252] simulating a Sertoli cell tumor with spindled growth. However, conventional areas of endometrioid carcinoma, often with squamous or mucinous metaplasia, typically merge with the Sertoli-like areas and foci of endometriosis may be seen. Finally, in contrast to Sertoli cell tumors, endometrioid carcinomas (including the sertoliform variant) are CK7 and EMA positive but negative for inhibin [144, 150]. Although FATWOs are usually sited in the broad ligament, these tumors can be centered in the ovary and may be composed of pseudotubular structures lined by cells exhibiting some resemblance to those seen in Sertoli cell tumor. However, other histologic patterns including slit-like and solid areas composed of small oval or spindle-shaped cells are also found. FATWO may stain for inhibin, but the staining is usually weak and focal [75, 253]. Other markers including CD99, CD10, and calretinin are not useful in this differential diagnosis as both tumors may show variable positivity. Carcinoid tumor may rarely be considered in the differential diagnosis of a Sertoli cell tumor, as they also grow in acini, cords, and trabeculae. However, carcinoids have cells with stippled chromatin and neuroendocrine granules, an often prominent fibromatous background and may have other associated teratomatous components. Chromogranin is the most discriminatory neuroendocrine marker, as NSE, synaptophysin and CD56 are also frequently positive in Sertoli cell tumors. Conversely, carcinoid tumors generally do not express inhibin or calretinin. Although rare Leydig cells may be found in Sertoli cell tumors, the presence of these cells should raise a strong suspicion for SLCT. Some Krukenberg tumors may be confused with Sertoli cell tumors as they are associated with a tubular morphology, luteinized stroma, and estrogenic manifestations. However, the presence of an extra-ovarian primary tumor, bilateral ovarian involvement, extra-ovarian spread, and the presence of signet ring cells containing cytoplasmic mucin should aid in the distinction of these tumors from Sertoli cell tumors.

Treatment and Clinical Outcome

Most Sertoli cell tumors are clinically benign and confined to the ovary at the time of diagnosis and therefore are adequately treated by salpingo-oophorectomy. Histologic parameters associated with adverse outcome include nuclear atypia and mitotic activity (>5/10HPF) [236].

Sex Cord Tumor with Annular Tubules

Prevalence and Clinical Presentation

Approximately one-third of women with sex cord tumors with annular tubules (SCTAT) have Peutz-Jeghers syndrome (PJS), an autosomal dominant condition characterized by mucocutaneous pigmentation, gastrointestinal hamartomatous polyposis, and predisposition to benign and malignant tumors of the gastrointestinal tract, breast, ovary, uterine cervix, and testis [215, 254, 255]. In this setting, patients are young with an average age of 27 years, usually have no associated hormonal manifestations, and the tumors are often an incidental finding [215, 255, 256]. Patients without PJS are slightly older (mean, 34 years), almost always present with a palpable mass [215, 257, 258] and may have associated estrogenic or progestogenic manifestations [259].

Macroscopic and Microscopic Features

Ovaries in patients with PJS frequently show bilateral, multiple, small yellowish nodules, some of which may not be grossly visible [256]. In contrast, tumors found in patients without PJS are typically unilateral with a solid yellow cut surface and range from 0.5 to 33 cm in maximum dimension. Microscopically, SCTATs are characterized by the presence of simple or complex annular tubules (Fig. 14.56). Simple tubules are ring shaped and are composed of cells with nuclei oriented peripherally and around a central hyaline core containing basement membrane-like material, with an intervening anucleate cytoplasmic zone (Fig. 14.57). The complex tubules are rounded structures composed of intercommunicating rings surrounding multiple rounded hyaline cores [260]. In the PJS, tumorlets of varying sizes are scattered within the ovarian stroma. In addition, some of them may show foci of diffuse growth, islands of lipid-rich sex cord cells as well as extensive calcification [256]. SCTATs not associated with PJS may show prominent hyalinization of the tubules and stroma as well as foci of typical Sertoli cell or granulosa cell tumor with a microfollicular pattern [215]. Typically, the cells have tall pale cytoplasm and round to oval nuclei with tiny nucleoli, and in most cases mitotic activity is scant. Only tumors not associated with the PJS may show brisk mitotic activity as well as marked nuclear atypia [261].

Immunohistochemical Profile

These tumors typically express inhibin and calretinin [75, 76, 79]. They are also positive for CD56 [262], WT-1, and cytokeratin [79] but do not stain for CD10 [12, 79], EMA [79], or Melan-A [263, 264].

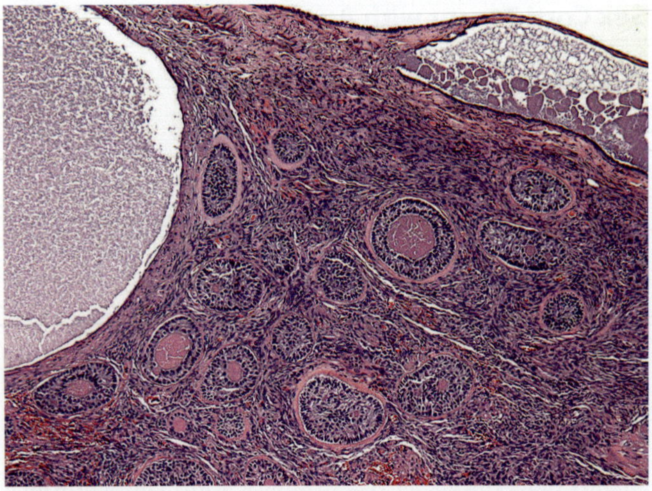

Fig. 14.56 Sex cord tumor with annular tubules. Multiple small nodules are composed of simple and complex annular tubules

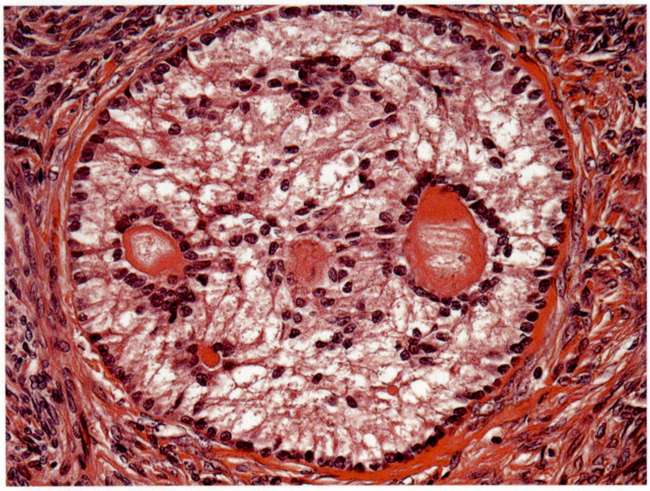

Fig. 14.57 Sex cord tumor with annular tubules. The cells have nuclei oriented peripherally and around a central hyaline core with an intervening anucleate, cytoplasmic zone

Electron Microscopy

Ultrastructurally, basal lamina is found surrounding the tubules and within them. Rarely, cells may contain Charcot-Böttcher crystals [265], raising the possibility of Sertoli cell differentiation, but they also may display ultrastructural features of granulosa cells [266], which has generated some controversy about the potential origin of these tumors.

Genotype

Germline mutations of the PJS gene (STK11/LKB1) at chromosome 19p13.3 have been reported in tumors of PJS patients, but similar somatic mutations have not been found in sporadic cases [267, 268].

Differential Diagnosis

The differential diagnosis includes gonadoblastoma as like SCTAT, this tumor also occurs in young women, has tubules formed by cells with polarized nuclei, and is often associated with calcifications that may be extensive. In contrast to SCTAT, gonadoblastomas typically occur in the setting of gonadal dysgenesis, contain tubules with a dual cell population (germ cells and Sertoli cells), and in some cases, may be associated with a second germ cell tumor, typically a dysgerminoma [269]. Sertoli cell tumor [238] and granulosa cell tumor [270] are also considered in the differential diagnosis, as both may show areas that overlap with a typical SCTAT. However, the presence of more conventional areas in these tumors will help to establish the correct diagnosis.

Treatment and Clinical Outcome

Surgery is the mainstay of treatment for these tumors. When they occur in the setting of PJS, patients typically follow a benign clinical course, although exceptionally rare malignant tumors have been reported [271, 272]. Approximately 20% of SCTATs not associated with PJS behave in a malignant fashion, with a propensity for lymphatic spread and late recurrences. Müllerian-inhibiting substance, alpha-inhibin, and progesterone may serve as useful serum markers to monitor for tumor recurrence [273, 274].

Stromal Luteoma

Prevalence and Clinical Presentation

Stromal luteomas account for 20% of all steroid cell tumors. They are presumed to arise from the ovarian stroma as most tumors lie within the ovarian parenchyma and 90% are associated with stromal hyperthecosis, either within the same or contralateral ovary [5, 275, 276]. Eighty percent of tumors occur in postmenopausal women [5] in whom the most common clinical presentation is vaginal bleeding (60%), secondary to hyperestrinism. In contrast to other steroid cell tumors that typically display a more androgenic profile, androgenic manifestations are seen in only 12% of stromal luteomas [5].

Macroscopic and Microscopic Features

The vast majority of tumors are unilateral and measure <3.0 cm in diameter. They are well demarcated with a gray-white to yellow solid cut surface but a third of tumors may have red or brown areas. Microscopically, a rounded nodule of lutein cells with small round nuclei, prominent nucleoli,

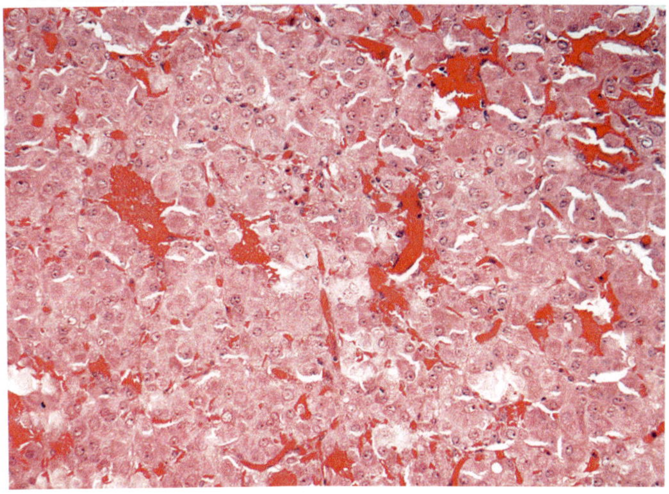

Fig. 14.58 Stromal luteoma. Lutein cells have small round nuclei, prominent nucleoli, and eosinophilic cytoplasm. They grow in tight nests and are separated by dilated pseudovascular spaces containing red blood cells

and eosinophilic cytoplasm is seen (Fig. 14.58). The cells may contain cytoplasmic lipochrome pigment, but there is usually a paucity of lipid content. The tumor cells may be arranged diffusely, in small nests or cords, and are typically surrounded by ovarian stroma, which may occasionally be hyalinized. Mitotic activity is rare. Degenerative changes with formation of pseudoglandular/pseudovascular spaces that may contain lipid-laden cells and chronic inflammatory cells occur in one-fifth of neoplasms [5].

Immunohistochemical Profile

The tumor cells express inhibin, calretinin, Melan-A, and vimentin, and may show focal immunoreactivity for keratin [75, 264, 277].

Differential Diagnosis

In some cases of stromal hyperthecosis, nests of lutein cells may form nodules, mimicking stromal luteoma. However, these nodules are usually multiple and measure <0.5 cm in diameter [2]. Like stromal luteoma, Leydig cell tumor of the non-hilar type and luteinized thecoma also occur in patients of the same age group. However, patients with Leydig cell tumors usually present with androgenic manifestations and tumor cells containing Reinke crystals are identified on microscopic examination, while a fibromatous stroma, lacking in stromal luteoma, is a feature of luteinized thecomas. Another differential to consider is pregnancy luteoma that may appear histologically similar to stromal luteoma. In contrast to the latter, these lesions occur in younger, pregnant women and are typically multifocal [34].

Treatment and Clinical Outcome

Stromal luteomas are benign neoplasms, warranting only surgical excision of the tumor or unilateral oophorectomy.

Leydig Cell Tumor

Prevalence and Clinical Presentation

Leydig cell tumors account for 20% of all steroid tumors and are characterized by the presence of cytoplasmic Reinke crystals, identified either on light or electron microscopy [278]. This finding is unique to Leydig cells and distinguishes these cells from lutein and cortical adrenal cells. Leydig cell tumors are divided into two subtypes [279]: hilus cell tumor and Leydig cell tumor non-hilar type, the latter being a rare entity. Hilus cell tumor arises from hilar Leydig cells and is located within the ovarian hilum whereas Leydig cell tumors of the non-hilar type are thought to originate from ovarian stromal cells [280] and found within the ovarian stroma, away from the hilum. These neoplasms commonly occur in postmenopausal women (mean, 58 years) who often present with androgenic manifestations secondary to testosterone production by the tumor cells [61, 281]. Estrogenic manifestations are rarely encountered [282].

Macroscopic and Microscopic Features

The tumors typically are unilateral, small (mean diameter of 2.4 cm), and circumscribed with a solid red-brown to yellow cut surface [61]. Rarely, bilateral tumors have been reported [283, 284]. Microscopically, these neoplasms are unencapsulated but well circumscribed showing a diffuse or lobular arrangement of the tumor cells separated by hyalinized or edematous stroma (Fig. 14.59). Clustering of cells separated by nucleus-free eosinophilic zones is a characteristic but not a universal feature [61] (Fig. 14.60). The tumor cells have rounded nuclei, small nucleoli, and moderate to abundant eosinophilic or vacuolated (lipid rich) cytoplasm that may contain Reinke crystals (refractile, cylindrical, or rhomboid structures) and lipochrome pigment. Some degree of cytologic atypia, intranuclear pseudoinclusions, bizarre nuclei, and multinucleated cells may be seen [61]. Mitotic figures are rare. One-third of tumors may show fibrinoid replacement of blood vessel walls and degenerative spaces (similar to that seen in stromal luteomas) can be found. In the absence of Reinke crystals, a diagnosis of Leydig cell tumor may still be favored if the neoplasm is located in the hilum around non-medullated nerve fibers, is associated with hilus cell hyperplasia, and shows fibrinoid necrosis of vessel walls as well as nuclear clustering.

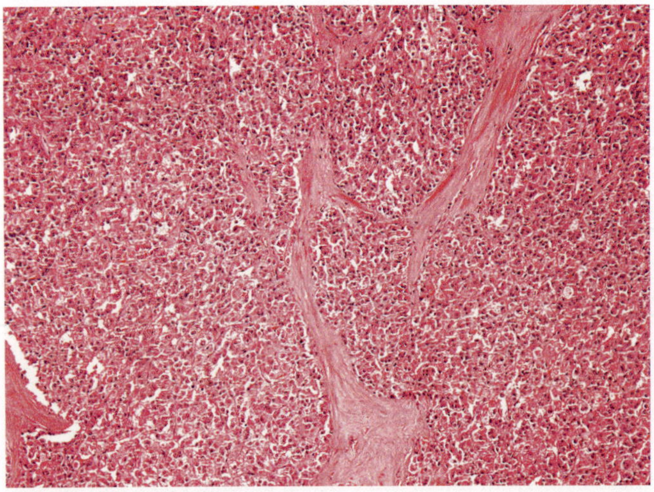

Fig. 14.59 Leydig cell tumor. The cells are arranged in lobules separated by hyalinized stroma

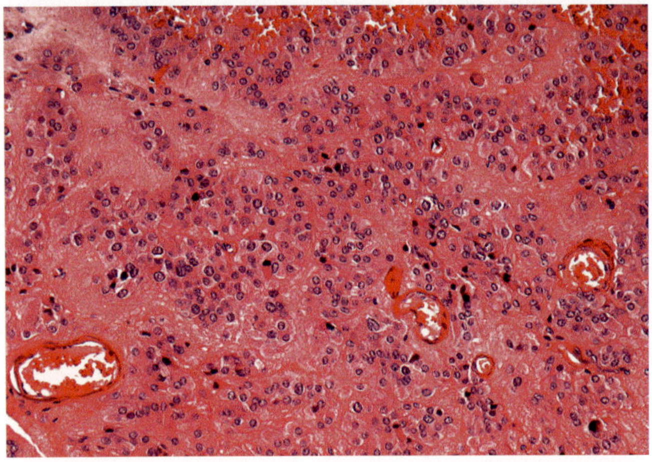

Fig. 14.60 Leydig cell tumor. Clustering of tumor cell nuclei with formation of anucleate zones is a typical but not constant feature of this tumor

Immunohistochemical Profile

The tumor cells stain for vimentin, inhibin, calretinin, CD56 [262], CD10 [13], Melan-A (Fig. 14.61), and galectin-3 [285], and some tumors may also show immunoreactivity for actin and keratin.

Electron Microscopy

Ultrastructural examination reveals abundant smooth endoplasmic reticulum, mitochondria, lysozymes containing lipochrome pigment, and lipid in the tumor cells. Reinke crystals appear rod shaped when cut longitudinally and hexagonal in cross section with a cross-hatched interior. Intracytoplasmic eosinophilic spheres may also be present [278].

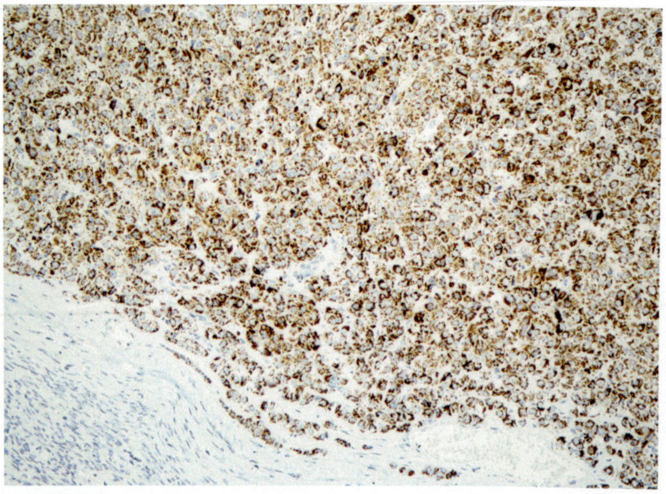

Fig. 14.61 Leydig cell tumor. The tumor cells show immunoreactivity for Melan-A

Differential Diagnosis

Leydig cell tumors, especially when present in the ovarian cortex, should be distinguished from other tumors containing cells with abundant eosinophilic cytoplasm, including luteinized granulosa cell tumor [137], stromal luteoma [5], and less likely, luteinized thecoma, stromal Leydig cell tumor (composed of intersecting fascicles of spindle cells associated with a collagenous stroma containing aggregates of luteinized cells with Reinke crystals) [80], and pregnancy luteoma [34]. The presence of Reinke crystals is diagnostic of a non-hilar Leydig cell tumor. In the hilum, the differential diagnosis includes Leydig cell hyperplasia, which does not form a tumorous mass and adrenal cortical adenoma that rarely may occur in the hilum. The latter is composed of clear cells containing abundant small vacuoles, but it is not associated with hilar cell hyperplasia and lacks Reinke crystals as well as other characteristic features of Leydig cell tumors [286].

Treatment and Clinical Outcome

The mainstay of therapy is surgery as no definite malignant tumors have been reported in literature. GnRH analogues may be considered for symptomatic relief in patients with hormonal manifestations [287].

Steroid Cell Tumor, NOS

Prevalence and Clinical Presentation

Steroid cell tumors, NOS account for 60% of all steroid cell tumors and are comprised of tumors that lack diagnostic

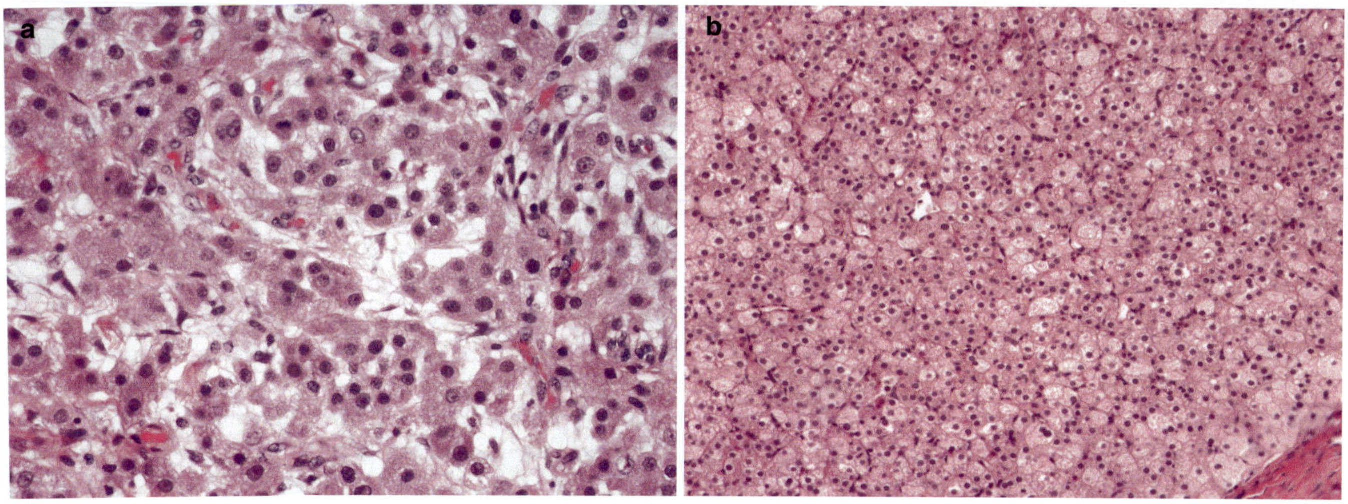

Fig. 14.62 Steroid cell tumor, not otherwise specified. Cells grow in sheets and have round to polygonal ample eosinophilic (**a**) or clear vacuolated (**b**) cytoplasm, and centrally placed nuclei with small prominent nucleoli

features of stromal luteoma or Leydig cell tumor. Even though they can occur at any age, these tumors are usually found in younger patients (mean, 43 years) [106] compared to other steroid cell tumors. Fifty percent of patients present with androgenic manifestations while estrogenic changes are only noted in 10%. Rarely, tumors may be associated with cortisol (with or without concurrent Cushing's syndrome) [288, 289] or aldosterone production [290], hypercalcemia, erythrocytosis, and ascites [105].

Macroscopic and Microscopic Features

Ninety-five percent of the tumors are unilateral and typically well circumscribed ranging in size from 1.2 to 45 (mean, 8.0) cm [105]. Most neoplasms are solid with an occasionally lobulated cut surface, but some may be solid and cystic or predominantly cystic. Depending on the amount of intracytoplasmic lipid content, the gross appearance of the tumors may range from yellow or orange (lipid rich) to red or brown (lipid poor). Tumors containing significant amount of lipochrome pigment may display a dark brown to black cut surface. Foci of hemorrhage and necrosis may be observed [105].

Microscopic examination frequently reveals a diffuse proliferation of tumor cells that on occasion, may grow in large or small nests and clusters, thin cords, or columns separated by a delicate and often conspicuous network of thin-walled blood vessels, reminiscent of the architectural patterns seen in the adrenal cortex. The stroma is typically scant but in a minority of tumors, may be abundant with a fibromatous, edematous, or even myxoid appearance. The tumor cells are round to polygonal and have centrally placed nuclei, prominent nucleoli, ample eosinophilic and granular to clear and vacuolated cytoplasm and distinct cellular borders (Fig. 14.62a, b) [105]. Cytoplasmic lipochrome pigment is found in less than 50% of neoplasms and Reinke crystals are absent. Infrequently, signet ring-type cells with a single large lipid-containing cytoplasmic vacuole may also be seen. Nuclear atypia and mitotic activity are usually absent or minimal; however 40% of tumors may exhibit variable degrees of cytologic atypia with increased mitotic rate (up to 15/10 HPFs) [105]. Foci of calcification, necrosis, and hemorrhage can be identified, with the latter being more frequent findings in tumors exhibiting atypical cytologic features.

Immunohistochemical Profile

By histochemistry, fat stain highlights the presence of cytoplasmic lipid in these tumors [105]. Steroid cell tumors, show immunoreactivity for vimentin, inhibin, and calretinin, although the latter has been reported to be a more sensitive marker than inhibin for these neoplasms [75, 76, 142, 144, 291]. Some tumors may stain for CD56 [262], CD10 [12], CD99 [143, 144, 292], Melan-A [263, 291], HMB45 [291], cytokeratin, and S100, but MART-1 is typically negative [277, 291]. Rarely, these tumors are EMA positive.

Differential Diagnosis

The differential diagnosis of steroid cell tumor, NOS include the other two tumors in the category of steroid cell tumors, stromal luteoma and Leydig cell tumor. The former is typically located within the ovarian stroma, occurs frequently in association with stromal hyperthecosis, and may show degenerative pseudovascular spaces containing red blood cells [275], while Leydig cell tumor has a hilar location, contains cells with cytoplasmic Reinke crystals, and is frequently associated with Leydig cell hyperplasia [61]. Even though pregnancy luteoma may resemble a lipid-poor steroid cell tumor microscopically, it is multifocal in 50% of cases, occurs bilaterally in one-third of patients, and tends to regress after pregnancy [34]. As steroid cell tumors may contain a predominance of oxyphilic cells, the differential diagnosis would also include oxyphilic endometrioid or clear cell carcinoma, hepatoid carcinoma, metastatic hepatocellular carcinoma, hepatoid yolk sac tumor, and rarely the oxyphilic variant of struma ovarii as well as metastatic or primary malignant melanoma. Oxyphilic endometrioid carcinoma shows true gland formation, squamous and/or mucinous differentiation, association with an adenofibromatous background or endometriosis, and diffuse immunoreactivity for EMA and CK7 [251]. Oxyphilic clear cell carcinoma often exhibits other features typical of conventional clear cell carcinoma including tubulocystic and/or papillary structures lined by cuboidal, hobnail, or flattened cells, nests and sheets of cells with abundant clear cytoplasm, and commonly an adenofibromatous component [293]. Hepatoid yolk sac tumor, hepatoid carcinoma, and metastatic hepatocellular carcinoma are rare tumors that frequently show a diffuse growth of oxyphilic cells. They are associated with elevated serum levels of alpha-fetoprotein [294], often exhibit focal gland formation, contain PAS-positive, diastase-resistant intracytoplasmic and extracellular hyaline globules, and stain for alpha-fetoprotein as well as HepPar1 [295]. Furthermore, hepatoid yolk sac tumor may show areas of conventional yolk sac neoplasia [224], and hepatoid carcinoma is frequently associated with another surface epithelial component [296], while in metastatic hepatocellular carcinoma, the clinical history or imaging studies may be helpful [297]. The presence of other teratomatous elements, colloid, and immunoreactivity for thyroglobulin and/or TTF-1 would distinguish an oxyphilic struma from a steroid cell tumor [133]. Primary [160] or metastatic [159, 161] melanomas can mimic steroid cell tumors as melanin pigment may be mistaken for lipochrome. In addition, melanoma cells also show immunoreactivity for Melan-A, HMB45 and S100. However, melanomas typically display overt malignant cytologic features and show staining for Mart-1 in contrast to steroid cell tumors.

When steroid cell tumors contain mostly clear cells, the differential diagnoses include primary ovarian clear cell carcinoma, metastatic renal cell and adrenocortical carcinomas, and lipid-rich Sertoli cell tumor. Clear cell carcinoma of the ovary is usually composed of a mixture of clear, hobnail cells, and eosinophilic cells arranged in tubulocystic, papillary, and solid patterns, and tumor cells show immunoreactivity for CK7 and EMA. Like steroid cell tumor, metastatic renal carcinoma may have a diffuse growth pattern, shows a delicate vascular network, expresses both CD10 and vimentin, and contains intracellular lipid. However, renal carcinoma cells also contain glycogen and they stain for RCC [298]. A history of a prior adrenal carcinoma and radiographic correlation would be invaluable in differentiating metastatic adrenocortical carcinoma from steroid cell tumor as both tumors may appear morphologically identical. Lipid-rich Sertoli cell tumor shares a similar morphology and immunoprofile with steroid cell tumor but the presence of estrogenic manifestations as well as a solid tubular growth pattern can aid in this distinction [236]. Rare steroid cell tumors with a nested growth pattern may be confused with paragangliomas [299]. The presence of sustentacular cells (highlighted by S100 immunostain) and immunoreactivity for neuroendocrine markers help to distinguish them from steroid cell tumors. It should be noted, however, that paragangliomas can show focal staining for inhibin and calretinin.

Treatment and Clinical Outcome

Surgery is mainstay of treatment for steroid cell tumors. As one-fifth of tumors show extra-ovarian extension at presentation and approximately 43% are clinically malignant, unilateral salpingo-oophorectomy with staging may be performed in younger patients with stage 1a tumors who want to preserve fertility. Otherwise, total hysterectomy with bilateral salpingo-oophorectomy and staging should be performed. In 10% of patients, tumor recurrence may occur >5 years after initial presentation. Clinicopathologic parameters that have been reported to correlate with adverse behavior include older age at presentation, size ≥ 7.0 cm, mitotic rate of ≥ 2/10HPFs, grade 2–3 nuclear atypia, necrosis, and hemorrhage [105]. However, occasional cytologically benign tumors may also behave aggressively.

Gynandroblastoma

Gynandroblastomas are tumors that contain both well-differentiated testicular (Sertoli) and ovarian (granulosa) type cells, each accounting for at least 10% of the tumor [300, 301]. These neoplasms usually occur in young adults but may be seen at any age and patients can present with hormonal (androgenic and estrogenic) manifestations. The differential diagnosis includes sex cord stromal tumors that contain a minor (<10%) component of another type. Most patients have stage I tumors when initially diagnosed and behave in a benign fashion.

Sex Cord Stromal Tumor, Unclassified

This group of neoplasms account for 10% of all sex cord stromal tumors. They occur in patients ranging in age from 12 to 83 (mean, 50) years [302, 303]. The clinical presentation is nonspecific and includes abdominal pain, abnormal uterine bleeding, ascites, and abdominal or pelvic masses. These tumors range from 4 to 27 cm in diameter and may be partially encapsulated with a solid and cystic cut surface. Histologically, they may consist predominantly of primitive spindle stromal cells without specific differentiation or cells with features suggestive of both granulosa and Sertoli-Leydig cell differentiation without fulfilling the diagnostic requirements for these neoplasms [302, 303]. The tumors show immunoreactivity for vimentin and may also stain for cytokeratin CAM 5.2 and AE1/3, but they generally do not express EMA [303]. Inhibin is not consistently positive [75].

This subset of tumors accounts for a significant number of neoplasms removed during gestation, especially when prominent intercellular edema, marked luteinization of the granulosa cells, or increased Leydig cell maturation obscures the true nature of the underlying tumor. Another tumor that can be included within this category is the rare sex cord stromal tumor reported in patients with Peutz-Jeghers syndrome, which can exhibit a variable solid, tubular, papillary, or cystic morphology and is composed of Sertoli and Leydig-like cells [303, 304]. In a study of 32 unclassified sex cord stromal tumors, it was reported that the behavior of these neoplasms appeared to be similar to that of granulosa and Sertoli-Leydig cell tumors, with a favorable prognosis when confined to the ovary [302].

References

1. Manieri C, Di Bisceglie C, Fornengo R, et al. Postmenopausal virilization in a woman with gonadotropin dependent ovarian hyperthecosis. J Endocrinol Invest. 1998;21(2):128–32.
2. Sasano H, Fukunaga M, Rojas M, Silverberg SG. Hyperthecosis of the ovary. Clinicopathologic study of 19 cases with immunohistochemical analysis of steroidogenic enzymes. Int J Gynecol Pathol. 1989;8(4):311–20.
3. Dunaif A, Hoffman AR, Scully RE, et al. Clinical, biochemical, and ovarian morphologic features in women with acanthosis nigricans and masculinization. Obstet Gynecol. 1985;66(4):545–52.
4. Mantzoros CS, Lawrence WD, Levy J. Insulin resistance in a patient with ovarian stromal hyperthecosis and the hyperandrogenism, insulin resistance and acanthosis nigricans syndrome. Report of a case with a possible endogenous ovarian factor. J Reprod Med. 1995;40(6):491–4.
5. Hayes MC, Scully RE. Stromal luteoma of the ovary: a clinicopathological analysis of 25 cases. Int J Gynecol Pathol. 1987;6(4): 313–21.
6. Adashi EY. Potential utility of gonadotropin-releasing hormone agonists in the management of ovarian hyperandrogenism. Fertil Steril. 1990;53(5):765–79.
7. Detre Z, Foldes E. Mucinous cystadenocarcinoma of ovary with Leydig cell hyperplasia. Pathol Res Pract. 1984;178(4):400–2.
8. Pascal RR, Grecco LA. Mucinous cystadenoma of the ovary with stromal luteinization and hilar cell hyperplasia during pregnancy. Hum Pathol. 1988;19(2):179–80.
9. Yoon BS, Seong SJ, Park CT, Park H, Shim JY, Kim JY. Cellular fibroma of the ovary containing Leydig cell hyperplasia: a case report. J Gynecol Oncol. 2010;21(1):56–8.
10. Delibasi T, Erdogan MF, Serinsoz E, Kaygusuz G, Erdogan G, Sertcelik A. Ovarian hilus-cell hyperplasia and high serum testosterone in a patient with postmenopausal virilization. Endocr Pract. 2007;13(5):472–5.
11. Bamberger AM, Ivell R, Balvers M, et al. Relaxin-like factor (RLF): a new specific marker for Leydig cells in the ovary. Int J Gynecol Pathol. 1999;18(2):163–8.
12. Oliva E, Garcia-Miralles N, Vu Q, Young RH. CD10 expression in pure stromal and sex cord-stromal tumors of the ovary: an immunohistochemical analysis of 101 cases. Int J Gynecol Pathol. 2007; 26(4):359–67.
13. Zafrakas M, Venizelos ID, Theodoridis TD, Zepiridis L, Agorastos T, Bontis JN. Virilizing ovarian hilus (Leydig) cell tumor with concurrent contralateral hilus cell hyperplasia: a rare diagnosis. Eur J Gynaecol Oncol. 2009;30(3):338–40.
14. Nogales FF, Martin-Sances L, Mendoza-Garcia E, Salamanca A, Gonzalez-Nunez MA, Pardo Mindan FJ. Massive ovarian oedema. Histopathology. 1996;28(3):229–34.
15. Young RH, Scully RE. Fibromatosis and massive edema of the ovary, possibly related entities: a report of 14 cases of fibromatosis and 11 cases of massive edema. Int J Gynecol Pathol. 1984;3(2): 153–78.
16. Natarajan A, Wales JK, Marven SS, Wright NP. Precocious puberty secondary to massive ovarian oedema in a 6-month-old girl. Eur J Endocrinol. 2004;150(2):119–23.
17. Siller BS, Gelder MS, Alvarez RD, Partridge EE. Massive edema of the ovary associated with androgenic manifestations. South Med J. 1995;88(11):1153–5.
18. Roth LM, Deaton RL, Sternberg WH. Massive ovarian edema. A clinicopathologic study of five cases including ultrastructural observations and review of the literature. Am J Surg Pathol. 1979; 3(1):11–21.
19. Bazot M, Detchev R, Cortez A, Uzan S, Darai E. Massive ovarian edema revealing gastric carcinoma: a case report. Gynecol Oncol. 2003;91(3):648–50.
20. Hameed A, Ying AJ, Keyhani-Rofagha S, Xie D, Copeland LJ. Ovarian mucinous cystadenoma associated with mural leiomyomatous nodule and massive ovarian edema. Gynecol Oncol. 1997;67(2): 226–9.
21. Khalbuss WE, Dipasquale B. Massive ovarian edema associated with ovarian serous cystadenoma: a case report and review of the literature. Int J Gynecol Cancer. 2006;16 Suppl 1:326–30.
22. Krasevic M, Haller H, Rupcic S, Behrem S. Massive edema of the ovary: a report of two cases due to lymphatic permeation by metastatic carcinoma from the uterine cervix. Gynecol Oncol. 2004; 93(2):564–7.
23. Levavi H, Sabah G, Heifetz M, Feinmesser M. Sclerosing peritonitis associated with bilateral luteinized thecoma, linked to anticonvulsant therapy. Eur J Gynaecol Oncol. 2009;30(6):695–700.
24. Staats PN, McCluggage WG, Clement PB, Young RH. Luteinized thecomas (thecomatosis) of the type typically associated with sclerosing peritonitis: a clinical, histopathologic, and immunohistochemical analysis of 27 cases. Am J Surg Pathol. 2008;32(9): 1273–90.
25. Bianco R, de Rosa G, Staibano S, Somma P, Bianco AR. Ovarian luteinized thecoma with sclerosing peritonitis in an adult woman treated with leuprolide and toremifene in complete remission at 5 years. Gynecol Oncol. 2005;96(3):846–9.
26. Schonman R, Klein Z, Edelstein E, Czernobilsky B, Fishman A. Luteinized thecoma associated with sclerosing peritonitis–conservative surgical approach followed by corticosteroid and GnRH agonist treatment–a case report. Gynecol Oncol. 2008; 111(3):540–3.
27. McCluggage WG, Young RH. Non-neoplastic granulosa cells within ovarian vascular channels: a rare potential diagnostic pitfall. J Clin Pathol. 2004;57(2):151–4.

28. Smith AE, Harmon JP, Deogracias FA, Pizer ES. Vascular embolization of benign granulosa cells. Gynecol Oncol. 2000;77(2):336–8.
29. Vydianath B, Ganesan R, McCluggage WG. Displaced granulosa cells in the fallopian tube mimicking small cell carcinoma. J Clin Pathol. 2008;61(12):1323–5.
30. Clement PB, Scully RE. Large solitary luteinized follicle cyst of pregnancy and puerperium: a clinicopathological analysis of eight cases. Am J Surg Pathol. 1980;4(5):431–8.
31. Medeiros F, Bell DA. Pseudoneoplastic lesions of the female genital tract. Arch Pathol Lab Med. 2010;134(3):393–403.
32. Choi CH, Pritchard JR. Large cystic granulosa cell tumor: case report. Am J Obstet Gynecol. 1990;163(1 Pt 1):74–6.
33. Mulvany NJ, Riley CB. Granulosa cell tumors of unilocular cystic type. Pathology. 1997;29(4):348–53.
34. Clement PB. Tumor-like lesions of the ovary associated with pregnancy. Int J Gynecol Pathol. 1993;12(2):108–15.
35. Yapar EG, Vural T, Ekici E, Kuscu E, Gokmen O. Hyperreactio luteinalis masquerading as an ovarian neoplasm in a triplet pregnancy. Eur J Obstet Gynecol Reprod Biol. 1996;65(2):177–80.
36. Kohorn EI. Theca lutein ovarian cyst may be pathognomonic for trophoblastic neoplasia. Obstet Gynecol. 1983;62(3 Suppl):80s–1.
37. Abdel-Aleem H, Al-Hussaini T, Karoush S. Hyperreactio luteinalis associated with normal singleton pregnancy. J Obstet Gynaecol. 2000;20(3):315.
38. Tulandi T, McInnes RA, Arronet GH. Ovarian hyperstimulation syndrome following ovulation induction with human menopausal gonadotropin. Int J Fertil. 1984;29(2):113–7.
39. Rajashekar L, Krishna D, Patil M. Polycystic ovaries and infertility: our experience. J Hum Reprod Sci. 2008;1(2):65–72.
40. Lussiana C, Guani B, Restagno G, et al. Ovarian hyper-stimulation syndrome after spontaneous conception. Gynecol Endocrinol. 2009;25(7):455–9.
41. Rodien P, Beau I, Vasseur C. Ovarian hyperstimulation syndrome (OHSS) due to mutations in the follicle-stimulating hormone receptor. Ann Endocrinol (Paris). 2010;71(3):206–9.
42. Smits G, Olatunbosun O, Delbaere A, Pierson R, Vassart G, Costagliola S. Ovarian hyperstimulation syndrome due to a mutation in the follicle-stimulating hormone receptor. N Engl J Med. 2003;349(8):760–6.
43. Upadhyaya G, Goswami A, Babu S. Bilateral theca lutein cysts: a rare cause of acute abdomen in pregnancy. Emerg Med Australas. 2004;16(5–6):476–7.
44. Angioni S, Portoghese E, Milano F, Melis GB, Fulghesu AM. Hirsutism and hyperandrogenism associated with hyperreactio luteinalis in a singleton pregnancy: a case report. Gynecol Endocrinol. 2007;23(5):248–51.
45. Beji O, Brahmi N, Thabet H, et al. Compressive pleural effusion after ovarian hyperstimulation syndrome–a case report and review. Fertil Steril. 2008;89(6):1826. e1821–e1823.
46. Onodera N, Kishi I, Tamaoka Y, Yamazaki K, Kamei K. A case of recurrent hyperreactio luteinalis. Am J Obstet Gynecol. 2008;198(5):e9–10.
47. Herr JC, Heidger Jr PM, Scott JR, Anderson JW, Curet LB, Mossman HW. Decidual cells in the human ovary at term. I. Incidence, gross anatomy and ultrastructural features of merocrine secretion. Am J Anat. 1978;152(1):7–27.
48. Israel SL, Rubenstone A, Meranze DR. The ovary at term. I. Decidua-like reaction and surface cell proliferation. Obstet Gynecol. 1954;3(4):399–407.
49. Ober WB, Grady HG, Schoenbucher AK. Ectopic ovarian decidua without pregnancy. Am J Pathol. 1957;33(2):199–217.
50. Kondi-Pafiti A, Grapsa D, Kontogianni-Katsarou K, Papadias K, Kairi-Vassilatou E. Ectopic decidua mimicking metastatic lesions–report of three cases and review of the literature. Eur J Gynaecol Oncol. 2005;26(4):459–61.
51. Suster S, Moran CA. Deciduosis of the appendix. Am J Gastroenterol. 1990;85(7):841–5.
52. Zaytsev P, Taxy JB. Pregnancy-associated ectopic decidua. Am J Surg Pathol. 1987;11(7):526–30.
53. Buttner A, Bassler R, Theele C. Pregnancy-associated ectopic decidua (deciduosis) of the greater omentum. An analysis of 60 biopsies with cases of fibrosing deciduosis and leiomyomatosis peritonealis disseminata. Pathol Res Pract. 1993;189(3):352–9.
54. Wu DC, Hirschowitz S, Natarajan S. Ectopic decidua of pelvic lymph nodes: a potential diagnostic pitfall. Arch Pathol Lab Med. 2005;129(5):e117–20.
55. Flieder DB, Moran CA, Travis WD, Koss MN, Mark EJ. Pleuro-pulmonary endometriosis and pulmonary ectopic deciduosis: a clinicopathologic and immunohistochemical study of 10 cases with emphasis on diagnostic pitfalls. Hum Pathol. 1998;29(12):1495–503.
56. O'Leary SM. Ectopic decidualization causing massive postpartum intraperitoneal hemorrhage. Obstet Gynecol. 2006;108(3 Pt 2):776–9.
57. Clement PB, Young RH, Scully RE. Ovarian granulosa cell proliferations of pregnancy: a report of nine cases. Hum Pathol. 1988;19(6):657–62.
58. McKay DG, Hertig AT, Hickey WF. The histogenesis of granulosa and theca cell tumors of the human ovary. Obstet Gynecol. 1953;1(2):125–36.
59. Wang YC, Su HY, Liu JY, Chang FW, Chen CH. Maternal and female fetal virilization caused by pregnancy luteomas. Fertil Steril. 2005;84(2):509.
60. Piana S, Nogales FF, Corrado S, Cardinale L, Gusolfino D, Rivasi F. Pregnancy luteoma with granulosa cell proliferation: an unusual hyperplastic lesion arising in pregnancy and mimicking an ovarian neoplasia. Pathol Res Pract. 1999;195(12):859–63.
61. Paraskevas M, Scully RE. Hilus cell tumor of the ovary. A clinicopathological analysis of 12 Reinke crystal-positive and nine crystal-negative cases. Int J Gynecol Pathol. 1989;8(4):299–310.
62. Dockerty MB, Masson JC. Ovarian fibromas: a clinical and pathologic study of 283 cases. Am J Obstet Gynecol. 1944;47:741–52.
63. Irving JA, Alkushi A, Young RH, Clement PB. Cellular fibromas of the ovary: a study of 75 cases including 40 mitotically active tumors emphasizing their distinction from fibrosarcoma. Am J Surg Pathol. 2006;30(8):929–38.
64. Siddiqui M, Toub DB. Cellular fibroma of the ovary with Meigs' syndrome and elevated CA-125. A case report. J Reprod Med. 1995;40(11):817–9.
65. Samanth KK, Black 3rd WC. Benign ovarian stromal tumors associated with free peritoneal fluid. Am J Obstet Gynecol. 1970;107(4):538–45.
66. Moran-Mendoza A, Alvarado-Luna G, Calderillo-Ruiz G, Serrano-Olvera A, Lopez-Graniel CM, Gallardo-Rincon D. Elevated CA125 level associated with Meigs' syndrome: case report and review of the literature. Int J Gynecol Cancer. 2006;16 Suppl 1:315–8.
67. Bale AE. The nevoid basal cell carcinoma syndrome: genetics and mechanism of carcinogenesis. Cancer Invest. 1997;15(2):180–6.
68. Johnson AD, Hebert AA, Esterly NB. Nevoid basal cell carcinoma syndrome: bilateral ovarian fibromas in a 3 1/2-year-old girl. J Am Acad Dermatol. 1986;14(2 Pt 2):371–4.
69. Tytle T, Rosin D. Bilateral calcified ovarian fibromas. South Med J. 1984;77(9):1178–80.
70. Young RH, Scully RE. Ovarian stromal tumors with minor sex cord elements: a report of seven cases. Int J Gynecol Pathol. 1983;2(3):227–34.
71. Tiltman AJ, Haffajee Z. Sclerosing stromal tumors, thecomas, and fibromas of the ovary: an immunohistochemical profile. Int J Gynecol Pathol. 1999;18(3):254–8.
72. He H, Luthringer DJ, Hui P, Lau SK, Weiss LM, Chu PG. Expression of CD56 and WT1 in ovarian stroma and ovarian stromal tumors. Am J Surg Pathol. 2008;32(6):884–90.

73. Lastarria D, Sachdev RK, Babury RA, Yu HM, Nuovo GJ. Immunohistochemical analysis for desmin in normal and neoplastic ovarian stromal tissue. Arch Pathol Lab Med. 1990;114(5):502–5.
74. Costa MJ, Ames PF, Walls J, Roth LM. Inhibin immunohistochemistry applied to ovarian neoplasms: a novel, effective, diagnostic tool. Hum Pathol. 1997;28(11):1247–54.
75. Kommoss F, Oliva E, Bhan AK, Young RH, Scully RE. Inhibin expression in ovarian tumors and tumor-like lesions: an immunohistochemical study. Mod Pathol. 1998;11(7):656–64.
76. Movahedi-Lankarani S, Kurman RJ. Calretinin, a more sensitive but less specific marker than alpha-inhibin for ovarian sex cord-stromal neoplasms: an immunohistochemical study of 215 cases. Am J Surg Pathol. 2002;26(11):1477–83.
77. Choi YL, Kim HS, Ahn G. Immunoexpression of inhibin alpha subunit, inhibin/activin betaA subunit and CD99 in ovarian tumors. Arch Pathol Lab Med. 2000;124(4):563–9.
78. Tsuji T, Catasus L, Prat J. Is loss of heterozygosity at 9q22.3 (PTCH gene) and 19p13.3 (STK11 gene) involved in the pathogenesis of ovarian stromal tumors? Hum Pathol. 2005;36(7):792–6.
79. Deavers MT, Malpica A, Liu J, Broaddus R, Silva EG. Ovarian sex cord-stromal tumors: an immunohistochemical study including a comparison of calretinin and inhibin. Mod Pathol. 2003;16(6): 584–90.
80. Zhang J, Young RH, Arseneau J, Scully RE. Ovarian stromal tumors containing lutein or Leydig cells (luteinized thecomas and stromal Leydig cell tumors)–a clinicopathological analysis of fifty cases. Int J Gynecol Pathol. 1982;1(3):270–85.
81. Chang KL, Crabtree GS, Lim-Tan SK, Kempson RL, Hendrickson MR. Primary extrauterine endometrial stromal neoplasms: a clinicopathologic study of 20 cases and a review of the literature. Int J Gynecol Pathol. 1993;12(4):282–96.
82. Lerwill MF, Sung R, Oliva E, Prat J, Young RH. Smooth muscle tumors of the ovary: a clinicopathologic study of 54 cases emphasizing prognostic criteria, histologic variants, and differential diagnosis. Am J Surg Pathol. 2004;28(11):1436–51.
83. Irving JA, Lerwill MF, Young RH. Gastrointestinal stromal tumors metastatic to the ovary: a report of five cases. Am J Surg Pathol. 2005;29(7):920–6.
84. Prat J, Scully RE. Cellular fibromas and fibrosarcomas of the ovary: a comparative clinicopathologic analysis of seventeen cases. Cancer. 1981;47(11):2663–70.
85. Kraemer BB, Silva EG, Sneige N. Fibrosarcoma of ovary. A new component in the nevoid basal-cell carcinoma syndrome. Am J Surg Pathol. 1984;8(3):231–6.
86. Christman JE, Ballon SC. Ovarian fibrosarcoma associated with Maffucci's syndrome. Gynecol Oncol. 1990;37(2):290–1.
87. Tsuji T, Kawauchi S, Utsunomiya T, Nagata Y, Tsuneyoshi M. Fibrosarcoma versus cellular fibroma of the ovary: a comparative study of their proliferative activity and chromosome aberrations using MIB-1 immunostaining, DNA flow cytometry, and fluorescence in situ hybridization. Am J Surg Pathol. 1997;21(1):52–9.
88. Klein WM, Kurman RJ. Lack of expression of c-kit protein (CD117) in mesenchymal tumors of the uterus and ovary. Int J Gynecol Pathol. 2003;22(2):181–4.
89. Celyk C, Gungor S, Gorkemly H, et al. Ovarian fibrosarcomas. Acta Obstet Gynecol Scand. 2002;81(4):375–6.
90. Shakfeh SM, Woodruff JD. Primary ovarian sarcomas: report of 46 cases and review of the literature. Obstet Gynecol Surv. 1987;42(6): 331–49.
91. Huang YC, Hsu KF, Chou CY, Dai YC, Tzeng CC. Ovarian fibrosarcoma with long-term survival: a case report. Int J Gynecol Cancer. 2001;11(4):331–3.
92. Bjorkholm E, Silfversward C. Theca-cell tumors. Clinical features and prognosis. Acta Radiol Oncol. 1980;19(4):241–4.
93. Aboud E. A review of granulosa cell tumours and thecomas of the ovary. Arch Gynecol Obstet. 1997;259(4):161–5.
94. Vieira SC, Santos LG, Cruz FJ, Goncalves WM, Rocha Sde F. Sclerosing peritonitis associated with luteinized thecoma and elevated serum CA 125 levels: case report. São Paulo Med J. 2008;126(2):123–5.
95. Banner EA, Dockerty MB. Theca cell tumors of the ovary. A clinical and pathological study of 23 cases (including 13 new cases) with a review. Surg Gynecol Obstet. 1945;81:234–42.
96. Geist RR, Rabinowitz R, Zuckerman B, et al. Massive edema of the ovary: a case report and review of the pertinent literature. J Pediatr Adolesc Gynecol. 2005;18(4):281–4.
97. Young RH, Clement PB, Scully RE. Calcified thecomas in young women. A report of four cases. Int J Gynecol Pathol. 1988;7(4): 343–50.
98. Young RH, Scully RE. Ovarian sex cord-stromal tumors with bizarre nuclei: a clinicopathologic analysis of 17 cases. Int J Gynecol Pathol. 1983;1(4):325–35.
99. Morizane M, Ohara N, Mori T, Murao S. Ossifying luteinized thecoma of the ovary. Arch Gynecol Obstet. 2003;267(3):167–9.
100. Paoletti M, Pridjian G, Okagaki T, Talerman A. A stromal Leydig cell tumor of the ovary occurring in a pregnant 15-year-old girl. Ultrastructural findings. Cancer. 1987;60(11):2806–10.
101. Park SH, Kim I. Histogenetic consideration of ovarian sex cord-stromal tumors analyzed by expression pattern of cytokeratins, vimentin, and laminin. Correlation studies with human gonads. Pathol Res Pract. 1994;190(5):449–56.
102. Klemi PJ, Gronroos M. An ultrastructural and clinical study of theca and granulosa cell tumors. Int J Gynaecol Obstet. 1979;17(3): 219–25.
103. Kim MS, Hur SY, Yoo NJ, Lee SH. Mutational analysis of FOXL2 codon 134 in granulosa cell tumour of ovary and other human cancers. J Pathol. 2010;221(2):147–52.
104. Shah SP, Kobel M, Senz J, et al. Mutation of FOXL2 in granulosa-cell tumors of the ovary. N Engl J Med. 2009;360(26):2719–29.
105. Hayes MC, Scully RE. Ovarian steroid cell tumors (not otherwise specified). A clinicopathological analysis of 63 cases. Am J Surg Pathol. 1987;11(11):835–45.
106. Dudzinski M, Cohen M, Ducatman B. Ovarian malignant luteinized thecoma–an unusual tumor in an adolescent. Gynecol Oncol. 1989;35(1):104–9.
107. Waxman M, Vuletin JC, Urcuyo R, Belling CG. Ovarian low-grade stromal sarcoma with thecomatous features: a critical reappraisal of the so-called "malignant thecoma". Cancer. 1979;44(6):2206–17.
108. Nocito AL, Sarancone S, Bacchi C, Tellez T. Ovarian thecoma: clinicopathological analysis of 50 cases. Ann Diagn Pathol. 2008;12(1):12–6.
109. Chalvardjian A, Scully RE. Sclerosing stromal tumors of the ovary. Cancer. 1973;31(3):664–70.
110. Duska LR, Flynn C, Goodman A. Masculinizing sclerosing stromal cell tumor in pregnancy: report of a case and review of the literature. Eur J Gynaecol Oncol. 1998;19(5):441–3.
111. Ismail SM, Walker SM. Bilateral virilizing sclerosing stromal tumours of the ovary in a pregnant woman with Gorlin's syndrome: implications for pathogenesis of ovarian stromal neoplasms. Histopathology. 1990;17(2):159–63.
112. Terauchi F, Onodera T, Nagashima T, et al. Sclerosing stromal tumor of the ovary with elevated CA125. J Obstet Gynaecol Res. 2005;31(5):432–5.
113. Bildirici K, Yalcin OT, Ozalp SS, Peker B, Ozden H. Sclerosing stromal tumor of the ovary associated with Meigs' syndrome: a case report. Eur J Gynaecol Oncol. 2004;25(4):528–9.
114. Hsu C, Ma L, Mak L. Sclerosing stromal tumor of the ovary: case report and review of the literature. Int J Gynecol Pathol. 1983;2(2):192–200.
115. Kawauchi S, Tsuji T, Kaku T, Kamura T, Nakano H, Tsuneyoshi M. Sclerosing stromal tumor of the ovary: a clinicopathologic, immunohistochemical, ultrastructural, and cytogenetic analysis

with special reference to its vasculature. Am J Surg Pathol. 1998;22(1):83–92.
116. Zamecnik M. Calcifying sclerosing tumor of the ovary: a late stage of sclerosing stromal tumor? Cesk Patol. 2002;38(3):121–4.
117. Kostopoulou E, Moulla A, Giakoustidis D, Leontsini M. Sclerosing stromal tumors of the ovary: a clinicopathologic, immunohistochemical and cytogenetic analysis of three cases. Eur J Gynaecol Oncol. 2004;25(2):257–60.
118. Saitoh A, Tsutsumi Y, Osamura RY, Watanabe K. Sclerosing stromal tumor of the ovary. Immunohistochemical and electron-microscopic demonstration of smooth-muscle differentiation. Arch Pathol Lab Med. 1989;113(4):372–6.
119. Shaw JA, Dabbs DJ, Geisinger KR. Sclerosing stromal tumor of the ovary: an ultrastructural and immunohistochemical analysis with histogenetic considerations. Ultrastruct Pathol. 1992;16(3):363–77.
120. Zekioglu O, Ozdemir N, Terek C, Ozsaran A, Dikmen Y. Clinicopathological and immunohistochemical analysis of sclerosing stromal tumours of the ovary. Arch Gynecol Obstet. 2010;282(6):671–6. Epub Feb 5, 2010.
121. Kiyokawa T, Young RH, Scully RE. Krukenberg tumors of the ovary: a clinicopathologic analysis of 120 cases with emphasis on their variable pathologic manifestations. Am J Surg Pathol. 2006;30(3):277–99.
122. Begum M, Katabuchi H, Tashiro H, Suenaga Y, Okamura H. A case of metastatic malignant hemangiopericytoma of the ovary: recurrence after a period of 17 years from intracranial tumor. Int J Gynecol Cancer. 2002;12(5):510–4.
123. Ramzy I. Signet-ring stromal tumor of ovary. Histochemical, light, and electron microscopic study. Cancer. 1976;38(1):166–72.
124. Matsumoto M, Hayashi Y, Ohtsuki Y, et al. Signet-ring stromal tumor of the ovary: an immunohistochemical and ultrastructural study with a review of the literature. Med Mol Morphol. 2008;41(3):165–70.
125. Dickersin GR, Young RH, Scully RE. Signet-ring stromal and related tumors of the ovary. Ultrastruct Pathol. 1995;19(5):401–19.
126. Vang R, Bague S, Tavassoli FA, Prat J. Signet-ring stromal tumor of the ovary: clinicopathologic analysis and comparison with Krukenberg tumor. Int J Gynecol Pathol. 2004;23(1):45–51.
127. Irving JA, Young RH. Microcystic stromal tumor of the ovary: report of 16 cases of a hitherto uncharacterized distinctive ovarian neoplasm. Am J Surg Pathol. 2009;33(3):367–75.
128. McCluggage WG, Young RH. Immunohistochemistry as a diagnostic aid in the evaluation of ovarian tumors. Semin Diagn Pathol. 2005;22(1):3–32.
129. Che M, Tornos C, Deavers MT, Malpica A, Gershenson DM, Silva EG. Ovarian mixed-epithelial carcinomas with a microcystic pattern and signet-ring cells. Int J Gynecol Pathol. 2001;20(4):323–8.
130. Scully RE, Young RH, Clement PB. Sex cord-stromal tumor. Tumors of the ovary, maldeveloped gonads, fallopian tube and broad ligament, Atlas of tumor pathology, third series. Fascicle 23. Washington, DC: Arm Forces Institute of Pathology; 1998. p. 169–225.
131. Devouassoux-Shisheboran M, Silver SA, Tavassoli FA. Wolffian adnexal tumor, so-called female adnexal tumor of probable Wolffian origin (FATWO): immunohistochemical evidence in support of a Wolffian origin. Hum Pathol. 1999;30(7):856–63.
132. Ordi J, Romagosa C, Tavassoli FA, et al. CD10 expression in epithelial tissues and tumors of the gynecologic tract: a useful marker in the diagnosis of mesonephric, trophoblastic, and clear cell tumors. Am J Surg Pathol. 2003;27(2):178–86.
133. Szyfelbein WM, Young RH, Scully RE. Struma ovarii simulating ovarian tumors of other types. A report of 30 cases. Am J Surg Pathol. 1995;19(1):21–9.
134. Evans 3rd AT, Gaffey TA, Malkasian Jr GD, Annegers JF. Clinicopathologic review of 118 granulosa and 82 theca cell tumors. Obstet Gynecol. 1980;55(2):231–8.
135. Fox H, Agrawal K, Langley FA. A clinicopathologic study of 92 cases of granulosa cell tumor of the ovary with special reference to the factors influencing prognosis. Cancer. 1975;35(1):231–41.
136. Nakashima N, Young RH, Scully RE. Androgenic granulosa cell tumors of the ovary. A clinicopathologic analysis of 17 cases and review of the literature. Arch Pathol Lab Med. 1984;108(10):786–91.
137. Young RH, Oliva E, Scully RE. Luteinized adult granulosa cell tumors of the ovary: a report of four cases. Int J Gynecol Pathol. 1994;13(4):302–10.
138. Nogales FF, Concha A, Plata C, Ruiz-Avila I. Granulosa cell tumor of the ovary with diffuse true hepatic differentiation simulating stromal luteinization. Am J Surg Pathol. 1993;17(1):85–90.
139. Susil BJ, Sumithran E. Sarcomatous change in granulosa cell tumor. Hum Pathol. 1987;18(4):397–9.
140. Hildebrandt RH, Rouse RV, Longacre TA. Value of inhibin in the identification of granulosa cell tumors of the ovary. Hum Pathol. 1997;28(12):1387–95.
141. McCluggage WG, Maxwell P. Immunohistochemical staining for calretinin is useful in the diagnosis of ovarian sex cord-stromal tumours. Histopathology. 2001;38(5):403–8.
142. Rishi M, Howard LN, Bratthauer GL, Tavassoli FA. Use of monoclonal antibody against human inhibin as a marker for sex cord-stromal tumors of the ovary. Am J Surg Pathol. 1997;21(5):583–9.
143. Loo KT, Leung AK, Chan JK. Immunohistochemical staining of ovarian granulosa cell tumours with MIC2 antibody. Histopathology. 1995;27(4):388–90.
144. Matias-Guiu X, Pons C, Prat J. Mullerian inhibiting substance, alpha-inhibin, and CD99 expression in sex cord-stromal tumors and endometrioid ovarian carcinomas resembling sex cord-stromal tumors. Hum Pathol. 1998;29(8):840–5.
145. Aguirre P, Thor AD, Scully RE. Ovarian endometrioid carcinomas resembling sex cord-stromal tumors. An immunohistochemical study. Int J Gynecol Pathol. 1989;8(4):364–73.
146. Costa MJ, DeRose PB, Roth LM, Brescia RJ, Zaloudek CJ, Cohen C. Immunohistochemical phenotype of ovarian granulosa cell tumors: absence of epithelial membrane antigen has diagnostic value. Hum Pathol. 1994;25(1):60–6.
147. Ahmed E, Young RH, Scully RE. Adult granulosa cell tumor of the ovary with foci of hepatic cell differentiation: a report of four cases and comparison with two cases of granulosa cell tumor with Leydig cells. Am J Surg Pathol. 1999;23(9):1089–93.
148. Kobel M, Gilks CB, Huntsman DG. Adult-type granulosa cell tumors and FOXL2 mutation. Cancer Res. 2009;69(24):9160–2.
149. Fujibayashi M, Aiba M, Iizuka E, Igarashi A, Muraoka M, Takagi K. Granulosa cell tumor-like variant of endometrioid carcinoma of the ovary exhibiting nuclear clearing with biotin activity: a subtype showing close macroscopic, cytologic, and histologic similarity to adult granulosa cell tumor. Arch Pathol Lab Med. 2005;129(10):1288–94.
150. Zhao C, Bratthauer GL, Barner R, Vang R. Comparative analysis of alternative and traditional immunohistochemical markers for the distinction of ovarian sertoli cell tumor from endometrioid tumors and carcinoid tumor: a study of 160 cases. Am J Surg Pathol. 2007;31(2):255–66.
151. Irving JA, Young RH. Granulosa cell tumors of the ovary with a pseudopapillary pattern: a study of 14 cases of an unusual morphologic variant emphasizing their distinction from transitional cell neoplasms and other papillary ovarian tumors. Am J Surg Pathol. 2008;32(4):581–6.
152. Eichhorn JH, Young RH. Transitional cell carcinoma of the ovary: a morphologic study of 100 cases with emphasis on differential diagnosis. Am J Surg Pathol. 2004;28(4):453–63.
153. Robboy SJ, Norris HJ, Scully RE. Insular carcinoid primary in the ovary. A clinicopathologic analysis of 48 cases. Cancer. 1975;36(2):404–18.

154. Robboy SJ, Scully RE, Norris HJ. Primary trabecular carcinoid of the ovary. Obstet Gynecol. 1977;49(2):202–7.
155. Young RH, Prat J, Scully RE. Endometrioid stromal sarcomas of the ovary. A clinicopathologic analysis of 23 cases. Cancer. 1984;53(5):1143–55.
156. Young RH, Scully RE. Sarcomas metastatic to the ovary: a report of 21 cases. Int J Gynecol Pathol. 1990;9(3):231–52.
157. McCluggage WG, Oliva E, Connolly LE, McBride HA, Young RH. An immunohistochemical analysis of ovarian small cell carcinoma of hypercalcemic type. Int J Gynecol Pathol. 2004;23(4):330–6.
158. Young RH, Oliva E, Scully RE. Small cell carcinoma of the ovary, hypercalcemic type. A clinicopathological analysis of 150 cases. Am J Surg Pathol. 1994;18(11):1102–16.
159. Gupta D, Deavers MT, Silva EG, Malpica A. Malignant melanoma involving the ovary: a clinicopathologic and immunohistochemical study of 23 cases. Am J Surg Pathol. 2004;28(6):771–80.
160. McCluggage WG, Bissonnette JP, Young RH. Primary malignant melanoma of the ovary: a report of 9 definite or probable cases with emphasis on their morphologic diversity and mimicry of other primary and secondary ovarian neoplasms. Int J Gynecol Pathol. 2006;25(4):321–9.
161. Young RH, Scully RE. Malignant melanoma metastatic to the ovary. A clinicopathologic analysis of 20 cases. Am J Surg Pathol. 1991;15(9):849–60.
162. Gagnon Y, Tetu B. Ovarian metastases of breast carcinoma. A clinicopathologic study of 59 cases. Cancer. 1989;64(4):892–8.
163. Monteagudo C, Merino MJ, LaPorte N, Neumann RD. Value of gross cystic disease fluid protein-15 in distinguishing metastatic breast carcinomas among poorly differentiated neoplasms involving the ovary. Hum Pathol. 1991;22(4):368–72.
164. Dubuc-Lissoir J, Berthiaume MJ, Boubez G, Van Nguyen T, Allaire G. Bone metastasis from a granulosa cell tumor of the ovary. Gynecol Oncol. 2001;83(2):400–4.
165. Lee YK, Park NH, Kim JW, Song YS, Kang SB, Lee HP. Characteristics of recurrence in adult-type granulosa cell tumor. Int J Gynecol Cancer. 2008;18(4):642–7.
166. Bjorkholm E, Silfversward C. Prognostic factors in granulosa-cell tumors. Gynecol Oncol. 1981;11(3):261–74.
167. Fox H. Pathologic prognostic factors in early stage adult-type granulosa cell tumors of the ovary. Int J Gynecol Cancer. 2003; 13(1):1–4.
168. Nosov V, Silva I, Tavassoli F, Adamyan L, Farias-Eisner R, Schwartz PE. Predictors of recurrence of ovarian granulosa cell tumors. Int J Gynecol Cancer. 2009;19(4):628–33.
169. King LA, Okagaki T, Gallup DG, Twiggs LB, Messing MJ, Carson LF. Mitotic count, nuclear atypia, and immunohistochemical determination of Ki-67, c-myc, p21-ras, c-erbB2, and p53 expression in granulosa cell tumors of the ovary: mitotic count and Ki-67 are indicators of poor prognosis. Gynecol Oncol. 1996;61(2):227–32.
170. Miller BE, Barron BA, Wan JY, Delmore JE, Silva EG, Gershenson DM. Prognostic factors in adult granulosa cell tumor of the ovary. Cancer. 1997;79(10):1951–5.
171. Fujimoto T, Sakuragi N, Okuyama K, et al. Histopathological prognostic factors of adult granulosa cell tumors of the ovary. Acta Obstet Gynecol Scand. 2001;80(11):1069–74.
172. Staibano S, Franco R, Mezza E, et al. Loss of oestrogen receptor beta, high PCNA and p53 expression and aneuploidy as markers of worse prognosis in ovarian granulosa cell tumours. Histopathology. 2003;43(3):254–62.
173. Korach J, Perri T, Beiner M, Davidzon T, Fridman E, Ben-Baruch G. Promising effect of aromatase inhibitors on recurrent granulosa cell tumors. Int J Gynecol Cancer. 2009;19(5):830–3.
174. Pautier P, Gutierrez-Bonnaire M, Rey A, et al. Combination of bleomycin, etoposide, and cisplatin for the treatment of advanced ovarian granulosa cell tumors. Int J Gynecol Cancer. 2008;18(3): 446–52.
175. Mom CH, Engelen MJ, Willemse PH, et al. Granulosa cell tumors of the ovary: the clinical value of serum inhibin A and B levels in a large single center cohort. Gynecol Oncol. 2007;105(2):365–72.
176. Young RH, Dickersin GR, Scully RE. Juvenile granulosa cell tumor of the ovary. A clinicopathological analysis of 125 cases. Am J Surg Pathol. 1984;8(8):575–96.
177. Zaloudek C, Norris HJ. Granulosa tumors of the ovary in children: a clinical and pathologic study of 32 cases. Am J Surg Pathol. 1982;6(6):503–12.
178. Calaminus G, Wessalowski R, Harms D, Gobel U. Juvenile granulosa cell tumors of the ovary in children and adolescents: results from 33 patients registered in a prospective cooperative study. Gynecol Oncol. 1997;65(3):447–52.
179. Lack EE, Perez-Atayde AR, Murthy AS, Goldstein DP, Crigler Jr JF, Vawter GF. Granulosa theca cell tumors in premenarchal girls: a clinical and pathologic study of ten cases. Cancer. 1981;48(8): 1846–54.
180. Kalfa N, Meduri G, Philibert P, et al. Unusual virilization in girls with juvenile granulosa cell tumors of the ovary is related to intratumoral aromatase deficiency. Horm Res Paediatr. 2010;74(2):83–91. Epub Apr 15, 2010.
181. Vassal G, Flamant F, Caillaud JM, Demeocq F, Nihoul-Fekete C, Lemerle J. Juvenile granulosa cell tumor of the ovary in children: a clinical study of 15 cases. J Clin Oncol. 1988;6(6):990–5.
182. Bouffet E, Basset T, Chetail N, et al. Juvenile granulosa cell tumor of the ovary in infants: a clinicopathologic study of three cases and review of the literature. J Pediatr Surg. 1997;32(5):762–5.
183. Gell JS, Stannard MW, Ramnani DM, Bradshaw KD. Juvenile granulosa cell tumor in a 13-year-old girl with enchondromatosis (Ollier's disease): a case report. J Pediatr Adolesc Gynecol. 1998;11(3):147–50.
184. Spaun E, Glavind K. Cystic juvenile granulosa cell tumor of the ovary. Acta Obstet Gynecol Scand. 1988;67(2):177–80.
185. Schmidt D, Kommoss F. Diagnosis and differential diagnosis of granulosa cell tumor. Pathologe. 2007;28(3):195–202.
186. Kondi-Pafiti A, Grapsa D, Kairi-Vassilatou E, et al. Granulosa cell tumors of the ovary: a clinicopathologic and immunohistochemical study of 21 cases. Eur J Gynaecol Oncol. 2010;31(1):94–8.
187. Ohishi Y, Kaku T, Oya M, Kobayashi H, Wake N, Tsuneyoshi M. CD56 expression in ovarian granulosa cell tumors, and its diagnostic utility and pitfalls. Gynecol Oncol. 2007;107(1):30–8.
188. Perez-Atayde AR, Joste N, Mulhern H. Juvenile granulosa cell tumor of the infantile testis. Evidence of a dual epithelial-smooth muscle differentiation. Am J Surg Pathol. 1996;20(1):72–9.
189. Hisaoka M, Horie A, Kajiwara Y. Juvenile granulosa cell tumor in a three-year-old infant. An immunohistochemical and ultrastructural study. Acta Pathol Jpn. 1990;40(8):616–21.
190. Pysher TJ, Hitch DC, Krous HF. Bilateral juvenile granulosa cell tumors in a 4-month-old dysmorphic infant. A clinical, histologic, and ultrastructural study. Am J Surg Pathol. 1981;5(8): 789–94.
191. Roth LM, Nicholas TR, Ehrlich CE. Juvenile granulosa cell tumor: a clinicopathologic study of three cases with ultrastructural observations. Cancer. 1979;44(6):2194–205.
192. Kalfa N, Philibert P, Patte C, et al. Extinction of FOXL2 expression in aggressive ovarian granulosa cell tumors in children. Fertil Steril. 2007;87(4):896–901.
193. Kalfa N, Philibert P, Patte C, et al. Juvenile granulosa-cell tumor: clinical and molecular expression. Gynécol Obstét Fertil. 2009;37(1):33–44.
194. Stenwig JT, Hazekamp JT, Beecham JB. Granulosa cell tumors of the ovary. A clinicopathological study of 118 cases with long-term follow-up. Gynecol Oncol. 1979;7(2):136–52.
195. Daubenton JD, Sinclair-Smith C. Severe hypercalcemia in association with a juvenile granulosa cell tumor of the ovary. Med Pediatr Oncol. 2000;34(4):301–3.

196. Piura B, Wiznitzer A, Shaco-Levy R. Juvenile granulosa cell tumor of the ovary associated with hypercalcemia. Arch Gynecol Obstet. 2008;277(3):257–62.
197. Crozier MA, Copeland LJ, Silva EG, Gershenson DM, Stringer CA. Clear cell carcinoma of the ovary: a study of 59 cases. Gynecol Oncol. 1989;35(2):199–203.
198. Vang R, Whitaker BP, Farhood AI, Silva EG, Ro JY, Deavers MT. Immunohistochemical analysis of clear cell carcinoma of the gynecologic tract. Int J Gynecol Pathol. 2001;20(3):252–9.
199. Silva EG, Tornos C, Bailey MA, Morris M. Undifferentiated carcinoma of the ovary. Arch Pathol Lab Med. 1991;115(4):377–81.
200. Roth LM, Talerman A. Recent advances in the pathology and classification of ovarian germ cell tumors. Int J Gynecol Pathol. 2006;25:305–20.
201. Plantaz D, Flamant F, Vassal G, et al. Granulosa cell tumors of the ovary in children and adolescents. Multicenter retrospective study in 40 patients aged 7 months to 22 years. Arch Fr Pédiatr. 1992;49(9):793–8.
202. Erdreich-Epstein A, Monforte HL, Lavey RS, Joshi S, Phillips JD, Villablanca JG. Successful multimodality therapy of recurrent multifocal juvenile granulosa cell tumor of the ovary. J Pediatr Hematol Oncol. 2002;24(3):229–33.
203. Powell JL, Connor GP, Henderson GS. Management of recurrent juvenile granulosa cell tumor of the ovary. Gynecol Oncol. 2001;81(1):113–6.
204. Jamieson S, Fuller PJ. Management of granulosa cell tumour of the ovary. Curr Opin Oncol. 2008;20(5):560–4.
205. Roth LM, Anderson MC, Govan AD, Langley FA, Gowing NF, Woodcock AS. Sertoli-Leydig cell tumors: a clinicopathologic study of 34 cases. Cancer. 1981;48(1):187–97.
206. Young RH, Scully RE. Ovarian Sertoli-Leydig cell tumors. A clinicopathological analysis of 207 cases. Am J Surg Pathol. 1985; 9(8):543–69.
207. Zaloudek C, Norris HJ. Sertoli-Leydig tumors of the ovary. A clinicopathologic study of 64 intermediate and poorly differentiated neoplasms. Am J Surg Pathol. 1984;8(6):405–18.
208. Young RH, Dudley AG, Scully RE. Granulosa cell, Sertoli-Leydig cell, and unclassified sex cord-stromal tumors associated with pregnancy: a clinicopathological analysis of thirty-six cases. Gynecol Oncol. 1984;18(2):181–205.
209. Young RH, Scully RE. Ovarian Sertoli-Leydig cell tumors with a retiform pattern: a problem in histopathologic diagnosis. A report of 25 cases. Am J Surg Pathol. 1983;7(8):755–71.
210. Lenhard M, Kuemper C, Ditsch N, et al. Use of novel serum markers in clinical follow-up of Sertoli-Leydig cell tumours. Clin Chem Lab Med. 2007;45(5):657–61.
211. Chadha S, Honnebier WJ, Schaberg A. Raised serum alpha-fetoprotein in Sertoli-Leydig cell tumor (androblastoma) of ovary: report of two cases. Int J Gynecol Pathol. 1987;6(1):82–8.
212. Gagnon S, Tetu B, Silva EG, McCaughey WT. Frequency of alpha-fetoprotein production by Sertoli-Leydig cell tumors of the ovary: an immunohistochemical study of eight cases. Mod Pathol. 1989;2(1):63–7.
213. Prat J, Young RH, Scully RE. Ovarian Sertoli-Leydig cell tumors with heterologous elements. II. Cartilage and skeletal muscle: a clinicopathologic analysis of twelve cases. Cancer. 1982;50(11): 2465–75.
214. Azuma A, Koyama T, Mikami Y, et al. A case of Sertoli-Leydig cell tumour of the ovary with a multilocular cystic appearance on CT and MR imaging. Pediatr Radiol. 2008;38(8):898–901.
215. Young RH, Prat J, Scully RE. Ovarian Sertoli-Leydig cell tumors with heterologous elements. I. Gastrointestinal epithelium and carcinoid: a clinicopathologic analysis of thirty-six cases. Cancer. 1982;50(11):2448–56.
216. Roth LM, Slayton RE, Brady LW, Blessing JA, Johnson G. Retiform differentiation in ovarian Sertoli-Leydig cell tumors. A clinicopathologic study of six cases from a Gynecologic Oncology Group study. Cancer. 1985;55(5):1093–8.
217. Mooney EE, Vaidya KP, Tavassoli FA. Ossifying well-differentiated Sertoli-Leydig cell tumor of the ovary. Ann Diagn Pathol. 2000;4(1):34–8.
218. Talerman A. Ovarian Sertoli-Leydig cell tumor (androblastoma) with retiform pattern. A clinicopathologic study. Cancer. 1987;60(12): 3056–64.
219. Reddick RL, Walton LA. Sertoli-Leydig cell tumor of the ovary with teratomatous differentiation: clinicopathologic considerations. Cancer. 1982;50(6):1171–6.
220. Waxman M, Damjanov I, Alpert L, Sardinsky T. Composite mucinous ovarian neoplasms associated with Sertoli-Leydig and carcinoid tumors. Cancer. 1981;47(8):2044–52.
221. Aguirre P, Scully RE, DeLellis RA. Ovarian heterologous Sertoli-Leydig cell tumors with gastrointestinal-type epithelium. An immunohistochemical analysis. Arch Pathol Lab Med. 1986; 110(6):528–33.
222. Virk R, Lu D. Mucinous adenocarcinoma as heterologous element in intermediately differentiated Sertoli-Leydig cell tumor of the ovary. Pathol Res Pract. 2010;206(7):489–92. Epub Aug 11, 2009.
223. Sweeney EC, Barry-Walsh C, Robinson A. Sertoli-Leydig cell tumor of the ovary with heterologous elements and carcinoid: an immunohistochemical and ultrastructural study. Ultrastruct Pathol. 1983;5(2–3):185–94.
224. Prat J, Bhan AK, Dickersin GR, Robboy SJ, Scully RE. Hepatoid yolk sac tumor of the ovary (endodermal sinus tumor with hepatoid differentiation): a light microscopic, ultrastructural and immunohistochemical study of seven cases. Cancer. 1982;50(11): 2355–68.
225. Mooney EE, Nogales FF, Tavassoli FA. Hepatocytic differentiation in retiform Sertoli-Leydig cell tumors: distinguishing a heterologous element from Leydig cells. Hum Pathol. 1999;30(6):611–7.
226. Young RH, Perez-Atayde AR, Scully RE. Ovarian Sertoli-Leydig cell tumor with retiform and heterologous components. Report of a case with hepatocytic differentiation and elevated serum alpha-fetoprotein. Am J Surg Pathol. 1984;8(9):709–18.
227. Cao QJ, Jones JG, Li M. Expression of calretinin in human ovary, testis, and ovarian sex cord-stromal tumors. Int J Gynecol Pathol. 2001;20(4):346–52.
228. Zhao C, Vinh TN, McManus K, Dabbs D, Barner R, Vang R. Identification of the most sensitive and robust immunohistochemical markers in different categories of ovarian sex cord-stromal tumors. Am J Surg Pathol. 2009;33(3):354–66.
229. Busam KJ, Iversen K, Coplan KA, et al. Immunoreactivity for A103, an antibody to melan-A (Mart-1), in adrenocortical and other steroid tumors. Am J Surg Pathol. 1998;22(1):57–63.
230. Kooijman CD, Straks W. Sertoli cell and sertoli-Leydig cell tumors of the ovary. A report of three cases with ultrastructural findings. Eur J Obstet Gynecol Reprod Biol. 1982;13(2):93–104.
231. Ordi J, Schammel DP, Rasekh L, Tavassoli FA. Sertoliform endometrioid carcinomas of the ovary: a clinicopathologic and immunohistochemical study of 13 cases. Mod Pathol. 1999;12(10): 933–40.
232. Roth LM, Liban E, Czernobilsky B. Ovarian endometrioid tumors mimicking Sertoli and Sertoli-Leydig cell tumors: sertoliform variant of endometrioid carcinoma. Cancer. 1982;50(7):1322–31.
233. McCluggage WG, Young RH. Ovarian sertoli-leydig cell tumors with pseudoendometrioid tubules (pseudoendometrioid sertoli-leydig cell tumors). Am J Surg Pathol. 2007;31(4):592–7.
234. Hughesdon PE. Ovarian tumours of Wolffian or allied nature: their place in ovarian oncology. J Clin Pathol. 1982;35(5):526–35.
235. Young RH. From krukenberg to today: the ever present problems posed by metastatic tumors in the ovary: part I. Historical perspective, general principles, mucinous tumors including the krukenberg tumor. Adv Anat Pathol. 2006;13(5):205–27.

236. Oliva E, Alvarez T, Young RH. Sertoli cell tumors of the ovary: a clinicopathologic and immunohistochemical study of 54 cases. Am J Surg Pathol. 2005;29(2):143–56.
237. Young RH, Scully RE. Ovarian Sertoli cell tumors: a report of 10 cases. Int J Gynecol Pathol. 1984;2(4):349–63.
238. Tavassoli FA, Norris HJ. Sertoli tumors of the ovary. A clinicopathologic study of 28 cases with ultrastructural observations. Cancer. 1980;46(10):2281–97.
239. Zung A, Shoham Z, Open M, Altman Y, Dgani R, Zadik Z. Sertoli cell tumor causing precocious puberty in a girl with Peutz-Jeghers syndrome. Gynecol Oncol. 1998;70(3):421–4.
240. Tracy SL, Askin FB, Reddick RL, Jackson B, Kurman RJ. Progesterone secreting Sertoli cell tumor of the ovary. Gynecol Oncol. 1985;22(1):85–96.
241. Ehrlich EN, Dominguez OV, Samuels LT, Lynch D, Oberhelman HJ, Warner NE. Aldosteronism and precocious puberty due to an ovarian androblastoma (Sertoli cell tumor). J Clin Endocrinol Metab. 1963;23(4):358–67.
242. Korzets A, Nouriel H, Steiner Z, et al. Resistant hypertension associated with a renin-producing ovarian Sertoli cell tumor. Am J Clin Pathol. 1986;85(2):242–7.
243. Ferry JA, Young RH, Engel G, Scully RE. Oxyphilic Sertoli cell tumor of the ovary: a report of three cases, two in patients with the Peutz-Jeghers syndrome. Int J Gynecol Pathol. 1994;13(3):259–66.
244. McCluggage WG. Recent advances in immunohistochemistry in gynaecological pathology. Histopathology. 2002;40(4):309–26.
245. Riopel MA, Perlman EJ, Seidman JD, Kurman RJ, Sherman ME. Inhibin and epithelial membrane antigen immunohistochemistry assist in the diagnosis of sex cord-stromal tumors and provide clues to the histogenesis of hypercalcemic small cell carcinomas. Int J Gynecol Pathol. 1998;17(1):46–53.
246. Stewart CJ, Jeffers MD, Kennedy A. Diagnostic value of inhibin immunoreactivity in ovarian gonadal stromal tumours and their histological mimics. Histopathology. 1997;31(1):67–74.
247. Oliva E, Vu Q, Young RH. CD10 Expression in Sex Cord Stromal Tumors (SCSTs) and Steroid Cell Tumors (StCTs) of the ovary. Mod Pathol. 2002;15:204a.
248. McCluggage WG, Shanks JH, Whiteside C, Maxwell P, Banerjee SS, Biggart JD. Immunohistochemical study of testicular sex cord-stromal tumors, including staining with anti-inhibin antibody. Am J Surg Pathol. 1998;22(5):615–9.
249. Ramzy I, Bos C. Sertoli cell tumors of ovary: light microscopic and ultrastructural study with histogenetic considerations. Cancer. 1976;38(6):2447–56.
250. Young RH, Prat J, Scully RE. Ovarian endometrioid carcinomas resembling sex cord-stromal tumors. A clinicopathological analysis of 13 cases. Am J Surg Pathol. 1982;6(6):513–22.
251. Pitman MB, Young RH, Clement PB, Dickersin GR, Scully RE. Endometrioid carcinoma of the ovary and endometrium, oxyphilic cell type: a report of nine cases. Int J Gynecol Pathol. 1994;13(4): 290–301.
252. Tornos C, Silva EG, Ordonez NG, Gershenson DM, Young RH, Scully RE. Endometrioid carcinoma of the ovary with a prominent spindle-cell component, a source of diagnostic confusion. A report of 14 cases. Am J Surg Pathol. 1995;19(12):1343–53.
253. Rahilly MA, Williams AR, Krausz T, al Nafussi A. Female adnexal tumour of probable Wolffian origin: a clinicopathological and immunohistochemical study of three cases. Histopathology. 1995;26(1):69–74.
254. McGowan L, Young RH, Scully RE. Peutz-Jeghers syndrome with "adenoma malignum" of the cervix. A report of two cases. Gynecol Oncol. 1980;10(2):125–33.
255. Srivatsa PJ, Keeney GL, Podratz KC. Disseminated cervical adenoma malignum and bilateral ovarian sex cord tumors with annular tubules associated with Peutz-Jeghers syndrome. Gynecol Oncol. 1994;53(2):256–64.
256. Scully RE. Sex cord tumor with annular tubules a distinctive ovarian tumor of the Peutz-Jeghers syndrome. Cancer. 1970;25(5): 1107–21.
257. Hart WR, Kumar N, Crissman JD. Ovarian neoplasms resembling sex cord tumors with annular tubules. Cancer. 1980;45(9):2352–63.
258. Shen K, Wu PC, Lang JH, Huang RL, Tang MT, Lian LJ. Ovarian sex cord tumor with annular tubules: a report of six cases. Gynecol Oncol. 1993;48(2):180–4.
259. Dolan J, Al-Timimi AH, Richards SM, et al. Does ovarian sex cord tumour with annular tubules produce progesterone? J Clin Pathol. 1986;39(1):29–35.
260. Young RH, Welch WR, Dickersin GR, Scully RE. Ovarian sex cord tumor with annular tubules: review of 74 cases including 27 with Peutz-Jeghers syndrome and four with adenoma malignum of the cervix. Cancer. 1982;50(7):1384–402.
261. Gloor E. Ovarian sex cord tumor with annular tubules. Clinicopathologic report of two benign and one malignant cases with long follow-ups. Virchows Arch A Pathol Anat Histol. 1979;384(2):185–93.
262. McCluggage WG, McKenna M, McBride HA. CD56 is a sensitive and diagnostically useful immunohistochemical marker of ovarian sex cord-stromal tumors. Int J Gynecol Pathol. 2007;26(3): 322–7.
263. Stewart CJ, Nandini CL, Richmond JA. Value of A103 (melan-A) immunostaining in the differential diagnosis of ovarian sex cord stromal tumours. J Clin Pathol. 2000;53(3):206–11.
264. Yao DX, Soslow RA, Hedvat CV, Leitao M, Baergen RN. Melan-A (A103) and inhibin expression in ovarian neoplasms. Appl Immunohistochem Mol Morphol. 2003;11(3):244–9.
265. Ahn GH, Chi JG, Lee SK. Ovarian sex cord tumor with annular tubules. Cancer. 1986;57(5):1066–73.
266. Crissman JD, Hart WR. Ovarian sex cord tumors with annular tubules. An ultrastructural study of three cases. Am J Clin Pathol. 1981;75(1):11–7.
267. Connolly DC, Katabuchi H, Cliby WA, Cho KR. Somatic mutations in the STK11/LKB1 gene are uncommon in rare gynecological tumor types associated with Peutz-Jegher's syndrome. Am J Pathol. 2000;156(1):339–45.
268. Kato N, Romero M, Catasus L, Prat J. The STK11/LKB1 Peutz-Jegher gene is not involved in the pathogenesis of sporadic sex cord-stromal tumors, although loss of heterozygosity at 19p13.3 indicates other gene alteration in these tumors. Hum Pathol. 2004;35(9):1101–4.
269. Susman M, Cooper C, Cooper M, Carter J, Russell P. Gonadoblastoma progressing to dysgerminoma in a 55-year-old woman with normal karyotype: follow-up studies. Pathology. 2008;40(6):645–6.
270. Matamala MF, Nogales FF, Lardelli P, Navarro N. Metastatic granulosa cell tumor with pattern of sex cord tumor with annular tubules. Int J Gynecol Pathol. 1987;6(2):185–93.
271. Barker D, Sharma R, McIndoe A, et al. An unusual case of sex cord tumor with annular tubules with malignant transformation in a patient with Peutz-Jeghers syndrome. Int J Gynecol Pathol. 2010;29(1):27–32.
272. Lele SM, Sawh RN, Zaharopoulos P, et al. Malignant ovarian sex cord tumor with annular tubules in a patient with Peutz-Jeghers syndrome: a case report. Mod Pathol. 2000;13(4):466–70.
273. Gustafson ML, Lee MM, Scully RE, et al. Mullerian inhibiting substance as a marker for ovarian sex-cord tumor. N Engl J Med. 1992;326(7):466–71.
274. Puls LE, Hamous J, Morrow MS, Schneyer A, MacLaughlin DT, Castracane VD. Recurrent ovarian sex cord tumor with annular tubules: tumor marker and chemotherapy experience. Gynecol Oncol. 1994;54(3):396–401.
275. Scully RE. Stromal luteoma of the ovary. A distinctive type of lipoid-cell tumor. Cancer. 1964;17:769–78.

276. Yamada S, Tanimoto A, Wang KY, Shimajiri S, Sasaguri Y. Stromal luteoma and nodular hyperthecosis of the bilateral ovaries associated with atypical endometrial hyperplasia of the uterus. Pathol Int. 2009;59(11):831–3.
277. Seidman JD, Abbondanzo SL, Bratthauer GL. Lipid cell (steroid cell) tumor of the ovary: immunophenotype with analysis of potential pitfall due to endogenous biotin-like activity. Int J Gynecol Pathol. 1995;14(4):331–8.
278. Schnoy N. Ultrastructure of a virilizing ovarian Leydig-cell-tumor. Hilar cell tumor. Virchows Arch A Pathol Anat Histol. 1982; 397(1):17–27.
279. Roth LM, Sternberg WH. Ovarian stromal tumors containing Leydig cells. II. Pure Leydig cell tumor, non-hilar type. Cancer. 1973;32(4):952–60.
280. Sternberg WH, Roth LM. Ovarian stromal tumors containing Leydig cells. Cancer. 1973;32(4):940–51.
281. Boivin Y, Richart RM. Hilus cell tumors of the ovary. A review with a report of 3 new cases. Cancer. 1965;18:231–40.
282. Ichinohasama R, Teshima S, Kishi K, et al. Leydig cell tumor of the ovary associated with endometrial carcinoma and containing 17 beta-hydroxysteroid dehydrogenase. Int J Gynecol Pathol. 1989;8(1):64–71.
283. Baramki TA, Leddy AL, Woodruff JD. Bilateral hilus cell tumors of the ovary. Obstet Gynecol. 1983;62(1):128–31.
284. Duun S. Bilateral virilizing hilus (Leydig) cell tumors of the ovary. Acta Obstet Gynecol Scand. 1994;73(1):76–7.
285. Devouassoux-Shisheboran M, Deschildre C, Mauduit C, et al. Expression of galectin-3 in gonads and gonadal sex cord stromal and germ cell tumors. Oncol Rep. 2006;16(2):335–40.
286. Adeyemi SD, Grange AO, Giwa-Osagie OF, Elesha SO. Adrenal rest tumour of the ovary associated with isosexual precocious pseudopuberty and cushingoid features. Eur J Pediatr. 1986;145(3): 236–8.
287. Klotz RK, Muller-Holzner E, Fessler S, et al. Leydig-cell-tumor of the ovary that responded to GnRH-analogue administration - case report and review of the literature. Exp Clin Endocrinol Diabetes. 2010;118(5):291–7.
288. Donovan JT, Otis CN, Powell JL, Cathcart HK. Cushing's syndrome secondary to malignant lipoid cell tumor of the ovary. Gynecol Oncol. 1993;50(2):249–53.
289. Young RH, Scully RE. Ovarian steroid cell tumors associated with Cushing's syndrome: a report of three cases. Int J Gynecol Pathol. 1987;6(1):40–8.
290. Kulkarni JN, Mistry RC, Kamat MR, Chinoy R, Lotlikar RG. Autonomous aldosterone-secreting ovarian tumor. Gynecol Oncol. 1990;37(2):284–9.
291. Deavers MT, Malpica A, Ordonez NG, Silva EG. Ovarian steroid cell tumors: an immunohistochemical study including a comparison of calretinin with inhibin. Int J Gynecol Pathol. 2003;22(2): 162–7.
292. Gordon MD, Corless C, Renshaw AA, Beckstead J. CD99, keratin, and vimentin staining of sex cord-stromal tumors, normal ovary, and testis. Mod Pathol. 1998;11(8):769–73.
293. Young RH, Scully RE. Oxyphilic clear cell carcinoma of the ovary. A report of nine cases. Am J Surg Pathol. 1987;11(9): 661–7.
294. Talerman A, Haije WG, Baggerman L. Serum alphafetoprotein (AFP) in diagnosis and management of endodermal sinus (yolk sac) tumor and mixed germ cell tumor of the ovary. Cancer. 1978;41(1):272–8.
295. Pitman MB, Triratanachat S, Young RH, Oliva E. Hepatocyte paraffin 1 antibody does not distinguish primary ovarian tumors with hepatoid differentiation from metastatic hepatocellular carcinoma. Int J Gynecol Pathol. 2004;23(1):58–64.
296. Ishikura H, Scully RE. Hepatoid carcinoma of the ovary. A newly described tumor. Cancer. 1987;60(11):2775–84.
297. Young RH, Gersell DJ, Clement PB, Scully RE. Hepatocellular carcinoma metastatic to the ovary: a report of three cases discovered during life with discussion of the differential diagnosis of hepatoid tumors of the ovary. Hum Pathol. 1992;23(5): 574–80.
298. Insabato L, De Rosa G, Franco R, D'Onofrio V, Di Vizio D. Ovarian metastasis from renal cell carcinoma: a report of three cases. Int J Surg Pathol. 2003;11(4):309–12.
299. McCluggage WG, Young RH. Paraganglioma of the ovary: report of three cases of a rare ovarian neoplasm, including two exhibiting inhibin positivity. Am J Surg Pathol. 2006;30(5):600–5.
300. Chalvardjian A, Derzko C. Gynandroblastoma: its ultrastructure. Cancer. 1982;50(4):710–21.
301. Neubecker RD, Breen SL. Gynandroblastoma. A report of five cases, with a discussion of the histogenesis and classification of ovarian tumors. Am J Clin Pathol. 1962;38:60–9.
302. Seidman JD. Unclassified ovarian gonadal stromal tumors. A clinicopathologic study of 32 cases. Am J Surg Pathol. 1996; 20(6):699–706.
303. Simpson JL, Michael H, Roth LM. Unclassified sex cord-stromal tumors of the ovary: a report of eight cases. Arch Pathol Lab Med. 1998;122(1):52–5.
304. Young RH, Dickersin GR, Scully RE. A distinctive ovarian sex cord-stromal tumor causing sexual precocity in the Peutz-Jeghers syndrome. Am J Surg Pathol. 1983;7(3):233–43.

15 Pathology of Mesenchymal and Hematopoietic Tumors

Esther Oliva

Keywords

Smooth muscle tumors • Endometrioid stromal sarcoma • Mesodermal adenosarcoma • Angiosarcoma • Rhabdomyosarcoma • Primitive neurectodermal tumor • Osteosarcoma • Pecoma • Lymphoma/Leukemia

Smooth Muscle Tumors

Prevalence and Clinical Presentation

In contrast to their uterine counterparts, smooth muscle tumors arising in the ovary are very rare, accounting for less than 1% of all ovarian neoplasms. Less than 100 leiomyomas have been reported in the literature and leiomyosarcomas are even more uncommon. In one study, leiomyomas were found in 0.5% of all ovaries evaluated over a 3.5-year period [1]. Patients with leiomyomas may present with signs and symptoms related to a pelvic mass or they may be incidental findings in middle-aged to postmenopausal patients that have simultaneous uterine leiomyomas. In rare instances, patients have presented with ascites or hydrothorax, obstructive symptoms (secondary to bowel obstruction or hydronephrosis), virilization associated with elevated testosterone serum levels [2], or even polymyositis [3]. Leiomyomas during pregnancy have been reported to rapidly grow and cause clinical concern for malignancy. Patients with leiomyosarcoma are slightly older than those with leiomyomas (mean age, 55 years). They typically present with signs and symptoms related to malignant pelvic mass and they may develop paraneoplastic syndromes [4].

E. Oliva
Department of Pathology, Massachusetts General Hospital, 55 Fruit St., Boston, MA 02114, USA and Harvard Medical School, Boston, MA, USA

Macroscopic and Microscopic Features

Benign tumors are typically unilateral and range widely in size, although most are 4–5 cm in greatest dimension [5]. Rarely, tumors may be bilateral and multiple [6]. They are well circumscribed and show a characteristic lobulated, white, whorled, firm, and homogeneous cut surface (Fig. 15.1). However, yellow cut surface (if cellular), calcification, edema, and cystic degeneration may be seen. Malignant tumors also range in size but they show a more variegated, often solid, fleshy, gray to yellow cut surface with more common areas of necrosis and/or hemorrhage when compared to benign tumors. They frequently extend to paraovarian soft tissues and are associated with adhesions. Myxoid tumors often display a gelatinous appearance [4].

Benign and malignant smooth muscle tumors are composed of interlacing fascicles of spindled cells with variable amount of eosinophilic to pale cytoplasm and fusiform "cigar-shaped" nuclei. Benign tumors typically have banal cytologic features and low mitotic counts (often <2/10 HPFs) (Fig. 15.2). In contrast, leiomyosarcomas frequently show moderate to severe cytologic atypia associated with brisk mitotic activity (Fig. 15.3) and, less frequently, tumor

R.A. Soslow and C. Tornos (eds.), *Diagnostic Pathology of Ovarian Tumors*,
DOI: 10.1007/978-1-4419-9751-7_15, © Springer Science+Business Media, LLC 2011

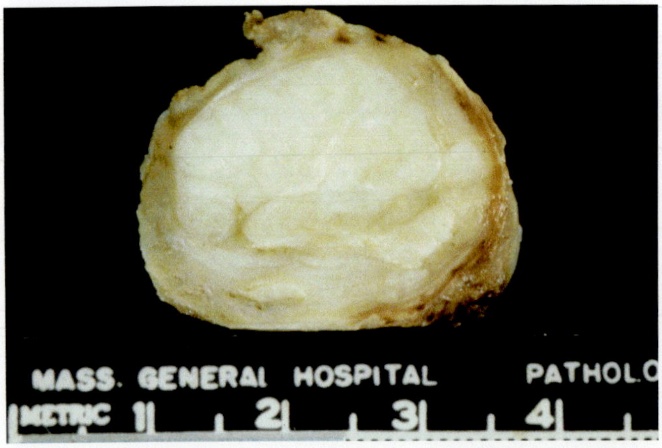

Fig. 15.1 Leiomyoma. The tumor is well-circumscribed and has a white and whorled cut surface

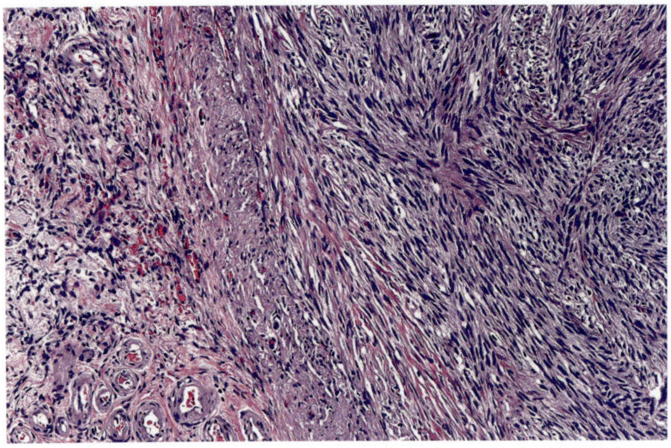

Fig. 15.2 Leiomyoma. The cells have elongated "cigar-shaped" nuclei and eosinophilic cytoplasm and are arranged in intersecting fascicles

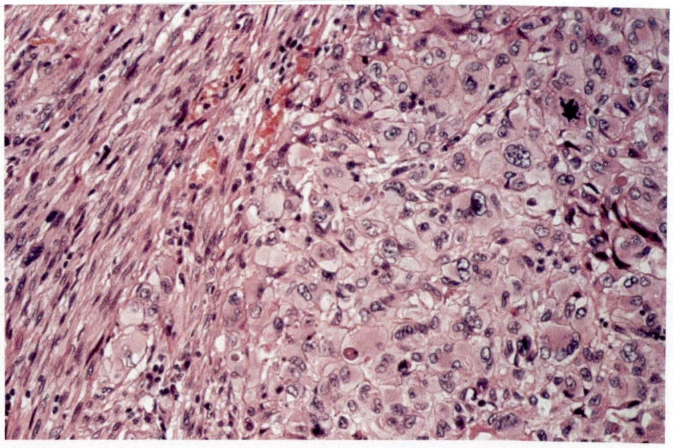

Fig. 15.3 Leiomyosarcoma. The tumor alternates spindle and epithelioid morphologies. The cells in both areas show cytologic atypia

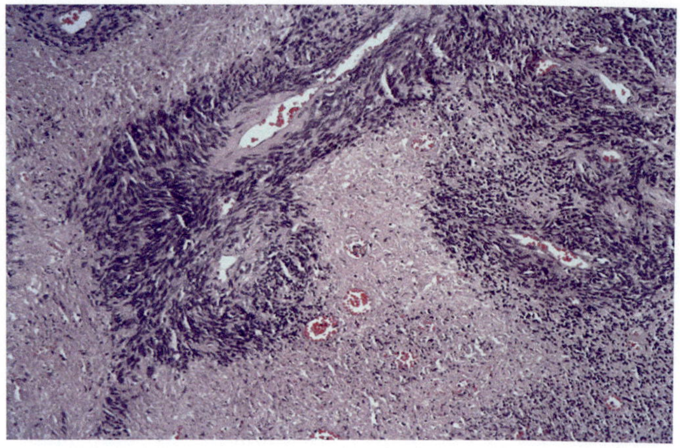

Fig. 15.4 Leiomyosarcoma. The tumor cells grow around vessels and there is an abrupt transition from the viable to the non-viable areas (so-called tumor cell necrosis)

cell necrosis (Fig. 15.4). Hyaline plaques, stromal hyalinization, edema, calcification, and infarct-type necrosis may be encountered in benign and malignant smooth muscle neoplasms. Tumors may show the same morphologic variations as seen in the uterus including cells with bizarre nuclei (leiomyoma with bizarre nuclei/symplastic leiomyoma), increased mitotic activity with no associated cytologic atypia (mitotically active leiomyoma), increased cellularity (cellular leiomyoma), admixture with fat (lipoleiomyoma), cells with abundant clear or eosinophilic cytoplasm (epithelioid leiomyoma/leiomyosarcoma), and abundant myxoid change (myxoid leiomyoma/leiomyosarcoma) [4, 5, 7]. The diagnostic criteria for leiomyosarcoma of the ovary differ slightly from those used in the uterus. Most leiomyosarcomas show at least two of the following: (a) moderate or severe cytologic atypia, (b) mitotic rate ≥10/10 HPFs, and (c) tumor cell necrosis. However, cytologically atypical tumors may demonstrate lesser mitotic activity (5–9/10 HPFs) in the absence of tumor cell necrosis but still be clinically malignant [4]. These tumors are thought to originate from the hilar vessels (Fig. 15.5) or smooth muscle metaplasia. Tumors have been reported to occur in association with endometriosis (some arising from the wall of an endometriotic cyst), dermoid cyst, and mucinous cystadenoma.

Immunohistochemical Profile

Smooth muscle tumors of the ovary stain for vimentin, smooth muscle actin, desmin, and other smooth muscle markers. They can also be positive for keratin, EMA, ER,

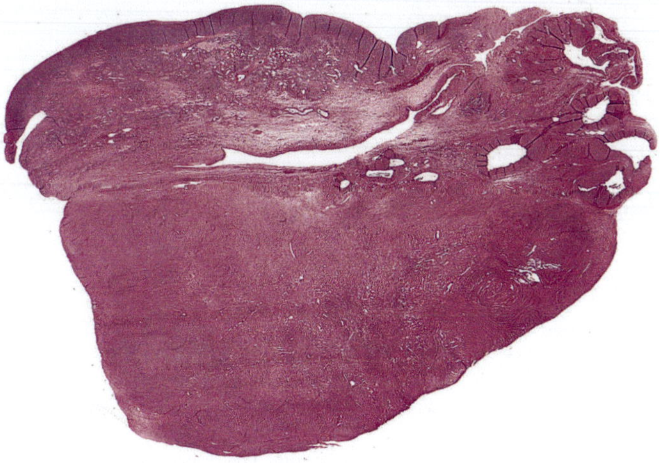

Fig. 15.5 Leiomyoma. The tumor appears to be centered in the hilum of the ovary raising the possibility of an origin from the hilar vessels

PR, and c-kit, as has been reported in the uterus; the frequency and degree of positivity are often lower in malignant tumors [4].

Differential Diagnosis

Before establishing a diagnosis of primary ovarian smooth muscle tumor, parasitic leiomyoma, intravenous leiomyomatosis, leiomyomatosis peritonealis disseminata, and metastatic leiomyosarcoma should be excluded as they may involve the ovary secondarily. Smooth muscle tumors with benign cytologic features may be confused with tumors in the fibroma/thecoma category while malignant tumors should be mostly distinguished from metastatic gastrointestinal stromal tumors, spindle cell carcinoma, and other rare malignant tumors with spindle morphology.

The fascicular growth of ovarian smooth muscle tumors overlaps with that seen in fibroma and its variants. However, fibromas much more frequently show a storiform growth and have abundant extracellular collagen while the cells display scant cytoplasm and wavy nuclei. Smooth muscle actin may be positive, but desmin and h-caldesmon (typically diffusely positive in smooth muscle tumors, especially if benign) are negative or only focally positive [8]. Thecomas may also be in the differential diagnosis of epithelioid smooth muscle tumors. The former, however, have a nodular growth, frequently display hyalinized/calcified plaques, and are positive for inhibin.

Gastrointestinal stromal tumors (GIST) may involve the ovary secondarily. When composed of spindled cells, they can be distinguished from smooth muscle tumors by the finding of shorter fascicles of pale eosinophilic cells with ovoid to spindle-shaped nuclei. Furthermore, in contrast to malignant smooth muscle tumors, these neoplasms rarely exhibit the same degree of cytologic atypia or number of atypical mitoses. A panel including c-kit, CD34, and desmin is very helpful in this differential diagnosis as GISTs typically are c-kit and CD34 positive and desmin negative and the opposite immunoprofile is characteristic of smooth muscle tumors (although leiomyosarcomas in the uterus have been reported to stain with c-kit without showing associated mutations) [9]. Rarely, spindle cell carcinoma, malignant mixed mesodermal tumors, or even endometrioid stromal sarcoma with extensive smooth muscle differentiation may enter in the differential diagnosis of leiomyosarcoma. Extensive sampling should allow for the identification of typical areas and immunohistochemistry may also be helpful.

Finally, benign and malignant myxoid smooth muscle tumors should be distinguished from myxoma. Even though myxoid leiomyosarcoma is hypocellular, it shows abundant intercellular matrix and may not be positive for desmin; in contrast to myxoma, it typically displays intersecting fascicles of spindled cells with "cigar-shaped nuclei" and at least some degree of cytologic atypia.

Treatment and Clinical Outcome

Leiomyomas (including leiomyoma variants) are typically associated with an excellent prognosis if completely excised either by oophorectomy or tumor enucleation. Even though patients with leiomyosarcoma are treated aggressively with hysterectomy, bilateral salpingo-oophorectomy and adjuvant therapy, 70% have tumor recurrences within a mean time interval of 19 months and at least 50% die of their disease within 2 years of the initial diagnosis. Furthermore, at the time of diagnosis, approximately 40% of the tumors have extraovarian extension [4, 10].

Endometrioid Stromal Sarcoma

Prevalence and Clinical Presentation

The ovary is the most common location for extrauterine endometrioid stromal sarcomas, although it is still rare [11, 12]. These tumors are thought to arise from preexistent endometriosis. Patients are peri- or menopausal and present with signs and non-specific symptoms related to a pelvic/abdominal mass.

Macroscopic and Microscopic Features

On gross examination, tumors range widely in size (average 10 cm). They are often solid or solid and cystic with the solid

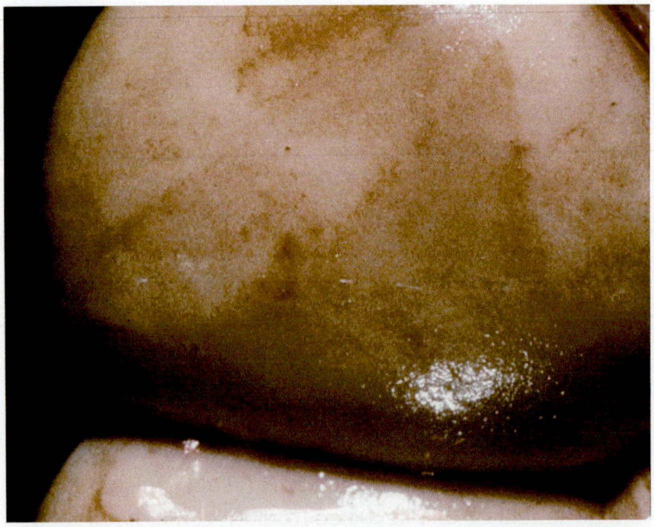

Fig. 15.6 Endometrioid stromal sarcoma. The tumor has a homogeneous tan cut surface

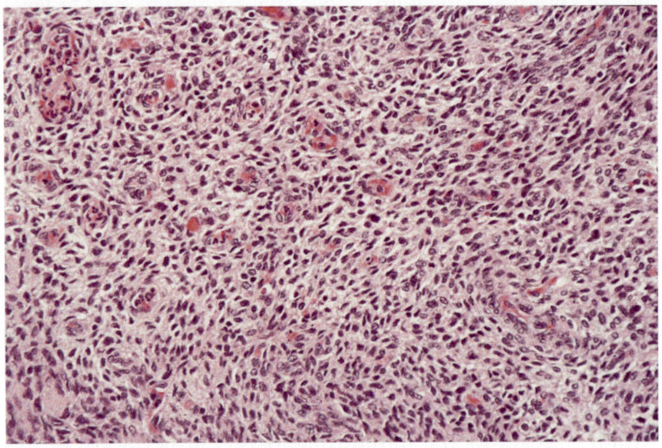

Fig. 15.7 Endometrioid stromal sarcoma. The neoplastic cells are small and uniform with scant cytoplasm and oval nuclei and they whorl around arterioles

areas having a tan or tan to yellow cut surface (Fig. 15.6). Rarely, tumors may be predominantly cystic with plaques or nodules present in the cyst wall. The cysts often contain hemorrhagic fluid. Tumor may be seen coating the ovarian surfaces and necrosis and/or hemorrhage are not infrequent [12, 13]. The morphologic appearance of endometrioid stromal sarcomas in the ovary greatly overlaps with that of their uterine counterpart. Tumors are composed of sheets of small "blue" cells with scant cytoplasm and oval to spindled nuclei with small nucleoli that may whorl around arterioles resembling the proliferative-phase endometrium (Fig. 15.7). In contrast to uterine tumors, a nodular growth is relatively common but the characteristic infiltrative irregular tongues can only be identified at the ovarian hilum and extraovarian soft tissues. These tumors may also show a fibromatous growth with a storiform arrangement of collagen bundles and

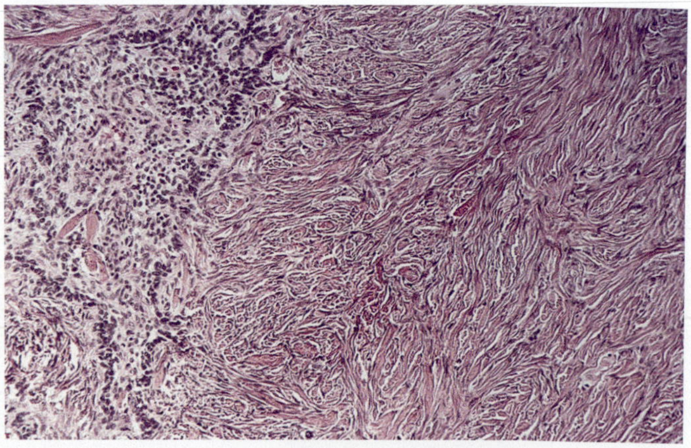

Fig. 15.8 Endometrioid stromal sarcoma. The tumor shows epithelial-like differentiation (*left*) as well as extensively collagenized areas reminiscent of an ovarian fibroma (*right*)

cells with elongated nuclei set in a prominent collagenous background, resembling an ovarian fibroma (Fig. 15.8). Smooth muscle, sex cord-like (Fig. 15.8), and glandular differentiation, as well as myxoid change, epithelioid appearance, and bizarre cells, can be seen [11–16]. Endometriosis is often seen in the vicinity of these tumors [15].

Immunohistochemical Profile

The immunohistochemical profile of endometrioid stromal sarcomas mirrors that reported in tumors originating in the uterus [17–26]. They are typically positive for vimentin, muscle-specific and smooth muscle actin, and frequently keratin. Most tumors stain for CD10, as in the uterus. However, as many tumors in the differential diagnosis also are CD10 positive, this antibody should not be used in isolation. Desmin, h-caldesmon, myosin, and HDCA8 may be positive, even more if areas of smooth muscle differentiation are present. Areas of sex cord-like differentiation may be positive for inhibin, calretinin, CD99, WT-1, and Melan-A as reported in the uterus [27]. Endometrioid stromal tumors frequently are ER and PR positive, although its presence is not specific to these tumors. As c-kit may be expressed in uterine tumors [28], it should be kept in mind that potentially ovarian tumors may be also positive.

Differential Diagnosis

The finding of bilateral ovarian involvement should raise suspicion for metastases from a uterine low-grade endometrial stromal sarcoma, as histology alone is not helpful to distinguish between primary or metastatic tumor them [29]. In general, features used to establish the diagnosis of metastases to the ovary can also be applied to metastatic uterine low-grade endometrial stromal sarcomas. If a concurrent or previous uterine tumor has been reasonably excluded, the diagnosis of primary endometrioid stromal sarcoma of the ovary can be rendered. A fibromatous appearance, one of the less common patterns seen in endometrioid stromal sarcoma, may cause diagnostic problems, particularly in the ovary as it closely mimics an ovarian fibroma. Other features that may be seen in both neoplasms include dense cellularity and mitoses. The tightly whorled storiform pattern of small spindled cells closely resembling normal ovarian cortex frequently seen in fibromas is not a usual feature of ovarian endometrioid stromal sarcoma [12, 13]. Furthermore, fibromas do not show the abundant and distinctive vasculature present in endometrial/oid stromal tumors and are not typically associated with endometriosis [30]. Fibrosarcoma may rarely enter in the differential diagnosis if an endometrioid stromal sarcoma has a prominent spindle growth as both may be large, are commonly associated with extraovarian spread, and have abundant mitoses. However, the long fascicles and cytologic atypia of fibrosarcoma are rarely seen in endometrioid stromal sarcomas [11, 27]. CD10 is helpful in the differential diagnosis of fibroma and fibrosarcoma from endometrioid stromal sarcoma, as both are negative for this marker [31].

In the ovary, when sex cord-like differentiation is extensive, these tumors may mimic a sex cord stromal tumor, particularly a granulosa cell tumor or intermediate-to-poorly differentiated Sertoli-Leydig cell tumor. However, the latter tumors are often confined to the ovary at the time of diagnosis, lack the typical vascular pattern of endometrioid stromal tumors, and may exhibit other classical patterns. Sex cord stromal tumors, including granulosa cell tumors, and endometrioid stromal sarcoma are CD10 positive [31]. Furthermore, areas of sex cord-like differentiation in endometrioid stromal sarcomas frequently show positivity for inhibin, calretinin, Melan-A, CD99, and CD56, as occurs in sex cord stromal tumors. Thus, it is very important to correlate areas of positivity with tumor morphology, as only typical areas of endometrioid stromal sarcoma will be negative for sex cord markers. Finally, hormonal manifestations are common in patients with sex cord stromal tumors in contrast to endometrioid stromal sarcoma.

Low-grade mesodermal adenosarcoma and endometrioid stromal sarcoma with glands share the following features: biphasic growth pattern with benign-appearing glands and low-grade endometrial stroma, and sex cord-like differentiation. A variety of Müllerian-type epithelium, phyllodes-like architecture, and characteristic periglandular stromal condensation are typical of low-grade mesodermal adenosarcoma in contrast to endometrioid stromal sarcoma [32]. In the very rare case of recurrent adenosarcoma, where only the sarcomatous component is present, the differential diagnosis may include a low-grade endometrioid stromal sarcoma. However, in this scenario, the sarcoma tends to be of higher grade without resemblance to endometrial stroma [33].

Treatment and Clinical Outcome

Although these tumors are considered to be low-grade from the histologic point of view, when originating from the ovary they often show surface involvement or extension to periovarian soft tissues and patients are more prone to recurrences. As in the uterus, recurrences may occur many years after the initial diagnosis. No differences in outcome have been noted between conventional tumors and their variants and histologic parameters, including mitotic index, cannot predict clinical behavior. In a large study, only size appeared to correlate with poor outcome. The mainstay of treatment is surgery that may be followed by hormonal treatment, chemotherapy, or radiation therapy [11].

Mesodermal Adenosarcoma

Prevalence and Clinical Presentation

Even though adenosarcoma is most common in the uterine corpus (low-grade Müllerian adenosarcoma), this neoplasm can also occur at extrauterine sites. The ovary is the most common location, followed by peritoneum, intestine, cervix, vagina, and others [34]. In the ovary, these tumors occur in younger patients (median, 50 years) when compared to their uterine counterpart [32, 35, 36]. Patients often present with a palpable mass on physical exam or may have symptoms and signs related to it, including abdominal pain or distension, but occasionally can present with irregular vaginal bleeding or may be asymptomatic. Patients may have elevated levels of serum CA-125 [37].

Macroscopic and Microscopic Features

On gross examination, these tumors are typically large, with a mean size of 14 cm. They can have polypoid or villous projections if they involve the ovarian surface, as they often rupture [32]. The cut section is frequently spongy with cystic spaces filled with watery, mucoid, or hemorrhagic fluid, surrounded by white to tan tissue; however, it may be mostly solid or predominantly cystic. On microscopic examination, there is typically a biphasic growth of low-grade malignant stroma and benign epithelium, sometimes with a prominent phyllodes architecture (Fig. 15.9). The stromal component often resembles the morphologic appearance of a low-grade endometrial stromal sarcoma or a cellular fibromatous neoplasm characteristically condensating around the glands and

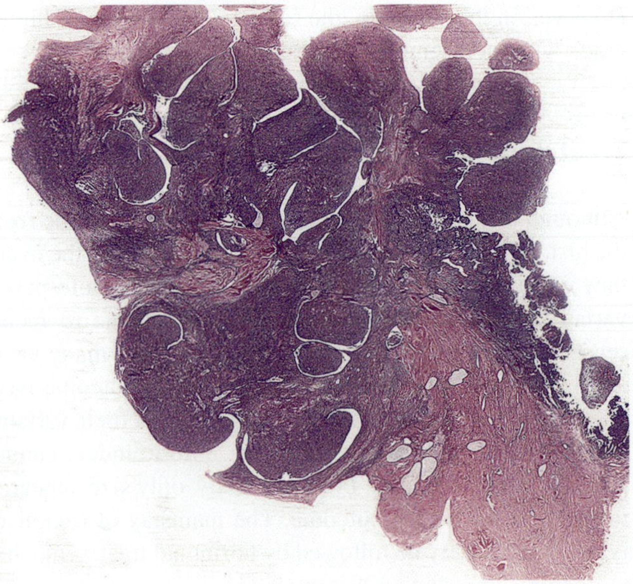

Fig. 15.9 Low-grade mesodermal adenosarcoma. At low-power the tumor has a "phyllodes-like" architecture

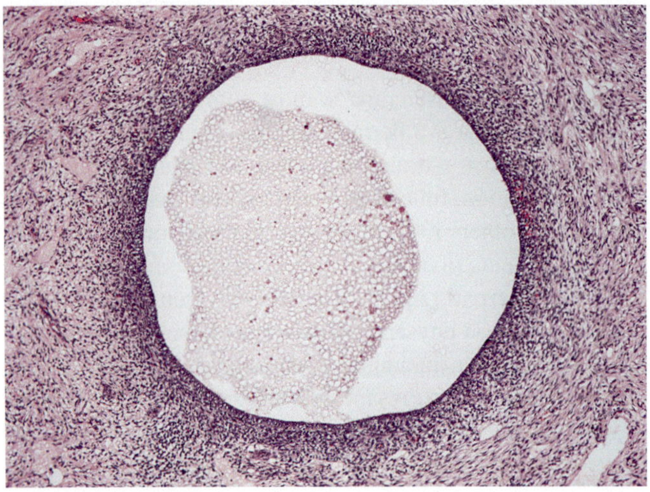

Fig. 15.10 Low-grade mesodermal adenosarcoma. The low-grade malignant stroma condensates around a dilated gland lined by flattened epithelium

cysts (Fig. 15.10) and forming intraluminal polypoid projections (Fig. 15.11). The stroma away from the epithelial component is often less cellular. The stromal cells typically exhibit mild to moderate cytologic atypia and the mitotic activity is usually >4/10 HPFs (Fig 15.12), but tumors can be diagnosed as adenosarcoma with >2 mitoses/10 high power fields if they show the characteristic low-power architecture. Heterologous elements including rhabdomyoblasts, fat and fetal-type cartilage, as well as sex cord-like elements can be seen. Other stromal findings include hyaline plaques, edema, hemorrhage, extensive myxoid change, smooth-muscle differentiation, multinucleated giant cells, ectopic decidua,

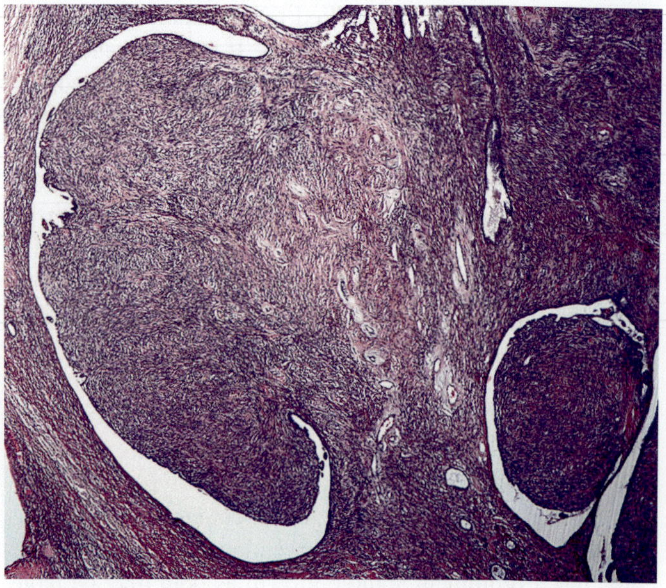

Fig. 15.11 Low-grade mesodermal adenosarcoma. The neoplastic stroma forms cellular intracellular polypoid projections into the glands

foamy macrophages, or luteinization [32]. Sarcomatous overgrowth, defined as an area of pure sarcoma occupying at least 25% of the tumor and most often but not always being high-grade, appears to occur more often in extrauterine adenosarcomas.

The epithelial component is present either as cystically dilated glands (Fig. 15.10) or broad polypoid fronds lined more often by proliferative-type epithelium. However, in occasional instances, cytologic atypia up to carcinoma in situ may be found. Other types of Müllerian epithelium including mucinous endocervical, ciliated, secretory, or indifferent, and non-keratinizing squamous epithelium may be encountered.

Immunohistochemical Profile

The stromal cells are typically positive for vimentin, WT-1, CD10, ER, and PR with variable cytokeratin, muscle actin, and androgen receptor expression [38]. The immunohistochemical profile overlaps with that seen in endometrial stromal tumors. Areas of sarcomatous overgrowth are typically negative for CD10, ER and PR [38, 39].

Differential Diagnosis

These tumors should be distinguished from adenofibroma, endometrioid stromal sarcoma with glandular differentiation and polypoid endometriosis. Adenofibroma may have a similar gross appearance; however, on microscopic examination, there is no stromal condensation of the stroma, no cytologic

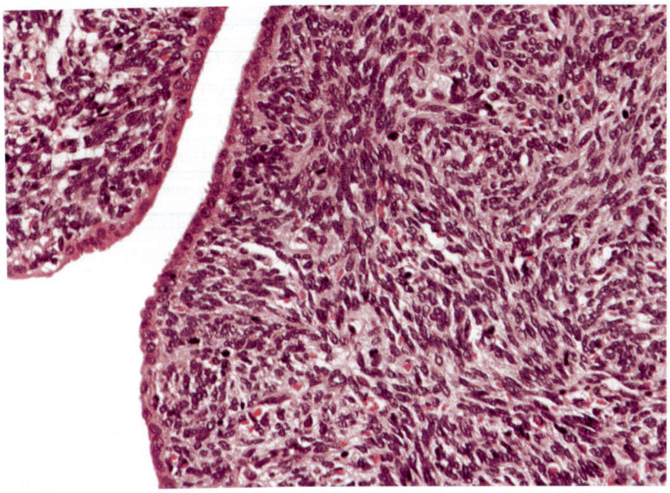

Fig. 15.12 Low-grade mesodermal adenosarcoma. The low-grade malignant stroma is composed of small hyperchromatic cells with brisk mitotic activity. The gland lining shows cuboidal cells with no cytologic atypia

atypia, and no mitotic activity. It has been shown that tumors with even ≤2 mitoses/10 high power fields in the stromal component may behave in a malignant fashion if they have the typical architecture of adenosarcoma [32]. Endometrioid stromal sarcoma is composed of a low-grade sarcoma and may have abundant glands and sex cord-like differentiation with similar morphologic appearance to that seen in adenosarcoma; however, it lacks condensation of the stroma around the glands as well as intraluminal stromal projections. Finally, although at low power, polypoid endometriosis may enter in the differential diagnosis of an adenosarcoma, it typically has the appearance of an endometrial polyp with collagenized stroma and thick blood vessels, lacking the epithelial/stromal relationship seen in adenosarcoma [40]. If tumors show extensive sex cord-like differentiation or heterologous elements, the differential diagnosis is more extensive and may include sex cord stromal tumors, immature teratoma, or even carcinosarcoma. If sarcomatous overgrowth is present, primary sarcomas of the ovary including rhabdomyosarcoma, leiomyosarcoma, fibrosarcoma, or sarcomas secondarily involving the ovary such as gastrointestinal stromal tumor, should be considered in the differential diagnosis.

Treatment and Clinical Outcome

Patients with ovarian tumors often present at a higher stage when compared to those with uterine tumors and they have a poorer prognosis. In the largest study up to date, 75% of patients had tumors stage IC or higher at the time of diagnosis [32]. Extraovarian spread as well as a high-grade stromal component are associated with high recurrence rates with the mean interval to a first recurrence being approximately 2 years. Even when considering tumors confined to the ovary, recurrence rate has been reported to be as high as 60%. Age <53 years, tumor rupture, high-grade stroma, and sarcomatous overgrowth in the primary tumor appear to be associated with recurrence, especially in stage I tumors [32, 36]. Patients have 5 and 10 year survival rates of 64% and 46%, respectively, being even lower at 15 years (30%). Recurrent tumors occur as pure sarcomas in most instances and have an intraabdominal or pulmonary location. Patients are treated with surgery with or without adjuvant chemotherapy, radiation or hormonal therapy; although no standard protocols exist due to the rarity of the tumor [32].

Angiosarcoma

Prevalence and Clinical Presentation

Pure malignant vascular tumors of the ovary are very rare [41–43] and may be seen in association with a mature cystic teratoma or surface epithelial tumors [41, 44–46]. Ovarian angiosarcomas frequently occur in young to middle-aged women in contrast to their vulvovaginal counterparts, which are typically seen in elderly women often after radiation therapy [47]. Patients present with abdominal pain, distention, and anemia if the tumor is large and associated with hemoperitoneum [48].

Macroscopic and Microscopic Features

Grossly, tumors range in size but show frequent cystic degeneration. They have a brown cut surface, are typically hemorrhagic, and frequently have necrosis [41, 42]. At low power, a lobulated architecture associated with cystic spaces, hemorrhage, and necrosis is often seen. The tumor cells tend to recapitulate to some extent the formation of vascular spaces; however, the morphologic appearance ranges from solid to spindled to reticular, slit-like vascular spaces to dilated and anastomosing channels (Fig. 15.13). A papillary growth can also be seen. The vascular spaces have irregular shapes and variable sizes and are often lined by spindled cells with eosinophilic cytoplasm and very atypical nuclei. In the epithelioid variant, tumor cells show abundant cytoplasm and vesicular nuclei and they frequently have a hobnail appearance (Fig. 15.14). Binucleated and multinucleated giant tumor cells may be present. Mitotic rate is typically brisk. Prominent stromal hyalinization may occur.

Immunohistochemical Profile

Tumor cells are positive for vimentin, smooth muscle actin, HHF-35, CD31, CD34, factor VIII, and/or Ulex europeus [42].

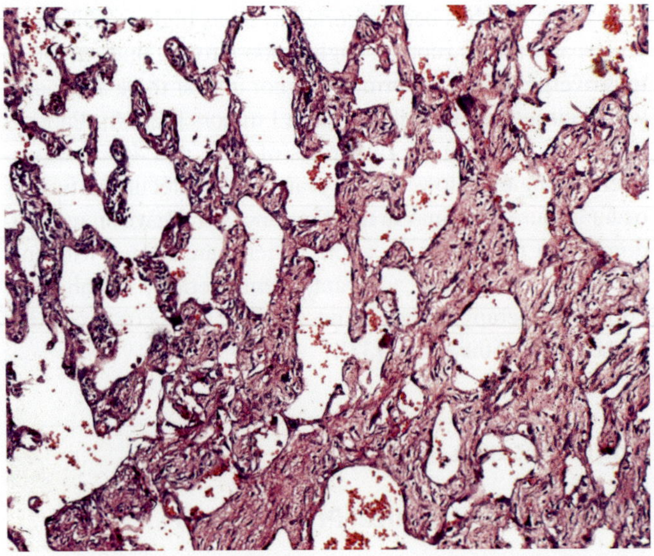

Fig. 15.13 Angiosarcoma, conventional. The tumor is composed of anastomosing channels lined by atypical endothelial cells

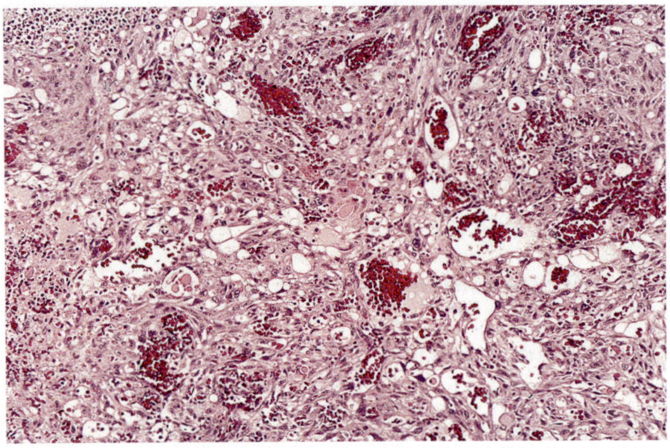

Fig. 15.14 Angiosarcoma, epithelioid. The neoplastic cells form sheets and vascular channels, display abundant cytoplasm as well as cytoplasmic vacuoles, some of them containing red blood cells

Cytokeratin may show focal strong staining, especially in epithelioid angiosarcoma.

Differential Diagnosis

Ovarian angiosarcoma should be distinguished from other high-grade sarcomas, especially spindled and/or epithelioid leiomyosarcoma. Finding of more typical morphologic areas as well as immunohistochemistry are helpful in establishing the correct diagnosis. Mesodermal adenosarcoma with sarcomatous overgrowth and carcinosarcoma may also enter in the differential diagnosis; however, the finding of any epithelial component rules out an angiosarcoma. Distinction from metastatic angiosarcoma is largely based on clinicopathologic correlation (bilateral ovarian involvement, surface deposits, lymphovascular invasion, and a history of a primary tumor elsewhere) [49]. Less frequently, benign vascular lesions may enter in this differential diagnosis. Other potential mimics include yolk sac tumor, clear cell carcinoma, and choriocarcinoma. The presence of vasoformative channels, absence of other typical morphological features of the above mentioned tumors, and immunohistochemical findings support the diagnosis of angiosarcoma.

Treatment and Clinical Outcome

In general, angiosarcoma is associated with a poor prognosis despite aggressive treatment [42]. However, in one study patients with stage I tumors were alive at 5.5 and 9 years after the initial diagnosis [41]. Most patients with angiosarcoma undergo surgery followed by a variety of chemotherapy regimens often based on doxorubicin plus ifosfamide or paclitaxel [50].

Rhabdomyosarcoma

Prevalence and Clinical Presentation

Pure rhabdomyosarcoma of the ovary is very rare with approximately 30 cases reported in the literature [51–55]. It more commonly arises in association with a malignant mixed mesodermal tumor, mesodermal adenosarcoma, teratoma, Sertoli-Leydig cell tumor, or it may coexist with a clear cell carcinoma [56–60]. They occur in patients within a wide age range (<2–86 years) that often present with abdominal pain, distension, and/or weight loss. Rarely, the initial presentation may be related to diffuse bone marrow involvement simulating leukemia [54].

Macroscopic and Microscopic Features

Grossly, tumors are typically large (average 15 cm) and have solid yellow to white to pink cut surfaces (Fig. 15.15) with frequent areas of hemorrhage and necrosis [54]. Embryonal rhabdomyosarcoma may have a gelatinous cut surface and it is the most common subtype of rhabdomyosarcoma in the ovary. It is characterized by alternating hyper- and hypocellular myxoid areas (Fig. 15.16) and may contain cystic spaces filled with eosinophilic fluid (Fig. 15.17). Most of the cells are oval to spindled and display scant eosinophilic cytoplasm, hyperchromatic nuclei, and brisk mitotic activity (Fig. 15.16). Strap- or tadpole cells with cross striations may be seen (Fig. 15.18) but are not common and they are not required to establish a diagnosis of rhabdomyosarcoma.

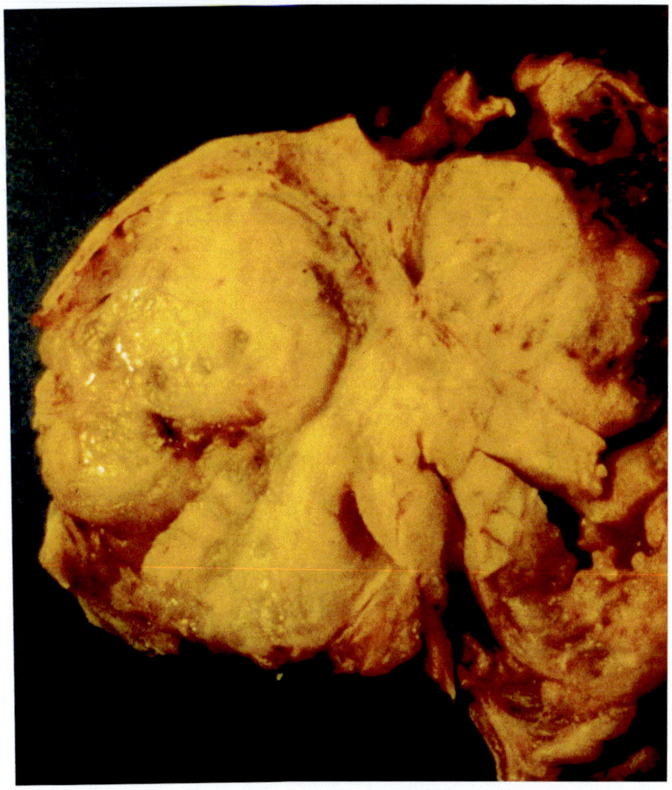

Fig. 15.15 Embryonal rhabdomyosarcoma. The tumor has a multinodular and yellow cut surface with some areas of hemorrhage

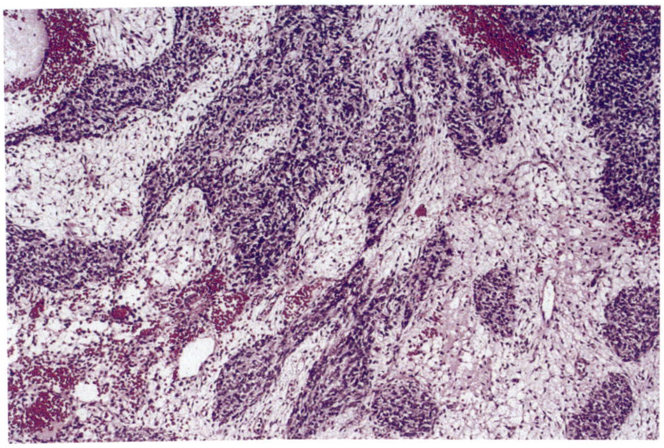

Fig. 15.16 Embryonal rhabdomyosarcoma. Alternating hypercellular and hypocellular/myxoid areas are seen. The hypercellular areas are composed of small hyperchromatic cells

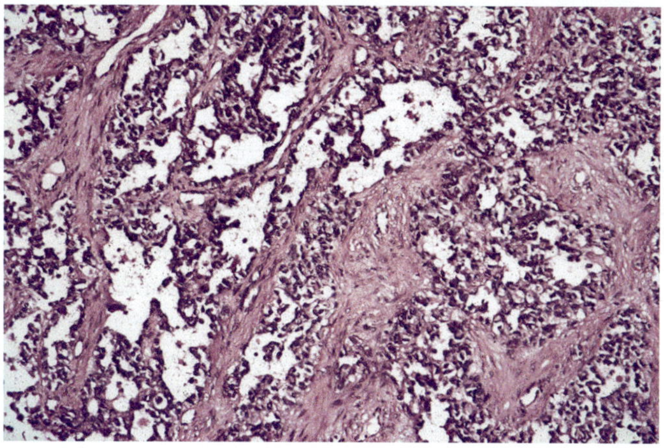

Fig. 15.17 Alveolar rhabdomyosarcoma. The tumor cells form irregular nests with pseudofollicular spaces

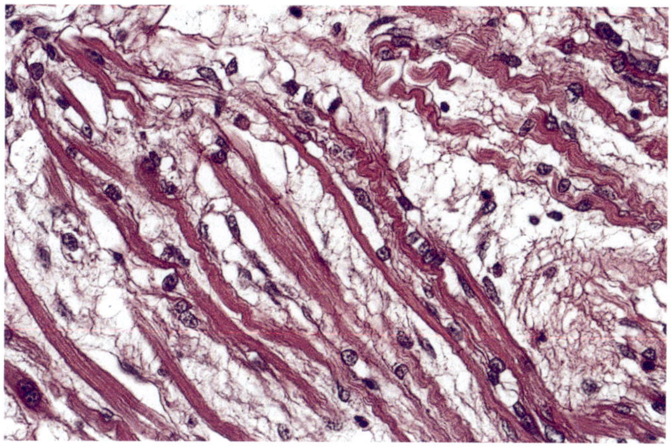

Fig. 15.18 Embryonal rhabdomyosarcoma. Rhabdomyoblasts with prominent cross-striations are seen

Chondroid differentiation resembling fetal cartilage may be seen [52, 54]. Much less common is alveolar rhabdomyosarcoma in which tumor cells are arranged in an alveolar pattern; they may not be cohesive, especially in the center of the alveoli. Fibrous septi of variable thickness may be present. The tumor cells have round to oval nuclei with prominent nucleoli and variable pale to deeply eosinophilic cytoplasm, and they are frequently admixed with wreath-like multinucleated giant cells. One spindled rhabdomyosarcoma characterized by eosinophilic spindle-shaped cells arranged in a fascicular or storiform pattern has been reported in the ovary.

Immunohistochemical Profile

Tumor cells are typically strongly positive for vimentin, muscle actin, desmin, myoglobin, myo-D1, as well as myogenin.

Cytogenetics

Alveolar rhabdomyosarcoma shows a typical t(2;13)(q35;q14) translocation while embryonal rhabdomyosarcoma frequently displays deletions in the long arm of chromosome 11.

Differential Diagnosis

Rhabdomyosarcoma from other gynecologic sites or soft tissues may metastasize to the ovary and, thus, should be considered in the differential diagnosis of a primary ovarian rhabdomyosarcoma. In these cases, the ovarian involvement is not always bilateral, the ovary may be the presenting site, and the morphology is identical to a primary tumor [61, 62]. Tumors composed of small cells, either primary or metastatic, should also be in the differential diagnosis when an embryonal rhabdomyosarcoma is under consideration. These include small cell carcinoma of the hypercalcemic and pulmonary types (primary and metastatic), lymphoma and leukemia, neuroblastoma, primitive neuroectodermal tumor (primary and metastatic), and others. However, in most cases, careful examination discloses the presence of cross striations which, coupled with the use of immunohistochemistry, will help to establish the diagnosis of rhabdomyosarcoma. As mesodermal adenosarcoma, carcinosarcoma, teratoma, and Sertoli-Leydig cell tumors can show extensive rhabdomyoblastic differentiation, these tumors should also be considered in the differential diagnosis. In these cases, extensive sampling is helpful to identify diagnostic areas. The rare alveolar rhabdomyosarcoma should be distinguished from its metastatic counterpart or even melanoma [54], while leiomyosarcoma should enter in the differential diagnosis of the spindle cell variant.

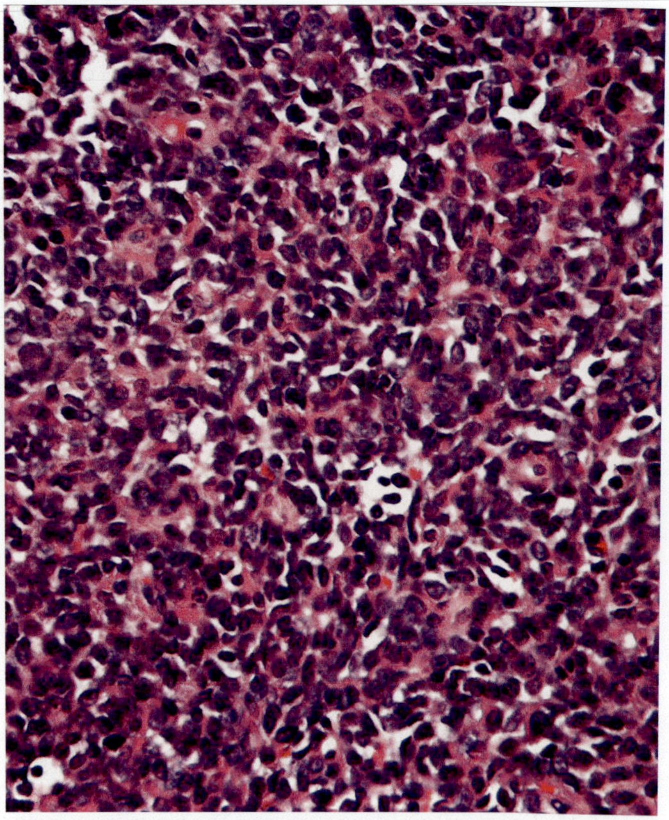

Fig. 15.19 Primitive neuroectodermal tumor, peripheral type. There is a diffuse growth of small hyperchromatic blue cells

Treatment and Clinical Outcome

In contrast to cervical rhabdomyosarcoma that is often associated with a good prognosis, ovarian tumors tend to be associated with a poor outcome [52, 54], especially when occurring in older women (>20 years) and when tumors have extraovarian spread [52]. In one recent study of rhabdomyosarcomas occurring in adults, the 5-year disease specific survival was 29%, in contrast to >85% 5-year overall survival in children with gynecologic tumors [52]. Patients are initially treated with surgery, which is often followed by chemotherapy.

Ewing Sarcoma/Primitive Neuroectodermal Tumor

Prevalence and Clinical Presentation

Extraskeletal Ewing sarcoma (ES) and primitive neuroectodermal tumor (pPNET) of the soft tissues as well as central-type PNETs have been rarely reported in the ovary [63–66]. Central-type PNETs are often seen in association with teratoma [65]. Most ES/pPNET and central-type PNET occur in young patients (<36 years). Common signs and symptoms are related to a rapidly growing abdominal/pelvic mass.

Macroscopic and Microscopic Features

Tumors are frequently large (>10 cm in maximum dimension) and have a multilobulated, soft, fleshy, and often tan to white and friable cut surface. Cystic degeneration, necrosis, and hemorrhage are commonly seen [65, 66]. On microscopic examination, neoplasms in both categories are characterized by a proliferation of small, round, primitive cells arranged in different patterns including diffuse (Fig. 15.19), lobulated, nested, trabecular, or cord-like. The cells typically have scant cytoplasm (but in some cases it may be more abundant depending on the amount of glycogen) and centrally located, irregularly shaped nuclei often with coarse chromatin, and brisk mitotic rate. The presence of a fibrillary background, Homer Wright rosettes, follicle-like spaces, and structures resembling primitive neural tubes such as perivascular pseudorosettes and ependymal and ependymoblastomatous rosettes, indicative of neuronal/neuroectodermal differentiation (Fig. 15.20) are helpful in distinguishing central type PNET from ES/pPNET tumors. As mentioned earlier, central-type PNET may be seen in association with a teratoma [65, 66] (see Chap. 13).

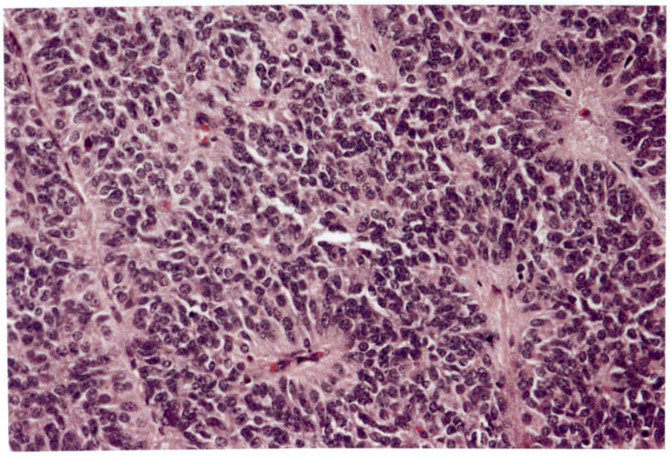

Fig. 15.20 Primitive neuroectodermal tumor, central type. The tumor cells are arranged around blood vessels (perivascular pseudorosettes) and are associated with a prominent fibrillary background

Immunohistochemical Profile

Both peripheral and central-type PNET show diffuse and strong vimentin, CD99, and FLI-1 (nuclear) expression. Glial Fibrillary Acid Protein (GFAP) indicative of glial differentiation (astrocytic or ependymal) appears to be the most discriminatory immunostain as positivity is suggestive of a central-type PNET [66]. Other markers, including neuroendocrine markers, neurofilaments, NSE, S100, desmin, and cytokeratin, show variable positivity.

Cytogenetics

Gene rearrangements involving 22q12 are characteristic of ES/pPNET, with t(11;22)(q24;q12) being the most common translocation. Thus, tumors lacking any degree of central nervous system differentiation should undergo molecular testing as they may represent ES/pPNET tumors.

Differential Diagnosis

As these tumors are rare, especially in a pure form, the differential diagnosis is broad and includes neoplasms composed of "small blue round" cells that can be encountered in the ovary including: small cell carcinoma of hypercalcemic type, small cell neuroendocrine carcinoma, immature teratoma with extensive neural differentiation, rhabdomyosarcoma, melanoma, desmoplastic round small cell tumor of the peritoneum and lymphoma/leukemia. Even though small cell carcinoma of the hypercalcemic type is CD99 positive, it is often associated with: (a) hypercalcemia (in two-third of patients), (b) large cell component or, rarely, mucinous differentiation; and (c) frequent positivity for keratin and less commonly for EMA [67, 68]. Small cell neuroendocrine carcinoma is typically associated with a surface epithelial carcinoma [69]. In immature teratoma showing extensive neural differentiation, no confluent growth is seen. Embryonal rhabdomyosarcoma may be CD99 and FLI-1 positive [70, 71] and in addition it may have t(11;22)(q24;q12). However, it also shows small blue cells admixed with spindle cells that may contain cross striations associated with a prominent myxoid background. These tumors are typically positive for myogenin [52]. Primary malignant melanoma can be composed of small cells but often will be associated with teratomatous components; it may show melanin pigment and more conventional areas; and it is typically positive for one or more melanocytic markers [72]. Although desmoplastic round small cell tumor is positive for CD99, its dual immunoreactivity for epithelial markers and desmin, as well as its characteristic histologic appearance with small blue cells forming irregular islands that may show central necrosis and are embedded in a desmoplastic stroma, help in establishing this diagnosis [73]. Lymphoma and leukemia when growing in a diffuse pattern still may show an admixture of lymphoid and myeloid cells in different stages of maturation but in many instances immunostains will be required to establish the correct diagnosis.

Treatment and Clinical Outcome

Stage at the time of diagnosis is the most important prognostic parameter [74]. Experience with these tumors in the ovary is very limited and although treatment of choice includes surgery often coupled with chemotherapy and/or radiation therapy, often they behave in an aggressive manner [64].

Osteosarcoma

Prevalence and Clinical Presentation

Pure primary osteosarcoma of the ovary is rare, with less than 10 cases reported in the literature [75, 76]. This tumor can also be seen in association with a mature cystic teratoma [77, 78]. Patients in both categories are often perimenopausal and may present with abdominal pain, pelvic mass, or signs and symptoms related to invasion of adjacent pelvic structures or distant metastases. Pain and pyrexia, commonly reported in patients with osteosarcoma of bone, have also been noted in patients with primary osteosarcoma of the ovary [79]. The histogenesis of pure primary tumors is unclear, although it has been hypothesized that most of these tumors may have initially originated from a teratoma which is secondarily overgrown by the high-grade sarcoma.

Macroscopic and Microscopic Features

On gross examination, tumors are typically large (often >10 cm) and may have a solid or solid and cystic cut surface with frequent areas of necrosis and hemorrhage [75]. Rarely, they may be extensively calcified [79]. On microscopic examination, they are often high grade and characterized by the presence of highly malignant cells associated with brisk mitotic rate growing in short fascicles or vaguely storiform sheets, reminiscent of a pleomorphic sarcoma. Typically, osteoid deposit is focal with brightly eosinophilic material forming irregular islands or trabeculae seen juxtaposed to malignant mononuclear cells [63].

Differential Diagnosis

Osteosarcoma may be a component of a malignant mixed mesodermal tumor or it may arise in association with a mature cystic teratoma, and osseous metaplasia may be present in endometrioid carcinomas with spindle cell features. Therefore, it is very important to extensively sample the tumor in order to identify a malignant epithelial component or benign elements of a teratoma before establishing a diagnosis of primary ovarian osteosarcoma. As osteosarcoma has been reported to metastasize to the ovary [29, 80] and patients with these tumors may present with bilateral ovarian involvement [29] or prior/concomitant metastases in other sites [80], awareness of the prior clinical history is of paramount importance in establishing the correct diagnosis as gross and microscopic appearances closely overlap with those seen in primary tumors.

Treatment and Clinical Outcome

Most ovarian osteosarcomas have a highly aggressive course. Patients frequently present with advanced stage disease (III or IV) and develop early recurrences and metastases despite aggressive treatment including chemotherapy. Patients with tumors confined to the ovary appear to have a good prognosis with approximately 2 years of follow-up [75], indicating that stage may be the single most important prognostic parameter in these tumors.

Cartilaginous Neoplasms

Prevalence and Clinical Presentation

Pure cartilaginous tumors are exceedingly rare in the female genital tract including the ovary. Only one primary chondroma [81] and one chondrosarcoma of the ovary have been reported in the literature [82]. In contrast, mature, immature, or malignant cartilage may be encountered in either benign or malignant ovarian tumors, including mature cystic teratoma, immature teratoma, Sertoli-Leydig cell tumor, mesodermal adenosarcoma, or malignant mixed mesodermal tumor. The clinical presentation is non-specific.

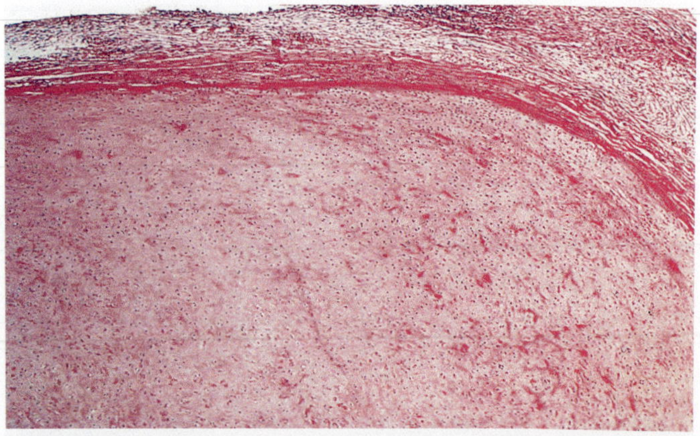

Fig. 15.21 Chondroma. The tumor is well circumscribed from the surrounding ovarian parenchyma and it is composed of small uniform cells embedded in a hyaline matrix

Macroscopic and Microscopic Features

The gross appearance overlaps with that seen in tumors within more typical locations and includes a characteristic lobulated and blue–gray to white cut surface. Cysts as well as areas of hemorrhage may be seen. On microscopic examination, benign tumors are hypocellular and have abundant hyaline matrix (Fig. 15.21). They are composed of evenly distributed and uniform chondrocytes that have eosinophilic granular cytoplasm and small round nuclei with a tiny nucleolus and only rare mitoses. The chondrocytes are free floating in lacunae. In contrast, chondrosarcomas are hypercellular, the chondrocytes vary in size and shape, they have enlarged, sometimes binucleated, hyperchromatic nuclei, and mitoses are easy to identify.

Immunohistochemical Profile

Cartilaginous tumors are positive for S100 and outside the female genital tract it has been shown that benign and malignant tumors may express ER and PR [83], with expression being lower in poorly differentiated neoplasms.

Differential Diagnosis

The presence of malignant cartilage in an ovarian tumor should raise the possibility of a malignant mixed mesodermal tumor or metastases. Extensive sampling of the tumor in order to identify other malignant mesenchymal, and epithelial components and inquiring about the past medical history of the patient are of paramount importance before establishing a diagnosis of primary chondrosarcoma of the ovary [29, 84, 85]. One case of chondrosarcoma from the rib metastatic to the ovary and scalp has been reported [85]. For chondromas, the differential diagnosis may include a mature cystic teratoma with extensive cartilaginous differentiation; however, typically other derivatives of the three embryonic layers are at least focally identified.

Treatment and Clinical Outcome

Experience with these tumors is extremely limited, but benign cartilaginous neoplasms should have an excellent prognosis after surgical excision while chondrosarcomas require aggressive and multimodal treatment and still are associated with very poor prognosis.

Nerve Sheath Tumors

Nerve sheath tumors, either benign or malignant, are exceedingly rare in the female genital tract including the ovary. Two plexiform neurofibromas involving the ovaries in patients with neurofibromatosis have been reported [86, 87]. One patient had neurofibromas also present in the cervix, endometrium, and myometrium [86]. On microscopic examination, these are poorly circumscribed tumors with a nodular configuration involving the hilum with or without cortical involvement. The neoplastic cells are elongated with tapered nuclei and dense eosinophilic cytoplasm and they stain for S100. Wagner-Meissner bodies can be seen. The stroma is frequently myxoid and mast cells may be admixed with the tumor cells. Rare malignant nerve sheath tumors of the ovary have also been reported, most of them associated with neurofibromatosis [88]. On microscopic examination, tumors tend to be hypercellular, they are composed of spindle-shaped cells with variable amounts of cytoplasm and wavy to oval pleomorphic nuclei with a fascicular or storiform growth. Mitotic activity is typically conspicuous.

Pecoma/Pecomatosis

Tumors composed of perivascular epithelioid (PEC) cells are very rare in the ovary. Three reports included a patient with a malignant uterine corpus PEComa associated with lymphangiomyomatosis involving multiple sites including the ovary [89], another patient with a cervical PEComa associated with a proliferation of PEC cells at multiple locations including the ovarian hilum [90], and a third patient with pulmonary and extrapulmonary lymphangiomyomatosis, the latter involving liver, periadrenal vessels, uterus, and ovaries [91]. On microscopic examination, clusters of epithelioid to spindle cells with clear to eosinophilic cytoplasm, often with a perivascular distribution, are seen. They may be associated with areas of hyalinization. Finally, a patient with an epithelioid angiomyolipoma, a tumor classically occurring in the kidney, was reported in the ovary. The tumor was composed of epithelioid cells, smooth muscle cells, thick blood vessels, and mature adipocytes [92]. Most of these patients had a history of tuberous sclerosis. The so-called PEC cells typically co-express melanocytic and smooth muscle markers.

Miscellaneous Mesenchymal Tumors

Other primary benign or malignant mesenchymal tumors of the ovary are exceedingly rare. Hemangiomas can be pure [93, 94] or arise from a mature cystic teratoma [95, 96]. They can be associated with ascites, elevated CA-125 serum levels causing concern for carcinoma, or hemangiomas at other sites (hemangiomatosis). Rare examples of glomus tumor (pure or associated with a dermoid cyst) also have been reported [97]. Synovial sarcoma occurs most commonly in the vulva, with only one biphasic example reported in the ovary [98]. Fibromyxoid sarcoma, a deceptively innocuous-appearing sarcoma, was recently reported in the ovary [99]. Rhabdoid tumor can also be seen in the ovary [100]; however, before establishing this diagnosis, tumors with a rhabdoid phenotype (epithelial, mesenchymal, stromal, melanoma, and hematopoietic) or rhabdoid component (endometrioid carcinomas with a dedifferentiated component) [101] should be excluded. Solitary fibrous tumor has been found in the broad ligament and paraovarian tissues but not arising from the ovary. Gastrointestinal stromal tumors can involve secondarily the ovary and they should be considered in the differential diagnosis when evaluating a tumor with spindle or epithelioid morphology [102].

Lymphoma and Leukemia

Prevalence and Clinical Presentation

Ovarian involvement by lymphoma or leukemia has been reported in approximately 25 and 10–20% of patients, respectively, in autopsy series, but it can be even higher [103–105]. In a recent large series, B-cell lymphomas, particularly diffuse large B-cell, were most common, followed by Burkitt lymphoma; the ovary was the most common location in the female genital tract [106]. Lymphoma as well as leukemia primarily occurring in the ovary are

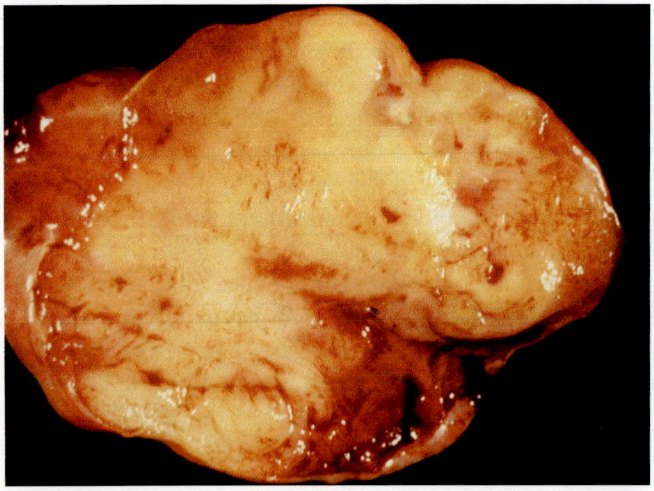

Fig. 15.22 Large B-cell lymphoma. The tumor replaces completely the ovarian parenchyma and has a homogeneous white to tan cut surface

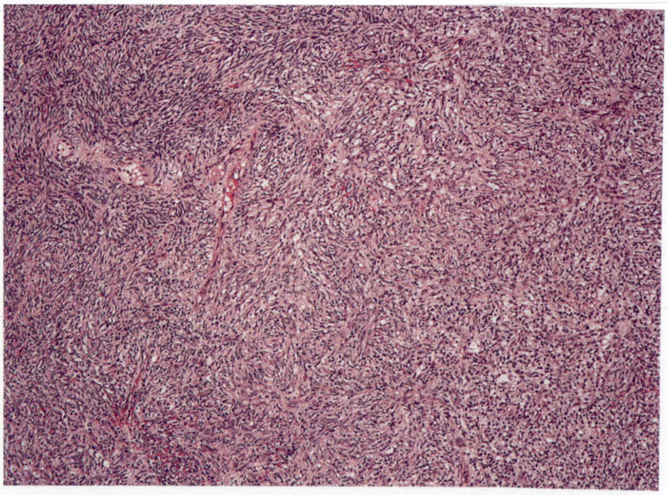

Fig. 15.23 Large B-cell lymphoma. The tumor cells grow in sheets but delicate collagenous bands are present focally imparting a "cord-like" appearance. A rim of ovarian parenchyma is seen (*left*)

exceedingly rare. Patients range widely in age but often are peri- or postmenopausal except for Burkitt lymphoma that occurs in young patients in endemic (in these cases, related to the Epstein–Barr virus) or non-endemic areas as well as in immunosuppressed patients [107]; the ovaries are involved more frequently in the endemic form. Patients present most commonly with pelvic pain, pelvic mass, symptoms related to compression of other organs or constitutional symptoms. Rarely, CA-125 may be elevated [108]. However, not infrequently, patients may be asymptomatic [109]. Patients with leukemia may have abnormal peripheral white blood counts. Granulocytic sarcoma may occur in patients with a prior history of myelodysplastic syndromes or chronic myelogenous leukemia with leukemic transformation or it may be the first manifestation of acute myelogenous leukemia [110].

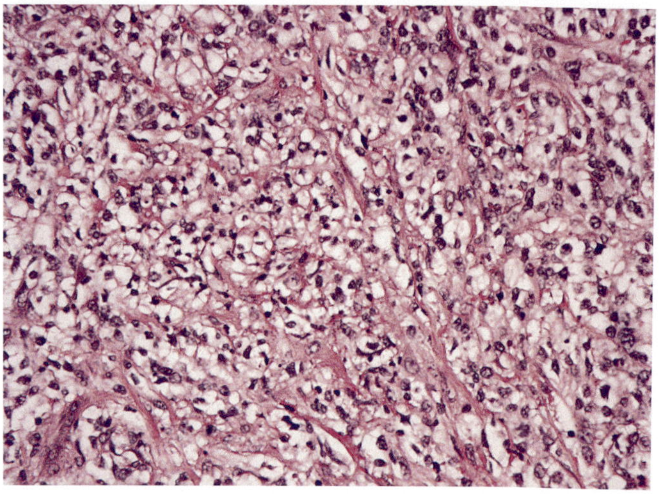

Fig. 15.24 Large B-cell lymphoma. The tumor cells are medium to large size and the nuclei have one or more nucleoli

Macroscopic and Microscopic Features

In lymphoma, both ovaries are frequently involved and range widely in size. There is often diffuse replacement of the ovarian parenchyma displaying a firm to soft or fleshy and white, gray to tan cut surface (Fig. 15.22). Areas of necrosis and hemorrhage may be present [109]. Although involvement by leukemia can occur, the finding of an ovarian tumor formed by neoplastic myeloid cells (chloroma, granulocytic sarcoma, myeloid sarcoma) is extremely rare. The latter is seen as a well-circumscribed, lobulated mass with a firm to fleshy cut surface, but sometimes it has a green color as a result of its myeloperoxidase contents [111]. Diffuse large B-cell lymphoma is the most common subtype of lymphoma involving the ovary independently if primary or secondary [106, 109]. Tumors typically show a diffuse growth (Fig. 15.23) that may be associated with areas showing a follicular architecture. Sclerosis is a common finding (Fig. 15.23). The diffuse areas consist of sheets of large cells with relatively abundant cytoplasm and large irregular nuclei, prominent nucleoli (resembling centroblasts or immunoblasts) (Fig. 15.24), and brisk mitotic activity. Follicular lymphoma has multiple follicles with poorly formed mantle zones and it is predominantly composed of small cleaved lymphoid cells [109]. Burkitt lymphoma is characterized by a diffuse growth of tumor cells and a striking starry-sky pattern due to multiple tingible body macrophages (that are clearing the high rate of spontaneous cell apoptosis). The tumor cells are uniform, have medium size, relatively abundant basophilic cytoplasm containing lipid vacuoles, and round nuclei with multiple nucleoli. Mitotic activity is brisk [112]. Other subtypes of B-cell as

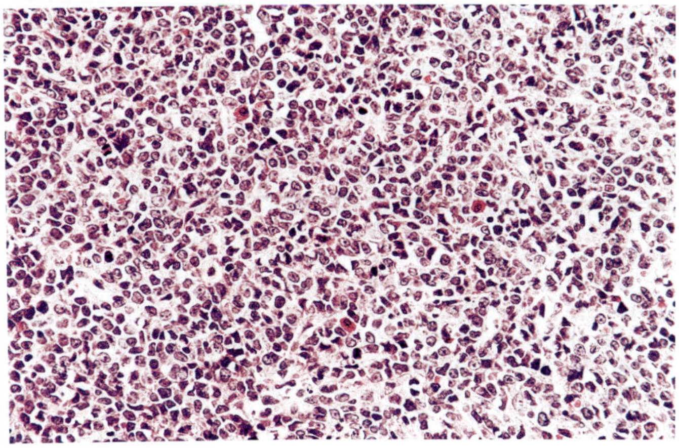

Fig. 15.25 Granulocytic sarcoma. There is a diffuse growth of non-cohesive, small to intermediate sized cells displaying scant cytoplasm, round to oval nuclei and tiny nucleoli. Some cells have brightly eosinophilic cytoplasm (myeloid differentiation)

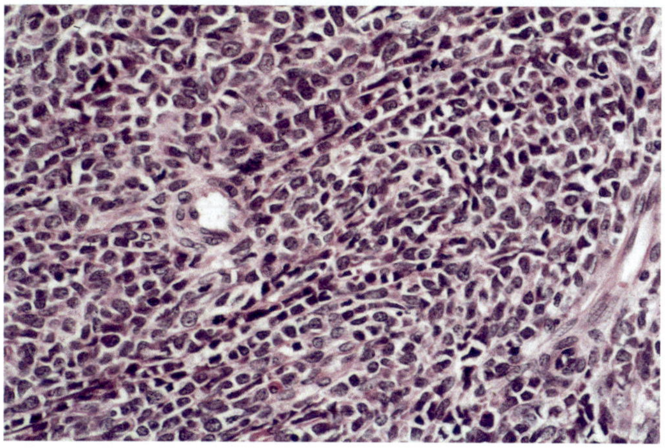

Fig. 15.26 Granulocytic sarcoma. The cells have a "cord-like" growth secondary to associated delicate sclerosis

well as T-cell lymphomas may occur but they are exceedingly rare [106, 109].

In granulocytic sarcoma, cells typically have a diffuse growth (Fig. 15.25), but in some instances, a cord-like growth associated with sclerosis (simulating breast carcinoma) (Fig. 15.26) or a pseudoalveolar growth can be seen. The cells are non-cohesive, small to intermediate in size with scant cytoplasm, round to oval nuclei with tiny nucleoli, and finely dispersed chromatin (Fig. 15.25). Some of the neoplastic cells may have myeloid differentiation displaying brightly eosinophilic cytoplasm. Cells with a signet ring-like morphology may be seen. Mitotic activity is often brisk. Lymphocytes, eosinophils, and rare megakaryocytes may be present admixed with the neoplastic cells [111].

Immunohistochemical Profile

Diffuse large B-cell lymphomas are typically positive for B-associated antigens including CD19, CD20, CD22, CD79a, and BCL-6 [109, 112]. They may be positive for CD5, CD10, and CD45 [112]. Cells in acute myelocytic leukemia are positive for Chloroacetate esterase, lysozyme, myeloperoxidase, and CD43. They may be positive for CD45 but are negative for L26 and CD3 [111, 113]. Burkitt lymphoma cells are positive for sIgM, CD19, CD20, CD22, and CD79a, they also express CD10 but are negative for CD5 and CD23 [114].

Differential Diagnosis

Lymphoid aggregates may occur in the hilum of the ovary, but they never have a confluent growth as seen in follicular or diffuse large cell lymphoma. Tumors in the differential diagnosis of lymphoma include dysgerminoma (mostly for its similar gross appearance), granulosa cell tumor, undifferentiated carcinoma, small cell carcinoma of the hypercalcemic type, and other tumors composed of small blue cells. In many of these cases, the morphologic features of the tumor should be coupled with the immunohistochemical as well as the clinical findings to establish the correct diagnosis. The main differential diagnosis of granulocytic sarcoma is with large B-cell lymphoma as both tumors share: (a) a diffuse growth and (b) non-cohesive tumor cells. In contrast to granulocytic sarcoma, lymphoma cells have slightly larger nuclei with more clumpy chromatin and lack myelocytic differentiation. Still, in some instances immunohistochemistry may be needed [111].

Treatment and Clinical Outcome

Patients with primary ovarian non-Hodgkin lymphoma are staged following the Ann Arbor rather than the FIGO classification and are regarded as having stage IE tumors (tumor confined to the ovary) [115]. Patients typically undergo surgery (as they are clinically thought to have an ovarian carcinoma) followed by chemotherapy and/or radiation therapy [109]. Complete remission rate for patients with non-Hodgkin lymphoma involving the ovary when treated with appropriate chemotherapy appears to be similar to that of patients with other nodal non-Hodgkin lymphomas [116], and survival rates oscillate between 80% and 100% at 5 years for stage I or IIE [109]. For granulocytic sarcoma, it has been suggested that patients who are treated aggressively with chemotherapy with or without radiation therapy, may have long survival without the development of acute leukemia [117].

References

1. Doss BJ, Wanek SM, Jacques SM, et al. Ovarian leiomyomas: clinicopathologic features in fifteen cases. Int J Gynecol Pathol. 1999;18:63–8.
2. Mallya S, MacFarlane IA, Taylor W, et al. Thecal cell reaction associated with an ovarian leiomyoma and presenting with virilization. Gynecol Endocrinol. 1990;4:271–6.
3. Van Winter JT, Stanhope CR. Giant ovarian leiomyoma associated with ascites and polymyositis. Obstet Gynecol. 1992;80:560–3.
4. Lerwill MF, Sung R, Oliva E, et al. Smooth muscle tumors of the ovary: a clinicopathologic study of 54 cases emphasizing prognostic criteria, histologic variants, and differential diagnosis. Am J Surg Pathol. 2004;28:1436–51.
5. Prayson RA, Hart WR. Primary smooth-muscle tumors of the ovary. A clinicopathologic study of four leiomyomas and two mitotically active leiomyomas. Arch Pathol Lab Med. 1992;116:1068–71.
6. Kandalaft PL, Esteban JM. Bilateral massive ovarian leiomyomata in a young woman: a case report with review of the literature. Mod Pathol. 1992;5:586–9.
7. Nogales FF, Ayala A, Ruiz-Avila I, et al. Myxoid leiomyosarcoma of the ovary: analysis of three cases. Hum Pathol. 1991;22:1268–73.
8. Irving JA, McCluggage WG. Ovarian spindle cell lesions: a review with emphasis on recent developments and differential diagnosis. Adv Anat Pathol. 2007;14:305–19.
9. Irving JA, Young RH. Lung carcinoma metastatic to the ovary: a clinicopathologic study of 32 cases emphasizing their morphologic spectrum and problems in differential diagnosis. Am J Surg Pathol. 2005;29:997–1006.
10. Monk BJ, Nieberg R, Berek JS. Primary leiomyosarcoma of the ovary in a perimenarchal female. Gynecol Oncol. 1993;48:389–93.
11. Chang KL, Crabtree GS, Lim-Tan SK, et al. Primary extrauterine endometrial stromal neoplasms: a clinicopathologic study of 20 cases and a review of the literature. Int J Gynecol Pathol. 1993;12:282–96.
12. Young RH, Prat J, Scully RE. Endometrioid stromal sarcomas of the ovary. A clinicopathologic analysis of 23 cases. Cancer. 1984;53:1143–55.
13. Oliva E, Young RH, Scully RE. Primary endometrioid stromal sarcomas of the ovary. A clinicopathologic analysis of 36 cases. Mod Pathol. 1998;11:110A.
14. Levine PH, Abou-Nassar S, Mittal K. Extrauterine low-grade endometrial stromal sarcoma with florid endometrioid glandular differentiation. Int J Gynecol Pathol. 2001;20:395–8.
15. Mitchard JR, Lott M, Afifi RA, et al. Low-grade endometrial stromal sarcoma with glandular differentiation arising in ovarian endometriosis. J Obstet Gynaecol. 2004;24:596–7.
16. Oliva E, Clement PB, Young RH. Epithelioid endometrial and endometrioid stromal tumors: a report of four cases emphasizing their distinction from epithelioid smooth muscle tumors and other oxyphilic uterine and extrauterine tumors. Int J Gynecol Pathol. 2002;21:48–55.
17. Baker P, Oliva E. Endometrial stromal tumours of the uterus: a practical approach using conventional morphology and ancillary techniques. J Clin Pathol. 2007;60:235–43.
18. Chu PG, Arber DA, Weiss LM, et al. Utility of CD10 in distinguishing between endometrial stromal sarcoma and uterine smooth muscle tumors: an immunohistochemical comparison of 34 cases. Mod Pathol. 2001;14:465–71.
19. Devaney K, Tavassoli FA. Immunohistochemistry as a diagnostic aid in the interpretation of unusual mesenchymal tumors of the uterus. Mod Pathol. 1991;4:225–31.
20. Farhood AI, Abrams J. Immunohistochemistry of endometrial stromal sarcoma. Hum Pathol. 1991;22:224–30.
21. Franquemont DW, Frierson Jr HF, Mills SE. An immunohistochemical study of normal endometrial stroma and endometrial stromal neoplasms. Evidence for smooth muscle differentiation. Am J Surg Pathol. 1991;15:861–70.
22. Lillemoe TJ, Perrone T, Norris HJ, et al. Myogenous phenotype of epithelial-like areas in endometrial stromal sarcomas. Arch Pathol Lab Med. 1991;115:215–9.
23. McCluggage WG, Maxwell P. Immunohistochemical staining for calretinin is useful in the diagnosis of ovarian sex cord-stromal tumours. Histopathology. 2001;38:403–8.
24. Nucci MR, O'Connell JT, Huettner PC, et al. h-Caldesmon expression effectively distinguishes endometrial stromal tumors from uterine smooth muscle tumors. Am J Surg Pathol. 2001;25:455–63.
25. Oliva E, Young RH, Amin MB, et al. An immunohistochemical analysis of endometrial stromal and smooth muscle tumors of the uterus: a study of 54 cases emphasizing the importance of using a panel because of overlap in immunoreactivity for individual antibodies. Am J Surg Pathol. 2002;26:403–12.
26. Rush DS, Tan J, Baergen RN, et al. h-Caldesmon, a novel smooth muscle-specific antibody, distinguishes between cellular leiomyoma and endometrial stromal sarcoma. Am J Surg Pathol. 2001;25:253–8.
27. Irving JA, Carinelli S, Prat J. Uterine tumors resembling ovarian sex cord tumors are polyphenotypic neoplasms with true sex cord differentiation. Mod Pathol. 2006;19:17–24.
28. Rushing RS, Shajahan S, Chendil D, et al. Uterine sarcomas express KIT protein but lack mutation(s) in exon 11 or 17 of c-KIT. Gynecol Oncol. 2003;91:9–14.
29. Young RH, Scully RE. Sarcomas metastatic to the ovary: a report of 21 cases. Int J Gynecol Pathol. 1990;9:231–52.
30. Sivanesaratnam V, Dutta R, Jayalakshmi P. Ovarian fibroma – clinical and histopathological characteristics. Int J Gynaecol Obstet. 1990;33:243–7.
31. Oliva E, Garcia-Miralles N, Vu Q, et al. CD10 expression in pure stromal and sex cord-stromal tumors of the ovary: an immunohistochemical analysis of 101 cases. Int J Gynecol Pathol. 2007;26:359–67.
32. Eichhorn JH, Young RH, Clement PB, et al. Mesodermal (mullerian) adenosarcoma of the ovary: a clinicopathologic analysis of 40 cases and a review of the literature. Am J Surg Pathol. 2002;26:1243–58.
33. Clement PB, Scully RE. Mullerian adenosarcoma of the uterus: a clinicopathologic analysis of 100 cases with a review of the literature. Hum Pathol. 1990;21:363–81.
34. McCluggage WG. Mullerian adenosarcoma of the female genital tract. Adv Anat Pathol. 2010;17:122–9.
35. Gallardo A, Prat J. Mullerian adenosarcoma: a clinicopathologic and immunohistochemical study of 55 cases challenging the existence of adenofibroma. Am J Surg Pathol. 2009;33:278–88.
36. Verschraegen CF, Vasuratna A, Edwards C, et al. Clinicopathologic analysis of mullerian adenosarcoma: the M.D. Anderson Cancer Center experience. Oncol Rep. 1998;5:939–44.
37. Recinos-Money E, Escobar-Alfaro G, Contreras J, et al. Ovarian adenosarcoma with elevated CA125 antigen. Case report and literature review. Cir Cir. 2008;76:71–5.
38. Soslow RA, Ali A, Oliva E. Mullerian adenosarcomas: an immunophenotypic analysis of 35 cases. Am J Surg Pathol. 2008;32:1013–21.
39. Van Mieghem T, Abeler VM, Moerman P, et al. CD10, estrogen and progesterone receptor expression in ovarian adenosarcoma. Gynecol Oncol. 2005;99:493–6.
40. Parker RL, Dadmanesh F, Young RH, et al. Polypoid endometriosis: a clinicopathologic analysis of 24 cases and a review of the literature. Am J Surg Pathol. 2004;28:285–97.
41. Nielsen GP, Young RH, Prat J, et al. Primary angiosarcoma of the ovary: a report of seven cases and review of the literature. Int J Gynecol Pathol. 1997;16:378–82.
42. Nucci MR, Krausz T, Lifschitz-Mercer B, et al. Angiosarcoma of the ovary: clinicopathologic and immunohistochemical analysis of

four cases with a broad morphologic spectrum. Am J Surg Pathol. 1998;22:620–30.
43. Platt JS, Rogers SJ, Flynn EA, et al. Primary angiosarcoma of the ovary: a case report and review of the literature. Gynecol Oncol. 1999;73:443–6.
44. Contreras AL, Malpica A. Angiosarcoma arising in mature cystic teratoma of the ovary: a case report and review of the literature. Int J Gynecol Pathol. 2009;28:453–7.
45. den Bakker MA, Ansink AC, Ewing-Graham PC. "Cutaneous-type" angiosarcoma arising in a mature cystic teratoma of the ovary. J Clin Pathol. 2006;59:658–60.
46. Jylling AM, Jorgensen L, Holund B. Mucinous cystadenocarcinoma in combination with hemangiosarcoma in the ovary. Pathol Oncol Res. 1999;5:318–9.
47. McAdam JA, Stewart F, Reid R. Vaginal epithelioid angiosarcoma. J Clin Pathol. 1998;51:928–30.
48. Bradford L, Swartz K, Rose S. Primary angiosarcoma of the ovary complicated by hemoperitoneum: a case report and review of the literature. Arch Gynecol Obstet. 2010;28(1):145–50.
49. Souza FF, Katkar A, den Abbeele AD, et al. Breast angiosarcoma metastatic to the ovary. Case Rep Med. 2009;2009:381015.
50. Serrano C, Garcia A, Brana I, et al. Angiosarcoma of the ovary: is it always a lethal disease? J Clin Oncol. 2010;28:e675–7.
51. Allende DS, Yang B. Primary ovarian rhabdomyosarcoma with heterologous elements: a case report. Int J Gynecol Pathol. 2008;27: 402–6.
52. Ferguson SE, Gerald W, Barakat RR, et al. Clinicopathologic features of rhabdomyosarcoma of gynecologic origin in adults. Am J Surg Pathol. 2007;31:382–9.
53. Guerard MJ, Arguelles MA, Ferenczy A. Rhabdomyosarcoma of the ovary: ultrastructural study of a case and review of literature. Gynecol Oncol. 1983;15:325–39.
54. Nielsen GP, Oliva E, Young RH, et al. Primary ovarian rhabdomyosarcoma: a report of 13 cases. Int J Gynecol Pathol. 1998;17:113–9.
55. Cribbs RK, Shehata BM, Ricketts RR. Primary ovarian rhabdomyosarcoma in children. Pediatr Surg Int. 2008;24:593–5.
56. Kefeli M, Kandemir B, Akpolat I, et al. Rhabdomyosarcoma arising in a mature cystic teratoma with contralateral serous carcinoma: case report and review of the literature. Int J Gynecol Pathol. 2009;28:372–5.
57. Prat J, Young RH, Scully RE. Ovarian Sertoli-Leydig cell tumors with heterologous elements. II. Cartilage and skeletal muscle: a clinicopathologic analysis of twelve cases. Cancer. 1982;50:2465–75.
58. Sant'Ambrogio S, Malpica A, Schroeder B, et al. Primary ovarian rhabdomyosarcoma associated with clear cell carcinoma of the ovary: a case report and review of the literature. Int J Gynecol Pathol. 2000;19:169–73.
59. Yanai H, Matsuura H, Kawasaki M, et al. Immature teratoma of the ovary with a minor rhabdomyosarcomatous component and fatal rhabdomyosarcomatous metastases: the first case in a child. Int J Gynecol Pathol. 2002;21:82–5.
60. Yasuoka H, Tsujimoto M, Fujita S, et al. Coexistence of a clear cell adenocarcinoma and an adenosarcoma with a heterologous rhabdomyosarcoma in an endometriotic cyst of the ovary: a case study. Int J Gynecol Pathol. 2009;28:362–6.
61. Young RH, Kozakewich HP, Scully RE. Metastatic ovarian tumors in children: a report of 14 cases and review of the literature. Int J Gynecol Pathol. 1993;12:8–19.
62. Young RH, Scully RE. Alveolar rhabdomyosarcoma metastatic to the ovary. A report of two cases and a discussion of the differential diagnosis of small cell malignant tumors of the ovary. Cancer. 1989;64:899–904.
63. Kawauchi S, Fukuda T, Miyamoto S, et al. Peripheral primitive neuroectodermal tumor of the ovary confirmed by CD99 immunostaining, karyotypic analysis, and RT-PCR for EWS/FLI-1 chimeric mRNA. Am J Surg Pathol. 1998;22:1417–22.
64. Kim KJ, Jang BW, Lee SK, et al. A case of peripheral primitive neuroectodermal tumor of the ovary. Int J Gynecol Cancer. 2004;14:370–2.
65. Kleinman GM, Young RH, Scully RE. Primary neuroectodermal tumors of the ovary. A report of 25 cases. Am J Surg Pathol. 1993;17:764–78.
66. Kojiro S, Quer A, Snuderl M, et al. Primitive Neuroectodermal Tumors (PNETs) of the Female Genital Tract (FGT): a morphologic, immunohistochemical, and molecular study of 18 cases. Mod Pathol. 2009;222a.
67. McCluggage WG, Oliva E, Connolly LE, et al. An immunohistochemical analysis of ovarian small cell carcinoma of hypercalcemic type. Int J Gynecol Pathol. 2004;23:330–6.
68. Young RH, Oliva E, Scully RE. Small cell carcinoma of the ovary, hypercalcemic type. A clinicopathological analysis of 150 cases. Am J Surg Pathol. 1994;18:1102–16.
69. Eichhorn JH, Young RH, Scully RE. Primary ovarian small cell carcinoma of pulmonary type. A clinicopathologic, immunohistologic, and flow cytometric analysis of 11 cases. Am J Surg Pathol. 1992;16:926–38.
70. Folpe AL, Hill CE, Parham DM, et al. Immunohistochemical detection of FLI-1 protein expression: a study of 132 round cell tumors with emphasis on CD99-positive mimics of Ewing's sarcoma/primitive neuroectodermal tumor. Am J Surg Pathol. 2000;24:1657–62.
71. Thorner P, Squire J, Chilton-MacNeil S, et al. Is the EWS/FLI-1 fusion transcript specific for Ewing sarcoma and peripheral primitive neuroectodermal tumor? A report of four cases showing this transcript in a wider range of tumor types. Am J Pathol. 1996;148:1125–38.
72. McCluggage WG, Bissonnette JP, Young RH. Primary malignant melanoma of the ovary: a report of 9 definite or probable cases with emphasis on their morphologic diversity and mimicry of other primary and secondary ovarian neoplasms. Int J Gynecol Pathol. 2006;25:321–9.
73. Slomovitz BM, Girotra M, Aledo A, et al. Desmoplastic small round cell tumor with primary ovarian involvement: case report and review. Gynecol Oncol. 2000;79:124–8.
74. Morovic A, Damjanov I. Neuroectodermal ovarian tumors: a brief overview. Histol Histopathol. 2008;23:765–71.
75. Fadare O, Bossuyt V, Martel M, et al. Primary osteosarcoma of the ovary: a case report and literature review. Int J Gynecol Pathol. 2007;26:21–5.
76. Nakayama M, Mitsuhashi T, Shimizu Y, et al. Immunohistochemical evaluation of KIT expression in sarcomas of the gynecologic region. Int J Gynecol Pathol. 2006;25:70–6.
77. Aygun B, Kimpo M, Lee T, et al. An adolescent with ovarian osteosarcoma arising in a cystic teratoma. J Pediatr Hematol Oncol. 2003;25:410–3.
78. Ngwalle KE, Hirakawa T, Tsuneyoshi M, et al. Osteosarcoma arising in a benign dermoid cyst of the ovary. Gynecol Oncol. 1990;37:143–7.
79. Hines JF, Compton DM, Stacy CC, et al. Pure primary osteosarcoma of the ovary presenting as an extensively calcified adnexal mass: a case report and review of the literature. Gynecol Oncol. 1990;39:259–63.
80. Eltabbakh GH, Belinson JL, Biscotti CV. Osteosarcoma metastatic to the ovary: a case report and review of the literature. Int J Gynecol Pathol. 1997;16:76–8.
81. Ramos Martínez E, Rodríguez Miguel L, Quijano Narezo M, et al. Primary chondroma of the ovary. Report of a case. Ginecol Obstet Méx. 1983;51:95–8.
82. Talerman A, Auerbach WM, van Meurs AJ. Primary chondrosarcoma of the ovary. Histopathology. 1981;5:319–24.
83. Grifone TJ, Haupt HM, Podolski V, et al. Immunohistochemical expression of estrogen receptors in chondrosarcomas and enchondromas. Int J Surg Pathol. 2008;16:31–7.

84. Piura B, Rabinovich A, Yanai-Inbar I, et al. Primary sarcoma of the ovary: report of five cases and review of the literature. Eur J Gynaecol Oncol. 1998;19:257–61.
85. Swift R, Jalleh R, Patel A, et al. Chondrosarcoma from a rib metastasizing to the ovary. Acta Orthop Scand. 1991;62:76.
86. Gordon MD, Weilert M, Ireland K. Plexiform neurofibromatosis involving the uterine cervix, endometrium, myometrium, and ovary. Obstet Gynecol. 1996;88:699–701.
87. Hegg CA, Flint A. Neurofibroma of the ovary. Gynecol Oncol. 1990;37:437–8.
88. Stone GC, Bell DA, Fuller A, et al. Malignant schwannoma of the ovary. Report of a case. Cancer. 1986;58:1575–82.
89. Liang SX, Pearl M, Liu J, et al. "Malignant" uterine perivascular epithelioid cell tumor, pelvic lymph node lymphangioleiomyomatosis, and gynecological pecomatosis in a patient with tuberous sclerosis: a case report and review of the literature. Int J Gynecol Pathol. 2008;27:86–90.
90. Fadare O, Parkash V, Yilmaz Y, et al. Perivascular epithelioid cell tumor (PEComa) of the uterine cervix associated with intraabdominal "PEComatosis": a clinicopathological study with comparative genomic hybridization analysis. World J Surg Oncol. 2004;2:35.
91. Lack EE, Dolan MF, Finisio J, et al. Pulmonary and extrapulmonary lymphangioleiomyomatosis. Report of a case with bilateral renal angiomyolipomas, multifocal lymphangioleiomyomatosis, and a glial polyp of the endocervix. Am J Surg Pathol. 1986;10:650–7.
92. Anderson AE, Yang X, Young RH. Epithelioid angiomyolipoma of the ovary: a case report and literature review. Int J Gynecol Pathol. 2002;21:69–73.
93. Gupta R, Singh S, Nigam S, et al. Benign vascular tumors of female genital tract. Int J Gynecol Cancer. 2006;16:1195–200.
94. Uppal S, Heller DS, Majmudar B. Ovarian hemangioma–report of three cases and review of the literature. Arch Gynecol Obstet. 2004;270:1–5.
95. Prus D, Rosenberg AE, Blumenfeld A, et al. Infantile hemangioendothelioma of the ovary: a monodermal teratoma or a neoplasm of ovarian somatic cells? Am J Surg Pathol. 1997;21:1231–5.
96. Itoh H, Wada T, Michikata K, et al. Ovarian teratoma showing a predominant hemangiomatous element with stromal luteinization: report of a case and review of the literature. Pathol Int. 2004;54:279–83.
97. Maeda D, Takazawa Y, Oda K, et al. Glomus tumor of the ovary: a case report. Int J Surg Pathol. 2010;18:557–60.
98. Smith CJ, Ferrier AJ, Russell P, et al. Primary synovial sarcoma of the ovary: first reported case. Pathology. 2005;37:385–7.
99. Winfield HL, De Las Casas LE, Greenfield WW, et al. Low-grade fibromyxoid sarcoma presenting clinically as a primary ovarian neoplasm: a case report. Int J Gynecol Pathol. 2007;26:173–6.
100. Leath III CA, Huh WK, Conner M, et al. Primary extrarenal rhabdoid tumor of the ovary. A case report. J Reprod Med. 2003;48:283–6.
101. Tafe LJ, Garg K, Chew I, et al. Endometrial and ovarian carcinomas with undifferentiated components: clinically aggressive and frequently underrecognized neoplasms. Mod Pathol. 2010;23:781–9.
102. Irving JA, Lerwill MF, Young RH. Gastrointestinal stromal tumors metastatic to the ovary: a report of five cases. Am J Surg Pathol. 2005;29:920–6.
103. Barcos M, Lane W, Gomez GA, et al. An autopsy study of 1206 acute and chronic leukemias (1958 to 1982). Cancer. 1987;60:827–37.
104. Paladugu RR, Bearman RM, Rappaport H. Malignant lymphoma with primary manifestation in the gonad: a clinicopathologic study of 38 patients. Cancer. 1980;45:561–71.
105. Osborne BM, Robboy SJ. Lymphomas or leukemia presenting as ovarian tumors. An analysis of 42 cases. Cancer. 1983;52:1933–43.
106. Kosari F, Daneshbod Y, Parwaresch R, et al. Lymphomas of the female genital tract: a study of 186 cases and review of the literature. Am J Surg Pathol. 2005;29:1512–20.
107. Neary B, Young SB, Reuter KL, et al. Ovarian Burkitt lymphoma: pelvic pain in a woman with AIDS. Obstet Gynecol. 1996;88:706–8.
108. Horger M, Muller-Schimpfle M, Yirkin I, et al. Extensive peritoneal and omental lymphomatosis with raised CA 125 mimicking carcinomatosis: CT and intraoperative findings. Br J Radiol. 2004;77:71–3.
109. Vang R, Medeiros LJ, Warnke RA, et al. Ovarian non-Hodgkin's lymphoma: a clinicopathologic study of eight primary cases. Mod Pathol. 2001;14:1093–9.
110. Meis JM, Butler JJ, Osborne BM, et al. Granulocytic sarcoma in nonleukemic patients. Cancer. 1986;58:2697–709.
111. Oliva E, Ferry JA, Young RH, et al. Granulocytic sarcoma of the female genital tract: a clinicopathologic study of 11 cases. Am J Surg Pathol. 1997;21:1156–65.
112. Roth LM, Vang R. Lymphomas and leukaemias. In: Tavassoli F, Devilee P, editors. WHO classification of tumours. Pathology and genetics of tumours of the breast and female genital organs. Lyons: IARC; 2003. p. 191–2.
113. Traweek ST, Arber DA, Rappaport H, et al. Extramedullary myeloid cell tumors. An immunohistochemical and morphologic study of 28 cases. Am J Surg Pathol. 1993;17:1011–9.
114. Garcia CF, Weiss LM, Warnke RA. Small noncleaved cell lymphoma: an immunophenotypic study of 18 cases and comparison with large cell lymphoma. Hum Pathol. 1986;17:454–61.
115. Vang R, Gown AM, Barry TS, et al. Cytokeratins 7 and 20 in primary and secondary mucinous tumors of the ovary: analysis of coordinate immunohistochemical expression profiles and staining distribution in 179 cases. Am J Surg Pathol. 2006;30:1130–9.
116. Dimopoulos MA, Daliani D, Pugh W, et al. Primary ovarian non-Hodgkin's lymphoma: outcome after treatment with combination chemotherapy. Gynecol Oncol. 1997;64:446–50.
117. Drinkard LC, Waggoner S, Stein RN, et al. Acute myelomonocytic leukemia with abnormal eosinophils presenting as an ovarian mass: a report of two cases and a review of the literature. Gynecol Oncol. 1995;56:307–11.

16 Tumors of Uncertain Histogenesis

Gkeok Stzuan Diana Lim and Esther Oliva

Keywords
Small cell carcinoma of hypercalcemic type • Tumor of Wolffian origin • Paraganglioma • Myxoma • Wilms tumor • Desmoplastic small round cell tumor

Small Cell Carcinoma of Hypercalcemic Type

Prevalence and Clinical Presentation

Small cell carcinoma of hypercalcemic type of the ovary (O-SCCHT) is a rare (<250 cases reported) but highly malignant neoplasm that affects adolescents and young adults, typically between 9 and 43 years of age, although occasionally it may occur in younger girls. Rare familial cases have been reported [1–5]. Nonspecific symptoms including abdominal pain and/or swelling are the most common clinical presentations. Although paraneoplastic hypercalcemia is detected in two-thirds of the patients, very few have related symptoms of fatigue, lethargy, polydipsia and polyuria, or even severe acute pancreatitis [5–7]. The underlying mechanism of the hypercalcemia in these tumors is unclear. It has been suggested that the parathormone-related peptide (PTHrp) – a hormone closely related to PTH with autocrine and paracrine functions, which binds to PTH receptors in bone and kidney – may be the cause of the hypercalcemia; however, this hypothesis has yet to be proven [8, 9].

Macroscopic and Microscopic Features

Tumors are typically large (average 15 cm) and mostly solid but may have a cystic component (Fig. 16.1) [5, 10]. They have a soft, gray to tan cut surface with frequent areas of hemorrhage and necrosis. The most common histologic pattern is a diffuse growth of closely packed cells (Fig. 16.2) [5]. Nests, cords, clusters of cells, and spindled cells may also be seen, and it is not infrequent to find a mixture of patterns (Fig. 16.3 and Fig. 16.4). Most tumors display follicle-like spaces that are typically few in number, have variable shapes, and often contain eosinophilic secretions (Fig. 16.5a and Fig. 16.5b). The tumor cells lining the spaces frequently lack an orderly arrangement at the periphery of the spaces and polarity of the inner layer. The majority of neoplasms are composed of small cells with scant cytoplasm and oval to round nucleus with one or two small nucleoli. Brisk mitotic activity is seen. Large cells are encountered in approximately 50% of tumors but may be the predominant or, rarely, the only component in a minority of SCCHT (so-called large

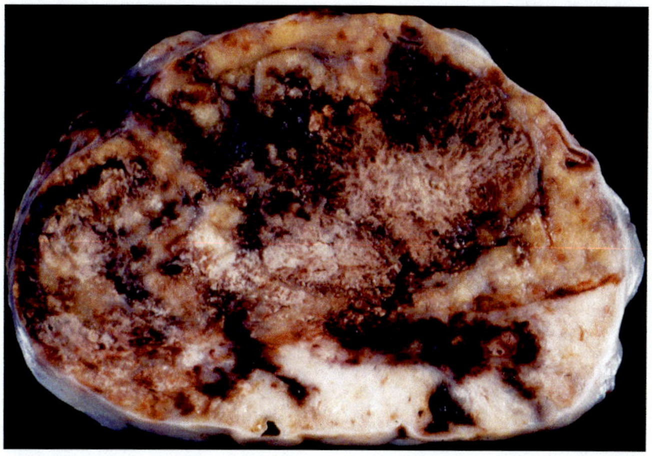

Fig. 16.1 Small cell carcinoma of hypercalcemic type. The tumor has a tan to gray cut surface with areas of necrosis and hemorrhage

E. Oliva (✉)
Department of Pathology, Massachusetts General Hospital, 55 Fruit St., Boston, MA 02114, USA and
Harvard Medical School, Boston, MA, USA

R.A. Soslow and C. Tornos (eds.), *Diagnostic Pathology of Ovarian Tumors*,
DOI: 10.1007/978-1-4419-9751-7_16, © Springer Science+Business Media, LLC 2011

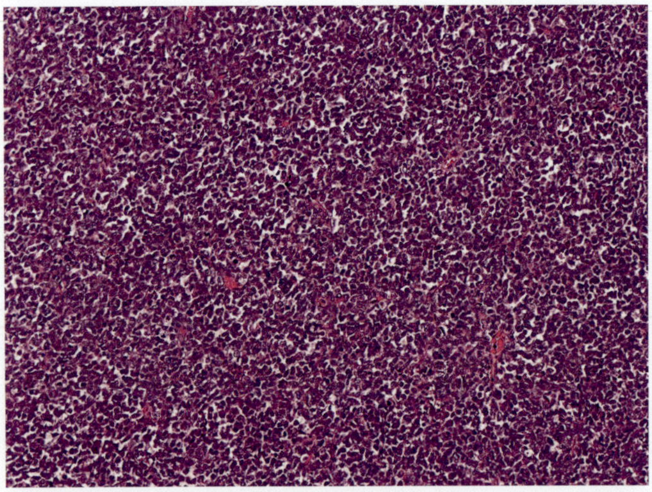

Fig. 16.2 Small cell carcinoma of hypercalcemic type. The tumor is composed of small blue cells with scanty cytoplasm growing in a diffuse pattern

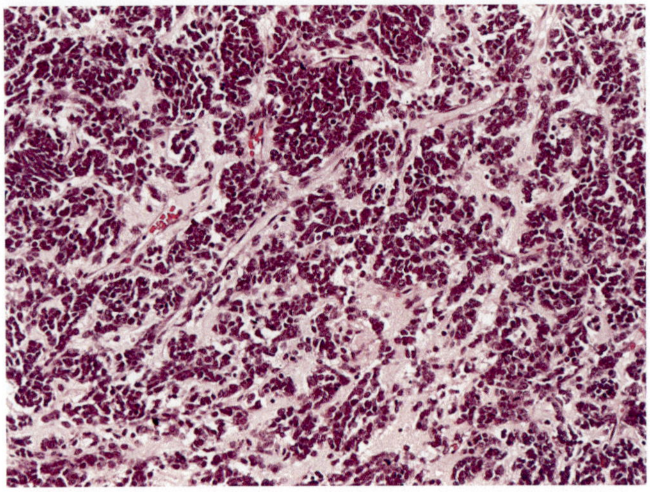

Fig. 16.3 Small cell carcinoma of hypercalcemic type. The tumor cells grow in nests and cords

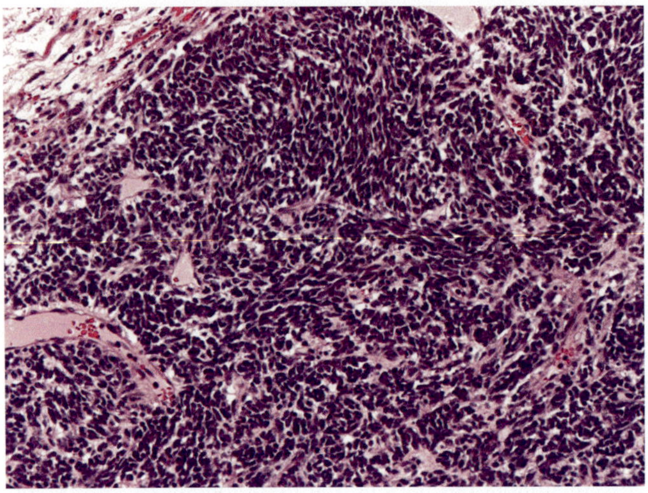

Fig. 16.4 Small cell carcinoma of hypercalcemic type. Neoplastic cells with focal spindle morphology (*top*) are seen next to tumor cells with a nested growth (*bottom*)

cell variant) [1, 3–5, 11–17]. In contrast to the small cells, large cells have moderate to abundant amounts of eosinophilic cytoplasm, which in more than half of the cases contain large hyaline globules that displace the nucleus to the side, imparting a rhabdoid appearance (Fig. 16.6) [5]. The nuclei are vesicular with prominent nucleoli. Mucinous differentiation seen as well-developed mucinous cells resembling endocervical cells lining small cysts (Fig. 16.7) or as signet-ring cells floating in myxoid stroma have been reported in approximately 8% of tumors [5]. There is typically scant stroma that may be myxoid, a finding more frequently associated with the large cell variant.

Immunohistochemical Profile

These tumors are typically positive for vimentin [4, 5, 11, 18], frequently express cytokeratin, and at least one to two-thirds stain for EMA [18–21]. Of note, more extensive staining for keratins and EMA has been noted in the "large cell variant," suggesting a greater degree of epithelial differentiation in these cells (Fig. 16.8) [15, 19]. The tumors are frequently positive for CD10, calretinin, and WT-1 (Fig. 16.9) [22, 23]; however, inhibin is consistently negative [19, 21]. They show variable staining for neuroendocrine markers, with NSE being the most frequently expressed marker [5, 11, 19, 20, 24]. SCCHT typically does not stain for CK5/6, α-fetoprotein, HCG, PLAP, OCT4, HPL, TTF-1 [1, 4, 10, 15, 16, 18, 19, 25–30], and c-kit [20, 26, 31]. These tumors show strong positivity for p53 [3, 19]. Cytometric studies have shown that the vast majority of the tumors are diploid [1, 27, 32, 33].

Electron Microscopy and Cytogenetics

By electron microscopy, the tumor cells typically disclose moderate to abundant amounts of dilated rough endoplasmic reticulum filled with medium-dense material. Other findings include free ribosomes and desmosomes. In some cases, tight junctions, primary and secondary lysosomes, and microfilaments forming paranuclear whorls are also found; the latter are typically seen in the large tumor cells [15, 34–36]. One cytogenetic study from a xenograft-derived monolayer cell line obtained from cells of a patient with O-SCCHT showed at (1;14) [37].

Differential Diagnosis

Small cell carcinoma of the hypercalcemic type may be grossly misdiagnosed as dysgerminoma because both tumors occur in young women. They may be associated with hypercalcemia (much less common in dysgerminoma),

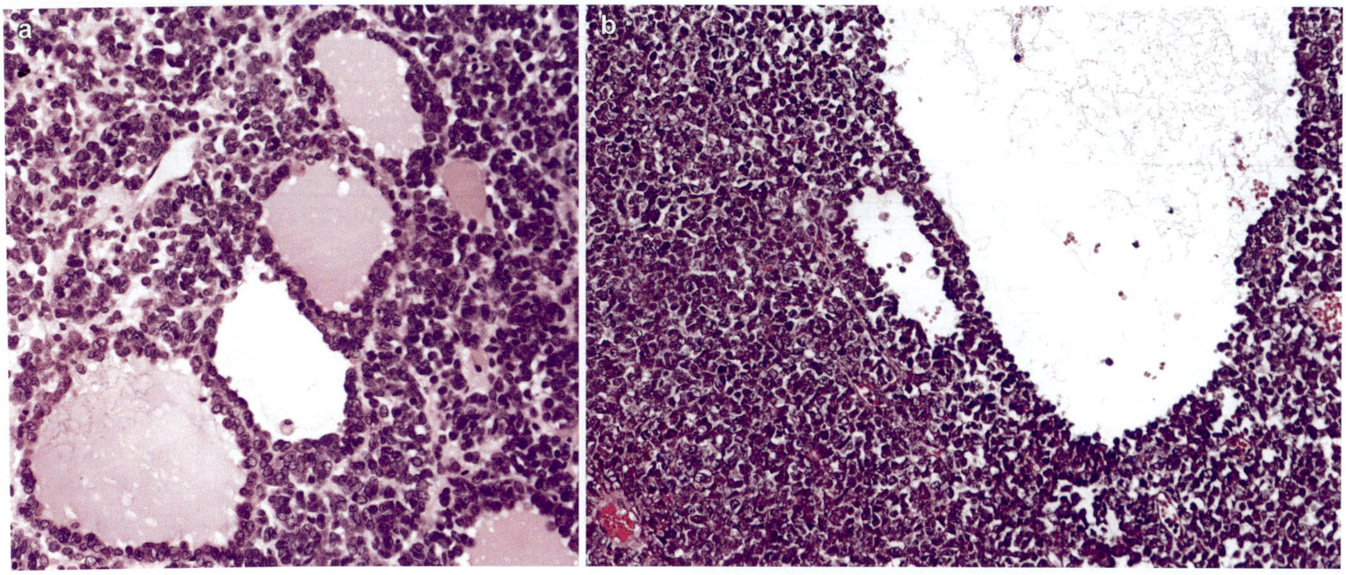

Fig. 16.5 Small cell carcinoma of hypercalcemic type. Follicle-like spaces of similar sizes contain abundant eosinophilic material (**a**). In other tumors, they vary widely in size but they are typically lined by non-cohesive neoplastic cells (**b**)

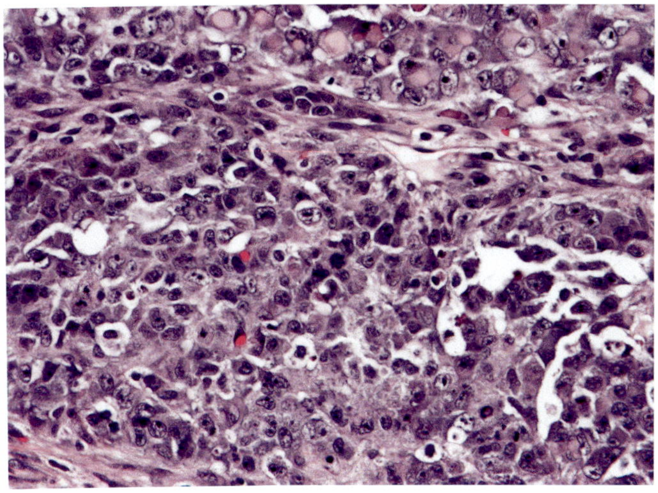

Fig. 16.6 Small cell carcinoma of hypercalcemic type, large cell variant. Large cells show round nuclei with prominent nucleoli and some have a rhabdoid appearance (*top*)

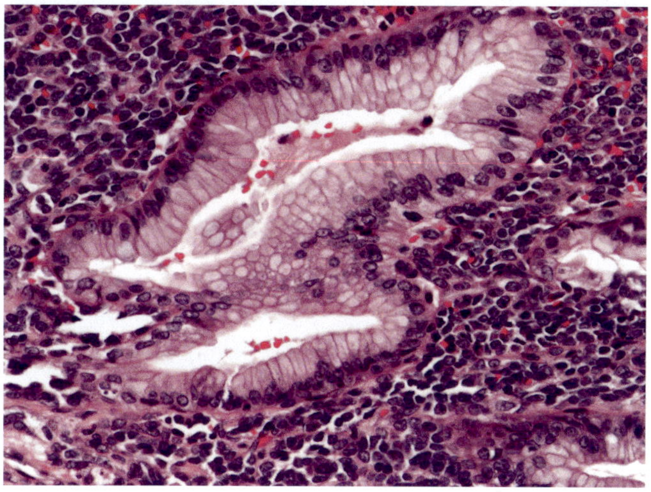

Fig. 16.7 Small cell carcinoma of hypercalcemic type. A well-formed mucinous gland lined by columnar epithelium is present in a background of small blue cells

and may have a solid, soft, gray cut surface. On microscopic examination, although dysgerminoma may have a nested growth, the finding of fibrous septi associated with a sprinkling of lymphocytes combined with cytologic features of the tumor cells (clear cytoplasm and "squared off nuclei") usually facilitate this distinction. From the immunohistochemical point of view, OCT4, PLAP, and c-kit are negative in O-SCCHT [26]. Microscopically, the most common pitfall is the misdiagnosis of O-SCCHT as granulosa cell tumor, more commonly of the juvenile type (JGCT), as both neoplasms typically occur in young women, display follicles, and are composed of mitotically active cells. Distinction between these two neoplasms is exceedingly important as most JGCTs are stage I at initial diagnosis and they are associated with an excellent prognosis in contrast to O-SCCHT, which is associated with a very poor survival (overall <10%). JGCT typically occurs during the first decade of life and is associated with estrogenic manifestations. Furthermore, the tumor cells have a diffuse or nodular growth, follicles are more numerous and irregular in shape when compared to O-SCCHT and have an associated fibrothecomatous component. The most helpful immunohistochemical marker in this differential diagnosis is inhibin, which is typically negative in O-SCCHT [21, 38]. Small cell

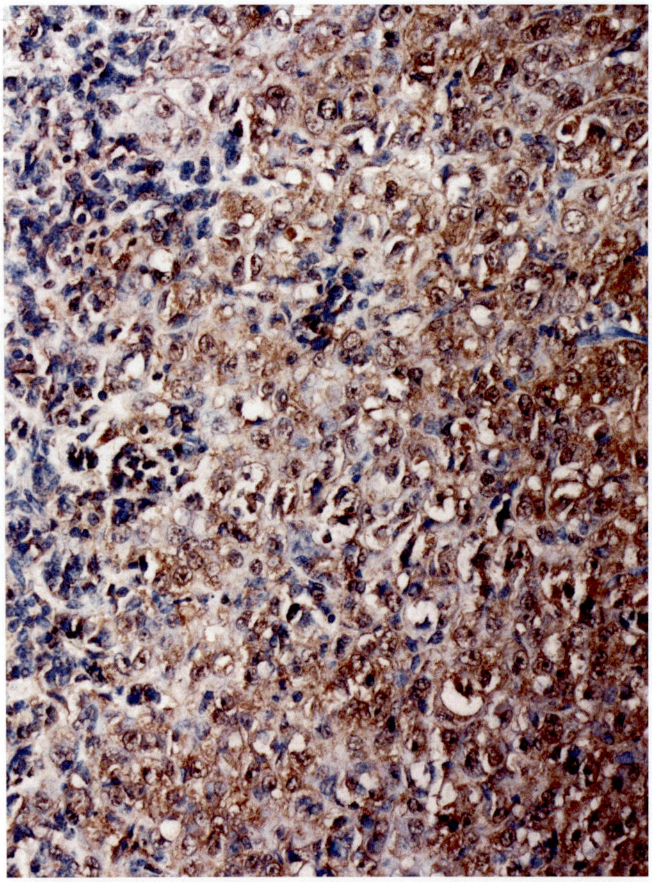

Fig. 16.8 Small cell carcinoma of hypercalcemic type, large cell variant. Most of the cells are positive for EMA

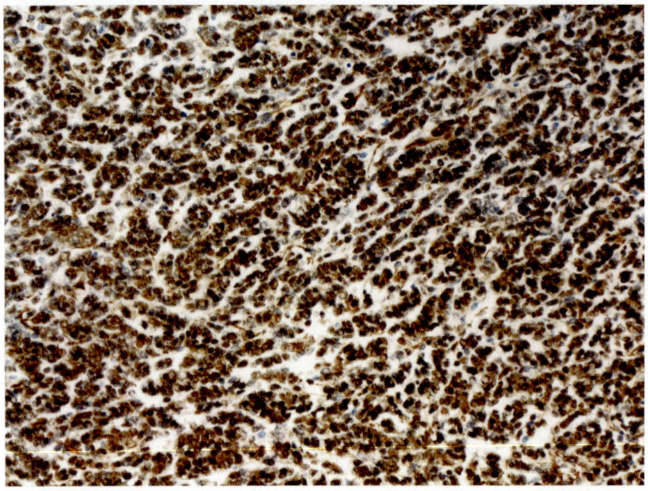

Fig. 16.9 Small cell carcinoma of hypercalcemic type. Diffuse WT-1 positivity is present in the tumor cells

carcinoma of the pulmonary type primary or metastatic to the ovary should also be included in this differential diagnosis. Primary small cell carcinoma of the pulmonary type typically occurs in older women, is not associated with hypercalcemia, and is frequently seen in association with another surface epithelial tumor (such as endometrioid carcinoma) [39]. From the immunohistochemical point of view, vimentin is negative in small cell carcinoma of the pulmonary type while it is positive in approximately 50% of O-SCCHT; in contrast, TTF-1 is positive in small cell carcinoma of the pulmonary type but negative in O-SCCHT [18, 39, 40]. Finally, small cell carcinoma of pulmonary type is frequently aneuploid, in contrast to O-SCCHT, which is typically diploid [39]. Metastatic small cell carcinoma of the pulmonary type to the ovary is rare. Previous clinical history, typical microscopic features seen in metastatic tumors, as well as immunoreactivity for TTF-1 are helpful in establishing this diagnosis [40]. Other tumors in this differential diagnosis include primitive neuroectodermal tumors, metastatic rhabdomyosarcoma, desmoplastic small round cell tumor, and lymphoma. Histologic features as well as immunohistochemical stains are helpful in these differential diagnoses [19, 41].

Treatment and Clinical Outcome

These are highly aggressive tumors associated with a dismal prognosis (<10% survival) despite aggressive treatment. Prognosis has been shown to be intimately related to tumor stage. In the largest study to date, only 33% of patients with stage I tumors were alive and free of disease at a mean follow-up time of 5.7 years. Factors that have been shown to influence prognosis in stage IA tumors include: (a) age >30 years, (b) normal preoperative calcium level, (c) tumor size <10 cm, and (d) absence of large cells [5]. In general, surgery is the mainstay of treatment as radiation and chemotherapy do not appear to improve long-term survival in these patients except in rare cases [5, 16, 17, 25, 42, 43].

Malignant Melanoma

Prevalence and Clinical Presentation

Primary malignant melanoma of the ovary is rare, whereas ovarian metastasis of melanoma is far more common. Thus, exclusion of a malignant melanoma elsewhere is mandatory. Most tumors originating in the ovary are associated with a mature cystic teratoma [44–49]. They occur in a wide age range and patients present with nonspecific signs and symptoms related to an adnexal mass [45, 48].

Macroscopic and Microscopic Features

The tumors range in size from <5 to 23 cm. Rarely, the tumor may have a black cut surface indicative of the presence of abundant melanin pigment [45]. Areas of necrosis and hemorrhage are common. On microscopic examination, these tumors are characterized by a great morphologic diversity; A diffuse growth pattern is most frequent (Fig. 16.10),

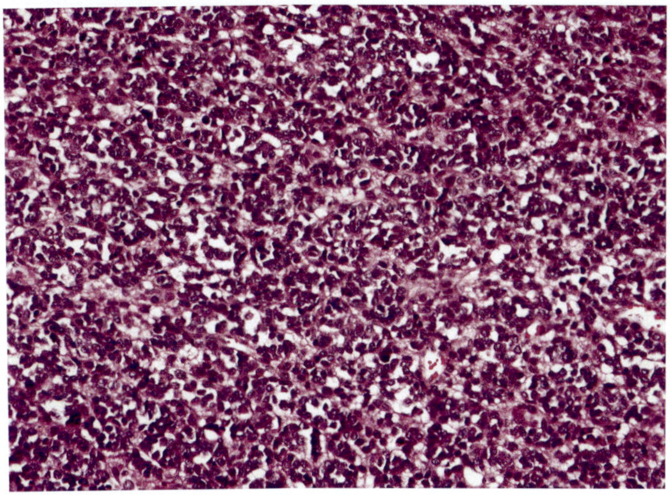

Fig. 16.10 Malignant melanoma. A diffuse growth of the tumor cells is seen

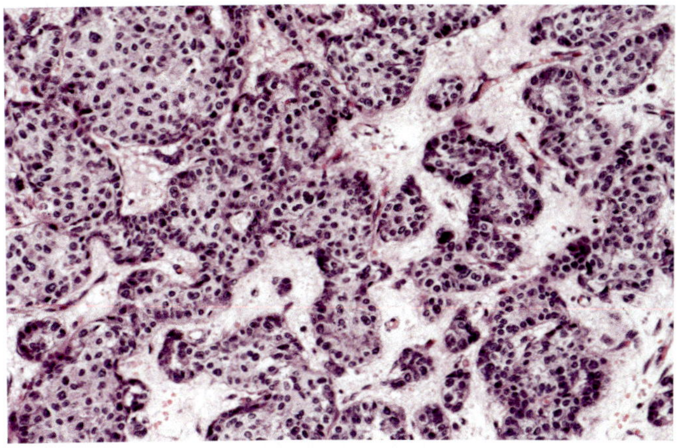

Fig. 16.11 Malignant melanoma. The tumor cells form nests and trabeculae surrounded by myxoid stroma

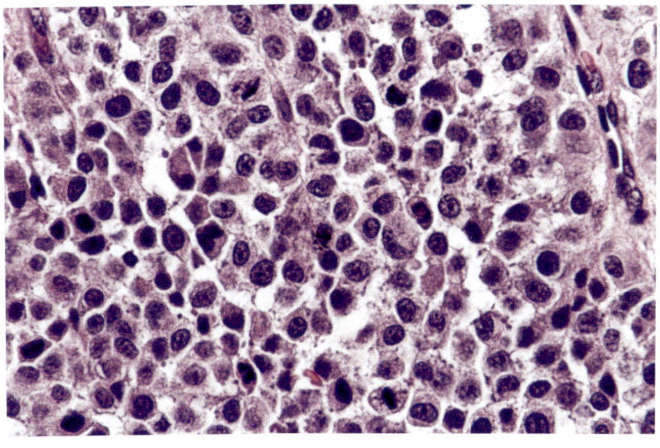

Fig. 16.12 Malignant melanoma. The cells display an epithelioid morphology with round nuclei, prominent nucleoli, and eosinophilic cytoplasm. No pigment is noted

although nested (Fig. 16.11) or fascicular growths as well as an admixture of patterns may be seen [45]. The tumor cells often have an epithelioid morphology displaying abundant

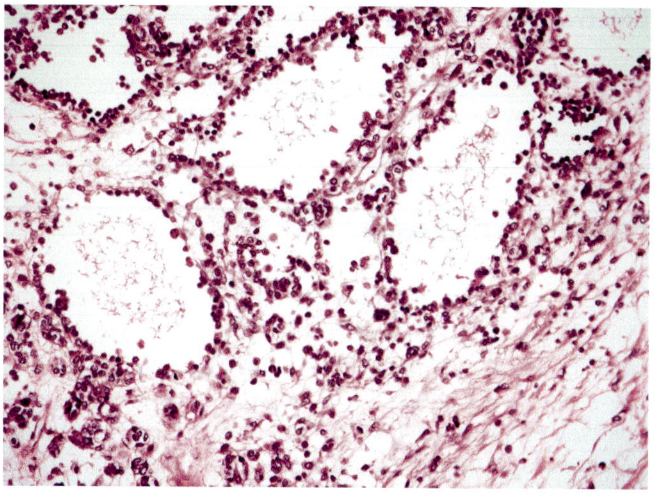

Fig. 16.13 Malignant melanoma. Striking follicle-like spaces are present

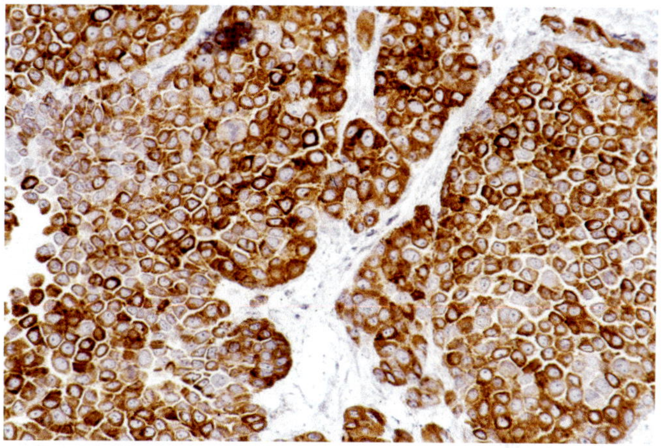

Fig. 16.14 Malignant melanoma. The cells show diffuse and strong immunoreactivity for Mart-1

eosinophilic and, less frequently, clear cytoplasm (Fig. 16.12). However, the tumor cells may be small (with scant cytoplasm) or spindled. Melanin pigment is often seen but may have a focal distribution. Nuclei are typically large and round and show prominent nucleoli and brisk mitotic activity (Fig. 16.12). Intranuclear pseudoinclusions and multinucleated tumor cells (some with wreath-like nuclei) may be seen. Unusual morphologic features include pseudopapillary architecture secondary to necrosis, follicle-like spaces (Fig. 16.13), and signet ring-like cells. The intervening stroma is scant, but abundant myxoid stroma may be appreciated in some tumors (Fig. 16.11) [45, 48]. As mentioned earlier, most melanomas are seen in association with a mature cystic teratoma; thus, hair and sebaceous material may be seen grossly [44–49].

Immunohistochemical Profile

These neoplasms stain for vimentin, S100, HMB45, Mart-1 (Fig. 16.14), tyrosinase, and TTF-1 [45, 48].

Electron Microscopy

Tumor cells characteristically contain melanosomes or premelanosomes and also show melanin pigment [48].

Differential Diagnosis

Primary malignant melanoma should first be distinguished from metastatic melanoma to the ovary [50–53]. As the histologic appearance and immunohistochemical profile show significant overlap, the findings of bilateral ovarian involvement, nodular growth, and a prior history of melanoma strongly point toward metastases. The most likely primary tumor of the ovary to be confused with malignant melanoma is a malignant steroid cell tumor because: (a) both neoplasms often show diffuse and nested growths, (b) tumor cells may have abundant eosinophilic cytoplasm and round nuclei with prominent nucleoli, (c) both neoplasms have overlapping immunohistochemical profile including positivity for Melan-A and S100, and (d) both tumors may be pigmented [54]. The finding of teratoma, melanin pigment, and positivity for HMB45, tyrosinase, and MITF support the diagnosis of malignant melanoma, while the findings of cells with vacuolated cytoplasm as well as positive staining for inhibin and calretinin will support the diagnosis of steroid cell tumor [22, 54–56]. Other differential diagnoses to consider include poorly differentiated/undifferentiated carcinoma, oxyphilic endometrioid and clear cell carcinoma, and hepatoid carcinoma when the tumor cells have abundant eosinophilic cytoplasm. The finding of focal clear-cut glandular differentiation and positive staining for epithelial markers including keratin and EMA favor a diagnosis of carcinoma [57]. When malignant melanoma is composed of small cells, the main differential diagnosis is with SSCHT, as primary melanoma may contain follicle-like spaces [45] and rarely can be associated with hypercalcemia [58]. However, SCCHT typically stains for WT-1 and cytokeratin but does not express melanocytic markers [19]. If the tumor has a predominant spindled morphology, the differential diagnosis includes a variety of mesenchymal neoplasms, most commonly leiomyosarcoma and fibrosarcoma.

Treatment and Clinical Outcome

Most primary melanomas of the ovary pursue an aggressive clinical course even when treated aggressively and are frequently associated with metastases at the time of initial diagnosis/ [45, 48].

Tumors Of Probable Wolffian Origin

Prevalence and Clinical Presentation

Tumors of probable Wolffian origin were first thought to originate from the broad ligament and around the fallopian tube [59], but they have also been reported to occur within the ovary [60, 61] and, rarely, along the lateral aspects of the uterus [62]. In the ovary, tumors arise from Wolffian remnants present in the hilum (rete ovarii). They occur in reproductive age or postmenopausal women (mean, 50 years) who may present with symptoms and signs related to an adnexal mass, vaginal bleeding, or may be asymptomatic.

Macroscopic and Microscopic Features

These neoplasms range widely in size (2–20 cm) and show either a solid or solid and cystic cut surface. The solid areas may have a lobulated appearance and are rubbery in consistency, with a white-gray to tan or yellow cut surface (Fig. 16.15). The cysts, which may be quite large, have a smooth inner lining and contain clear to hemorrhagic fluid. On microscopic examination, three growth patterns can be identified and they often coexist in the same neoplasm. They include: (a) closely packed tubules with compressed pseudolumens lined by cells that lack consistent polarity toward the luminal surface (commonly the predominant pattern) (Fig. 16.16); (b) cysts of different sizes and shapes, some of which contain eosinophilic secretions (sieve-like pattern) (Fig. 16.17); (c) sheets of cells (least predominant) (Fig. 16.18). Exceptionally, a diffuse prominent spindled growth may be seen [63]. The neoplastic cells typically have scant cytoplasm, but those lining tubules may be columnar with clear, occasionally vacuolated cytoplasm.

Immunohistochemical Profile

These tumors typically show luminal/cytoplasmic positivity for CD10 (Fig. 16.19) [64]. They are immunoreactive for vimentin and keratins (AE1/3, CK1, CAM 5. 2, CK7), and commonly stain for inhibin (although focal and weak) and calretinin [38, 65, 66]. They show variable immunoreactivity for ER, PR, and androgen receptors, and only <15% of tumors stain for EMA [65]. They are negative for carcinoembryonic antigen and cytokeratin 20 [65, 67, 68]. They can express C-kit although they do not typically harbor the associated mutation [69, 70] and rare patients have been reported to benefit from treatment with c-kit inhibitors [70].

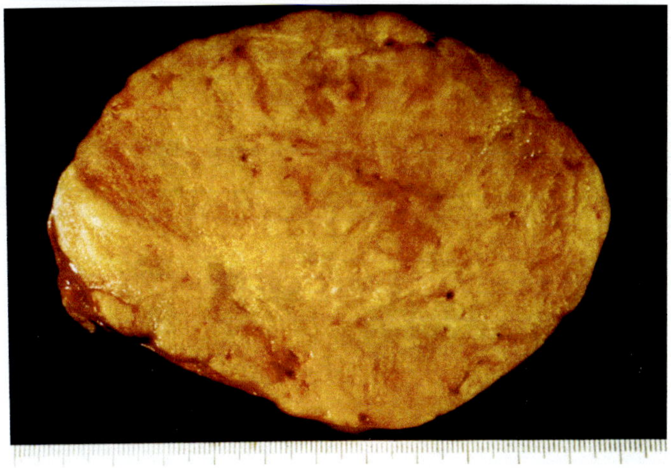

Fig. 16.15 Female adnexal tumor of probable Wolffian origin. The tumor replaces the ovary and has a homogenous solid yellow cut surface

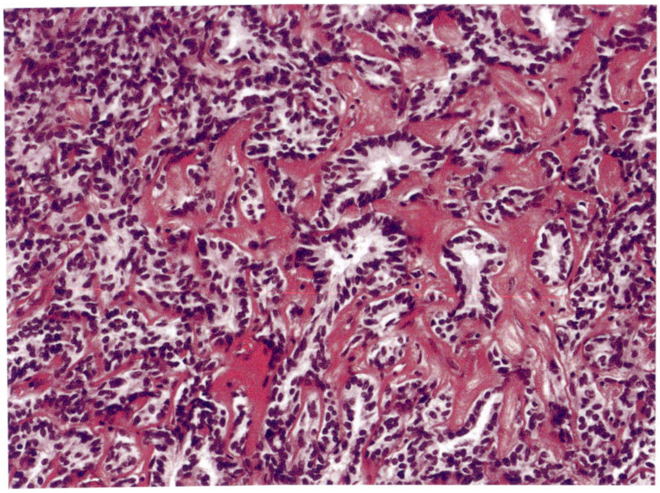

Fig. 16.16 Female adnexal tumor of probable Wolffian origin. Irregularly shaped tubular-like structures, some with compressed lumens are lined tumor cells with bland cytologic features

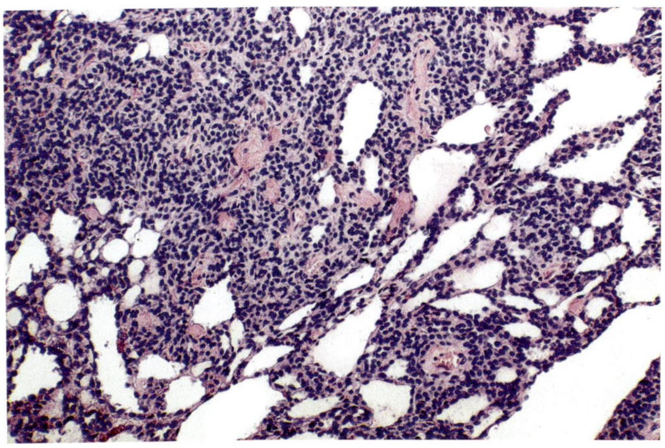

Fig. 16.17 Female adnexal tumor of probable Wolffian origin. Cystic spaces of varying caliber impart a sieve-like pattern

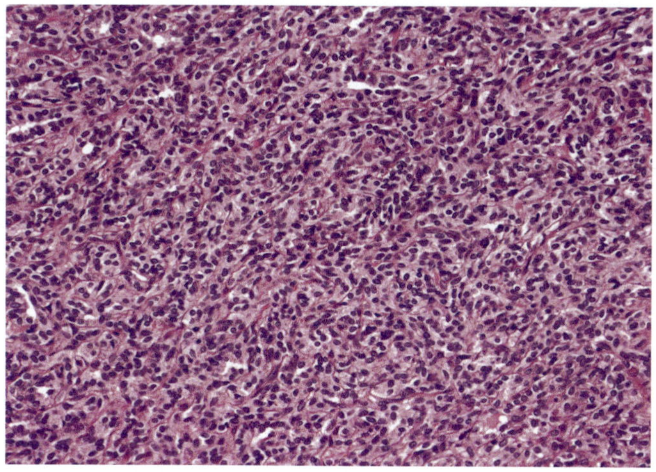

Fig. 16.18 Female adnexal tumor of probable Wolffian origin. Cells grow as compact nests

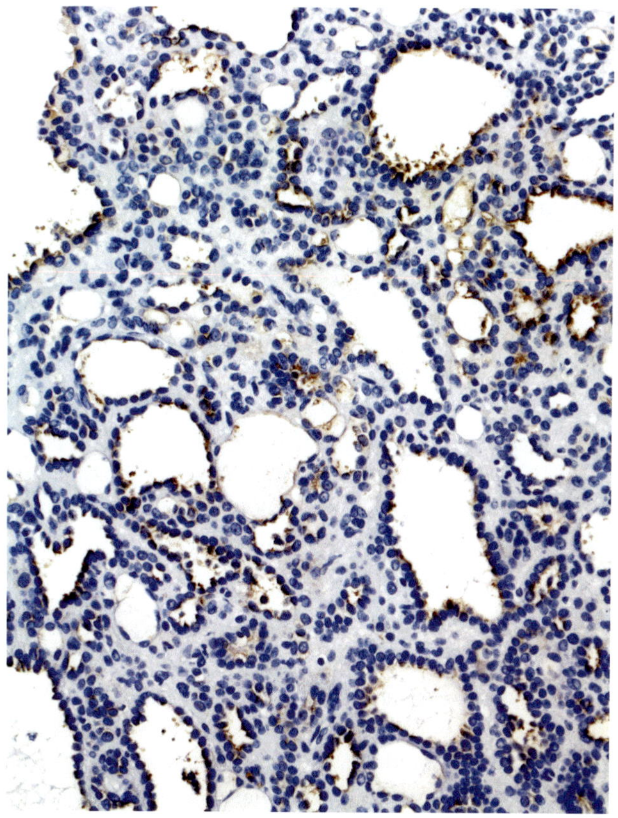

Fig. 16.19 Female adnexal tumor of probable Wolffian origin. The neoplastic cells show luminal staining for CD10

Differential Diagnosis

When the tumors are composed predominantly of spindle cells, the main differential diagnosis is endometrioid carcinoma with a spindled component as both tumors share a low-power lobulated architecture and may contain cystic areas, hollow tubules, and a spindle component [71]. Furthermore, Wolffian tumors may contain endometrioid-like glands [61]. However, squamous metaplasia as well as intraluminal mucin are lacking in these tumors and the degree of cytologic atypia is also less prominent than that seen in endometrioid carcinomas [61]. Another differential diagnosis is Sertoli cell tumor, as it also forms tubular structures and is composed of cells with minimal cytologic atypia that stain for keratin and vimentin, but not for EMA. However, in Sertoli cell tumor, the tubules have a luminal border, and the cells have at least moderate amount of cytoplasm. In addition, they lack the cystic and spindle areas seen in the adnexal tumor of probable Wolffian origin, and they stain more extensively and strongly with inhibin [72].

Treatment and Clinical Outcome

Most of these tumors are confined to the ovary at the time of diagnosis and they behave in a benign fashion. Investigators have recently shown that some tumors with aggressive behavior may be treated with c-kit inhibitors [70].

Paraganglioma

Prevalence and Clinical Presentation

These neoplasms typically arise in the adrenal medulla or extra-adrenal paraganglia, but very rarely they have been reported to be primary in the ovary [57, 73, 74]. Paraganglioma can occur at any age and patients may present with hypertension [57, 73, 74], as commonly described in tumors at other locations, or with nonspecific symptoms including abdominal discomfort or an adnexal mass [57, 74].

Macroscopic and Microscopic Features

These tumors are typically large solid masses with a tan-yellow to brown cut surface (with the exception for a 1.2 cm gangliocytic paraganglioma that arose in the wall of a mature cystic teratoma) [74]. On microscopic examination, paragangliomas are well circumscribed but unencapsulated. They are composed of closely packed nests of tumor cells resulting in a characteristic "Zellballen" appearance. Some nests may show central degeneration. The cells are often polygonal and display abundant granular and, much less frequently, clear cytoplasm, round to oval nuclei with specked chromatin, and tiny nucleoli (Fig. 16.20). Scattered tumor cells with bizarre nuclei may be seen but they are not striking. In one case, cells resembling ganglion cells have been reported (gangliocytic paraganglioma) [74]. Mitotic activity is low (0–2 mitoses/10 HPFs).

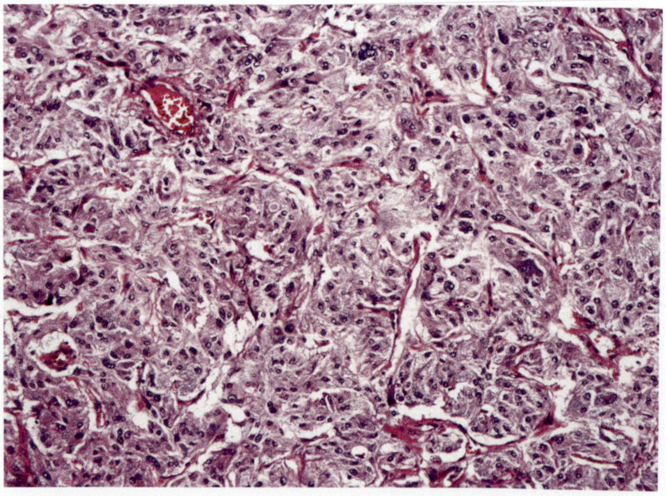

Fig. 16.20 Paraganglioma. Nests of cells with round to oval nuclei and small nucleoli are seen, imparting a "Zellballen" appearance

The intervening stroma is scant, but it may be hyalinized, and is typically rich in capillary-like vessels.

Immunohistochemical Profile

The tumor cells characteristically show cytoplasmic positivity for chromogranin and membranous staining for CD56. S-100 positivity has been described in two tumors (in keeping with sustentacular cells in one) [57, 74]. Two paragangliomas have also been reported to stain for inhibin and one focally for calretinin [57]. The tumors are negative for keratin and EMA [57, 74].

Electron Microscopy

As seen in other more typical locations, these tumors contain numerous dense core neurosecretory granules and are devoid of desmosomes [57].

Differential Diagnosis

Paraganglioma can be confused with sex cord stromal tumors of the ovary, especially with luteinized granulosa cell tumor and steroid cell tumor due to similar gross appearance, nested architecture, oxyphilic morphology of the tumor cells, presence of bizarre nuclei, and positivity for inhibin. However, luteinized granulosa cell tumors typically lack a vascular stroma, do not contain cells with granular cytoplasm or a speckled chromatin pattern, and, in some cases, exhibit focal areas of typical adult granulosa cell tumor with nuclear grooves [75]. Even though steroid cell tumors may show a nested growth, the nests are frequently larger and more irregular, often alternating with cords and trabeculae. The stroma is inconspicuous and the tumor cells have prominent nucleoli and vacuolated cytoplasm, which may contain lipochrome pigment [54]. It is important to note that rare paragangliomas may show immunoreactivity for inhibin and calretinin [57]; thus, a battery of markers should be used in this differential diagnosis. Malignant melanoma should also be considered in the differential diagnosis as it can show a nested growth and contain oxyphilic cells. However, the finding of other growth patterns, tumor cells with large nuclei and prominent nucleoli, nuclear pseudoinclusions, melanin pigment and positivity for HMB45 and Mart-1, as well as a clinical history of a prior melanoma, will help to establish the diagnosis of malignant melanoma [57]. Other differentials that should be considered include primary or metastatic carcinomas with oxyphilic cells and epithelioid smooth muscle tumor.

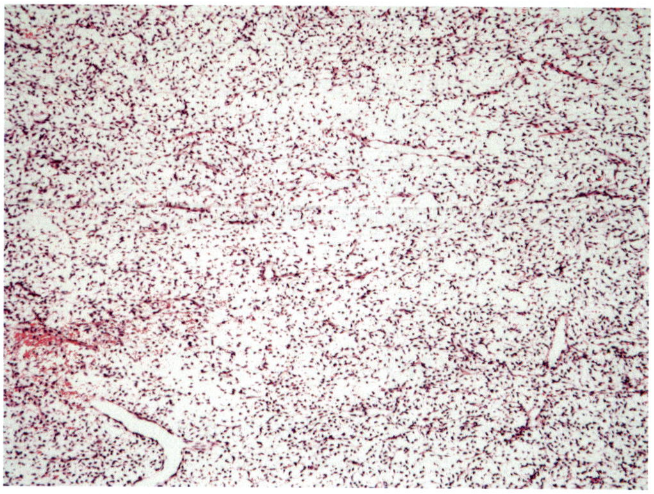

Fig. 16.21 Myxoma. The tumor is hypocellular with a striking vasculature of small capillary-sized blood vessels

Treatment and Clinical Outcome

As one of the five reported cases in the literature had extraovarian disease at the time of the initial surgery [57], these tumors should be considered potentially aggressive and long-term follow-up is advised.

Myxoma

Prevalence and Clinical Presentation

These are uncommon neoplasms of the ovary that typically occur in reproductive age women. They often present with symptoms and signs related to an ovarian mass including abdominal pressure, pain, or weight gain.

Macroscopic and Microscopic Features

On gross examination, these tumors are frequently large and have a thin capsule. They are typically cystic and contain gelatinous material, although some cysts may be filled with hemorrhagic contents. On microscopic examination, the tumors are well-circumscribed but not encapsulated. They are paucicellular (Fig. 16.21); however, the overall cellularity is greater when compared to intramuscular myxomas [76, 77]. The tumor cells are randomly distributed either singly (more common) or arranged in rows. They are spindled to stellate (with long tapering cytoplasmic processes) with oval and sometimes indented nuclei that may contain nuclear pseudoinclusions and small nucleoli (Fig. 16.22). Binucleated cells may be seen. Mitotic figures are rare to absent. The intercellular matrix contains pale blue mucoid material (positive for alcian blue and colloidal iron) that may be admixed with collagen bands and variable numbers of lymphocytes and mast cells. A striking vasculature is often seen, typically composed of capillary-sized vessels that at least focally may display a plexiform arrangement (Fig. 16.21) [77].

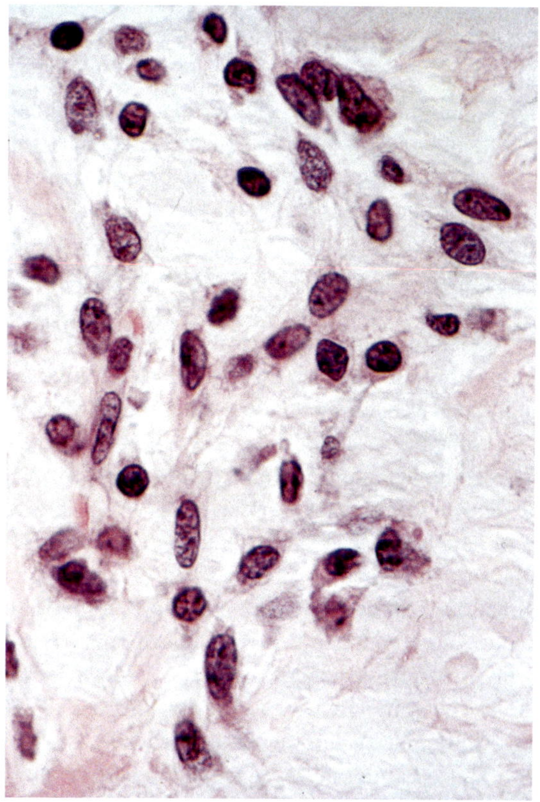

Fig. 16.22 Myxoma. The tumor cells have ovoid to elongated nuclei, small nucleoli, and inconspicuous cytoplasmic borders

Immunohistochemical Profile

These tumors typically stain for vimentin, and often for muscle actin and/or smooth muscle actin, but not for desmin, S100 [77], keratins, EMA, or factor VIII [78]. Based on their immunoprofile, some authors have postulated that ovarian myxomas may constitute part of the spectrum of the fibroma–thecoma category of tumors [79].

Electron Microscopy

The cells have variably sized plasmalemmal vesicles and peripheral aggregates of parallel thin filaments with focal densities indicative of myofibroblastic differentiation. Rare micropinocytic vesicles and intermediate-type junctions may be seen [78].

Differential Diagnosis

Myxomas should be distinguished from a variety of neoplastic and nonneoplastic lesions. Massive edema typically occurs in young women who frequently present with nonspecific symptoms, but occasionally may have androgenic manifestations, in contrast to myxoma. Other distinguishing features include entrapment of preexisting follicular structures, the finding of scattered luteinized cells, and a predominantly edematous but not myxoid background [80]. An edematous fibroma may also enter in the differential diagnosis, but it tends to occur in an older age group. In addition, it lacks the delicate vasculature commonly seen in myxoma, and areas more typical of a fibroma are often appreciated. Rarely, neurofibromas may be seen in the ovary, more commonly in patients with neurofibromatosis [81, 82]. Myxoid sarcomas that should be considered in the differential include embryonal rhabdomyosarcoma, myxoid leiomyosarcoma, fibromyxoid sarcoma, and myxoid liposarcoma. Embryonal rhabdomyosarcoma frequently shows extensive myxoid stroma; however, the cells show varying degrees of skeletal muscle differentiation and some tumors may also exhibit chondroid differentiation [83, 84]. Even though myxoid leiomyosarcoma is hypocellular, shows abundant intercellular matrix, and may not stain for desmin, it typically displays intersecting fascicles of spindled cells with "cigar-shaped nuclei" and at least some degree of cytologic atypia, in contrast to myxomas [85]. One fibromyxoid sarcoma involving the ovary was recently reported [86]. Even though like myxoma, the tumor had low-grade cytologic features, a delicate vasculature, and staineds focally for actin, the myxoid areas showed transition to extensively hyalinized areas with a variety of growth patterns (storiform, whorled, and fibromatosis-like) and no stellate cells were seen [86]. Primary myxoid liposarcoma of the female genital tract has been reported to occur in the uterus [87], vulva [88], and fallopian tube [89], but also recently, in the ovary [90]. Rarely, liposarcoma may present as a heterologous component of a malignant mixed mesodermal tumor [91].

Treatment and Clinical Outcome

Ovarian myxomas are considered low-grade neoplasms and patients may experience long-term recurrences [77].

Wilms Tumor

Prevalence and Clinical Presentation

Extrarenal Wilms tumors are exceedingly rare and, in the female genital tract, have been reported to occur in the uterine corpus [92], uterine cervix [93], and in the ovary [94]. The histogenesis of these tumors is poorly understood, but almost 50% of extrarenal Wilms tumors appear to arise from a teratoma [95, 96]. Patients present with nonspecific symptoms related to an adnexal mass.

Macroscopic and Microscopic Features

The tumors are large with solid pink to gray cut surfaces and variable areas of necrosis. On microscopic examination, they are composed of an admixture of sheets of closely packed small cells (blastema), small well-formed tubules, or poorly formed glomeruloid structures present in varying proportions. The cells typically have scant cytoplasm, oval primitive nuclei, and brisk mitotic activity. A stromal component with spindle cells may also be seen. Features of anaplasia or atypical mitoses, or the presence of a heterologous component (rhabdomyosarcoma) as described in kidney tumors, have not been reported in the ovary [94].

Immunohistochemical Profile

The epithelial elements (tubules and glomeruloid structures) but not the blastema or stromal cells stain for keratin. Vimentin is positive in the spindle cells [94].

Differential Diagnosis

In a background of teratoma, Wilms tumor should be distinguished from the neuroepithelial component of a teratoma or a PNET arising in it [97]. Less likely, the differential diagnosis may include a malignant mixed mesodermal tumor or even a Sertoli-Leydig cell tumor.

Treatment and Clinical Outcome

Experience with these tumors is limited and close follow-up is advised.

Desmoplastic Small Round Cell Tumor

Prevalence and Clinical Presentation

This is a highly aggressive malignant tumor of uncertain histogenesis with divergent differentiation first described in the abdomen of young males (4:1) [98, 99]. Uncommon sites of involvement include, but are not limited to, the paratesticular region, pleura, ovary, posterior cranium, bones and soft tissue of the hand and sinonasal tissues. In women, ovarian involvement frequently represents the initial manifestation of the disease frequently associated with widespread peritoneal tumor. Thus, the presenting symptoms or signs may be related to an ovarian mass or more often to the peritoneal disease including abdominal pain, constipation, weight loss, abdominal distension and jaundice [100]. Patients may also have increase CA-125 serum levels [101–103].

Macroscopic and Microscopic Features

These tumors typically are seen as multiple nodules that may be very large, have smooth bosselated outer surfaces with gray-white, firm cut surfaces not infrequently accompanied by necrosis. On microscopic examination, they show well-defined small to relatively large irregular aggregates of small blue cells surrounded by desmoplastic stroma (Fig. 16.23). The nests of tumor cells frequently show central necrosis as well as peripheral palisading of basaloid cells. Gland-like spaces, papillary structures, trabeculae, cords of single cells, rosettes, and spindle cells may be seen [100–105]. The cells are uniform with scant cytoplasm and indistinct cell borders. However, larger cells, cells with cytoplasmic inclusions and eccentric nuclei or even spindle cells may be seen. The nuclei are small to medium-sized, hyperchromatic with a small nucleoli and brisk mitotic activity.

Immunohistochemical Profile

As desmoplastic small round cell tumors are thought to have divergent differentiation, they typically express keratins and EMA as well as vimentin and desmin (Fig. 16.24) (both paranuclear dot-like, especially when rhabdoid). They are also positive for NSE and WT-1 and they rarely can express muscle actin (2%), chromogranin, CD99, CD15, S100, and CA125 [100, 106].

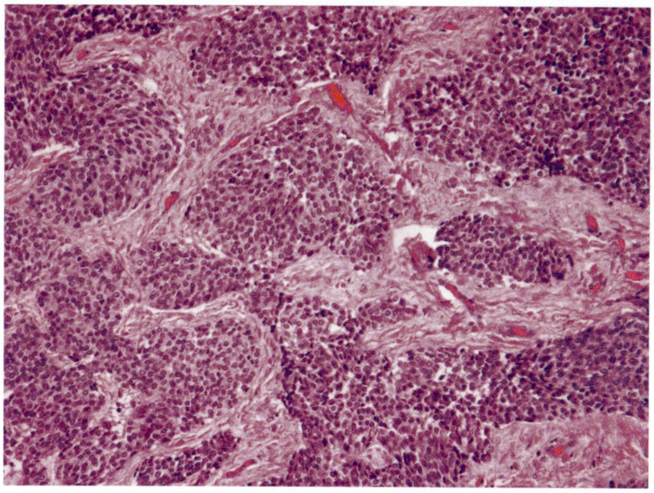

Fig. 16.23 Desmoplastic small round cell tumor. The tumor is composed of irregular nests of small cells separated by abundant desmoplastic stroma

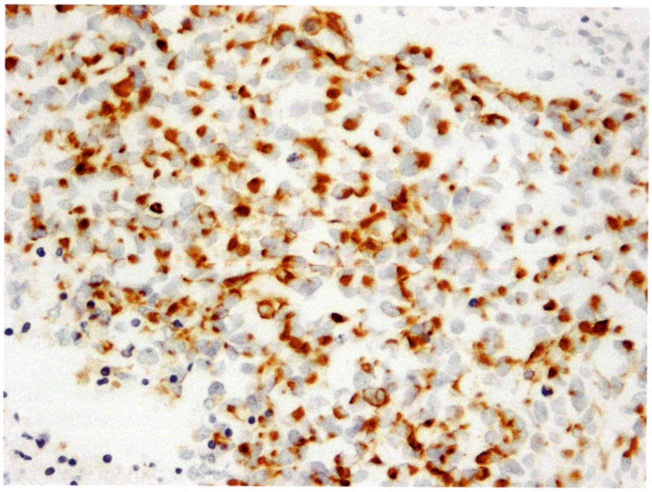

Fig. 16.24 Desmoplastic small round cell tumor. The neoplastic cells are strongly desmin positive

Cytogenetics

These tumors show a specific translocation between chromosomes 11 and 22 [t(11;22)(p13;q12)] with fusion of the EWS and Wilms tumor suppressor genes (WT-1) [100, 107, 108].

Differential Diagnosis

Tumors that may enter in the differential diagnosis are those composed mostly of small blue cells involving either the peritoneum or ovary that may be seen in young women and include small cell carcinoma of hypercalcemic type (often hypercalcemia and follicle-like spaces, much less often positive for epithelial markers and negative for desmin) [2, 8], dedifferentiated carcinoma (in older patients with an associated well-differentiated carcinoma) [109, 110], sex cord stromal tumors (mostly granulosa cell tumor and poorly differentiated Sertoli-Leydig cell tumor, typically endocrine manifestations, inhibin and calretinin positive), rhabdomyosarcoma (myogenin positive) [111], malignant neuroectodermal tumor (often associated with dermoid cyst if central type; if peripheral type similar t(11;22) but different immunoprofile) [56, 112], lymphoma (CD30 positive if B-type, CD43, myeloperoxidase and muramidase positive if leukemia) [113, 114], endometrioid stromal sarcoma (frequent endometriosis, diffuse growth, small arterioles, and CD10 positivity) [115, 116], and mesothelioma (very rarely only composed of small cells). The frequent co-expression of epithelial (keratin and EMA), smooth muscle (desmin), and neuroendocrine markers (NSE) and WT-1 along with the t(11;22) is quite unique to this tumor and is helpful in this extensive differential diagnosis.

Treatment and Clinical Outcome

This tumor is associated with a very poor prognosis with early and frequent widespread metastases. Sites of metastasis have included liver, spleen, lung and lymph nodes. Three-year survivals up to 55% have been reported in in patients treated with surgery, chemotherapy and radiotherapy [117].

References

1. Lamovec J, Bracko M, Cerar O. Familial occurrence of small-cell carcinoma of the ovary. Arch Pathol Lab Med. 1995;119:523–7.
2. Peccatori F, Bonazzi C, Lucchini V, et al. Primary ovarian small cell carcinoma: four more cases. Gynecol Oncol. 1993;49:95–9.
3. Seidman JD. Small cell carcinoma of the ovary of the hypercalcemic type: p53 protein accumulation and clinicopathologic features. Gynecol Oncol. 1995;59:283–7.
4. Ulbright TM, Roth LM, Stehman FB, et al. Poorly differentiated (small cell) carcinoma of the ovary in young women: evidence supporting a germ cell origin. Hum Pathol. 1987;18:175–84.
5. Young RH, Oliva E, Scully RE. Small cell carcinoma of the ovary, hypercalcemic type. A clinicopathological analysis of 150 cases. Am J Surg Pathol. 1994;18:1102–16.
6. Bourgain A, Acker O, Lambaudie E, et al. Small cell carcinoma of the ovary of the hypercalcemic type revealed by a severe acute pancreatitis: about one case. Gynécol Obstét Fertil. 2005;33:35–8.
7. Wynn D, Everett GD, Boothby RA. Small cell carcinoma of the ovary with hypercalcemia causes severe pancreatitis and altered mental status. Gynecol Oncol. 2004;95:716–8.
8. Matias-Guiu X, Prat J, Young RH, et al. Human parathyroid hormone-related protein in ovarian small cell carcinoma. An immunohistochemical study. Cancer. 1994;73:1878–81.
9. Walt H, Hornung R, Fink D, et al. Hypercalcemic-type of small cell carcinoma of the ovary: characterization of a new tumor line. Anticancer Res. 2001;21:3253–9.
10. Jensen ML, Rasmussen KL, Jacobsen M. Ovarian small cell carcinoma: a case report with histologic, immunohistochemical and ultrastructural findings. APMIS Suppl. 1991;23:126–31.
11. Abeler V, Kjorstad KE, Nesland JM. Small cell carcinoma of the ovary. A report of six cases. Int J Gynecol Pathol. 1988;7:315–29.
12. Dharan M. Intraoperative cytology of small cell carcinoma of the ovary with hypercalcemia. A case report. Acta Cytol. 1993;37:61–6.
13. Ferlicot S, Bessoud B, Martin V, et al. Large cell variant of small cell carcinoma of the ovary with hypercalcemia. Ann Pathol. 1998;18:197–200.
14. Fukunaga M, Miyazawa Y, Ushigome S. Endometrial low-grade stromal sarcoma with ovarian sex cord-like differentiation: report of two cases with an immunohistochemical and flow cytometric study. Pathol Int. 1997;47:412–5.
15. Lifschitz-Mercer B, David R, Dharan M, et al. Small cell carcinoma of the ovary: an immunohistochemical and ultrastructural study with a review of the literature. Virchows Arch A Pathol Anat Histopathol. 1992;421:263–70.
16. Powell JL, McAfee RD, McCoy RC, et al. Uterine and ovarian conservation in advanced small cell carcinoma of the ovary. Obstet Gynecol. 1998;91:846–8.
17. Rana S, Warren BK, Yamada SD. Stage IIIC small cell carcinoma of the ovary: survival with conservative surgery and chemotherapy. Obstet Gynecol. 2004;103:1120–3.
18. Aguirre P, Thor AD, Scully RE. Ovarian small cell carcinoma. Histogenetic considerations based on immunohistochemical and other findings. Am J Clin Pathol. 1989;92:140–9.
19. McCluggage WG, Oliva E, Connolly LE, et al. An immunohistochemical analysis of ovarian small cell carcinoma of hypercalcemic type. Int J Gynecol Pathol. 2004;23:330–6.
20. Neto AG, Deavers MT, Silva EG, et al. Immunohistochemical profile of ovarian small cell carcinoma, hypercalcemic type. Mod Pathol. 2003;16:202A.
21. Riopel MA, Perlman EJ, Seidman JD, et al. Inhibin and epithelial membrane antigen immunohistochemistry assist in the diagnosis of sex cord-stromal tumors and provide clues to the histogenesis of hypercalcemic small cell carcinomas. Int J Gynecol Pathol. 1998; 17:46–53.
22. Movahedi-Lankarani S, Kurman RJ. Calretinin, a more sensitive but less specific marker than alpha-inhibin for ovarian sex cord-stromal neoplasms: an immunohistochemical study of 215 cases. Am J Surg Pathol. 2002;26:1477–83.
23. Oliva E. CD10 expression in the female genital tract: does it have useful diagnostic applications? Adv Anat Pathol. 2004;11:310–5.
24. Aguirre P, Thor AD, Scully RE. Ovarian endometrioid carcinomas resembling sex cord-stromal tumors. An immunohistochemical study. Int J Gynecol Pathol. 1989;8:364–73.
25. Benrubi GI, Pitel P, Lammert N. Small cell carcinoma of the ovary with hypercalcemia responsive to sequencing chemotherapy. South Med J. 1993;86:247–8.
26. Carlson JW, Nucci MR, Brodsky J, et al. WT-1, TTF-1 and HPV testing can distinguish between small cell carcinomas of the ovary, cervix and lung. Mod Pathol. 2005;18:179A.
27. Forster C, Ostertag H, Schmitt J, et al. Small cell carcinoma of the ovary, hypercalcemic type. A case report with immunohistochemical, ultrastructural and cytophotometric analysis and review of the literature. Gen Diagn Pathol. 1997;142:365–70.
28. Hamilton S, Beattie GJ, Williams AR. Small cell carcinoma of the ovary: a report of three cases and review of the literature. J Obstet Gynaecol. 2004;24:169–72.
29. Idei Y, Kitazawa S, Fujimori T, et al. Ovarian small cell carcinoma with K-ras mutation: a case report with genetic analysis. Hum Pathol. 1996;27:77–9.
30. Longy M, Toulouse C, Mage P, et al. Familial cluster of ovarian small cell carcinoma: a new mendelian entity? J Med Genet. 1996; 33:333–5.
31. Chen L, Dinh TA, Haque A. Small cell carcinoma of the ovary with hypercalcemia and ectopic parathyroid hormone production. Arch Pathol Lab Med. 2005;129:531–3.
32. Buono JP, Dechelotte P, Glowaczower E, et al. Ploidy analysis in a case of ovarian small cell carcinoma with hypercalcemia. Ann Pathol. 1995;15:134–7.

33. Eichhorn JH, Bell DA, Young RH, et al. DNA content and proliferative activity in ovarian small cell carcinomas of the hypercalcemic type. Implications for diagnosis, prognosis, and histogenesis. Am J Clin Pathol. 1992;98:579–86.
34. Dickersin GR, Kline IW, Scully RE. Small cell carcinoma of the ovary with hypercalcemia: a report of eleven cases. Cancer. 1982;49:188–97.
35. Dickersin GR, Scully RE. An update on the electron microscopy of small cell carcinoma of the ovary with hypercalcemia. Ultrastruct Pathol. 1993;17:411–22.
36. McMahon JT, Hart WR. Ultrastructural analysis of small cell carcinomas of the ovary. Am J Clin Pathol. 1988;90:523–9.
37. Hanson JA, Parry EM, Gibbs AR, et al. Characterization of highly radiosensitive cell lines from a human ovarian small-cell cancer. Gynecol Oncol. 1997;67:147–53.
38. Kommoss F, Oliva E, Bhan AK, et al. Inhibin expression in ovarian tumors and tumor-like lesions: an immunohistochemical study. Mod Pathol. 1998;11:656–64.
39. Eichhorn JH, Young RH, Scully RE. Primary ovarian small cell carcinoma of pulmonary type. A clinicopathologic, immunohistologic, and flow cytometric analysis of 11 cases. Am J Surg Pathol. 1992;16:926–38.
40. Irving JA, Young RH. Lung carcinoma metastatic to the ovary: a clinicopathologic study of 32 cases emphasizing their morphologic spectrum and problems in differential diagnosis. Am J Surg Pathol. 2005;29:997–1006.
41. Clement PB. Selected miscellaneous ovarian lesions: small cell carcinomas, mesothelial lesions, mesenchymal and mixed neoplasms, and non-neoplastic lesions. Mod Pathol. 2005;18 Suppl 2:S113–29.
42. Sholler GL, Luks F, Mangray S, et al. Advanced small cell carcinoma of the ovary in a pediatric patient with long-term survival and review of the literature. J Pediatr Hematol Oncol. 2005;27:169–72.
43. Tewari K, Brewer C, Cappuccini F, et al. Advanced-stage small cell carcinoma of the ovary in pregnancy: long-term survival after surgical debulking and multiagent chemotherapy. Gynecol Oncol. 1997;66:531–4.
44. Davis GL. Malignant melanoma arising in mature ovarian cystic teratoma (dermoid cyst). Report of two cases and literature analysis. Int J Gynecol Pathol. 1996;15:356–62.
45. McCluggage WG, Bissonnette JP, Young RH. Primary malignant melanoma of the ovary: a report of 9 definite or probable cases with emphasis on their morphologic diversity and mimicry of other primary and secondary ovarian neoplasms. Int J Gynecol Pathol. 2006;25:321–9.
46. O'Gorman T, Olaitan A. Primary malignant melanoma arising in an ovarian cystic teratoma. Eur J Gynaecol Oncol. 2009;30:88–9.
47. O'Leary AJ, Tejura H. Malignant melanoma arising in a cystic teratoma of the ovary. J Obstet Gynaecol. 2005;25:626–7.
48. Ueng SH, Pinto MM, Alvarado-Cabrero I, et al. Ovarian malignant melanoma: a clinicopathologic study of 5 cases. Int J Surg Pathol. 2010;18:184–92.
49. Zarbo R, Scibilia G, Conoscenti G, et al. Ovarian cystic teratoma with primary epithelial cell melanoma. Eur J Gynaecol Oncol. 2005;26:71–4.
50. Fitzgibbons PL, Martin SE, Simmons TJ. Malignant melanoma metastatic to the ovary. Am J Surg Pathol. 1987;11:959–64.
51. Gupta D, Deavers MT, Silva EG, et al. Malignant melanoma involving the ovary: a clinicopathologic and immunohistochemical study of 23 cases. Am J Surg Pathol. 2004;28:771–80.
52. McCluggage WG, Wilkinson N. Metastatic neoplasms involving the ovary: a review with an emphasis on morphological and immunohistochemical features. Histopathology. 2005;47:231–47.
53. Young RH, Scully RE. Malignant melanoma metastatic to the ovary. A clinicopathologic analysis of 20 cases. Am J Surg Pathol. 1991;15:849–60.
54. Hayes MC, Scully RE. Ovarian steroid cell tumors (not otherwise specified). A clinicopathological analysis of 63 cases. Am J Surg Pathol. 1987;11:835–45.
55. McCluggage WG. Value of inhibin staining in gynecological pathology. Int J Gynecol Pathol. 2001;20:79–85.
56. McCluggage WG, Maxwell P. Immunohistochemical staining for calretinin is useful in the diagnosis of ovarian sex cord-stromal tumours. Histopathology. 2001;38:403–8.
57. McCluggage WG, Young RH. Paraganglioma of the ovary: report of three cases of a rare ovarian neoplasm, including two exhibiting inhibin positivity. Am J Surg Pathol. 2006;30:600–5.
58. Levy I, Feun L. Hypercalcemia and malignant melanoma. Am J Clin Oncol. 1990;13:524–6.
59. Kariminejad MH, Scully RE. Female adnexal tumor of probable Wolffian origin. A distinctive pathologic entity. Cancer. 1973;31:671–7.
60. Delaloye JF, Ruzicka J, De Grandi P. An ovarian tumor of probable Wolffian origin. Acta Obstet Gynecol Scand. 1993;72:314–6.
61. Young RH, Scully RE. Ovarian tumors of probable wolffian origin. A report of 11 cases. Am J Surg Pathol. 1983;7:125–35.
62. Ordi J, Nogales FF, Palacin A, et al. Mesonephric adenocarcinoma of the uterine corpus: CD10 expression as evidence of mesonephric differentiation. Am J Surg Pathol. 2001;25:1540–5.
63. Fanghong L, Szallasi A, Young RH. Wolffian tumor of the ovary with a prominent spindle cell component: report of a case with brief discussion of unusual problems in differential diagnosis, and literature review. Int J Surg Pathol. 2008;16:222–5.
64. Ordi J, Romagosa C, Tavassoli FA, et al. CD10 expression in epithelial tissues and tumors of the gynecologic tract: a useful marker in the diagnosis of mesonephric, trophoblastic, and clear cell tumors. Am J Surg Pathol. 2003;27:178–86.
65. Devouassoux-Shisheboran M, Silver SA, Tavassoli FA. Wolffian adnexal tumor, so-called female adnexal tumor of probable Wolffian origin (FATWO): immunohistochemical evidence in support of a Wolffian origin. Hum Pathol. 1999;30:856–63.
66. Portugal R, Oliva E. Calretinin: diagnostic utility in the female genital tract. Adv Anat Pathol. 2009;16:118–24.
67. Rahilly MA, Williams AR, Krausz T, et al. Female adnexal tumour of probable Wolffian origin: a clinicopathological and immunohistochemical study of three cases. Histopathology. 1995;26:69–74.
68. Tiltman AJ, Allard U. Female adnexal tumours of probable Wolffian origin: an immunohistochemical study comparing tumours, mesonephric remnants and paramesonephric derivatives. Histopathology. 2001;38:237–42.
69. Harada O, Ota H, Takagi K, et al. Female adnexal tumor of probable wolffian origin: morphological, immunohistochemical, and ultrastructural study with c-kit gene analysis. Pathol Int. 2006;56:95–100.
70. Steed H, Oza A, Chapman WB, et al. Female adnexal tumor of probable wolffian origin: a clinicopathological case report and a possible new treatment. Int J Gynecol Cancer. 2004;14:546–50.
71. Tornos C, Silva EG, Ordonez NG, et al. Endometrioid carcinoma of the ovary with a prominent spindle-cell component, a source of diagnostic confusion. A report of 14 cases. Am J Surg Pathol. 1995;19:1343–53.
72. Oliva E, Alvarez T, Young RH. Sertoli cell tumors of the ovary: a clinicopathologic and immunohistochemical study of 54 cases. Am J Surg Pathol. 2005;29:143–56.
73. Fawcett FJ, Kimbell NK. Phaeochromocytoma of the ovary. J Obstet Gynaecol Br Commonw. 1971;78:458–9.
74. Mahdavi A, Silberberg B, Malviya VK, et al. Gangliocytic paraganglioma arising from mature cystic teratoma of the ovary. Gynecol Oncol. 2003;90:482–5.
75. Young RH, Oliva E, Scully RE. Luteinized adult granulosa cell tumors of the ovary: a report of four cases. Int J Gynecol Pathol. 1994;13:302–10.

76. Brady K, Page DV, Benn LE, et al. Ovarian myxoma. Am J Obstet Gynecol. 1987;156:1240–2.
77. Eichhorn JH, Scully RE. Ovarian myxoma: clinicopathologic and immunocytologic analysis of five cases and a review of the literature. Int J Gynecol Pathol. 1991;10:156–69.
78. Costa MJ, Thomas W, Majmudar B, et al. Ovarian myxoma: ultrastructural and immunohistochemical findings. Ultrastruct Pathol. 1992;16:429–38.
79. Costa MJ, Morris R, DeRose PB, et al. Histologic and immunohistochemical evidence for considering ovarian myxoma as a variant of the thecoma-fibroma group of ovarian stromal tumors. Arch Pathol Lab Med. 1993;117:802–8.
80. Young RH, Scully RE. Fibromatosis and massive edema of the ovary, possibly related entities: a report of 14 cases of fibromatosis and 11 cases of massive edema. Int J Gynecol Pathol. 1984;3:153–78.
81. Gordon MD, Weilert M, Ireland K. Plexiform neurofibromatosis involving the uterine cervix, endometrium, myometrium, and ovary. Obstet Gynecol. 1996;88:699–701.
82. Hegg CA, Flint A. Neurofibroma of the ovary. Gynecol Oncol. 1990;37:437–8.
83. Ferguson SE, Gerald W, Barakat RR, et al. Clinicopathologic features of rhabdomyosarcoma of gynecologic origin in adults. Am J Surg Pathol. 2007;31:382–9.
84. Nielsen GP, Oliva E, Young RH, et al. Primary ovarian rhabdomyosarcoma: a report of 13 cases. Int J Gynecol Pathol. 1998;17:113–9.
85. Lerwill MF, Sung R, Oliva E, et al. Smooth muscle tumors of the ovary: a clinicopathologic study of 54 cases emphasizing prognostic criteria, histologic variants, and differential diagnosis. Am J Surg Pathol. 2004;28:1436–51.
86. Winfield HL, De Las Casas LE, Greenfield WW, et al. Low-grade fibromyxoid sarcoma presenting clinically as a primary ovarian neoplasm: a case report. Int J Gynecol Pathol. 2007;26:173–6.
87. Sosnik H, Jelen M, Sosnik K, et al. Liposarcoma of the uterine corpus coexisting with preinvasive cervical cancer–a case report. Pol J Pathol. 2006;57:171–3.
88. Nucci MR, Fletcher CD. Liposarcoma (atypical lipomatous tumors) of the vulva: a clinicopathologic study of six cases. Int J Gynecol Pathol. 1998;17:17–23.
89. Oishi H, Moriyama S, Kotera K, et al. First case of liposarcoma arising from the fallopian tube: case report and review of the literature. J Obstet Gynaecol Res. 2008;34:713–6.
90. Tirabosco R, Arora R, Ye H, et al. Primary myxoid liposarcoma of the ovary in an adolescent girl: a case report. Int J Gynecol Pathol. 2010;29:256–9.
91. Costa MJ, Khan R, Judd R. Carcinoma (malignant mixed mullerian [mesodermal] tumor) of the uterus and ovary. Correlation of clinical, pathologic, and immunohistochemical features in 29 cases. Arch Pathol Lab Med. 1991;115:583–90.
92. Bittencourt AL, Britto JF, Fonseca Jr LE. Wilms' tumor of the uterus: the first report of the literature. Cancer. 1981;47:2496–9.
93. Bell DA, Shimm DS, Gang DL. Wilms' tumor of the endocervix. Arch Pathol Lab Med. 1985;109:371–3.
94. Sahin A, Benda JA. Primary ovarian Wilms' tumor. Cancer. 1988;61:1460–3.
95. Malik TK, Malik GB, Diesh G. Retroperitoneal teratoma with nephroblastic tissue as the main component. Int Surg. 1967;47: 246–9.
96. Ward SP, Dehner LP. Sacrococcygeal teratoma with nephroblastoma (Wilm's tumor): a variant of extragonadal teratoma in childhood. A histologic and ultrastructural study. Cancer. 1974;33: 1355–63.
97. Kleinman GM, Young RH, Scully RE. Primary neuroectodermal tumors of the ovary. A report of 25 cases. Am J Surg Pathol. 1993;17:764–78.
98. Ordonez NG, Zirkin R, Bloom RE. Malignant small-cell epithelial tumor of the peritoneum coexpressing mesenchymal-type intermediate filaments. Am J Surg Pathol. 1989;13:413–21.
99. Gerald WL, Rosai J. Case 2. Desmoplastic small cell tumor with divergent differentiation. Pediatr Pathol. 1989;9:177–83.
100. Gerald WL, Ladanyi M, de Alava E, et al. Clinical, pathologic, and molecular spectrum of tumors associated with t(11;22)(p13;q12): desmoplastic small round-cell tumor and its variants. J Clin Oncol. 1998;16:3028–36.
101. Ordonez NG, Sahin AA. CA 125 production in desmoplastic small round cell tumor: report of a case with elevated serum levels and prominent signet ring morphology. Hum Pathol. 1998;29: 294–9.
102. Parker LP, Duong JL, Wharton JT, et al. Desmoplastic small round cell tumor: report of a case presenting as a primary ovarian neoplasm. Eur J Gynaecol Oncol. 2002;23:199–202.
103. Young RH, Eichhorn JH, Dickersin GR, et al. Ovarian involvement by the intra-abdominal desmoplastic small round cell tumor with divergent differentiation: a report of three cases. Hum Pathol. 1992;23:454–64.
104. Ordonez NG, el-Naggar AK, Ro JY, et al. Intra-abdominal desmoplastic small cell tumor: a light microscopic, immunocytochemical, ultrastructural, and flow cytometric study. Hum Pathol. 1993;24:850–65.
105. Slomovitz BM, Girotra M, Aledo A, et al. Desmoplastic small round cell tumor with primary ovarian involvement: case report and review. Gynecol Oncol. 2000;79:124–8.
106. Lae ME, Roche PC, Jin L, et al. Desmoplastic small round cell tumor: a clinicopathologic, immunohistochemical, and molecular study of 32 tumors. Am J Surg Pathol. 2002;26:823–35.
107. Antonescu CR, Gerald WL, Magid MS, et al. Molecular variants of the EWS-WT1 gene fusion in desmoplastic small round cell tumor. Diagn Mol Pathol. 1998;7:24–8.
108. Rodriguez E, Sreekantaiah C, Gerald W, et al. A recurring translocation, t(11;22)(p13;q11.2), characterizes intra-abdominal desmoplastic small round-cell tumors. Cancer Genet Cytogenet. 1993;69:17–21.
109. Silva EG, Deavers MT, Bodurka DC, et al. Association of low-grade endometrioid carcinoma of the uterus and ovary with undifferentiated carcinoma: a new type of dedifferentiated carcinoma? Int J Gynecol Pathol. 2006;25:52–8.
110. Tafe LJ, Garg K, Chew I, et al. Endometrial and ovarian carcinomas with undifferentiated components: clinically aggressive and frequently underrecognized neoplasms. Mod Pathol. 2010;23: 781–9.
111. Ferguson SE, Gerald W, Barakat RR, et al. Clinicopathologic features of rhabdomyosarcoma of gynecologic origin in adults. Am J Surg Pathol. 2007;31:382–9.
112. Kawauchi S, Fukuda T, Miyamoto S, et al. Peripheral primitive neuroectodermal tumor of the ovary confirmed by CD99 immunostaining, karyotypic analysis, and RT-PCR for EWS/FLI-1 chimeric mRNA. Am J Surg Pathol. 1998;22:1417–22.
113. Oliva E, Ferry JA, Young RH, et al. Granulocytic sarcoma of the female genital tract: a clinicopathologic study of 11 cases. Am J Surg Pathol. 1997;21:1156–65.
114. Vang R, Medeiros LJ, Warnke RA, et al. Ovarian non-Hodgkin's lymphoma: a clinicopathologic study of eight primary cases. Mod Pathol. 2001;14:1093–9.
115. Chang KL, Crabtree GS, Lim-Tan SK, et al. Primary extrauterine endometrial stromal neoplasms: a clinicopathologic study of 20 cases and a review of the literature. Int J Gynecol Pathol. 1993;12:282–96.
116. Young RH, Prat J, Scully RE. Endometrioid stromal sarcomas of the ovary. A clinicopathologic analysis of 23 cases. Cancer. 1984;53:1143–55.
117. Lal DR, Su WT, Wolden SL, et al. Results of multimodal treatment for desmoplastic small round cell tumors. J Pediatr Surg. 2005;40:251–5.

Fallopian Tube and Peritoneum

17

Terence J. Colgan and Martin C. Chang

Keywords
Fallopian tube carcinoma • Peritoneal serous carcinoma • Mesothelioma • Psammoma-carcinoma • Staging of tubal carcinoma

Introduction

The majority of primary carcinomas of the ovary, fallopian tube, and peritoneum are of the serous type. It is well understood that primary Müllerian-type epithelial tumors of the fallopian tube and peritoneum (including the uterine ligaments) are phenotypically indistinguishable from ovarian carcinoma. Assigning a primary tubal or peritoneal site by WHO criteria (discussed in respective sections below) requires minimal or no ovarian involvement and depends on the location(s) of the dominant tumor mass(es). Based on these criteria, the lifetime risk of fallopian tube carcinoma is estimated to be 0.3–1.1%, and that of primary peritoneal carcinoma to be 1 per 500 (0.2%) [1].

Along with the increased risk of ovarian cancer, *BRCA1* and *BRCA2* mutations are also associated with increased risk of both fallopian tube and primary peritoneal carcinoma of the high-grade serous type. As discussed in Chap. 18, thorough sectioning of specimens from risk-reducing salpingo-oophorectomy specimens (RRSOs) shows otherwise occult carcinomas in about 10% of *BRCA* mutation carriers, with the prevalence ranging from 8% to 25% (Table 18.6) [2–9]. A clear difference in the prevalence of these occult carcinomas between *BRCA1* and *BRCA2* mutation carriers has not been demonstrated (Table 17.7). The risk of peritoneal carcinoma in *BRCA* mutation positive women subsequent to RRSO is about 1%. In one large study, 2 of 259 (0.77%) *BRCA*-carrier women who underwent RRSO subsequently developed primary peritoneal serous carcinoma after intervals of 3.8 and 8.6 years [10].

The current system of assigning primary site is based on the classical paradigm that müllerian-type serous carcinomas – even those that are multifocal/disseminated at diagnosis – are of a unifocal ovarian surface epithelial-stromal origin unless proven otherwise, and that "primary peritoneal" carcinomas are distinct from ovarian primaries because of a presumed multifocal de novo origin from the peritoneal lining (i.e., from metaplastic mesothelium). This model is supported by evidence that ovarian serous carcinomas with concomitant omental masses demonstrate less genetic heterogeneity (e.g., with respect to allelic loss and *TP53* mutations) than primary peritoneal carcinomas at different intraperitoneal sites [11].

However, recent work in serous carcinomas of *BRCA* mutation carriers, including characterization of lesions in RRSO specimens, has demonstrated a spectrum of previously undetected precursor lesions in the distal fallopian tube (see Chap. 18). Although initial reports focused on *BRCA* mutation carriers, it is now apparent that – irrespective of family history – concomitant early cancers of the distal fallopian tube can be found in one half of cases designated as ovarian or primary peritoneal carcinomas by WHO criteria [12]. This realization raises the question: "how many of the multifocal tumors characterized prior to the adoption of fimbrial sectioning were classified correctly with respect to the location of precursor lesions?"

Although the WHO criteria still represent the standard in assigning a primary site for diagnostic reporting, understanding serous carcinogenesis must take into account the importance of the distal fallopian tube as a source for pelvic serous carcinomas (see section below, and Table 17.1) [13, 14].

T.J. Colgan (✉)
Pathology and Laboratory Medicine, Mount Sinai Hospital, 600 University Avenue, Toronto, ON M5G 1X5, Canada

R.A. Soslow and C. Tornos (eds.), *Diagnostic Pathology of Ovarian Tumors*,
DOI: 10.1007/978-1-4419-9751-7_17, © Springer Science+Business Media, LLC 2011

Table 17.1 Conceptual schema for carcinogenesis of the ovary, fallopian tube, and peritoneum

	Histologic pattern(s)	Supporting evidence
Ovarian origin (unifocal)	• Low-grade serous • Endometrioid • Mucinous • Clear cell • Borderline tumors	• Intracystic or surface-epithelial ovarian precursors, including borderline tumors • Frequent *BRAF*, *KRAS* mutations
Primary peritoneal origin (multifocal)[a]	• Low-grade serous • Serous borderline tumors	• Continuity/concurrence with peritoneal endosalpingiosis
Tubal fimbrial origin[b] (unifocal origin; variable final distribution)	• "Pelvic serous carcinoma": high-grade serous carcinoma with dominant disease in ovary, tube, peritoneum[c], arising in association with STIC/tubal cancer[d]	• High frequency of STIC in RRSOs • High rate of concurrent STIC in sporadic and familial pelvic cancers • Specificity of p53 signatures
Undefined pelvic origin (possibly multifocal peritoneal)[a]	• "Pelvic serous carcinoma": high-grade serous carcinoma with dominant disease of the ovary or peritoneum, no evidence of tubal disease[a]	• Small number of primary peritoneal carcinomas arise post-RRSO • Concurrent tubal lesions not found in every case of pelvic cancer
Serous uterine-endometrial origin	• Serous carcinoma of the endometrium, with peritoneal involvement but no concurrent STIC[c]	• Cases in which a large extent of disease is in the endometrium, with intra-abdominal spread but no tubal lesion found

References: [12, 18, 20, 21, 23]
STIC serous tubal intraepithelial carcinoma
[a]Low-grade serous carcinomas may arise from ovarian or peritoneal endosalpingiosis and there are unpublished data that suggest possible origin in the fallopian tube.
[b]This category appears to represent the majority of cases.
[c]Uterine-endometrial serous carcinoma with concurrent STIC/tubal cancer may also represent a distinct variant of pelvic serous carcinoma.
[d]Pelvic serous carcinomas in which the fallopian tube is obliterated by carcinoma are also included in this category.

In this chapter, the terms "ovarian cancer/carcinoma," "fallopian tube carcinoma," and "primary peritoneal" carcinoma refer to the WHO-defined entities.

Borderline and Low-Grade Tumors of the Fallopian Tube and Peritoneum

A classification of fallopian tube neoplasms is shown in Table 17.2. Every type of surface-epithelial stromal tumor seen in the ovary has been reported as a fallopian tube primary. Müllerian-type borderline tumors and low-grade adenocarcinomas are rare in the fallopian tube [1]. Morphologically, these tumors are the same as their ovarian counterparts and may be seen in association with a benign precursor (e.g., adenofibroma, endometriosis). Non-serous primary peritoneal borderline tumors and low grade carcinomas are very rare, and are also thought to arise from endometriosis or other benign müllerian epithelial rests [15].

In contrast, primary peritoneal low-grade serous neoplasms involving the tube are more commonly seen, and often present as multiple implants throughout the peritoneum. As with low-grade serous ovarian neoplasms, the distinction between a primary peritoneal serous borderline tumor (PPSBT) and low-grade serous carcinoma (PPLGSC) may be difficult [16]. Both PPSBT and PPLGSC may be

Table 17.2 Classification of fallopian tube neoplasms

1. Epithelial
   - Carcinomas of all types, including intraepithelial carcinoma, including STICs
   - Low malignant potential
   - Benign – cystadenomas (including papillary cystadenoma of von Hippel Lindau) [34–36], metaplastic papillary tumor, endometrioid polyp
   - Female adnexal tumor of probable Wolffian origin (FATWO)
2. Mixed epithelial-mesenchymal tumors
   - Adenofibroma and cystadenofibroma
   - Carcinosarcoma (malignant mixed mesodermal tumor)
   - Adenosarcoma
3. Mesenchymal tumor – includes leiomyoma and sarcomas
4. Mesothelial tumor
   - Benign adenomatoid tumor
   - Mesothelial inclusion cyst (multicystic mesothelioma)
   - Well-differentiated papillary mesothelioma
   - Malignant mesothelioma
5. Germ cell tumor– includes teratomas, and ependymomas
6. Trophoblastic tumors
   - Choriocarcinoma
   - Placental site trophoblastic tumor
   - Hydatidiform mole
   - Placental site nodule
7. Hematopoietic

Modified from the WHO classification

associated with endosalpingiosis, may involve lymph nodes and the omentum, and may have numerous psammoma bodies. The major criterion for this differential diagnosis is the presence of invasion.

Carcinoma

Defining Primary Tubal Cancer

Traditional Definition

The classical (WHO) definition of primary fallopian tube carcinoma requires a macroscopic lesion within the fallopian tube (including isthmus, ampulla, or fimbria), no concurrent carcinoma of the same type within the ovary or uterine corpus, and a transition between benign and malignant tubal epithelium [17]. Based on these stringent criteria, fallopian tube carcinoma is a rare entity, far outnumbered by its ovarian counterpart. Insights gained from more detailed pathologic examination of the tube in both RRSO and non-RRSO specimens significantly challenge the biologic basis for this model. For example, there has been a striking and disproportionately high number of early high-grade serous carcinomas (both macroscopic and microscopic) found in the distal fallopian tube of RRSO specimens. Moreover, the earliest detectable and unequivocally clonal precursor to high-grade serous carcinoma (the benign "p53 signature") is highly specific to the fallopian tube [18, 19].

New Proposals for Defining Tubal Cancer

The evidence for a unified model of high-grade serous carcinogenesis, with the tubal fimbriae as the primary site in the majority of cases, is substantial. The spectrum of otherwise occult lesions found in RRSO specimens is described in Chap. 18. RRSO specimens may show the full range of precursors to serous carcinoma and occur frequently in the fallopian tube, both alone and concurrent to ovarian and primary peritoneal cancers [12, 18]. Moreover, although these precursor lesions are often isolated findings in the fallopian tube, they are almost never seen in the müllerian epithelium of the ovaries and peritoneum.

Nevertheless, not every high-grade serous carcinoma of the female genital tract can be attributed to a fallopian tube origin because even with thorough sectioning, a tubal precursor is not found in every case of high-grade serous carcinoma. Rarely, high-grade serous carcinoma can arise from the low-grade serous pathway [20, 21]. In cases with a concurrent uterine serous carcinoma, it may be difficult to demonstrate or exclude a single source of the tumor. Although most uterine serous carcinomas progress through myometrial invasion and lymphovascular spread, a subset appear to undergo transtubal spread onto the fimbrial, ovarian, and peritoneal surfaces. Recently, 11 concurrent early serous endometrial and tubal carcinomas have been reported [22], two of which shared genetic alterations. In the absence of larger studies, it remains controversial whether peritoneal involvement by uterine serous carcinoma may actually originate in the distal fallopian tube.

The WHO/FIGO reporting systems for high-grade serous carcinoma are well established and cannot be abandoned without significant impact on disease classification and treatment. Nevertheless, it is important to recognize that these established methods of assigning a primary site to high-grade serous carcinomas of the upper genital tract are, by necessity, arbitrary. The alternative simple and encompassing term "pelvic serous carcinoma" has been advocated; it suggests that disseminated abdominal serous carcinomas originate in the tubal fimbriae, regardless of the final disease distribution [23].

Clinical Features

Fallopian tube carcinoma is clinically similar to its corresponding primary peritoneal and ovarian carcinomas. Clinical signs and symptoms are often nonspecific, and not clearly attributable to the difference in anatomic site. Vaginal bleeding/discharge, pain, and abdominal distension are the three most common presenting complaints of tubal carcinoma [24]. It is thought that abdominal pain occurs more frequently than in ovarian cancer, and is colicky owing to accumulation and discharge of fluid through a blocked fallopian tube lumen ("hydrops tubae profluens"). Nevertheless, this presentation is uncommon and the vast majority of cases (>90%) are unsuspected preoperatively.

Microscopic Features

Types

The morphologic features of these tumors have been described in previous chapters. The vast majority of tubal carcinomas (~80%) are high-grade serous carcinomas (Fig. 17.1). Fallopian tube transitional cell carcinomas also occur relatively more frequently, presumably arising from metaplastic transitional epithelium, which is frequently encountered in the tubal mucosa. Transitional cell elements can also be identified in the more common serous carcinomas, as well. A substantial proportion of tubal carcinomas are endometrioid in type, with a more glandular appearance than the papillary architecture of the serous adenocarcinomas. Clear cell carcinomas of the fallopian tube are very uncommon, but resemble clear cell carcinomas seen elsewhere in the gynecologic tract. Squamous carcinomas of the fallopian tube have been reported, but

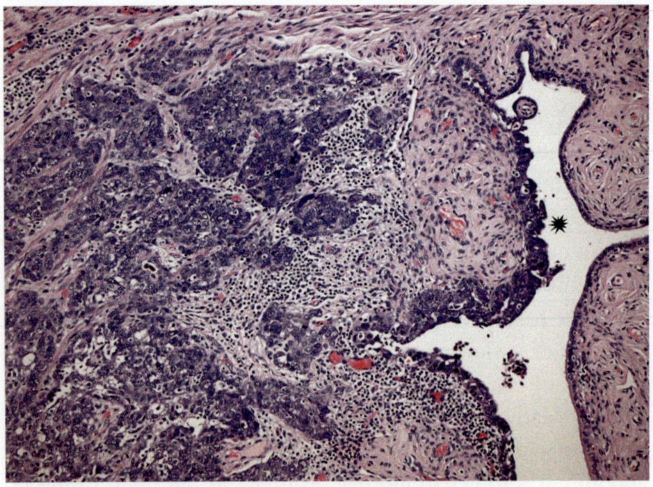

Fig. 17.1 Invasive serous carcinoma of the fallopian tube. Serous tubal intraepithelial carcinoma (STIC) compose part of the tubal lining (*asterisk*), with an adjacent invasive carcinoma (*center* and *left*)

must be distinguished from tubal involvement by ascending squamous intraepithelial lesion from the cervix via the endometrium. Primary carcinosarcoma, that is, a mixed tumor of both malignant glands and stroma, has been rarely described in the fallopian tube and must be distinguished from the more common endometrial and ovarian primary origins. The mesenchymal elements in carcinosarcomas may be composed of elements normally found in the fallopian tube (homologous) or of elements foreign to the tube, such as striated muscle and cartilage (heterologous). The fallopian tube is also a frequent site of involvement for metastatic endometrial (non-hematogenous route) and breast carcinoma (hematogenous route); however, most secondary tumors in the fallopian tube are classically considered metastases from a high-grade serous ovarian primary. Although a substantial portion of these likely originate in the fallopian tube fimbriae (see above), a tumor that forms a mass involving both ovary and tube is currently reported as an ovarian primary with tubal involvement.

Precursors

Detailed histologic evaluation of the fallopian tube of RRSO specimens through complete embedding was first proposed by Colgan [25]. Subsequently, a specific grossing protocol was developed by the group at Brigham and Women's Hospital in Boston, the SEE-FIM protocol, which stressed the longitudinal sectioning of the fimbriae [14]. Use of the SEE-FIM protocol in the fallopian tube of RRSO specimens is discussed in Chap. 18. The SEE-FIM protocol may also be useful in non-RRSO pathology specimens (Table 17.3). In cases of suspected early-stage ovarian carcinomas should also be completely submitted using the SEE-FIM protocol, with special attention paid to the fimbriae for macroscopic or microscopic lesions since tubal findings may upstage the disease. Use of the SEE-FIM protocol in advanced ovarian and peritoneal carcinoma is of limited clinical utility. The role of fallopian tube evaluation in staging uterine-endometrial adenocarcinoma is less defined; however, use of the SEE-FIM protocol has been recommended for cases of uterine serous carcinoma [22].

Table 17.3 Value of the SEE-FIM protocol in surgical pathology

1. Situations where SEE-FIM is of proven clinical value:
   - All RRSO specimens
   - Ovarian carcinomas initially found to be Stage 1a or b
   - Positive peritoneal cytology in the absence of a known primary
2. Situations where SEE-FIM is of theoretical clinical value (staging and/or classification):
   - Uterine serous carcinoma
   - Peritoneal serous carcinoma
3. Situations in which SEE-FIM is of investigational value only, and not routinely necessary:
   - Benign hysterectomy and salpingo-oophorectomy specimens
   - Non-serous uterine carcinoma
   - Advanced ovarian carcinoma

RRSO risk reducing salpingo-oophorectomy

Primary fallopian tube carcinoma forms a mass lesion, which may involve the tubal lumen or expand a tubal fimbria. Most tubal tumors demonstrate high-grade serous histology (Fig. 17.1), with varying degrees of invasion into the endosalpinx, muscularis, mesosalpinx, and serosa. The staging of tubal carcinoma is discussed below. Serous tubal intraepithelial carcinoma (STIC, also known as tubal adenocarcinoma in situ) is usually not apparent macroscopically. Although not an invasive carcinoma, STIC is a lesion of high-grade neoplastic cells with access to the peritoneal cavity, and has a demonstrated potential to disseminate. It must be distinguished from benign mimics or "dysplastic" lesions of uncertain significance. The differential diagnosis of STIC is discussed in detail in Chap. 18 (Table 18.5). To summarize, the major criteria for STIC are as follows: high nuclear-to-cytoplasmic ratio, nuclear atypia (irregular contours, rounded), lack of cilia, prominent nucleoli, high mitotic activity, and loss of polarity.

Differential Diagnosis

Distinguishing ovarian carcinoma from fallopian tubal carcinoma is the principal differential diagnostic challenge in the diagnosis of tubal carcinoma. Contiguous spread of endometrial carcinoma to the proximal tube mucosa and/or muscularis can occur and should not be mistaken for primary tubal carcinoma. Similarly, fallopian tubal serosa and mucosa can be sites of metastatic spread of carcinomas from both gynecologic and non-gynecologic sites. Rarely, tubal endometriotic foci may give rise to carcinomas. Finally, hyperplasia of the tubal epithelium can occur in a number of inflammatory conditions, including tuberculous salpingitis. The mucosal complexity and epithelial proliferation should not be mistaken for adenocarcinoma in situ.

Staging

The staging system for fallopian tube carcinoma is surgical-pathologic, not clinical, and is based upon the concept that the tube is a hollow viscus, and that the documentation of the spread of tumor is analogous to bowel. The current International Federation of Obstetricians and Gynecologists (FIGO) staging system builds upon this concept (Table 17.4) [26]. In using the FIGO staging system two caveats should be borne in mind. Diligent surgical pathologic examination and documentation is critical to correct staging of tubal carcinoma, particularly in the early stages of the disease. Nevertheless, surgical pathologic examination alone is not sufficient to correctly stage all carcinomas, since staging requires the integration of other non-surgical pathologic findings. For example, cytopathologic examination of pelvic washings may indicate a more advanced stage than surgical pathologic examination alone. Similarly, knowledge of the presence of extrapelvic peritoneal or regional lymph nodal disease, particularly if unresected, would be critical in correctly designating the FIGO stage. Secondly, the current FIGO system is suboptimal for the staging of early and fimbrial tubal carcinomas, and amendments have been proposed (Table 17.5 and Fig. 17.2) [27–29]. These amendments may be particularly useful in the reporting of tubal carcinomas seen in RRSO specimens.

Table 17.4 FIGO staging system for fallopian tube carcinoma [26]

Tis – carcinoma in situ
T1 – tumor limited to tubes
T1A – tumor limited to tube, without penetrating serosa, no ascites
T1B – tumor limited to tubes, without penetrating serosa, no ascites
T1C – tumor extends onto or through serosa or positive washings
T2 – tumor limited to one or both tubes, with pelvic extension
T2A – tumor extends to pelvis: uterus and/or ovaries
T2B – tumor extends to pelvis: other structures than uterus/ovaries
T2C – tumor extends to pelvis, and positive washings
T3 – tumor involves tube(s) with peritoneal implants beyond pelvis and/or positive regional lymph nodes
T3A – microscopic tumor implants beyond pelvis
T3B – macroscopic tumor implants ≤2 cm beyond pelvis
T3C – macroscopic tumor implants >2 cm beyond pelvic and/or positive nodes
TX – primary tumor cannot be assessed
T0 – no evidence of primary tumor

Table 17.5 Proposed amendments to FIGO staging system for tubal carcinoma

Stage 0 – "carcinoma in situ" (limited to tubal epithelium), including STIC
Stage Iai – "growth limited to one tube with invasion of the lamina propria only, no ascites"
Stage Iaii – "growth limited to one tube with invasion of the muscularis propria only, no ascites"
Stage IF – "growth limited to one tubal fimbria with invasion of the lamina propria"

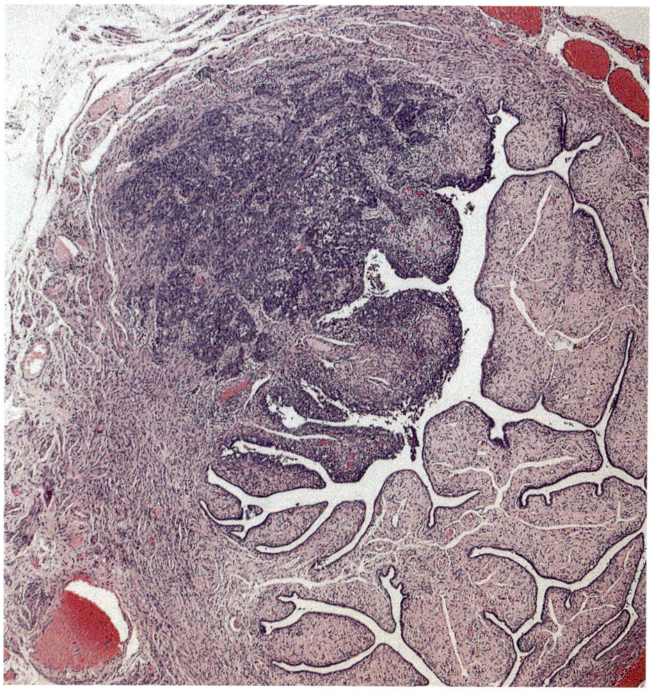

Fig. 17.2 An early-stage invasive carcinoma of the fallopian tube (*upper left*) involves only a portion of the tubal circumference, and is less than full thickness

Other Tubal Neoplasms

The most common benign fallopian tube neoplasm is adenomatoid tumor, a benign tumor of mesothelial origin that can be mistaken for signet ring cell adenocarcinoma and lipoleiomyoma. The tumor is composed of irregular to round, small spaces formed by cuboidal to attenuated cells, dissecting throughout the muscularis. These cells are strongly positive for keratin and calretinin, and negative for vascular markers. A variety of other benign tumors have been described in the fallopian tube or mesosalpinx including serous cystadenoma, adenofibromas, serous tumors of low malignant potential, lipomas, leiomyomas, and teratomas (Table 17.2). Fimbrial adenofibromas can present as pedun-

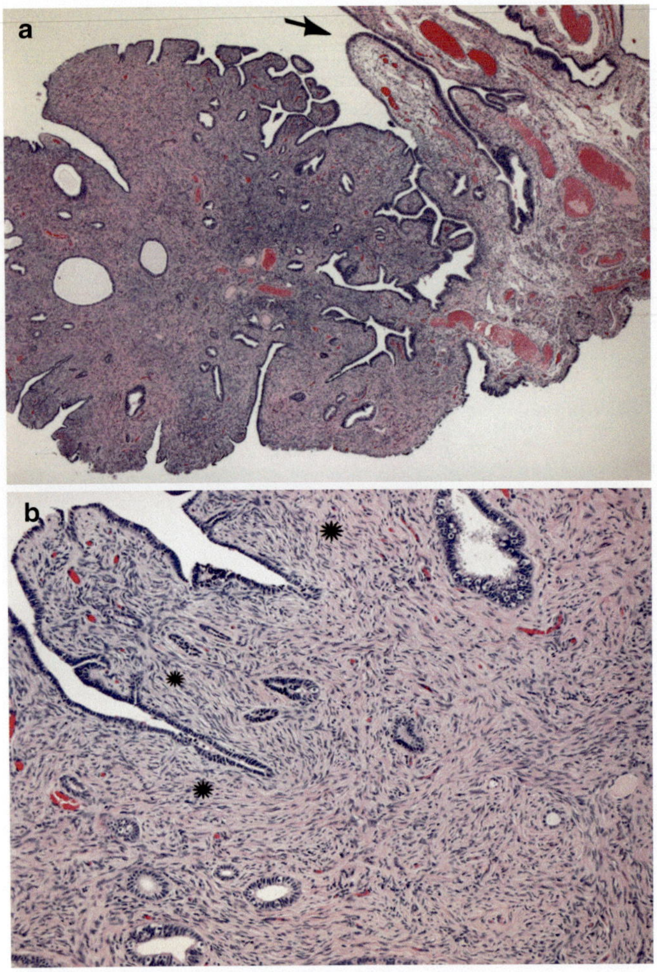

Fig. 17.3 (**a** and **b**). Adenofibroma of the fallopian tube. This benign mixed tumor arises from the tubal fimbria (*arrow*, Fig. 17.3a) and is composed of cellular fibromatous stoma (*asterisks*, Fig. 17.3b), and bland epithelium covering the papillae and forming glands

culated small solid masses, and have now been reported in RRSO specimens; they must be distinguished from early fimbrial adenocarcinomas (Fig. 17.3a, b).

Peritoneal Carcinoma, Serous Type

Although many peritoneal neoplasms have been described (e.g., adenomatoid tumors), this section discusses peritoneal serous carcinomas and related neoplasms only, since these tumors have been historically distinguished from ovarian carcinomas. A significant minority of patients with seemingly advanced stage ovarian carcinoma have normal, or only slightly enlarged ovaries, yet widespread serous carcinoma throughout the peritoneum. In these cases, a diagnosis of "primary peritoneal serous carcinoma" is warranted, provided all five criteria are fulfilled as outlined in Table 17.6. The fifth criterion, the exclusion of any invasive or in situ serous carcinoma of the endometrium and fallopian tube, is suggested since it is now clear that both of these extraperitoneal sites can give rise to disseminated peritoneal disease with only minimal disease at the primary site. If this additional criterion is accepted, then the definition of primary peritoneal serous carcinoma is more restricted, and the separation of tubal and endometrial cases from true peritoneal primaries may facilitate our understanding of these diseases in the future. Immunohistochemistry does not permit separation of ovarian from peritoneal serous carcinomas.

Table 17.6 Diagnostic criteria for primary peritoneal serous carcinoma

1. The malignancy is predominantly serous in type
2. Normal sized ovaries, or enlarged only by a benign process
3. Extra-ovarian disease exceeds any ovarian surface disease
4. Ovaries are not involved, or involvement is restricted to serosa and superficial cortex, with no nodule measuring larger than 5×5 mm
5. Exclusion of invasive or in situ serous carcinoma of the endometrium or tube

Modified after original SGO criteria

The clinical presentation and behavior of primary peritoneal serous carcinoma is identical or similar to advanced ovarian carcinoma. Consequently, surgical and chemotherapeutic treatment is also identical or similar.

Differential Diagnosis

The morphologic diagnosis of primary peritoneal serous carcinoma is usually straightforward, although small biopsies or cores with minimal disease can present a challenge. Nevertheless, there are a number of entities that should be considered and excluded before settling on a diagnosis of primary peritoneal serous carcinoma (Table 17.7).

As described above, the definitive exclusion of ovarian, tubal, and endometrial origins requires histologic examination of these organs. In most cases, this will require complete histologic embedding of the uterus, tubes, and ovaries if deemed clinically appropriate (Table 17.6). The major microscopic differential diagnoses are mesothelioma, metastatic carcinoma, and occasionally mesothelial hyperplasia. Metastatic carcinoma to peritoneum from both abdominal

Table 17.7 Differential diagnosis of primary peritoneal serous carcinoma

- Serous carcinoma of ovary, tube, and endometrium
- Malignant mesothelioma
- Serous borderline tumor of the peritoneum
- Psammocarcinoma
- Low-grade serous carcinoma of the peritoneum
- Well-differentiated papillary mesothelioma of the peritoneum
- Metastatic carcinoma
- Mesothelial hyperplasia

and extra-abdominal sites can also present as peritoneal plaques and masses.

Distinguishing mesothelioma from primary peritoneal serous carcinoma may have important management and medicolegal implications. The clinical presentation of peritoneal mesothelioma is similar or identical to peritoneal carcinoma: ascites and abdominal distension. The macroscopic findings of peritoneal plaques and nodules is common to both diseases. Consequently, distinguishing mesothelioma from primary peritoneal serous carcinoma is dependent upon histopathologic features. Although biphasic mesotheliomas with both epithelial and sarcomatous elements, and sarcomatoid mesotheliomas, may occur, most cases of peritoneal mesothelioma are epithelioid in type. Usually, the typical histologic appearance of serous carcinoma is distinctive from that of mesothelioma. By definition, mesotheliomas show invasion of underlying stroma and structures, but may assume a number of different patterns. Papillary, solid, tubular, and tubulopapillary patterns are common in epithelial mesothelioma, and the constituent cells retain a mesothelial-like appearance with a cuboidal or low columnar cell shape and moderate amounts of eosinophilic cytoplasm, and round to ovoid nuclei with a small nucleolus (Fig. 17.4). In contrast, serous carcinomas have complex papillae with tufting, PAS positive diastase-resistant mucin positivity, and marked nuclear anaplasia. Nevertheless, the microscopic appearance of mesothelioma can be variable, and foci of more marked pleomorphism resembling adenocarcinoma may occur. Psammoma bodies have been described in malignant mesothelioma, so their presence should not lead to a conclusion that primary peritoneal serous carcinoma is present. Consequently, an immunohistochemical panel may be useful or necessary to separate malignant mesothelioma from primary peritoneal serous carcinoma. Both mesothelioma and serous carcinoma exhibit immunopositivity for keratins, epithelial membrane antigen, CA125 (OC125), and WT-1. Immunopositivity for calretinin, CK 5/6, D2-40, HBME-1, and thrombomodulin are characteristic of mesothelioma (Fig. 17.5). Some caution is warranted in interpreting CK 5/6 in suspected peritoneal mesotheliomas, since some abdominal adenocarcinomas may also be positive [30]. In contrast, immunopositivity for Ber-EP4, B72.3, MOC-31, CA-19-9, Leu M1, ER, and PLAP is characteristic of serous carcinoma [31].

The rare deciduoid mesothelioma is characterized by sheets of eosinophilic cells with well-defined membranes, resembling decidua. This type of mesothelioma occurs not only in the peritoneum of young women [32], but also in the pleura and in men [33].

Less common primary mesothelial neoplasms may also have a papillary architecture (Table 17.7). Serous borderline tumors of the peritoneum lack evidence of stromal invasion. Psammocarcinoma has long been recognized, but remains a very rare entity if strict criteria are used. Criteria for the diagnosis of low-grade serous carcinoma of the peritoneum have been defined, but have not been validated and adopted widely [16]. The rare well-differentiated papillary mesothelioma has a rather unique appearance, but since it has a predominantly papillary architecture it could be mistaken for primary peritoneal serous carcinoma. It lacks nuclear atypia, epithelial stratification, mitotic activity, and stromal invasion. Proliferative areas of mesothelial hyperplasia can be striking and need to be distinguished from serous carcinoma.

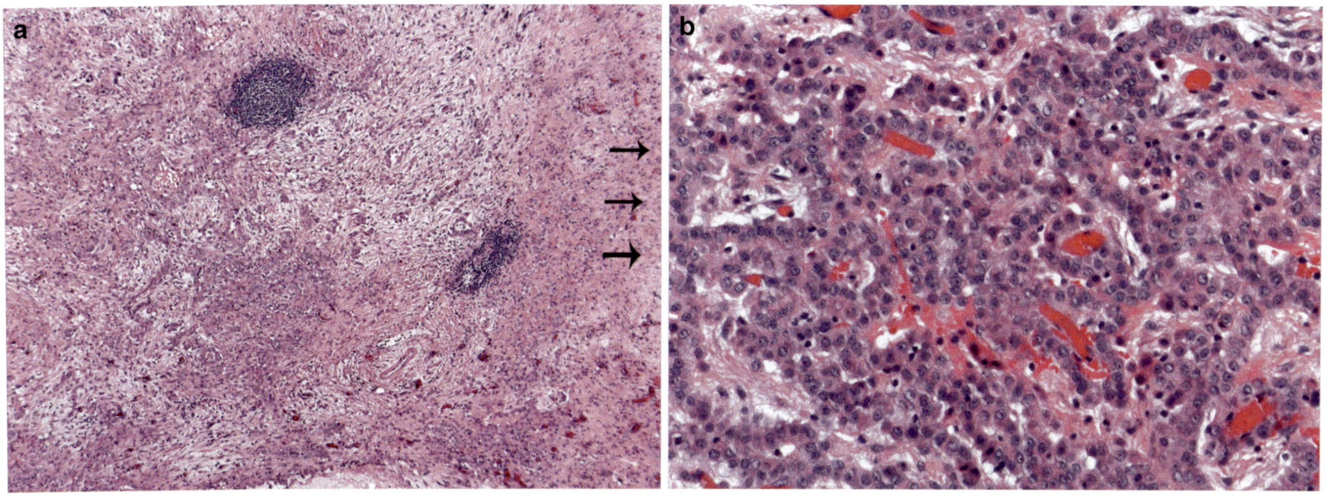

Fig. 17.4 (**a** and **b**). Diffuse malignant mesothelioma, epithelial type, in omentum. The extensive infiltration of the omentum beneath its surface (*arrows*) is key to the diagnosis of malignancy (Fig. 17.4a, low power). The constituent cells have abundant eosinophilic cytoplasm, resembling hyperplastic mesothelium, and only mild to moderate nuclear atypia (Fig. 17.4b, high power)

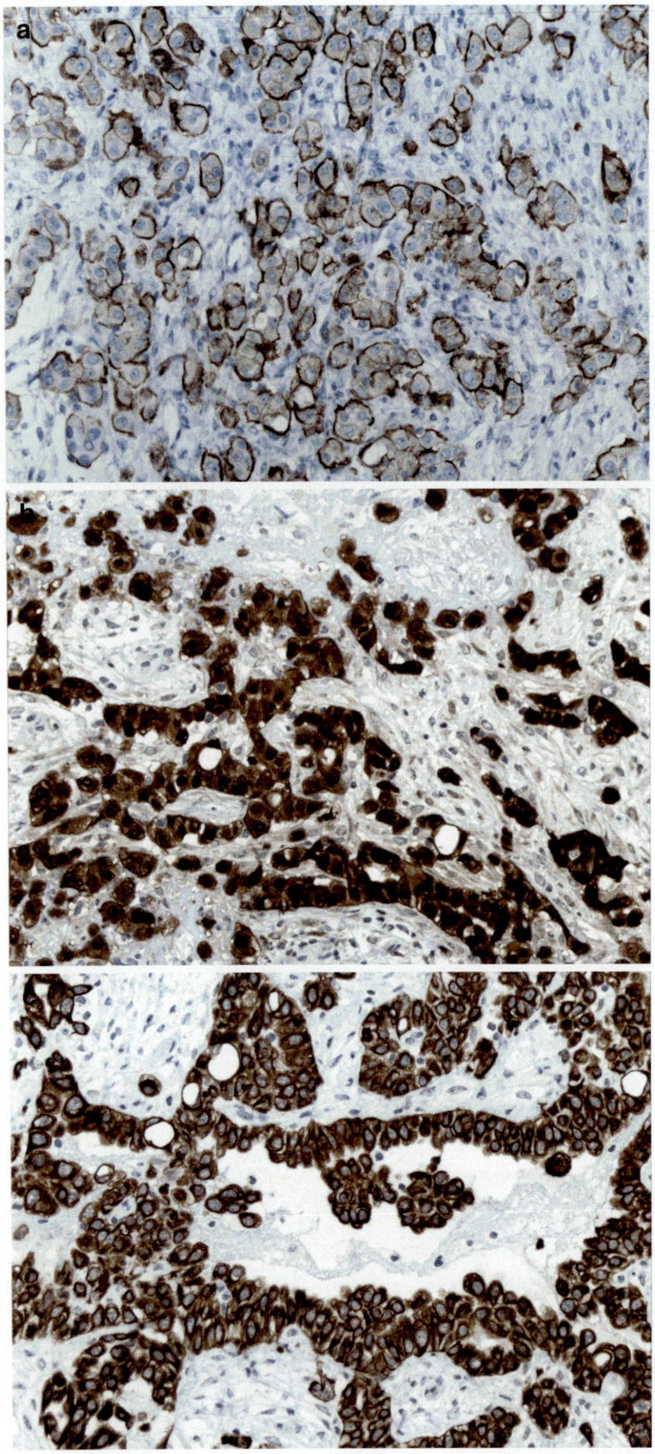

Fig. 17.5 (a–c). Immunohistochemistry (IHC) in diffuse malignant mesothelioma, epithelial type. IHC staining for CA125 is often positive and does not serve to distinguish mesothelioma from serous carcinoma (Fig. 17.5a). Strong nuclear and cytoplasmic IHC staining for calretinin (Fig. 17.5b) and cytoplasmic staining for CK5 (Fig. 17.5c) are typical of mesothelioma

References

1. Tavassoli FA, Devilee P, editors. Tumours of the breast and female genital organs. Lyon: IARC; 2003.
2. Olivier RI, van Beurden M, Lubsen MA, Rookus MA, Mooij TM, van de Vijver MJ, et al. Clinical outcome of prophylactic oophorectomy in BRCA1/BRCA2 mutation carriers and events during follow-up. Br J Cancer. 2004;90(8):1492–7.
3. Lu KH, Garber JE, Cramer DW, Welch WR, Niloff J, Schrag D, et al. Occult ovarian tumors in women with BRCA1 or BRCA2 mutations undergoing prophylactic oophorectomy. J Clin Oncol. 2000;18(14):2728–32.
4. Leeper K, Garcia R, Swisher E, Goff B, Greer B, Paley P. Pathologic findings in prophylactic oophorectomy specimens in high-risk women. Gynecol Oncol. 2002;87(1):52–6.
5. Powell CB, Kenley E, Chen LM, Crawford B, McLennan J, Zaloudek C, et al. Risk-reducing salpingo-oophorectomy in BRCA mutation carriers: role of serial sectioning in the detection of occult malignancy. J Clin Oncol. 2005;23(1):127–32.
6. Carcangiu ML, Peissel B, Pasini B, Spatti G, Radice P, Manoukian S. Incidental carcinomas in prophylactic specimens in BRCA1 and BRCA2 germ-line mutation carriers, with emphasis on fallopian tube lesions: report of 6 cases and review of the literature. Am J Surg Pathol. 2006;30(10):1222–30.
7. Shaw PA, Rouzbahman M, Pizer ES, Pintilie M, Begley H. Candidate serous cancer precursors in fallopian tube epithelium of BRCA1/2 mutation carriers. Mod Pathol. 2009;22(9):1133–8.
8. Lamb JD, Garcia RL, Goff BA, Paley PJ, Swisher EM. Predictors of occult neoplasia in women undergoing risk-reducing salpingo-oophorectomy. Am J Obstet Gynecol. 2006;194(6):1702–9.
9. Hirst JE, Gard GB, McIllroy K, Nevell D, Field M. High rates of occult fallopian tube cancer diagnosed at prophylactic bilateral salpingo-oophorectomy. Int J Gynecol Cancer. 2009;19(5):826–9.
10. Rebbeck TR, Lynch HT, Neuhausen SL, Narod SA, Van't Veer L, Garber JE, et al. Prophylactic oophorectomy in carriers of BRCA1 or BRCA2 mutations. N Engl J Med. 2002;346(21):1616–22.
11. Lu KH, Bell DA, Welch WR, Berkowitz RS, Mok SC. Evidence for the multifocal origin of bilateral and advanced human serous borderline ovarian tumors. Cancer Res. 1998;58(11):2328–30.
12. Kindelberger DW, Lee Y, Miron A, Hirsch MS, Feltmate C, Medeiros F, et al. Intraepithelial carcinoma of the fimbria and pelvic serous carcinoma: evidence for a causal relationship. Am J Surg Pathol. 2007;31(2):161–9.
13. Longacre TA, Oliva E, Soslow RA. Recommendations for the reporting of fallopian tube neoplasms. Hum Pathol. 2007;38(8):1160–3.
14. Medeiros F, Muto MG, Lee Y, Elvin JA, Callahan MJ, Feltmate C, et al. The tubal fimbria is a preferred site for early adenocarcinoma in women with familial ovarian cancer syndrome. Am J Surg Pathol. 2006;30(2):230–6.
15. Quddus MR, Sung CJ, Lauchlan SC. Benign and malignant serous and endometrioid epithelium in the omentum. Gynecol Oncol. 1999;75(2):227–32.
16. Weir MM, Bell DA, Young RH. Grade 1 peritoneal serous carcinomas: a report of 14 cases and comparison with 7 peritoneal serous psammocarcinomas and 19 peritoneal serous borderline tumors. Am J Surg Pathol. 1998;22(7):849–62.
17. Hu CY, Taymor ML, Hertig AT. Primary carcinoma of the fallopian tube. Am J Obstet Gynecol. 1950;59(1):58–67.
18. Jarboe E, Folkins A, Nucci MR, Kindelberger D, Drapkin R, Miron A, et al. Serous carcinogenesis in the fallopian tube: a descriptive classification. Int J Gynecol Pathol. 2008;27(1):1–9.

19. Folkins AK, Jarboe EA, Saleemuddin A, Lee Y, Callahan MJ, Drapkin R, et al. A candidate precursor to pelvic serous cancer (p53 signature) and its prevalence in ovaries and fallopian tubes from women with BRCA mutations. Gynecol Oncol. 2008;109(2):168–73.
20. Vang R, Shih Ie M, Kurman RJ. Ovarian low-grade and high-grade serous carcinoma: pathogenesis, clinicopathologic and molecular biologic features, and diagnostic problems. Adv Anat Pathol. 2009;16(5):267–82.
21. Dehari R, Kurman RJ, Logani S, Shih Ie M. The development of high-grade serous carcinoma from atypical proliferative (borderline) serous tumors and low-grade micropapillary serous carcinoma: a morphologic and molecular genetic analysis. Am J Surg Pathol. 2007;31(7):1007–12.
22. Jarboe EA, Miron A, Carlson JW, Hirsch MS, Kindelberger D, Mutter GL, et al. Coexisting intraepithelial serous carcinomas of the endometrium and fallopian tube: frequency and potential significance. Int J Gynecol Pathol. 2009;28(4):308–15.
23. Crum CP. Intercepting pelvic cancer in the distal fallopian tube: theories and realities. Mol Oncol. 2009;3(2):165–70.
24. Nordin AJ. Primary carcinoma of the fallopian tube: a 20-year literature review. Obstet Gynecol Surv. 1994;49(5):349–61.
25. Colgan TJ, Murphy J, Cole DE, Narod S, Rosen B. Occult carcinoma in prophylactic oophorectomy specimens: prevalence and association with BRCA germline mutation status. Am J Surg Pathol. 2001;25(10):1283–9.
26. Edge S, Byrd D, Compton C, Fritz A, Greene F, Trotti A, editors. American joint committee on cancer: cancer staging, handbook/cervix uteri. 7th ed. New York: Springer; 2010.
27. Colgan TJ. Challenges in the early diagnosis and staging of fallopian-tube carcinomas associated with BRCA mutations. Int J Gynecol Pathol. 2003;22(2):109–20.
28. Alvarado-Cabrero I, Navani SS, Young RG, Scully RE. Tumor of the fimbriated end of the fallopian tube: a clinicopathologic analysis of 20 cases, including nine carcinomas. Int J Gynecol Pathol. 1997;16:189–96.
29. Alvarado-Cabrero I, Young RH, Vamvakas EC, Scully RE. Carcinoma of the fallopian tube: a clinicopathologic study of 105 cases with observations on staging and prognostic factors. Gynecol Oncol. 1999;72:367–79.
30. Chu PG, Weiss LM. Expression of cytokeratin 5/6 in epithelial neoplasms: an immunohistochemical study of 509 cases. Mod Pathol. 2002;15(1):6–10.
31. Ordonez NG. Role of immunohistochemistry in distinguishing epithelial peritoneal mesotheliomas from peritoneal and ovarian serous carcinomas. Am J Surg Pathol. 1998;22(10):1203–14.
32. Nascimento AG, Keeney GL, Fletcher CD. Deciduoid peritoneal mesothelioma. An unusual phenotype affecting young females. Am J Surg Pathol. 1994;18(5):439–45.
33. Shanks JH, Harris M, Banerjee SS, Eyden BP, Joglekar VM, Nicol A, et al. Mesotheliomas with deciduoid morphology: a morphologic spectrum and a variant not confined to young females. Am J Surg Pathol. 2000;24(2):285–94.
34. Funk KC, Heiken JP. Papillary cystadenoma of the broad ligament in a patient with von Hippel-Lindau disease. AJR Am J Roentgenol. 1989;153(3):527–8.
35. Gaffey MJ, Mills SE, Boyd JC. Aggressive papillary tumor of middle ear/temporal bone and adnexal papillary cystadenoma. Manifestations of von Hippel-Lindau disease. Am J Surg Pathol. 1994;18(12):1254–60.
36. Gersell DJ, King TC. Papillary cystadenoma of the mesosalpinx in von Hippel-Lindau disease. Am J Surg Pathol. 1988;12(2):145–9.

18 Familial Cancer and Prophylactic Surgery

Terence J. Colgan and Martin C. Chang

Keywords
Risk-reducing salpingo-oophorectomy • *BRCA* and fallopian tube • Serous tubal intraepithelial carcinoma • p53 signature • Fimbrial carcinoma • HNPCC • Lynch syndrome

Familial Ovarian Cancer

Prevalence and Epidemiologic Overview

The American Cancer Society estimates that 21,550 new cases of ovarian cancer will be diagnosed in the United States in 2009, representing about 3% of all cancers and over 5% of cancer deaths in women. The lifetime incidence of ovarian cancer in the general population is between 1% and 2%, and over 80% of cases are associated with extra-ovarian spread at diagnosis. Heritable risk of ovarian cancer has been well documented [1], with large meta-analyses in the pre-genetic testing era demonstrating a two-to-fourfold increase in risk among first-degree relatives of women affected.

Familial ovarian cancers account for approximately 10% of all ovarian cancers, of which the vast majority arises in the setting of well-defined genetic syndromes [1]. Detectable mutations in the *BRCA1* and *BRCA2* genes account for 90% of ovarian cancers in the Hereditary Breast and Ovarian Cancer syndrome, and up to 85% of all familial ovarian cancers. *BRCA1* mutations markedly elevate ovarian cancer risk; the lifetime ovarian-cancer risk in affected individuals is between 40% and 60%. *BRCA2* mutations, depending on location within the gene, confer a more moderate ovarian cancer risk increase, with a lifetime risk of up to 25%.

T.J. Colgan (✉)
Pathology and Laboratory Medicine, Mount Sinai Hospital,
600 University Avenue, Toronto, ON M5G 1X5, Canada

BRCA1 and *BRCA2*, and Role of Other Genes

BRCA1 and *BRCA2* are distinct tumor-suppressor genes, both of which maintain genomic stability by playing roles in DNA repair. Mutations throughout the coding regions of both genes have been observed, although a subset of mutations have a particularly high prevalence owing to geographic and ethnic founder effects. Table 18.1 summarizes the genetic features and clinical syndromes associated with *BRCA1* and *BRCA2*.

The syndromes associated with mutations of both genes are phenotypically heterogeneous with respect to penetrance and the ratio of breast-to-ovarian cancer risk. A clinically significant correlation between genotype and phenotype has not been determined, but moderate differences in cancer risk have been attributed to mutation position within *BRCA1* and *BRCA2*. For example, *BRCA1* mutations in the 3′ third of the gene are associated with relatively fewer ovarian cancers, whereas *BRCA2* mutations in an area of exon 11 confer relatively high ovarian cancer risk. Genetic variation in other genes related to the function of *BRCA1* (e.g., HRAS1 variants favor ovarian cancer) and *BRCA2* (e.g., RAD51 variants favor breast cancer), as well as endocrine-signaling genes (e.g., NCOA3 variants favor breast cancer) also modify cancer risk [1].

R.A. Soslow and C. Tornos (eds.), *Diagnostic Pathology of Ovarian Tumors*,
DOI: 10.1007/978-1-4419-9751-7_18, © Springer Science+Business Media, LLC 2011

Table 18.1 Overview of *BRCA1* and *BRCA2* syndromes

BRCA1	*BRCA2*
Chromosome 17q21	Chromosome 13q12.13
Prevalence: 0.1% (1% in Ashkenazi)	Prevalence: 0.1% (1.5% in Ashkenazi, 0.6% of Icelanders)
Involved in DNA repair and cell-cycle control through incompletely characterized protein interactions	Involved in DNA repair and cell-cycle control through formation of a *BRCA2*-RAD51 complex
	Mutation in exon 11 region ("Ovarian Cancer Cluster Region") results in moderate increase of ovarian cancers relative to breast cancers
Breast cancers are usually "triple negative"	Breast cancers are usually ER/PR positive, Her2/neu negative
	Higher risk of male breast cancer
Very high lifetime risk of ovarian cancer (up to 60%)	Moderate lifetime risk of ovarian cancer (~25%)
Younger cancer onset than *BRCA2*	
Ovarian cancers are usually high-grade serous	Ovarian cancers not well characterized; appear similar to those seen in *BRCA1*[a]

[a]Although statistically, hereditary ovarian cancers will be predominantly *BRCA1*, there has been no reported phenotypic difference between *BRCA1*- and *BRCA2*-associated ovarian cancers. Most studies of serous carcinoma precursors in the high-risk population do not distinguish between *BRCA1* and *BRCA2* mutation carriers

Aside from *BRCA1* and *BRCA2*, two other distinct syndromes are considered to increase ovarian cancer risk. Lynch syndrome (Hereditary Nonpolyposis Colon Cancer Syndrome, HNPCC), arises from inherited defects in DNA mismatch repair, and the resulting microsatellite instability is specifically associated with an increased risk of colon, endometrial, and ovarian cancers. The significance of ovarian cancer in Lynch syndrome will be discussed briefly at the end of this chapter. The Peutz–Jeghers syndrome arises from a mutation of a serine-threonine kinase coded on Chromosome 19, and results in an increase in a number of characteristic tumors, especially hamartomatous polyps of the gastrointestinal tract and, in the ovary, sex-cord stromal tumors such as the sex-cord tumor with annular tubules (discussed in Chap. 14).

The Untested Patient

For a woman who has not been diagnosed with either breast or ovarian cancer, the most important familial risk factors are number of family members affected, early onset in affected family members, high-prevalence ethnicity (e.g., Ashkenazi, Icelandic, French-Canadian), and breast cancer in a male relative (more common in *BRCA2*).

Histologic Types in Familial Ovarian Cancer

The majority of familial ovarian cancers are serous carcinomas, occurring with a somewhat higher relative frequency than other histologic types within sporadic cases. Some have reported an increased overall incidence of endometrioid and undifferentiated carcinomas [1], but this has not been validated with immunohistochemistry. Mucinous carcinomas occur less frequently in *BRCA1* mutation carriers. Owing to the lower ovarian cancer risk, similar data for *BRCA2* carriers are scarce. Although statistically, hereditary ovarian cancers are predominantly *BRCA1*-associated, there has been no reported phenotypic difference between *BRCA1*- and *BRCA2*-associated ovarian cancers. *BRCA1* or *BRCA2* mutations do not appear to increase the risk for borderline tumors, sex-cord stromal tumors, or germ cell tumors (see Table 18.2).

Table 18.2 *BRCA* mutation positivity and associated ovarian tumors

Increased incidence with *BRCA1* mutation[a]	No increase with *BRCA1* mutation[a]
• High-grade serous carcinoma[b]	• Mucinous carcinoma • Clear-cell carcinoma • Transitional cell • Low-grade serous carcinoma • Epithelial borderline tumors • Germ cell tumors • Sex-cord stromal tumors

Limited data report occasional cases of endometrioid carcinoma, undifferentiated carcinoma, and carcinosarcoma
[a]Not systematically studied for *BRCA2* mutations
[b]Increase reported for ovarian, fallopian tube, and primary peritoneal carcinomas; seen in both *BRCA1* and *BRCA2* mutation carriers

In the remainder of the chapter, no distinction will be made between carriers of *BRCA1* and *BRCA2* with respect to the microscopic findings in surgical specimens.

Fallopian Tube as a Source of "Serous Pelvic Carcinoma"

The strikingly high cancer risk conferred by *BRCA* mutations has provided insights into ovarian carcinogenesis by allowing the detection of the earliest disease precursors. It is ironic that a syndrome originally dubbed the "Hereditary Breast Ovarian Cancer" syndrome has led to the identification of a likely non-ovarian source for serous carcinoma in general – the distal fallopian tube – as discussed in Chap. 17. In this new paradigm of tubal-fimbrial origin ovarian, tubal, and primary peritoneal carcinomas classically defined by WHO criteria are unified as pelvic serous carcinoma. Unequivocal ovarian high-grade serous carcinomas arising in the ovary proper, for example, from low-grade serous neoplasms, have only been reported rarely [2]. In the remainder of the chapter, "ovarian cancer" refers to a defined clinical syndrome, or a specific WHO-classified entity. The term "pelvic serous carcinoma," although not used in the WHO system, is applied to the discussion of issues of carcinogenesis and precursor lesions. (Excluded from the discussion are serous carcinomas that arise in the pancreas and rarely involve the peritoneum.)

Pelvic Surgery for *BRCA1/2* Carriers

Terminology

Given the high lifetime risk of developing serous carcinoma of the pelvis, many women will choose to undergo prophylactic resection of tissues at risk. This surgery is often labeled "prophylactic salpingo-oophorectomy," but since there does remain a small residual risk for the development of peritoneal serous carcinoma, a preferred term is "risk-reducing salpingo-oophorectomy (RRSO)"

Selection of Patients for Risk-Reducing Surgery

For women with a *BRCA1* mutation, the risk of serous carcinoma begins to rise in the late 30s and early 40s, whereas this rise is delayed for about a decade in women with a *BRCA2* mutation. Risk assessment of women with a personal and/or family history suggestive of an inherited predisposition, but who has not had genetic testing or has not had a *BRCA* mutation identified, is challenging. RRSO is usually offered to *BRCA1* + women upon completion of childbearing, and deferred after early 40s only with disclosure of risks. Although delaying RRSO in *BRCA2* women could be considered, such a deferral may lead to a substantial loss of the protective benefit against breast carcinoma that RRSO achieves. Some pre-menopausal women with a history of breast carcinoma may elect to undergo RRSO not only for its protective effects against serous carcinoma, but for the surgical induction of menopause.

Nature of Surgery

Risk-reducing surgery which consists of oophorectomy only is inadequate since there continues to be a substantial risk of subsequent serous carcinoma. As a minimum, both fallopian tubes and ovaries (RRSO) are resected. RRSO may be performed as an outpatient laparoscopic procedure. Amputation of the tubes as close as possible to the uterine cornua has been recommended. Post-RRSO ovarian remnants should be avoided. It is unclear if salpingectomy only, with retention of the ovaries, could deliver the same risk reduction. Peritoneal washings for cytology and selective peritoneal biopsies may be performed at the time of surgery, since these can assist in the staging of disease in the event that occult disease is detected.

Whether hysterectomy with its increased risk of perioperative morbidity should also be performed to reduce the risk of uterine-endometrial serous carcinoma (USC), and facilitate post-operative hormonal therapy, is still controversial. USCs have been found in *BRCA1* and *BRCA2* mutation carriers, in some reports with high frequency of association [3]; however, a causative role has not been proven. It is theoretically possible that these cases represent drop metastases from an occult tubal primary, manifesting clinically as endometrial cancer. RRSO has the effect of closing the continuity between endometrium and peritoneum. In most cases, the endometrium is amenable to clinical follow-up and biopsy; as such, a routine role for prophylactic hysterectomy is doubtful. Nevertheless, patients undergoing RRSO frequently have concurrent indications for hysterectomy (fibroids, post-menopausal bleeding). In such cases, the pathologic examination of the uterine corpus would proceed as with routine cases.

Intra-operative consultation ("frozen section") should be restricted to those cases in which a mass lesion is identified. In the event carcinoma is detected, a complete staging procedure can be completed at the time of the RRSO.

Grossing Procedures of the Specimen

The entire RRSO specimen should be examined microscopically. The detection of occult tubal and ovarian malignancies in RRSO specimens that have not been subjected to such complete sectioning is much reduced (Table 18.3). Sections through the short axis of the ovary maximize the amount of serosal surface available for examination. Since most of the tubal carcinomas are distal or fimbrial in location (see below), transverse sections of the fallopian tubes, in combination with longitudinal sections of the fimbriae (dubbed the "SEE-FIM" protocol at its originating institution, Brigham and Women's Hospital) maximize sensitivity in detecting occult malignancy [4]. Routine "levels" of the RRSO specimen have not been shown to be useful. Exceptionally, such sectioning should be undertaken in the presence of a positive peritoneal wash, and may reveal a putative primary focus. The use of the SEE-FIM grossing protocol in non-RRSO specimens is discussed in Chap. 17.

Microscopic Findings of Specimens

Definition of Lesions

Thorough microscopic evaluation of RRSO specimens has demonstrated otherwise occult malignancies within the

Table 18.3 Prevalence of occult tubal and ovarian malignancies in RRSO specimens: findings in completely embedded vs incompletely embedded specimens

Study (Ref. number)	Presence of occult malignancy in tube/ovary		Proportion of malignancy in tube
	BRCA +	*BRCA* unknown /untested	
Complete histologic embedding			
Lu et al. [16] $n=33$	4/23 (17%)	0/11 (0%)	100%
Colgan et al. [17] $n=60$	5/39 (13%)	0/21 (0%)	80%
Leeper et al. [18] $n=30$	4/17 (24%)	0/13	75%
Olivier et al. [12] $n=90$	5/65 (8%)	0/25 (0%)	60%
Powell et al. [19] $n=20$	5/20 (25%)		80%
Carcangiu et al. [20] $n=50$	6/50 (12%)		67%
Shaw et al. [21] $n=176$	15/176 (8.5%)		100%
Lamb et al. [22] $n=113$	6/62 (10%)	0/51 (0%)	83%
Hirst et al. [23] $n=45$	5/15 (11%)	0/30 (0%)	100%
Total:	**55/467 (11.7%)**	**0/151 (0%)**	**47/55 (85%)**
Incomplete histologic embedding			
Kauff et al. [24] $n=98$	3/98 (3%)		NA
Olivier et al. [12] $n=38$ Oophorectomy	0/29 (0%)	0/9 (0%)	NA
Powell et al.[a] [19] $n=47$	2/47 (14%)		NA
Callahan et al.[a] [13] $n=122$	7/120 (6%)	0/2 (0%)	NA
Rebbeck et al. [25] $n=259$	6/259 (2%)		NA
Total:	**18/553 (3.2%)**	**0/11 (0%)**	

BRCA +, mutation positive in *BRCA1* and/or *BRCA2*; *BRCA* unknown, "High-risk" but no, or inconclusive, *BRCA* testing[a]. Some specimens examined in toto

adnexa in a large proportion (from 8% to 38%) of high-risk patients. Overall, about 10% of completely embedded *BRCA*+RRSO specimens will show occult tubal and/or ovarian malignancy upon histologic examination (Table 18.3). This prevalence may provide a useful benchmark for laboratories undertaking examination of this specimen type. Most of these occult malignancies are serous tubal intraepithelial carcinomas. In comparison, occult malignancy in RRSO specimens from *BRCA* unknown/untested patients seems to be exceedingly rare. Studies of risk-reducing ("prophylactic") surgical specimens from the high-risk population have not identified a convincing difference with respect to occult lesions between *BRCA1* and *BRCA2* mutation carriers (Table 18.4). The recognition and correct classifications of both occult malignancy and precursors have provided new insights into the origins of pelvic serous carcinoma (see also Chap. 17).

Ovarian Epithelial Precursor Lesions and Rarity of Serous Precursors

Primary ovarian carcinomas ("surface-epithelial stromal tumors") are classically thought to arise from ovarian surface epithelium or its derivatives, namely, cortical (mesothelial) inclusion cysts. It has historically been difficult to detect unequivocal premalignant/precursor lesions within these ovarian surface-epithelial derivatives. Thus, the biologic basis of progression from the epithelium of cortical inclusion cysts to the diverse histologic types of ovarian epithelial neoplasia remains elusive, although it has been studied recently [5]. The principal challenge has been that ovarian neoplasia is typically only diagnosed when clinically apparent.

As noted, with the recent studies of RRSO specimens from high-risk patients, the detection of pre-clinical ovarian malignancy has become common (Table 18.3). Indeed, a spectrum of occult ovarian disease has been found in RRSO

Table 18.4 Prevalence of occult tubal/ovarian malignancies in completely embedded RRSO specimens: *BRCA1* mutation + vs *BRCA2* mutation +

Study (Ref. number) n = *BRCA* + cases	*BRCA1* mutation +	*BRCA2* mutation +	*BRCA1* mutation + and *BRCA2* mutation +
Lu et al. [16] $n = 22$	3/20 (15%)	1/2 (50%)	
Colgan et al. [17] $n = 39$	5/27 (13%)	0/12 (0%)	
Leeper et al. [18] $n = 17$	3/14 (21%)	1/3 (33%)	
Olivier et al. [12] $n = 65$	5/58 (6%)	0/6 (0%)	0/1 (0%)
Carcangiu et al. [20] $n = 50$	6/37 (16%)	0/13 (0%)	
Lamb et al. [22] $n = 62$	5/40 (13%)	1/22 (5%)	
Shaw et al. [21] $n = 176$	8/103 (8%)	7/73 (10%)	
Total:	**35/299 (11.7%)**	**10/121 (8.3%)**	

specimens, ranging from metastatic carcinoma (e.g., from a breast primary) and advanced serous carcinoma to microscopic intracystic borderline tumors. Close evaluation of RRSO specimens across multiple studies underscores the separation of two distinct categories of precursor lesions: [1] ovarian cortical inclusion cysts and related intracystic tumors, and [2] high-grade serous tubal intraepithelial carcinoma and related lesions. As discussed in earlier chapters, the former is typically associated with alterations in *KRAS* and *BRAF*, and the latter with mutations in *TP53*. True ovarian-surface serous intraepithelial carcinomas have been described, but are extremely rare relative to lesions in the tubal fimbriae (see below). Invasive primary carcinomas of the fallopian tube are discussed in Chap. 17.

Ovarian cortical inclusion cysts are thought to arise from invaginations of the surface mesothelium as a healing response to injury (e.g., post-ovulation). It has been suggested recently that ciliated epithelium found in inclusion cysts may derive from the fallopian tube [6]. The mesothelium may undergo metaplasia to müllerian epithelium, and progress with variable amounts of admixed stromal growth to become large cystadenomas, intracystic borderline tumors, or rarely, invasive (cyst) adenocarcinomas (of the low grade serous, endometrioid, mucinous, and/or clear-cell types). Within RRSO specimens from the *BRCA*-carrier population – despite the markedly increased risk for ovarian carcinoma – the prevalence of cortical inclusion cysts and borderline tumors is not increased relative to the general population.

Normality of Tubal Mucosa and Hyperplasia

Fallopian tube epithelium is predominantly non-stratified and lined with a mixture of ciliated and secretory cells. A small subset of secretory cells ("intercalated/peg cells") may have a more elongated nucleus. The ratio of ciliated to secretory cells varies significantly with hormonal factors and location in the tube. Mitoses can be found in normal tubal epithelium, and should not be taken as *prima facie* evidence of neoplasia (Fig. 18.1). There is considerable variability in the appearance of normal tubal epithelium (Fig. 18.2). Larger contiguous regions of secretory cells may appear hyperplastic, displaying variations in density, epithelial stratification, and nuclear shape. These changes are mostly physiologic and should not be interpreted as neoplastic.

Serous Tubal Intraepithelial Carcinoma (STIC)

Serous tubal intraepithelial carcinoma (STIC) or adenocarcinoma in situ is an early high-grade serous carcinoma characterized by an intraepithelial proliferation of highly atypical non-ciliated tubal epithelial cells, with loss of polarity and varying degrees of stratification (including a single layer) that displace the normal mucosal epithelium (Fig. 18.3) The importance of STICs in high-grade serous carcinogenesis has been underscored by studies sequencing the *TP53* gene that have shown mutations in almost all STICs examined which are concordant with concomitant mutations of associated primary serous carcinoma of any pelvic site.

The cytologic features of the lesional epithelium are of key importance in the diagnosis of STIC [1, 7]. Malignant cells within STICs lack cilia, demonstrate elevated nuclear-to-cytoplasmic ratio, and a high mitotic index. Nuclei are also atypical, demonstrating a less elongated shape and more prominent nucleoli (Fig. 18.4). Although nuclear pleomorphism is traditionally cited in STICs, it has been noted that normal fallopian tube epithelial cells vary considerably in nuclear morphology – a fact that must be considered to avoid diagnostic pitfalls [7]. The key architectural feature in STIC is loss of polarity, which can be appreciated even in a single layer of malignant cells. Disorganized stratification of the epithelium is a more common pattern. Irregular fracturing of the stratified epithelium with surface exfoliation is a common architectural pattern but is not invariably present (Fig. 18.5). Many mutations result in the accumulation of

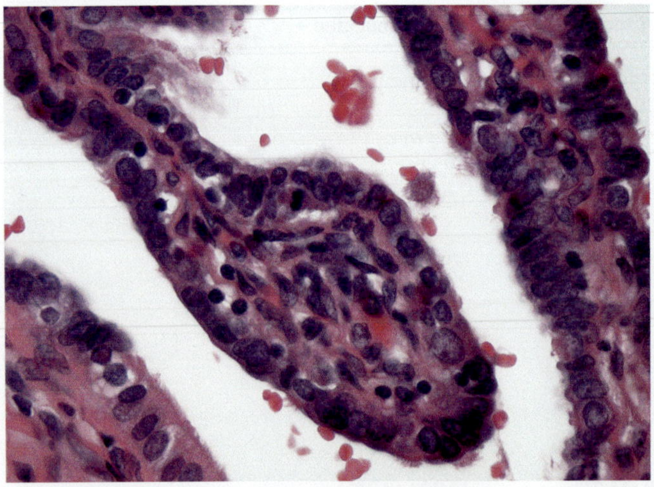

Fig. 18.1 Normal tubal mucosal plicae (H&E stain). Note mitosis (*center*). Mitotic activity is not necessarily an indicator of malignancy

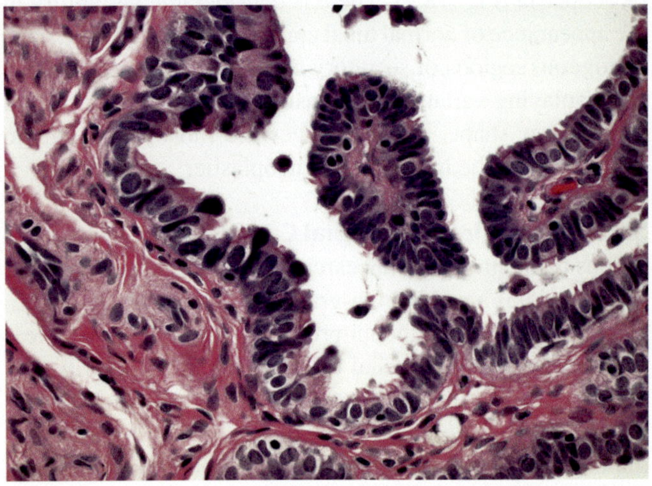

Fig. 18.2 Normal tubal epithelium (H&E stain). Even normal tubal epithelium can demonstrate considerable stratification and nuclear hyperchromasia and variability

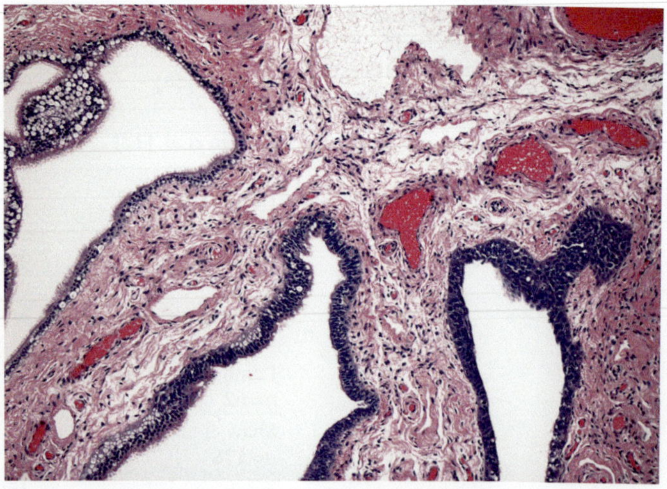

Fig. 18.3 Tubal mucosa (H&E stain) showing both normal mucosa (*left*) and serous intraepithelial carcinoma, STIC, (*right*). STIC can often be suspected even at such low power assessment in many cases

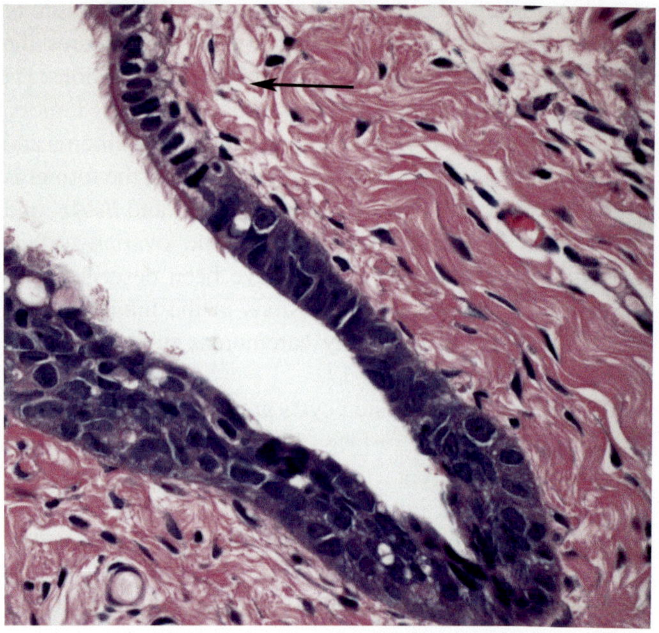

Fig. 18.4 Tubal epithelium (H&E stain) showing the junction between normal, ciliated serous epithelium (*arrow*), and serous intraepithelial carcinoma (STIC)

non-functioning protein in the nucleus, and thus strong p53 immunoreactivity. Although STICs frequently demonstrate strong p53 and Ki-67 immunoreactivity (Figs. 18.6 and 18.7), the diagnosis is made on morphologic grounds. Reliance on the p53 immunostain to make the diagnosis of STIC, in the absence of the major diagnostic criteria, is to be discouraged. A significant proportion (10–20%) of serous cancers and precursors has underlying null *TP53* mutations; moreover, there is a spectrum of benign changes that may also be associated with p53 immunoreactivity (see below).

These morphologic criteria are important in distinguishing STIC from its benign mimics (Table 18.5). Tangential sectioning may result in a stratified or pseudostratified appearance within the tubal mucosa (Fig. 18.8a, b). Thermal damage can give the superficial appearance of an epithelial abnormality (Fig. 18.9). Areas of true secretory-cell hyperplasia with epithelial stratification are frequently found, some of which feature nuclear hyperchromasia and size variation. It is believed that this pattern is favored by the hormonal environment of the luteal phase [8]. These areas may typically be distinguished from STIC by the predominance of elongated nuclei and an organized or polarized pattern of growth. The reactive epithelial changes seen in acute salpingitis can be striking (Fig. 18.10). The surface epithelium associated with a tubal adenofibroma may have reparative-type atypia,

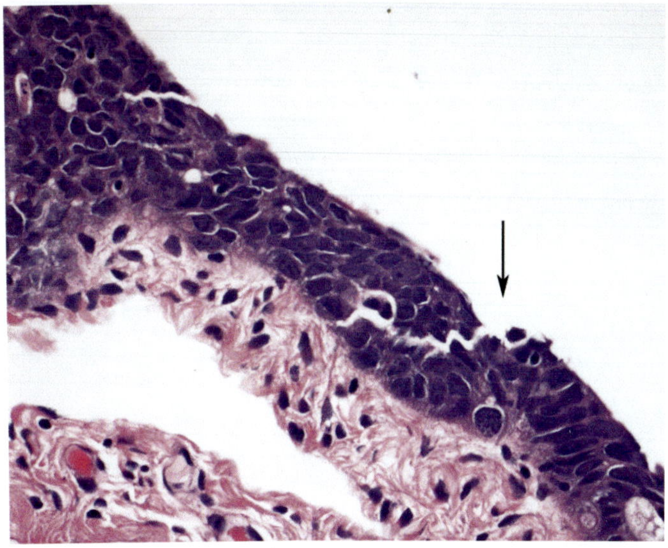

Fig. 18.5 Serous intraepithelial carcinoma, STIC (H&E stain). Note the stratified and marked atypia of the epithelium with "cracking artifact" (*arrow*)

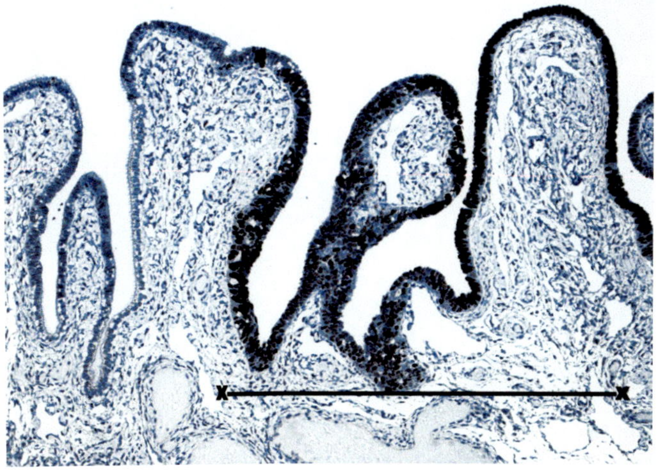

Fig. 18.6 Immunohistochemical stain for p53 of tubal mucosa. Serous intraepithelial carcinoma, STIC shows strong, uniform staining (*above line*), while non-neoplastic epithelium shows minimal/absent staining (*left of center*). p53 IHC is a useful adjunct to the diagnosis of STIC, but is not specific for this lesion

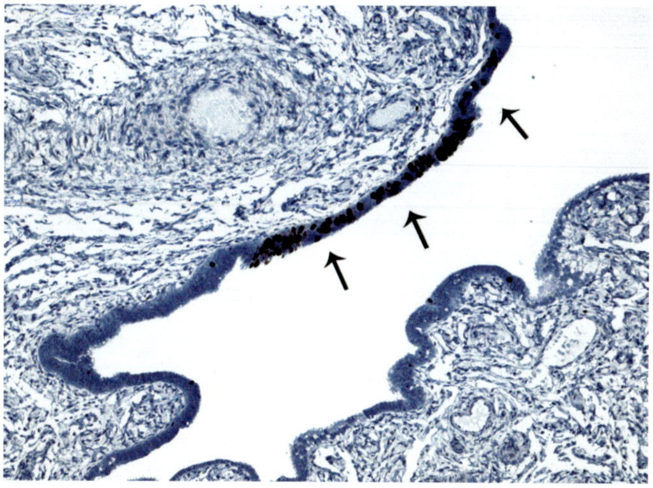

Fig. 18.7 MIB1 immunohistochemical stain of tubal mucosa. Serous intraepithelial carcinoma, STIC, shows strong uniform staining (*arrows*), while non-neoplastic epithelium shows absent staining (*below*). MIB1 IHC can also be a useful adjunct to the diagnosis of STIC

distinguishable from serous neoplasia by the presence of degenerated hyperchromatic nuclei and dense eosinophilic cytoplasm (Fig. 18.11). Transitional-cell metaplasia is a frequent benign finding in the fallopian tube and, when prominent, may also mimic STIC [9] (Fig. 18.12). Tubal transitional cells often grow in a nested pattern, cytologically resembling benign transitional cells of other sites, and lacking both nuclear atypia and mitotic activity. Oncocytic metaplasia of the tubal epithelium may also occur (Fig. 18.13). Finally, STIC must be distinguished from early invasive tubal carcinoma (discussed below and in Chap. 17).

Table 18.5 Serous intraepithelial carcinoma (STIC): criteria and differential diagnosis

Major criteria
• High N/C ratio
• Nuclear atypia (irregular contours, rounded)
• Lack of cilia
• Prominent nucleoli
• High mitotic activity
• Loss of polarity
Other helpful features
• Disorganized stratification
• Epithelial "cracks"/exfoliation
Differential diagnosis
• Tangential sectioning
• Secretory-cell hyperplasia
• "Atypical hyperplasia"/mild-to-moderate "dysplasia"
• Surface reparative changes (e.g., over an adenofibroma)
• Transitional cell metaplasia
• Early invasive tubal carcinoma

Dysplasia and Atypical Hyperplasia

Cytologic atypia not considered fully diagnostic of STIC ("atypical hyperplasia"/"dysplasia") has been described; however, the significance of these lesions is unclear [7, 10]. Rare cases of STIC with disseminated carcinoma on follow-up have been reported, but none of the "earlier" lesions, including "*p53 signatures*," have been shown to be of clinical significance. Cases with varying degrees of cytologic atypia pose the most diagnostic uncertainty, since some but not all of the criteria for STIC are met.

The term "atypical hyperplasia" is used inconsistently, to describe a pattern with stratification of mucosal cells, at times papillary or cribriform epithelial projections, nuclear

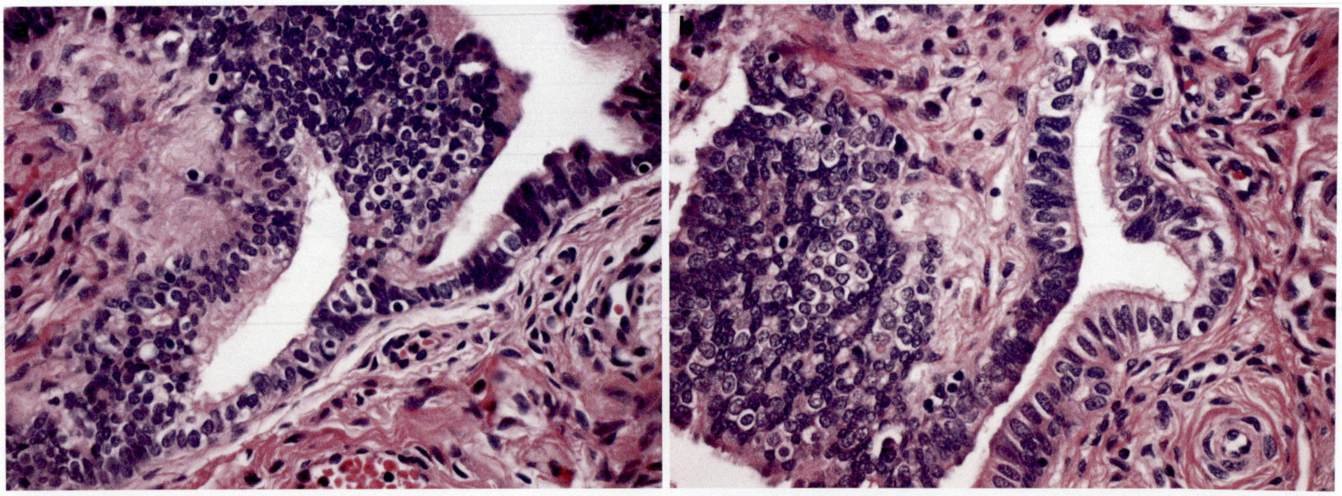

Fig. 18.8 (**a** and **b**) Normal tubal epithelium (H&E stain) which is presented in tangential sections should not be mistaken for serous intraepithelial carcinoma, STIC

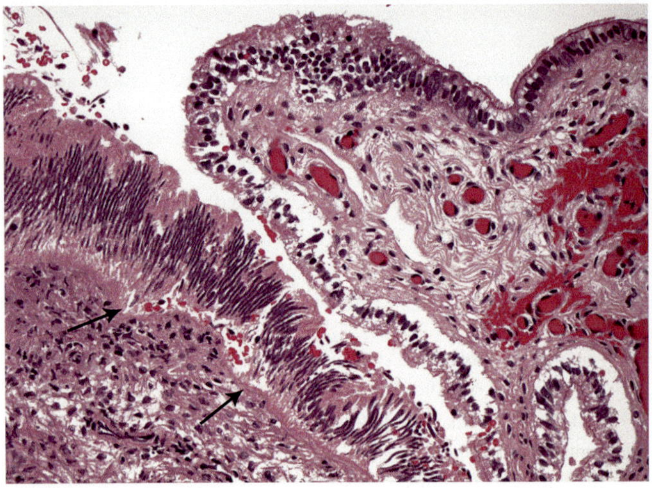

Fig. 18.9 Thermal artifact in tubal epithelium (H&E stain, arrows) can superficially resemble serous intraepithelial carcinoma

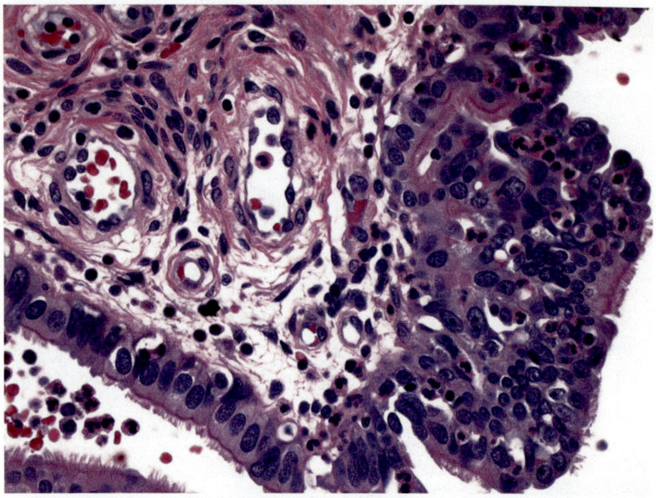

Fig. 18.10 Acute salpingitis (H&E stain). The reactive serous epithelium can develop considerable stratification and cytologic atypia

hyperchromasia, and size variation [10]. In these areas, however, the nuclei demonstrate a lack of mitotic activity, preserved polarity, even spacing, and a linear or radial arrangement. This is thought to represent a benign reactive mimic of neoplasia, and usually demonstrates low p53 immunoreactivity and Ki-67 proliferative activity.

Cases exhibiting discrete and focal mild to moderate nuclear atypia, with variable amounts of disorganization and epithelial displacement along a single or pseudostratified layer have been labelled as "dysplasia" by some. These areas stand out as being different from the surrounding mucosa; however, the number of atypical cells may be small. Immunoreactivity for p53 may be seen, and may even be strong within a continuous stretch of cells. Nevertheless, the full criteria for STIC are typically absent. Mitoses are usually not seen, and Ki-67 immunoreactivity may be low, even in atypical cells. The nuclei may be poorly preserved, with

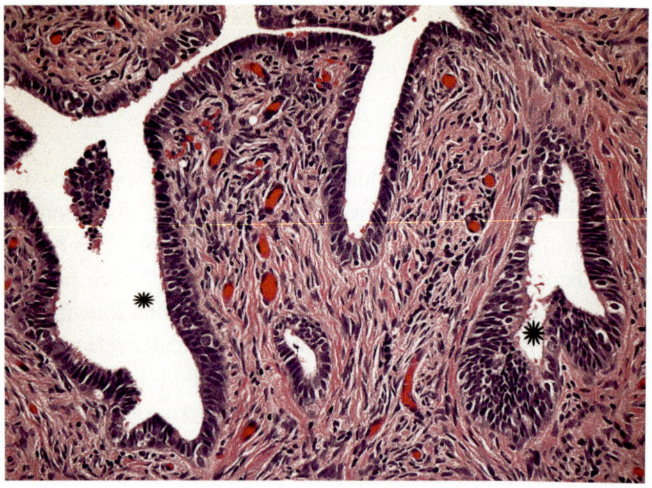

Fig. 18.11 Adenofibroma of the tubal fimbria (H&E stain). The serous epithelium can be stratified and cellular (*asterisks*)

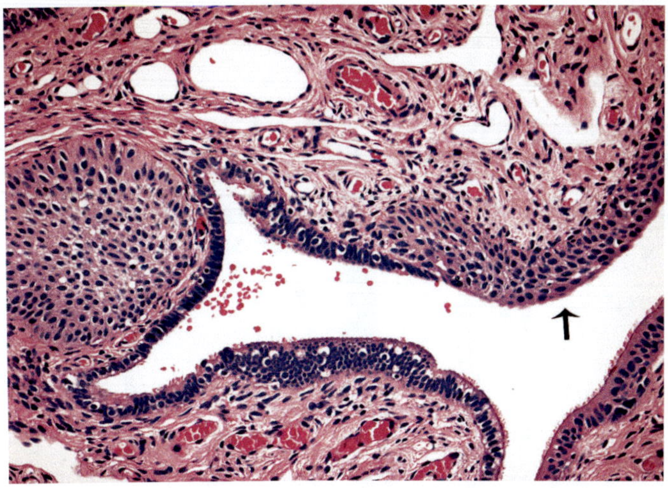

Fig. 18.12 Transitional cell metaplasia of fimbrial epithelim (H&E stain). The transitional epithelium may replace the surface tubal epithelium (*arrow*) and present as nests in the submucosa

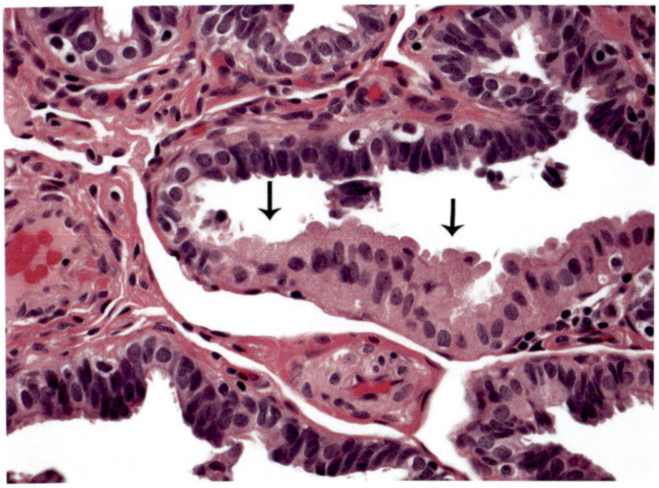

Fig. 18.13 Oncocytic metaplasia of tubal epithelium (H&E stain). The distinctive appearance of this metaplasia is unlikely to be confused with serous tubal intraepithelial carcinoma, STIC

indistinct nucleoli; evidence of degenerative changes may also be seen within the cells, including shedding of cilia (Fig. 18.14). There may be an admixture of ciliated cells, suggesting that the atypia is not associated with a continuous clonal process. If highly atypical cells displace normal epithelium over a single intramucosal layer this finding should be considered diagnostic for STIC. The combination of strong p53 immunoreactivity and high Ki-67 proliferative activity may support a morphologic diagnosis of STIC, but morphologically non-carcinomatous p53 and Ki67 immunoreactive proliferations have been described (see "tubal intraepithelial lesion in transition" discussion below). Since the diagnosis of both "atypical hyperplasia" and "tubal dysplasia" lacks both precision and any known clinical significance, *they should not be used in routine reporting*.

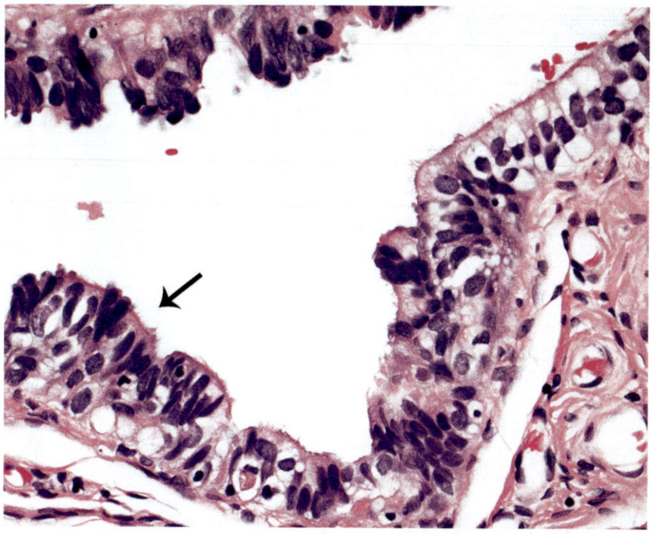

Fig. 18.14 Tubal epithelial atypia (H&E stain). An area of the tubal epithelium shows some nuclear atypia in comparison to the adjacent normal epithelium (*right*), but ciliated cells are retained. Some pathologists might interpret this as a "dysplastic" lesion, but atypia in ciliated cells is better interpreted as a variant of normal

Early Invasive Tubal Carcinoma

The fallopian tube mucosa overlies the endosalpingeal stroma, lined by a basement membrane. By definition, early tubal invasive carcinoma is distinguished from STIC by the presence of invasion into the stroma. In high-risk patients, examination of RRSO specimens may reveal a primary occult invasive carcinoma of the fallopian tube (by WHO criteria) in some cases (Table 18.3). A majority of these are associated with concomitant STIC. Staging and management of invasive tubal carcinoma is discussed in Chap. 17.

p53 Signature and Tubal Intraepithelial Lesions in Transition (TILT) and Immunohistochemistry

Although the relative numbers of secretory and ciliated cells in normal tubal epithelium vary, continuous stretches (at least 12 cells) of benign tubal secretory cells that demonstrated clonality, strong p53 immunoreactivity, and co-localizing evidence of DNA damage can be identified in RRSO specimens from *BRCA* mutation carriers [11]. In contrast to STICs, these areas (called "p53 signatures") do not have cytologic atypia and do not demonstrate proliferative activity on Ki-67 immunohistochemistry (Fig. 18.15). Thus, p53 signatures are benign but share molecular features with serous carcinoma, consistent with an early ("latent") precursor to STIC (see Table 18.4).

Interestingly, p53 signatures occur in *BRCA* carriers with equal frequency (~30–40%) compared to the general population, unlike isolated STICs which are extremely rare in non-*BRCA* carriers. Moreover, STICs concomitant with sporadic ovarian and primary peritoneal carcinomas are common and indistinguishable from their counterparts in *BRCA* carriers.

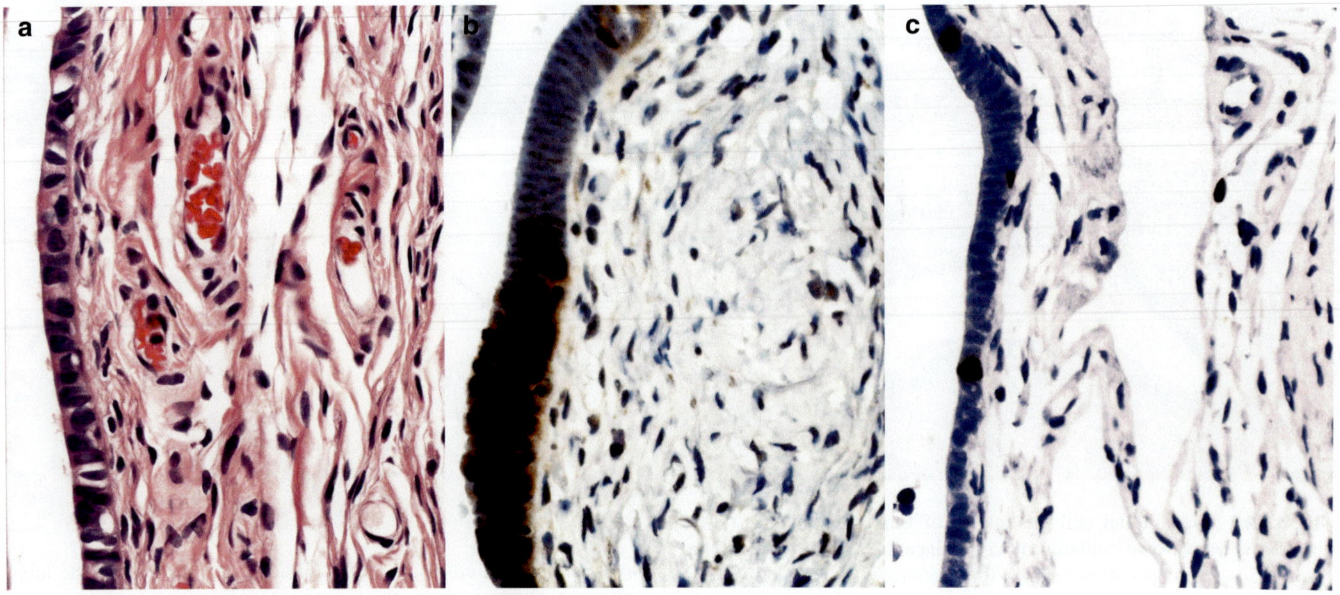

Fig. 18.15 "p53 signature" (H&E and p53, Mib-1 immunostains). This area of the tubal fimbrial epithelium shows no cellular atypia (**a**), but a discrete continuous area of strong nuclear p53 immunoreactivity is present, (**b**, *bottom left*) with proliferative activity within the normal range (**c**)

Table 18.6 A comparison of the features, biologic significance, and clinical importance of tubal precursor and carcinoma lesions

Finding at RRSO	Features	Biologic significance	Clinical utility and reporting
"p53 signature"[b]	More than 12 continuous benign non-ciliated cells immunoreactive for p53[a], but not Ki-67[a]	Benign non-obligate precursor lesion	None[a]
"Tubal intraepithelial lesion in transition"[b]	Stratified proliferation of benign-appearing non-ciliated cells immunoreactive for p53[a] and Ki-67[a]	Proposed benign intermediate precursor lesion	None[a]
Tubal epithelial "dysplasia"[b]	Discrete areas of atypia, but falls short of diagnostic criteria for STIC	Unclear. (Some may represent benign reactive changes)	None
STIC	Meets morphologic criteria for intraepithelial carcinoma	Cancer precursor with risk of theoretical dissemination	Yes
Invasive tubal serous carcinoma	Meets morphologic criteria for primary invasive carcinoma	Risk of dissemination	Yes

[a]Routine p53 or Ki-67 immunohistochemistry is *not* recommended in the evaluation of RRSO specimens, but may be used adjunctively
[b]Terminology not recommended for diagnostic reporting

These findings suggest that (1) the existence of a p53 signature is likely not associated with cancer risk beyond the general population, (2) a role for the p53-signature-STIC pathway cannot be excluded in sporadic as well as familial serous cancers, and (3) mutations in *BRCA1* and *BRCA2* play a key as-yet-undiscovered role in the progression of p53 signatures to STIC and serous carcinoma (Table 18.6).

Lesions with features intermediate between p53 signatures and STICs have been described, consisting of continuous stretches of variably stratified tubal-secretory cells that demonstrate strong p53 and Ki-67 immunoreactivity but no cytologic atypia. To date, only two of these "tubal epithelial lesions in transition" (TILT) have been reported, and their clinical significance is unknown. Nevertheless, their existence points to a degree of continuity within the high-grade serous carcinogenic pathway.

The descriptions of a p53 signature and tubal intraepithelial lesion in transition has provided a conceptual framework of tubal neoplasia, but use of these morphologic categories should be restricted to the research setting only. Their use in routine diagnostic reports is of no clinical value, and may lead to adverse consequences for both the *BRCA*+and normal woman.

Occult Malignancies of the Non-serous Types

Although serous carcinoma accounts for over two thirds of ovarian carcinomas associated with *BRCA1* mutation carriers [1], there is both a relative and absolute increase among *BRCA1* carriers of ovarian carcinomas diagnosed as endometrioid and undifferentiated. At least two cases of occult tubal endometrioid carcinoma from a *BRCA1* carrier at RRSO have been reported [12, 13]. While there is evidence for a specific role of *BRCA1* (and *BRCA2*) mutations in the serous carcinogenesis pathway, their role in ovarian cancers of these other histologic types is unclear. It should be noted that diagnostic criteria used to separate endometrioid and undifferentiated carcinomas from serous carcinomas are in the process of revision (see Chaps. 4 and 5); it is at least theoretically possible that some pathologists regard these as variants of high-grade serous carcinoma.

Sporadic high-grade serous carcinomas are associated with a similar spectrum of precursor lesions with site-specificity for the distal fallopian tube. *BRCA1* and *BRCA2* germline mutations both lead to genomic instability. Although this may give rise to a disproportionate number of early *TP53* mutations and thus high-grade serous carcinoma, a role for *BRCA1* and *BRCA2* (and indeed *TP53*) in endometrioid and undifferentiated types of ovarian carcinoma cannot be excluded at present.

Sites of Tubal Carcinoma: Fimbrial Predominance

In numerous studies, the tubal fimbriae are the preferred site for p53 signatures, STICs, and early invasive carcinomas within the *BRCA*-carrier population. Concomitant STICs in cases of sporadic ovarian and primary peritoneal serous carcinomas also predominate within the fimbriae. It is unclear whether this site-specificity arises from a simple difference in epithelial surface area, and to what extent intrinsic properties of the fimbrial epithelial cells and the local microenvironment (e.g., genotoxic stress related to ovulation) play a role. The tubal fimbria may represent a "transition zone" between epithelium of müllerian differentiation and the mesothelium of the peritoneal cavity [14], much like in the esophagus and uterine cervix. This junctional epithelium may be subject to cyclic injurious stimuli and impart an underlying vulnerability to neoplastic transformation. The importance of histologic evaluation of the tubal fimbrial epithelium in RRSO specimens from *BRCA* carriers is shown by Table 18.7, which summarizes studies' findings of the prevalence of these morphologic lesions in both these specimens and the general at-risk population.

Table 18.7 A comparison of ovarian and tubal lesions in general versus *BRCA*+RRSO populations

Lesion	Prevalence in *BRCA*+(based on RRSO)	Prevalence general population (based on evaluations of BSO)
Ovarian CICs	~40%	~40%
Intracystic borderline tumors	Rare	Rare
P53 signature	30–40%	30–40%
"TILT"	Rare (only 2 cases reported)	Not yet described
STIC, no concomitant cancer	2–9%	<1–3%
STIC, concomitant serous ca	50–60%	50–60%
Primary tubal serous ca (WHO)	Up to 6%	<1%

Outcome and Management: Following RRSO

Risk of Peritoneal Malignancy Following RRSO

BRCA+women who have undergone RRSO still have a residual 1–6% lifetime risk of developing peritoneal carcinoma. Consequently, the most efficacious method of following these women has yet to be determined.

Outcome of STICs and Early Tubal Carcinoma

The natural history of STIC is unclear, but STIC does have metastatic potential since there have been reports of disseminated peritoneal carcinoma after a diagnosis of STIC in a salpingectomy specimen. Management options include follow-up with or without staging, or post-operative adjuvant chemotherapy. Early invasive tubal carcinomas should be managed as for other tubal carcinomas of similar stage.

Established Tubal and Ovarian Carcinoma in *BRCA* Carriers

There is evidence that these tumors are relatively prognostically favorable, as compared to carcinomas that arise in a sporadic setting. Details are provided in Chap. 5.

HNPCC/Lynch Syndrome

Hereditary nonpolyposis colorectal carcinoma (HNPCC/Lynch syndrome) is an autosomal-dominant inherited cancer-susceptibility syndrome associated with defects in one or more DNA mismatch repair (MMR) genes (MLH1, MSH2, MSH6, and PMS2). When not checked by the function of a normal allele, the mutations result in tumorigenesis through the accumulation of repeated microsatellite sequences. Although first characterized with respect to colon cancer risk, MMR mutations in female carriers result in an even higher lifetime risk of endometrial cancer. Women with MMR mutations also have an increased risk of ovarian cancer: 12–15% lifetime risk in several large familial studies [15].

In contrast to *BRCA1*- and *BRCA2*-associated tumors, ovarian cancers associated with microsatellite instability are overwhelmingly of the endometrioid and clear cell histologic types [15]. As discussed in previous chapters (Chaps. 7 and 8), the tumorigenesis of endometrioid and clear cell carcinomas of the ovary is thought to parallel that of endometrial adenocarcinoma. In both types, transformation within a benign precursor (epithelial inclusion cyst, endometriosis) has been implicated. With Lynch syndrome, there appears to be an interaction between the MMR defects of ovarian ectopic endometrioid epithelium, the influence of unopposed estrogens, and loss of function in the *PTEN* tumor suppressor gene. These alterations, and the resulting cancer risk, parallel those in the endometrium proper.

A full discussion of the clinical management of Lynch syndrome is outside the scope of this monograph. The dominant risk resides in the endometrium. Although both ovarian and endometrial cancer risk in this population can be reduced with hormonal therapy, the current evidence is that there is a benefit for prophylactic hysterectomy and bilateral salpingo-oophorectomy in those MMR-mutation carriers past childbearing years. In the pathologic specimens from these surgeries, complete sampling of the endometrium, ovaries, and fallopian tubes is indicated. The high-grade serous lesions described earlier in this chapter are not presumably more prevalent in MMR-mutation carriers than in the general population. Nevertheless, endometriosis and endometrioid neoplasia – the relevant precursors in this population – commonly involve the fallopian tube. As with all specimens, signs of endometriosis should be sought carefully on gross examination of serosal surfaces.

References

1. Tavassoli FA, Devilee P, editors. Tumours of the breast and female genital organs. Lyon: IARC; 2003.
2. Vang R, Shih IeM, Kurman RJ. Ovarian low-grade and high-grade serous carcinoma: pathogenesis, clinicopathologic and molecular biologic features, and diagnostic problems. Anat Pathol. 2009;16(5): 267–82.
3. Lavie O, Neuman M, Rosemann E, Diamant Y. Serous papillary adenocarcinoma of the endometrium. A unique entity with a grave prognosis. Case report and review of the literature. Eur J Gynaecol Oncol. 1993;14(1):46–50.
4. Medeiros F, Muto MG, Lee Y, Elvin JA, Callahan MJ, Feltmate C, et al. The tubal fimbria is a preferred site for early adenocarcinoma in women with familial ovarian cancer syndrome. Am J Surg Pathol. 2006;30:230–6.
5. Pothuri B, Leitao M, Levine DA, et al. Genetic analysis of the early natural history of epithelial ovarian carcinoma. PLoS ONE. 2010;26:e10358.
6. Kurman RJ, Shih IeM. The origin and pathogenesis of epithelial ovarian cancer: a proposed unifying theory. Am J Surg Pathol. 2010;34:433–43.
7. Jarboe E, Folkins A, Nucci MR, et al. Serous carcinogenesis in the fallopian tube: a descriptive classification. Int J Gynecol Pathol. 2008;27(1):1–9.
8. Hendrickson MR, Kempson RL. Normal histology of the uterus and fallopian tubes. In: Sternberg SS, editor. Histology for pathologists. 2nd ed. Philadelphia: Lippincott Williams & Wilkins; 1997. p. 879–928.
9. Rabban JT, Crawford B, Chen LM, Powell CB, Zaloudek CJ. Transitional cell metaplasia of fallopian tube fimbriae: a potential mimic of early tubal carcinoma in risk reduction salpingo-oophorectomies from women With BRCA mutations. Am J Surg Pathol. 2009;33(1):111–9.
10. Stern J, Buscema J, Parmley T, Woodruff JD, Rosenshein NB. Atypical epithelial proliferations in the fallopian tube. Am J Obstet Gynecol. 1981;140(3):309–12.
11. Lee Y, Miron A, Drapkin R, et al. A candidate precursor to serous carcinoma that originates in the distal fallopian tube. J Pathol. 2007;211(1):26–35.
12. Olivier RI, van Beurden M, Lubsen MA, et al. Clinical outcome of prophylactic oophorectomy in BRCA1/BRCA2 mutation carriers and events during follow-up. Br J Cancer. 2004;90(8):1492–7.
13. Callahan MJ, Crum CP, Medeiros F, et al. Primary fallopian tube malignancies in BRCA-positive women undergoing surgery for ovarian cancer risk reduction. J Clin Oncol. 2007;25(25):3985–90.
14. Auersperg N, Woo MM, Gilks CB. The origin of ovarian carcinomas: a developmental view. Gynecol Oncol. 2008;110(3):452–4.
15. Lynch HT, Casey MJ, Snyder CL, Bewtra C, Lynch JF, Butts H, et al. Hereditary ovarian carcinoma: heterogeneity, molecular genetics, pathology, and management. Mol Oncol. 2009;3:97–137.
16. Lu KH, Garber JE, Cramer DW, et al. Occult ovarian tumors in women with BRCA1 or BRCA2 mutations undergoing prophylactic oophorectomy. J Clin Oncol. 2000;18(14):2728–32.
17. Colgan TJ, Murphy J, Cole DE, Narod S, Rosen B. Occult carcinoma in prophylactic oophorectomy specimens: prevalence and association with BRCA germline mutation status. Am J Surg Pathol. 2001;25(10):1283–9.
18. Leeper K, Garcia R, Swisher E, Goff B, Greer B, Paley P. Pathologic findings in prophylactic oophorectomy specimens in high-risk women. Gynecol Oncol. 2002;87(1):52–6.
19. Powell CB, Kenley E, Chen LM, et al. Risk-reducing salpingo-oophorectomy in BRCA mutation carriers: role of serial sectioning in the detection of occult malignancy. J Clin Oncol. 2005;23(1):127–32.
20. Carcangiu ML, Peissel B, Pasini B, Spatti G, Radice P, Manoukian S. Incidental carcinomas in prophylactic specimens in BRCA1 and BRCA2 germ-line mutation carriers, with emphasis on fallopian tube lesions: report of 6 cases and review of the literature. Am J Surg Pathol. 2006;30(10):1222–30.
21. Shaw PA, Rouzbahman M, Pizer ES, Pintilie M, Begley H. Candidate serous cancer precursors in fallopian tube epithelium of BRCA1/2 mutation carriers. Mod Pathol. 2009;22(9):1133–8.
22. Lamb JD, Garcia RL, Goff BA, Paley PJ, Swisher EM. Predictors of occult neoplasia in women undergoing risk-reducing salpingo-oophorectomy. Am J Obstet Gynecol. 2006;194(6):1702–9.
23. Hirst JE, Gard GB, McIllroy K, Nevell D, Field M. High rates of occult fallopian tube cancer diagnosed at prophylactic bilateral salpingo-oophorectomy. Int J Gynecol Cancer. 2009;19:826–9.
24. Kauff ND, Satagopan JM, Robson ME, et al. Risk-reducing salpingo-oophorectomy in women with a BRCA1 or BRCA2 mutation. N Engl J Med. 2002;346(21):1609–15.
25. Rebbeck TR, Lynch HT, Neuhausen SL, et al. Prevention and observation of surgical end points study group prophylactic oophorectomy in carriers of BRCA1 or BRCA2 mutations. N Engl J Med. 2002;346(21):1616–22.

19 Peritoneal Cytology

Andre L. Moreira and Oscar Lin

Keywords

Cytology • Effusion • Pelvic washing • Peritoneal washing - Ovarian tumors • Peritoneum

Peritoneal washing cytology has been used in the evaluation of gynecologic malignancies for many decades. The International Federation of Gynecology and Obstetrics (FIGO) introduced peritoneal washing as part of the staging of ovarian carcinomas in 1975. The presence of malignant cells in serous effusions indicates spread of disease beyond the organ of origin and is associated with significant therapeutic and prognostic implications. The prototypic well-differentiated ovarian endometrioid carcinoma, FIGO stage IA, has an expected 5 year survival rate that exceeds 95%. In contrast, tumors that involve the ovarian surface and/or have positive peritoneal washings, many of which are non-endometrioid, are assigned FIGO stage IC and have significantly reduced 5 year survivals [1]. Peritoneal cytology is also a prognostic marker regardless of disease stage in patients with ovarian carcinoma [2, 3].

Unfortunately, the morphologic examination of cytologic specimens has not proven to be a sensitive or specific diagnostic tool. The false-positive rate of cytology in the evaluation of serous effusions has been reported to be as high as 4.5% [4]. The main reason for a false-positive diagnosis is the misinterpretation of reactive mesothelial cells as malignant epithelial cells. Reactive mesothelial cells can mimic a malignant cell by showing nuclear enlargement and hyperchromasia. Other potential pitfalls include interpretative errors caused by the presence of psammoma bodies or other epithelial clusters originating from endometriosis and endosalpingiosis [4–6]. The false-negative rate for cytology in serous effusions is approximately 20%. Factors that contribute to this relatively high false-negative rate include scant numbers of malignant cells in peritoneal washings, infrequent exfoliation of malignant cells, and interpretive errors. Additionally, peritoneal washings are sometimes performed before a specific tissue diagnosis is rendered, which may lead to discordant findings between cytology and surgical material. This is more commonly seen in benign or low-grade tumors of the female genital tract or in cases of carcinoma arising in the fallopian tube [7].

Cytological evaluation should ideally be done in correlation with the surgical pathology specimen. Even in the setting of a prophylactic salpingo-oophorectomy, it is best to finalize the cytology report after reviewing the surgical pathology slides when atypical cells are seen.

Since the presence of malignant cells in a peritoneal washing leads to treatment with chemotherapy in many instances, diagnostic accuracy is essential. In this chapter, we will discuss the cytomorphological criteria for the diagnosis of ovarian tumors in liquid cytology, potential pitfalls, and ancillary methods that may be helpful in characterizing cell types.

Cytological Evaluation

Mesothelial Cells

Any evaluation of a cytologic preparation such as peritoneal washing and peritoneal effusion should start with the accurate identification of mesothelial cells, which are very common in these specimens. Mesothelial cells can be seen in peritoneal washings as single cells or in clusters. The cells have abundant dense cytoplasm that may show peripheral basophilia referred by some authors as "skirts" or biphasic cytoplasm (Fig. 19.1). This biphasic appearance is the result

A.L. Moreira (✉)
Department of Pathology, Memorial Sloan-Kettering Cancer Center, 1275 York Avenue, New York, NY, USA

R.A. Soslow and C. Tornos (eds.), *Diagnostic Pathology of Ovarian Tumors*,
DOI: 10.1007/978-1-4419-9751-7_19, © Springer Science+Business Media, LLC 2011

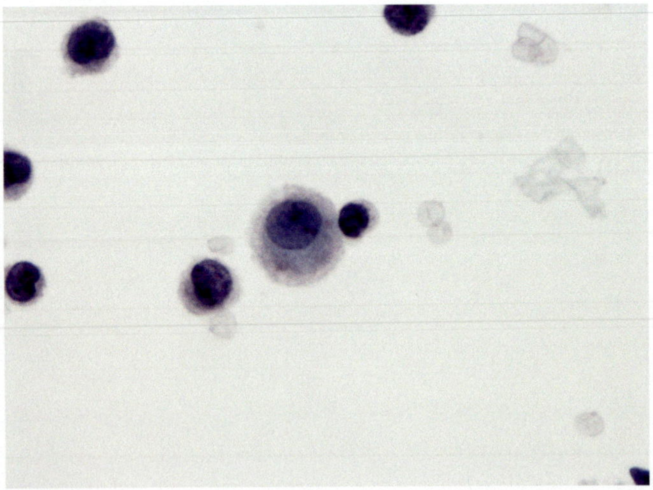

Fig. 19.1 Benign mesothelial cells showing biphasic cytoplasm. Note the denser cytoplasm around the nucleus and lighter cytoplasm at the periphery

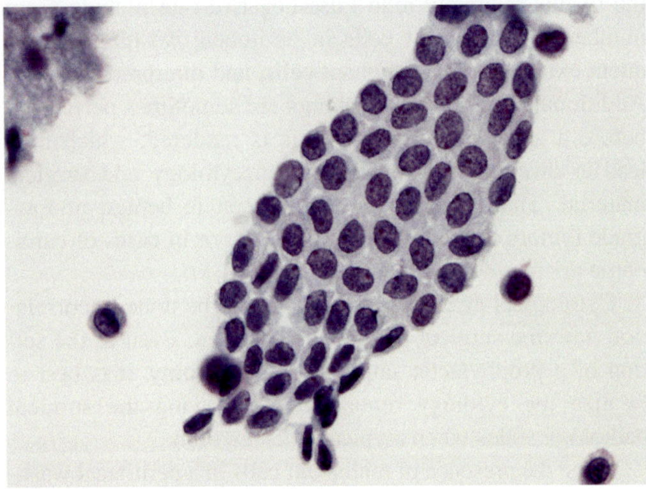

Fig. 19.2 Benign mesothelial cells arranged in flat sheets. Note the presence of intercellular spaces (*windows*) within the cells

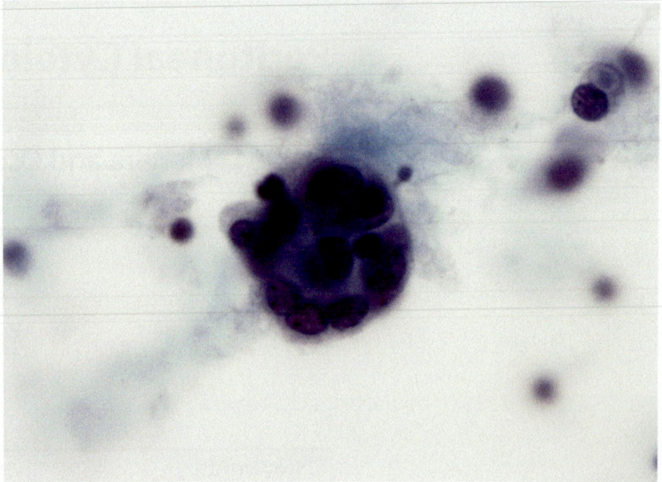

Fig. 19.3 Cluster of reactive mesothelial cells in a three-dimensional arrangement. The cluster can mimic adenocarcinoma by showing hyperchromasia and nuclear pleomorphism

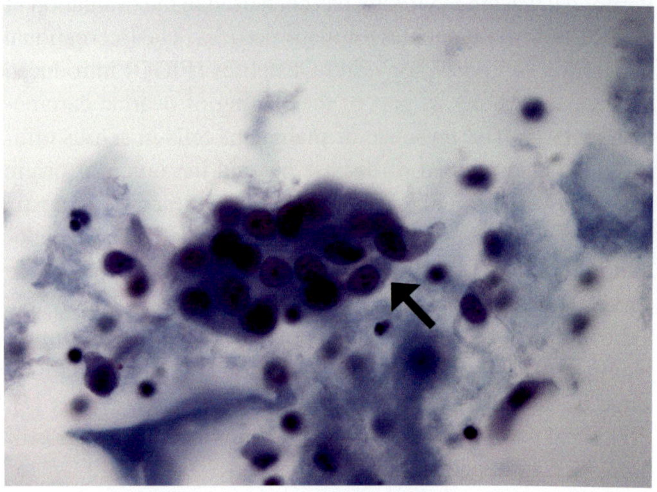

Fig. 19.4 Cluster of reactive mesothelial cells in a three-dimensional arrangement. Note the presence of intercellular spaces (*windows*) within the cell clusters (*arrow*)

of a denser concentration of intermediate filaments around the nucleus. The nuclei of mesothelial cells are often single, but multinucleation and size variation might occur. The cells can have single or multiple nucleoli. Despite nuclear variation, the cells share morphological similarities. When present in clusters, mesothelial cells are often arranged in flat sheets (Fig. 19.2) and show small spaces separating the cells, a phenomenon also known as "windows."

Reactive mesothelial cells, in addition to flat sheets and single cells, can present as tridimensional clusters with occasional cell ball or papillary cluster formation (Fig. 19.3), thus mimicking epithelial clusters, which represents a potential pitfall. The presence of cellular "windows" might help to identify the cells as mesothelial in origin (Fig. 19.4). Reactive mesothelial cells show cellular enlargement, dense cytoplasm and large nuclei with increased nucleus to cytoplasm (N/C) ratio, and may contain prominent nucleoli. Nuclear pleomorphism is commonly seen as well as binucleation and multinucleation (Fig. 19.5). Mitotic figures can be present (Fig. 19.6). Occasionally, the cells might be vacuolated with single large cytoplasmic vacuole that may displace the nucleus, thus mimicking signet-ring adenocarcinoma (Fig. 19.7). Vacuolization as a result of endocytosis of surrounding fluid is commonly seen in chronic effusions. Clusters of reactive mesothelial cells very often have the appearance of scalloped borders (Fig. 19.8), a feature that although not diagnostic may be helpful in the distinction

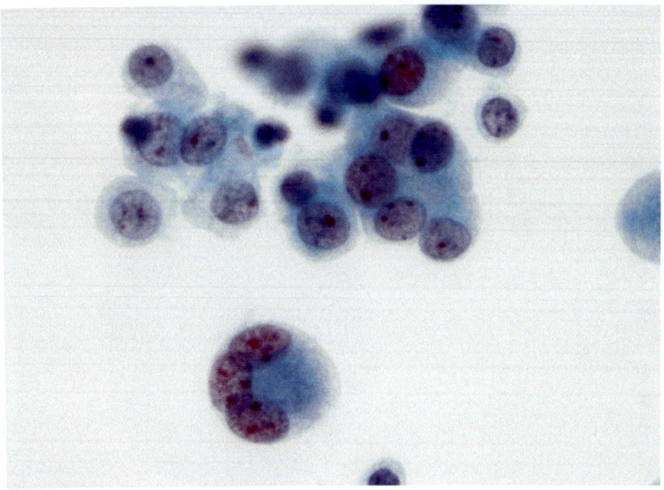

Fig. 19.5 Reactive mesothelial cells showing multinucleation and binucleation. Note the presence of biphasic cytoplasm in the isolated multinucleated cell and the presence of windows in a small cluster with scalloped borders

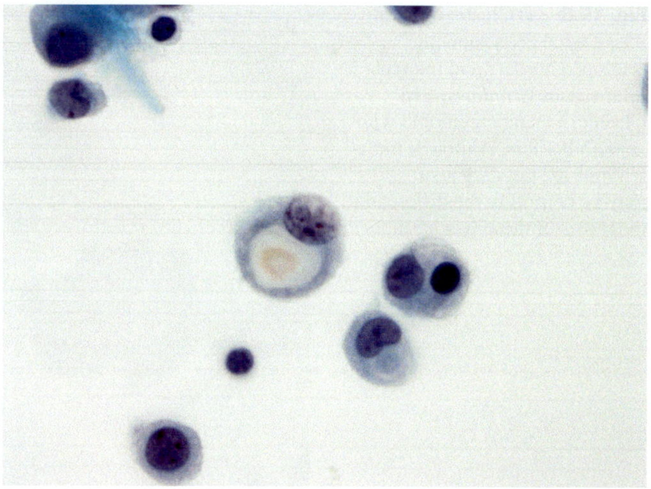

Fig. 19.7 Isolated reactive mesothelial cells showing vacuolization of the cytoplasm, thus mimicking a signet-ring cell adenocarcinoma. Vacuolization can be seen in chronic effusions or as an artifact in peritoneal washings. Note that the nucleus of the vacuolated cell is similar to the other recognizable mesothelial cells in the background

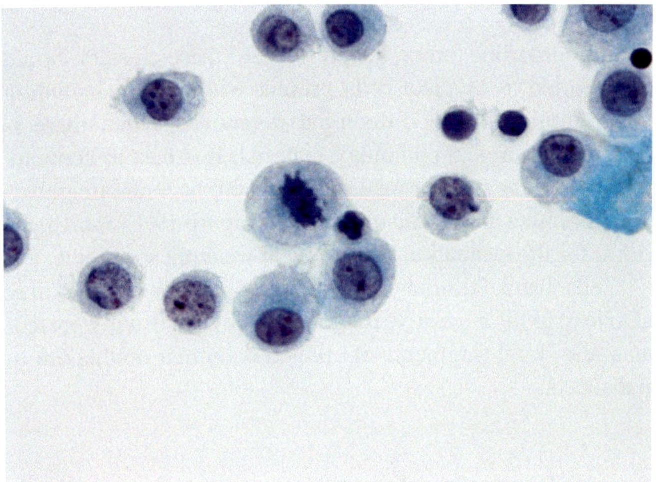

Fig. 19.6 Reactive mesothelial cells showing a mitotic feature

between mesothelial cells and carcinoma. The latter tends to form clusters with smooth borders.

The presence of a single population of cells with recognizable characteristics of mesothelial cells is the most important single criterion for the diagnosis of a negative peritoneal washing. The exception to this rule is malignant mesothelioma, but the distinction between reactive mesothelial cell and mesothelioma is not in the scope of this chapter. The presence of two distinct cell populations in a cytologic preparation of peritoneal washing should raise the possibility of a neoplastic process. Table 19.1 illustrates the most common mimickers of adenocarcinoma in peritoneal washing cytology and clues to the correct diagnosis.

Endosalpingiosis and Endometriosis

The presence of clusters of benign epithelial cells in peritoneal washings can be very challenging. Very often, these clusters are few and clearly defy the rule of "one population of cells" as a hallmark of benign cytology (Fig. 19.9). The origin of these clusters of benign epithelial cells include endosalpingiosis and endometriosis, which frequently contribute to diagnostic difficulties. Endosalpingiosis is characterized by organized, tight clusters with occasional non-branching papillary formations (Fig. 19.10). The cells are uniform with scant basophilic cytoplasm. The nuclei have smooth nuclear membranes, a fine chromatin pattern, and small nucleoli. It is important to remember that psammoma bodies might also be associated with endosalpingiosis; therefore, psammoma bodies per se are not diagnostic of a malignant process (Fig. 19.11). In a well-preserved specimen, the presence of cilia or terminal bars is diagnostic of this entity (Fig. 19.10). In general, the cells lack characteristics of malignancy, but the nuclear features of endosalpingosis might be nearly impossible to distinguish from well-differentiated serous neoplasms such as serous borderline tumor and low-grade serous carcinoma. Distinction of endosalpingiosis from serous borderline tumor is based on the presence of large papillary clusters with architectural disorganization in the latter. These findings are better appreciated in cell block preparation.

Endometriosis is another potential pitfall in peritoneal washings. It is characterized by the presence of round to oval cells arranged in three-dimensional clusters, tubular structures, and flat sheets, similar to what is described in vaginal–

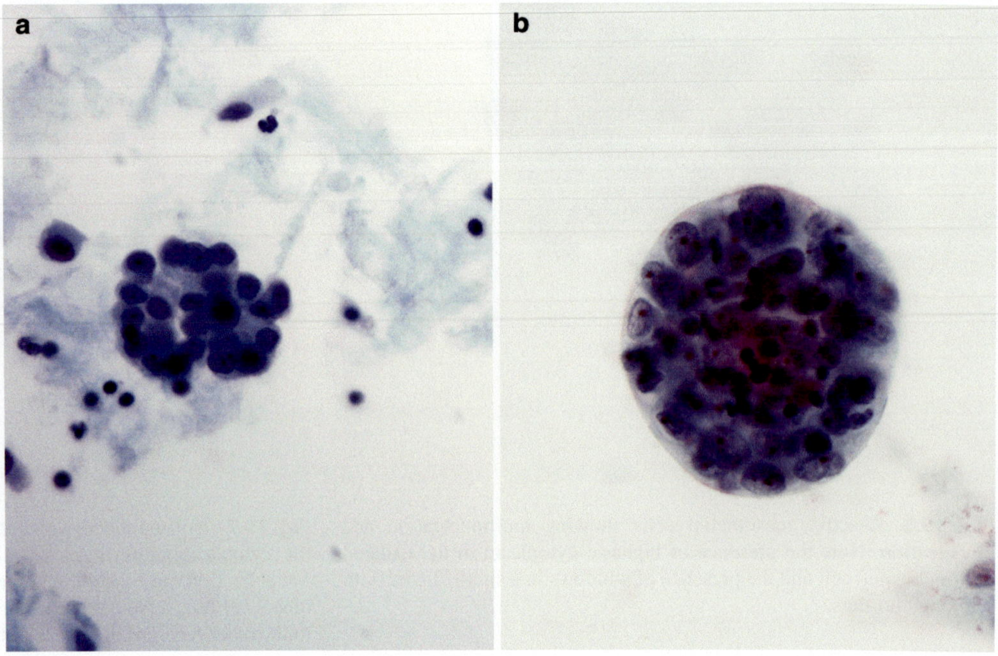

Fig. 19.8 (**a**) Cluster of reactive mesothelial cells showing scalloped borders and intercellular spaces (*windows*). (**b**) Cluster of adenocarcinoma with smooth borders. Although these features are not diagnostic, they may be helpful in the differential diagnosis of these two entities

Table 19.1 Mimickers of adenocarcinoma in peritoneal washings

	Clues
Reactive mesothelial cells	Presence of "windows"[a]
Endometriosis	Bland nuclei and associated hemosiderin-laden macrophages[b]
Endosalpingiosis	Bland nuclei and presence of cilia
Fallopian tube epithelium	Flat clusters, bland nuclei, and presence of cilia
Borderline tumor	Atypical papillary clusters of mildly atypical epithelial cells[c]
Psammoma bodies	Benign cytologic features

[a]If the diagnosis of adenocarcinoma is still suspected. Immunostains should be performed to exclude an epithelial nature of the clusters
[b]Best seen in a cell block preparation
[c]If a borderline tumor is suspected, comparison with the surgical material is recommended

cervical specimens (Fig. 19.12). The nuclei are round or bean shaped with fine chromatin and rare nucleoli. The cytoplasm is scant and vacuolated. Diagnostic stromal cells are extremely rare in washings. The presence of hemosiderin-laden macrophages is not helpful in the diagnosis because these cells can be seen in other conditions such as inflammation and metastatic carcinomas. Similar to endosalpingiosis, the main morphological difference between endometriosis and borderline tumor is the presence of papillary clusters seen in the latter.

In most cases, a cell block preparation is helpful to evaluate the presence of scant epithelial cells. Endosalpingiosis and endometriosis should be reported as benign glandular cells. In cases where the diagnosis is less definite due to degenerated cells, or if the patient has a clinical history of serous borderline tumor, the presence of these clusters should be reported as atypical cells present with a note, including these entities in the differential diagnosis. When there is concurrent surgical pathology material, it is best to compare that with the cytology material to ensure accurate diagnosis and minimize diagnostic uncertainty. Figure 19.13 is an algorithm for the evaluation of peritoneal washing specimen.

Cells from fallopian tubes in peritoneal washings may also lead to false-positive diagnosis. They are usually present as a few small fragments of ciliated columnar epithelium in flat sheets.

Serous Borderline Tumor

Serous borderline tumors shed into the peritoneal washings as tightly cohesive clusters of epithelial cells with rare single cells. The clusters may be large with papillary formation and smooth borders, and the cells have increased nucleus to cytoplasm ratio (Fig. 19.14). Nuclear molding can be seen (Fig. 19.15). Most specimens have few clusters of neoplastic cells. Frequently, the patient has a frozen section diagnosis of a serous borderline tumor and the peritoneal washings show the presence of atypical cells. The pathologist might elect to release the report describing the presence of neoplastic cells along with the differential diagnosis, which includes serous borderline tumor and carcinoma. Alternatively, the pathologist could wait for the evaluation of the surgical pathology specimen before releasing the final diagnosis, thus reducing the chance of making a mistaken diagnosis of carcinoma, which might lead to unnecessary additional therapy.

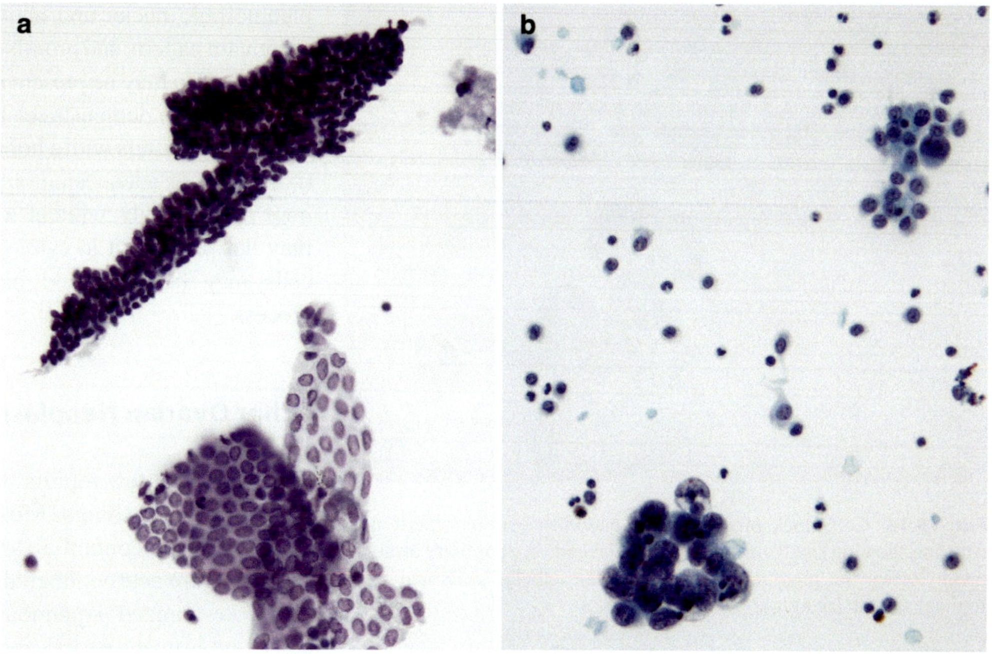

Fig. 19.9 (**a**) Low-power view of a pelvic wash showing a flat sheet of mesothelial cells together with a cluster of epithelial cells in a papillary arrangement. At high magnification, cilia could be seen within this cluster, indicating endosalpingiosis. (**b**) Low-power view of a peritoneal washing showing two distinct cell populations. In the upper corner, there is a loosely arranged cluster of mesothelial cells with scalloped borders. In the lower corner is a three-dimensional cluster of adenocarcinoma. Note large cell size and nuclear hyperchromasia

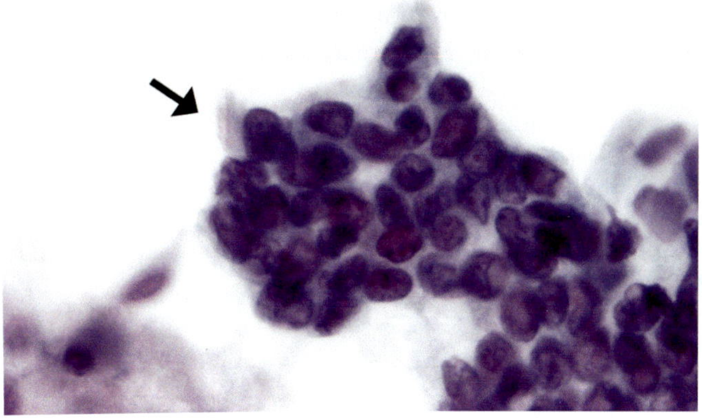

Fig. 19.10 Epithelial cells in a peritoneal washing. Note the disorganization of the cluster with irregular nuclei in the presence of a terminal bar with cilia (*arrow*). These features are indicative of a benign diagnosis, endosalpingiosis

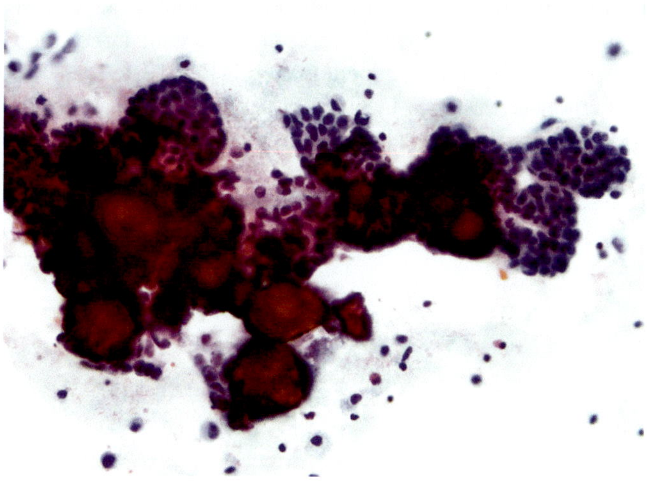

Fig. 19.11 Psammoma bodies in a peritoneal washing. Here, psammoma bodies are associated with papillary branching clusters of neoplastic cells, diagnostic of a serous borderline tumor. Psammoma bodies are not specific and can be seen in association with endosalpingiosis, reactive mesothelial cells, and carcinoma

Ovarian Carcinomas

Malignant effusions typically show two populations of cells: reactive mesothelial and neoplastic cells. Similar to other conditions, the malignant cells may be few. A cell block preparation might be helpful in these cases to render an accurate diagnosis as ancillary studies can be performed in these difficult cases. Malignant cells may be present in a pelvic wash or effusion, singly (Fig. 19.16) or in clusters (Fig. 19.17). The latter are often large and have tridimensional morphology with a cannonball appearance including smooth edges, although scalloped borders such as seen in reactive mesothelial cells can also be present. Rarely, acinar formations are identified. Like any adenocarcinoma, the cells are large, have increased nucleus to cytoplasm ratio, nuclear pleomorphism, hyperchromasia, and irregular nuclear contours. Delicate and vacuolated cytoplasm is often present.

Although it is difficult and unnecessary to accurately classify ovarian carcinomas in a cytology specimen, there

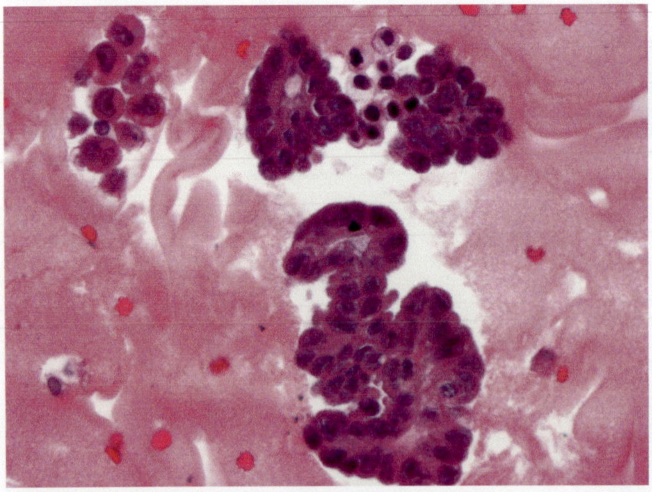

Fig. 19.12 Cell block preparation of endometriosis in a peritoneal washing showing small clusters of epithelial cells in a papillary arrangement. Notice the presence of clusters of mesothelial cells and macrophages in the background

are clues to their specific histological type. Serous carcinoma usually presents as cellular specimens containing single cells or poorly cohesive irregular cell clusters with large, pleomorphic nuclei and prominent nucleoli (Fig. 19.18).

Peritoneal washings involved by endometrioid carcinomas display loose, three-dimensional clusters of cells with eccentric, pleomorphic nuclei and abundant delicate cytoplasm, coarse chromatin pattern, and prominent nucleoli (Fig. 19.19). Papillary configuration may be present as well as acinar formation. Clear cell carcinoma demonstrates large cells, clear cytoplasm, large nuclei, and clusters with a hobnail pattern. Mucinous neoplasms (borderline or adenocarcinoma) may be very difficult to diagnose cytologically without a specific clinical history. Mucin may not be evident in cytological preparations and neoplastic cells may be scant and bland, thus mimicking a benign process.

Other Ovarian Neoplasms

There are very few reports of cases of germ cell tumor cells in peritoneal washings. Most reports are from ascitic fluid specimens that contain a variety of neoplastic cells, including immature neuroepithelial cells forming rosette-like structures, keratinized squamous cells, squamoid metaplastic cells, and immature glial-appearing cells [8].

Sex cord stromal tumors in most cases are confined to the ovary, but involvement of the ovarian surface or ovarian rupture is seen in a small number of cases. Therefore, it is very rare to encounter these tumor cells in a peritoneal washing. Nevertheless, when present, the neoplastic cells may have a variety of morphological presentation similar to the histological counterpart. Sertoli cell tumors with

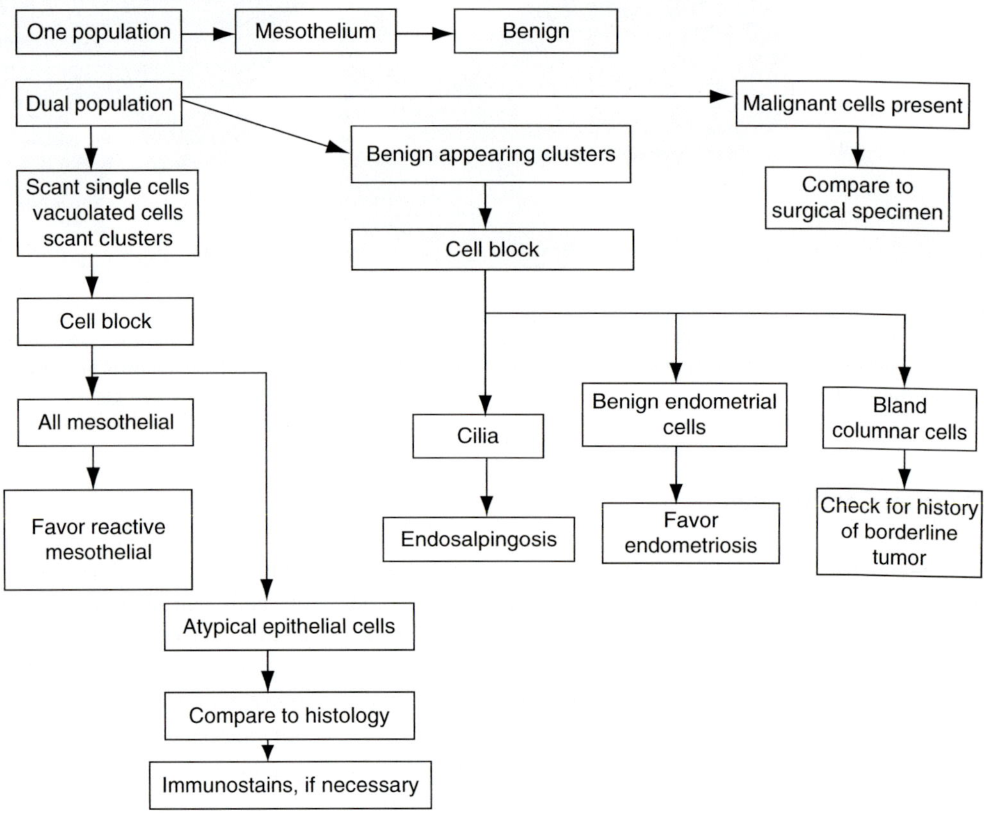

Fig. 19.13 Diagram illustrating an algorithm for the evaluation of peritoneal washing. This highlights the importance of cell block preparation and comparison of the cytologic material to the surgical specimen to avoid interpretative errors

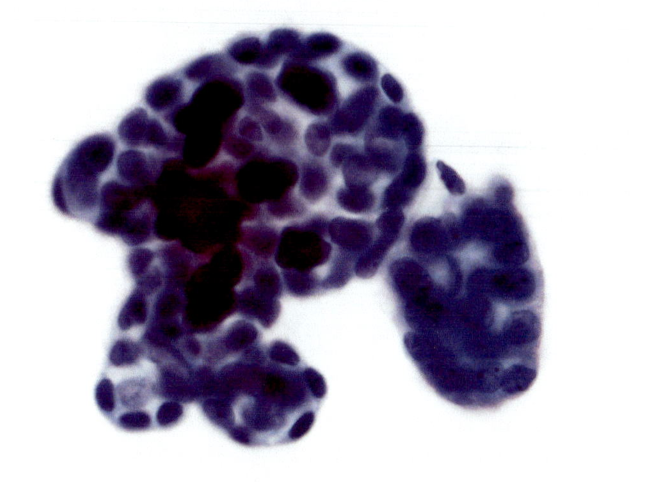

Fig. 19.14 A cluster of atypical cells with psammoma bodies in a papillary arrangement. Surgical specimen revealed the presence of an ovarian serous borderline tumor

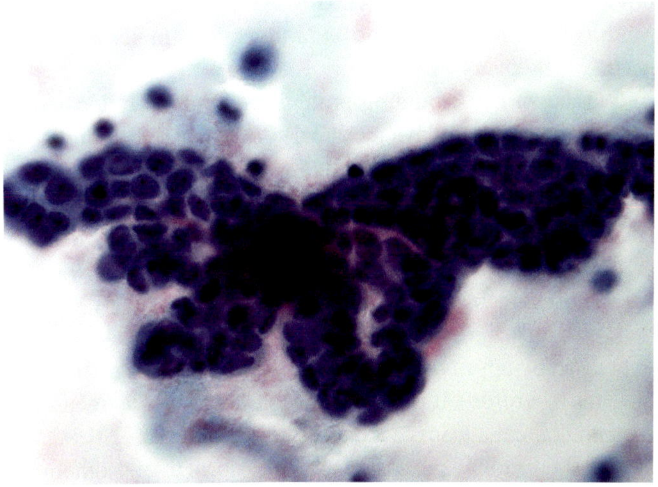

Fig. 19.15 Branched papillary cluster of atypical cells. Note nuclear hyperchromasia and nuclear molding. This is representative of a serous borderline tumor in a peritoneal washing

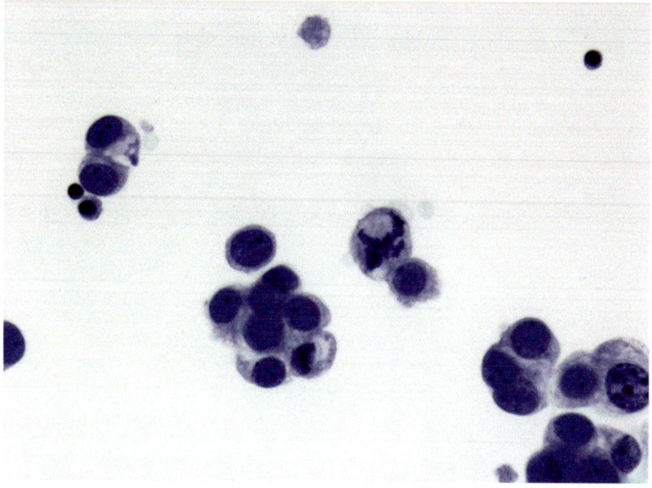

Fig. 19.16 High-grade ovarian serous carcinoma. Malignant cells can be present singly or in small clusters. Malignant cells have eccentric nuclei, finely vacuolated cytoplasm, and high nucleus to cytoplasm ratio. An atypical mitotic figure is also present

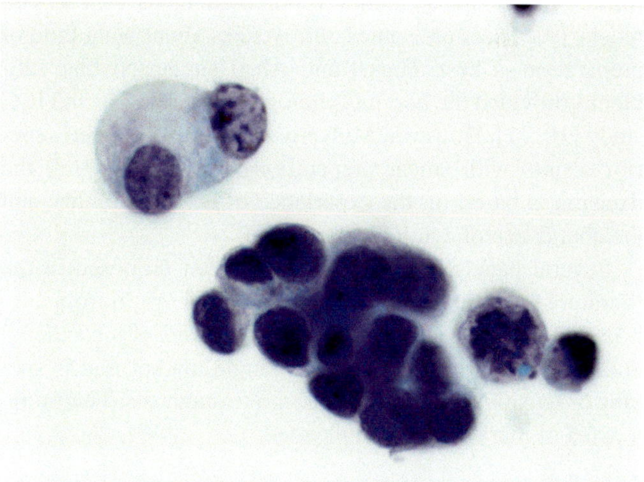

Fig. 19.17 Cluster of malignant cells. Notice that these clusters can also have scalloped borders, mimicking a mesothelial cell. But the high nuclear to cytoplasmic ratio, nuclear atypia, and hyperchromasia are diagnostic of a malignant process. An atypical mitotic figure is also seen

lipidized cytoplasm may present in a peritoneal washing as stripped oval and bland nuclei (Fig. 19.20). Sex cord tumor with annular tubules may present as acinar structures with large and atypical nuclei, thus mimicking a carcinoma (Fig. 19.21).

When confronted with a positive peritoneal washing, it is imperative to correlate the findings with clinical history and histological diagnosis. This will avoid confusion when different terms are used for the diagnosis. As mentioned before, in cases where the surgical diagnosis is pending, the pathologist may elect to wait and sign out the case after the surgical specimen has been reviewed and correctly classified. This approach may prevent unnecessary therapy.

Ancillary Studies

In order to improve the accuracy of cytology in peritoneal washings and serous effusions, ancillary studies can be performed. Ancillary studies are recommended in the presence of atypical cells, when it is difficult to distinguish reactive mesothelial cells from adenocarcinoma (Fig. 19.13) or when it is necessary to differentiate adenocarcinoma from mesothelioma. The most common ancillary tests are immunocytochemical studies (ICC), although there is a limited number of studies reporting the use of flow cytometry and molecular studies. Immunocytochemical studies can be per-

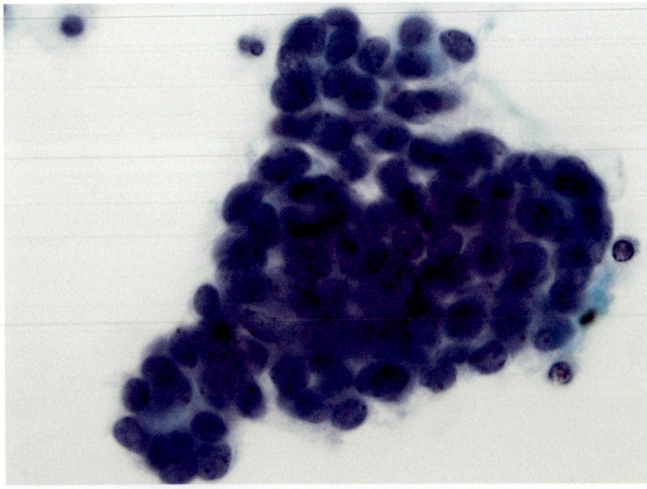

Fig. 19.18 A cluster of highly atypical cells present in a three-dimensional arrangement. Note the presence of nuclear molding. The differential diagnosis includes carcinoma or serous borderline tumor. Oophorectomy showed a low-grade serous carcinoma

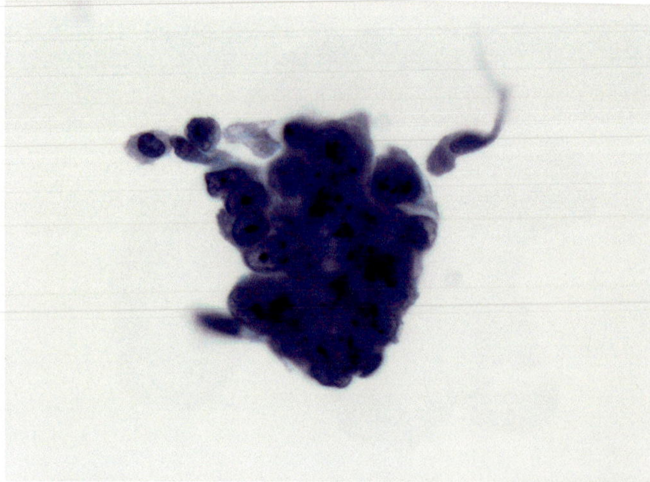

Fig. 19.19 A cluster of malignant epithelial cells arranged in a three-dimensional cluster. Prominent nucleoli are evident. Surgical excision reveals an ovarian endometrioid adenocarcinoma

formed in cytospins; ThinPrep preparations as well as cell blocks [9]. There are some controversies about what kind of preparation is best. Fetsch and Abati suggested that cell-blocks provided the best morphologic interpretation and ICC study [10, 11]. However, Mulvany reportedly obtained superior results with smear preparations [12]. In general, the decision is based on the experience of each laboratory and the abundance of cytological material.

Several panels of antibodies have been proposed in the literature to distinguish cells of mesothelial origin from adenocarcinoma. The antibodies include markers of mesothelial and epithelial origin. There is no single marker that is specific or sensitive enough for an accurate diagnosis. Therefore, a panel of markers is recommended.

Mesothelial Markers

Mesothelial markers previously evaluated include D2-40, calretinin, mesothelin, CK5/6, WT-1, and HBME-1. D2-40, a monoclonal antibody directed against oncofetal antigen M2A is present on cell membrane; it has been described as a useful marker of mesothelial cells. It has been shown to have good sensitivity and specificity to distinguish epithelioid malignant mesothelioma and reactive mesothelial cells from adenocarcinoma [13–15]. However, D2-40 positivity has been reported in approximately 20% of ovarian carcinoma cases [16] and it is well known that WT-1 is expressed in both mesothelial and serous cells, as discussed subsequently.

Calretinin, a calcium-binding protein widely expressed throughout the central and peripheral nervous systems, is also expressed by mesothelial cells in the cytoplasm and frequently in the nucleus. Calretinin expression has also been demonstrated in a small fraction of adenocarcinomas from other organs [11–15, 17–21]. However, positivity for calretinin has been reported to be as high as 40% in serous carcinoma of the ovary [15, 22].

WT-1 is a DNA-binding protein predominantly located in the nucleus, which plays a critical role in the development of the genitourinary tract. In adult tissues, it is expressed by mesangial cells of the kidney, Sertoli cells of the testis, ovarian stromal cells and surface epithelium, mesothelial cells, and some other stromal cells in the female genital tract. WT-1 is also expressed by malignant mesotheliomas and by tumors derived from the ovarian surface epithelium. WT-1 can be expressed by approximately 70% of ovarian/pelvic serous carcinoma. Therefore, it is not a good marker in the distinction of mesothelial cells from ovarian serous carcinomas [23].

Cytokeratin 5 is expressed by normal mesothelium, squamous and transitional epithelia, and myoepithelial cells in the cytoplasm. Unfortunately, a significant proportion of adenocarcinomas (breast, lung, pancreas, and ovary) also express CK5, precluding its use in differential diagnosis [24].

Mesothelin is a surface protein that may be involved in cell–cell adhesion. Its expression is seen in mesothelial cells and surface ovarian epithelial cells, and pancreatic ductal carcinomas [24]. HBME-1 is a mouse monoclonal antibody prepared from a human malignant epithelial mesothelioma; however, its specificity is not very high for mesothelial cells [10, 11, 20, 24–27].

The markers discussed above might be useful to characterize cells of mesothelial origin and are particularly useful when the cells in question are abundant in the specimen. However, they are not particularly useful in cases in which it is necessary to characterize a small population of atypical cells in an effusion specimen containing abundant mesothe-

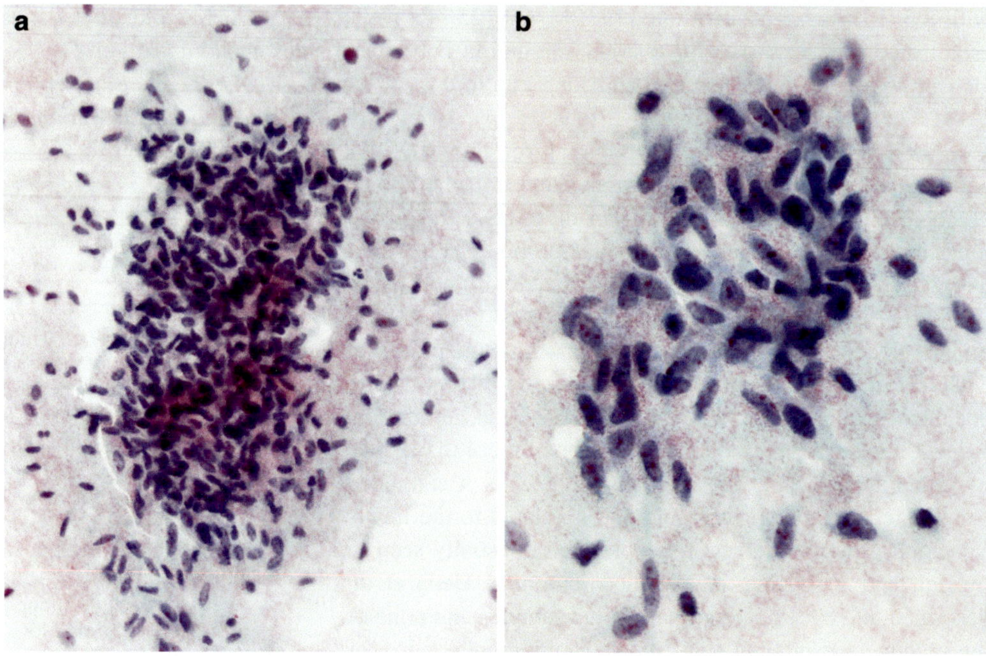

Fig. 19.20 (**a**) Presence of spindle-shaped cells, devoid of cytoplasm, seen in a proteinaceous background in a peritoneal washing. (**b**) On higher magnification of the same cells, note delicate chromatin and small nucleoli. Surgical excision showed a lipidized Sertoli cell tumor of the ovary

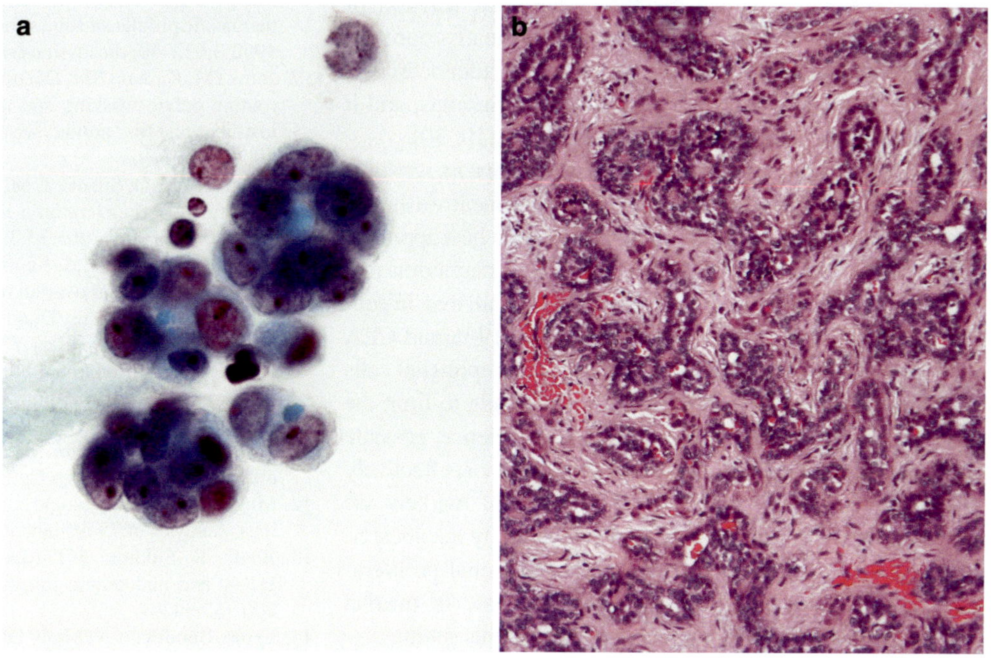

Fig. 19.21 (**a**) Cluster of highly atypical cells with an acinar arrangement, prominent nucleoli, and dense hyalinized material within gland-like formation. (**b**) Surgical excision of the ovarian mass was diagnostic of sex cord tumor with annular tubules

lial cells. False-negative staining can occur due to lack of sensitivity or technical reasons. Therefore, the use of epithelial-associated markers, which can positively identify epithelial cells, is preferred.

Epithelial-Associated Markers

Antibodies for the detection of epithelial cells include CEA, CA125, MOC-31, Ber-EP4, CA19-9, and B72.3. CEA was one of the first antibodies used to distinguish epithelial cells from mesothelial cells. Immunoreactivity varies considerably, depending on the antibody clone and the range of tumor types used. Mesothelial cells are generally negative for CEA staining, as are most carcinomas derived from gynecologic sites; this limits its utility in the evaluation of peritoneal washings from patients thought to have ovarian or endometrial carcinoma [27]. Several investigators have also reported that CEA stains macrophages and other inflammatory cells owing to nonspecific cross-reactivity, which has been associated with a 5–15% false-positive rate [19, 26]. CA125 is a well-established tumor marker for ovarian carcinoma but it can also be expressed in pancreatic and lung carcinomas [28]. CA125 expression has been shown to have low specificity and low sensitivity in effusion samples [26]. The MOC-31 antibody recognizes an epithelial-associated transmembrane glycoprotein of unknown function in the GLS-1 small cell lung carcinoma cell line and in epithelial tumors. The Ber-EP4 antibody is generated from mice immunized with cells from the MCF-7 breast carcinoma cell line and two non-covalently bound glycopeptides. Most studies have documented that MOC-31 and Ber-EP4 are highly effective in distinguishing adenocarcinoma from reactive mesothelial cells, although a few authors have reported low expression of these markers in mesothelial cells [14, 15, 24, 29]. CA19-9 is a sialylated lacto-N-fucopentose II molecule that is related to the Lewis (a) blood group substance. Its expression is usually seen in ovarian and gastrointestinal adenocarcinomas. Ueda et al. described that 89% of serous carcinoma samples and none of the reactive mesothelial cells showed immunoreactivity for the CA19-9 antibody [15, 26]. The monoclonal antibody B72.3 is generated with a membrane-enriched fraction of human mammary carcinoma and recognizes the glycoprotein TAG-72 that is expressed in a wide variety of adenocarcinomas. B72.3 is expressed in 70% of adenocarcinomas, and it lacks affinity for reactive mesothelial cells [15, 19, 30].

In summary, there is no single marker that appears sensitive and specific enough to correctly diagnose peritoneal washings. Therefore, a panel of antibodies appears to be the best approach in the discrimination between gynecologic adenocarcinoma and reactive mesothelial cells. We agree with Ueda et al. that, in general, a combination of MOC-31, Ber-EP4, CA19-9, and CEA antibodies is a suitable panel to detect glandular epithelial cells in peritoneal fluids [26], although it is worthwhile to limit the number of stains in any given case. In our experience, possible adenocarcinoma cases with limited material are most effectively evaluated with B72.3 and BerEp4. Mesothelial markers are indicated to distinguish peritoneal involvement by adenocarcinoma from mesotheliomas and atypical mesothelial proliferations. Due to their relatively high positivity rate in ovarian tumors, the use of WT1 is not recommended in this setting.

Flow Cytometry and Molecular Studies

The evaluation of flow cytometry and molecular studies in effusion specimens has been limited. Dong et al. have shown that it is possible to analyze epithelial and mesothelial markers (Ber-EP4 and EMA), CD138, and integrin subunits with a four-color flow cytometer [31]. Holloway et al. applied a real-time PCR-based assay involving 17 genes to 13 independent cytologic samples from biopsy-proven malignant mesothelioma and lung adenocarcinomas. They used microarray data for 190 lung adenocarcinomas and 33 malignant mesotheliomas to identify genes differentially expressed between the two diseases. Their test was able to distinguish between lung adenocarcinoma and mesothelioma in cells collected from pleural effusions [32]. However; there are no reports of molecular techniques applied to the diagnosis of ovarian carcinoma in peritoneal washing or abdominal effusion.

References

1. Tauchi PS, Caraway N, Truong LD, Kaplan AL, Ramzy I. Serous surface carcinoma of the peritoneum: useful role of cytology in differential diagnosis and follow-up. Acta Cytol. 1996;40:429–36.
2. Zuna RE, Behrens A. Peritoneal washing cytology in gynecologic cancers: long-term follow-up of 355 patients. J Natl Cancer Inst. 1996;88:980–7.
3. Mathew S, Erozan YS. Significance of peritoneal washings in gynecologic oncology. The experience with 901 intraoperative washings at an academic medical center. Arch Pathol Lab Med. 1997;121:604–6.
4. Fanning J, Markuly SN, Hindman TL, et al. False positive malignant peritoneal cytology and psammoma bodies in benign gynecologic disease. J Reprod Med. 1996;41:504–8.
5. Ravinsky E. Cytology of peritoneal washings in gynecologic patients. Diagnostic criteria and pitfalls. Acta Cytol. 1986;30:8–16.
6. Sneige N, Fanning CV. Peritoneal washing cytology in women: diagnostic pitfalls and clues for correct diagnosis. Diagn Cytopathol. 1992;8:632–40; discussion 640–2.
7. Jorns JM, Knoepp SM. Occult fallopian tube carcinoma detected in routine pelvic washing specimens submitted for staging: another justification for pelvic washing cytology? Diagn Cytopathol. 2009;37:923–9.
8. Ikeda K, Tate G, Suzuki T, Mitsuya T. Cytomorphologic features of immature ovarian teratoma in peritoneal effusion: a case report. Diagn Cytopathol. 2005;33:39–42.
9. Gammon R, Hameed A, Keyhani-Rofagha S. Peritoneal washing in borderline epithelial ovarian tumors in women under 25: the use of cell block preparations. Diagn Cytopathol. 1998;18:212–4.
10. Fetsch PA, Abati A, Hijazi YM. Utility of the antibodies CA 19–9, HBME-1, and thrombomodulin in the diagnosis of malignant mesothelioma and adenocarcinoma in cytology. Cancer. 1998;84:101–8.
11. Fetsch PA, Simsir A, Abati A. Comparison of antibodies to HBME-1 and calretinin for the detection of mesothelial cells in effusion cytology. Diagn Cytopathol. 2001;25:158–61.
12. Mulvany N. Cytohistologic correlation in malignant peritoneal washings. Analysis of 75 malignant fluids. Acta Cytol. 1996;40:1231–9.
13. Bhalla R, Siddiqui MT, Mandich D, et al. Diagnostic utility of D2-40 and podoplanin in effusion cell blocks. Diagn Cytopathol. 2007;35:342–7.
14. Lyons-Boudreaux V, Mody DR, Zhai J, Coffey D. Cytologic malignancy versus benignancy: how useful are the "newer" markers in body fluid cytology? Arch Pathol Lab Med. 2008;132:23–8.
15. Ordonez NG. The diagnostic utility of immunohistochemistry and electron microscopy in distinguishing between peritoneal mesotheliomas and serous carcinomas: a comparative study. Mod Pathol. 2006;19:34–48.
16. Nofech-Mozes S, Khalifa MA, Ismiil N, et al. Immunophenotyping of serous carcinoma of the female genital tract. Mod Pathol. 2008;21:1147–55.
17. Chhieng DC, Yee H, Schaefer D, et al. Calretinin staining pattern aids in the differentiation of mesothelioma from adenocarcinoma in serous effusions. Cancer. 2000;90:194–200.
18. Comin CECE, Saieva CC, Messerini LL. h-caldesmon, calretinin, estrogen receptor, and Ber-EP4: a useful combination of immunohistochemical markers for differentiating epithelioid peritoneal

mesothelioma from serous papillary carcinoma of the ovary. Am J Surg Pathol. 2007;31:1139–48.
19. Ko EC, Jhala NC, Shultz JJ, Chhieng DC. Use of a panel of markers in the differential diagnosis of adenocarcinoma and reactive mesothelial cells in fluid cytology. Am J Clin Pathol. 2001;116:709–15.
20. Politi E, Kandaraki C, Apostolopoulou C, Kyritsi T, Koutselini H. Immunocytochemical panel for distinguishing between carcinoma and reactive mesothelial cells in body cavity fluids. Diagn Cytopathol. 2005;32:151–5.
21. Wieczorek TJ, Krane JF. Diagnostic utility of calretinin immunohistochemistry in cytologic cell block preparations. Cancer. 2000;90:312–9.
22. Takeshima Y, Amatya VJ, Kushitani K, Inai K. A useful antibody panel for differential diagnosis between peritoneal mesothelioma and ovarian serous carcinoma in Japanese cases. Am J Clin Pathol. 2008;130:771–9.
23. Hwang H, Quenneville L, Yaziji H, Gown AM. Wilms tumor gene product: sensitive and contextually specific marker of serous carcinomas of ovarian surface epithelial origin. Appl Immunohistochem Mol Morphol. 2004;12:122–6.
24. Yaziji H, Battifora H, Barry TS, et al. Evaluation of 12 antibodies for distinguishing epithelioid mesothelioma from adenocarcinoma: identification of a three-antibody immunohistochemical panel with maximal sensitivity and specificity. Mod Pathol. 2006;19:514–23.
25. Ascoli V, Carnovale-Scalzo C, Taccogna S, Nardi F. Utility of HBME-1 immunostaining in serous effusions. Cytopathology. 1997;8:328–35.
26. Ueda J, Iwata T, Ono M, Takahashi M. Comparison of three cytologic preparation methods and immunocytochemistries to distinguish adenocarcinoma cells from reactive mesothelial cells in serous effusion. Diagn Cytopathol. 2006;34:6–10.
27. Reid-Nicholson M, Iyengar P, Hummer AJ, Linkov I, Asher M, Soslow RA. Immunophenotypic diversity of endometrial adenocarcinomas: implications for differential diagnosis. Mod Pathol. 2006;19:1091–100.
28. Nap M. Immunohistochemistry of CA 125. Unusual expression in normal tissues, distribution in the human fetus and questions around its application in diagnostic pathology. Int J Biol Markers. 1998;13:210–5.
29. Hecht JL, Pinkus JL, Pinkus GS. Monoclonal antibody MOC-31 reactivity as a marker for adenocarcinoma in cytologic preparations. Cancer. 2006;108:56–9.
30. Selvaggi SM. Diagnostic pitfalls of peritoneal washing cytology and the role of cell blocks in their diagnosis. Diagn Cytopathol. 2003;28:335–41.
31. Dong HP, Holth A, Berner A, Davidson B, Risberg B. Flow cytometric immunphenotyping of epithelial cancer cells in effusions–technical considerations and pitfalls. Cytometry B Clin Cytom. 2007;72:332–43.
32. Holloway AJ, Diyagama DS, Opeskin K, et al. A molecular diagnostic test for distinguishing lung adenocarcinoma from malignant mesothelioma using cells collected from pleural effusions. Clin Cancer Res. 2006;12:5129–35.

Index

R.A. Soslow and C. Tornos (eds.), *Diagnostic Pathology of Ovarian Tumors*,
DOI: 10.1007/978-1-4419-9751-7, © Springer Science+Business Media, LLC 2011

Q

R

S

T

W

Y

MIX
Papier aus verantwortungsvollen Quellen
Paper from responsible sources
FSC® C105338

If you have any concerns about our products,
you can contact us on
ProductSafety@springernature.com

In case Publisher is established outside the EU,
the EU authorized representative is:
Springer Nature Customer Service Center GmbH
Europaplatz 3, 69115 Heidelberg, Germany

Printed by Libri Plureos GmbH
in Hamburg, Germany